AF719112

DESCRIPTIVE COLLEGE PHYSICS

The San Onofre Nuclear Generating Station, located near San Clemente, Calif., generates 430 megowatts of electricity—enough to meet the needs of a city of more than a half-million population.

DESCRIPTIVE COLLEGE PHYSICS

Third Edition

HARVEY E. WHITE, Ph. D., Sc. D.
University of California, Berkeley

D. Van Nostrand Company
New York • Cincinnati • Toronto • London • Melbourne

D. Van Nostrand Company Regional Offices:
New York Cincinnati
D. Van Nostrand Company International Offices:
London Toronto Melbourne

Copyright © 1971 by Litton Educational Publishing, Inc.

Library of Congress Catalog Card Number: 70-102047
ISBN: 0-442-09416-7

All rights reserved. Certain portions of this work copyright © 1955, © 1963 by Litton Educational Publishing, Inc. No part of this work covered by the copyrights hereon may be reproduced or used in any form or by any means—graphic, electronic, or mechanical, including photocopying, recording, taping, or information storage and retrieval systems—without written permission of the publisher. Manufactured in the United States of America.

PUBLISHED BY D. VAN NOSTRAND COMPANY
450 West 33rd Street, New York, N.Y. 10001

10 9 8 7 6

PREFACE

This textbook is designed primarily for use in a one-semester course in college physics for liberal arts students. These are the students for whom mathematics beyond the simplest of algebra would reduce the class size to a fraction of the desirable enrollment.

Although the first and second editions of this book were also written for a one-semester course, physics instructors in some colleges and universities in the United States, as well as in foreign countries, have been using the text for a one-year course.

Many instructors have adopted this text because of the large amount of space devoted to atomic and nuclear physics. Others have selected it because of the many diagrams and explanations of easy-to-perform lecture demonstration experiments.

For many years the author has taught a one-semester physics survey course at the University of California, Berkeley. Presented as a nonmathematical course it is attended largely by students majoring in social sciences, economics, history, political science, philosophy, etc. Although simple algebraic equations are given in the lectures to illustrate principles and laws of nature, and frequent numerical problems are solved by the instructor, the students are not required to work problems. Believing that some exposure by liberal arts students to classical and modern physics is better than no physics at all, the author has adopted such a nonmathematical treatment. He has found that students' interests can be aroused by the lecturer performing numerous demonstrations.

The reasons for including the solutions to a number of problems within the text material proper, as well as problems at the ends of chapters, are twofold: (1) many students develop a sufficient interest in physics to go beyond the work required of them by solving problems; and (2) some physics instructors prefer to include numerical examples in the students' homework and examinations. To encourage such students, answers are given to the even-numbered problems at the ends of the chapters.

The author has attempted to update all subjects where recent discoveries in scientific research and technical developments have been achieved. New sections have been written on such subjects as lasers, conservation laws for nuclear reactions, and world-line diagrams. In the few cases where the *cgs* system of units was still used, conversion to the *MKSA* system has been accomplished. A considerable number of example problems with their step-by-step solutions have been added to the text material. A large number of biographies of famous men of science have been added to those already given in the first and second editions. These should add considerably to the interest of liberal arts students in the history of science.

The author wishes to take this opportunity to thank the many physics instructors and students who have written to the author and included corrections and suggestions, as well as encouragement to write again.

January, 1971
Berkeley, California

HARVEY E. WHITE

CONTENTS

SOUND

LIGHT

ELECTRICITY AND MAGNETISM

ATOMIC PHYSICS

NUCLEAR PHYSICS

INTRODUCTION · OPTICAL ILLUSIONS

Physics is a natural science that historically has no beginning and no end. As a science it is a continuous process by which, as the years go by, man learns more and more of the seemingly endless detail of the physical world about him.

In an attempt to establish an origin for the basic sciences, many historians go back more than two thousand years to the era of the great philosophers Aristotle, Copernicus, Archimedes, Hippocrates, etc. These men are considered great because in their time they had a keen sense of reasoning, and they applied their powers to the observation and explanation of many natural phenomena. While many of the concepts these men formulated showed remarkable insight, progress was relatively slow because they overlooked one important factor. They failed to recognize the importance of experimentation. Herein lies the secret to the rapid advancement of science in this the twentieth century.

To advance science, whether it is one of the life sciences or one of the physical sciences, well-planned experiments must be performed, observations must be made, and measurements must be taken and recorded. It is through a study of these recorded measurements, and their correlation and mathematical formulation, that fundamental laws of nature are established, verified, and then extended to include other natural phenomena. *Physics may well be defined, therefore, as that branch of knowledge and experimentation that treats of the inanimate world and its phenomena.*

1 PHYSICS AS AN OBJECTIVE METHOD It has long been known that when experiments are to be performed the observer cannot rely entirely upon the human senses of touch, sight, hearing, etc., to make accurate observations. Methods of measurement which rely upon the senses alone are called *subjective methods.* Methods which make use of scientific instruments are generally called *objective methods.*

In the early history of science, laws were frequently discovered by the use of subjective methods. Progress was slow, however, until such methods were replaced by objective methods using measuring instruments devised to give greater and greater precision.

It is true that many scientific discoveries have been made in the past with what we now would call the crudest of apparatus and equipment. It is the development of precision instruments and apparatus, however, which has, particularly within the last several decades, led to discoveries which are far-reaching in their theoretical implications and are of extreme practical importance to the advancement of civilization.

As an introduction to the subject of physics we will first consider a number of experiments illustrating the false impressions so easily arrived at from the use of subjective methods of observation. Although these experiments are of the nature of an entertainment, they do have more serious aspects, for they demonstrate the necessity for using objective methods in advancing science.

2 SUBJECTIVE METHODS If someone asks you to determine the temperature of a pan of water, your first impulse, if the water is not too hot, is to use your hand or your fingertips and not to bother looking for a thermometer. To illustrate the gross inaccuracy of the touch in determining temperature, consider the three pans of water as shown in Fig. A. If the hand is first held

Fig. A. Experiment illustrating the uncertainty of subjective methods of measurement.

for some little time in the pan containing *cold* water and then plunged into the *warm* water, the senses tell you it is hot. If, however, the hand is first held in the *hot* water and then plunged into the *warm* water, your senses tell you it is cold. Your conclusion in either case is thus influenced by your experience immediately preceding your determination of the temperature of the middle pan. When a thermometer is used in this experiment, the same temperature will be arrived at in either case. Although this latter would be called an objective method of measurement, one still relies upon the senses to obtain a reading of the thermometer scale.

If the length and breadth of a table top are to be measured, a *foot rule*, a *yardstick*, or a *meter stick* should be used and not the *span of a hand*. In a similar way the time that it takes a sprinter to run the "*one-hundred yard dash*" is measured by a *clock*, a *watch*, or a *chronometer* and not by the *heartbeat* or *pulse*.

3 THE EYE In making many scientific measurements the eye is considered as the most useful of all recording instruments. In some instances, however, the eye is not and should not be used directly in making observations, since it cannot be relied upon to observe what is really there. To illustrate how unreliable the sense of vision can be in some cases we will consider in the next section a number of examples commonly referred to as "optical illusions."

Despite its many and sometimes serious imperfections and limitations, the human eye is a marvelous optical instrument. It is nature's priceless gift to man, enabling him to enjoy the beauties of form, color, and motion made possible by light. Optically the eye is like an exceptionally fine camera with an elaborate lens system on the one side and a sensitive screen or photographic film, called the *retina*, on the other. See Fig. B. The refracting media of the eye consist of the *cornea*, the *aqueous humor*, the *crystalline lens*, and *vitreous humor*, and its function is to focus an image of the objects to be seen on the retina. Like a camera, the eye contains an *iris diaphragm* which opens wider for faint light and closes down to a bare pinhole opening for very bright sunlight. It is this iris that contains the pigment determining the color of the eye.

In the retina of the eye the light pulses are received by tiny *cones* and *rods* whose function it seems to be to change the light into electrical impulses. Each cone and rod is connected with an individual nerve which conducts these impulses through the nerve canal to the brain. Just how these electrical impulses are produced by the cell-like structures, the cones and rods, and how they are interpreted by the brain as vision is still but vaguely understood by scientists. Experiments seem to indicate that the cones respond only to bright light and are particularly

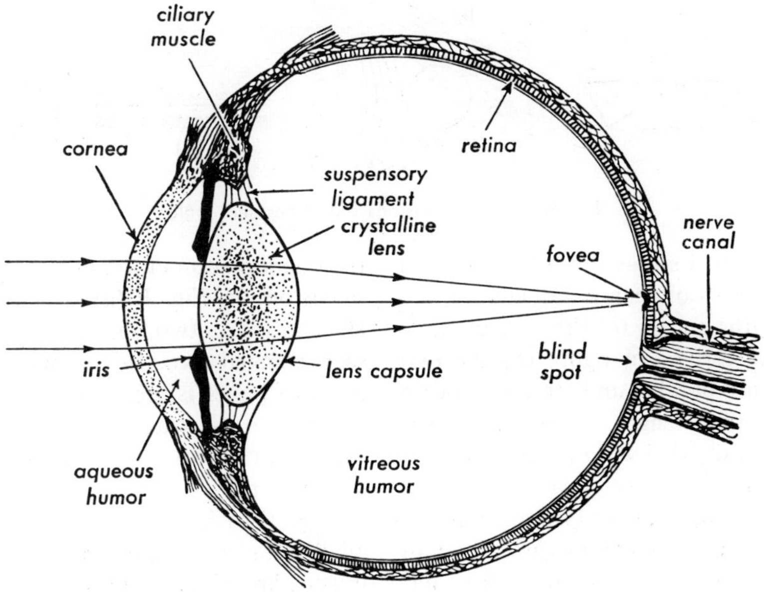

Fig. B. Cross-section diagram of the human eye.

responsible for the detection and distinction of color, whereas the rods are sensitive to very feeble light, to motion, and to slight variations in intensity.

At the very center of the retina is a small yellowish-looking spot called the *fovea*. This small region contains a large number of cones, but no rods. It is on this spot in each eye that the words and letters of this page are focused one after the other when reading. Note when scrutinizing one word or particularly a single letter in a word that the rest of the page and even the words and letters close by appear indistinct.

It is customary to divide all sensory data that contribute to sight perception of any object into two parts: first, the formation of the retinal image by the light coming from the object, and second, the integrative property of the brain to interpret this image.

4 THE BLIND SPOT Not far from the fovea on the retina of the eye is a small region called the *blind spot*. This spot, which is insensitive to light, is where the nerve canal joins the eyeball. The existence of the blind spot can be demonstrated by closing the right eye, and holding the book at arm's length, looking continuously at the center of the circle of Fig. C with the left eye.

Fig. C. Experiment illustrating the blind spot of the eye.

Both the circle and square will be seen from this distance. If the book is now moved slowly toward the eye, still fixing the eye upon the circle, a position (about 8 to 10 in. from the eye) will be reached where the square disappears. When both eyes are open no position will be found where either the cross or the square disappears. One eye always sees that part of an object to which the other eye is blind. A similar experiment with the right eye focused on the square will cause the circle to disappear.

5 OPTICAL ILLUSIONS Of the hundreds of well-known optical illusions, only a few of the most interesting ones will be presented here. In Fig. D is a group of six figures classified as illusions due to *lines* and *angles*. In (a), the first figure, the

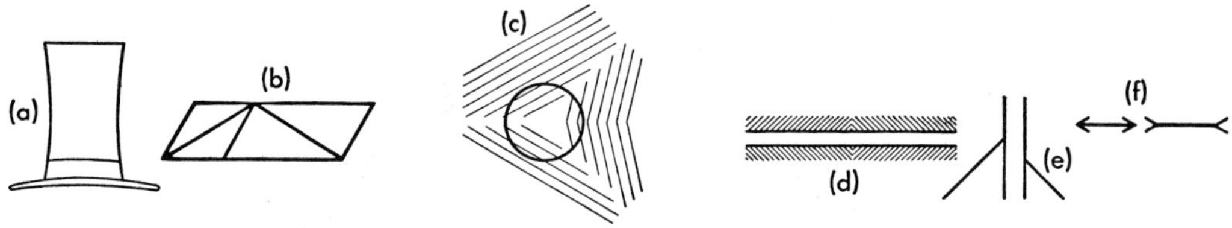

Fig. D. Optical illusions with lines and angles.

brim of the hat is as long as the hat is high; in (b) the diagonal lines of each parallelogram are of the same length; and in (c) the perfect circle appears to be distorted. In figure (d) the two horizontal lines are parallel and straight, and in (f) they are of equal length. In (e) the lower right-hand line if extended will intersect the left-hand line where it joins the vertical.

Fig. E is an example of *perspective*, an illusion suggesting depth to the picture when in reality it is flat. Actually this is just a line diagram enclosing three figures of equal height. By means of slanting lines these figures are made to appear to have different heights. Experiences from early childhood have trained us to interpret the slanting lines as depth.

The next set of illusions, shown in Fig. F, are classified as *equivocal figures*. These illustrate the phenomenon of the *fluctuation* of the process of vision. In Fig. F (a) six cubes may be seen stacked three, two, one, or seven cubes may be seen stacked two, three, two. In (b) a folded sheet of paper is seen opening either toward or away from the reader. In (c) is a flight of steps seen from above looking down, or from below looking up.

Fig. F(d) is one of the most interesting of all illusions. To fully appreciate the effect one must himself perform the experiment with a small wire cube about one inch in size. The cube is held by a small handle at one corner and viewed with one eye at a distance of from 1 to 2 ft. By the principle of fluctuation the observer next tries to make the farthest corner of the cube appear as the nearest corner. When this condition is attained the cube upon being turned

Fig. E. Which figure is tallest? Measure them.

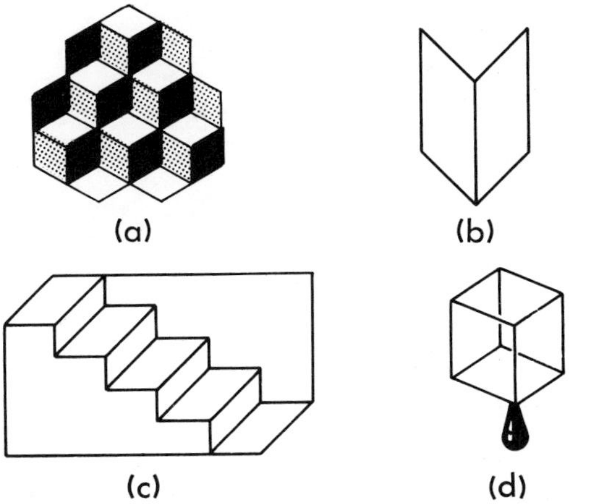

Fig. F. Optical illusions illustrating fluctuation of the attention.

about a horizontal or vertical axis will appear to turn in the opposite direction. A little practice in the fluctuation of the visual senses is required in this experiment and it is well worth performing.

The diagram in Fig. G illustrates imperfec-

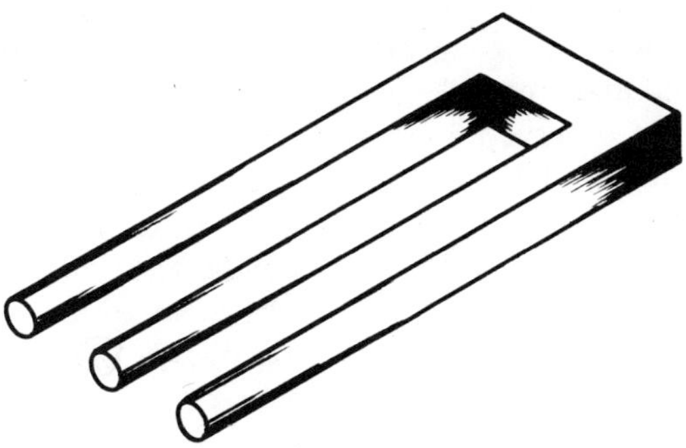

Fig. G. An illusion in perspective and dimensions.

tions in our mental processes of interpreting a two-dimensional drawing as a three-dimensional object. As a drawing there is nothing wrong; as a real three-dimensional object the drawing represents an impossibility. Cover over either end of the drawing and what you see looks perfectly feasible and a possible reality.

In Fig. H are two pairs of similar figures of equal area. The slanting lines at the ends make the lower figure in each case appear to be larger than the one immediately above. Such figures

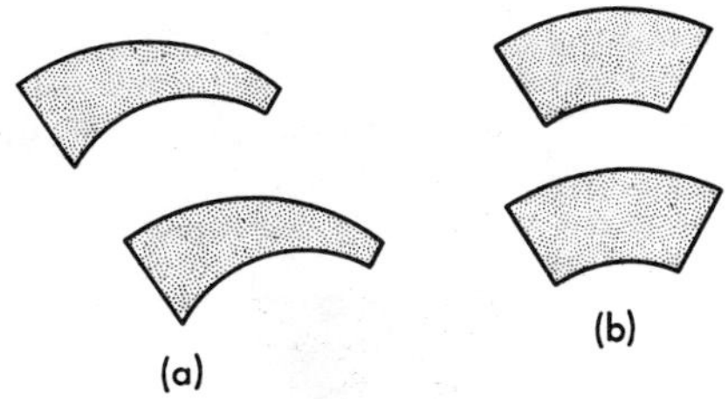

Fig. H. Optical illusions of area.

should be cut from white cardboard and held one above the other. When the upper figure is interchanged with the corresponding lower figure, one figure seems to grow and the other to shrink before your eyes.

In Fig. I(a) are two small squares of equal

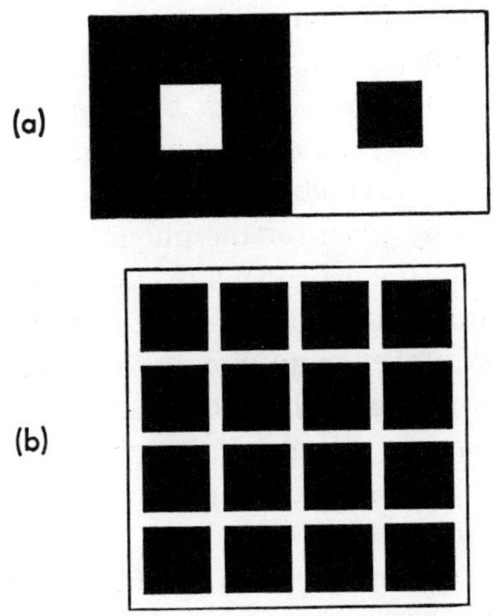

Fig. I. Light and dark objects illustrating the phenomenon of irradiation.

size: a white square on a black background, and a black square on a white background. When an image of this is formed on the retina of the eye, the cones and rods just beyond the white edges are stimulated by those nearby, thus causing the white square to be larger than the black one. This phenomenon is called *irradiation* or *brightness contrast*. A similar phenomenon is illustrated in Fig. I(b) where gray spots are seen at the intersections of the white lines.

6 FATIGUE AND COLOR ILLUSION All of the illusions described above have been confined to figures in black and white. There exists a large

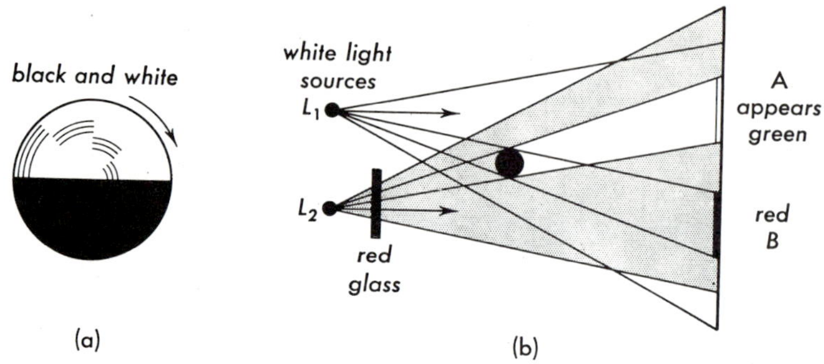

Fig. J. Diagrams of experiments demonstrating color illusions.

number of illusions which are classified as *color illusions.* Two of these are diagrammed in Fig. J. A disk painted black and white as shown in (a) will appear to be colored when set rotating at a relatively low speed. The colors to be seen are faint pastel shades of violet, blue, green, yellow, and pink. The speed of the wheel should be from about 4 to 15 revolutions per second. The explanation usually given for the phenomenon is that the retina of the eye responds more quickly to some colors than to others. Since the white image of the disk moves around on the retina and since white light contains all the colors of the rainbow, some colors are perceived at each given spot on the retina sooner than others and the effect of color is produced.

The second diagram (b) in Fig. J illustrates the appearance of color by virtue of contrast. If a white patch of light, for example, is seen on a background of red it will appear to be pale green. If, on the other hand, a white patch is seen on a background of green it will appear to be pink. The experiment may be performed with two similar arc lights producing white light. Each of these is made to cast a shadow of the same rod R on a white screen. If a piece of red glass is placed in front of light L_2 as shown in the figure, the white patch of light at A will appear to be pale green. If a green glass is inserted in its place the region A will appear pink. In each case, A receives light from L_1 only and must therefore really be white. Red and blue-green, or cyan, are complementary colors, and, when they are added together, they produce white light. The subject of the mixing of colors will be taken up in detail in Chap. 22.

7 COMPLEMENTARY IMAGES When the eyes are subjected to bright light for some little time, the retina seems to show tiring or *fatigue.* Furthermore, continued subjection of any part of the retina to one particular color causes only those cones sensitive to that color to tire. When the same retinal area is subsequently subjected to white light, the previously inactive cones respond more strongly than those originally stimulated and a complementary color is seen.

To observe these colored images, fix the attention on the black star in the lower right-hand corner of the field of the flag in Fig. K and keep it there for about 15 to 20 seconds. Then turn the eyes toward a white wall of the room, or toward the open sky, and in one or two seconds the American flag will appear in all of its true colors. Similar effects can be observed with other color photographs.

Delayed images of this kind are always complementary in color to the original pictures: black becomes white, yellow becomes blue, green becomes magenta, magenta becomes green, etc. (For an explanation of complementary colors see Chap. 22.)

8 THE STROBOSCOPIC EFFECT In motion pictures, when a wagon with spoked wheels is coming to a stop the wheels are often noticed to stand still, then turn backward, stop, turn forward, and then stop again. This phenomenon,

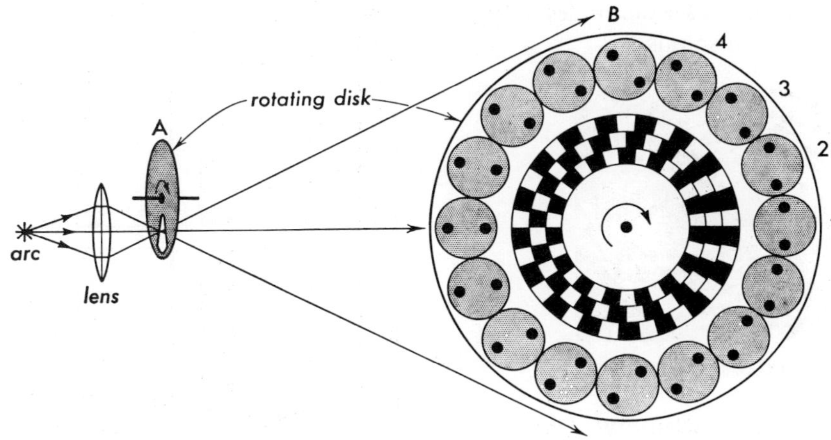

Fig. L. Experimental arrangement for demonstrating the stroboscopic effect.

known as the "stroboscopic effect," is due to interrupted illumination of the motion-picture screen and can be illustrated in many ways. An interesting experiment illustrating the phenomenon is shown in Fig. L. Two disks are mounted on the shafts of two separate motors. The smaller disk *A* with a narrow slot is used to interrupt the light beam illuminating the large disk. The disk *B* is white with black circles and dots arranged exactly as shown. Suppose now that disk *A* makes sixteen revolutions per second, thus illuminating disk *B* with sixteen short flashes of light per second. Suppose also that *B* makes only one revolution per second, and that one flash of light comes when the disk has the position shown in the figure. Confining the attention to the circle at position (1), the two enclosed dots are one above the other. When the second flash of light appears, the circle (2) will be in position (1) and the two dots will appear to have shifted slightly clockwise. When the next flash of light comes, the circle (3) will be in position (1) and the two dots will have shifted still farther. This process continued shows that the circles will appear to stand still and the dots to rotate within them.

If the light flashes in any experiment as the one just described are slower than 16 per second, the illuminated object will appear to flicker badly. If, however, the flashes come at an increasingly higher rate, the flicker will soon disappear entirely and the illumination will seem to be steady. The reason for this is that each retinal image is somehow retained by the vision mechanism for about one sixteenth of a second. This is called the *persistence of vision.*

9 CIRCLES AND SPIRALS If this page of the book is held about one foot in front of the observer's eyes and the book moved rapidly in a circle about 2 to 3 in. in diameter, the spiral in Fig. M will appear to rotate in the direction of motion. A set of alternately dark and light concentric circles will show the same effect, the apparent rotation being due to persistence of vision.

Fig. M. Illusion of rotation.

We have seen in the previous illustrations how some optical illusions break down or diminish under critical inspection. There are others, however, that persist. No amount of staring or thought will teach you to see the circles of Fig. N as anything other than spirals.

Fig. N. Circles appear to be spirals.

10 THE TRAPEZOIDAL WINDOW One of the most striking optical illusions in perspective is that of a slowly rotating window having the design shown in Fig. O. This device is cut from a single sheet of $\frac{1}{4}$ in. plywood that is painted white, gray, and black on both sides,* and is mounted on a slowly rotating motor driven shaft.

Viewed in a darkened room with front and side illumination of the window by two lights shielded from the audience, the window appears to rotate first in one direction, then the other.

An added effect is obtained by hanging a brightly colored ball at the upper corner of the small end. The ball appears to continue around in the same direction, whereas the window appears to oscillate back and forth.

* In constructing such a window for classroom use, make all dimensions proportional to the reproduction in Fig. O, with the maximum height about 16 in.

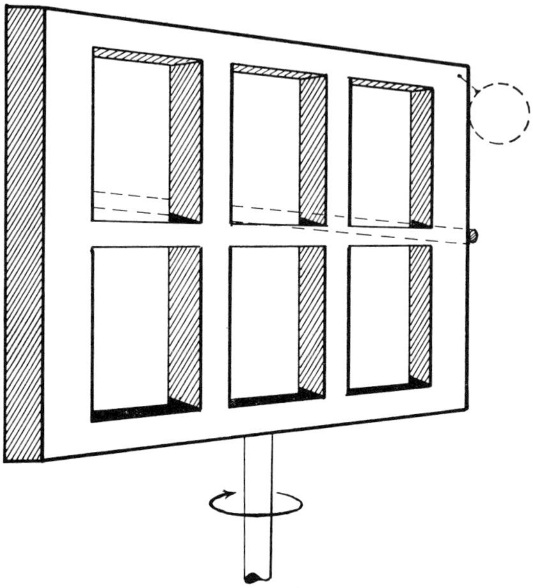

Fig. O. Rotating slowly and continuously in one direction this trapezoidal-shaped window appears to rotate alternately one way, then the other.

Another striking effect is to place a colored rod through the upper center window pane, as shown dotted in the figure. The rod's apparent behavior must be observed to be appreciated.

QUESTIONS

1. What in the human eye is the function of (a) the cornea, (b) the crystalline lens, (c) the iris, (d) the retina, and (e) the fovea? (See also Chap. 23.)

2. Before diving into a swimming pool the first time on any particular day, does putting your toes into the water give a good indication of the temperature of the water?

3. Briefly explain how the subjective method enters into the measurement of the overall length of an automobile even though you use a good tape measure. How would you measure the length of an automobile to an accuracy of one-eighth of an inch?

4. Explain why the wheels of a wagon may appear to turn backward on a motion picture screen when the wagon is moving forward.

5. Briefly explain why motion pictures flicker when the projector runs too slowly, but do not flicker when it runs fast.

6. What is the stroboscopic effect? Briefly explain.

1 / UNITS OF MEASUREMENTS

We are living today in a world of science and technology. To be well educated, and to feel at home in this world, everyone should at some time study the subjects of physics, chemistry, and biology.

Physics may be defined as that branch of knowledge dealing with the inanimate world and its phenomena. The subject includes:

mechanics
properties of matter
heat
sound
electricity
magnetism
X rays
atomic structure
radioactivity
cosmic rays
nuclear structure
atomic energy

Physics is called an exact science because experiments are performed and accurate observations are made with measuring instruments. Laws and theories are formulated from the measured results of these experiments and then used to predict the results of new experiments. If these new experimental results do not agree with theory, the theory is either modified and brought into agreement or it is discarded for a new and better theory. Those who perform new experiments are called "experimental physicists," and those who formulate theories based upon experiments are called "theoretical physicists." The nature of physics with its objective methods of experimentation justifies its title as one of the most exact of all the sciences.

1.1 FUNDAMENTAL AND DERIVED UNITS Since physics is a science based upon exact measurement, it is essential that the student first become familiar with several of the more commonly used measuring devices and the units into which each is usually divided. Every measurement, whether it be of distance, weight, an interval of time, or of anything else, requires *two* things: first, a *number;* and second, a *unit.* One might, for example, obtain as the result of the measurement of different distances, *20 feet*, *5 miles*, *3 rods*, or as the result of the measurement of different weights, *6 pounds*, *25 tons*, *4½ ounces*, or as the result of the measurement of different time intervals, *7 hours*, *26 seconds*, etc. As the result of some experiment or the reading of certain instruments one might obtain the measurements, *10.7 calories, 90 horsepower, 6 volts, 12 kilowatts*, etc. In each case the *unit* is just as essential as the *number* expressing the amount.

Although there are numerous different units, each one can be expressed in not more than three special units. These three, called *fundamental units*, are *length, mass*, and *time*. All other units are called *derived units* since, as we shall see later, they can always be written as some combination of the three fundamental units.

There are in general two widely used systems of fundamental units: (a) the *metric system*, and (b) the *English system*. Throughout the civilized world scientific observations are nearly always expressed in the metric system. This system employs the *standard meter* as the unit of length, the *standard kilogram* as the unit of mass, and the *second* as the unit of time. The English system, sometimes called *the Engineers' system*, employs the *foot* as the unit of length, the *pound* as the unit of weight, and the *second* as the unit of time.

1.2 THE STANDARD METER AND YARD Until 1960, the standard meter was taken to be a certain platinum-iridium bar kept in the vaults of the International Bureau of Weights and Measures in France. Three facsimiles of this bar are kept at the United States Bureau of Standards in Washington, D. C. Each of these duplicate copies was referred to as an *International Prototype Meter* and legally defined as the standard of length in the United States. From these prototypes all other measuring rods and tapes were standardized.

In 1960, at an international meeting of the General Conference on Weights and Measures in Paris, France, the standard meter adopted by all countries was defined in terms of the wavelength of light. The light selected was the orange light of wavelength 6057.80×10^{-7} meters, produced by krypton gas in an electrical discharge tube. The standard meter is defined today (see Sec. 24.4) as

$$1 \text{ meter} = 1{,}650{,}763.73 \text{ wavelengths (orange light of krypton)}$$

When the standard meter was first devised it was intended that it have a length equal to one ten-millionth part of the distance from one of the earth's poles to the equator. More recent measurements of the earth's dimensions have shown that the distance from pole to equator is about 10,000,880 *standard meters*.

The standard meter is usually divided into one hundred equal parts. Each of these parts is called the *centimeter*:

$$1 \text{ meter} = 100 \text{ centimeters}$$

or, abbreviated,

$$1 \text{ m} = 100 \text{ cm}$$

The centimeter is further divided into ten equal parts. Each of these parts is called the *millimeter*. See Fig. 1A.

$$1 \text{ centimeter (cm)} = 10 \text{ millimeters (mm)}$$

$$1000 \text{ millimeters (mm)} = 1 \text{ meter (m)}$$

In civil life in the United States the *yard* is used as the standard unit of length. By an act of Congress in 1866 the standard yard to be used legally in the United States was defined as

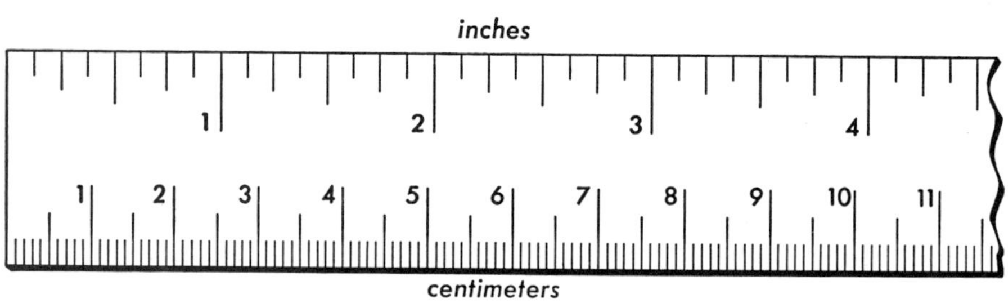

Fig. 1A. Diagram comparing the centimeter scale (lower) with the inch scale (upper).

3600/3937 part of a standard meter. Since the yard is divided into thirty-six inches,

$$1 \text{ meter (m)} = 39.37 \text{ inches (in.)}$$

With twelve inches to one foot,

$$3 \text{ ft} = 1 \text{ yd}$$

$$1 \text{ ft} = 30.48 \text{ cm}$$

and

$$1 \text{ in.} = 2.54 \text{ cm}$$

The sizes of the inch and fractions of an inch, relative to the centimeter and millimeter, are illustrated in Fig. 1A.

When large distances are to be measured it is convenient as well as customary to use large units of length. Such units are the *kilometer* in the metric system and the *mile* in the English system. One kilometer is equivalent to one-thousand meters, and one mile is equivalent to five-thousand-two-hundred-and-eighty feet.

$$1000 \text{ meters (m)} = 1 \text{ kilometer (Km)}$$

$$1 \text{ mile (mi)} = 5280 \text{ feet (ft)}$$

The relations between units of length in the metric system as against the English system are given in Table 1A.

EXAMPLE 1. How many meters are there in 150 ft?

SOLUTION. Referring to Table 1A, we find that

$$1 \text{ ft} = 0.30480 \text{ m}$$

When we write 150 ft we mean 150×1 ft. Therefore, in place of 1 ft we can write its equivalent of 0.30480 m. Hence we get

$$150 \text{ ft} = 150 \times 0.30480 \text{ m}$$

$$150 \text{ ft} = 45.72 \text{ m}$$

which to three significant figures is 45.7 m.

1.3 THE STANDARD KILOGRAM AND POUND

The standard unit of mass is the *kilogram*, a block of platinum also preserved at the International Bureau of Weights and Measures near Paris. Two copies of this kilogram (which may be called International Prototype Kilograms) are kept in the vaults of the U. S. Bureau of Standards. The kilogram is divided into one-thousand equal parts called *grams*.

$$1000 \text{ grams (g)} = 1 \text{ kilogram (Kg)}$$

The original intent was to base the standard kilogram upon the gram, the gram being the mass of one cubic centimeter of pure water taken at a temperature of four degrees centigrade.

The standard *pound* is defined in terms of the standard kilogram by the relation that its mass shall equal 0.4536 kilogram. From this we obtain the following relations:

There are 0.4536 Kg in 1 lb
453.6 g in 1 lb
28.35 g in 1 oz
2.2045 lb in 1 Kg
16 oz in 1 lb
2000 lb in 1 ton

TABLE 1A. Table of Conversion Factors for Units of Length

	Km	m	cm	in.	ft	mi
1 kilometer =	1	1000	100,000	39,370	3280.83	0.62140
1 meter =	0.00100	1	100	39.370	3.28083	6.21×10^{-4}
1 centimeter =	1.0×10^{-5}	0.0100	1	0.39370	0.032808	6.21×10^{-6}
1 inch =	2.54×10^{-5}	0.02540	2.5400	1	0.08333	1.58×10^{-5}
1 foot =	3.05×10^{-4}	0.30480	30.480	12	1	1.89×10^{-4}
1 mile =	1.60934	1609.34	160,934	63,360	5280	1

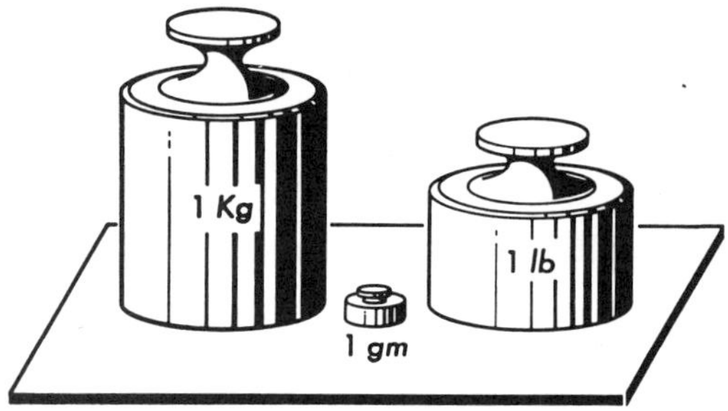

Fig. 1B. The mass of a standard kilogram is 2.2 times the mass of a standard pound weight.

The beginning student should become familiar with the above numbers and the names of the various units. Fig. 1B illustrates the relative sizes of standard weights and masses.

EXAMPLE 2. How many kilograms are there in one ton?

SOLUTION. Since there are 2000 lb in 1 ton, and 0.4536 Kg in 1 lb, we can multiply the two numbers to obtain

$$2000 \text{ lb} \times 0.4536 \frac{\text{Kg}}{\text{lb}} = 907.2 \text{ Kg}$$

which, to three significant figures, is 907 Kg.

1.4 HISTORICAL TIMEPIECES Instruments for the measurement of time go back historically to the ancient Egyptians of at least eight centuries B.C. The earliest timepieces on record were chiefly water clocks, some of very simple design and others of more elaborate design. These clocks were based upon the elementary principle that it takes the same time for equal amounts of water to flow through a small opening. The *hour glass*, employing the same principle and using sand instead of water, is an outgrowth of the water clock and dates back to medieval times.

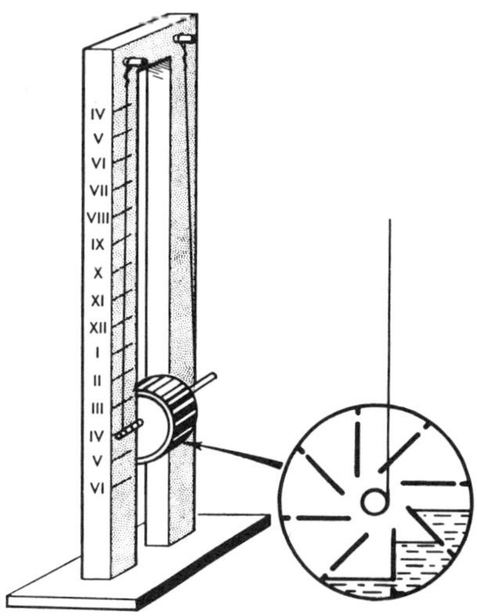

Fig. 1C. Diagram of an early form of water clock.

A water clock of moderately simple design is shown in Fig. 1C. Small holes at the edge of the vanes (shown in the detail diagram) allow the water to flow from one compartment to the other. This permits the cylinder to turn slowly, thus unwinding the suspending cords.

The *sundial* dates back to the Chaldean astronomer Berosus who lived about the time of Alexander the Great, 300 B.C. Today, similar instruments serve as ornaments in many of our public parks. A sundial of common design, as shown in Fig. 1D, consists essentially of a *pin*

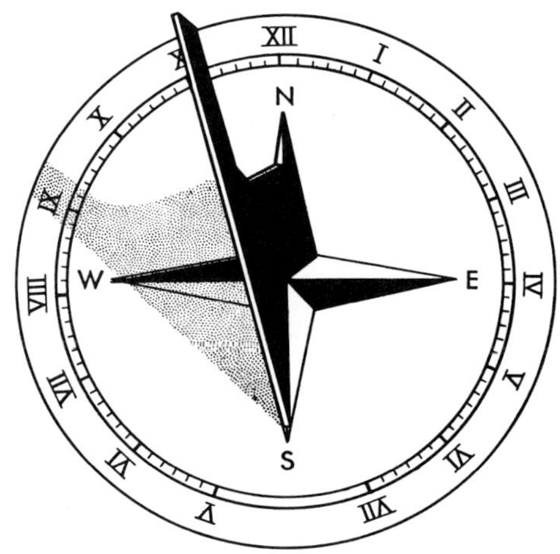

Fig. 1D. Diagram of a sundial.

called a gnomon mounted at an angle on a circular plate called the *dial*. The gnomon is mounted in the vertical North and South plane. The edge of the gnomon is parallel to the earth's rotational axis and its purpose is to cast a shadow on the dial which is marked with the hours of the day. Due in part to the apparent yearly precessional motion of the earth's axis, small corrections must be made to the time as shown by the shadow. These corrections amounting to several minutes are usually engraved upon every sundial. The sundial on the University of California campus at Berkeley, for example, has the following corrections:

Jan. 10, + 17 min		July 19, + 15 min	
Feb. 9, + 23	"	Aug. 18, + 18	"
Mar. 11, + 19	"	Sept. 17, + 4	"
Apr. 10, + 8	"	Oct. 17, − 5	"
May 20, + 5	"	Nov. 16, − 6	"
June 19, + 10	"	Dec. 16, + 5	"

Modern clocks depend for their regulation upon the swinging of a pendulum or the oscillation of a balance wheel. Examples of these are the grandfather clock and the modern wrist watch. Such devices will be discussed later under the heading of vibrations and waves, Chap. 18. Probably the most accurate watch made today employs the vibrations of a tuning fork. Electric clocks commonly found in the city homes of today are run by tiny electric motors. The speed of these motors is controlled at the city's electric power plant by controlling the frequency of the alternating current supplied to the power lines leading to the house. The master clock at the power house is frequently a pendulum clock.

1.5 TIME AND THE MEAN SOLAR DAY Three kinds of time are recognized by astronomers: first, *sidereal time;* second, *apparent solar time;* and third, *mean solar time*. The last is the time used in civil life. If at any given point on the earth's surface we adjust the gnomon of a sundial to lie in the North and South vertical plane, the time interval between two successive transits of the sun's shadow over the 12 o'clock mark is called the *apparent solar day*. For several reasons, one being that the earth's orbit around the sun is elliptical, this interval of time varies slightly from day to day. An apparent solar day in December is about one minute longer than an apparent solar day in September. It is clear, therefore, why in this day of accurate timepieces we do not regulate our clocks to apparent solar time.

The *average* length of all apparent solar days throughout a solar year is called the *mean solar day*. Since there are 3600 seconds in one hour, 24 hours in one day, and 365.241 days in one tropical year, there are approximately 3600 × 24 × 365.241, or 31,556,822 seconds in one year. The *tropical year* is defined as the time between two successive passages of the vernal equinox by the sun. Since tropical years vary slightly, the *second* is defined as*

$$\frac{1}{31{,}556{,}825.9747} \text{ part of tropical year 1900}$$

The second is abbreviated s.

For astronomical purposes a different time scale known as sidereal time is used. There is one more sidereal day in one solar year than there are mean solar days. One solar year equals 366.2421 sidereal days. The reason for the additional day is that in making one complete turn around the sun in its orbit the earth has actually made 366.2421 revolutions with respect to the fixed stars. The sidereal second as ticked off by an astronomical clock is therefore slightly shorter than the second given by an ordinary clock keeping mean solar time.

1.6 METRIC UNITS Nearly all scientific experiments, in the United States as well as abroad, are performed using metric units. In these units distance is usually measured in millimeters, centimeters, meters, or kilometers; mass is measured in grams or kilograms; and time is measured in seconds, minutes, or hours. In metric units the abbreviation *cgs* means *centimeter, gram, second*, and *mks* means *meter, kilogram, second.*

The English system, as already referred to, uses the foot, yard, and mile as units of length; the ounce, pound, and ton as units from which to calculate mass; and the second as the unit of time. We Americans inherited the English system with all of its cumbersome fractions, and

* International standard adopted on October 14, 1960, by the General Conference on Weights and Measures, at Paris, France.

because it is so firmly ingrained in our civil life it will be a difficult task to ever change completely over to the simpler metric system. There is a strong movement today, however, advocating such a change.

The chief advantage of the metric system over the English system is that all units are divided into ten or one hundred parts. This enables fractional distances and masses to be expressed as decimals. Decimals, it is well known, are easier to manipulate in the addition, subtraction, multiplication, and division of two or more quantities.

PROBLEMS

1. How many meters are there in 12 ft? [Ans. 3.658 m.]*

2. Express the distance to the moon in centimeters, if the distance is 239,000 mi.

3. An automobile is 14 ft 4 in. in length. Find the length in meters. [Ans. 4.37 m.]

4. Find the height of a man in meters if he says he is 6 ft 3 in. tall.

5. Find the number of square meters in one square mile. [Ans. 2,590,000 m^2.]

6. How many square meters are there in 5 acres? One square mile contains 640 acres.

7. An 85-Kg man weighs how many pounds? [Ans. 187.5 lb.]

8. An automobile weighs 3600 lb. What is the equivalent mass in Kg?

9. How many meters are there in 50 mi? [Ans. 80,500 m.]

10. How many miles are there in 32 Km?

11. A gallon of water weighs 8 lb. What is the equivalent mass in grams? [Ans. 3629 g.]

12. What weight in pounds is equivalent to a mass of 60 Kg?

13. What mass in Kg is equivalent to a 25-lb weight? [Ans. 11.34 Kg.]

14. How many seconds are there in one full day?

15. How many seconds are there in a 31-day month? [Ans. 2,678,000 s.]

16. Find the number of seconds in 6 hr and 30 min.

17. A jet passenger plane weighs 276,000 lb. What is its mass in Kg? [Ans. 125,200 Kg.]

18. What is the height of a man in millimeters if he is 5 ft 10 in. tall?

19. Find the number of millimeters in a fathom. [Ans. 1829.]

20. What is the circumference in meters of a ball 9 in. in diameter?

* All numerical answers beginning with 1, 2, or 3 should be given to four significant figures. Answers beginning with 4 through 9 should be given to three significant figures unless otherwise specified. See Appendix XIX.

2 | VELOCITY, SPEED, AND ACCELERATION

The first ten chapters of this book deal with many natural phenomena classified as *mechanics.* To launch a rocket, to dive to the bottom of the deepest ocean, to set a broken bone, to study the structure of atoms and molecules, to build a legal case involving an automobile accident, or to vote intelligently on government issues, such as those involving atomic energy, one must have some knowledge of the subject of mechanics.

Mechanics is defined as that branch of physics dealing with the motions or states of material bodies. It is generally divided into three parts: *kinematics*, *dynamics*, and *statics.* Some of the concepts and principles to be described and classified under each of these headings are as follows:

Kinematics
- *speed*
- *velocity*
- *acceleration*
- *angular velocity*

Dynamics
- *force*
- *impulse*
- *momentum*
- *kinetic energy*

Statics
- *equilibrium*
- *balanced forces*
- *composition of forces*
- *resolution of forces*

Kinematics is concerned with kinds of motion, but not with the properties of the object that is moving. Dynamics is concerned with causes for changes in motion, and it involves the masses of bodies and how they react to forces acting upon them. Statics deals with bodies in equilibrium. As an introduction to the kinematics of motion, the elementary concepts of *speed* and *velocity* will be considered first.

2.1 SPEED AND VELOCITY *Speed and velocity are both defined as the rate of change of position.* Since by *change in position* of a body is meant the distance traveled, the definition of average speed can be written

$$\textbf{av. speed} = \frac{\textbf{distance traveled}}{\textbf{time}} \tag{2a}$$

As an algebraic equation

$$\boxed{v = \frac{x}{t}} \tag{2b}$$

where v is the average speed, x the distance traveled, and t the elapsed time.

Change of position is illustrated in Fig. 2A. A car traveling with uniform speed along a straight line passes the point A at one instant and the point B at some later instant of time. In substituting in Eq. (2b) the distance traveled is AB, and the time interval between A and B is t seconds.

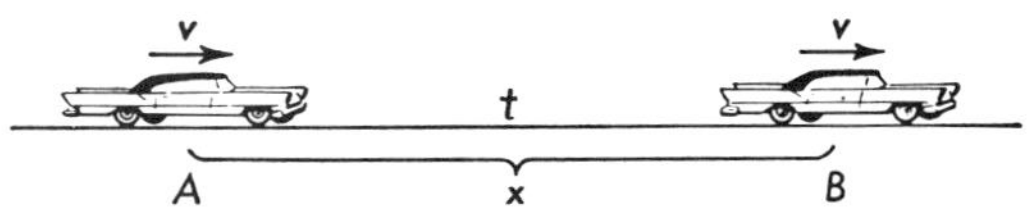

Fig. 2A. Diagram of a body moving with constant velocity.

When a body moves along a straight line it is proper to specify its direction and speak of its *velocity* along that line. Frequently the *direction* is implied but not explicitly specified. Motion along a straight or curved path, with no reference being made to direction, can be referred to as *speed.* The words speed and velocity are commonly used interchangeably.

The magnitude of any velocity measurement is expressed in derived units. When, for example, one says that *the speed or velocity of a car is 35 miles per hour*, the magnitude may be written

$$v = 35\,\frac{\text{mi}}{\text{hr}}$$

Miles in the numerator and *hours* in the denominator represent fundamental units of *distance* and *time*, respectively. Mathematically we can write

$$v = 35\,\frac{1\text{ mi}}{1\text{ hr}} \tag{2c}$$

which means that v is equal to 35 times 1 mile, divided by 1 hour. The values of unity are seldom written in front of the units, but when needed for clarity they may be inserted as in Eq. (2c).

EXAMPLE 1. A man takes two hours to drive his car to a distant city 120 kilometers due east. What is his average velocity?

SOLUTION. Here the distance traveled $x = 120$ Km, and the time of travel $t = 2$ hr. The velocity, therefore, is

$$v = \frac{120\text{ Km}}{2\text{ hr}} = 60\,\frac{\text{Km}}{\text{hr}}$$

The answer is 60 Km/hr due east, and is read *sixty kilometers per hour due east.* The units are just as important as the numbers and must be included in the answer.

If in the answer to the above example the distance of 120 Km is expressed in meters, and the time of 2 hr is changed to seconds, we may write,

$$v = \frac{120{,}000\text{ m}}{7{,}200\text{ s}} = 16.67\,\frac{\text{m}}{\text{s}} \tag{2d}$$

If in the answer to the above example the fundamental unit of 1 Km is changed to meters, and the fundamental unit of 1 hr is changed to seconds, we may write

$$v = 60\,\frac{1\text{ Km}}{1\text{ hr}} = 60\,\frac{1000\text{ m}}{3600\text{ s}}$$

$$v = 16.67\,\frac{\text{m}}{\text{s}}$$

EXAMPLE 2. A car is traveling along a straight and level road with a speed of 60 mi/hr. Find the equivalent speed in ft/s.

SOLUTION. The distance of 1 mi in the numerator is replaced by 5280 ft, and the time of 1 hr in the denominator is replaced by 3600 s.

$$v = 60\,\frac{\text{mi}}{\text{hr}} = 60\,\frac{5280\text{ ft}}{3600\text{ s}} = 88\,\frac{\text{ft}}{\text{s}}$$

This result, that

$$60\,\frac{\text{mi}}{\text{hr}} = 88\,\frac{\text{ft}}{\text{s}} \tag{2e}$$

should be remembered as it is very useful in the solving of many practical problems. A speed of 120 mi/hr, for example, is equivalent to 2×88 of 176 ft/s, while a speed of 30 mi/hr is equivalent to $\frac{1}{2} \times 88$ or 44 ft/s.

EXAMPLE 3. A toy train running along a straight track at constant speed requires 8 s to travel a distance of 20 m. Find the speed.

SOLUTION. Since $x = 20$ m and $t = 8$ s, the velocity

$$v = \frac{20\text{ m}}{8\text{ s}} = 2.5\,\frac{\text{m}}{\text{s}}$$

The answer is read *two point five meters per second.* To change this answer to centimeters per second the unit 1 m in the numerator is changed to 100 cm, and the answer becomes

$$v = 2.5 \frac{\text{m}}{\text{s}} = 2.5 \frac{100 \text{ cm}}{\text{s}} = 250 \frac{\text{cm}}{\text{s}}$$

2.2 DISTANCE TRAVELED If the speed of a body is known, the distance traveled can be calculated for any given interval of time. For such problems Eq. (2b) is conveniently changed by solving for x. Multiplying both sides of the equation by t does not alter the equality.

$$vt = \frac{xt}{t}$$

From this we obtain $vt = x$, or

$$\boxed{x = vt} \qquad (2f)$$

EXAMPLE 4. If a body moves with a speed of 45 cm/s how far will it travel in 2 min?

SOLUTION. By using Eq. (2f), we obtain

$$x = 45 \frac{\text{cm}}{\text{s}} \times 2 \text{ min} = 90 \frac{\text{cm min}}{\text{s}}$$

In order to eliminate *time* units in this answer they must both be expressed in the same units. To do this, the minutes may be changed to seconds as follows:

$$x = 45 \frac{\text{cm}}{\text{s}} \times 120 \text{ s} = 5400 \text{ cm}$$

Note that with s in the numerator and s in the denominator $\frac{s}{s} = 1$, leaving *cm* in the answer as the unit of length. This illustrates a common practice that is usually followed in the solving of all problems. Like quantities are expressed in the same units.

Dividing both sides of Eq. (2f) by v and setting $\frac{v}{v} = 1$ on the right-hand side gives

$$t = \frac{x}{v} \qquad (2g)$$

an equation for the time of travel in terms of x and t.

EXAMPLE 5. If a car travels with an average speed of 30 mi/hr, how long will it take to go 175 mi?

SOLUTION. By using Eq. (2g), we obtain

$$t = \frac{x}{v} = \frac{175 \text{ mi}}{30 \text{ mi/hr}} = 5.83 \text{ hr}$$

2.3 VECTORS AND SCALARS Nearly all physical measurements, whether they are made with the simplest of instruments or with the most complex of apparatus, may be classified as *vector* or *scalar* quantities. *Measurable quantities that have magnitude and direction are called vectors.* Examples of vector quantities are *displacement, velocity, acceleration,* and *force. Measurable quantities that have magnitude only are called scalars.* Examples of scalar quantities are *volume, area, density,* and *mass.*

The importance of this seemingly trivial distinction between quantities that have direction and those that do not is realized when, in solving certain problems, the simple process of the addition of two or more like quantities becomes necessary.

No difficulty is generally encountered with scalars since such quantities are added algebraically. For example, in the addition of volumes, the sum of 2 gallons and 3 gallons is 5 gallons. The addition of two vectors, on the other hand, is more complicated and requires a special process called *vector addition,* and will be treated in detail in Chap. 5.

2.4 SPEED AND VELOCITY Although the terms *speed* and *velocity* are often used synonymously, strictly speaking *speed is a scalar quantity* and *velocity is a vector quantity.* In the last section it was explained that vector quantities have magnitude and direction, whereas scalars have magnitude only.

Speed is a term applied only to the magnitude of velocity and does not specify the direction of motion. When an object is moving along a straight line, its *speed* and *velocity* are numerically equal. If, however, the speed along a curved path is constant, the velocity is not considered to be constant because of its changing direction.

TABLE 2A. Conversion Factors for Speed and Velocity

Velocity	m/s	ft/s	Km/hr	mi/hr	knots
1 m/s =	1	3.281	3.600	2.240	1.940
1 ft/s =	0.30480	1	1.0973	0.6818	0.5921
1 Km/hr =	0.27778	0.9113	1	0.6214	0.5396
1 mi/hr =	0.44704	1.4667	1.6093	1	0.8684
1 knot =	0.51480	1.689	1.853	1.152	1

Speed and velocity always have the *dimensions of length divided by time.*

EXAMPLE 6. Change 30 mi/hr to kilometers per hour.

SOLUTION. From the table, 1 mi/hr in the left-hand column is (read across to the fourth column) equal to 1.6093 Km/hr. Therefore,

$$30 \times 1.6093 = 48.279 \frac{\text{Km}}{\text{hr}}$$

By dropping off the last two figures the answer to three significant figures is

$$30 \frac{\text{mi}}{\text{hr}} = 48.3 \frac{\text{Km}}{\text{hr}}$$

The *knot* is a nautical unit of speed, about 15% larger than speed in miles per hour. It is *not* correct to say the speed of a ship is 10 knots/hr; it is correct to say the speed is 10 knots.

2.5 UNIFORM ACCELERATION

Whenever the speed or velocity of a body changes, the body is said to be accelerated. *Acceleration is defined as the rate of change of velocity.* A car "picking up speed" has a *positive acceleration*, while another slowing down has a *negative acceleration*. Standing still or moving with constant velocity a car has zero acceleration.

Consider as an illustration of accelerated motion the car shown in Fig. 2B. Due to a constantly acting force, exerted by the motor through the drive wheels, this car is constantly accelerated as it moves along the straight line *AB*. As it passes *A* it has a relatively low velocity v_0, while farther along its path at a point *B* it is moving faster and has a velocity v. The *initial velocity* is called v_0 and the *final velocity* is called v.

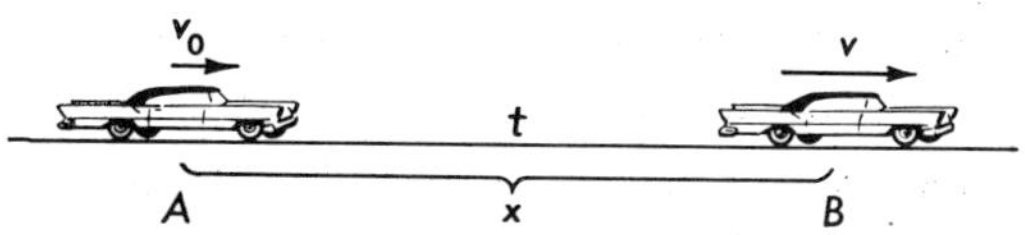

Fig. 2B. A car is accelerated for a period of t seconds.

If the time required to go from *A* to *B* is t seconds, the acceleration, by the definition given above, is schematically written

$$\textbf{acceleration} = \frac{\textbf{final vel.} - \textbf{initial vel.}}{\textbf{time}}$$

or algebraically as

$$\boxed{a = \frac{v - v_0}{t}} \qquad \text{(2h)}$$

EXAMPLE 7. Suppose at A, in Fig. 2B, that the velocity of the car is 5 m/s, that at B it has increased to 20 m/s, and that it takes 5 s to go from A to B. What is the acceleration?

SOLUTION. By direct substitution in Eq. (2h), we obtain

$$a = \frac{20 \text{ m/s} - 5 \text{ m/s}}{5 \text{ s}} = \frac{15 \text{ m/s}}{5 \text{ s}} = 3 \frac{\text{m}}{\text{s}^2}$$

The answer is read *three meters per second per second* and means that the velocity increases 3

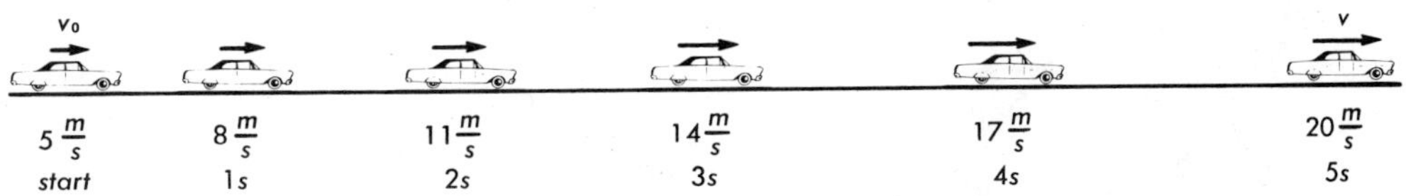

Fig. 2C. Diagram of accelerated car in Example 7.

m/s every second of time. Initially the velocity is 5 m/s. An increase of 3 m/s means that at the end of one second the velocity is 8 m/s, at the end of 2 s it is 11 m/s, at the end of 3 s it is 14 m/s, at the end of 4 s it is 17 m/s, and at the end of 5 s it is 20 m/s. See Fig. 2C. A constant acceleration is, therefore, one in which *the velocity changes by equal amounts each second.*

When a body is slowing down, the initial velocity is greater than the final velocity, and the acceleration, as given by Eq. (2h), is negative.

EXAMPLE 8. In going up a long steep hill a car slows down from 30 m/s to 10 m/s in 2 min. Find the acceleration.

SOLUTION. By direct substitution in Eq. (2h), we obtain,

$$a = \frac{10\ \text{m/s} - 30\ \text{m/s}}{2\ \text{min}} = -\frac{20\ \text{m/s}}{2\ \text{min}} = -10\,\frac{\text{m/s}}{\text{min}}$$

The minus sign means that the velocity decreases by 10 m/s each minute of time. During the first minute it drops from 30 m/s to 20 m/s, and during the next minute from 20 m/s to 10 m/s.

2.6 STARTING FROM REST When a body starts from rest and undergoes a constant acceleration, the initial velocity v_0 as given in Eq. (2h) is zero, i.e., $v_0 = 0$. The acceleration a is then given by the special equation

$$\boxed{a = \frac{v}{t}} \tag{2i}$$

When transposed, this takes the form

$$v = at \tag{2j}$$

EXAMPLE 9. An airplane starting from rest at one end of a runway acquires its takeoff speed of 60 mi/hr in 8 s. What is its acceleration?

SOLUTION. By applying Eq. (2i), we obtain

$$a = \frac{60\ \text{mi/hr}}{8\ \text{s}} = 7.5\,\frac{\text{mi/hr}}{\text{s}}$$

The answer is read *seven point five miles per hour per second.*

Remembering that 60 mi/hr is equivalent to 88 ft/s, this same acceleration can be expressed as

$$a = \frac{88\ \text{ft/s}}{8\ \text{s}} = 11\,\frac{\text{ft}}{\text{s}^2}$$

The answer is read *eleven feet per second per second*, or *eleven feet per second squared.*

2.7 ACCELERATION DOWN AN INCLINED PLANE A study of acceleration, and an experimental proof of the formulas given above, may be made with an inclined plane. In Fig. 2D a marble is shown rolling down a groove in the top of a long inclined plane. As the marble moves down with ever-increasing speed, its position is noted at each tick of a *metronome*, or *seconds clock*. In the particular experiment, diagrammed here, the angle of the plane has been adjusted so that the distance traveled in the first second is 20 cm. After two seconds the total distance traveled is then measured and found to be 80 cm. After 3 s, it is 180 cm, etc. These measurements, when tabulated for the first 5 s, appear as in the first two columns of Table 2B.

To find the velocity attained at the end of each second, a shorter horizontal track is placed first in position 1, then at 2, 3, 4, etc. In each of these positions the distance traveled along the horizontal track in 1 s is a direct measure of the velocity acquired from the incline. Thus, at the end of 1 s the velocity is 40 cm/s, at the end of 2 s the velocity is 80 cm/s, etc. These measured values from the experiment are tabulated in column three.

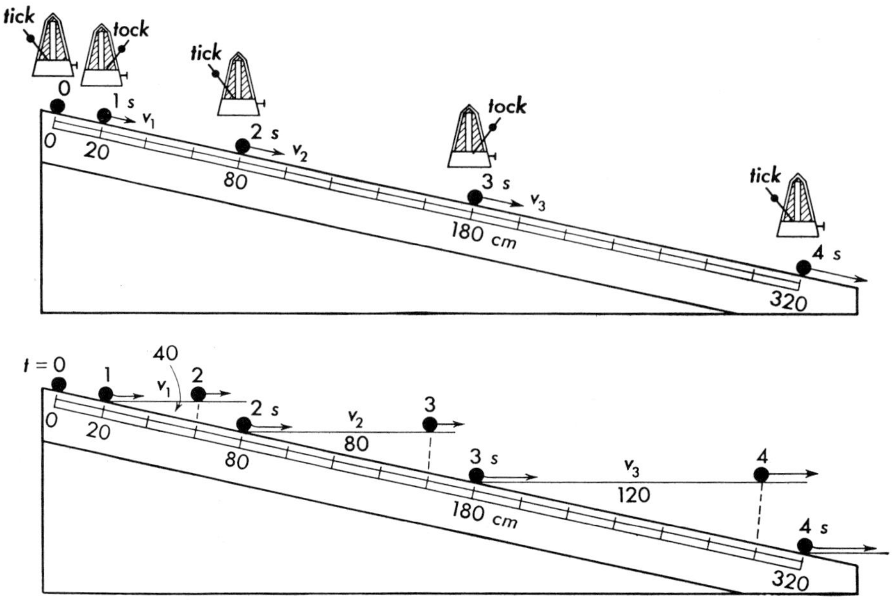

Fig. 2D. Diagram of an inclined-plane experiment showing the numerical distances traveled by a marble for each second of time as it rolls down the plane.

A careful study of the first and third columns shows that the velocity v is directly proportional to the time t.

$$v \propto t$$

To write this as an algebraic equation, the proportionality sign is replaced by a constant k.

$$v = kt \qquad (2k)$$

The value of k can be determined from the experiment as follows. Divide each value of v in the third column by the corresponding time t in the first column and obtain 40 cm/s² as a common answer. Therefore, $k = 40$ cm/s² and

$$v = 40t \qquad (2l)$$

By referring to Eq. (2j) it is seen that the constant 40 is none other than the *acceleration*. In other words it is the increase in velocity per second of time. Each second, the marble on the incline increases its velocity by 40 cm/s. By replacing 40 in Eq. (2l) by a, we obtain

$$v = at$$

the Eq. (2j).

To find a relation for the distance traveled along the inclined plane, it is observed that x in

TABLE 2B. Experimental and Calculated Values Determined from the Inclined Plane Experiment Shown in Fig. 2D

(1) Time t s	(2) Distance x cm	(3) Velocity v cm/s	(4) $\frac{v}{t}$	(5) at	(6) $\frac{x}{t^2}$	(7) t^2	(8) $\frac{1}{2}at^2$
0	0	0	—	0	—	0	0
1	20	40	40	40	20	1	20
2	80	80	40	80	20	4	80
3	180	120	40	120	20	9	180
4	320	160	40	160	20	16	320
5	500	200	40	200	20	25	500

the second column is proportional to the square of t listed in the seventh column.

$$s \propto t^2, \quad \text{or} \quad s = kt^2$$

To find the constant of proportionality k, each distance x when divided by the corresponding value of t^2 gives 20 as shown in column six. Since this is just half of the acceleration, we may write $\frac{1}{2}a$ for the constant k and obtain

$$\boxed{x = \tfrac{1}{2}at^2} \qquad (2m)$$

QUESTIONS

1. Briefly explain the difference between speed and velocity.

2. Define each of the following: (a) speed, (b) velocity, (c) scalar, (d) vector, (e) acceleration, and (f) knot.

PROBLEMS

1. A car travels 320 Km in 4 hr and 30 min. Find the average speed. [Ans. 71.1 Km/hr.]

2. A train makes the trip from Chicago to New York, a distance of 1500 Km in 16 hr. Find the average speed in (a) Km/hr, and (b) m/s.

3. A racing car travels 620 Km/hr on a straight and level road. What is its speed in (a) mi/hr, and (b) ft/s? [Ans. (a) 385.3 mi/hr, (b) 565 ft/s.]

4. A DC-8 jet airliner has a cruising speed of 980 Km/hr. Find its cruising speed in (a) m/s, (b) mi/hr, and (c) ft/s.

5. A ship at sea maintains a speed of 22 knots. What is its speed in Km/hr, and (b) mi/hr? [Ans. (a) 40.8 Km/hr, (b) 25.34 mi/hr.]

6. If a ship has a speed of 24 knots, find its speed in (a) m/s, and (b) ft/s.

7. A plane flies 560 mi in 1 hr and 40 min. Find the average speed in (a) mi/hr, and (b) Km/hr. [Ans. (a) 336.0 mi/hr, (b) 541 Km/hr.]

8. A ship making 18 knots travels east for 24 hr. How far has it traveled in (a) mi, and (b) Km?

9. How long will it take a plane cruising at 620 mi/hr to fly from Honolulu, Hawaii, to San Francisco, California? The distance is 2200 mi. [Ans. 3.548 hr = 3 hr and 33 min.]

10. How long will it take a car averaging 48 mi/hr to travel from New York to California, a distance of 3000 mi?

11. How long will it take a man to walk 50 Km if his average speed is 4.5 Km/hr? [Ans. 11 hr and 6.6 min.]

12. A car starting from rest acquires a speed of 50 Km/hr in 8 s. What is its acceleration in m/s^2?

13. Starting from rest at the end of a runway a plane undergoes a constant acceleration of 5 m/s^2. Find its velocity at the end of 15 s. [Ans. 75 m/s.]

14. A jet plane starting from rest at the end of a runway acquires a speed of 260 Km/hr in 32 s. Find the acceleration in m/s.

15. A train moving with a speed of 2 m/s is given a constant acceleration. If at the end of 3 min it has a speed of 26 m/s, what is its acceleration? [Ans. 0.1333 m/s^2.]

16. A passenger plane starts from rest at the end of a runway and maintains a constant acceleration of 1.4 m/s^2. What will be its speed at the end of 30 s?

17. Starting from rest, a racing car acquires a speed of 120 Km/hr in 1 min. Find (a) the acceleration, and (b) the distance traveled. [Ans. (a) 0.556 m/s^2, (b) 1000 m.]

18. A motor boat starting from rest acquires a speed of 50 Km/hr in 12 s. Find (a) the acceleration, and (b) the distance traveled.

19. A car traveling at 80 Km/hr has the brakes applied, bringing the car to rest in 5 s. Find (a) the acceleration, and (b) the distance traveled. [Ans. (a) -4.44 m/s^2, (b) 55.6 m.]

20. A train traveling at 100 Km/hr has its brakes applied, bringing it to a stop in 1 min 10 s. Find (a) the acceleration, and (b) the distance traveled.

3 / NEWTON'S LAWS OF MOTION

The laws and formulas that describe the *speed*, *velocity*, and *acceleration* of moving bodies, are classified as that branch of mechanics called *kinematics*. These laws of motion consider neither the size nor weight of a body, nor the material or structural details of the system.

In this chapter we are concerned with the causes for the motion of a body. In order to go beyond simple kinematics we shall introduce two new concepts, *mass* and *force*.

To Isaac Newton * goes the credit of having been the first to systematically introduce these concepts into mechanics and to formulate the fundamental laws governing all motion. These laws constitute the fundamental principles of the branch of mechanics called "dynamics" and resolve themselves into three laws commonly referred to as "Newton's Laws of Motion."

3.1 NEWTON'S FIRST LAW OF MOTION *A body at rest or in uniform motion will remain at rest or in uniform motion unless some external*

* SIR ISAAC NEWTON (1642–1727), English physicist and mathematician, was born in England on Christmas Day, 1642. He obtained his education at Trinity College, Cambridge, where in 1665 he was awarded the Master of Arts degree. At just this time the prevalence of the Black Plague forced him into retirement at his old home in Woolsthorpe where in the two years 1665 and 1666 his genius became manifest. In this period he invented the calculus, discovered the composition of white light, and conceived the idea of universal gravitation. In the years that followed, he published much of his work on optics and developed his ideas on gravitation, which were published in 1687 in his *Principia*. At the age of fifty he suffered from a nervous breakdown, and never again did any extensive scientific work, but devoted his time to theology. He became very absent minded and slovenly in his personal appearance. It is needless to say, therefore, that he never married. His *Principia* is considered to be one of the greatest monuments of the human intellect. In it, Newton lays the foundations of mechanics which are broad enough to include all future developments, and these he applies to the motions of heavenly bodies under the law of gravitation. He was elected to Parliament, was president of the Royal Society for twenty-five years, and was knighted by Queen Anne in 1705. The greatness of this modest man is illustrated by his last words: "If I have seen farther than others, it is by standing on the shoulders of giants."

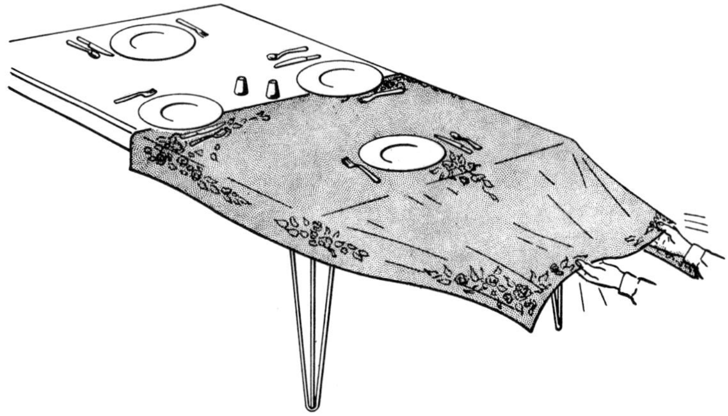

Fig. 3A. A tablecloth can be pulled from a table without dislodging the dishes.

force is applied to it. This law can be demonstrated by many simple experiments. In Fig. 3A a tablecloth is shown being jerked from under the dishes and silverware on a table without disturbing their original setting. In Fig. 3B a

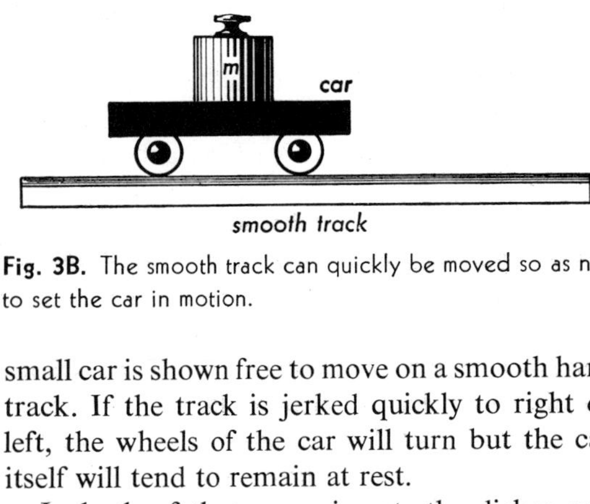

Fig. 3B. The smooth track can quickly be moved so as not to set the car in motion.

small car is shown free to move on a smooth hard track. If the track is jerked quickly to right or left, the wheels of the car will turn but the car itself will tend to remain at rest.

In both of these experiments the dishes and the car are at rest. They tend to remain at rest because the sudden motion of the objects on which they are resting exert no large force on them for any appreciable length of time. Actually, the dishes and the car do move slightly due to small frictional forces between the moving parts in contact. The tendency for each body to remain at rest is due to that property, common to all material bodies, called *inertia. The inertia of a body may be defined as that property of a body which tends to resist a change in its state of rest or motion. Mass is defined as a quantitative measure of inertia.* In the metric system, mass is measured in grams or kilograms.

A third experiment illustrating inertia, and Newton's First Law, is illustrated in Fig. 3C. A

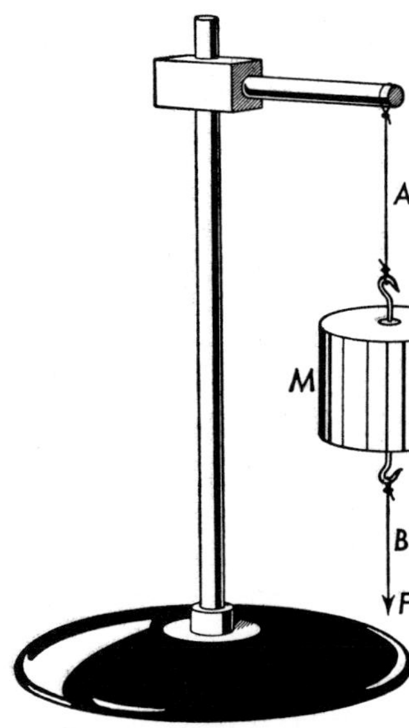

Fig. 3C. A slow, steady pull at F breaks the thread at A, while a quick pull at F breaks the thread at B.

small mass M of 1000 g is suspended by a fine thread A, then pulled downward by another piece of the same thread B. If the force F is a slow steady pull, the thread will always break at A; whereas if it is a sudden jerk, it will always break at B. In the first case, the tension in the upper thread is greatest and is equivalent to the force F plus the weight of the mass M. In the second case the force F is momentarily very large, causing the thread to break before the mass M has had time to move down far enough to stretch and break the upper thread. It is the inertia of M that permits the very large force F to be momentarily applied to the lower thread only.

If the car in Fig. 3B is started rolling along the track, Newton's First Law states that it should keep moving with the same velocity. The law, of course, neglects friction, for we know that friction will eventually bring the car to rest. The greater the friction, the sooner it will stop. The smaller the friction, the longer it will move. If friction could be entirely eliminated, the *inertia* of the car would keep it moving indefinitely with constant velocity.

3.2 NEWTON'S SECOND LAW OF MOTION

Newton's First Law, concerning bodies at rest or moving with constant velocity, assumes that no forces are acting to change their state. Newton's Second Law, however, assumes that such a force is acting and describes the resulting *change in motion*. In Fig. 3D, for example, a small car of mass

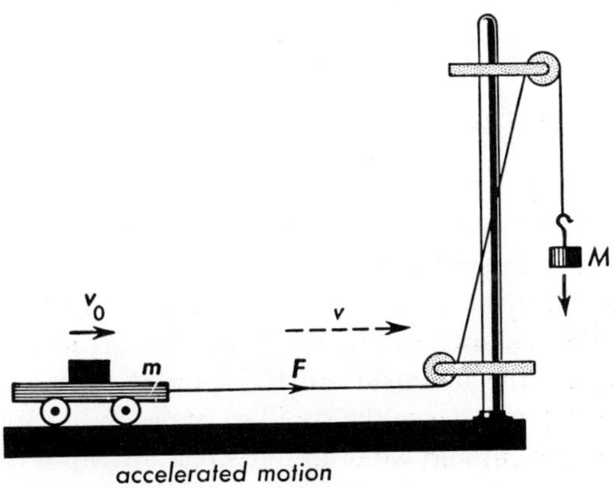

Fig. 3D. A small car being accelerated by a constant force.

m is acted upon by a constant force F. The force is produced by the pull of gravity on the mass M, and is transmitted to the car by a cord passing over pulleys as indicated. If the car is initially at rest, this force will start it moving; if it is already moving with a velocity v_0, the force will increase its velocity. Thus the car is accelerated.

Newton's Second Law of Motion may be stated as follows: *When a body is acted upon by a constant force, its resulting acceleration is proportional to the force and inversely proportional to the mass.*

Symbolically,

$$a \propto \frac{F}{m}$$

This law is stated as a proportionality because it holds regardless of the units in which each of the three quantities are measured. If the units are properly chosen, an equal-sign may be inserted and the law written as an equation,

$$a = \frac{F}{m}$$

By transposing m to the other side of the equation, we obtain the so-called "force equation," which forms the basis of so many principles in mechanics.

$$F = ma \qquad \text{(3a)}$$

force = mass × acceleration

To the above statement of Newton's Second Law should be added the statement that *the acceleration takes place in the direction of the acting force.*

EXAMPLE 1. Neglecting friction, find what constant force will give a mass of 50 Kg an acceleration of 5 m/sec^2?

SOLUTION. By substituting directly into the force equation, Eq. (3a), we obtain

$$F = 50 \text{ Kg} \times 5 \frac{\text{m}}{\text{s}^2} = 250 \frac{\text{Kg m}}{\text{s}^2}$$

The answer is a force of 250 Kg m/s^2. Thus force is not as simple a concept as it might seem at first hand; it involves all three of the fundamental units, *length*, *mass*, and *time*.

According to the *MKS* system (meter-kilogram-second system), unit force is called the newton in honor of Sir Isaac Newton. *The newton is defined as that force which, applied to a mass of 1 Kg, will give it an acceleration of 1 m/s².*

$$1 \text{ newton} = 1 \text{ Kg} \times 1 \frac{\text{m}}{\text{s}^2} \qquad (3b)$$

As a unit of force, the newton is represented by the capital letter N.

A second unit of force also common in physics is the dyne. *The dyne is a force which acting on a 1-g mass will give it an acceleration of 1 cm/s².* In the *cgs* system (*centimeter-gram-second* system), the units of Eq. (3a) become

$$1 \text{ dyne} = 1 \text{ g} \times 1 \frac{\text{cm}}{\text{s}^2}$$

There is a growing preference among scientists and teachers of physics to use the kilogram and meter in place of the gram and centimeter as units of mass and length.

EXAMPLE 2. Neglecting friction, find what constant force in newtons will give a mass of 4 Kg an acceleration of 3.8 m/s²?

SOLUTION. By applying the force equation, $F = ma$, we obtain

$$F = 4 \text{ Kg} \times 3.8 \frac{\text{m}}{\text{s}^2} = 15.2 \frac{\text{Kg m}}{\text{s}^2} = 15.2 \text{ N}$$

The answer is a constant force of 15.2 N.

Dynes and *newtons* are absolute units of force. They arise from the force equation when the absolute units of *mass* and *time* are used as they are above.

Since 1 Kg = 1000 g, and 1 m = 100 cm,

$$1 \text{ newton} = 100{,}000 \text{ dynes}$$

3.3 THE ENGINEERING SYSTEM OF UNITS The engineer seldom uses the metric system of units described above; he finds it more convenient to measure *force* in *pounds*. To apply Newton's Second Law of Motion, as expressed by the force equation $F = ma$, it is necessary to introduce a unit of mass called the *slug*. See Fig. 3E. *One slug is defined as that mass which, when acted upon by a force of 1 lb, is given an acceleration of 1 ft/s².*

$$1 \text{ lb} = 1 \text{ slug} \times 1 \frac{\text{ft}}{\text{s}^2} \qquad (3c)$$

The mass of a body in slugs may be obtained by dividing the weight of that body in pounds by 32.

We will see in Sec. 4.3 that the number 32 is derived from the acceleration of freely falling bodies.

The difference between the mass of an object and its weight will be treated in detail in Sec. 6.1.

EXAMPLE 3. What horizontal force in pounds applied to the tongue of a small wagon weighing 176 lb will give it an acceleration of 6 ft/s²?

SOLUTION. By applying Newton's Second Law, Eq. (3a), we obtain

$$F = \frac{176}{32} \text{ slugs} \times 6 \frac{\text{ft}}{\text{s}^2}$$

$$F = 5.5 \text{ slugs} \times 6 \frac{\text{ft}}{\text{s}^2} = 33 \text{ lb}$$

The answer is a constant force of 33 lb.

As given in Sec. 1.3, 1 Kg = the mass of a 2.2046 lb weight, and since the weight of a 1 Kg mass is mg = 9.80 N, we obtain the relation,

$$9.80 \text{ N} = 2.2046 \text{ lb}$$

or

$$1 \text{ N} = 0.22496 \text{ lb}$$

To four figures

$$1 \text{ N} = 0.2250 \text{ lb} \qquad (3d)$$

To illustrate how Newton's Second Law may always be relied upon, consider the following paradox. A spool of thread is placed upon the table as shown in Fig. 3F. With the thread leading off the underside of the spool a horizontal force F will cause the spool to move. Will it roll to the right and wind up the thread or will it roll to the left and unwind? The performance and

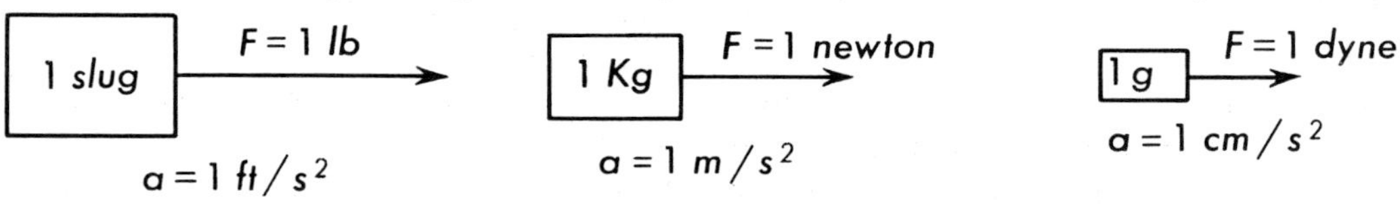

Fig. 3E. Unit force acting on unit mass produces unit acceleration.

Fig. 3F. If the thread is pulled to the right, which way will the spool roll?

explanation of this experiment are left as a problem for the student.

3.4 NEWTON'S THIRD LAW OF MOTION

Of Newton's three laws of motion, the third is perhaps the least understood. This is probably because it is seldom used in solving problems, and often when it is used it is incorrectly applied. The law states that *to every action force there is an equal and opposite reaction force.*

The principle of action and reaction may be illustrated by a bat striking a ball, Fig. 3G. During impact the bat exerts a force *F* on the ball, and the ball exerts an equal but opposite force *G* on the bat. The force *F* being exerted on the ball gives it an acceleration to the right, while the force *G* being exerted on the bat gives it an acceleration to the left. The ball speeds up during the impact and acquires a high velocity while the bat in the same time interval slows down to a lower velocity.

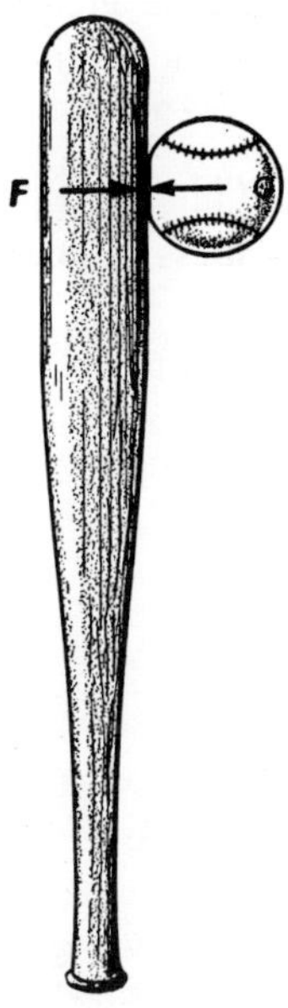

Fig. 3G. A bat at all times exerts a force on the ball equal in magnitude to the force the ball exerts on the bat.

Consider the second example of a block hanging by a cord as illustrated in Fig. 3H. The weight of the block *W* is the force with which the earth pulls downward on the block, while the equal and opposite force *X* is the upward force exerted by the block on the earth.

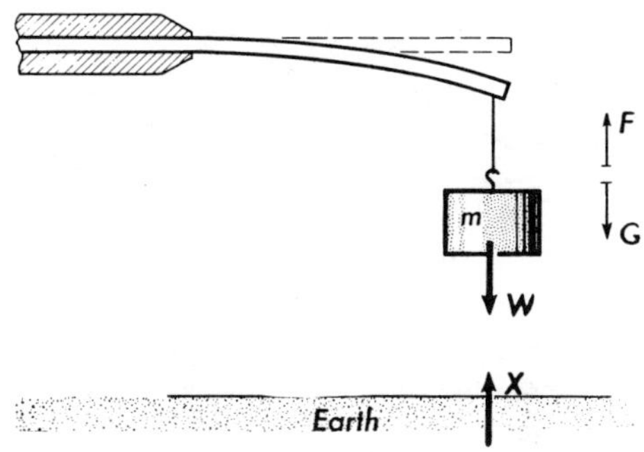

Fig. 3H. Diagram illustrating Newton's Third Law of Motion. Forces always exist in pairs.

In addition to this pair of forces, the block exerts a downward force *G* on the cord, while the cord pulls upward with the reaction force *F*. Although to many people these forces may seem confusing, it should be pointed out that Newton himself had some difficulty in applying his third law to certain problems. The difficulty arises from trying to apply action and reaction forces to the same body when in reality they apply to different ones. It is important to note that *the action force and the reaction force in Newton's Third Law of Motion act on different bodies.*

Whether a body is at rest or in motion, the state of that body depends upon the forces acting on it and not upon the forces it exerts on something else. So far as the body is concerned, the latter do not determine its motion.

3.5 THE TRAIN-AND-TRACK EXPERIMENT

Another illustration of Newton's Third Law is that of a train on a track. The drive wheels push back on the track with a force *B*, and the track pushes forward on the wheels with an equal and opposite force F. These two form an action and reaction pair. The train-and-track experiment is designed to show that both of these forces exist.

In Fig. 3I, the track is mounted on a large wheel with its axis of rotation vertical. With the track free to move, both forces of the pair are seen to be real, the track moves backward and the train moves forward. The track moves backward because the wheels exert a force B upon it in that direction, and the train goes forward because the track exerts a force upon it in that direction. If, when the train acquires a certain speed, the power is shut off, the force F vanishes—so also does B—and the train and track continue to move with constant speed.

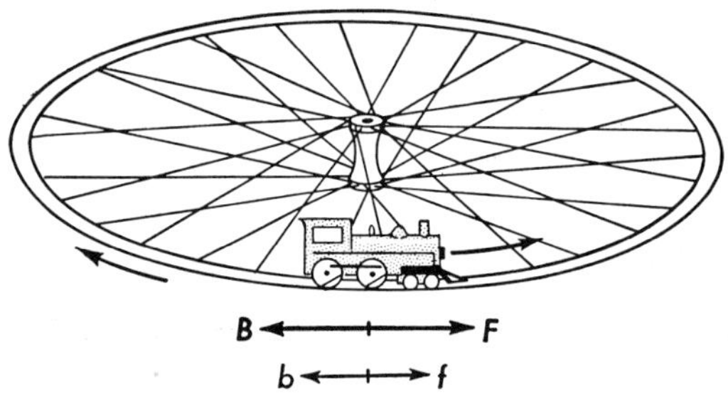

Fig. 3I. Demonstration of Newton's Third Law of Motion. The train moves forward, and the track, if free to move, moves backward.

In a practical case, however, the track is not only fastened down, but there is some frictional resistance to motion. Because of this friction, although the train is coasting, the track pushes backward on the wheels with a force b, and the wheels push forward on the track with an equal but opposite force f.

In order to keep a train moving with constant speed, a minimum force B great enough to overcome friction must continually be supplied to the locomotive drive wheels. The two forces acting on the train then are F and b, and these have a zero resultant since they are equal and opposite. There being no resultant force, there is no acceleration and the train continues to move with constant velocity.

To start the train moving, and to maintain an acceleration, F must be greater than b. Under these conditions the acceleration, by Newton's Second Law, is

$$F - b = ma$$

force = mass × acceleration

QUESTIONS

1. State Newton's First Law of Motion, and describe an experiment that could be performed to illustrate the law.

2. State Newton's Second Law of Motion. Write down the algebraic equation representing the law and indicate the meaning of each symbol.

3. State Newton's Third Law of Motion. Give an example and briefly explain it in terms of the forces involved.

4. Define or briefly explain each of the following: (a) newton, (b) dyne, (c) slug, (d) inertia, and (e) mass.

PROBLEMS

1. A 12-Kg mass is given an acceleration of 4.6 m/s^2. Calculate the applied force in newtons. [Ans. 55.2 N.]

2. A mass of 2.8 Kg is given an acceleration of 1.25 m/s^2. Find the applied force in (a) newtons, and (b) dynes.

3. A car weighing 4000 lb can accelerate at 3.6 ft/s^2. Find the effective driving force in pounds. [Ans. 450 lb.]

4. A force of 220 lb is applied to a car weighing 2000 lb. Find (a) the mass of the car in slugs, and (b) the acceleration.

5. A passenger plane weighing 480 tons is capable of an acceleration of 2.2 ft/s^2. Calculate (a) the mass of the plane in slugs, and (b) the force developed in pounds. [Ans. (a) 3×10^4 slugs, (b) 6.60×10^4 lb.]

6. A racing car has a mass of 2450 Kg. What force will give it an acceleration of 0.4 m/s^2?

7. A force of 30 newtons acts on a mass of 5.4 Kg. Find the acceleration. [Ans. 5.56 m/s^2.]

8. A 2.5 Kg mass is allowed to fall freely under the pull of gravity. Find the acceleration.

9. A 1000 Kg car starting from rest acquires a speed of 90 Km/hr in 1 min. Calculate (a) the acceleration, and (b) the force exerted on the pavement by the drive wheels. [Ans. (a) 0.417 m/s^2, (b) 417 N.]

10. A force of 40 newtons is applied to a mass of 5 Kg. Find (a) the acceleration, (b) the distance traveled in 10 s, and (c) the velocity at the end of 10 s.

11. A force of 50 N is applied to a mass of 8 Kg. Find (a) the acceleration, (b) the distance traveled in 12 s, and (c) the velocity acquired at the end of 12 s. [Ans. (a) 6.25 m/s^2, (b) 450 m, (c) 75 m/s.]

12. Starting from rest at the end of a runway a 2400 ton plane acquires its take-off speed of 180 mi/hr in 45 s. Find (a) the acceleration, (b) the total thrust of the motors, and (c) the distance traveled before it became airborne.

13. A 2200 Kg car starting from rest acquires a speed of 90 Km/hr in 8 s. Find (a) the acceleration, (b) the force applied through the wheels, and (c) the distance traveled. [Ans. (a) 3.125 m/s^2, (b) 6880 N, (c) 100 m.]

14. A locomotive weighing 400 tons starts from rest and acquires a speed of 60 mi/hr in 20 s. Find (a) the acceleration, (b) the force, and (c) the distance traveled.

15. A force of 360 N applied to a mass produces an acceleration of 5m/s^2. Find (a) the mass, (b) the velocity at the end of 6 s, and (c) the distance traveled in 20 s. [Ans. (a) 72 Kg, (b) 30 m/s, (c) 1000 m.]

16. A 650 Kg mass undergoes an acceleration of 2.75 m/s^2. Find (a) the applied force, (b) the velocity acquired in the first 10 s, and (c) the distance traveled the first 10 s.

4 FALLING BODIES, PROJECTILES, AND NEWTON'S LAW OF GRAVITATION

The newborn child learns many things during the first year of his life. One is that he always falls down, and never up. He also discovers that every thing else also falls down when dropped or released. If he runs too fast he may stumble and fall, and if he drops his glass of milk, he may look down and smile because it hit the floor.

As he grows older he continues to learn more about gravitation and the way different kinds of objects fall when they are set free.

When the physicist refers to a law applicable to a specified group of phenomena, he is referring actually to a formal description of the behavior of nature under prescribed conditions.

Neglecting friction, all bodies, large and small, fall with the same acceleration. This, the law of falling bodies, is a physical paradox for it contradicts the conclusion the average person might derive *a priori*. This is not to be wondered at, because centuries ago the great philosopher Aristotle* (384–322 B.C.) taught that heavy bodies fall proportionately faster than lighter ones.

It took the world nearly two thousand years to produce a challenger of Aristotle's scientific teachings. In the year 1590 Galileo† was pondering over the question of falling bodies and found

* ARISTOTLE (384–322 B.C.), famous Greek philosopher, logician, moralist, political thinker, biologist, and founder of literary criticism, spent his early years as a student and fellow worker with Plato. While practically all of Aristotle's teachings concerning physical principles are now known to be erroneous, his contributions to other fields of knowledge have placed him high among the great men of ancient Greece.

† GALILEO GALILEI (1564–1642), Italian mathematician, astronomer, and experimental physicist. At the early age of twenty-four Galileo wrote a treatise on the center of gravity of solids. This led the following year to his appointment as professor of mathematics at the University of Pisa. A rumor that a Dutch lens grinder had observed that two lenses used together make distant objects appear close at hand led Galileo to construct a telescope. Successful telescopes of greater and greater magnification enabled him eventually to observe, for the first time, the mountains on the moon, the major satellites of Jupiter, and sunspots. While at Pisa, Galileo carried out many experiments and public demonstrations of principles which laid the foundations of mechanics and the laws of projectiles and falling bodies.

apparent inconsistencies with Aristotle's teachings.

On one occasion, Galileo is alleged to have attracted a large crowd to the leaning tower, where he climbed the spiral staircase to the bell chamber at the top and there through an open archway dropped two stones, one large and one small. These two bodies fell side by side and struck the ground together, thus sounding the death knell of an old hypothesis and the birth of a new era in science.

Although this particular incident is apparently legend, the importance of Galileo's many authentic experiments lies not in the fact that they demonstrated the fallacy of Aristotle's reasoning, but that they presented to the world a new and more reliable scientific method, the method of experimentation.

4.1 GRAVITATION The principle that all objects fall with the same acceleration can be demonstrated in various ways. One of these is illustrated in Fig. 4A where two steel balls, one large and one small, are supported in the groove of a wooden block 10 or 20 ft above the ground. When the block is tipped by pulling the cord, both balls fall together and strike the ground together. Dropped from a height of 16 ft, the time of fall is just 1 sec. The shaded circles in the figure show the position of the two bodies at the end of each quarter second.

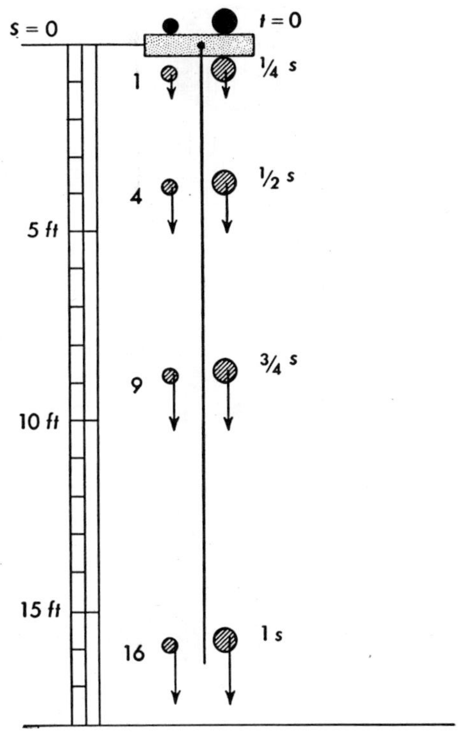

Fig. 4A. All bodies falling freely under the constant pull of gravity fall a distance of 16 ft in the first second.

If the balls in the experiment are replaced by two marbles of the same size, one steel and the other wood, they too will fall side by side and strike the ground together. In this case, the steel marble weighs fifteen times as much as the wood (density of steel, 7600 Kg/m^3; density of wood, 500 Kg/m^3).

The question of air friction usually arises in this latter experiment, for careful observation will show that the wooden ball lags ever so slightly behind the steel ball. This lagging due to air friction increases the farther they fall, and is even more pronounced when a still lighter object like a feather or leaf is allowed to fall at the same time. Due to its large surface area, a feather or leaf flutters to the ground, being held back by the inertia of the large amount of air that must be pushed aside to let it by.

In the absence of air, even a feather will fall with the acceleration of a solid steel ball. An experiment illustrating just this is shown in Fig. 4B. A long glass cylinder containing a feather and silver coin is connected by a flexible rubber tube to a vacuum pump. If, after evacuation, the cylinder is turned upside down, the feather and the coin will be observed to fall together. When the air is once more admitted to the cylinder, the feather will again flutter slowly to the bottom.

In the absence of air friction, all bodies fall with the same acceleration.

In the treatment of falling bodies given in the remainder of this chapter, air friction is entirely neglected. The formulas presented and used in working problems are known to hold only approximately. In most practical cases, however, the calculated results are so nearly realized experimentally that corrections for air friction need

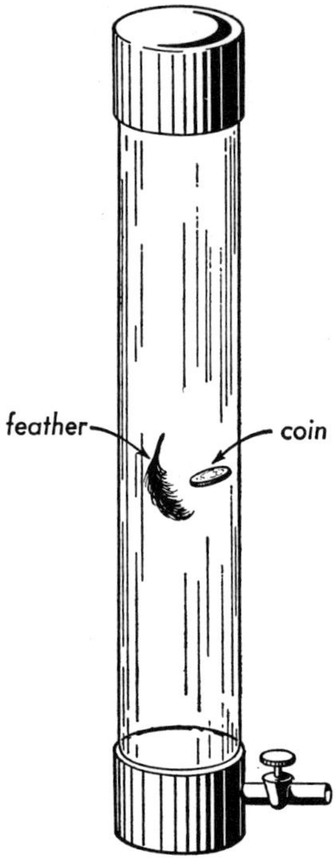

Fig. 4B. In a vacuum a feather and coin fall with the same acceleration and strike the bottom together.

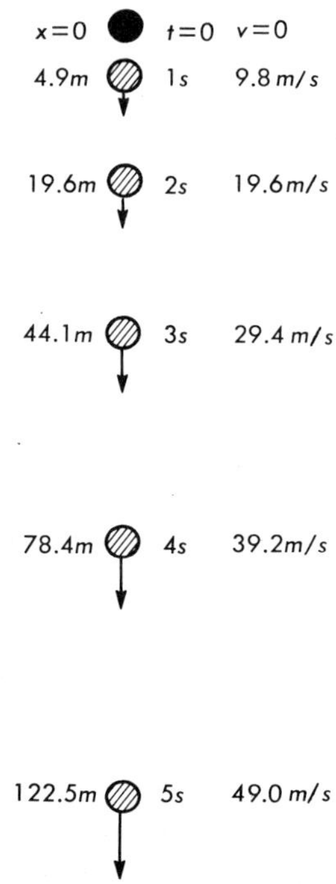

Fig. 4C. Illustration of the distance and velocity of freely falling bodies at the end of each of the first 5 s.

only be made where the distances and velocities involved are large.

4.2 FREE FALL Many laboratory experiments can be performed to demonstrate the well-established laws of falling bodies. One of these is the inclined plane experiment described in Chap. 2. If the angle the plane makes with the horizontal is increased, the acceleration of the ball down the plane will also increase. The velocities as well as the distances traveled will increase in the proper proportions to show that all of the equations summarized at the end of Chap. 2 are valid. This is true even for the limiting angle of 90° where the inclined plane becomes straight up and down and the steel ball falls freely under the full force of gravity.

As indicated in Fig. 4C, the measured distance a body falls is 4.9 m in the first second; four times sixteen or 19.6 m in 2 seconds, nine times 4.9 m or 44.1 m in 3 seconds, etc. Inserting these distances in Eq. (2m), the constant of acceleration is computed to be $a = 9.8$ m/s². If the distances are measured in centimeters, the same formula gives the equivalent acceleration as 980 cm/s² and when measured in feet gives 32 ft/s².

4.3 THE ACCELERATION DUE TO GRAVITY Experiments carried on at many points over the earth show that the acceleration due to gravity is not everywhere the same; there are slight variations. Although these variations are small and are not of any consequence in most practical problems, they do exist and should be mentioned.

Over the earth's surface the values of the acceleration due to gravity lie between a minimum of 32.09 ft/s² or 9.7804 m/s² at the equator and a maximum of 32.26 ft/s² or 9.8321 m/s² at the North and South poles. Reference here to the equator and the poles is only a generalization, for not all points on the equator have the same values as quoted above, nor do all points on any one latitude have the same value. Irregularities of the earth's structure give rise to minute random differences.

The International Committee on Weights and Measures has adopted as a standard or accepted value, 9.80665 m/s^2 or 32.174 ft/s^2. For practical purposes, however, it is customary to use the simpler values 9.80 m/s^2 and 32 ft/s^2, and in formulas for free fall to use the small letter g in place of a as given in Chap. 2. For freely falling bodies then

$$g = 9.80 \text{ m/s}^2, \quad g = 980 \text{ cm/s}^2,$$

or

$$g = 32 \text{ ft/s}^2$$

and Eq. (2i) becomes

$$\boxed{v = gt} \tag{4a}$$

while Eq. (2m) becomes

$$\boxed{x = \tfrac{1}{2}gt^2} \tag{4b}$$

EXAMPLE 1. A boy standing on a high bridge drops a stone into the water below. By looking at his watch he notes that it takes just 3 s for the stone to fall. Calculate (a) the speed of the stone as it hits the water, and (b) the height of the bridge in meters.

SOLUTION. To find the speed, it is convenient to use Eq. (4a) and substitute the known quantities, $g = 9.8$ m/s^2, and $t = 3$ s.

$$v = 9.80 \text{ m/s}^2 \times 3 \text{ s} = 29.40 \text{ m/s}.$$

To find the height, use Eq. (4b) and substitute the same quantities as follows:

$$x = \frac{1}{2} 9.8 \frac{\text{m}}{\text{s}^2} \times 9 \text{ s}^2 = 44.1 \text{ m}$$

The answers are (a) the stone hits the water with a speed of 29.40 m/s, and (b) the height of the bridge is 44.1 m.

4.4 PROJECTION STRAIGHT UPWARD When a body is projected straight upward its speed will rapidly diminish until at some point it comes momentarily to rest and then falls back toward the earth, acquiring again at the ground the same speed it had upon projection. Experiment shows that the time taken to rise to the highest point of its trajectory is equal to the time taken to fall from there to the ground. This implies that the upward motions are just the same as the downward motions, but in reverse, and that the time and speed for any point along the path are given by the same equations for free fall, Eqs. (4a) and (4b).

In Fig. 4D, a particle is shown projected upward with a velocity of 39.2 m/s. After each second's time, its speed on the way up is shown to be the same as its speed at the same level on the way down.

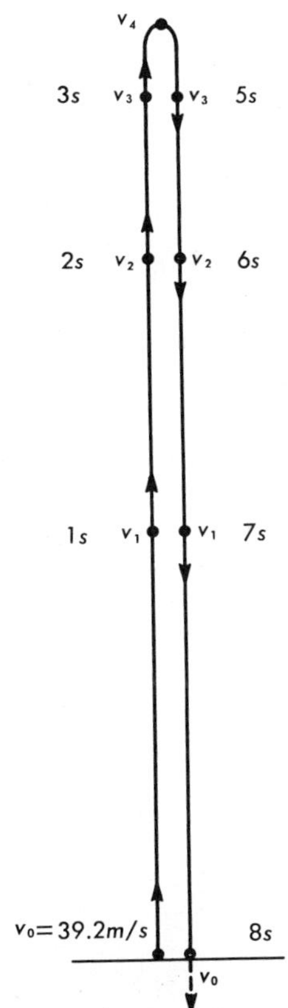

Fig. 4D. Neglecting air friction, the upward motion of a body is the same as the downward motion in reverse. A stone thrown upward returns to the ground with the same speed.

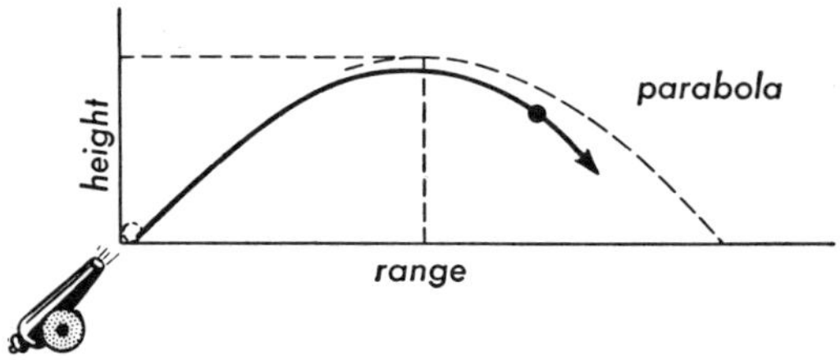

Fig. 4E. Projectiles tend to follow a parabolic path. Because of air friction they fall short.

EXAMPLE 2. A ball is thrown straight upward with a speed of 39.2 m/s. Neglecting friction, find (a) the time taken to reach the top of its flight, and (b) the maximum height reached.

SOLUTION. Since the time to rise equals the time to fall, we can apply Eq. (4a) for free fall. By substituting directly into the equation $v = gt$, we obtain

$$39.2 \frac{\text{m}}{\text{s}} = 9.8 \frac{\text{m}}{\text{s}^2} \times t$$

By solving for t, we find

$$t = \frac{39.2}{9.8} \times \frac{\text{m}}{\text{s}} \times \frac{\text{s}^2}{\text{m}} = 4.0 \text{ s}$$

By applying Eq. (4b), we obtain by direct substitution

$$x = \frac{1}{2} \times 9.8 \frac{\text{m}}{\text{s}^2} \times (4 \text{ s})^2 = 78.4 \text{ m}$$

4.5 PROJECTILES Every object when projected into the air follows the path of a parabola. Such is the case only when air friction is negligible. In practical cases, air friction may be considered negligible only for slowly moving objects with high densities, like a large stone, a block of metal, or a solid ball. For high-speed projectiles the air continually slows down the projectile and the path departs from a parabola somewhat in the fashion shown in Fig. 4E.

The interesting and important information to be obtained from a study of projectiles is the *maximum height*, the *range*, and the *time of flight*.

The maximum height is defined as the greatest vertical distance reached by the projectile as measured from the starting point. The range is defined as the horizontal distance from the point of projection to the point where the projectile returns again to the same level. The time of flight is defined as the time required for the projectile to again reach the level of its initial projection. Experiment shows that all three of these factors will depend upon two things: first, on the *initial velocity* of the projectile, and second, on its *angle of projection*. The latter is always measured from the horizontal and is called the *elevation angle*.

The paths of several projectiles all having the same initial velocity but different elevation angles are shown in Fig. 4F. The greatest possible height is obtained when the projection is straight up and the maximum range when the elevation angle is 45°. Experimentally this can be illustrated by a small stream of water from a flexible tube and nozzle. Each drop of water represents a projectile.

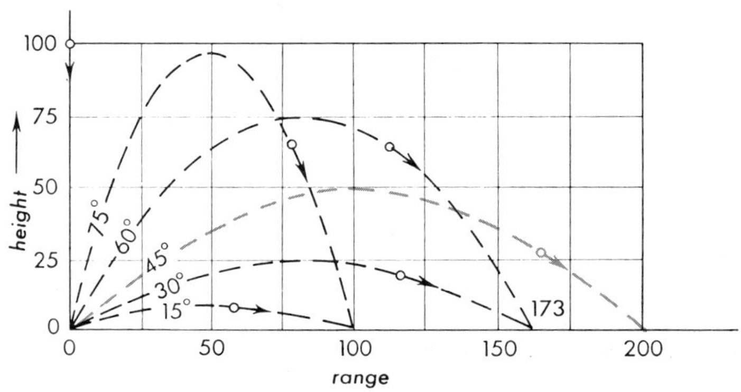

Fig. 4F. Diagram illustrating the shape of the trajectories of objects projected at different elevation angles. The vertical and horizontal scales are for the special case where the velocity of projection is 80 ft/s.

For maximum range with high-speed bullets, because of air friction, the elevation angle should be a little greater than 45°. For slowly moving objects like a shot-put, hammer, javelin, or broad jumper on the athletic field, the angle of 45° will give the maximum distance.

If v is the velocity of projection and θ is the elevation angle, the maximum height, range, and time of flight are given by the formulas in Table 4A.

TABLE 4A. Formulas for Projectiles

Elevation Angle θ	Maximum Height h	Range x	Time of Flight t
0°	0	0	0
30°	$v^2/8g$	$0.866v^2/g$	v/g
45°	$2v^2/8g$	v^2/g	$1.414v/g$
60°	$3v^2/8g$	$0.866v^2/g$	$1.732v/g$
90°	$4v^2/8g$	0	$2v/g$

In Fig. 4F the horizontal and vertical distances given are calculated for projectiles having an initial velocity of 80 ft/s. For different initial velocities the paths of the projectiles will have exactly the same shape as those given in the figure but the distances will be different.

If an object is projected with a velocity v it will, neglecting air friction, rise to its maximum height in the same time it takes to fall from that height to the ground. As might be suspected from Fig. 4F, it will arrive at the ground with the same speed it had on projection.

EXAMPLE 3. In the 1948 Olympic Games held in London, the 16-lb shot-put was won by Wilbur Thompson of the United States with a toss of 17.12 m. Assuming a 45° elevation angle for the toss, calculate the initial velocity.

SOLUTION. Since the elevation angle is given as 45° and the range is known, the third equation in the third column of Table 4A is selected, namely, $x = v^2/g$. Multiplying both sides of this equation by g gives

$$v^2 = gx$$

By substituting the known quantities, we obtain

$$v^2 = 9.80\,\frac{\text{m}}{\text{s}^2} + 17.12\text{ m} = 168\,\frac{\text{m}^2}{\text{s}^2}$$

By taking the square root of both sides, we finally obtain

$$v = 13.0\,\frac{\text{m}}{\text{s}}$$

4.6 MONKEY AND HUNTER EXPERIMENT A hunter aims and shoots an arrow at a monkey in a tree. At the instant the arrow leaves the bow the monkey drops from the branch upon which he has been sitting. The two should meet in mid-air regardless of the speed of the arrow. If gravity could be eliminated, the arrow, as shown in Fig. 4G, would travel the straight line path AM and the monkey would stay at M and be hit there. With gravity acting, however, the arrow travels the path ABC and the monkey drops from M to C. During each fraction of a second, indicated by t = 1, 2, 3, and 4, both fall the same distance from their gravity-free positions and collide at C. The greater the speed of the missile the shorter will be the distance MC.*

4.7 NEWTON'S LAW OF GRAVITATION Nearly everyone has heard the story of how young Isaac Newton, while sitting under an apple tree one day, was struck on the head by a falling apple. This incident set Newton to thinking about falling bodies and led him, at the early age of twenty-three, to the discovery of the law of gravitation.

* The monkey and hunter experiment may be performed by blowing a small wooden marble through a tube about a foot long. Another ball representing the monkey can be released at M by a small electromagnet. Two fine copper wires completing an electric circuit are crossed just in front of the tube at A. When the projectile passes this point the circuit is broken, releasing M. The body M can be made of iron, hollow or solid, and in any shape, or it can be a small stuffed toy monkey with a small piece of iron sewed on the top of his head for the magnetic attraction by the electromagnet.

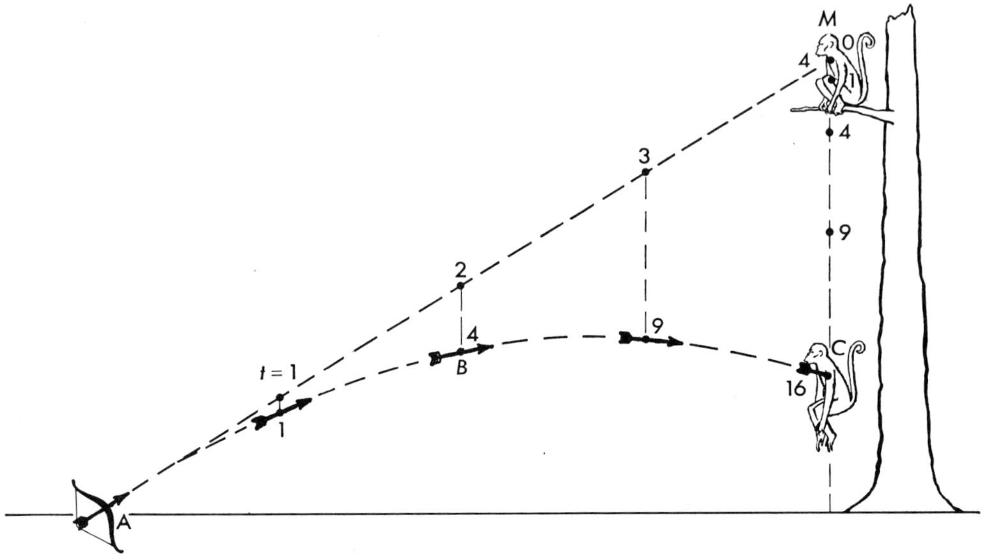

Fig. 4G. Diagram illustrating the monkey and hunter experiment.

It has often been said incorrectly that Newton discovered gravity. What Newton discovered was the *universal law of gravitation. Any two bodies attract each other with a force proportional to the product of their masses and inversely proportional to the square of the distance between them.* Written in algebraic symbols,

$$F \propto \frac{m_1 m_2}{d^2}$$

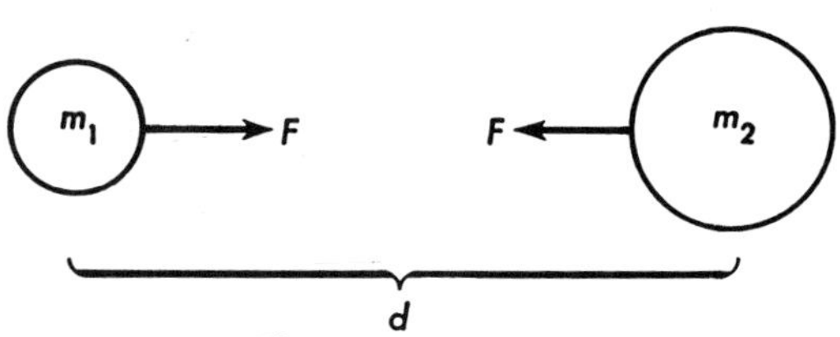

Fig. 4H. The gravitational attraction of one body of mass m_1 for another of mass m_2.

As illustrated in Fig. 4H, F is the force of attraction, m_1 and m_2 are the two masses, and d is the distance between them. Mass m_1 pulls on m_2 with a force F to the left and m_2 pulls on m_1 with an equal force F to the right. To make an equation of this symbolism it is only necessary to replace the proportionality symbol above by an equal sign and insert a constant on either side of the equality.

$$\boxed{F = G\frac{m_1 m_2}{d^2}} \tag{4c}$$

Experiment shows that if F is measured in newtons, m_1 and m_2 in kilograms, and d in meters, the "Newtonian Constant of Gravitation" G has the value

$$G = 0.000{,}000{,}000{,}0667 \frac{\text{m}^3}{\text{Kg s}^2}$$

or in short-hand notation, see Appendix VIII,

$$G = 6.67 \times 10^{-11} \frac{\text{m}^3}{\text{Kg s}^2} \tag{4d}$$

If F is measured in pounds, m_1 and m_2 in slugs, and d in feet,

$$G = 3.419 \times 10^{-8} \frac{\text{ft}^3}{\text{slug s}^2} \tag{4e}$$

EXAMPLE 4. Two locomotives, weighing 64 tons each, stand beside each other with their centers 10 ft apart. Calculate the gravitational attractive force between them.

SOLUTION. First find the mass of each locomotive in slugs as follows:

$$m = \frac{64 \times 2000}{32} = 4000 \text{ slugs}$$

Direct substitution in Eq. (4c) now gives

$$F = 3.419 \times 10^{-8} \frac{4000 \times 4000}{10^2} = 0.00547 \text{ lb}$$

This force of 0.00547 lb is extremely small and would be difficult to detect, much less measure.

If we now consider the attraction between a very large object, like the earth, and another object like our own body, the force becomes one that is quite measurable for it is none other than our own weight. It is the force that keeps all of us on the earth.

To go one step further, gravitational attraction is the force that keeps the moon in its orbit around the earth and the earth in its orbit around the sun. Such forces have magnitudes of billions of billions of tons.

The masses and some of the distances and dimensions of the sun, the moon, and the earth are given in Fig. 4I.

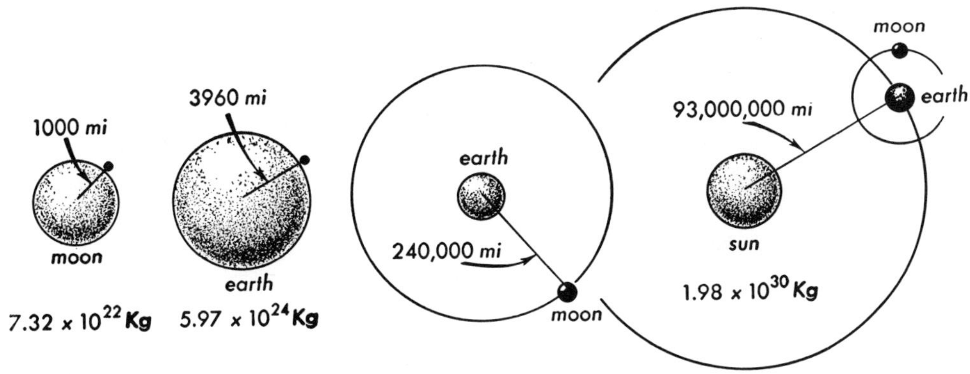

Fig. 4I. Gravitational attraction keeps the moon in its orbit around the earth and the earth in its orbit around the sun. Equivalents in meters: 1000 mi = 1609 Km; 3960 mi = 6373 Km; 240,000 mi = 386,000 Km; 93,000,000 mi = 150,000,000 Km.

PROBLEMS

1. A stone dropped from a bridge strikes the water below after 5.2 s. Find the height of the bridge in meters. [Ans. 132.5 m.]

2. A flowerpot falls from a window sill and strikes the ground below after 2.6 s. Find the height of the window sill in meters.

3. A stone is dropped from a bridge 220 m above the water. (a) How long will it take to fall, and (b) with what speed will it hit the water? [Ans. (a) 6.70 s, (b) 65.7 m/s.]

4. A baseball thrown straight upward reaches a maximum height of 120 m. (a) How long will it be in the air? (b) With what speed will it strike the ground?

5. An arrow shot vertically upward reaches a height of 480 ft. Find (a) the total time of flight, and (b) the velocity with which it strikes the ground. [Ans. (a) 10.95 s. (b) 175.2 ft/s.]

6. An arrow is shot vertically upward with an initial velocity of 42 m/s. Find (a) the maximum height reached, and (b) the total time of flight.

7. A baseball hit straight upward is caught 12 s later by the catcher. Find

(a) the maximum height reached, and (b) the initial upward velocity. [Ans. (a) 176.4 m, (b) 58.8 m/s.]

8. A javelin is thrown at an elevation angle of 45° for a record distance of 85.3 m. Find (a) the initial velocity, and (b) the maximum height reached.

9. A shot-put is thrown at an elevation angle of 45° for a record distance of 22.8 m. Find (a) the initial velocity, and (b) the time of flight. [Ans. (a) 14.95 m/s, (b) 2.157 s.]

10. An arrow is shot vertically upward with an initial velocity of 29.4 m/s. Two seconds later another arrow is shot upward with the same velocity. At what height will the two arrows pass each other? (Note: Make a graph by plotting height vertically and time horizontally.)

11. A stone is thrown with a velocity of 30 m/s at an elevation angle of 30°. Find (a) the maximum height reached, (b) the range, and (c) the time of flight. [Ans. (a) 11.48 m, (b) 79.5 m, (c) 3.061 s.]

12. In the 1956 Olympic Games the 16-lb hammer throw was won by Harold Connolly of the United States with a throw of 63.19 m. Assuming a 45° elevation angle for the toss, calculate (a) the maximum height reached, (b) the initial velocity, and (c) the time of flight.

13. In the 1960 Olympic Games the javelin throw was won by Viktor Tsibulenko of Russia with a throw of 84.64 m. Assuming a 45° elevation angle for the toss, calculate (a) the maximum height reached, (b) the initial velocity, and (c) the time of flight. [Ans. (a) 21.16 m, (b) 28.80 m/s, (c) 4.16 s.]

14. In the 1964 Olympic Games the 16-lb shot-put was won by Dallas Long of the United States with a toss of 20.33 m. Find (a) the greatest height reached, (b) the time of flight, and (c) the initial velocity. Assume an elevation angle of 45°, and projection from the ground level.

15. The javelin throw in the 1964 Olympic Games was won by Pauli Nevala of Finland with a toss of 271 ft $2\frac{1}{4}$ in. Find (a) the maximum height reached, (b) the time of flight, and (c) the velocity of projection. Assume an elevation angle of 45°, and projection from the ground level. [Ans. (a) 67.8 ft, (b) 4.12 s, (c) 93.2 ft/s.]

16. A baseball thrown at an elevation angle of 60° reaches a height of 80 m. Find (a) the initial velocity, (b) the range, and (c) the time of flight.

17. Explain why a 10-Kg stone will fall with the same acceleration as a 1-Kg stone. Use Newton's Second Law of Motion as well as the Law of Gravitation. $\left[\text{Ans. } F = G\dfrac{m_1 m_2}{d^2} = m_2 g, \text{ so } g = G\dfrac{m_1}{d^2}, \text{ independent of } m_2.\right]$

18. A world's record in the javelin throw was made in June 1968 by Janis Lusis of Latvia. His record of 301 ft $9\frac{1}{4}$ in. was well over previous records. Motion pictures show the javelin leaving his hand inclined at an angle of 45° with the horizontal. Since the tail of the javelin starts close to the ground, and the spearhead strikes the ground, assume the start and finish of the trajectory to be at the ground level. Find (a) the maximum height reached, (b) the time of flight, and (c) the projection velocity.

5 FORCES, VECTORS, AND CENTER OF GRAVITY

In many of the laws of nature, the physical concept we call *force* plays a most prominent role. In order to gain a better understanding of all laws it is important that we gain a clear understanding of forces, their origin, their interaction with one another, and the consequences of their interaction with matter.

Everyone knows that when he weighs himself he is measuring the downward force he exerts on the foot board of the scales, and that this force causes some mechanism within the scales to indicate his weight. The greater the downward force, the greater is the indicated weight. See Fig. 5A. We are not interested here in the system of levers, weights, or springs within the scales, but rather with the downward force we call our *weight*.

Weight, as explained in the last chapter, is due to the gravitational attraction of the earth for all bodies.

As illustrated in Fig. 5B, gravitational forces always act in the direction of a line joining the body and the center of the earth and are, for all practical purposes, perpendicular to the earth's surface at the body.

The term *force* is not confined to weight alone but to the action of any one body upon another. For example, in towing an automobile as shown in Fig. 5C, there are two forces acting: (1) a downward force due to gravity, and (2) a horizontal force due to the pull of the tow line. The latter force is supplied by some external object or machine. In pushing a lawn mower there are also two forces: (1) the downward force due to gravity, and (2) a diagonal force due to some person pushing on the lawn-mower handle.

Fig. 5A. Weight is a downward force. The earth attracts all bodies toward its center

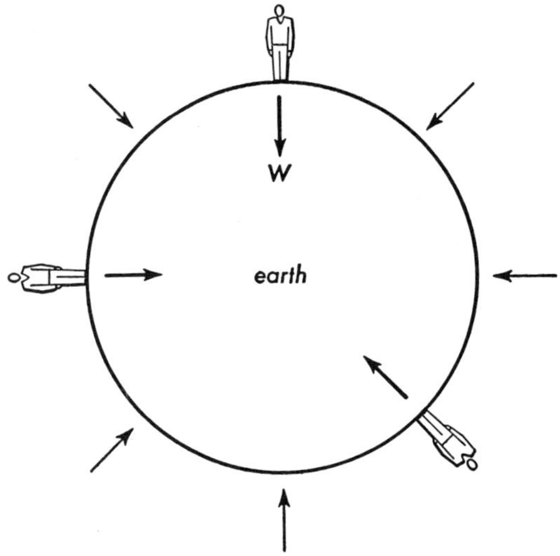

Fig. 5B. Weight is a force due to gravitational attraction, and gravitational force acts in the direction of a line joining the body and the center of the earth.

Fig. 5D. Two forces acting at an angle to each other are equivalent to a single force acting in a direction between them.

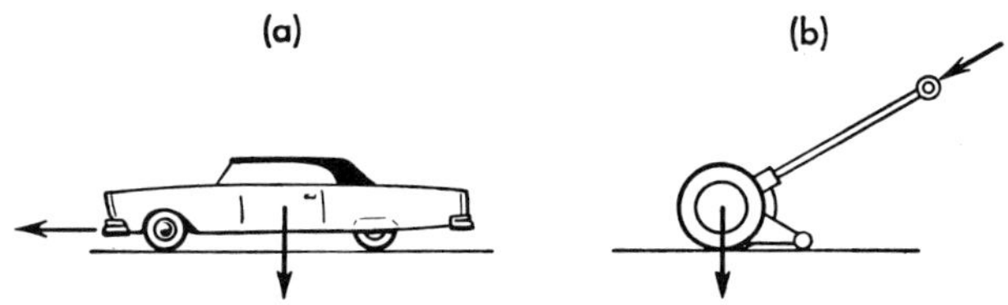

Fig. 5C. Illustration of two independent forces acting on the same body.

Regardless of the direction in which a force may act, its magnitude may be expressed in *newtons*, *dynes*, or *pounds*.

5.1 FORCES ARE ADDED VECTORIALLY Since forces have both magnitude and direction, they are vector quantities and, therefore, subject to the rules of *vector addition*. Consider the diagram in Fig. 5D, illustrating a heavy trunk being pulled along the floor by two ropes. With steady pulls of 40 N and 25 N exerted in directions at 90° from each other, the trunk moves in a direction indicated by the dotted arrow. By *vector addition a resultant force* can be found which, upon taking the place of the two forces shown, will produce the same motion.

When this *resultant* is determined by vector addition, it is found to have a magnitude of 47.2 N and a direction making an angle of 32° with the 40 N force.

To illustrate the common methods employed in the vector addition of two forces, consider the illustrations shown in Fig. 5E where two forces of 3 N and 5 N respectively, are applied to a body at the common point *B*. The relative direction of the 3-N force is different in each of the three cases. There are in general two graphical methods for finding the resultant: first, *the parallelogram method;* and second, *the triangle method.*

The parallelogram method of vector addition is shown in Fig. 5F. To illustrate the method consider diagram (a). First a horizontal line *OA* is

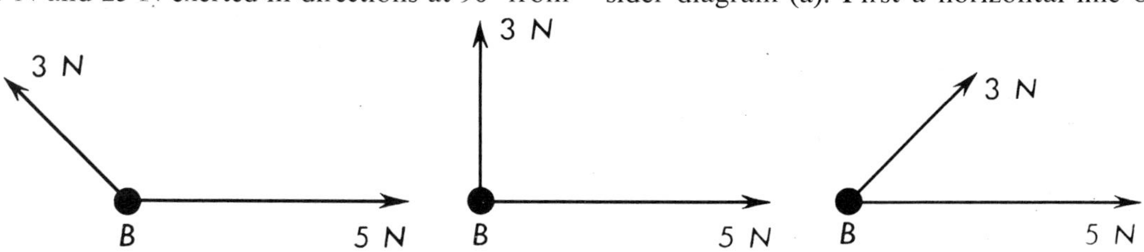

Fig. 5E. Force diagrams showing two forces acting on the same body but in different relative directions.

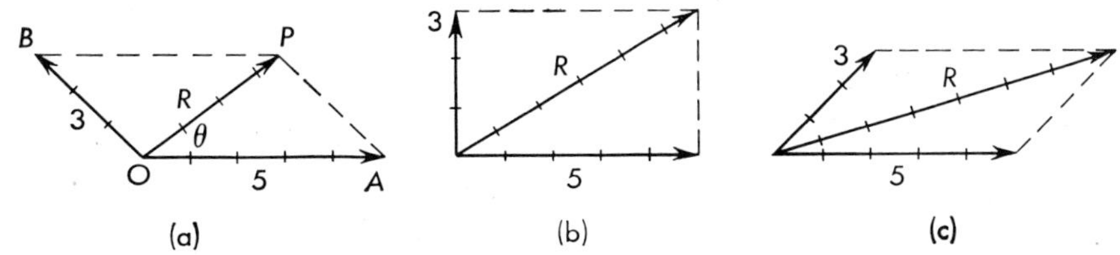

Fig. 5F. Illustration of the parallelogram method of vector addition.

drawn to represent the 5-N force. The length of the line is made to be 5 cm to represent the magnitude of the force, and the arrowhead is inserted at the right end to indicate its direction. In a similar way the line OB is drawn 3 cm long and in the proper direction to represent the 3-N force.

From the head of the vector representing the 5-N force, a dotted line is drawn parallel to the 3-N vector; and, from the head of the 3-N vector, a dotted line is drawn parallel to the 5-N vector. The resultant R is drawn from the origin O to the point P where these two dotted lines intersect. Using the same scale that was used to mark off the 3-N and 5-N vectors, the length of R can be read directly as 3.5 N, and the angle θ it makes with the horizontal can be read in degrees with a protractor as 36°. In other words, the length of R gives the magnitude of the resultant force and the direction of R gives the direction of the force.

This resultant force if acting on the body will produce exactly the same effect and is equivalent to the two original forces. It is to be noted that when the forces are in the same direction or in opposite directions, the resultant R is equal to the arithmetical sum and difference respectively. Thus the magnitude of R may have any value from the arithmetic difference, 2 N, up to the arithmetic sum, 8 N, depending solely upon the relative directions of the two original forces.

The triangle method of vector addition, as shown in Fig. 5G, follows directly from the parallelogram method just described. Only half of the parallelogram need be drawn. In diagram (a), for example, the 5-N vector is first drawn to some scale. Second, the 3-N vector is started at the arrowhead of the 5-N vector and drawn in its true direction and to the same scale. The resultant is finally drawn starting at the origin O and ending at P, the second arrowhead. For convenience alone, this triangle method is to be preferred over the parallelogram method.

5.2 FORCE POLYGON When three or more forces act simultaneously upon a body, a single force called their resultant can be found which, acting alone upon that body, will produce the same result. To find such a resultant force, the *polygon method* of vector addition is often employed. In principle, this is an extension of the triangle method and consists of placing the tail of one vector at the head of the one preceding it, and continuing this process until all vectors have been added.

An illustration of the polygon method applied to five forces is given in Fig. 5H. The space diagram at the left shows the forces acting on a body at P, while the *vector diagram* at the right shows the vector addition and resultant force R. Starting at A as origin, vector AB is drawn 8 cm long (or 8 units of any convenient scale) and parallel to the 8-N vector in diagram (a). Vector BC is next drawn 7 cm long and parallel to the 7-N vector. These are followed in succession by vectors CD, DE, and EF, respectively. With all five

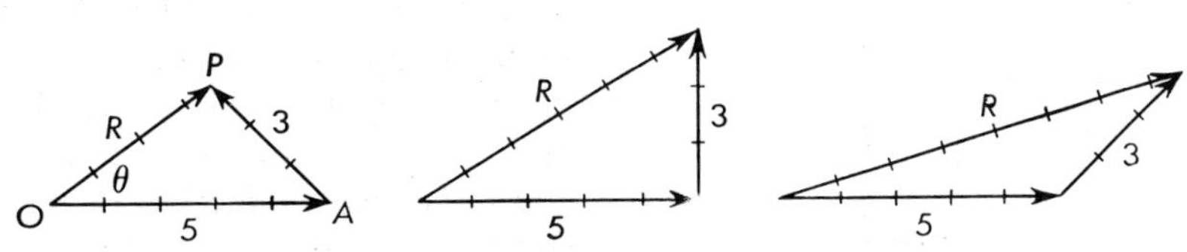

Fig. 5G. Illustration of the triangle method of vector addition.

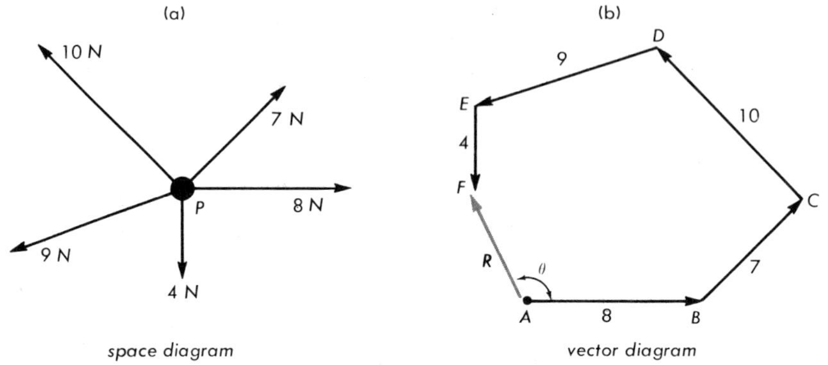

Fig. 5H. Illustration of the graphical addition of five forces to find their resultant. (The polygon method.)

vectors added, the resultant R is found by joining the last arrowhead F with the origin A.

Drawn to scale, the measured length of R will give the magnitude of the resultant force and the measured angle θ will give the direction in which it acts.

5.3 CONDITIONS FOR EQUILIBRIUM When one or more forces act upon a body at rest, and their resultant sum is not zero, the body will be set into motion. Under such conditions there is an *unbalanced force* acting and this force alone accounts for the acceleration. If, however, the vector sum of all the forces acting is zero, the body is in equilibrium and the body will remain at rest. To turn this statement around is to say that "*any object remaining at rest is in equilibrium and the resultant of all forces acting upon it is zero.*"

If two and only two forces act upon a body in equilibrium, a little study will show that they must be equal in magnitude and opposite in direction. A book lying on the table or a lamp hanging from the ceiling are good examples of dual forces in equilibrium. See Figs. 5I and 5J.

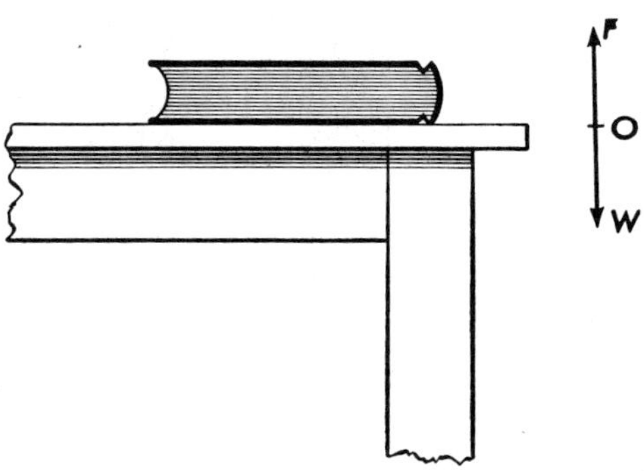

Fig. 5I. A book lying on a table is in equilibrium.

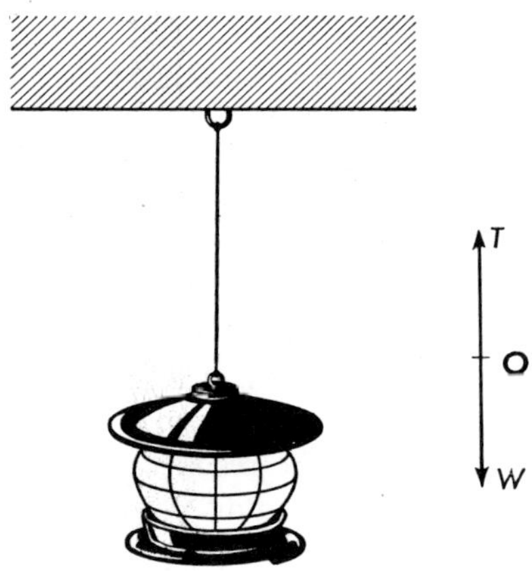

Fig. 5J. A lamp hanging from the ceiling is in equilibrium.

The two forces acting on the book are W, the downward pull of the earth, called the *weight*, and F, the upward thrust of the table. Since the book is in equilibrium, the force F is equal in magnitude to the weight W. For the lamp at the right the downward force or weight is counterbalanced by the upward tension in the cord. Here again, the forces are equal in magnitude and opposite in direction. A body moving with constant velocity is in equilibrium: since there is no acceleration, there is no unbalanced force.

In the game of tug-of-war, when two opposing teams are pulling with equal but opposite forces at the ends of a rope, a condition of equilibrium exists. As illustrated in Fig. 5K, the force F of

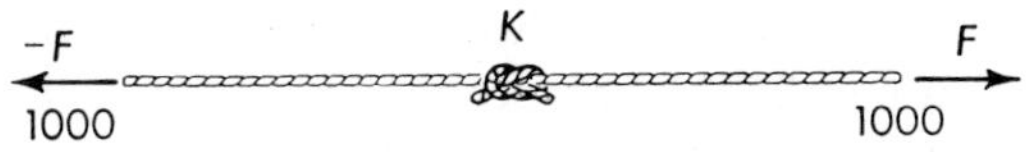

Fig. 5K. The tension in the rope is 1000 N.

1000 N acting to pull the knot K to the right is counterbalanced by an equal but opposite force $-F$ of 1000 N pulling it to the left. If the two forces become unequal, equilibrium will no longer exist and the knot K will accelerate in the direction of the greater force. It should be noted in the equilibrium case that the tension in the rope is 1000 N and not 2000 N. This apparent paradox can be explained away by supposing that one team ties their end of the rope to a post. The other team still pulling with their 1000 N maintains the same equilibrium conditions as before and in so doing maintains the tension of 1000 N. One team can be looked upon as holding the rope so the other team can pull.

5.4 THREE FORCES IN EQUILIBRIUM When, as the result of the action of three forces, a body is in equilibrium, the *resultant* of all three forces must be zero. In other words, *to be in equilibrium, the force polygon must close*. With three forces, such a polygon would have only three sides, i.e., it would be a triangle. As an illustration, consider the street light suspended from two poles as shown in Fig. 5L.

The three forces acting through the common point O are W the weight of the lamp 5 lb acting straight downward, F_1 the pull of one rope at 45° up and to the left, and F_2 the pull of the other rope at 30° up and to the right. The force polygon is shown in diagram (b), where vectorially

$$\vec{W} + \vec{F_2} + \vec{F_1} = 0 \tag{5a}$$

In constructing this diagram, the conditions of equilibrium are imposed as a means of determining the magnitude of the forces F_1 and F_2. The graphical procedure is as follows: a vector of length 5 units is first drawn straight downward to represent W, the weight of the light. To the head of this vector at B, a dotted line BC is drawn parallel to the rope exerting the force F_2. From A, another dotted line AD is drawn parallel to the rope exerting the force F_1. At E, where these two lines intersect, the vectors F_1 and F_2 are terminated and arrowheads inserted in the directions shown. The solid lines AE and BE, when measured, are found to have lengths of 4.48 and 3.66 units and represent the forces F_1 = 4.48 lb and F_2 = 3.66 lb, respectively.

5.5 RESOLUTION OF A FORCE INTO COMPONENTS Many of the force problems in mechanics are most easily solved by the so-called "*method of components*." To apply this method to typical problems it is necessary that we first see how a

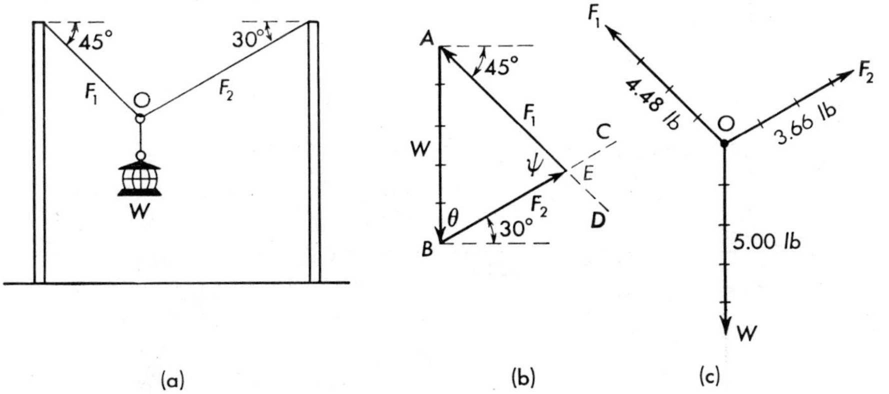

Fig. 5L. The lamp hanging from two cords is in equilibrium.

single vector may be resolved into two components. Consider as an illustration the known force F, making an angle of θ degrees with the x axis as shown in Fig. 5M.

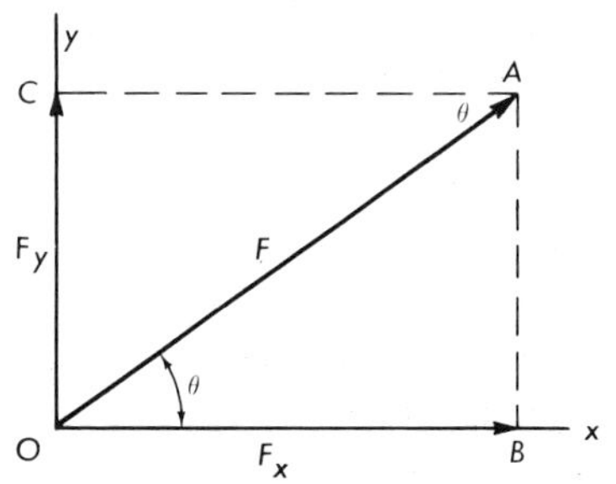

Fig. 5M. The resolution of a vector into two rectangular components.

By dropping lines from A, perpendicular to the x and y axes, the component forces F_x and F_y are obtained. These two components are equivalent to the original force F, since by adding them vectorially they give F as a resultant.

With F_x and F_y perpendicular to each other, triangles OAB and OAC are congruent right triangles with corresponding sides equal. F_y = BA and F_x = CA.

Consider the following example of the resolution of a force into two rectangular components. A boy is pulling a small 40-lb wagon up a 30% grade as illustrated in Fig. 5N. A 30% grade means that for every 100 ft traveled in the horizontal direction there is a vertical rise of 30 ft. (This is not equivalent to an angle of 30°.)

To find how hard the boy must pull on the wagon, the weight of the wagon W is resolved into two components, one parallel to the incline and the other perpendicular to it. As shown in diagram (b), the 40-lb vector W is first drawn vertically downward. Two dotted lines are next drawn parallel and perpendicular to the incline, one from one end of W and the other from the other. Where these two lines intersect at Q the vector components F_x and F_y are terminated. When the magnitudes of these two forces are measured they are found to have the values F_x = 11.5 lb and F_y = 38.3 lb. The force F_y is the force the wheels of the wagon exert against the incline, and, being perpendicular to it, neither aids nor hinders the motion. To pull the wagon the boy must exert a force P equal to or greater than 11.5 lb, the magnitude of the component F_x down the incline.

To illustrate how this same problem can be solved mathematically, we first note that the *force diagram* at the right and inclined plane at the left are similar triangles. (*The sides of one are perpendicular respectively to the corresponding sides of the other.*) We therefore make use of the elementary proposition in geometry that corresponding sides of similar triangles are proportional. We may therefore write

$$F_x : W = BC : AB, \quad \text{or} \quad \frac{F_x}{W} = \frac{BC}{AB}$$

To calculate F_x we must know the length of the side AB. Since triangle ABC is a right triangle the square of the hypotenuse is equal to the sum

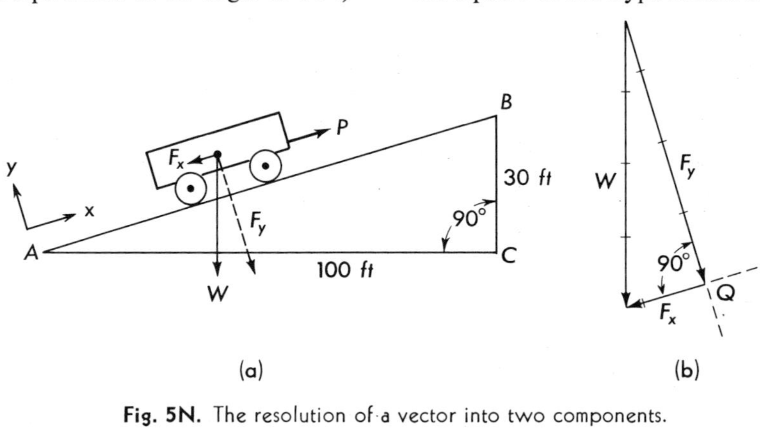

Fig. 5N. The resolution of a vector into two components.

of the squares of the other two sides. Squaring both sides, adding, and taking the square root gives $AB = 104.4$ ft. Inserting the known values in the preceding equation we obtain as a result,

$$\frac{F_x}{40} = \frac{30}{104.4} \quad \text{or} \quad F_x = \frac{40 \times 30}{104.4} = 11.5 \text{ lb}$$

By a similar proportion the component F_y can be calculated: it is found to be 38.3 lb.

5.6 THE SAILBOAT A problem that puzzles many people, particularly those more or less familiar with sailboats, is that of sailing across the water into the wind. This phenomenon, commonly known as *tacking*, is another illustration of the resolution of a force into rectangular components.

As shown in Fig. 5O, the wind is from the

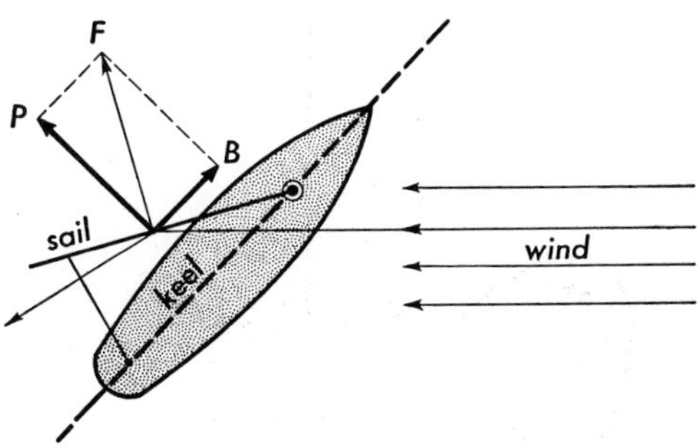

Fig. 5O. A boat sailing into the wind. An example of the resolution of a force ***F*** into two rectangular components ***P*** and ***B***.

east and the boat is headed NE. When the sail is properly set, the wind, in blowing across the canvas, is deflected away in such a way that it exerts a force F normal to the surface as shown. By resolving this force into two rectangular components, one parallel and the other perpendicular to the boat, the force B, responsible for the boat's motion, is found.

The other component, P, has little effect upon the boat since it is perpendicular to the motion. It is a useless force which tends to tip the boat and move it to leeward. To prevent tipping or being pushed sideways, sailboats are equipped with a deep heavy keel. By increasing the angle between the sail and the wind, the force F will increase but the forward component will decrease. If the boat is headed more directly into the wind, without changing the relative position of the sail, the useful component B will again decrease. Most rapid progress upwind is attained when the wind and keel make an angle of 45° and the sails are so rigged that the rudder is parallel to the keel.

5.7 CENTER OF MASS *The center of mass* (abbr. *c of m*) *of any given body, or system of bodies, is a point such that if any plane is passed through it the mass moments on one side of the plane are equal to the mass moments on the other.* Consider, for example, two spheres of mass m_1 and m_2 as shown in Fig. 5P. The *c of m P* lies on a line

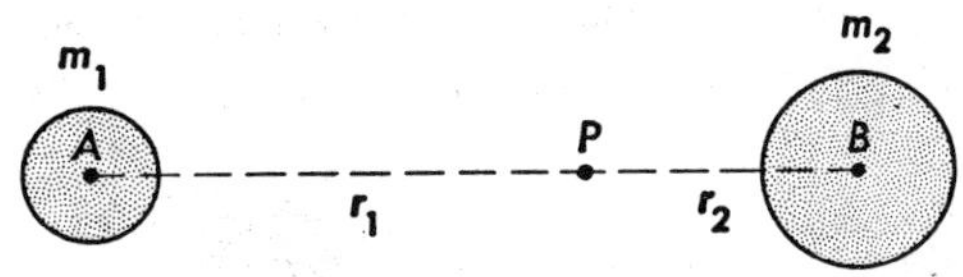

Fig. 5P. The center of mass of two bodies is located at some point on a line joining their centers of mass.

connecting the centers of the two bodies and in such a position that

$$\boxed{m_1 r_1 = m_2 r_2} \qquad (5b)$$

For a vertical plane through P perpendicular to the plane of the page, $m_1 r_1$ is the mass moment of m_1 and $m_2 r_2$ is the mass moment of m_2.

EXAMPLE 1. Find the *c of m* of two bodies $m_1 = 2$ g and $m_2 = 5$ g placed 14 cm apart.

SOLUTION. Since the distance $r_1 + r_2 = 14$ cm, we obtain

$$r_2 = 14 - r_1 \qquad (5c)$$

By substituting all known quantities in Eq. (5b), we obtain

$$2r_1 = 5(14 - r_1) \quad \text{or} \quad 2r_1 = 70 - 5r_1$$

giving

$$7r_1 = 70 \quad \text{or} \quad r_1 = 10 \text{ cm}$$

By substituting this value of r_1 in Eq. (5c), we find

$$r_2 = 4 \text{ cm}$$

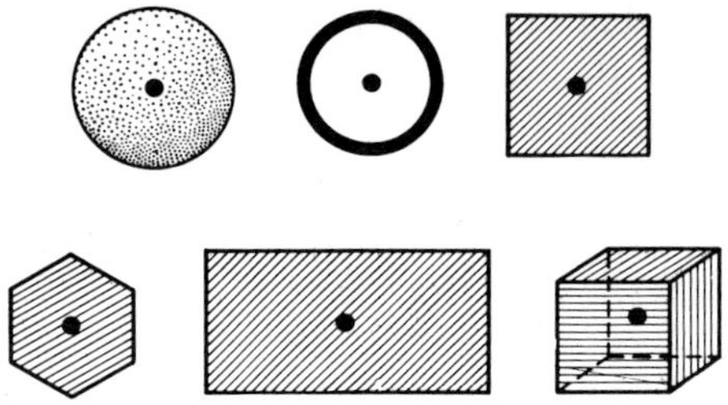

Fig. 5Q. Illustration of the center of mass of regularly shaped objects.

The *c of m* of all regularly shaped bodies like those shown in Fig. 5Q is at their geometrical center. A plane passed through the center of any of these figures will divide the body into two equal parts.

5.8 ROTATION ABOUT THE CENTER OF MASS

In Fig. 5R two masses m_1 and m_2 are shown

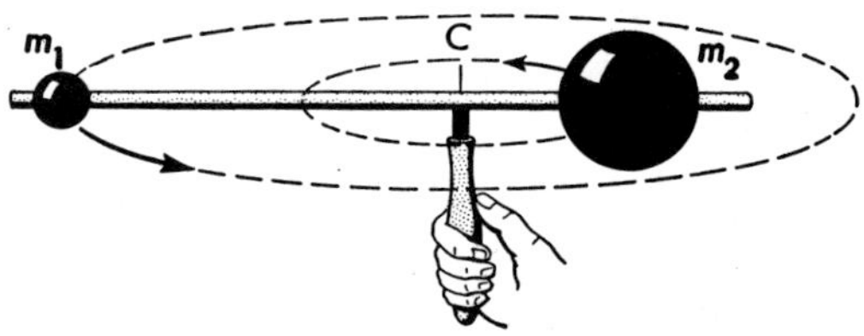

Fig. 5R. Diagram illustrating the smooth rotation of two bodies around their center of mass.

supported at the ends of a thin rod and rotating smoothly around a pin through the *c of m*. If the pin is located at any other point—for example, halfway between the two masses—the experimenter will experience an unbalanced force on his hand, tending to make it "wobble."

Should the two-body system be thrown spinning into the air, it will be observed to rotate about its *c of m* while the *c of m* traces out the smooth trajectory of a projectile. Rotation is smooth around this point because the centrifugal forces of the two bodies are counterbalanced there. The two centrifugal forces are equal in magnitude but opposite in direction. Being in opposite directions they exert no resultant force on the pivot.

5.9 CENTER OF GRAVITY The *c of m* of the two bodies in Fig. 5P is the one and only point about which the two bodies will, if pivoted, balance under the earth's gravitational pull. (Here we assume they are mounted on a rod of negligibly small mass.) Furthermore, a single upward force applied at *P*, equal in magnitude to the weight of the two bodies, will maintain equilibrium; the system will not tend to move in any direction nor will it tend to rotate.

The *c of m* is therefore a point at which all of the weight can be considered as concentrated. For this reason the *c of m* is often called the center of gravity (abbr. *c of g*).

The *c of g* of a regular or irregular shaped body of uniform or nonuniform density can be found by suspending it from any one point and then another as shown in Fig. 5S. With each

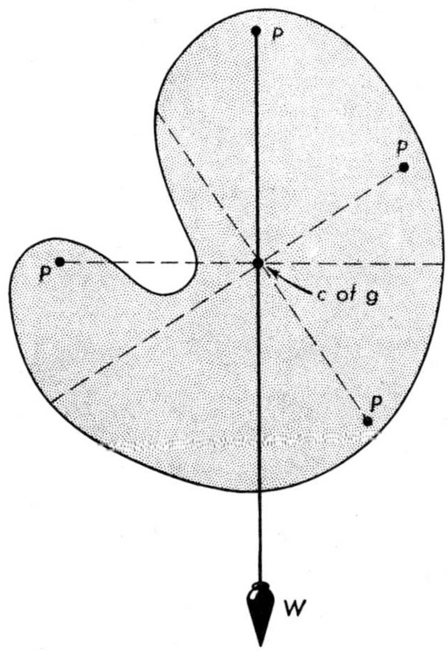

Fig. 5S. The center of gravity of any freely suspended object lies at or directly beneath the point of suspension.

suspension from a point *P* near the periphery, the body will hang with its *c of g* directly under that point. Lines drawn along the string supporting the plumb bob for each suspension will all cross at the common point, the *c of g*.

If pivoted at this point and set turning, or thrown spinning into the air, the rotation will be smooth about the *c of g*. (*Note:* Because the force of gravity decreases with altitude, the *c of g* of a body is not always at exactly the same point as the *c of m*. The lower part of a mass, for example, is closer to the earth's center than the upper part and therefore has a greater weight per unit mass. For all practical purposes, however, the two terms *c of g* and *c of m* are considered synonymous.)

PROBLEMS

In solving all vector problems in this chapter the graphical method should be used. Use a ruler and protractor, and draw all vectors to a scale and measure all unknown quantities asked for in the problem. Use a sharp pencil and measure with care.

1. Two forces of 30 N and 45 N, respectively, are applied to a single body. If the angle between their two directions is 90°, find (a) the magnitude, and (b) the direction of the resultant. [Ans. (a) 54.1 N, (b) 33.7° with the 45-N force.]

2. Two forces of 56 N and 34 N, respectively, are exerted simultaneously on a large object. If these two forces are at right angles to each other, what is the magnitude and direction of their resultant? Find the angle using a protractor.

3. Two forces of 10 N and 30 N, respectively, act at right angles to each other. Find their resultant force and specify its direction. [Ans. 31.60 N at 18.4° with the larger force.]

4. A vertical force of 55 N and a horizontal force of 35 N act on a heavy stone. Find the magnitude and direction of the resultant force.

5. Two forces act on the same body. The forces are 20 N in a direction straight west, and 30 N in a direction 30° east of North. Find (a) the resultant force, and (b) specify the direction in which it acts. [Ans. (a) 26.46 N, (b) 10.9° west of North.]

6. A boy pulls a 60-lb sled up an icy slope, the slope making an angle of 12° with the horizontal. What is the force he must exert parallel to the slope?

7. A large box on wheels, weighing 2200 N, is pulled up a plank onto a truck. If the plank makes an angle of 25° with the horizontal, find the required force parallel to the plank. [Ans. 930 N.]

8. Three forces act simultaneously on the same body. These forces are (1) 6 N in a direction 30° east of north, (2) 5 N in a direction 45° west of north, and (3) 8 N in a direction 35° south of east. Find the resultant force graphically by the polygon method and specify the direction of the resultant.

9. The three forces acting on a body are, (1) 8 N in a direction 30° east of north, (2) 12 N in a direction 45° north of west, and (3) 6 N in a direction 30° south of west. Find the resultant force graphically by the polygon method and specify the direction. [Ans. 15.74 N at 38° west of north.]

10. If the lamp in Fig. 5L weighs 60 lb and the angles the ropes make with the horizontal are 40° and 50°, in place of 45° and 30°, respectively, find the magnitudes of F_1 and F_2.

11. If the lamp in Fig. 5L weighs 250 N, and the angles the ropes make with the horizontal are 48° and 35° in place of 45° and 30°, respectively, find the magnitudes of F_1 and F_2. [Ans. 206.2 N, and 168.5 N.]

12. A box weighing 150 lb rests on a plank inclined at an angle of 28° with the horizontal. Find the two components of force exerted (a) parallel, and (b) perpendicular, respectively, to the plank.

13. A man pushes with a force of 250 N along the handle of a 220 Kg roller. The handle makes an angle of 35° with the ground. Find (a) the horizontal, and (b) the vertical components of this force. Find (c) the total vertically downward force exerted by the roller on the ground. [Ans. (a) 204.8 N, (b) 143.5 N, (c) 2299 N.]

14. A sailboat is tacking at 45° into the wind. If the resultant force exerted by the wind on the sail is 320 N, and the boom makes an angle of 32° with the keel, find the forward thrust on the boat.

15. The wind exerts a resultant force of 150 lb on the sail of a small boat. If the angle between the wind and the keel is 45°, and the angle between the boom and the keel is 28°, find the forward thrust on the boat. [Ans. 70.5 lb.]

16. Two large stones of 250 Kg and 600 Kg, respectively, are located with their centers of mass 1.2 m apart. Find their common center of mass.

17. Two weights of 15 Kg and 28 Kg are located with their centers of gravity 50 cm apart. Find their center of gravity. [Ans. 32.56 cm from the smaller mass.]

18. Three equal masses of 25 Kg each are located at the corners of an equilateral triangle 1 m on a side. Find the center of mass.

6 WORK, ENERGY, AND POWER

It is well known that there are many forms of energy—heat, sound, light, mechanical, nuclear, chemical, electrical, etc. Each form plays an important role in our everyday life. In this chapter we are concerned with mechanical energy only, and in particular with kinetic energy and potential energy.

There is little doubt that the most important concept in all nature is energy. It is important because it represents a fundamental entity common to all forms of matter in all parts of the known physical universe. Closely associated with energy is another concept-*work*, a term used in civil life to describe the expenditure of one's stored up bodily energy. Because energy is most easily described in terms of work, this latter will first be treated in detail. In defining work we must first make a clear distinction between *weight* and *mass*.

6.1 WEIGHT AND MASS When a mass m is allowed to fall freely, it is the constant downward force of gravity on the mass that gives rise to its constant acceleration. If Newton's Second Law is applied to this motion, see Eq. (3a), the force F is none other than the weight F_g of the body, and the acceleration a is the acceleration g due to gravity. For falling bodies the force equation, $F = ma$, is written in different symbols.

weight = mass × acceleration

$$\boxed{F_g = mg} \tag{6a}$$

As we have seen in the preceding chapter, *weight* and *force* have both magnitude and direction and are therefore vector quantities. Mass, on the other hand, is a scalar quantity since it has only magnitude. The distinction between *weight* and *mass* is illustrated by imagining a given body to be carried out into free space far removed from other bodies. There it is free of all gravitational attraction. There, a body at rest will still have its mass but it will have no weight. That a body under these circumstances has its mass would be demonstrated if another mass were to bump into it. The smaller the mass of the incoming body, the less would be the recoil of the first mass from the impact.

Weight here on the earth is due to the gravitational attraction of the earth upon a mass at its surface. In the equation $F_g = mg$, we may define g as the *weight per unit mass*. F_g is the *mass* times the *weight per unit mass*.

EXAMPLE 1. Calculate the weight of a body having a mass of 1 Kg.

SOLUTION. By direct substitution in Eq. (6a), we obtain

$$F_g = 1\text{ Kg} \times 9.80\,\frac{\text{m}}{\text{s}^2} = 9.80\,\frac{\text{Kg m}}{\text{s}^2} = 9.80\text{ N}$$

This answer shows that to lift a mass of 1 Kg requires an upward force of 9.80 newtons and that weight and mass differ from each other numerically by the acceleration due to gravity. By similar calculations to the above, the weight of a 1-g mass is found to be 980 dynes, and the weight of a 1-slug mass is found to be 32 lb. See Fig. 6A.

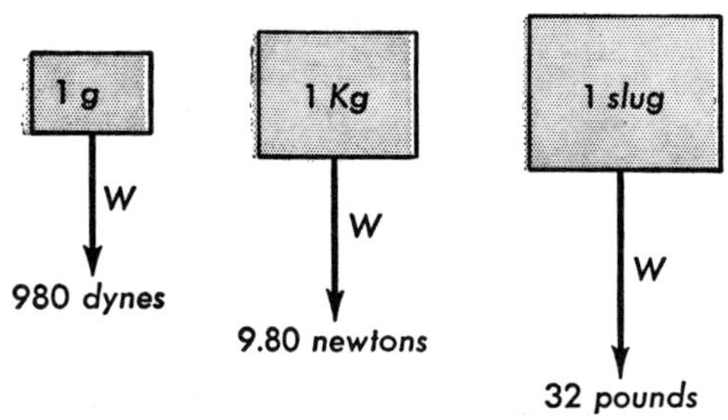

Fig. 6A. Weight is a force equal in magnitude to the mass multiplied by the acceleration due to gravity. $F_g = mg$.

It frequently becomes necessary to convert known masses, weights, and forces from the metric system to the engineering system or vice versa. For this purpose the following conversion factors are useful:

$$1\text{ newton} = 0.224\text{ lb}$$
$$1\text{ dyne} = 0.00000224\text{ lb}$$
$$1\text{ slug} = 14.6\text{ Kg}$$

The first two equations involve forces, while the last one involves masses.

6.2 WORK In its simplest mechanical form *work is defined as the force times the distance through which the force acts.*

work = force × distance

Algebraically,

$$\boxed{W = F \times h} \tag{6b}$$

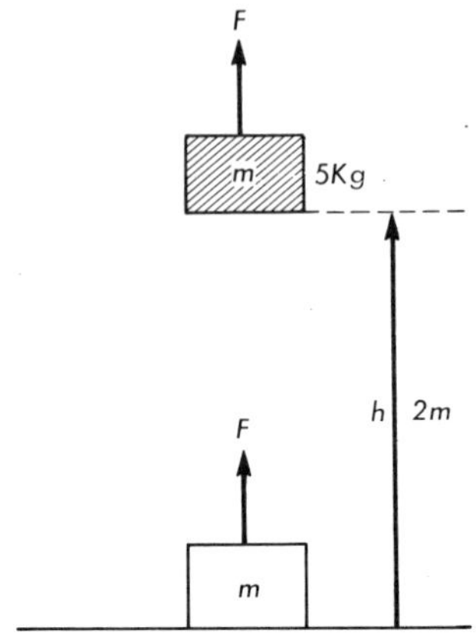

Fig. 6B. To raise a body vertically, work must be done against the pull due to gravity.

Consider the general problem of calculating the work done in lifting a mass m to a height h above the ground. See Fig. 6B.

By Eq. (6a) the force required to lift any mass m is equal to its own weight:

$$F_g = mg$$

By substituting the weight mg for F in Eq. (6b), we obtain

$$W = mg \times h \tag{6c}$$

To give numerical values, assume that a mass of 5 Kg is lifted vertically a distance of 2 m. By direct substitution in Eq. (6c), we obtain

$$W = 5\text{ Kg} \times 9.8\,\frac{\text{m}}{\text{s}^2} \times 2\text{ m} = 98\,\frac{\text{Kg m}^2}{\text{s}^2}$$

Since *force, in newtons*, has the units Kg m/s^2, the answer can also be written

$$W = 98\text{ newton meters} \tag{6d}$$

Work in the *MKS* system is seen to have the absolute units Kg m^2/s^2, which are equal to the derived units *newton meters*. In the *cgs* system the corresponding absolute units are *g cm^2/s^2*, which are equal to the derived units *dyne centimeters*. In the English or engineering system, the units of work are *foot-pounds*, abbreviated *ft-lb*.

EXAMPLE 2. Find the work done in lifting a weight of 5 lb to a height of 10 ft.

SOLUTION. By substituting directly in Eq. (6b), we obtain

$$W = 5 \text{ lb} \times 10 \text{ ft} = 50 \text{ ft-lb}$$

6.3 JOULES AND ERGS

In the *MKS* system of units, a force of *1 newton* acting through a distance of *1 meter* in the same direction performs an amount of work equivalent to *1 joule*.*

$$\boxed{1 \text{ newton meter} = 1 \text{ joule}} \qquad (6e)$$

The joule as a unit of work is abbreviated J. In the *cgs* system the *dyne cm* as a unit of work is called the *erg*.

$$1 \text{ dyne cm} = 1 \text{ erg}$$

A force of 1 dyne acting through a distance of 1 cm in the same direction does 1 erg of work.

Since the *newton* as a unit of force = $1\text{Kg} \times 1 \text{ m/s}^2 = 1000 \text{ g} \times 100 \text{ cm/s}^2 = 10^5$ dynes, the newton meter = 10^5 dynes $\times$ 100 cm = 10^7 dyne cm. In other words,

$$1 \text{ joule} = 10^7 \text{ ergs} \qquad (6f)$$

The *joule* as a unit of work is, therefore, much larger than the *erg* and in many practical problems is to be preferred because of the smaller numbers involved in calculations.

EXAMPLE 3. Calculate the work done in lifting a mass of 4 Kg to a height of 2.50 m.

SOLUTION. The known quantities are $m = 4$ Kg, $h = 2.5$ m, and $g = 9.80 \text{ m/s}^2$. By direct substitution in Eq. (6c), we obtain

$$W = 4 \text{ Kg} \times 9.80 \frac{\text{m}}{\text{s}^2} \times 2.5 \text{ m}$$

$$W = 98.0 \frac{\text{Kg m}^2}{\text{s}^2}$$

$$W = 98.0 \text{ J}$$

The answer is read ninety-eight point zero joules.

* JAMES PRESCOTT JOULE (1818–1889), English physicist, born December 24, 1818, near Manchester. He owned a large brewery but devoted his life to scientific research. In 1840, he stated the law of the electrical equivalent of heat, and later the first law of thermodynamics. He determined the mechanical equivalent of heat in four different ways. With Lord Kelvin he performed work on the compression of gases and the heat changes that take place when the gases escape through small holes. His numerous scientific papers were published by the Royal Society of London in 1884.

6.4 WORK DONE AGAINST FRICTION In sliding a mass of 5 Kg along a horizontal plane a distance of 2 m, the work done *will not* in general be as great as that required to lift the same mass 2 m vertically.

As shown by an example in Fig. 6C, the

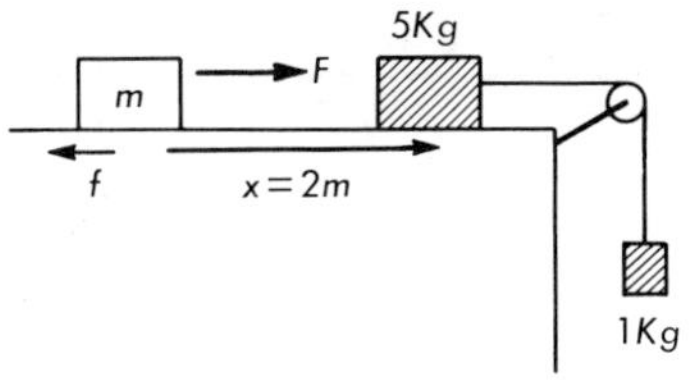

Fig. 6C. To slide a body along a level plane, work must be done against friction.

weight of a 1-Kg mass is large enough to overcome friction and slide the 5-Kg mass along the table. Here too the work is defined as force times distance, and we write

$$W = F \times x \qquad (6g)$$

where x is the distance traveled. Direct substitution in Eq. (6g) shows that the work done

$$W = 1 \text{ Kg} \times 9.8 \frac{\text{m}}{\text{s}^2} \times 2 \text{ m}$$

$$W = 19.6 \frac{\text{Kg m}^2}{\text{s}^2}$$

or

$$W = 19.6 \text{ J}$$

This is only one fifth as much work as that required to lift the same 5-Kg mass an equal vertical distance of 2 m. See Eq. (6d). By reducing the friction between the block and plane, the force F can be reduced still further. Such a reduction can be accomplished by smoothing and lubricating the sliding surfaces, or better by mounting the block on wheels. Could the friction be eliminated entirely, the work done in moving any object in a horizontal direction would be practically zero, for once started it would continue moving with constant velocity. A vertical lift, however, requires at least an amount of work equal to the weight mg times the height h.

When a force is applied to an object such that it makes an angle with the direction of motion, only the component of the force in the direction of motion is effective in doing work. This is illustrated in Fig. 6D where a force F is applied

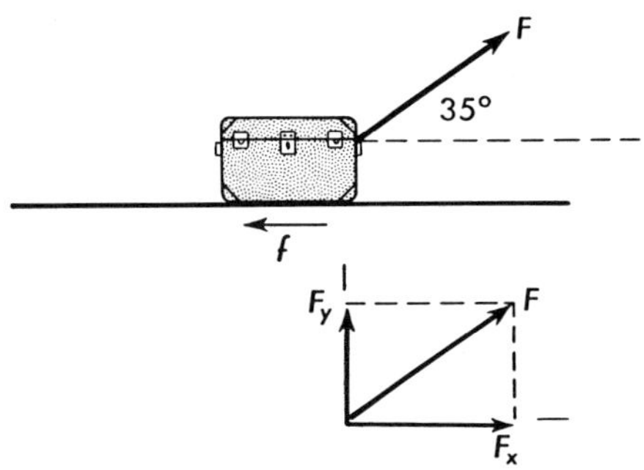

Fig. 6D. The force pulling a heavy trunk across a floor is resolved into components.

to a heavy trunk to pull it across the floor. In the right-hand diagram, F is shown resolved into vertical and horizontal components F_y and F_x. The upward force F_y has a lifting effect and helps to reduce friction, while the horizontal force F_x is the one that does the work. This force multiplied by the horizontal distance moved is equal to the work done.

$$W = F_x \times x$$

In calculating work done, the *force* and *distance* are always measured in the same direction. If the distance is horizontal it is customary to use the letter x as in Eq. (6g), and if the distance moved is vertical to use h as in Eq. (6b).

6.5 POTENTIAL ENERGY Mechanical energy is divided into two categories, *potential energy* and *kinetic energy*. A system *is said to have potential energy if by virtue of its configuration or state it is able to do work*. Water at the top of a fall or a wound clock spring is an example of a system with potential energy. The clock spring may keep a clock running for a certain length of time, and the water may, by falling, turn a paddle wheel. Potential energy is measured by the amount of work that is available. It is therefore measured in *joules*, *ergs*, or *foot-pounds*.

If a given mass m is raised to a specified height h, as illustrated in Fig. 6E, it then has potential

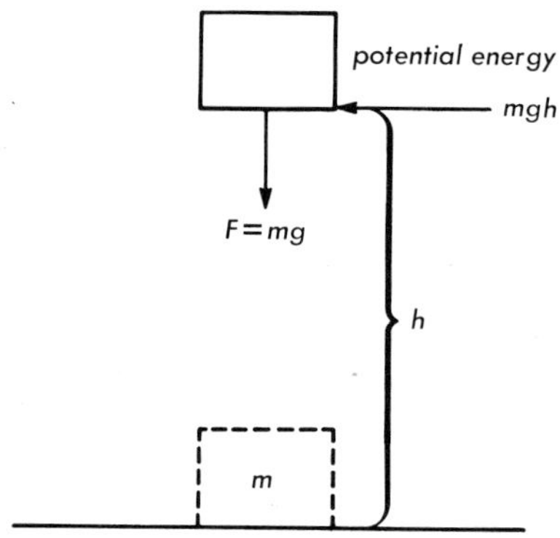

Fig. 6E. A body has potential energy by virtue of its position.

energy $F \times h$ by virtue of its position above the ground level from which it has been lifted. The *work done* in lifting it has been stored up as potential energy in the block. This energy can be regained by dropping the mass back to the ground, for in so doing it can be made to perform some kind of work. By definition,

$$\text{potential energy} = F \times h \qquad (6h)$$

or, expressing F in weight units, Eq. (6a),

$$\textbf{P.E.} = mg \times h$$

EXAMPLE 4. A mass of 2.50 Kg is raised to a height of 5 m above the ground. Calculate its potential energy.

SOLUTION. By substituting the known quantities in Eq. (6h), we obtain

$$\textbf{P.E.} = 2.50 \text{ Kg} \times 9.80 \frac{\text{m}}{\text{s}^2} \times 5 \text{ Kg} = 122.5 \text{ J}$$

If a body is lifted straight upward, carried up a staircase, or pulled up an inclined plane, the potential energy acquired is given by the *weight* × *vertical height* to which it is raised.

The meaning of *positive*, *zero*, or *negative potential energy* is illustrated in Fig. 6F. Located

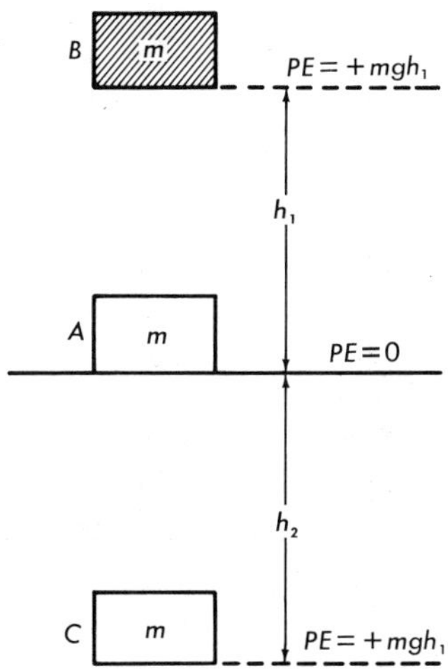

Fig. 6F. Potential energy with respect to a base plane may be plus or minus.

at any point above the *base plane* a body has positive potential energy, while at points below that line it has negative potential energy. To lift the mass *m* from A to B, work is done and the mass acquires potential energy to the amount of mgh_1.

In returning from B to A the mass loses potential energy, performing *work* = mgh_1 on some other body. Similarly, in going from A to C, the body loses energy and ends up at C with mgh_2 less energy than it had at A. To raise it again to A an equivalent amount of work mgh_2 will have to be done on the body.

The choosing of a *base plane* as a zero energy level is a purely arbitrary selection. In most practical applications it is customary to select the lowest point to be reached by a body as the zero level so that all displacements from there will be positive in sign. In the engineering systems of units *potential energy*, like *work*, is expressed in *foot-pounds*, i.e., *pounds* × *vertical distance in feet*.

6.6 KINETIC ENERGY *The kinetic energy of a moving body is defined as its ability to do work by virtue of its motion.* A car moving along the highway has kinetic energy of translation, and a rotating wheel on a machine has kinetic energy of rotation. See Fig. 6G. For a given mass *m*,

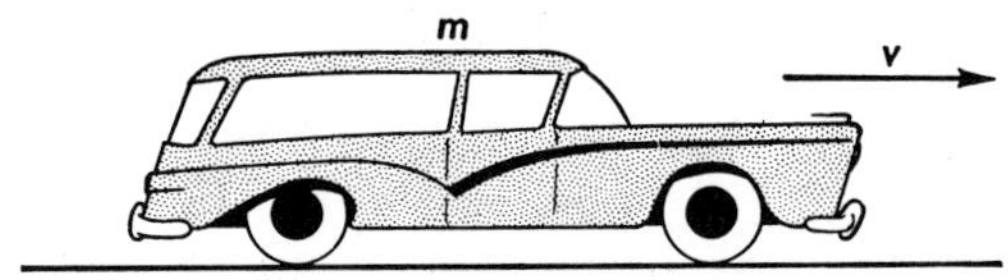

Fig. 6G. A moving body has kinetic energy $\frac{1}{2}mv^2$.

moving in a straight line with constant velocity *v*, the kinetic energy is given by

$$\text{kinetic energy} = \tfrac{1}{2}mv^2 \tag{6i}$$

EXAMPLE 5. Calculate the kinetic energy of a 20-Kg mass moving with a velocity of 4 m/s.

SOLUTION. By direct substitution in Eq. (6i), we obtain

$$\text{K.E.} = \tfrac{1}{2} \times 20\ \text{Kg} \times \left(4\ \frac{\text{m}}{\text{s}}\right)^2 = 160\ \frac{\text{Kg m}^2}{\text{s}^2}$$

This answer has exactly the dimensions of *work*, and *potential energy*, and can be written in the same derived units.

$$\text{K.E.} = 160\ \text{joules}$$

A moving body has energy because in being brought to rest it must exert a force *F* on some other object, and this force acting through a distance *x* does work. In other words, work can be done by a moving body. Conversely, by applying a constant horizontal force *F* on a body of mass *m*, for a distance *x*, it will be given a kinetic energy $\frac{1}{2}mv^2$. See Fig. 6H.

$$F \times x = \tfrac{1}{2}mv^2 \tag{6j}$$

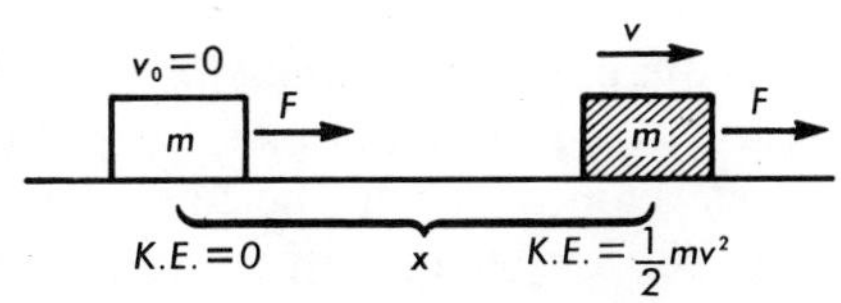

Fig. 6H. A body has kinetic energy by virtue of its motion.

This is known as the "work equation." In it friction is entirely neglected, and the body is presumed to start from rest.

EXAMPLE 6. A constant horizontal force of 40 N acts for a distance of 20 m on a 500 Kg small racing car. If friction is neglected and the car starts from rest, what is its velocity?

SOLUTION. The known quantities are $F =$ 40N, $x =$ 20 m, and $m =$ 500 Kg. In applying Eq. (6j) to this problem, all factors but v^2 are transposed to the other side of the equality and then substitutions made as follows:

$$v^2 = \frac{2(F \times x)}{m}$$

$$v^2 = \frac{2(40 \text{ N} \times 20 \text{ m})}{500 \text{ Kg}}$$

$$v^2 = 3.200 \frac{\text{m}^2}{\text{s}^2}$$

$$v = 1.789 \frac{\text{m}}{\text{s}}$$

6.7 POWER *Power is defined as the rate of doing work, or the rate at which work is being done.*

$$\textbf{Power} = \frac{\textbf{work}}{\textbf{time}}$$

$$\boxed{P = \frac{F \times x}{t}} \qquad (6k)$$

The faster a given amount of work is done the greater is the power. In other words, the smaller the time t in the above equation the greater is the fraction $F \times x/t$ and the power P.

A good illustration of power is given in Fig. 6l. An elevator car with a mass m is raised a given distance h, in a specified time t. If all three of these quantities are known, their values substituted in Eq. (6k) will give the power that is being developed during the time it is rising.

In the metric system, with work measured in *ergs* or *joules*, power is expressed either in *ergs per second* or in *joules per second*. One joule per second is called the *watt*,* a unit of power.

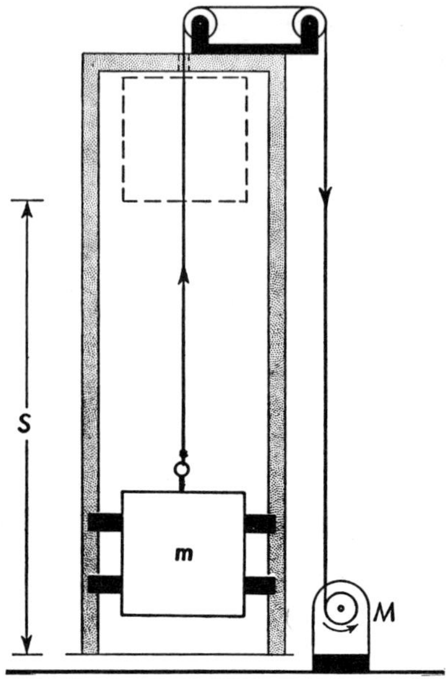

Fig. 6l. Demonstration of a concrete hoist. See Examples 5 and 15.

$$1 \frac{\text{joule}}{\text{second}} = 1 \text{ watt} \qquad (6l)$$

The *kilowatt* is another unit of power and is equal to *1000 watts.*

In the engineering system, with work measured in foot-pounds, power is expressed in *foot-pounds per second*, and in *horsepower* (abbr. hp).

$$1 \text{ hp} = 550 \text{ ft-lb/s} \qquad (6m)$$

EXAMPLE 7. Find the power of an engine capable of lifting 200 Kg to a height of 65 m in 8 s.

* JAMES WATT (1736–1819), Scottish engineer and inventor for whom the watt is named. He is best known for his long and painstaking work of improving the efficiency of the steam engine. He also invented the centrifugal governor and the steam pump. In spite of his life-long work on the steam engine, he took a dim view of its use in locomotives.

SOLUTION. By direct substitution in Eq. (6k), we obtain

$$P = \frac{200 \text{ Kg} \times 9.80 \text{ m/s}^2 \times 65 \text{ m}}{8 \text{ s}}$$

$$P = 15{,}925 \frac{\text{Joule}}{\text{second}}$$

To four significant figures†

$$P = 15{,}920 \text{ watts}$$

or

$$P = 15.92 \text{ KW}$$

† When the last digit to be dropped from any number is 5, (a) leave the next-to-last digit unchanged if it is even, or (b) raise the next-to-last digit by one if it is odd.

QUESTIONS

1. Define or briefly explain each of the following: (a) kinetic energy, (b) work, (c) watt, (d) erg, and (e) newton meter.

2. Define or briefly explain each of the following: (a) potential energy, (b) power, (c) horsepower, (d) joule, and (e) dyne centimeter.

3. Clearly explain the difference between (a) energy and power, (b) joule and watt, and (c) foot-pound and horsepower.

PROBLEMS

1. A 250-Kg load is lifted 32 m by an elevator. Find the work done. [Ans. 78,400 J.]

2. A concrete hoist raises a load of 680 Kg to a height of 260 m. Find the work done.

3. Find the work done when a hiker whose mass is 82 Kg climbs a mountain from sea level to a height of 4220 m. [Ans. 3.391×10^6 J.]

4. An elevator car with a load of people has a total mass of 2500 Kg. Find the work done in raising this car from the first floor to the twentieth floor if the average distance between floors is 4.5 m.

5. A man weighing 60 Kg, on a motorcycle of 125 Kg, is traveling at 90 Km/hr. Find the total kinetic energy. [Ans. 1.031×10^5 J.]

6. A car with a mass of 1200 Kg travels along a straight and level road at 80 Km/hr. Find the kinetic energy.

7. A car with a mass of 1500 Kg has a kinetic energy of 1.4×10^6 J. What is its velocity? [Ans. 43.2 m/s.]

8. A jet plane with a total mass of 1.2×10^5 Kg has a cruising speed of 900 Km/hr. Find its kinetic energy.

9. A bullet of 25 g is shot from a hunting rifle with a velocity of 620 m/s. Find the kinetic energy in (a) joules, and (b) ergs. [Ans. (a) 4800 J, (b) 4.80×10^{10} ergs.]

10. A 2400-lb car is traveling along the road at 60 mi/hr. Find its kinetic energy.

11. A small plane weighing 3600 lb has a cruising speed of 120 mi/hr. What is its kinetic energy? [Ans. 1.742×10^6 ft lb.]

12. A 12-Kg mass falls freely from a height of 160 m. Find (a) the potential energy just before it is released, and (b) the kinetic energy just before it hits the ground.

13. An 8-Kg stone is dropped from a height of 42 m. Find (a) the potential energy before it is released, and (b) the kinetic energy just before it hits the ground. [Ans. (a) 3293 J, (b) 3293 J.]

14. A loaded elevator with a total mass of 3200 Kg rises 60 m in 12 s. Find the power developed in KW.

15. A 5000-Kg truck in going up a 5 per cent grade climbs a vertical height of 400 m in 16 min. Find (a) the potential energy stored, and (b) the average power developed. [Ans. (a) 1.960×10^7 J, (b) 20.42 KW.]

16. A concrete hoist raises a 250 Kg load a vertical distance of 90 m in 18 s. Find the power expended in KW.

17. A truck with a total mass of 4200 Kg requires a force of 1000 N to keep it moving along a straight and level road at 12 m/s. Find (a) the kinetic energy, and (b) the power developed. [Ans. (a) 3.024×10^5 J, (b) 12 KW.]

18. Water is pumped from a river into a tank, a total lift of 135 m, at a rate of 6 m^3/hr. What is the power developed? One cubic meter of water has a mass of 1000 Kg.

19. How many cubic meters of water can be pumped in one hour from a river to a reservoir by a 3-KW pump? Assume a lift of 75 m. One cubic meter of water has a mass of 1000 Kg. [Ans. 14.69 m^3.]

7 CONSERVATION OF ENERGY AND MOMENTUM

Over the past two hundred years many principles and laws have been formulated from controlled laboratory experiments to describe the physical behavior of nature. Some of these laws hold accurately within certain boundary conditions but deviate from ideal behavior when extreme experimental conditions are imposed.

There are a few laws, however, which have been found to hold under all conditions. They seem to hold from the smallest known particles of atomic structure, to the largest dynamic systems in the universe, the galaxies themselves. One of these is the law of conservation of energy.

7.1 CONSERVATION OF ENERGY Most important of all the laws of nature is the law of the conservation of energy. Although the law has been stated in almost as many different ways as there are books written on the subject, they all have in reality the same meaning. The following three examples are typical statements: (1) *in transforming energy from one form to another, energy is always conserved,* (2) *energy is never created nor destroyed, or* (3) *the sum total of all energy in the universe remains constant.*

In this chapter we are concerned with the law of conservation of energy only as it applies to the two forms of mechanical energy, *potential* and *kinetic*. The law will again be encountered in connection with the other forms in the chapters on heat, electricity, and atomic structure.

Consider the energy involved in a waterfall as shown in Fig. 7A. The water at the top of the

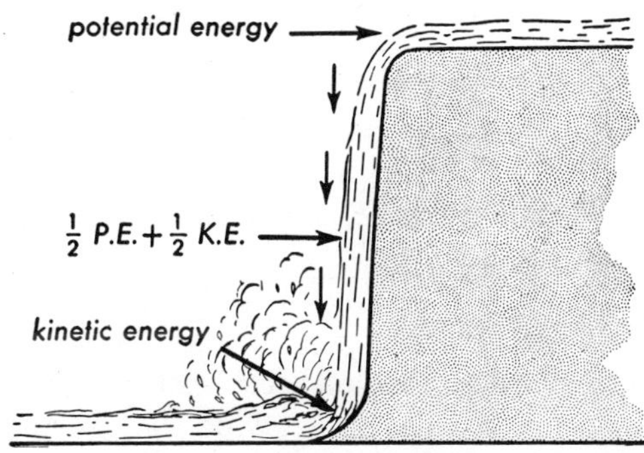

Fig. 7A. All the available energy at the top of a waterfall is potential. At the bottom it is kinetic.

fall has potential energy by virtue of its position above the base. As it falls over and then downward with ever-increasing speed, the kinetic energy $\frac{1}{2}mv^2$ increases while the potential energy decreases. At the bottom of the fall the potential energy approaches zero and the kinetic energy approaches its maximum value. At the top the energy was practically all potential while near the bottom it is mostly kinetic. Assuming the water starts from rest at the top and no energy is lost in falling, the P.E. at the top of the falls equals the K.E. at the bottom.

P.E. at top = K.E. at bottom

$$F \times h = \tfrac{1}{2} mv^2 \qquad (7a)$$

or

$$mgh = \tfrac{1}{2} mv^2 \qquad (7b)$$

Solving for v,

$$v^2 = 2gh \quad \text{or} \quad v = \sqrt{2gh} \qquad (7c)$$

This special equation for the speed of freely falling bodies can also be derived from Eqs. (4a) and (4b). Its derivation is left as an exercise for the student. See Prob. 9.

EXAMPLE 1. A mass of 25 Kg is dropped from a height of 5 m. Find the kinetic energy and velocity just as it reaches the ground.

SOLUTION. Since the P.E. at the top is equivalent to the K.E. at the bottom,

$$\text{P.E.} = 25\ \text{Kg} \times 9.80\ \frac{\text{m}}{\text{s}^2} \times 5\ \text{m}$$

$$= 1225\ \text{joules} = \text{K.E.}$$

The velocity is found by Eq. (7c) to be

$$v = \sqrt{2 \times 9.8 \times 5} = 9.9\ \text{m/s}$$

When the falling body in the above problem is part way down, it has some P.E. and some K.E. Its total energy E is therefore of two kinds

$$\boxed{E = \tfrac{1}{2} mv^2 + mgh} \qquad (7d)$$

At the instant the body reaches the ground it is suddenly stopped and all of the energy is quickly transformed into heat. The transformation of mechanical energy into heat is often demonstrated in the physics laboratory by an experiment in which a quantity of lead-shot is dropped from a height of several feet and its temperature measured before and after falling. By raising the shot and dropping it many times, the rise in temperature amounts to several degrees.

Consider next the case of a body sliding down an incline. Starting from rest at the top of an incline a mass m has potential energy mgh and no kinetic energy. See Fig. 7B. Upon reaching the

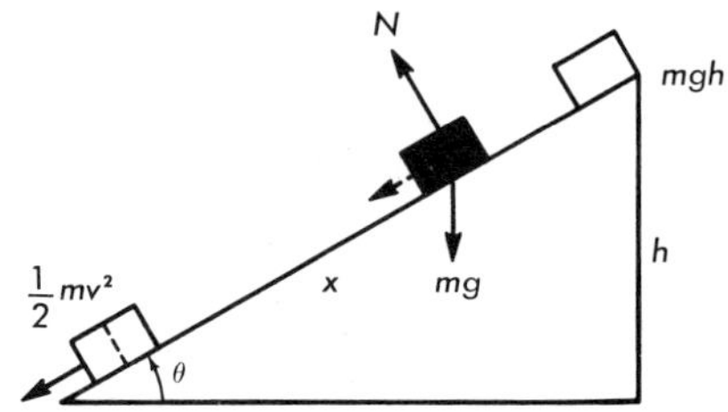

Fig. 7B. In sliding downhill, potential energy may be transformed into kinetic energy.

bottom the potential energy has been reduced to zero and the energy is all kinetic. By conservation the total energy is at all times constant, so that

$$\text{total E at top} = \text{total E at bottom}$$

$$\text{P.E.} + \text{K.E.} = \text{P.E.} + \text{K.E.} \qquad (7e)$$

$$mgh + 0 = 0 + \tfrac{1}{2} mv^2$$

from which

$$v = \sqrt{2gh} \qquad (7f)$$

This is just the velocity acquired by a body falling freely from a height h. See Eq. (7c). Such a result is interesting because it shows that the velocity acquired by a body sliding without friction down an inclined plane of any angle θ is equal in magnitude to the velocity acquired by a body falling freely from the same height h. It should be noted, however, that while the final speeds are the same, the accelerations and the times of descent are different.

7.2 THE SIMPLE PENDULUM A similar energy treatment can be given for the swinging of a simple pendulum. At the extreme ends of each swing, see Fig. 7C, the bob comes momentarily to rest and the energy E is all potential and equal to mgh. At the bottom of the swing the energy E is all kinetic and equal to $\frac{1}{2} mv^2$.

The motion of the pendulum bob is like that of a body sliding, without friction, down an inclined plane of changing angle. The kinetic

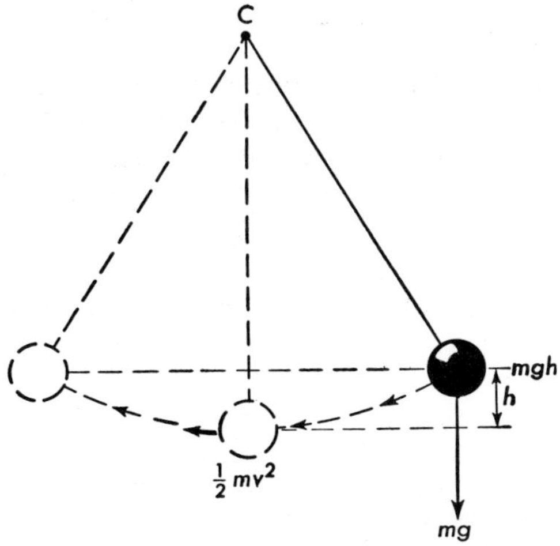

Fig. 7C. Potential energy of a pendulum changes to kinetic energy and back again.

energy acquired in going down one side is just sufficient to carry it up to an equal height on the other. If a pin is placed at P as shown in Fig. 7D the bob will rise to the same height h as before.

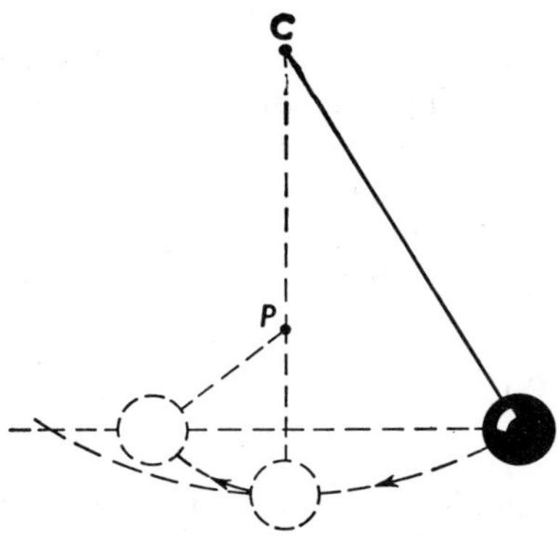

Fig. 7D. The pendulum bob rises to the same height.

7.3 CONSERVATION OF MOMENTUM When two or more bodies collide with each other, momentum is conserved. This law, and the law of conservation of energy, constitute the two most basic laws in all dynamics.

Any moving object has *momentum*, a measurable quantity that is given by the following product:

$$\textbf{momentum} = mv$$

where m is the mass of the body, and v is its velocity. Like several other quantities in mechanics, momentum is a vector quantity; it has a magnitude equal to mv, and a direction the same as that of the velocity.

The law of conservation of momentum applies to all collision phenomena and states that the total momentum before impact equals the total momentum after impact. Consider as an example the "head on" collision of two balls as shown in Fig. 7E. Before impact the mass m_1 is moving

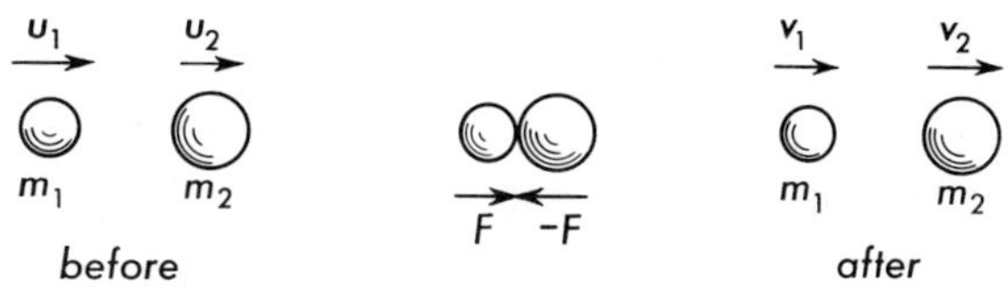

Fig. 7E. The total momentum of two bodies before impact is equal to the total momentum after impact.

with a velocity u_1 and has a momentum m_1u_1, while m_2 is moving with a velocity u_2 and has a momentum m_2u_2. The total momentum before impact is therefore equal to the sum of the two momenta, $m_1u_1 + m_2u_2$.

By similar reasoning it is clear that after impact, m_1 and m_2, with their new velocities v_1 and v_2, have a total momentum $m_1v_1 + m_2v_2$. The law of conservation of momentum requires that

total momentum before impact

= total momentum after impact

$$\boxed{m_1u_1 + m_2u_2 = m_1v_1 + m_2v_2} \tag{7g}$$

During impact two equal but opposite forces are set up between the bodies, one the force exerted by m_1 on m_2 and the other the force exerted by m_2 on m_1. These two equal but opposite forces are an action and reaction pair explained by Newton's Third Law of Motion. Each force acts for the same short interval of time giving equal changes of momenta to both bodies. One body gains as much momentum as the other loses. In other words, the total momentum remains constant.

EXAMPLE 2. An ivory ball of mass 5 Kg moving with a velocity of 20 m/s collides with another ivory ball of mass 10 Kg moving in the same direction along the same line with a velocity of 10 m/s. After impact the first mass is still moving in the same direction but with a velocity of only 8 m/s. Calculate the velocity of the second mass after impact.

SOLUTION. By direct substitution in Eq. (7g), we obtain

$$(5 \times 20) + (10 \times 10) = (5 \times 8) + (10 \times v^2)$$
$$200 = 40 + 10v_2$$
$$10v_2 = 160$$
$$v_2 = 16 \text{ m/s}$$

After impact the second mass has a velocity of 16 m/s.

If, in the preceding example, the total kinetic energy after impact is calculated and compared with the total kinetic energy before impact the two will not be found equal. Employing the equation K.E. $= \frac{1}{2} mv^2$,

$$\text{K.E.} = \tfrac{1}{2}(5 \times 20^2) + \tfrac{1}{2}(10 \times 10^2)$$
$$= 1500 \text{ joules, before impact}$$

$$\text{K.E.} = \tfrac{1}{2}(5 \times 8^2) + \tfrac{1}{2}(10 \times 16^2)$$
$$= 1440 \text{ joules, after impact}$$

The difference in energy, to the amount of 60 joules, has disappeared as mechanical energy and gone into heat. During impact both masses were slightly deformed in shape as a result of the mutually acting forces, and a small amount of heat was generated internally. This heat goes to raise the temperature of the two colliding bodies. It is only by including this heat energy of 60 joules with the mechanical energy after impact that the law of conservation of energy is upheld.

This is just another way of stating that such collisions are not perfectly elastic. If they were perfectly elastic, conservation of mechanical energy would hold as well as conservation of momentum. Perfectly elastic collisions are known to occur between the *ultramicroscopic* atoms and molecules of a gas but not with the *macroscopic* bodies encountered in everyday life. The more inelastic the colliding bodies, the more energy is transformed into heat. A treatment of elasticity and its application to collision problems will be given in Chap. 11.

It is important to note from the above discussion of Example 2, that *for most impact problems the law of conservation of momentum should be applied and not the law of conservation of energy.*

7.4 EXPERIMENTS An interesting experiment illustrating conservation of momentum is shown in Fig. 7F. Two small cars of equal mass,

$$m_1 = m_2,$$

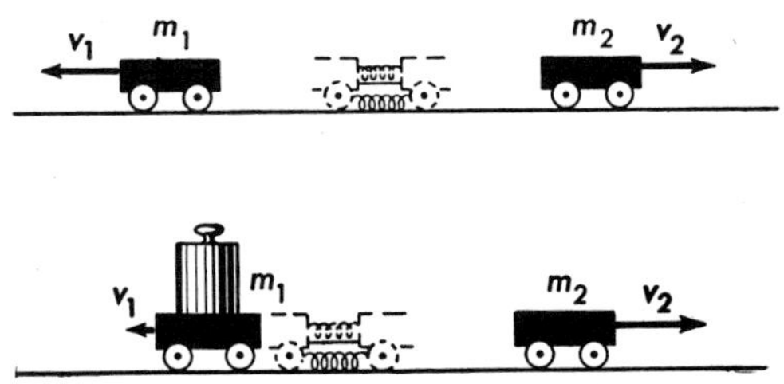

Fig. 7F. Demonstration of conservation of momentum.

are tied together with a compressed spring between them. When the cord tie is burned with a match, releasing the spring, the two cars fly apart with equal velocities. Before the spring is released the cars are at rest and the total momentum is zero. After the spring is released the total momentum is still zero since the two velocities are oppositely directed. Momentum being a vector quantity,

$$m_1v_1 + m_2v_2 = 0 \tag{7h}$$

With motions to the right taken as positive, v_2 is positive and v_1 negative, and the two momenta cancel.

If this same experiment is repeated with one of the cars heavily loaded as shown in diagram (b), the two fly apart as before but with unequal speeds. The lighter mass moves away with a high speed while the heavier mass recoils with a low speed. The product $m_1 \times v_1$, however, is equal in magnitude to the product $m_2 \times v_2$, and the sum of the two momenta is zero.

EXAMPLE 3. A shell weighing 60 Kg is shot with an initial velocity of 500 m/s, from a gun weighing 2000 Kg. What is the initial velocity with which the gun recoils?

SOLUTION. By applying Eq. (7h), and substituting directly the known quantities, we obtain

$$60 \text{ Kg} \times 500 \text{ m/s} + 2000 \text{ Kg} \times v_2 = 0$$
$$30{,}000 \text{ Kg m/s} + 2000\, v_2 \text{ Kg} = 0$$
$$v_2 = \frac{-30{,}000 \text{ Kg m/s}}{2000 \text{ Kg}} = -\ 15 \text{ m/s}$$

The gun recoils with a velocity of 15 m/s.

An experiment illustrating the recoil velocity of a heavy mass like that in Example 3 is shown by two pendulums in Fig. 7G. A U-shaped spring,

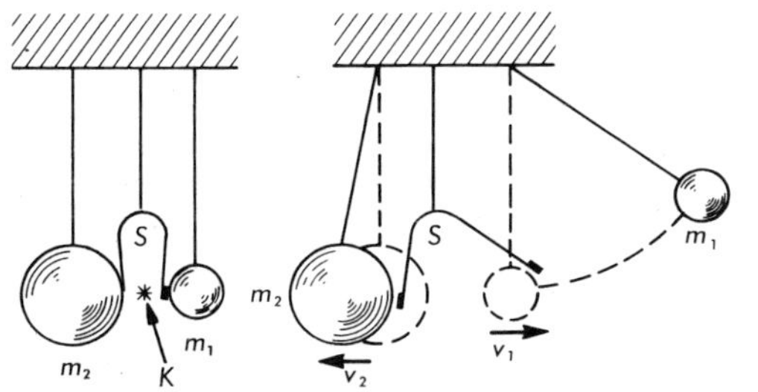

Fig. 7G. Experiment with a light and heavy pendulum, illustrating the law of conservation of momentum.

compressed and tied with a cord K, rests against a heavy iron ball m_2 and a small wooden ball m_1. When the cord K is burned through by a lighted match, the spring is released and the two masses fly apart with equal but opposite speeds and momenta. The small mass acquires a high velocity and swings up to a considerable height while the large mass is barely perceived to move.

Another interesting experiment illustrating conservation of momentum may be performed with eight or nine duck-pin balls and a grooved board as shown in Fig. 7H. When one ball is rolled up to the others, it will be stopped by collision with the end ball, and the ball on the opposite end will roll out with almost the same speed. If two balls are rolled up, as indicated in the diagram, two will roll out on the other end, and if three are rolled up, three will roll out, etc. Glass or steel marbles work best in this experiment as they are highly elastic; however, duck-pin balls or billiard balls produce a more spectacular effect.

QUESTION. When the two balls are rolled up, why doesn't just one ball roll off on the other side with twice the velocity, thus conserving momentum?

ANSWER. The answer to this question involves the conservation of energy as well as momentum. If only one ball came off with twice the velocity, its kinetic energy would be twice the energy available from the two incident balls.

A spectacular experiment of a similar nature may be performed by suspending four or five large bowling balls from the ceiling of the room as shown in Fig. 7I. When one ball is pulled back

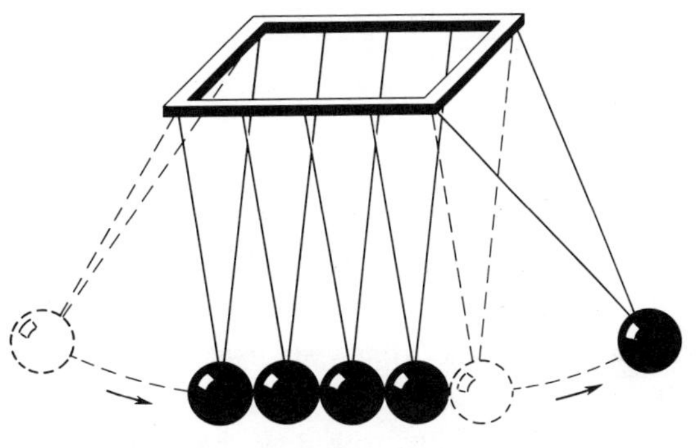

Fig. 7I. Experiment with bowling balls, illustrating the law of conservation of momentum.

and released, it will upon impact with the others be brought to rest and at the same time cause the ball at the opposite end to recoil outward with approximately the same speed. Two balls re-

Fig. 7H. Experiment with the duckpin balls, illustrating the law of conservation of momentum.

leased will cause two to move out on the other side, etc.

The propelling force of a jet plane or rocket is derived from the principle of momentum. See Fig. 7J.

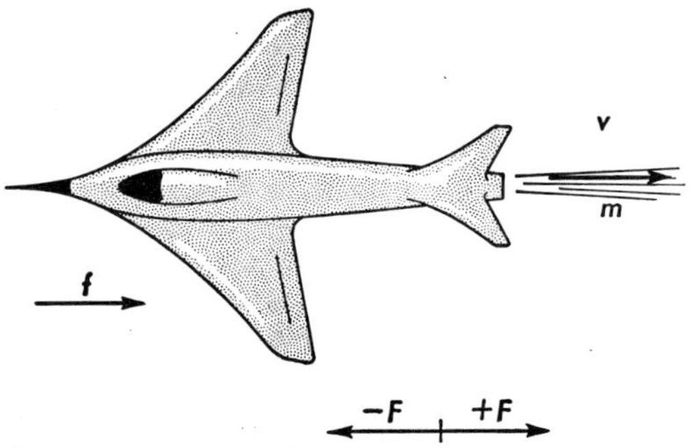

Fig. 7J. Conservation of momentum accounts for the forward thrust on a rocket, plane, or missile.

As the gases are burned within the combustion chambers of the engine, they exert a large forward force $-F$ on the plane and an equal and opposite force $+F$ on the exhausting gases. As a result of the backward force the gases acquire a very high velocity v and a momentum mv. When the plane is flying at constant speed, the forward thrust $-F$ just balances the frictional resistance f of the air. To relate the thrust with the exhaust gas momentum, we write down the force equation:

$$F = ma$$

For uniformly accelerated motion we also have

$$v = at$$

or

$$a = v/t$$

Direct substitution in the force equation gives

$$F = m\frac{v}{t}$$

or

$$\boxed{Ft = mv} \tag{7i}$$

To apply this, the *impulse equation*, to the jet or rocket engine, m represents the mass of gas exhausted in any chosen number of seconds t. The impulse $F \times t$ exerted on the gases is just equal and opposite to the impulse $-F \times t$ exerted on the plane.

QUESTIONS

1. State the law of conservation of momentum. Give an example by making a diagram and explaining it in detail.

2. State the law of conservation of energy in your own words. Give an example in which potential energy is transformed into kinetic energy.

3. Give an example from your own experience where kinetic energy is transformed into potential energy.

4. At what point in its path does the bob of a swinging pendulum have (a) its maximum speed, (b) zero speed, (c) maximum acceleration, and (d) zero acceleration?

5. When a car goes down a hill and the brakes are applied to keep its speed at 20 mi/hr, what becomes of the potential energy? Explain.

6. A small rocket engine is mounted on the stern of a small boat so it fires its exhaust blast toward the rear. Explain the forces involved, and the motions expected from the firing.

PROBLEMS

1. A 12-Kg stone is dropped from a height of 35 m. Find (a) its potential energy before being dropped, (b) its kinetic energy 10 m above the ground, and (c) the kinetic energy with which it arrives at the ground. [Ans. (a) 4120 J, (b) 2940 J, (c) 4120 J.]

2. A 4-Kg weight is dropped from a height of 18 m. Find (a) its initial potential energy, (b) its maximum kinetic energy, and (c) its maximum velocity.

3. A boy on a bicycle coasts for 160 m down an 8% grade. What will be his speed at the bottom of the hill? [Ans. 15.81 m/s.]

4. A 50-g bullet with a velocity of 650 m/s enters and becomes embedded in a block of wood with a mass of 4.80 Kg. With what velocity will the block recoil if it was initially at rest?

5. A bowling ball with a mass of 4.50 Kg and moving with a velocity of 7.20 m/s collides with another ball of the same mass and moving in the same direction with a velocity of 1.80 m/s. After impact the first mass is still moving in the same direction but with a velocity of 2.90 m/s. Calculate the velocity of the second mass. [Ans. 6.10 m/s.]

6. A ball of 4 Kg and moving with a velocity of 6.80 m/s collides head-on with another ball with a mass of 2.50 Kg moving in the same direction with a velocity of 2.20 m/s. After impact the second mass is moving in the same direction with a velocity of 7.30 m/s. Calculate the velocity of the first mass.

7. A 640-Kg gun, mounted on wheels, fires a 4-Kg shell with a velocity of 620 m/s. Calculate the recoil velocity of the gun. [Ans. −3.875 m/s.]

8. A 50-g bullet with a velocity of 615 m/s passes through a 4.50 Kg block of wood initially at rest. If the bullet emerges from the other side with a velocity of 240 m/s, what is the recoil velocity of the block of wood?

9. A 5-Kg block of wood lying on a fence post is hit by a 6-g bullet. If the bullet enters the block with a velocity of 680 m/s and leaves the opposite side with a velocity of 120 m/s, find (a) the recoil velocity of the block, (b) the energy of the bullet before impact, (c) the energy of the bullet after impact, (d) the energy of the recoiling block, and (e) the energy dissipated in the form of heat in the wood. [Ans. (a) 0.672 m/s, (b) 1387 J, (c) 43.20 J, (d) 1.129 J, (e) 1343 J.]

10. A rifle aimed vertically upward has a 200-g block of wood placed on the muzzle. When the bullet of 1.8 g is fired into the block, the combined mass rises to a height of 1.2 m. Find (a) the velocity of the recoiling block, and (b) the muzzle velocity of the bullet.

11. A 3-Kg block of wood hangs as a pendulum by a long wire. When a 5-g bullet is fired horizontally into the wood and becomes embedded there, the block swings to a height 12 cm above its rest position. Find (a) the recoil speed of the block, and (b) the muzzle velocity of the bullet. [Ans. (a) 1.534 m/s, (b) 922 m/s.]

12. A jet plane with its cargo has a mass of 1×10^5 Kg. It takes off and climbs to a height of 5 Km, where it cruises at 800 Km/hr while in straight and level flight. Find (a) its potential energy, (b) its kinetic energy, and (c) its total energy.

8 | CIRCULAR MOTION

The world in which we live is a dynamic system of moving bodies. From the tiniest of atomic particles on the one hand, to the largest galaxies on the other, we can say that everything is in a state of motion. It can be shown that any motion we can impose upon any rigid body can be resolved into two components. One component is a simple *translation* without *rotation*, and the other is a *rotation* without *translation*.

Try moving any object from one specified position and orientation to some other position and orientation. You will find that this same change can be brought about by a rotation about some axis followed by a translation, or the reverse.

While the preceding chapters have been confined to motion along a line, this chapter is devoted to rotation only, and is called *circular motion*.

When a rigid body is set into rotation its motion is described with reference to the axis about which it revolves. The *axis of rotation*, as it is generally called, is sometimes fixed within the body and sometimes beyond its outermost boundary. In the case of most wheels of machinery, for example, the axes of rotation are lines through the geometrical centers perpendicular to the planes of rotation. For a stone whirled on the end of a string, however, the axis is at the opposite end of the string remote from the stone itself.

8.1 ANGULAR SPEED AND ANGULAR VELOCITY The speed with which a body rotates is called its *angular speed. This may be measured by the frequency of rotation* which refers to the number of revolutions a body makes in unit time and is designated by the letter f.

$$f = \textbf{number of revolutions per second} \qquad (8a)$$

A flywheel, for example, might be said to have an angular speed of 10 revolutions per second (*abbr. 10 rps*). This is equivalent to an angular speed of 600 revolutions per minute (*abbr. 600 rpm*), and to an angular speed of 36,000 revolutions per hour (*abbr. 36,000 rph*).

For formulating the laws of mechanics it will be found convenient to express all rotation in radians, not in degrees or revolutions. *The radian* (rad) *is a unit of angular measure* just as the meter is a unit of linear measure. It is defined as the angle subtended by the arc of a circle whose length is equal to the radius of the same circle. In Fig. 8A the distance x measured along the arc is equal to the radius r, and the angle $\theta = 1$ radian.

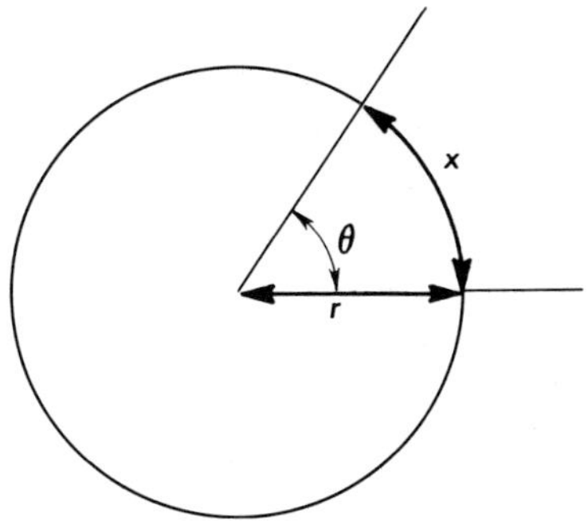

Fig. 8A. The radian is a unit of angular measure. When the arc equals the radius r, the angle θ equals one radian.

Since the entire circumference of a circle is just 2π times the radius r, there are 2π radians in one complete circle.

$$2\pi \text{ radians} = 360°$$

Since $\pi = 3.1416$,

$$1 \text{ radian} = 57.3°$$

It follows from the above relations that the angle θ in radians between any two points on the circumference of a circle is given by x, the length of the arc between the two points, divided by the radius r. In other words,

$$\text{angle in radians} = \frac{\text{arc length}}{\text{radius}}$$

or, in algebraic symbols,

$$\theta = \frac{x}{r} \tag{8a}$$

The reason for measuring angles in radians is that it simplifies all formulas for rotary motion. As an illustration, consider the speed of a stone being whirled on the end of a string as shown in Fig. 8B. The angular speed of the motion is defined as the angle turned through, divided by the elapsed time.

$$\textbf{angular speed} = \frac{\textbf{angle turned through}}{\textbf{time}}$$

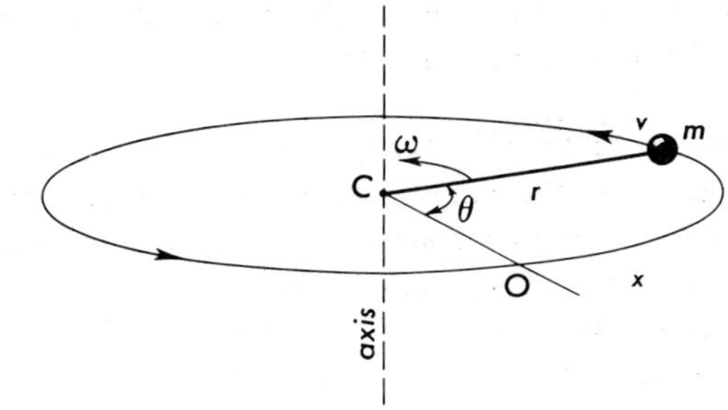

Fig. 8B. An illustration of circular motion.

In algebraic symbols,

$$\omega = \theta/t \tag{8b}$$

and is to be compared with the corresponding definition of linear speed,

$$v = x/t \tag{8c}$$

Since v, the speed of the body along its path, is defined as the distance traveled per unit time, the relation between v and f is given by

$$v = 2\pi rf \tag{8d}$$

The distance traveled in one second is equal to the circumference of the circle, $2\pi r$, multiplied by the number of revolutions per sec, f. If r is in meters, the speed v is in *meters per second;* and if r is in feet, the speed v is in feet per second.

In order to apply the term *velocity** to uniform circular motion, one refers to the *instantaneous velocity*. Such a velocity v has the same magnitude as the speed v and a direction which at any point has the direction of the tangent to the circle. If the speed is constant, the magnitude of the velocity is constant but the direction is continually changing.

As in linear motion we often use the terms *angular speed* and *angular velocity* interchangeably. Angular velocity as given by Eq. (8b) is a vector quantity if the direction of rotation is specified. To represent angular velocity by a vector, the magnitude is taken to be that of ω, and the direction is taken as that given by the *right-hand rule*. According to this rule, when the axis is grasped with the fingers pointing in

* Here we are speaking of the usual, or linear velocity, and not the angular velocity.

the direction of rotation, the thumb then points in the direction of the vector.

8.2 CENTRIPETAL FORCE When a stone is whirling on the end of a string there is an inward force exerted by the string on the ball. This force is called the *centripetal force*. By Newton's Third Law of Motion the ball exerts an equal but opposite force on the string. This is called the *centrifugal force*. Both forces are illustrated in Fig. 8C. Since the only force acting on the ball

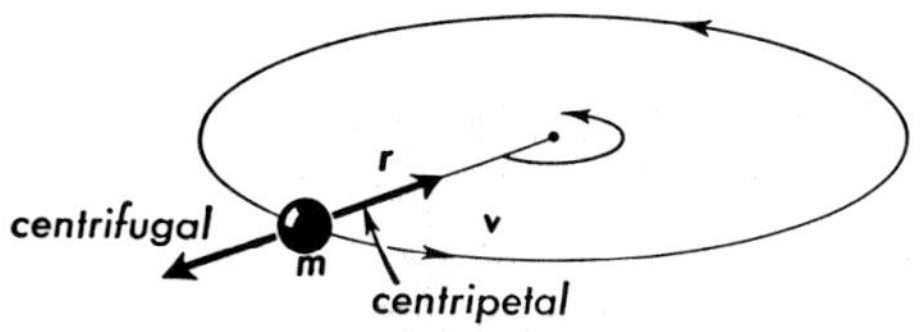

Fig. 8C. A mass m, moving in a circle, experiences an acceleration toward the center.

is inward, the ball is not in equilibrium but is being continually accelerated in the direction of the force, i.e., toward the center.

This appears to be a physical paradox, for here is a body moving with constant speed in a circle and yet being accelerated toward the center of the circle without getting any closer to it. If the string were to break suddenly, the ball would fly off on a tangent to the circle and move with constant velocity according to Newton's First Law.

To obtain a clearer picture of centripetal force and acceleration toward the center, motion in a circle as illustrated in Fig. 8D (c) is to be compared with the motion of a projectile accelerated downward by the pull of gravity as shown in Fig. 4G. Because of the earth's attraction for all bodies, a projectile is continually accelerated downward away from the straight line of its original projection. In circular motion the mass is continually accelerated toward the center, always at right angles to its instantaneous velocity and away from any straight-line-tangent along which it would travel if suddenly released.

In diagram (a) of Fig. 8D the instantaneous velocity is shown at two points, A and B. The velocity as indicated by the vectors v is seen to be changing in direction but not in magnitude. Diagram (b) is a velocity diagram showing v' as the change in velocity that occurs in going from A to B. Since this velocity triangle is similar to triangle ABC in diagram (a), corresponding sides are proportional to each other, and the following can be written:

$$\frac{x}{r} = \frac{v'}{v}$$

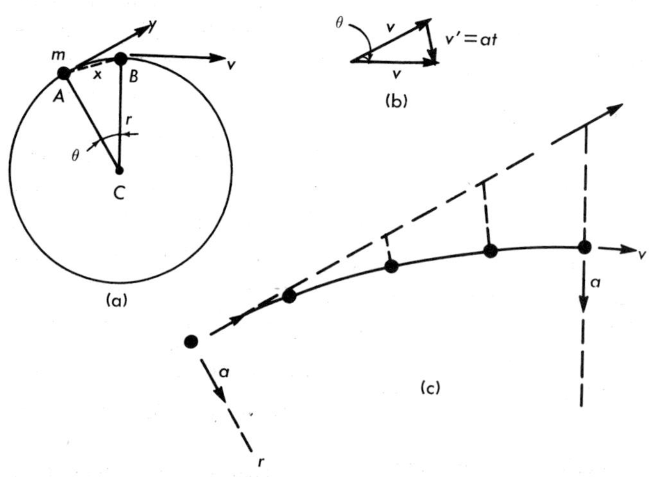

Fig. 8D. A body moving in a circle is continually being accelerated toward the center.

Since v', the change in velocity, is due to an acceleration, it can, by the equation $v' = at$, be replaced by at. During the time t the body moves from A to B a distance $x = vt$. For small angles θ the distance measured along the arc AB is approximately equal to the chord x so that to a close approximation x can be replaced by vt. These two substitutions made in the equation above give

$$\frac{vt}{r} = \frac{at}{v}$$

Canceling t and transposing v, we obtain the relation

$$a = \frac{v^2}{r} \tag{8e}$$

Thus, the *centripetal acceleration* is given by v^2/r.

As the angle θ in Fig. 8D is made smaller and smaller the arc distance x becomes more and more nearly equal to the chord, while the change in velocity v' which gives the direction of the acceleration a, becomes more nearly perpendicular to v. In the limit when θ becomes zero Eq. (8e) holds true exactly and the acceleration is perpendicular to v.

Centripetal force is defined as that force of constant magnitude which acting continuously at right angles to the motion of a particle causes it to move in a circle with constant speed. Since by Newton's Second Law of Motion, $F = ma$, centripetal force is given by

$$F = m\frac{v^2}{r} \tag{8f}$$

EXAMPLE 1. A mass of 5 Kg is moving with a speed of 6 m/s in a circle of 2-m radius. Find the centripetal force.

SOLUTION. The known quantities can be substituted directly into Eq. (8f).

$$F = 5\text{ Kg}\,\frac{(6\text{ m/s})^2}{2\text{m}} = 90\,\frac{\text{Kg m}}{\text{s}^2}$$

or

$$F = 90\text{ N}$$

8.3 EXPERIMENTS DEMONSTRATING CENTRIPETAL FORCE Many interesting experiments can be performed to illustrate centripetal force. In Fig. 8E mercury and water have been placed

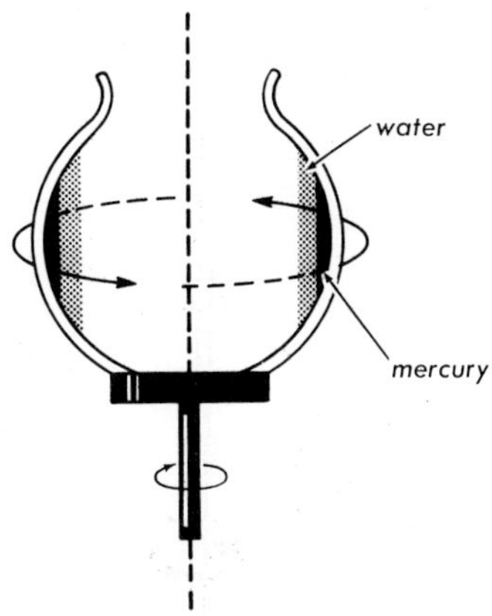

Fig. 8E. Mercury and water rotate in a dish. The water is inside the mercury. Centripetal force, like gravitational force, is greater for the denser substance.

in a dish and the dish set rotating rapidly about a vertical axis. From Eq. (8f) it is observed that r and v will be the same for both water and mercury and that m is different. Since mercury's mass is 13.6 times greater than an equal volume of water, the centripetal force F is 13.6 times greater for mercury. The mercury therefore takes the outermost position in the dish.

When the dish stops, the mercury goes to the bottom since there the dish can exert the larger upward force upon it. Because of its greater density, mercury sinks in water.

Although the earth is often said to be spherical, it is in reality an oblate spheroid, i.e., a slightly flattened sphere. Accurate measurements show that the earth's diameter is 28 mi greater through the equator than it is through the poles. The cause for this flattening is illustrated in Fig. 8F by two circular metal strips. Diagram (a) shows the strips round when at rest while (b) shows the flattening due to rapid rotation. The flattening of the earth is due to its own rotation of 2π radians every 24 hr. It is the enormous size of the earth and its lack of greater rigidity that

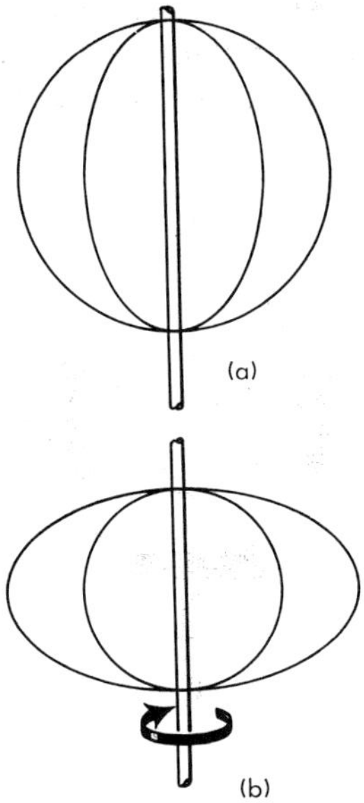

Fig. 8F. The flattening of the earth is due to its rotation about the polar axis.

make it behave as though it were soft and semi-plastic.

Mud or water clings to an automobile tire until the speed becomes too high and then it flies off on a tangent as shown in Fig. 8G. To

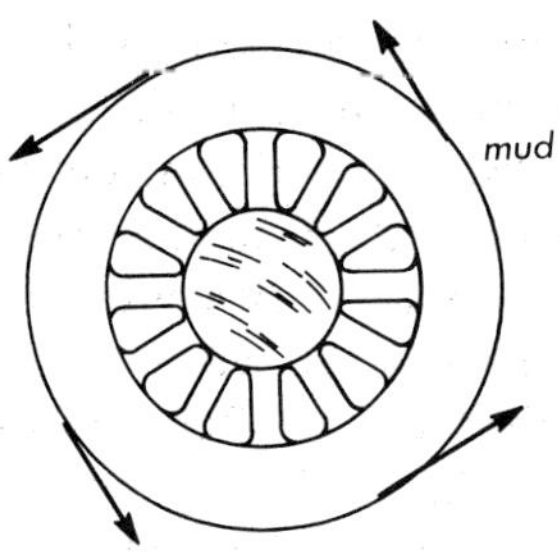

Fig. 8G. Mud or water on a fast turning wheel flies off on a tangent.

remain with the tire the required centripetal force cannot exceed the force of adhesion.

The opening of the loop in a lariat as whirled and thrown by a cowboy is due to centripetal force. See Fig. 8H. Due to rotation each small section of the rope, acting as an individual mass

Fig. 8H. A lariat takes a circular form because each small part tries to fly off on a tangent, thus getting as far from the center as possible. Centripetal force is responsible for keeping it in a circle.

m, tends to fly off on a tangent and thus get as far from the center of rotation as possible. The average distance from the center of all sections of the rope is a maximum when the loop takes the form of a circle rotating about an axis perpendicular to the plane of the loop.

When a small chain, as shown in Fig. 8I, is set

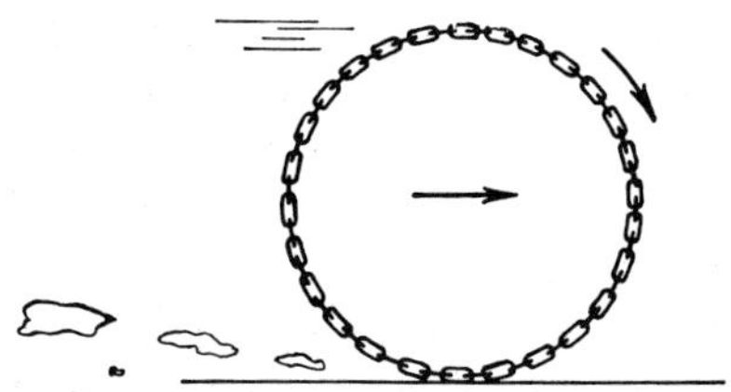

Fig. 8I. A chain turning at high speed will roll along the level as a rigid wheel.

rotating at high speed by an electric motor and then set free, it will roll along the floor as though it were a rigid metal hoop. Upon bumping into an obstacle, it will bounce into the air and, upon coming down, will retain its circular form as it rolls on. The rigidity of the hoop is due to the enormous centripetal force attained at high angular speeds.

8.4 KINETIC ENERGY OF ROTATION When a body is in a state of rotation it possesses kinetic energy. It possesses this energy because in being brought to rest it may be made to do work. To calculate the exact amount of energy stored up in a rotating body, however, it is necessary to take into account not only the speed of rotation but the distribution of the mass with respect to the axis of rotation. In other words, the formula must take into account the shape of the body. When this is done the resultant equations closely resemble the formula for the kinetic energy in linear motion, $\frac{1}{2}mv^2$.

8.5 ANGULAR MOMENTUM Just as a mass m, moving along a straight line with a velocity v, has momentum mv, so a mass m moving with a speed v in a circle of radius r has an angular momentum mvr. See Fig. 8J.

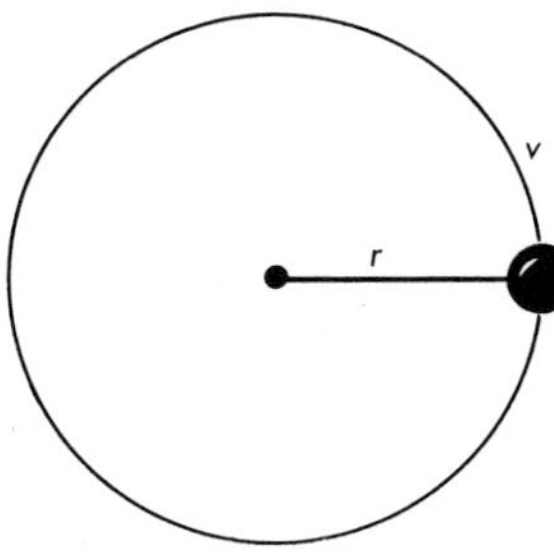

Fig. 8J. A ball of mass m whirling in a circle at the end of a string has an angular momentum mvr.

$$\boxed{\text{angular momentum} = mvr} \qquad (8g)$$

EXAMPLE 2. A mass of 3 Kg at the end of a cord 0.25 m long is set rotating in a circle as shown in Fig. 8J. If the angular speed is 4 rev/s, find the angular momentum.

SOLUTION. We first find the speed v by using Eq. (8d).

$$v = 2\pi rf = 2 \times 3.142 \times 0.25 \text{ m} \times 4 \frac{1}{\text{s}}$$

$$v = 6.28 \text{ m/s}$$

By substituting in Eq. (8g), we obtain

$$\text{ang. mom.} = mvr = 3 \text{ Kg} \times 6.28 \frac{\text{m}}{\text{s}} \times 0.25 \text{ m}$$

$$= 4.71 \frac{\text{Kg m}^2}{\text{s}}$$

The units are just as important and necessary in this answer as are the numbers in front of them, for if the units were changed the numbers would have to be changed accordingly.

Angular momentum, like angular velocity, is a vector quantity. The magnitude of the vector is given by the product mvr, while the direction is given by the right hand rule. See end of Sec. 8.1.

8.6 CONSERVATION OF ANGULAR MOMENTUM Just as there is a law of *conservation of momentum* for bodies moving in a straight line so also is there a law of the *conservation of angular momentum* for bodies in rotation. This law can best be illustrated by an experiment as shown in Fig. 8K. Two equal masses m are mounted on a

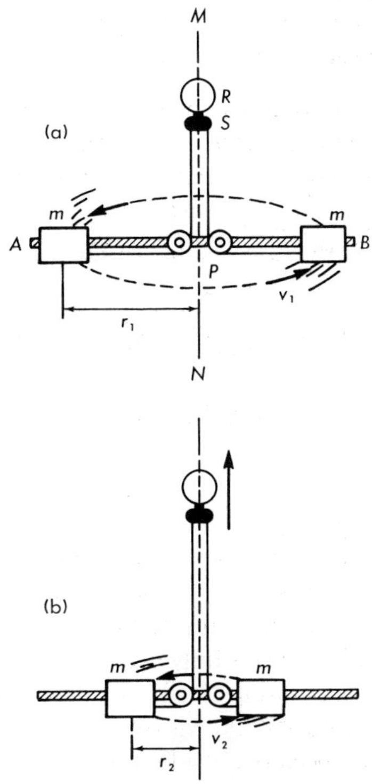

Fig. 8K. Experimental demonstration of the conservation of angular momentum.

rod AB capable of being rotated about a vertical axis MN. Cords fastened to each mass and leading over pulleys at P to the ring R enable the radial distance to be changed, from r_1 in (a), to

r_2 in (b), by simply pulling up on R. The swivel S prevents the cords from twisting. When the system is first set rotating as in (a) with a speed of v_1 the angular momentum of each mass is mv_1r_1. On pulling up on the ring R the radius decreases to r_2 and the speed changes to v_2, giving an angular momentum of mv_2r_2.

Conservation of angular momentum requires that

$$mv_1r_1 = mv_2r_2 \tag{8h}$$

Since the mass is not altered in value, conservation of angular momentum requires that any decrease in r must be compensated for by an increase in speed. This is necessary to keep both sides of the equation equal to each other.

Eq. (8f) shows, for example, that if r is reduced to half value, the velocity v must double. With v doubled and the circle only half as large, the angular speed increases fourfold.

EXAMPLE 3. Suppose in Fig. 8K that $m = 10$ g, $v_1 = 20$ cm/s, and $r_1 = 16$ cm. What will be the new speed if the radius r_1 is decreased to half value, i.e., $r_2 = 8$ cm?

SOLUTION. By direct substitution in Eq. (8h), we obtain

$$10\text{ g} \times 20\,\frac{\text{cm}}{\text{s}} \times 16\text{ cm} = 10\text{ g} \times v_2 \times 8\text{ cm}$$

$$3200\,\frac{\text{cm}}{\text{s}} = 80\,v_2$$

$$v_2 = 40\,\frac{\text{cm}}{\text{s}}$$

An interesting experiment illustrating the same principle is diagrammed in Fig. 8L. An observer stands on a turntable with weights in each hand. With arms fully extended horizontally he is first set rotating slowly. Upon drawing the hands and weights in toward the chest as shown, the angular speed is considerably increased. This experiment is best appreciated by the turning observer who feels himself speeded up by what seems to be a mysterious force.

The kinetic energy of rotation in each of these two experiments increases as the masses are pulled in closer to the center of rotation. This increase is due to the fact that work, equal to force times distance, must be done to pull them in.

Fig. 8L. Experiments illustrating conservation of angular momentum.

This principle is used by expert figure skaters on the ice. They start into a whirl with their arms, and perhaps one leg extended, and then upon drawing the arms and leg in, obtain a greatly increased angular speed.

8.7 GYROSCOPIC STABILITY If a balanced gyroscope wheel is mounted in gimbal rings as shown in Fig. 8M, it will, when set spinning at high speed, exhibit a property called "gyroscopic

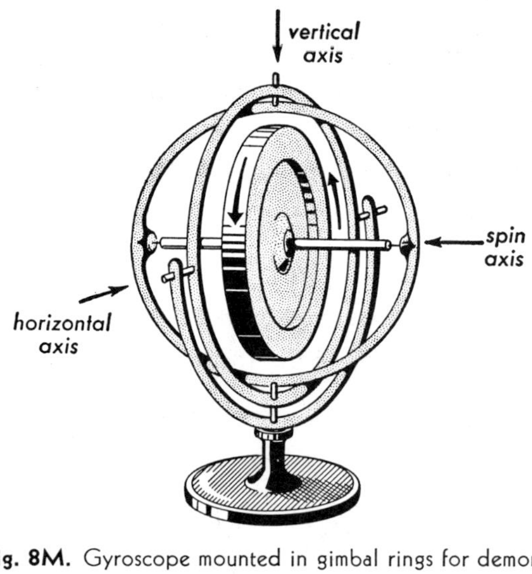

Fig. 8M. Gyroscope mounted in gimbal rings for demonstrating gyroscopic stability.

stability." When the gyro is picked up and carried about, the base can be turned in any direction without altering the direction of the spin axis relative to the earth. In other words, the plane of the gyro wheel seems to have assumed a rigidity in space.

To change the direction of the "spin axis" a torque must be exerted upon it. It is the function of the gimbal ring mounting to allow the base support to be turned in any way without exerting any torque whatever upon the wheel.

Newton's First Law of Motion, involving the inertia of a body, has its rotational counterpart in the above experiment.

A body in rotation about some fixed axis will continue to rotate about that same axis with constant speed unless it is acted upon by an unbalanced torque.

8.8 THE SPINNING TOP A common top, set spinning like the one shown in Fig. 8N, is a good illustration of angular momentum. It is but one of the many forms of gyroscopes. Spinning about its axis the top precesses about its peg or pivot point, the line AB describing an inverted cone about the vertical line AC. If when looking down from above, such a top is spinning in a clockwise direction the precession is clockwise; spinning counterclockwise the precession is counterclockwise.

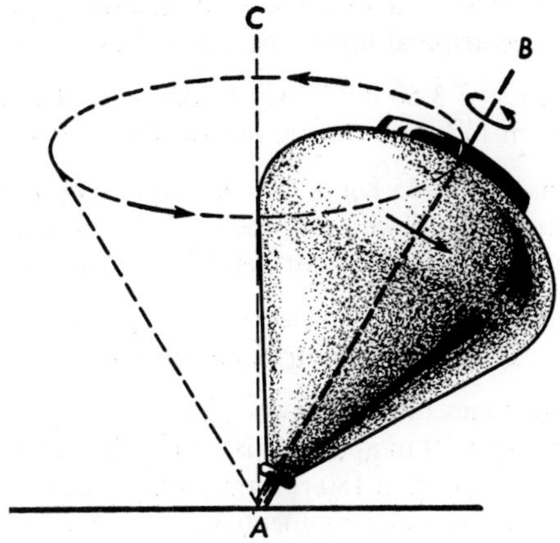

Fig. 8N. A spinning top precesses around a vertical axis.

QUESTIONS

1. Define or briefly explain each of the following: (a) frequency of rotation, (b) instantaneous velocity, (c) centripetal force, (d) centrifugal force, (e) centripetal acceleration, (f) angular momentum, and (g) rotational kinetic energy.

2. Since a lariat stays open in a more or less circular loop, as shown in Fig. 8H, what has become of the centripetal force? Explain.

3. Explain how an earth-circling satellite can be accelerated continually toward the earth and still remain at the same distance above the surface.

4. Why doesn't a spinning top, like the one shown in Fig. 8N, fall over? Explain.

PROBLEMS

1. A stone on the end of a string 1.30 m long is whirled in a circle with a speed of 4.50 rev/s. Calculate the centripetal acceleration. [Ans. 1039 m/s^2.]

2. A ball on the end of a string 60 cm long is whirled in a circle with a speed of 2.6 rev/s. If the ball has a mass of 425 g, find (a) the centripetal acceleration, and (b) the centripetal force.

3. A 2-Kg mass is fastened to the end of a wire and whirled in a circle of radius 0.75 m at a speed of 36 m/s. Find (a) the centripetal acceleration, and (b) centripetal force. [Ans. (a) 1728 m/s^2, (b) 3456 N.]

4. A mass of 3.50 Kg is tied to one end of a wire and whirled in a circle of 1.40 m radius with a speed of 4 m/s. Calculate the centripetal force.

5. A mass of 1.25 Kg is fastened to a wire and whirled in a circle 1.80 m in diameter, and with a speed of 3.50 rev/s. Calculate (a) the centripetal acceleration, and (b) centripetal force. [Ans. (a) 436 m/s^2, (b) 544 N.]

6. A 500-g mass is whirled in a circle of radius 1.20 m with a speed of 2 rev/s. Find (a) the centripetal force, and (b) the angular momentum.

7. Two stones of 2.50 Kg each are tied to opposite ends of a wire so their centers are 1.20 m apart. This system is thrown whirling into the air with an angular speed of 180 rev/min. Calculate (a) the tension in the wire, and (b) the total angular momentum. [Ans. (a) 533 N, (b) 33.93 Km^2/s.]

8. The driver of a car rounds a curve in the road at 60 mi/hr. If the radius of the curve is 450 ft, and the driver weighs 160 lb, find (a) the centripetal force exerted by the car seat on his body, and (b) his angular momentum.

9. Calculate the acceleration of a body toward the earth if it is released from a space ship 10,000 Km above the earth's surface. Assume the earth's radius and mass to be 6400 Km and 6×10^{24} Kg. [Ans. 1.482 m/s^2.]

10. Calculate the acceleration of the moon toward the earth. Assume the distance from the earth's center is 3.86×10^5 Km, and the moon takes 28 days to circle the earth.

11. Two masses of 0.500 Kg each, fastened to opposite ends of a wire, are thrown whirling into the air. If they rotate at a speed of 1.80 rev/s, and their centers are 2.50 m apart, find (a) the tension in the wire, and (b) the total angular momentum of the system. [Ans. (a) 79.9 N, (b) 17.67 Kg m^2/s.]

12. A boy whirls a stone over his head in a sling like that used by David when he slew Goliath. If the 80-g stone moves in a circle 2 m in diameter, with a frequency of 2 rps, find (a) the speed of the stone, and (b) the centripetal force.

9 PLANETARY MOTION AND SATELLITES

The landing of the Apollo 11 astronauts on the moon fulfilled an ancient dream of mankind. The realization of this history-making event was made possible through the teamwork of thousands of people—scientists, engineers, astronomers, physicians, technicians, and others. The calculations needed to carry out the complicated mission were based upon the orbital motions of the moon, the earth, and the spacecraft.

Man's realization of the planetary motion goes back in history to the time of the ancient Greeks. According to astronomical history it was the early Greek philosopher, Pythagoras (530 B.C.), who said, "The world is round and hangs in space." The earth, he said, "does not stand still but revolves around a central fire, called Hestia. This fire is not the sun, for the sun is illuminated, as are the planets, by reflection from Hestia."

This idea lay dormant for two thousand years before Copernicus* at the beginning of the sixteenth century said, "The sun stands still, and the earth and planets move in orbits around it." The observational and mathematical proof that all the planets move in elliptical orbits was first

* COPERNICUS (1473–1543), was born as Mikolj Kopernik in Poland. He studied mathematics at the University of Cracow and, after graduation, traveled first to Bologna, then to Rome, and finally to Ferrara, where he obtained a Doctor's degree in Canon Law. Later at Padua he studied medicine and practiced his skills on the poor as well as the rich. Yet he found time during all his work to study his great love, astronomy.

Copernicus assumed that the earth is a planet, like Venus, Mars, and Jupiter, and that all the planets move in circles around the sun. He visualized the earth rotating on an axis to account for the apparent daily motion of the sun, moon, and stars, and said that the latter are too far away for any motion of the earth to show an observable change in their relative positions. He listed the order of the known planets from the sun outward, and in his honor the solar system is frequently called the *Copernican System*.

This man cared little for fame, but he wanted to know the truth, and when he found it he wanted to pass it on to others. His great powers of observation, and his remarkable insight into complex problems, are set forth in his new theory of the universe in a book entitled *De Revolutionibus Orbium Coelestium* published near the end of his life.

presented in 1609 when Johannes Kepler† published a book containing two of his three laws now known as Kepler's Laws of Planetary Motion. Kepler established his laws by a detailed study of the thousands of precision measurements of the planets made by his contemporary and close friend Tycho Brahe.**

† JOHANNES KEPLER (1571–1630), German astronomer and philosopher. Born in Weil of a poor but noble family, Kepler was educated at the University of Maulbroon. It was in his position as professor of astronomy at Gratz that he first became interested in the planets. Hearing that Tycho Brahe had recorded great quantities of data on the motions of hundreds of stellar objects Kepler went to Prague. There he became a close and devoted friend of Tycho Brahe, and promised this grand old man that he would tabulate and publish the recorded observations. In 1609 Kepler published his "Commentaries on Mars" in which his first two laws of planetary motion are to be found. The third law came a little later. Kepler, a religious man, was sickly for most of his 59 years. He was twice married but had no children of his own. He died penniless.

** TYCHO BRAHE (1546–1601), noted Danish astronomer, born of a noble family, entered the University of Copenhagen at the age of 13. A year later he was deeply impressed by an eclipse of the Sun, which was visible from Copenhagen. What struck him most was that it had been predicted with extreme accuracy. Tycho resolved then and there to study astronomy, and later accepting a position as Director of the Observatory of Uraniborg, Denmark, spent some 20 years observing the planets, the moon, and the stars. Without a telescope, he made remarkably accurate observations of the planets and catalogued thousands of star positions and their time of observation.

Fortunately, there came into his life, not long before his death, a brilliant young mathematician named Johannes Kepler. Upon his deathbed Tycho left all his tables of observations to Kepler, and made the young man promise to publish them as the *Rudolphine Tables*. Although Tycho vehemently rejected the Copernican system, Kepler kept his word and, using the massive tables, proved beyond a shadow of a doubt that the planets move in orbits around a stationary and massive Sun.

9.1 KEPLER'S FIRST LAW *The planets move in elliptical orbits with the sun at one of the foci.*

An ellipse can be constructed by fastening the two ends of a piece of string to two pins F_1 and F_2 as shown in Fig. 9A. By keeping the string

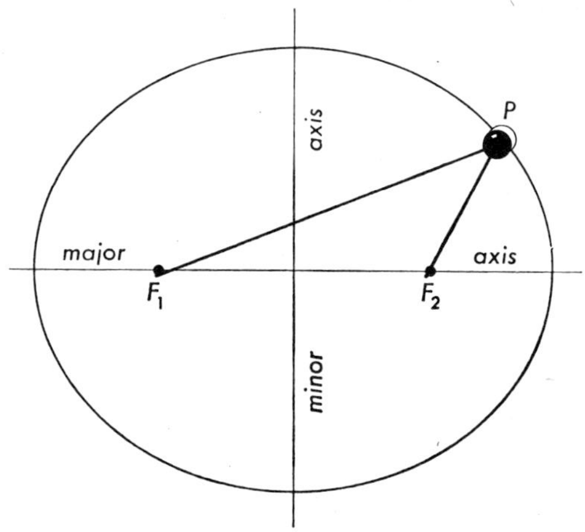

Fig. 9A. An ellipse can be drawn with two pins, a string, and a pencil.

tight, and with a pencil at P, the complete arc can be swung around much the same as one draws a circle with a compass.

If the string length remains unchanged, and the foci F_1 and F_2 are brought closer and closer together, the major axis AB and the minor axis CD become more and more nearly equal, and in the limit when the foci coincide the axes are equal and the ellipse becomes a circle. The real orbits of the planets are so nearly circular that if they were drawn with an ordinary compass, the ellipse would differ from the circle by less than the thickness of the line.

The eccentricity e of an ellipse (see Fig. 9B) is defined as the ratio of the distances SQ and AQ:

$$e = \frac{SQ}{AQ} \tag{9a}$$

where AQ is the semimajor axis a, and SQ is equal to ae. With the sun at one focus, the distance of closest approach AS is called perihelion, and the greatest distance BS is called aphelion.

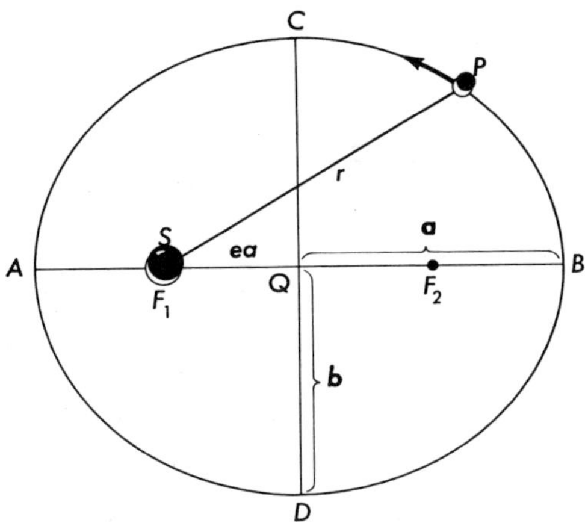

Fig. 9B. Elliptical orbit with an eccentricity e = 0.5.

A little study of Fig. 9B will enable the reader to find that

$$\text{aphelion} = a(1 + e) \tag{9b}$$

$$\text{perihelion} = a(1 - e) \tag{9c}$$

$$\text{semiminor axis} = a\sqrt{1 - e^2} \tag{9d}$$

9.2 KEPLER'S SECOND LAW *The straight line joining the sun and any planet sweeps out equal areas in equal intervals of time.*

As shown in Fig. 9C, the straight line referred to, called the radius vector, varies in length from a minimum at perihelion to a maximum at aphelion. Although the earth's orbit is nearly circular, the numbers 1, 2, 3, 4, etc., correspond to the earth's position at the beginning of each of twelve equal months.

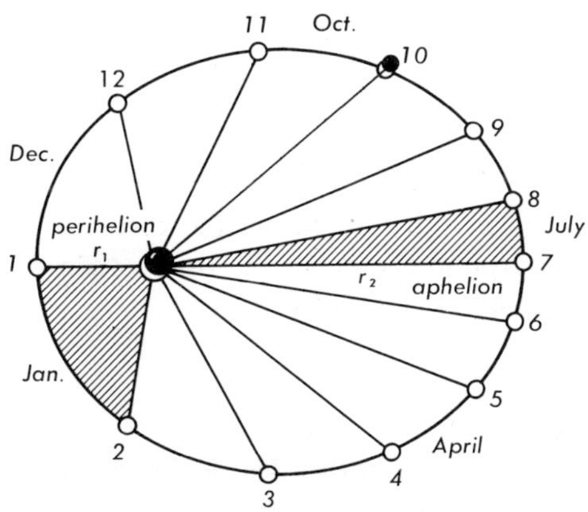

Fig. 9C. Elliptical orbit of the earth around the sun, showing equal areas swept out by the radius vector in equal intervals of time. (The eccentricity is exaggerated.)

To cover these unequal orbital distances in equal intervals of time, the speed must be a maximum at perihelion and a minimum six months later at aphelion. During the periods 1 to 2 and 7 to 8, for example, the areas swept out are equal.

As the earth moves along its orbit in September, October, November, etc., the attractive force of the sun causes it to speed up. Upon reaching perihelion at the end of December, its speed is a maximum and is too fast to remain at this distance r_1 from the sun. During the months of March, April, May, etc., the earth is receding from the sun and the attractive force of the sun slows it down. Upon reaching aphelion at the end of June, its speed is a minimum and is too slow to keep it at this greater distance r_2 from the sun. The average distance from the sun is 92,900,000 mi, whereas the average orbital speed of the earth is 18.5 mi/sec.

That point of a satellite's orbit closest to the earth, moon, or Jupiter is called perigee, perilune, and perijove, respectively, while the orbital point farthest away is called apogee, apolune, and apojove, respectively.

9.3 KEPLER'S THIRD LAW *The squares of the orbital periods of the planets are proportional to the cubes of their mean distances from the sun.*

The period T of a planet, or satellite, is defined as the time required to make one complete trip around its orbit, and the mean distance r is defined as the average distance away from the sun. Important data on the eight major planets of the solar system are given in Table 9A. The constant ratios T^2/r^3 in column four verify Kepler's Third Law.

While Kepler's laws were originally derived from Tycho Brahe's careful observations, they are readily derived from the basic laws of classical mechanics.

TABLE 9A. Measured Characteristics of the Planets

Name	Period T (years)	Mean Distance (mi $\times 10^6$)	Mean Distance (Km $\times 10^6$)	$\frac{T^2}{r^3}$	Mean Radius (mi)	Mean Radius (Km)	Mass (Kg $\times 10^{24}$)
Mercury	0.241	36.0	57.9	1.245	1504.3	2421.1	0.3244
Venus	0.615	67.1	108.1	1.252	3828.2	6161.0	4.861
Earth	1.000	92.9	149.5	1.247	3958.9	6371.0	5.975
Mars	1.881	141.5	227.8	1.249	2070.5	3332.1	0.6387
Jupiter	11.862	483.3	777.8	1.246	43429.0	69892.0	1902.1
Saturn	29.458	886.1	1426.0	1.247	35748.5	57532.0	569.4
Uranus	84.015	1783.0	2869.0	1.245	14727.0	23701.0	87.1
Neptune	164.790	2793.0	4496.0	1.246	13381.0	21535.0	103.1
Pluto	247.700	3665.0	5899.0	1.246	1781.4	2867.0	0.5 ?

For simplicity we will assume that the earth's orbit is circular as shown in Fig. 9D. In this diagram M is the mass of the sun, m is the mass of a planet like the earth, and r is the distance between their centers. The centripetal force F, as given by Eq. (8d), is just the force of gravitational attraction given by Eq. (4c). These are:

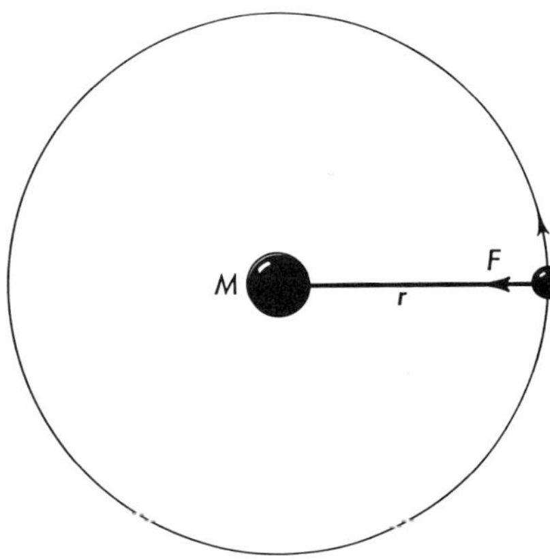

Fig. 9D. The gravitational force of attraction F is the centripetal force that keeps the earth in its nearly circular orbit around the sun.

$$F = m\frac{v^2}{r} \qquad (9e)$$

Centripetal force

$$F = G\frac{Mm}{r^2} \qquad (9f)$$

Newton's law of gravitation

where

$$G = 6.67 \times 10^{-11}\frac{m^3}{\text{Kg s}^2}$$

Since these two equations are different expressions for the same force F, the right-hand sides may be placed equal to each other, which gives

$$G\frac{Mm}{r^2} = m\frac{v^2}{r}$$

Upon simplifying, this equation becomes

$$\frac{GM}{r} = v^2 \qquad (9g)$$

The orbital velocity of any satellite moving in a circular orbit around any mass M in the universe is, therefore, given by the simple relation

$$v = \sqrt{\frac{GM}{r}} \qquad (9h)$$

In mechanics the velocity of a body is given by $v = x/t$. If we choose the distance x to be once around the orbit, the time t then becomes the period T, and we obtain

$$v = \frac{2\pi r}{T} \qquad (13i)$$

By squaring both sides of this equation, and substituting the right-hand side for v^2 in Eq. (13g), we can write

$$\frac{GM}{r} = \frac{4\pi^2 r^2}{T^2}$$

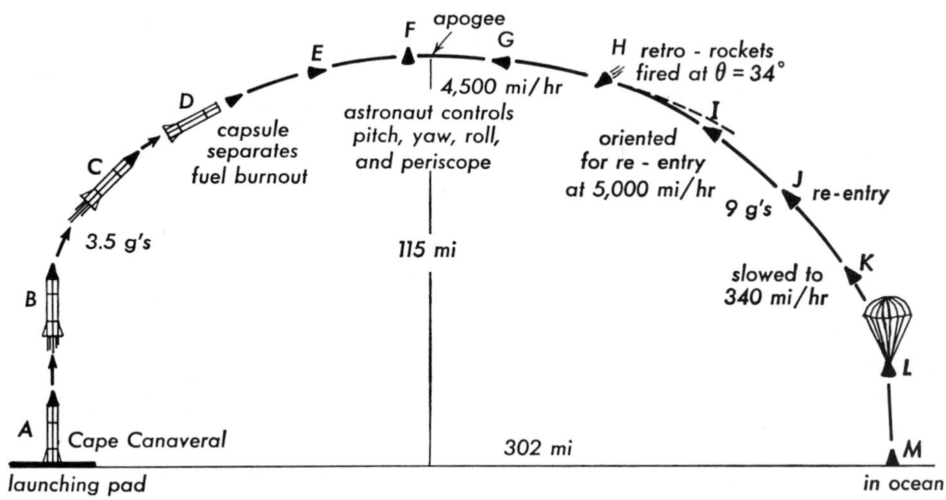

Fig. 9E. Diagram of the flight path taken by the first American astronaut to be launched into space, May 5, 1961.

or

$$T^2 = \left(\frac{4\pi^2}{GM}\right) r^3 \qquad (13j)$$

Since all quantities in parentheses are constants, $T^2 \propto r^3$, and Kepler's third law is consistent with the laws of classical mechanics.

9.4 SPACE FLIGHT When a large rocket takes off from the ground to carry a vehicle into space, its initial take-off direction is vertically upward as shown in Fig. 9E. As the rocket gains height, control fins or jets are set to make it turn slowly toward a horizontal trajectory.

The first American to be launched into space by a rocket was carried aloft by a Redstone missile from Cape Canaveral (now Cape Kennedy), Florida, May 5, 1961. The astronaut, Alan Shepard, piloted the pressurized space capsule in which he was jettisoned. Let us analyze some of the forces, velocities, and accelerations to which the rocket and astronaut were subjected at several points of the flight. The following cases are typical of space flight in general, and the equations can be applied equally well to orbiting satellites and flights to the moon and planets.

Since the value of g in Shepard's flight diminished by less than 5% at an altitude of 115 mi, and its direction changed by less than 5 degrees over the 302-mi range, it will be assumed that, to a first approximation, g remains constant over the entire flight path.

Case A. Before Vertical Launch. When the entire rocket assembly is at rest on the launching pad, just prior to take-off all velocities and accelerations are zero, and all forces are in equilibrium. See Fig. 9F. At this time there are two

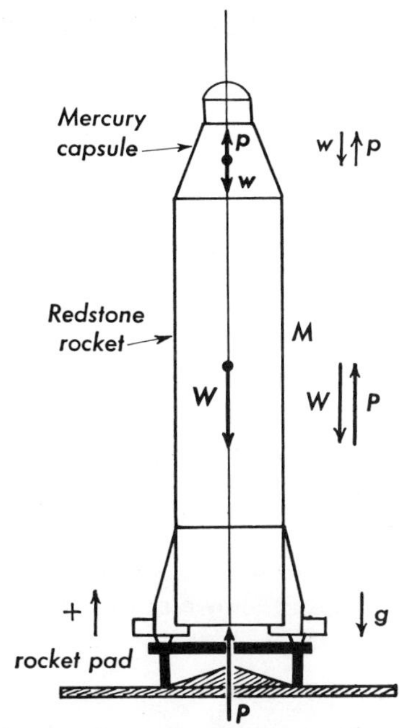

Fig. 9F. Diagram of a rocket and its manned space capsule at rest on its launching pad and ready for vertical take-off.

equal forces acting on the rocket, W the downward pull of the earth, and P the upward push of the launching pad. The downward force is given by

$$W = Mg$$

where M is the total mass of the rocket and its payload.

Assuming the upward direction is positive, g the acceleration due to gravity is negative, and so is W. Since the upward force P on the total mass M must be equal to W in magnitude,

$$P = Mg \tag{9k}$$

In a similar way the two equal but opposite forces acting on the astronaut are w, the downward pull of the earth, and p, the upward thrust of his supporting cradle. These forces are given by

$$w = mg \tag{9l}$$

$$p = mg \tag{9m}$$

Note that for each kilogram mass of the rocket structure, or its contents, the force p is equal to 9.8 newtons, and for each slug of mass it is 32 lb. Under these conditions the rocket and the astronaut are said to be withstanding "*1 g*."

Case B. Vertical Climb. When the rocket is accelerating vertically upward after launching (see Fig. 9G) the two forces acting on the total mass M are P the upward thrust of the rocket engines, and W the downward pull of the earth. The resultant force F acting on the rocket is therefore given by

$$F = P - W \tag{9n}$$

To lift the rocket from the ground, P must be greater than W, and their vector resultant F is the force that produces the upward acceleration a.

$$F = Ma \tag{9o}$$

Applying these same principles to the astronaut in the capsule shown at the top in Fig. 9G, we observe that

$$f = p - w \tag{9p}$$

where p is the thrust of the cradle on the astronaut's body and w is the downward pull of the earth on the same mass, the resultant f is the force that gives the astronaut the same upward acceleration as the rocket.

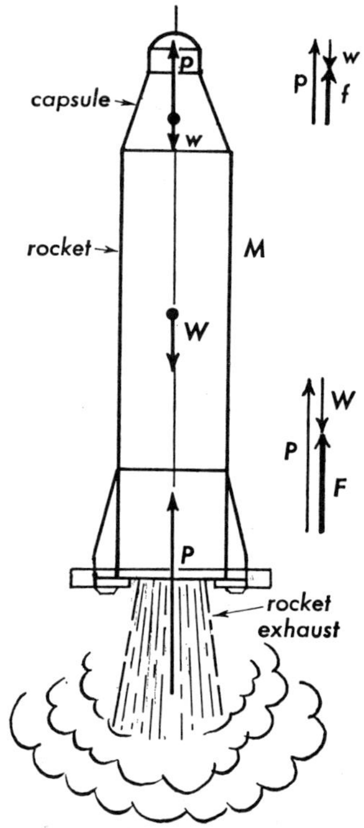

Fig. 9G. Force diagram for a rocket and its astronaut pilot during vertical flight under rocket thrust.

$$f = ma \tag{9q}$$

EXAMPLE 1. A 25-ton rocket with an astronaut of 160 lb aboard is climbing vertically upward with an acceleration of 96 ft/s². Find the total force of the cradle on the man's body (a) in g's and (b) in pounds.

SOLUTION. Transposing Eq. (9p), we obtain for the thrust

$$p = w + f$$

and substituting mg for w from Eq. (9l), and ma for f from Eq. (9q), we obtain

$$p = mg + ma \tag{9r}$$

Assuming $g = 32$ ft/s², the astronaut's upward acceleration of 96 ft/s² is just $3 \times g$:

$$a = 3g \tag{9s}$$

Direct substitution of 3 g for a in Eq. (9r) gives for the thrust

$$p = mg + m(3g)$$

or

$$p = m(4g) \qquad (9t)$$

Ans. (a) The astronaut is said to be "taking 4 g's." This means that all parts of his body are subjected to a force of four times their normal weight.

To find the thrust in pounds, direct substitution in Eq. (9t) gives

$$p = \frac{160 \text{ lb}}{32 \text{ ft/s}^2}\left(4 \times 32 \frac{\text{ft}}{\text{s}^2}\right)$$

$$p = 5 \text{ slugs} \times 128 \text{ ft/s}^2$$

$$p = 640 \text{ lb}$$

Ans. (b) The astronaut's apparent weight is four times normal.

Case C. Free Jettisoned Space Capsule. After the space capsule carrying the astronaut has been jettisoned from the rocket as shown at D in Fig. 9E, the vehicle becomes a free projectile. From point D to point H, where the retro-rockets are fired, the downward pull of the earth W is the only force acting on the body. This force gives rise to a downward acceleration g and is responsible for the continually changing direction of the missile as it travels along its flight path.

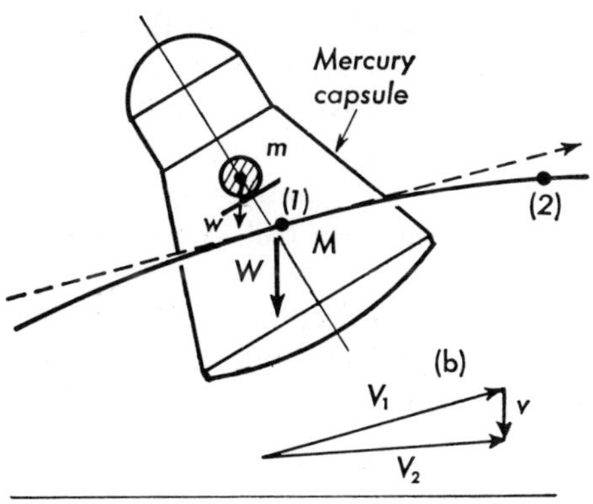

Fig. 9H. Diagram of a free jettisoned space capsule as it travels without rocket thrust along its flight path.

In the velocity diagram, Fig. 9H(b), V_1 represents the instantaneous velocity of the capsule as it passes point (1), v is the change in velocity imparted by W during the time it takes to reach point (2), and V_2 is the instantaneous velocity as it passes point (2).

$$v = gt$$

Since there is no thrust P from rocket motors for this part of the flight, the cradle exerts no force on the astronaut, and the astronaut exerts no force on the cradle. While the capsule, the astronaut, and the cradle fall with the same acceleration g, weightlessness prevails throughout the moving system.

9.5 SATELLITES If the velocity with which a space capsule is launched into space is great enough, the vehicle will orbit the earth as a satellite. To find the velocity required to circle the earth consider the details of Fig. 9I.

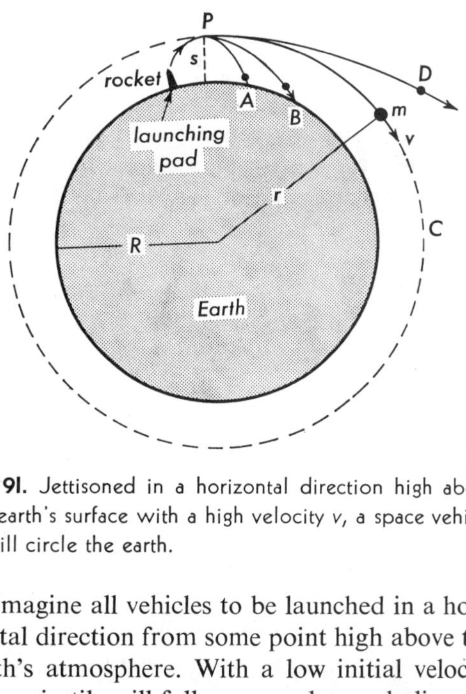

Fig. 9I. Jettisoned in a horizontal direction high above the earth's surface with a high velocity v, a space vehicle m will circle the earth.

Imagine all vehicles to be launched in a horizontal direction from some point high above the earth's atmosphere. With a low initial velocity the projectile will follow a nearly parabolic path as shown at A. At a somewhat higher velocity the trajectory will be that of path B. At a still higher velocity the projectile in falling toward the earth will follow a circular path of radius r. This particular velocity is called the orbiting velocity. At still higher velocities, such as shown at D, the projectile will follow an elliptical path or escape from the earth completely.

By Newton's Law of Gravitation, Eq. (4c), the force F exerted by the earth on a satellite of mass m is inversely proportional to r^2, and by Newton's Second Law of Motion, Eq. (3a), the force is proportional to a. It follows, therefore, that the acceleration of a mass m toward the earth is inversely proportional to the square of the distance from the earth's center.

Converting this relation into algebraic symbols we can write the following inverse square proportionality:

$$\frac{a}{g} = \frac{R^2}{r^2} \tag{9u}$$

where a is the inward acceleration of mass m at any distance r, and g is its inward acceleration at the earth's surface R. Transposing g, we obtain

$$a = g\frac{R^2}{r^2} \tag{9v}$$

This inward acceleration for an orbiting satellite is just the centripetal acceleration required to keep it in orbit and prevent it from flying off on a tangent. By Eq. (8c) we have

$$a = \frac{v^2}{r}$$

Equating these two relations, we obtain

$$\frac{v^2}{r} = g\frac{R^2}{r^2}$$

which upon solving for v, becomes

$$\boxed{v = \sqrt{g\frac{R^2}{r}}} \tag{9w}$$

orbiting velocity

A comparison of Eq. (9w) with Eq. (9h) shows that $GM/r = gR^2/r$, from which we obtain

$$GM = gR^2$$

If the orbit is very close to the earth's surface, we can write, to a first approximation, $r = R$, and Eq. (9w) becomes

$$v = \sqrt{gR} \tag{9x}$$

(The acceleration due to gravity, $g = 9.80$ m/s^2,is equivalent to 78,920 mi/hr^2.)

For a satellite to escape from the earth and never return it must be launched with a velocity about 40% greater than that required to make it orbit. The escape velocity is given by

$$\boxed{v = \sqrt{2gR}} \tag{9y}$$

escape velocity

EXAMPLE 2. If the earth's radius is 6360 Km, what horizontal speed must be imparted to a satellite to cause it to orbit in a circle 800 Km (about 500 mi) above the earth's surface?

SOLUTION. Since the acceleration due to gravity $g = 9.8$ m/s^2, direct substitution in Eq. (9w) gives

$$v = \sqrt{\frac{9.8 \times (6{,}360{,}000)^2}{7{,}160{,}000}\,\frac{\text{m}^2}{\text{s}^2}}$$

$$v = \sqrt{55.4 \times 10^6\,\frac{\text{m}^2}{\text{s}^2}}$$

$$v = 7450\,\frac{\text{m}}{\text{s}}$$

This is equivalent to 16,700 mi/hr.

9.6 MECHANICAL WELL MODEL The preceding sections of this chapter show that figuratively speaking, we on the surface of the earth are living at the bottom of a gravitational well thousands of miles deep. To reach the moon, the planets, or other worlds we must climb out of this well onto a horizontal plane we call gravitationless free space.

We have seen in Sec. 6.5 that the work done in raising a body of mass m to a height x is given by the product of force times distance.

$$W = F \times x$$

where the force is given by Newton's Second Law, $F = mg$,

$$W = mgx \tag{9z}$$

Furthermore, by virtue of its position above its starting point, a body has gained potential energy mgx.

$$\text{P.E.} = mgx \tag{9aa}$$

In applying this formula to the raising of a body it is assumed that the required force remains constant. Such an assumption is not justified, however, at large distances, since the pull of the earth diminishes as the square of the distance. See Eq. (9f).

If we use Newton's Gravitational Formula, Eq. (9f), for the force and calculate by means of the calculus the potential energy of a mass m at increasing distances above the earth, we obtain

$$\text{P.E.} = -G\frac{Mm}{r} \tag{9bb}$$

where M is the earth's mass, and r is the distance from the earth's center.

In deriving this equation it is assumed that the zero energy, or base plane, from which potential energy is measured is gravitationless free space hundreds of millions of miles (infinity) away from the earth. The minus sign means that the potential energy is negative with respect to the base plane.

To see why we speak of a well, we plot Eq. (9bb) and obtain a potential energy graph as shown in Fig. 9J. The potential energy P.E. is plotted vertically and the distance from the center of the earth r is plotted horizontally. (For r = 4000 mi GMm is arbitrarily set at 64,000.)

For a distance r that is infinitely far away, P.E. = 0. At smaller and smaller distances the magnitude of P.E. increases, but the P.E. is negative. The colored curve in the diagram comes down to the earth's surface where r has the value of the earth's radius R.

A mechanical well model for demonstrating satellite orbits is obtained by rotating this potential energy graph around the vertical axis. In so doing the curve describes a cone-like surface as shown in Fig. 9K. If we make a solid body of some hard material with the shape of the shaded section, a small steel or glass marble can be rolled around in the cone to describe satellite orbits.

A great variety of orbits can be generated simply by varying the marble's initial velocity. When viewed from directly overhead, most orbits are elliptical or circular. Orbital distortion arising from rolling friction can be held to a minimum by using hard materials with smooth surfaces.

If orbits including the moon are desirable, a small potential crater, far out on the periphery of the earth model, can be made. See Fig. 9L. The rolling of a marble toward this crater, away from the earth, is analogous to the path of an earth-based satellite on its projected trip to or around the moon. At the *saddle point* Q the gravitational pulls of the earth and moon are equal and opposite.

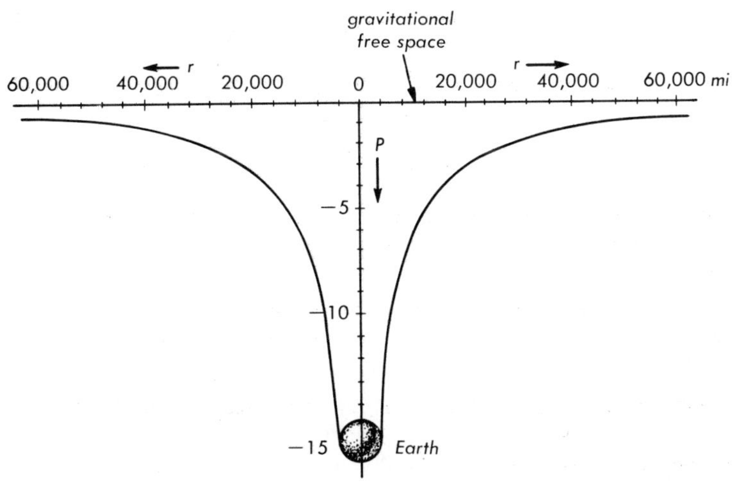

Fig. 9J. Potential energy graph illustrating the "well" analogy of the earth's gravitational field.

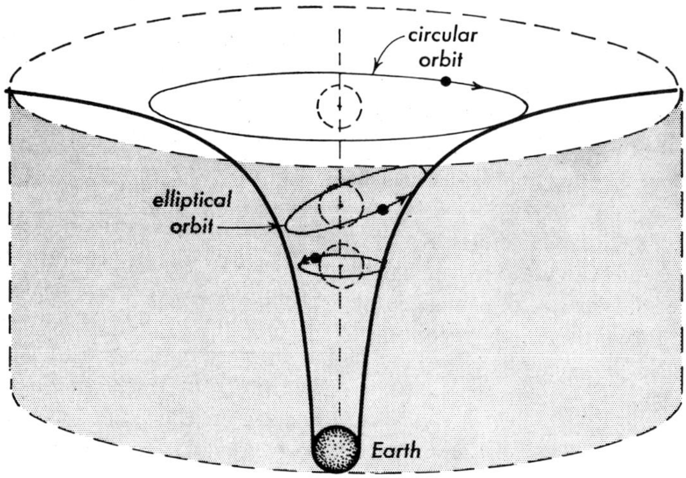

Fig. 9K. Mechanical well model for demonstrating elliptical and circular orbits.

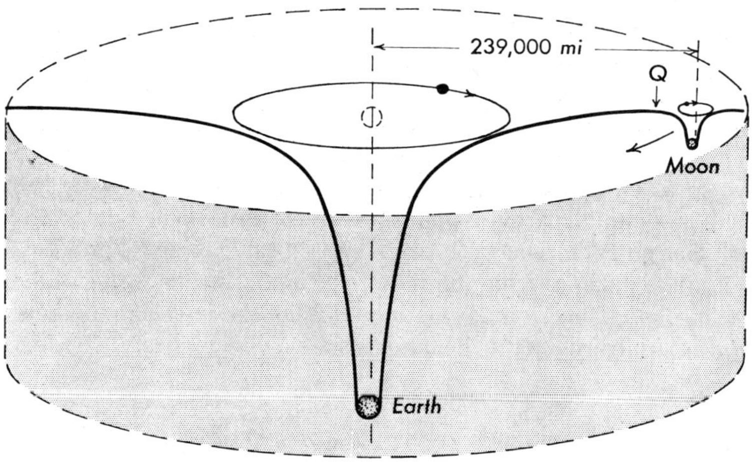

Fig. 9L. Mechanical well model for the earth and moon.

QUESTIONS

1. State Kepler's first law of planetary motion. Draw an ellipse with two pins and a piece of string, and label the major axis, the minor axis, the foci, perihelion, and aphelion.

2. State Kepler's second law of planetary motion. Draw an ellipse with two pins and a piece of string, and show the meaning of the second law by shaded areas.

3. State Kepler's third law of planetary motion. Draw three ellipses with two pins and a piece of string. Use the same string length for all three, but change the distance between pins. What can you say about the eccentricities?

PROBLEMS

1. A space capsule moves in an orbit around the earth having a semimajor axis of 12,000 Km, and a semiminor axis of 4,000 Km. Find (a) the eccentricity, (b) apogee, and (c) perigee. [Ans. (a) 0.943, (b) 23,310 Km, (c) 684 Km.]

2. A satellite orbit has a semimajor axis of 15,000 Km, and a semiminor axis of 7,500 Km. Find (a) the eccentricity, (b) apogee, and (c) perigee.

3. A 20,000-Kg rocket with a 75-Kg astronaut aboard is climbing vertically upward with an acceleration of 39.2 m/s². Find (a) the rocket thrust in N, and the force on the astronaut's body in *g*'s, and (c) N. [Ans. (a) 9.84×10^5 N, (b) 5 *g*'s, (c) 3675 N.]

4. Make a diagram to scale, showing a space vehicle circling the earth 3,000 Km above the surface. If retro-rockets are fired to slightly reduce speed, what will be the shape of the new orbit? If the rockets are fired to slightly increase the speed, what will be the shape of the new orbit? Draw in such modified orbits.

5. A 24-ton rocket with a 160-lb astronaut aboard is climbing vertically upward with an acceleration of 144 ft/s². Find (a) the rocket thrust in lb, and the force on the astronaut's body in (b) *g*'s, and (c) lb. [Ans. (a) 2.649×10^3 lb, (b) 5.50 *g*'s, (c) 880 lb.]

6. A 32,000-Kg rocket with a 60-Kg astronaut aboard is climbing vertically upward with an acceleration of 34.3 m/s². Find (a) the thrust of the rocket motors, and the force on the astronaut's body in (b) *g*'s, and (c) N.

7. A satellite is to orbit the earth 629 Km above the surface. Calculate (a) its speed in m/s, and (b) its period of revolution in s and min. [Ans. (a) 7.55×10^3 m/s, (b) 5.83×10^3 s, (c) 97.2 min.]

8. A 40-ton rocket with a 192-lb astronaut aboard is climbing vertically upward. If the astronaut experiences an apparent weight of six times normal, find (a) the vertical acceleration in ft/s², (b) the thrust of the rocket motors in lb, and (c) the astronauts apparent weight in lb.

9. (a) What must be the speed of a satellite if it is to orbit the earth 1629 Km above the surface? (b) What will be its period in min? [Ans. (a) 7.06×10^3 m/s, (b) 118.8 min.]

10. A satellite orbits the planet Mars 1668 Km above the surface. Find (a) its speed, and (b) its period in min.

11. A 20,000-Kg rocket is to be put into a circular orbit 500 Km above the surface of the earth. Calculate (a) its kinetic energy, (b) its potential energy when at the earth's surface, (c) its potential energy when in orbit, (d) the potential energy imparted to the satellite, and (e) the total energy expended to put the satellite in orbit. [Ans. (a) 5.80×10^{11} J, (b) -12.51×10^{11} J, (c) -11.60×10^{11} J, (d) 9.1×10^{10} J, (e) 6.70×10^{11} J.]

12. How much energy must be expended to carry a 10,000-Kg rocket from the earth's surface out into free space?

10 THE ATOMIC THEORY OF MATTER

All matter in the universe is composed of atoms, molecules, and elementary particles—electrons, proton, neutrons, mesons, pions, etc. Whether we are talking about atoms in free space, the molecules of which our earth is composed, or what goes on in the interior of a star like our sun, all atoms, molecules, and particles are in a rapid state of motion. The nature of this motion and its activity depend upon the temperature and the state of the matter in question, as well as the kinds of atomic particles of which it is composed.

In treating the physical properties of matter it is convenient to divide substances into five forms or states: (1) *the solid state*, (2) *the liquid state*, (3) *the gaseous state*, (4) *the plasma state*, and (5) *the nucleon state*. Most substances may be made to take on any one of these states simply by altering the temperature and pressure.

10.1 KINDS OF ATOMS Although there are thousands of different substances known to the scientific world they all, when broken down into their smallest component parts, are found to be composed of one or more kinds of atoms. A substance which contains atoms of one kind only is called an *element*, while those containing more than one kind are called *compounds* or *mixtures*. Iron, copper, aluminum, platinum, mercury, hydrogen, and helium are examples of elements; whereas water, salt, brass, wood, and air are examples of compounds and mixtures.

The technical names and chemical abbreviations of a few of the more commonly known elements are given in Table 10A. A complete table of the known elements is given in the Appendix.

With each element it is customary to associate two numbers, one called the *atomic number*, the other the *atomic weight*. The atomic number, given at the left in the tables, specifies the position that element always occupies with respect to all the others, while the atomic weight on the right gives the average weight of one atom of that element relative to the average weight of an oxygen atom as sixteen. On this basis the lightest known element, hydrogen, has atoms of an average weight of approximately unity.

To illustrate the minuteness of individual atoms, the actual masses in kilograms and approximate diameters in meters of the lightest element, hydrogen, and the very heavy element, plutonium, are as follows:

TABLE 10A. Some of the Chemical Elements

Atomic No.	Element	Symbol	Atomic Weight
1..........	hydrogen	H	1.0078
2..........	helium	He	4.004
3..........	lithium	Li	6.940
4..........	beryllium	Be	9.02
6..........	carbon	C	12.01
7..........	nitrogen	N	14.01
8..........	oxygen	O	16.000
10..........	neon	Ne	20.183
13..........	aluminum	Al	26.97
26..........	iron	Fe	55.84
29..........	copper	Cu	63.57
47..........	silver	Ag	107.88
50..........	tin	Sn	118.70
78..........	platinum	Pt	195.23
79..........	gold	Au	197.2
80..........	mercury	Hg	200.61
82..........	lead	Pb	207.18
88..........	radium	Ra	225.95
92..........	uranium	U	238.17
94..........	plutonium	Pu	239.18

$$\text{1 hydrogen} \begin{cases} \text{mass} = 1.67 \times 10^{-27}\ \text{Kg} \\ \text{diameter} = 1 \times 10^{-10}\ \text{m} \end{cases}$$

$$\text{94 plutonium} \begin{cases} \text{mass} = 3.9 \times 10^{-25}\ \text{Kg} \\ \text{diameter} = 6 \times 10^{-10}\ \text{m} \end{cases}$$

The actual mass of any atom in kilograms can be obtained by multiplying the atomic weight of that element by the unit atomic mass 1.67×10^{-27} Kg. (*Note:* For the power-of-ten number system used here, see Appendix VIII.)

Although the intricate structure of each atom plays an important part in its physical and chemical behavior, we will neglect this detailed structure for the time being and think only of each atom as being a tiny sphere-like particle with a very small mass. Later in other chapters where it is pertinent to do so the structure of individual atoms will be considered in detail.

10.2 MOLECULES One of the most important properties of atoms is their ability to act upon one another at a distance. Some atoms, when they come close together, attract each other, while others exhibit a force of repulsion. When, at the close approach of two or more atoms, attraction occurs, the atoms may combine to form a molecule. Once a molecule has formed it will move about and behave as a unit particle under various physical conditions.

Molecules in general may contain almost any number of atoms. Those having but one atom are called *monatomic molecules*, those with two are called *diatomic molecules*, and those with three *triatomic molecules*. In the free state of a gas some atoms like helium, neon, and krypton prefer to exist alone, whereas others like hydrogen, nitrogen, and oxygen prefer to combine and move about in pairs.

Examples of monatomic molecules are helium (He), neon (Ne), and krypton (Kr); of diatomic molecules are hydrogen (H_2), nitrogen (N_2), oxygen (O_2), and carbon monoxide (CO); and of triatomic molecules are ozone (O_3), carbon dioxide (CO_2), water (H_2O), and hydrocyanic acid (HCN). See Fig. 10A. Besides these simplest atomic aggregates there are molecules known to contain many atoms. Along with triatomic molecules they are called polyatomic molecules.

It is clear from the diagrams in Fig. 10A that the atoms of a molecule may be of the same kind or different. The question as to why some atoms cling together in pairs and others do not is a subject involving the structure of the atoms themselves. If the individual atoms of a molecule are brought much closer together than their normal separation they repel each other and are pushed apart. If they are pulled farther apart the forces become attractive, pulling them together. In other words, they act as though they were connected by springs as shown in Fig. 10B. Pushed closer together or pulled farther apart they tend to move back to some equilibrium distance. In terms of energy they occupy a position of *minimum potential energy*. To push them closer together or to separate them requires work.

At large distances all atomic forces become very weak so that if by some means or other the atoms of a molecule are pulled far enough apart they become completely separated as free atoms. A graph of the forces between atoms is shown in

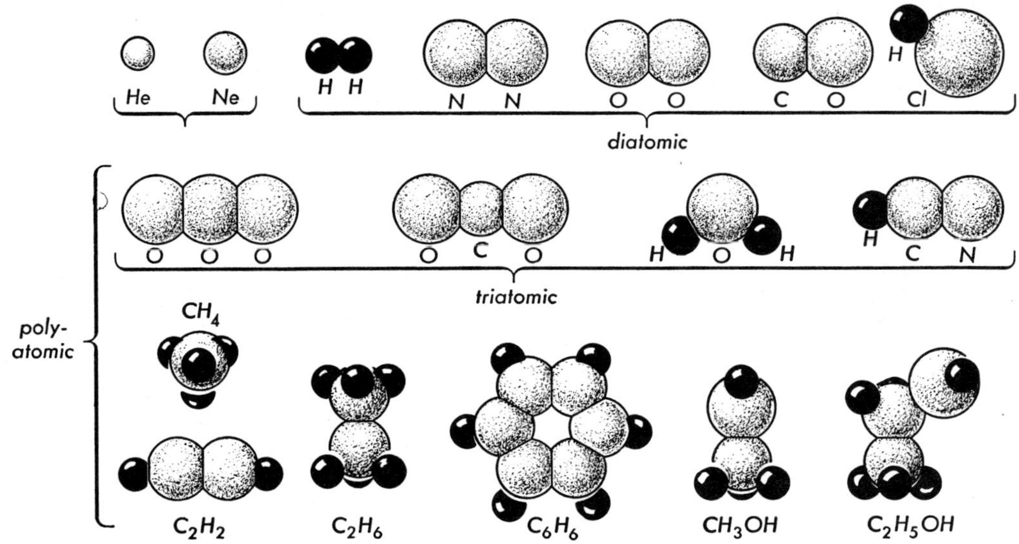

Fig. 10A. Schematic diagrams of a few common molecules. First Row. Helium, neon, hydrogen, nitrogen, oxygen, carbon monoxide, hydrochloric acid. Second Row. Ozone, carbon dioxide, water, hydrocyanic acid. Third Row. Methane, acetylene, ethane, benzine, methyl alcohol, ethyl alcohol.

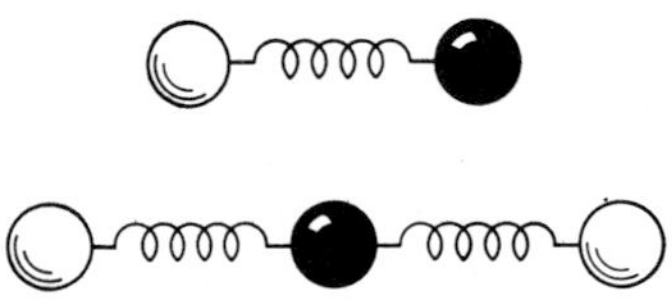

Fig. 10B. The forces between atoms in a molecule behave like springs.

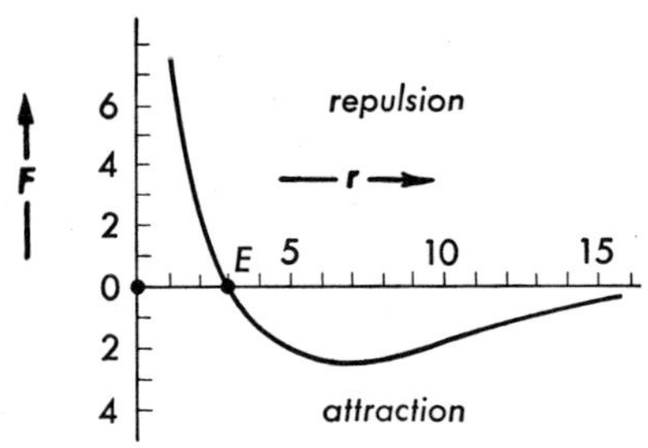

Fig. 10C. Typical graph of the force between the two atoms of a diatomic molecule.

Fig. 10C. The horizontal scale giving the distance r between atoms will be slightly different for different atoms, but the equilibrium position E is approximately 3×10^{-10} meter.

10.3 MOLECULAR WEIGHT The molecular weight of a substance is defined as the sum of the atomic weights of the atoms which make up one molecule of that substance. A carbon dioxide molecule, for example, has two oxygen atoms of weight 16 and one carbon atom of weight 12. The molecular weight of carbon dioxide is therefore $16 + 16 + 12 = 44$. Similarly the molecular weight of nitrogen is 28, oxygen is 32, and helium is 4. To find the mass of a molecule in grams, its molecular weight should be multiplied by unit atomic mass 1.67×10^{-27} Kg.

10.4 STATES OF MATTER The three common states of matter are (1) the solid state, (2) the liquid state, and (3) the gaseous state. Two additional states are now recognized: (4) the plasma state, and (5) the nucleon state. If a solid is heated sufficiently it can be made to melt or liquefy, and by continued heating can be boiled or vaporized. As a vapor it is in the gaseous state. If, on the other hand, a gas is cooled sufficiently it will condense and become a liquid. The continued cooling of a liquid will cause it to solidify or freeze. In the case of water, nature performs all these changes of state: ice is melted to become water, and water is vaporized to become steam; water vapor or clouds condense to become rain, and rain freezes to become ice or hail. Although it may sometimes require extreme heat or extreme cold, all substances can be transformed from any one state to another.

10.5 THE GASEOUS STATE When a substance is in the gaseous state it is in an extremely rarefied condition. Most of the atoms are grouped together into molecules which on the average are very far apart. These molecules are not at rest but are moving about with extremely high velocities, bumping into each other and into the walls of the container. It is the bumping of many millions of molecules against the walls of the containing vessel that gives rise to what is called gas pressure.

A good example of gas pressure is to be found where air has been pumped into an automobile tire or into a toy balloon. Since there are so many more air molecules bombarding the rubber walls inside than outside, the walls are held out by greater bombardment. In addition to moving linearly, a gas molecule, made up of two or more atoms, also vibrates and rotates about its center of mass. As the temperature of the gas is raised all of these motions increase in speed, causing an increase in pressure. As the temperature is decreased the atomic motions slow down, decreasing the pressure.

Molecular motion can be illustrated by means of a mechanical model as shown in Fig. 10D(a). Small steel marbles are placed between two parallel glass plates and set into motion by means of vibrating metal strips around the sides. Each strip *V* is mounted at the end of a short strip of spring steel and is set into vibration by means of small electromagnets *S*. As the steel balls bump into these strips they are bounced off with high speed. On the average, the small steel marbles move considerably faster than the larger ones. This is characteristic of the different sized molecules in a mixture of two different gases like helium and neon. By increasing the vibrations of the strips the steel marbles move faster. This is analogous to the heating of a gas to a higher temperature.

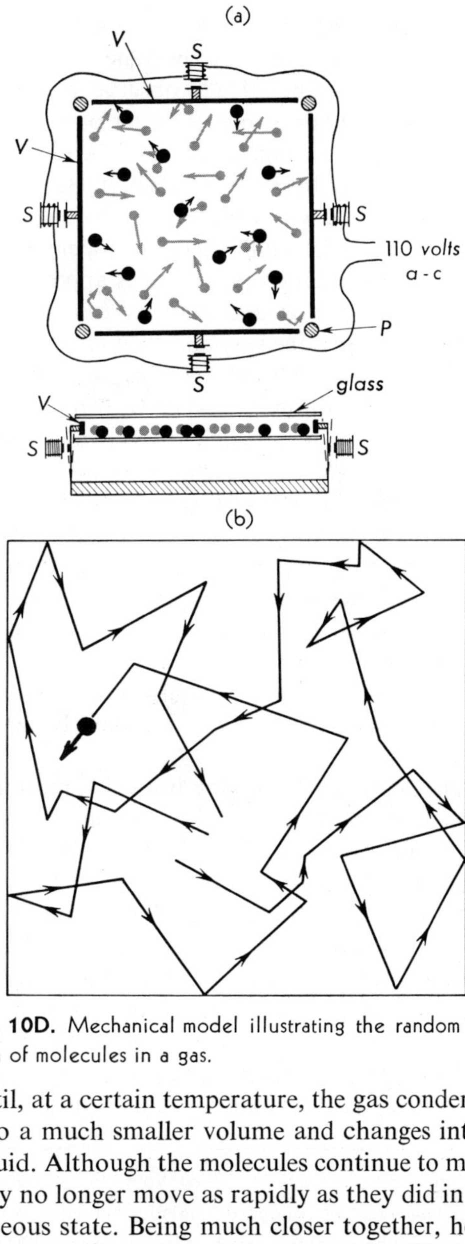

Fig. 10D. Mechanical model illustrating the random motion of molecules in a gas.

10.6 THE LIQUID STATE When a gas is continually cooled the molecular motion slows down until, at a certain temperature, the gas condenses into a much smaller volume and changes into a liquid. Although the molecules continue to move they no longer move as rapidly as they did in the gaseous state. Being much closer together, however, they now attract each other with sufficiently strong forces to cause them to move in closely packed swarms.

The swarming of honeybees as they fly through the air is comparable to the molecules of a gas. Their subsequent collection on the branch of a tree corresponds closely to condensation into the liquid state.

10.7 BROWNIAN MOTION Although no one has ever observed directly the random motions of molecules, it is possible to observe in a microscope the resultant recoils of larger particles under their continual bombardment. The effect was first discovered in 1827 by Robert Brown,* a British botanist, who observed the irregular but lifelike motions of small particles suspended in a liquid. These microscopic particles appear to be continually agitated and make a succession of quick jumps first one way and then another. The path of a single particle is illustrated in Fig. 10D(b). Such motions are called Brownian movements, after the name of their discoverer. These curious Brownian movements were carefully studied and verified by Wiener in 1863, and by Sir William Ramsey in 1879, and may be observed in either liquids or gases. The invisibly small molecules of air or water, as the case may be, move at relatively high velocities, and bombarding the larger visible particles vigorously from all sides make them dart here and there. The larger the particles the slower their Brownian movement.

One method of observing Brownian motion in a gas is illustrated in Fig. 10E. Smoke from the tip of a match, just extinguished, is drawn into a small box by squeezing and releasing the rubber bulb. A strong beam of light from an arc light, entering the box through a glass lens in the side, illuminates the smoke particles, enabling them to be seen from above with a high-power microscope. The tiny smoke particles appear as bright starlike points darting first one way then another.

To observe Brownian movement in a liquid,

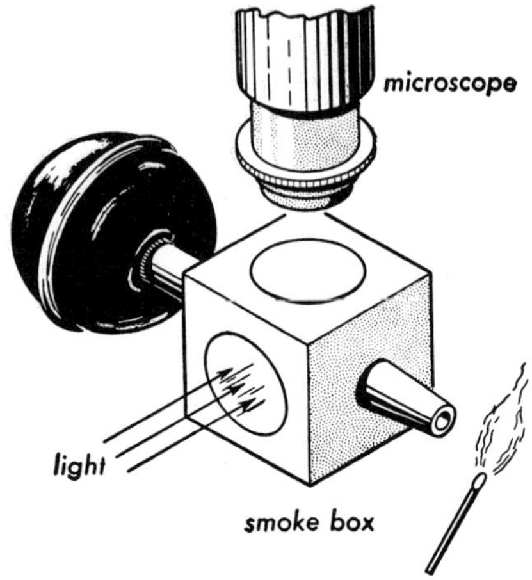

Fig. 10E. Experimental arrangement for observing the Brownian motion of smoke particles.

a small amount of powdered gamboge (an orange-yellow gum resin) is first put into some distilled water, and one drop of this solution put on a microscope slide. By illuminating the slide with a strong light, the microscopic gamboge particles will be seen dancing about as they are continually being hit by water molecules. Remembering that the gamboge particles are thousands of times heavier than the water molecules, it will be realized that the latter must be moving with very high speeds to cause such visible recoils.

10.8 THE SOLID STATE As the temperature of a liquid is lowered the molecular activity decreases. This permits the molecules to pack a little more closely together and accounts for the slight contraction of a liquid on cooling, and conversely for its expansion on heating. As they come closer and closer together the tendency of each molecule to wander through the liquid decreases. If the temperature is lowered still further a point is ultimately reached where the liquid freezes and becomes a solid.

In the solid state each molecule is confined to a definite small space between neighboring molecules. This is illustrated in Fig. 10F by atomic models of ultramicroscopic crystals. The first model illustrates a cubic lattice, a simple type of structure in which the atoms take positions at the

* ROBERT BROWN (1773–1858), noted British botanist. After two years of college in England, he went to Australia where he spent four years collecting over 4,000 new plant specimens. In 1810, he published the first of his great four-volume work on plants and their classification. In 1827, he put forward his idea that the dancing motion of microscopic inanimate particles of matter in water or gas is due to molecular motion of the fluid molecules and not some biological organism. It is for this discovery that he is best known in physics.

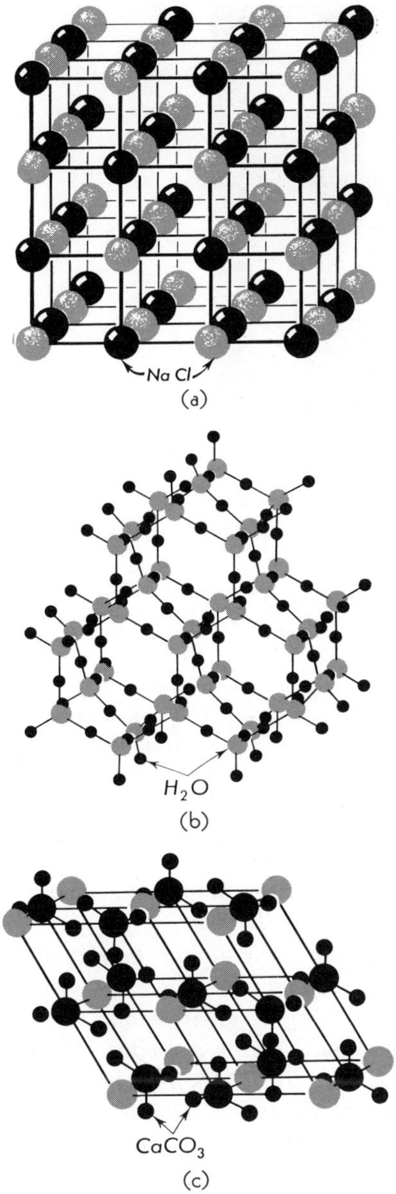

Fig. 10F. Atomic models of common crystal forms. (a) Common salt, (b) ice, and (c) calcite.

corners of cubes. Common table salt with its two kinds of atoms, sodium and chlorine, always forms such a cubic lattice, the individual atoms alternating in kind in each of the three directions, Na, Cl, Na, Cl, Na, etc.

The second crystal is of the hexagonal lattice form in which the principal structure presents parallel hexagonal "holes" through the crystal. Water, in freezing to form ice or snowflakes, takes on this form. See Fig. 10G. Note that within the atomic lattice each oxygen atom is bound by connecting links between four hydrogen atoms, while each hydrogen is linked between two oxygens. The silicon and oxygen atoms of quartz, chemically SiO_2, take on a somewhat similar structure and in the natural state also exhibit its hexagonal structure. See Fig. 10H.

The third model in Fig. 10F shows the atomic structure for a crystal of calcite. Calcite, which chemically is calcium carbonate ($CaCO_3$), is a clear transparent crystal found in nature. Note how the calcium and carbon atoms form the corners of parallelograms with each carbon atom surrounded by three oxygen atoms.

X-ray studies of crystalline forms of matter show that practically all solids belong to one of seven basic forms. Each of these seven fundamental geometrical patterns of the crystal lattice may be illustrated by the smallest section which contains a representative portion of the crystal structure. Such portions are called *unit cells.*

The repetition of the unit cell structure in all directions will reproduce nearly all of the known solids. The subject dealing with the properties of these crystal forms is known as solid state physics.

While the dimensions of unit cells differ between substances, and depend directly upon the kinds of atoms that go to make up the structure, most of them are parallelepipeds, in which the sizes and shapes are defined by the lengths of the three axes a, b, and c, and the angles between them α, β, and γ. See Fig. 10I and Table 10B.

TABLE 10B. Types of Unit Cells

Unit Cell	Axes	Angles
Cubic	$a = b = c$	$\alpha = \beta = \gamma = 90°$
Tetragonal	$a = b \neq c$	$\alpha = \beta = \gamma = 90°$
Orthorhombic	$a \neq b \neq c$	$\alpha = \beta = \gamma = 90°$
Monoclinic	$a \neq b \neq c$	$\alpha = \beta = 90°; \gamma \neq 90°$
Triclinic	$a \neq b \neq c$	$\alpha \neq \beta \neq \gamma \neq 90°$
Hexagonal	$a = b \neq c$	$\alpha = \beta = 90°; \gamma = 120°$
Rhombohedral	$a = b = c$	$\alpha = \beta = \gamma \neq 90°$

Fig. 10G. Photographs of snow crystals exhibiting hexagonal structure.

Fig. 10H. Natural crystals of quartz and calcite.

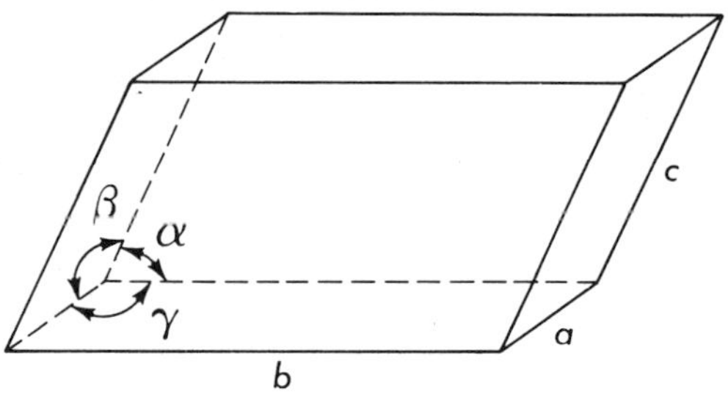

Fig. 10I. General diagram of a unit cell.

Variations of the seven types of unit cell give rise to thirty-two known crystal lattices. All thirty-two are based upon the fourteen unit cells shown in Fig. 10J. The atom sizes in this drawing have been scaled down with respect to the axis distances between them, in order to see the structure. Properly scaled models of three unit cells, made up of atoms of the same kind, are shown in Fig. 10K.

A crystal lattice containing several unit cells of the compound NaCl, sodium chloride, is shown in Fig. 10L. This simple cubic structure is a more exact representation of the solid state lattice in general than is the open type of drawing in Fig. 10F(a).

It should be noted that in representing the solid state of matter by unit cells the number of atoms per unit cell in the extended crystal lattice is not the same as shown in Fig. 10J. Some of the atoms of a single unit cell are shared by neighboring unit cells. A little study of the extended crystal lattice shows that the following rules can be applied to find the relative numbers of atoms that go to make up a solid.

1. *An atom lying completely within a unit cell belongs wholly to that cell.*
2. *An atom lying in the face of a unit cell belongs equally to two unit cells, and therefore counts as one-half an atom for that particular cell.*
3. *An atom lying on one edge of a unit cell is shared equally by four unit cells, and therefore counts as one-quarter of an atom per unit cell.*
4. *An atom lying at a corner of a unit cell is shared equally by eight unit cells, and therefore counts for one-eighth of an atom.*

A number of pure chemical elements that form face-centered or body-centered cubic structures in the solid state are listed in Table 10C. The size of the structure is specified in each case by the dimension *a* between cube corners, and is given in angstroms: 1 angstrom $= 1 \times 10^{-10}$ m.

While some elements or compounds always seem to form the same crystal pattern on solidi-

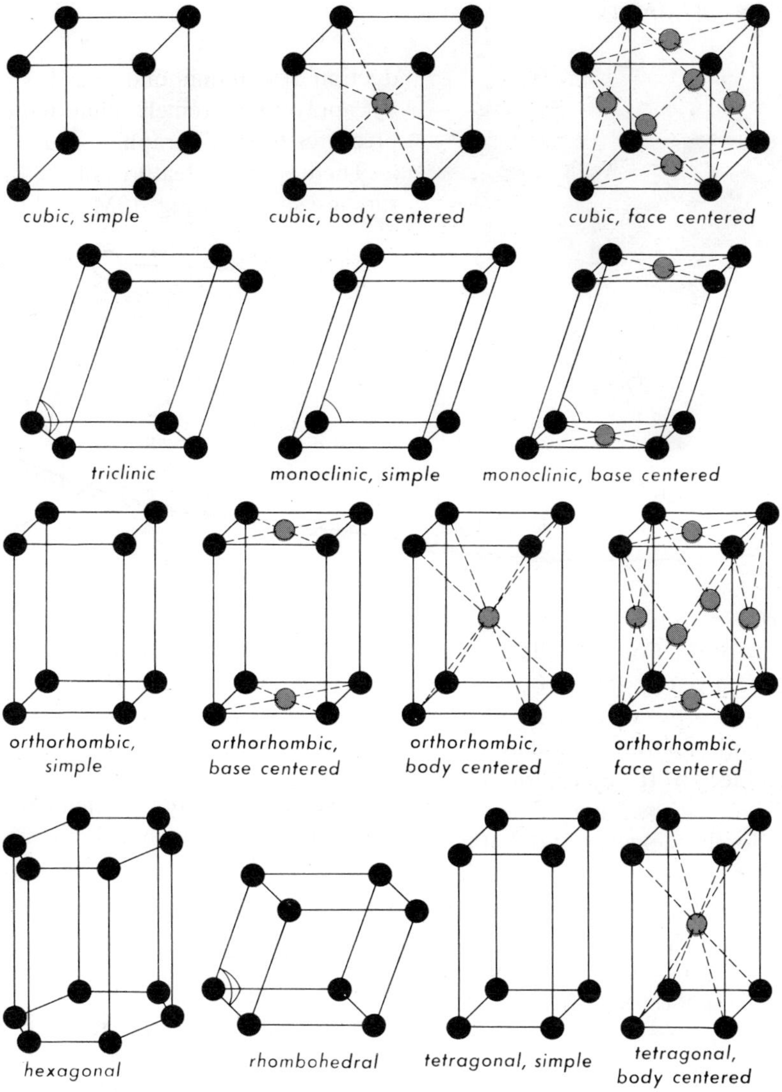

Fig. 10J. Fourteen variations of the seven types of unit cells which go to make up the crystal lattices of solids.

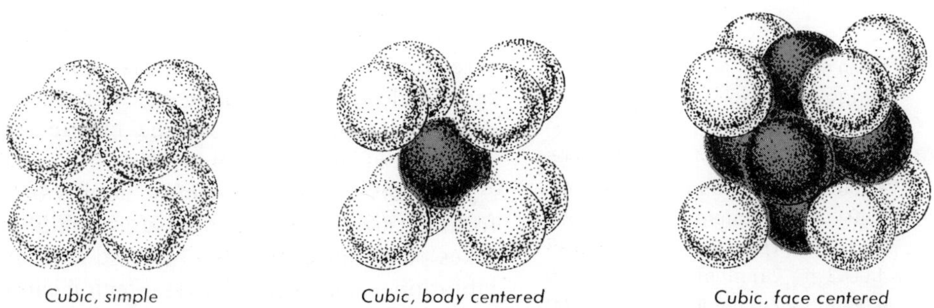

Fig. 10K. Unit cells showing the close packing of atoms in three variations of the cubic crystal lattice.

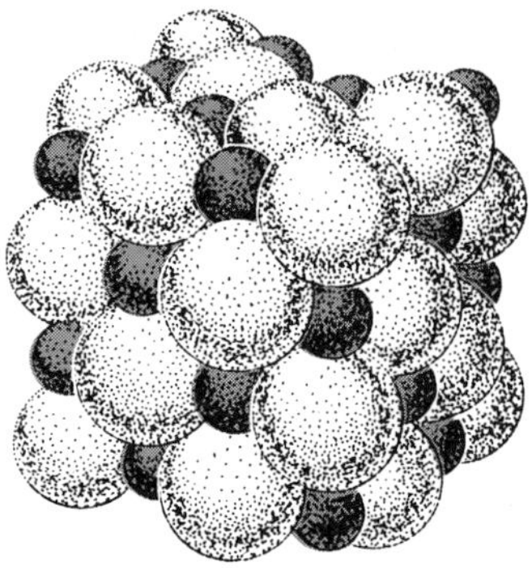

Fig. 10L. The face-centered cubic structure of sodium chloride, NaCl. Colored spheres represent Na^+ and gray spheres Cl^-.

fying, others are known to take on any one of a number of different forms. For example, two of the forms of iron are given in Table 10C. While most of the different forms of crystals known today have been found in nature, there are some that have been produced in the laboratory only. Diamond, one of several known crystal forms of carbon atoms, is a closely packed crystal that has, until recently, defied laboratory reproduction. Small diamonds for use in the manufacturing of high-speed cutting tools are now produced in quantities by the General Electric Company. It was in the company's laboratories that the production of real diamonds was first accomplished by applying extremely high temperatures and pressures to small graphite slugs.

TABLE 10C. Some Elements and Their Cubic Crystal Structures

Face-centered		Body-centered	
Element	a (Å)	Element	a (Å)
Aluminum	4.0493	Barium	5.0250
Copper	3.6149	α Iron	2.8665
Gold	4.0786	Rubidium	5.6300
γ Iron	3.5910	Sodium	4.2906
Lead	4.9505	Strontium	4.8500
Lithium	4.404	Titanium	3.3060
Nickel	3.5239	Tungsten	3.1147
Platinum	4.0862	Uranium	3.4740
Silver	4.0862	Zirconium	3.6200

The important features of the diamond structure are shown in Fig. 10M, and this is the characteristic structure for the elements C, Si, Ge, Sn, and Pb in the fourth column of the Periodic Table (see Appendix II). Note that while the outermost atoms form a large cube, all atoms are bound by chemical bonds to their four nearest neighbors. Furthermore, each set of four nearest neighbors lies at corners of a cube, with atoms diagonally opposite each other on the sides.

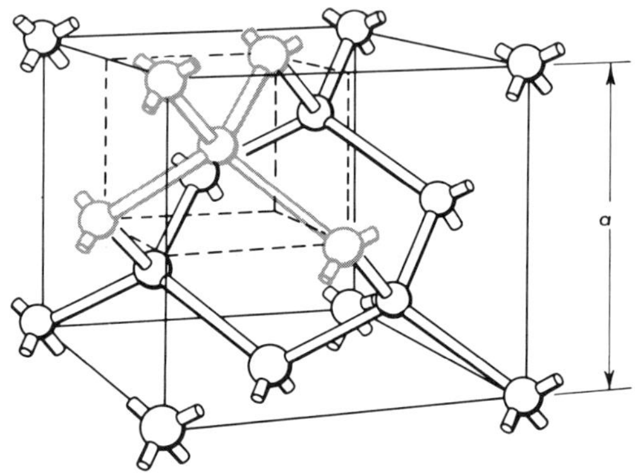

Fig. 10M. Diamond crystal structure. For carbon, silicon, and germanium a = 3.56, 5.43 and 5.66 Å, respectively. Nearest-neighbor spacing is $a\sqrt{3}/4$, and maintains tetragonal symmetry. (After Shockley.)

For C, Si, and Ge, a = 3.56, 5.43, and 5.66 Å, respectively, while the nearest-neighbor spacing is $a\sqrt{3}/4$.

Metals in general, when they cool down from the molten state, solidify into ultra-microscopic crystals that pack closely together to form a three-dimensional mosaic. This is well illustrated by the electron-microscope photographs reproduced in Fig. 10N. Note the clear-cut cubic structure of pure aluminum.

Actually the atoms in most liquids begin to form localized crystal arrays before solidification takes place but, due to the rapid state of atomic vibration, each localized crystal region can move with respect to another. A liquid may therefore

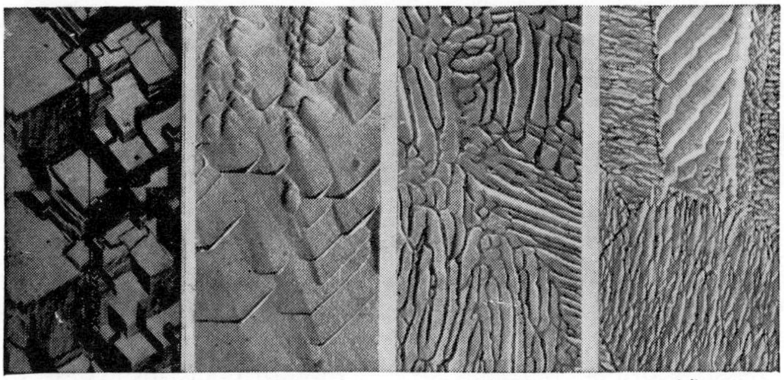

Fig. 10N. Electron microscope photographs showing the crystalline-like structure of metals: (a) pure aluminum 5600×, (b) magnesium-aluminum alloy 13,000×, (c) polished steel 14,000×, and (d) polished copper 14,000×.

be looked upon as a transition state between the gas where individual molecules exist and the solid where individual atoms become part of a crystal structure and can no longer be associated with any particular molecule. The average distance between atoms in all solids, crystal or amorphous, is of the order of 2×10^{-10} m.

A photograph showing how uniform spheres tend to pack together naturally in the hexagonal form is given in Fig. 10O.

While each atom or molecule in a solid is confined to a definite space within the body lattice, it is in a state of motion within that space. As the temperature decreases, this motion becomes slower and slower until, at absolute zero, −273°C, all molecular motion ceases. By molecular motion is meant motion of the molecule as a whole. At absolute zero, however, the atoms of certain solids are still vibrating. This residual energy of vibration is known to be an inherent property of some solids, which cannot be utilized or taken away.

10.9 THE PLASMA STATE. For many years scientists have classified all matter as belonging to one of three states, *solids*, *liquids* or *gases*. In recent years laboratory experiments and astronomical observations indicate there are at least two additional states of matter that should be included, the *plasma state* and the *nucleon state*.

At extremely high temperatures that exist deep within the sun, and millions of stars, the atoms are stripped of many or all of their electrons. Any region of this plasma state is a conglomeration of fast flying atomic particles, electrons, protons, neutrons, mesons, photons, neutrons, etc.—conditions far different from those on the earth.

Studies of plasmas of various kinds have been

Fig. 10O. Electron microscope photograph of polystyrene molecules, showing natural hexagonal packing. Diameter 5×10^{-3} cm.

carried out in the laboratory, however, by sending high voltage discharges through gases confined to special containers with magnetic fields as walls. These experiments have been carried out at low and high pressures, and at extremely high temperatures.

10.10 THE NUCLEON STATE The nucleon state, not yet produced in the laboratory, but believed to exist in the high density stars, have neutrons, protons, and other massive atomic particles so closely packed together that their average densities are thousands of times those of the densest metals found on the earth.

While little is known about this new state of matter some progress is being made to learn about it through laboratory studies of the hundred and more kinds of atomic particles now known to exist and from astronomical observations. The pulsating star recently discovered in the Crab Nebula, having a frequency of thirty pulses per second, is believed to be composed mainly of nucleons. These subjects will be taken up in more detail in the last chapters of this book.

QUESTIONS

1. Define or briefly explain each of the following: (a) atomic number, (b) atomic weight, (c) molecular weight, (d) monatomic molecule, and (e) diatomic molecule.

2. What is Brownian motion? How can it be observed?

3. Make a diagram and explain the atomic and molecular motions within a solid. How do these motions change with a rise in temperature?

4. Briefly explain the atomic and molecular motions within a liquid. What effect does temperature have upon these motions?

5. What are the three common states of matter? How do these states differ from each other? Which, in general, is the most compact?

6. How many known elements are there? What constitutes an element?

7. Give an example of a (a) monatomic molecule, (b) diatomic molecule, (c) triatomic molecule, and (d) polyatomic molecule.

8. What general treatment of most solids will change their state to the liquid or gas?

9. What can you say about the arrangement of atoms in a solid, like iron? In a solid like common table salt?

11 | PROPERTIES OF SOLIDS

The contents of this chapter will be confined to some of the more important physical properties of solids, and succeeding chapters will be confined to similar treatments of liquids and gases. One of the most important properties of solids is the one known as *elasticity*. There are several different aspects of elasticity—namely, *stretching*, *bending*, *compressing*, and *twisting*.

11.1 STRETCHING OF A SPRING If a vertically mounted rod, wire, or spring is supported rigidly at its upper end and weights are added to its lower end, the amount by which it is stretched is found to be proportional to the weight applied. This relation is a simple illustration of the behavior of solid bodies first studied by Hooke, and the basis of the law named after him. Hooke's law states that within certain limits *the deformation of a solid body is proportional to the force applied to it*. The stretching of a spring is illustrated in Fig. 11A. Due to an added weight W, the spring is stretched a distance x. If a second equal weight is added, the total distance stretched will be twice that of the first one. If a third weight is added the total distance stretched will be three times that of the first one, etc. This is illustrated by the graph shown at the right in Fig. 11A. Each value of x is plotted vertically and the corresponding loads W are plotted horizontally.

To be more specific, when the first weight of 5 N is added, the stretch or elongation is 2 cm. With two 5 N weights, the total elongation is 4 cm; and with three 5 N weights, $x = 6$ cm, etc. A continuation of this shows, as does the graph at the right in Fig. 11A, that each weight of 5 N produces an added elongation of 2 cm. To make an equation of this we write

$$\boxed{W = kx} \qquad (11a)$$

where k is a constant and equal in this experiment to 2.5 N/cm, or 250 N/m. Each value of x multiplied by 2.5 gives the corresponding weight

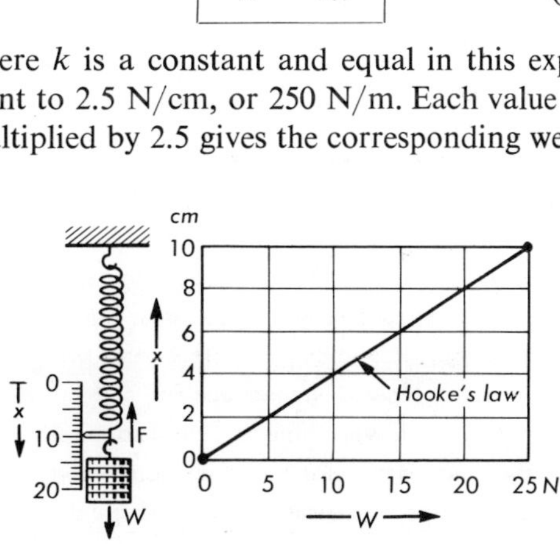

Fig. 11A. Experiment illustrating Hooke's law.

W. When the spring in Fig. 11A is stretched a distance x, the spring itself exerts an upward force F equal but opposite in direction to W. For the spring, then,

$$F = -kx \tag{11b}$$

The minus sign indicates that x and F are in opposite directions. This equation is often referred to as Hooke's law.*

In this experiment the proportionality constant k is a measure of the stiffness of the spring. Its value 250 N/m means that a force of 250 N is required to stretch the spring unit distance, i.e., 1 m. The larger the force required to stretch a spring unit distance, the stiffer the spring.

EXAMPLE 1. A steel spring 2 m long hangs from the ceiling. If a mass of 5 Kg is suspended from the lower end, the spring stretches 25 cm. What would be the additional stretch of the spring if 10 Kg more is added?

SOLUTION. Here we apply Hooke's law: as given by Eq. (11a), first to find the spring constant k. Solving the equation for the unknown k, we have,

$$k = \frac{W}{x}$$

Substituting the known values, along with the relation $W = mg$, we have,

$$k = \frac{mg}{x} = \frac{5 \text{ Kg} \times 9.80 \text{ m/s}^2}{0.25 \text{ m}}$$

$$k = 196.0 \text{ Kg/s}^2$$

We can now use this spring constant in Eq. (11a) to find the stretch for the total mass of 15 Kg. Solving the equation for the unknown x, we obtain,

$$x = \frac{W}{k} = \frac{mg}{k}$$

$$x = \frac{15 \text{ Kg} \times 9.8 \text{ m/s}^2}{196.0 \text{ Kg/s}^2}$$

$$x = 0.75 \text{ m} \quad \text{or} \quad x = 75 \text{ cm}$$

The spring stretches 50 cm more, or 5 cm per Kg.

* ROBERT HOOKE (1635–1703), English experimental physicist, known principally for his contributions to the wave theory of light, universal gravitation, and atmospheric pressure. He originated many physical ideas but perfected few of them. Hooke's scientific achievements would undoubtedly have received greater acclaim had his efforts been confined to fewer subjects. He had an irritable temper. He made many virulent attacks on Newton and other men of science, claiming that work published by them was due to him.

11.2 THE STRETCHING OF A WIRE Because a wire or rod will not stretch very far before reaching the breaking point, one must, in order to check Hooke's law, resort to some method of measuring extremely small changes in length. This is frequently done by means of a device known as the optical lever. As shown by the experiment diagrammed in Fig. 11B, a beam of light

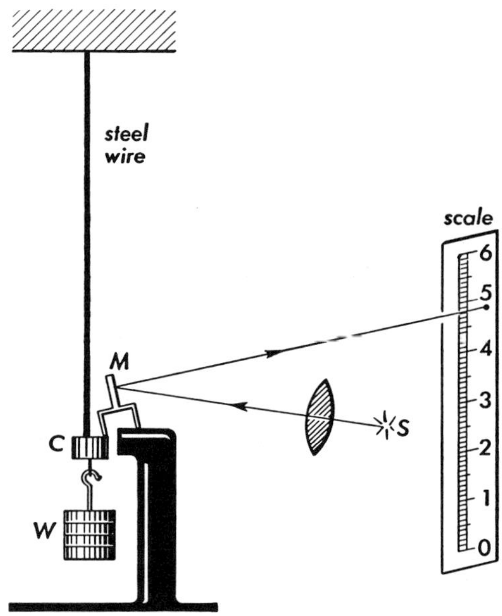

Fig. 11B. The stretching of a wire under increasing tension can be measured by an optical lever.

is reflected from a small mirror M mounted on a small three-legged stool, two legs of which rest on a stationary platform, as shown, and the third on a small clamp C at the lower end of the wire. As the wire stretches under an added weight W the mirror tips back and the light beam is reflected up a measurable amount on the distant scale.

Like the stretching of a spring described in the preceding section, the stretching of a wire obeys Hooke's law. The amount stretched is directly proportional to the force applied and is illustrated by the straight part of the graph *AB* in Fig. 11C. If the weights are removed, the wire

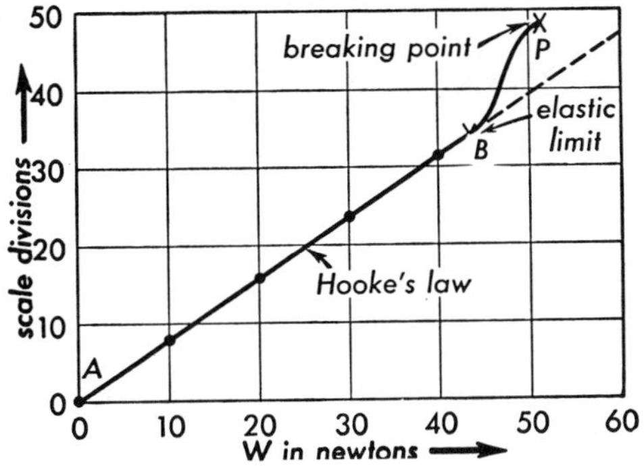

Fig. 11C. A graph of the stretching of a wire showing Hooke's law, the elastic limit, and the breaking point.

will return to its original length. If weights are continually added the forces applied will eventually become too great, and Hooke's law will no longer hold as the elongation will increase too rapidly. This is the region BP on the graph. Carried too far in this direction, the wire will break. The point B at which Hooke's law ceases to hold is called the *elastic limit*. If the wire is stretched beyond this point it will be permanently stretched and will not return to its original length when the weights are removed.

11.3 STRESS AND STRAIN When a force of any magnitude is applied to a solid body, the body becomes distorted. Whether the distortion is large or small, some portion of the body is moved with respect to some neighboring portion. As a result of this displacement, atomic forces of attraction set up restoring forces which resist the alteration and tend to restore the body to its original shape. The greater the applied force the greater will be the deformation necessitating the setting up of atomic restoring forces that will bring about equilibrium.

It is common engineering practice to describe the restoring forces in a distorted body as a stress and to give to this term the quantitative definition of *force per unit area*. The actual deformation of the body produced by an applied force involves a change in geometrical form called *strain*. Strain is defined as *a quantitative measure of deformation*.

11.4 COMPRESSION When a load F is applied to the ends of a rod to compress it as shown in Fig. 11D, the decrease in length l is the same in

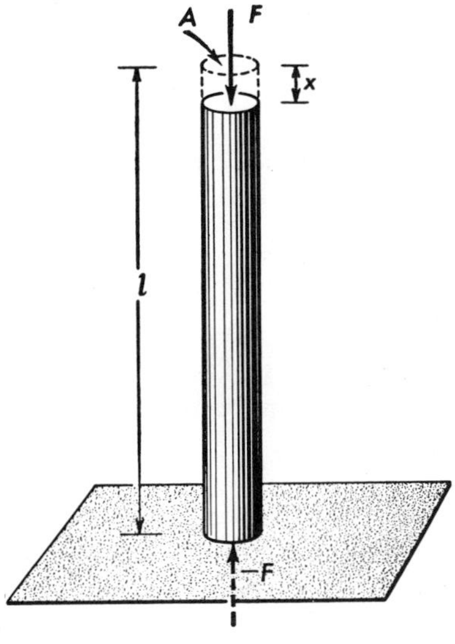

Fig. 11D. Young's modulus for the compression of a rod is given by Fl/Ae.

amount as the elongation it would acquire when the same load is applied as a tension. In other words, Hooke's law applies to compression.

11.5 BENDING When a rod or beam is subjected to a force tending to bend it, the amount of bending is proportional to the force applied. This is illustrated in Fig. 11E where, in diagram (a), a rod is shown clamped at one end with a weight hanging at the other. Diagram (b) is an illustration of a similar rod supported by pivots P at both ends with a load applied at the middle. The amount of bending, as measured in each case by the distance x, is found to be proportional to W.

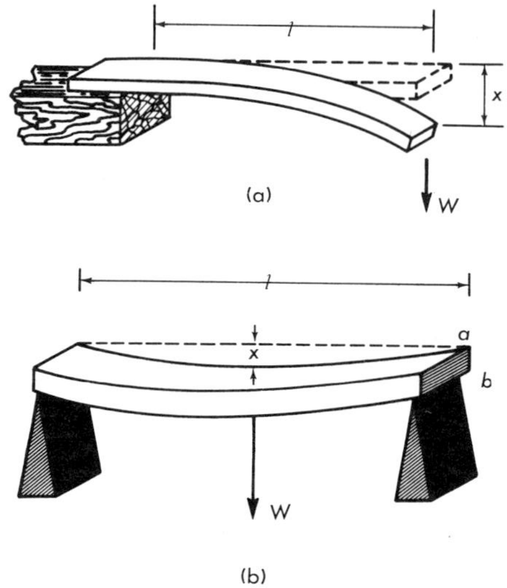

Fig. 11E. (a) The bending of a springboard obeys Hooke's law. (b) The bending of a beam follows Hooke's law.

By plotting W against x, one obtains a straight line just as in the case of stretching shown in Fig. 11A. Such a graph shows that Hooke's law applies to the bending of a rod or beam and that the phenomenon can be represented by a formula like Eq. (11a).

11.6 TORSION A detailed knowledge of torsional stresses and strains has considerable practical application in engineering design. When a rod or bar is clamped at one end and a torque is applied at the other, it is twisted as illustrated in Fig. 11F. Within the elastic limit the angular displacement θ is proportional to the applied torque L, where $L = F \times r$.

For example, if a torque of 20 Nm will twist the end of a rod through an angle of 5°, a torque twice as large will twist the end of the same rod through 2 times 5°, or 10°, and three times the original torque will twist it through 3 times 5°, or 15°. The proportionality constant k, in this case, will be 4 Nm per degree.

Fig. 11G illustrates an application of torsional

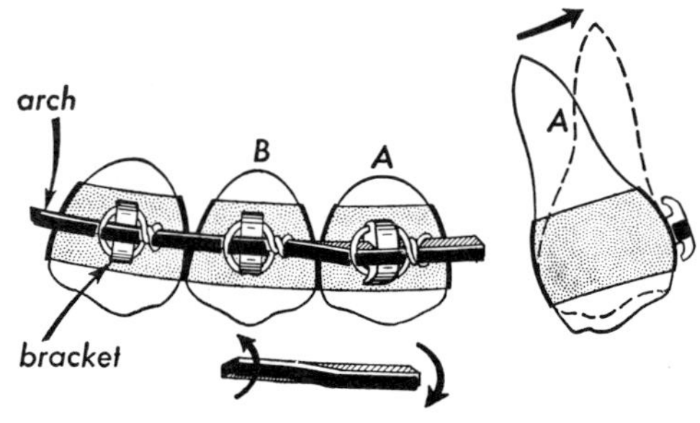

Fig. 11G. Illustration of an application of torsional stress to the straightening of teeth by the orthodontist.

stress in the profession of *orthodontics*. Orthodontia, or dental orthopedics, is that branch of dentistry dealing principally with the straightening of teeth. In one technique it is customary to fit each tooth with a wide metal band containing an arch bracket. Each bracket has a rectangular slot into which a square arch wire is tied.

To produce a rotation of tooth A as indicated at the right, a permanent twist is put into the square arch wire between that tooth and its neighbor. A torsional stress is thereby produced upon insertion of the arch wire, and over a period of several weeks the stress is gradually relieved by the turning of the tooth. The possible movement of the roots of teeth in any direction through the mandible as well as the upper jawbone is well known to every dentist.

11.7 IMPACT OF ELASTIC BODIES When two bodies collide with each other, the law of con-

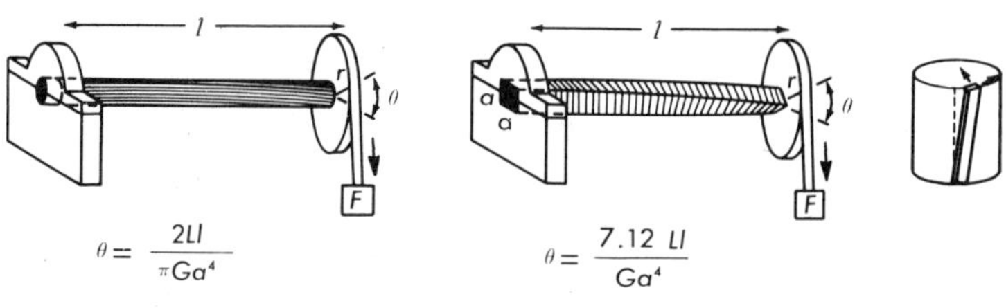

Fig. 11F. The twisting of a rod obeys Hooke's law.

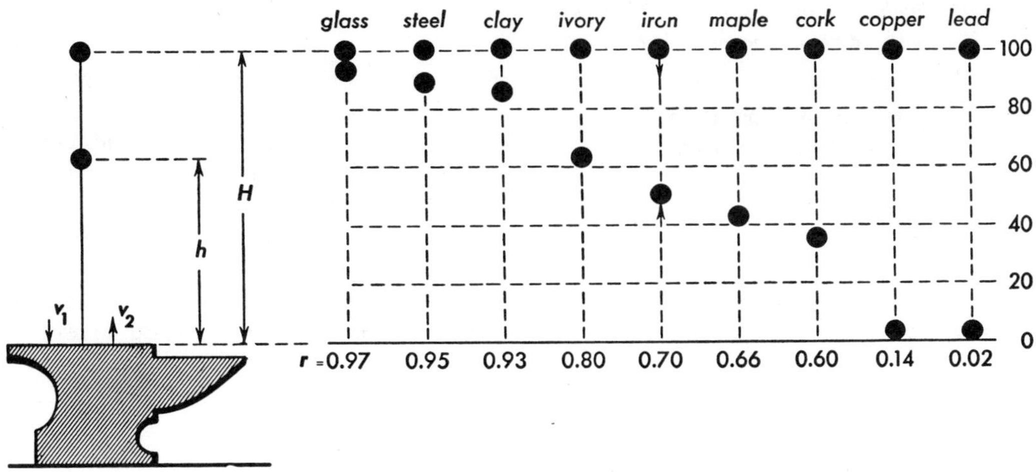

Fig. 11H. The bouncing marble experiment illustrating the resilience of different substances.

servation of momentum states that the total momentum before impact is equal to the total momentum after impact. This law is not sufficient, however, to determine what the individual velocities of each of the two bodies will be. See Secs. 7.3 and 7.4. Different kinds of material behave differently at impact and will move apart with different velocities. As an illustration, consider the experiment diagrammed in Fig. 11H.

Spheres of different substances are dropped successively, all from the same height, onto the smooth top surface of a large anvil and allowed to bounce to their various heights. Contrary to one's preconceived ideas of elasticity, a glass or steel marble will bounce to a greater height than will a ball made of the best Pará or India rubber. A lead ball or marble, on the other hand, hardly bounces at all.

In this particular experiment the two bodies in collision are the marble and the anvil. Because of its very great mass the recoiling velocity of the anvil is negligibly small.

As illustrations of the compression of a solid when in collision with another solid, note the deformation of the two balls in Fig. 11I. These are instantaneous photographs of (a) a golf ball being struck by a golf club, and (b) a tennis ball being struck by a tennis racket. They illustrate what is called the *resilience* of matter. *Resilience is defined as the ability of a body to undergo a compression, or rapid deformation, without the development of permanent deformation.* Resilience is the opposite of brittleness.

11.8 COEFFICIENT OF RESTITUTION *The coefficient of restitution is defined as the ratio of the*

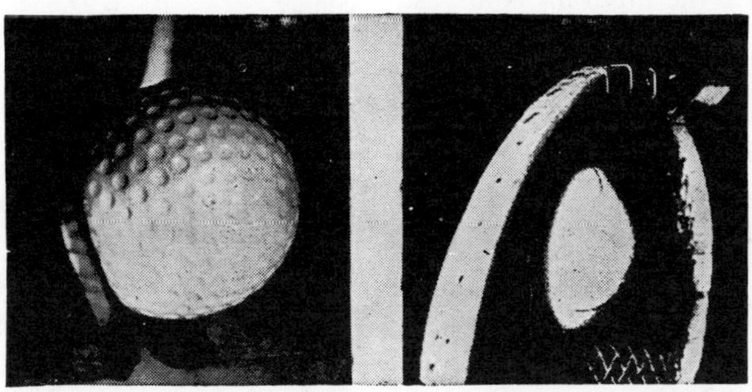

(a) (b)

Fig. 11I. Photographs of (a) the impact between a golf club and ball, and (b) the impact between a tennis racket and ball.

velocity with which two bodies separate after collision to the velocity of their approach before collision.

$$r = \frac{\textbf{velocity of separation}}{\textbf{velocity of approach}} \tag{11c}$$

or, as a formula,

$$r = \frac{v_2 - v_1}{u_1 - u_2} \tag{11d}$$

The meaning of each term is shown in Fig. 11J. Two spheres are shown there in head-on collision. Mass m_1 moving with a velocity u_1 collides head-on with another mass m_2 moving with a smaller velocity u_2. During impact, each mass exerts a force on the other. At every instant these forces are equal in magnitude and opposite in direction, action and reaction. After impact, mass m_2 has a velocity v_2 and mass m_1 a velocity v_1.

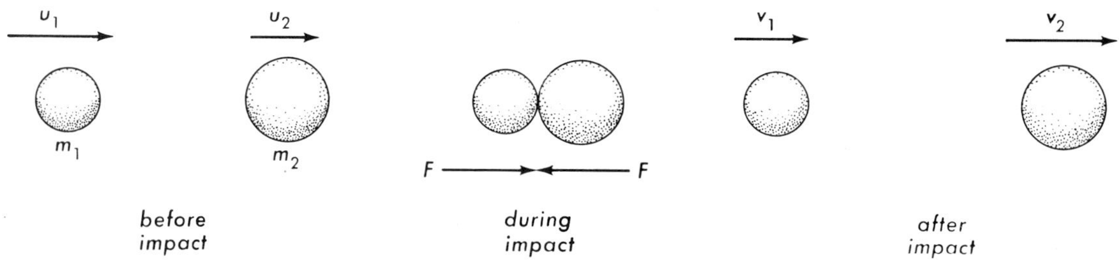

Fig. 11J. The velocity of approach is $u_1 - u_2$, and the velocity of separation is $v_2 - v_1$.

Values of the constant r for different substances may be obtained from the above described experiment by dropping marbles and determining the velocity u_1 of a marble just before it strikes the anvil, and the velocity v_1 just as it leaves on rebound. Since the anvil m_2 is at rest, u_2 and v_2 are both zero. Rather than measure the velocities directly it is more convenient to make use of the laws of falling bodies and to calculate the velocities from the heights to which the marbles are carried. Since $v = \sqrt{2gh}$ for falling bodies [see Eq. (7c)], $\sqrt{2gH}$ can be written for u_1 and $\sqrt{2gh}$ for v_1, to give

$$r = \frac{\sqrt{2gh}}{\sqrt{2gH}}$$

or

$$\boxed{r = \sqrt{\frac{h}{H}}} \tag{11e}$$

As illustrated in Fig. 11H, H is the height from which a marble falls and h is the height to which it rebounds. For a very elastic substance, like glass or steel, colliding with steel, r has a value of 0.95 or better, whereas for a very inelastic substance, like lead, colliding with steel, r is extremely small. It is seen from Eq. (11e) that the smallest value r can have is zero, whereas the largest value is unity.

Approximate values of the coefficient of restitution for the impact at moderate velocities of pairs of spheres of the same material are given in Table 11A.

TABLE 11A. Approximate Values of the Coefficient of Restitution

Material	r	Material	r
brass	0.30	iron	0.67
bronze	0.60	ivory	0.78
copper	0.22	lead	0.16
cork	0.60	maple	0.65
elm	0.48	rubber	0.75
glass	0.96	steel	0.95

11.9 HEAD-ON COLLISIONS We have seen in Sec. 7.3 that when two bodies collide head-on the law of conservation of momentum as given by Eq. (7g) applies. This law is not sufficient, however, to determine the recoil velocities of each of the two bodies after impact. Different kinds of materials have different elastic properties and do not behave alike during impact, and for this reason will move apart with different velocities.

Combining Eq. (7g) with the coefficient of restitution, as given by Eq. (11d), provides a means for determining the recoil velocities of all objects.

Suppose for example that the masses of two bodies, their initial velocities, and the coefficient of restitution are known. With two equations and two unknowns, the recoil velocities of the two bodies can be calculated.

By solving Eq. (11d) for r_2, and substituting for v_2 in Eq. (7g), we obtain the equation

$$v_1 = u_1 - \frac{m_2}{m_1 + m_2}(u_1 - u_2)(1 + r) \quad (11f)$$

By solving Eq. (11d) for v_1 and substituting for v_1 in Eq. (7g), we obtain the equation

$$v_2 = u_2 - \frac{m_1}{m_1 + m_2}(u_2 - u_1)(1 + r) \quad (11g)$$

EXAMPLE 2. A 2-Kg steel ball moving with a velocity of 8 m/s collides head-on with a 5-Kg steel ball moving in the same direction and along the same straight line at 1 m/s. If the coefficient of restitution is 0.90, find the recoil velocity of the two balls.

SOLUTION. By direct substitution in Eq. (11f), we obtain

$$v_1 = 8\,\frac{\text{m}}{\text{s}} - \frac{5\text{ Kg}}{7\text{ Kg}}\left(8\,\frac{\text{m}}{\text{s}} - 1\,\frac{\text{m}}{\text{s}}\right)(1 + 0.90)$$

$$v_1 = -1.5\,\frac{\text{m}}{\text{s}}$$

By direct substitution in Eq. (11g), we obtain

$$v_2 = 1\,\frac{\text{m}}{\text{s}} - \frac{2\text{ Kg}}{7\text{ Kg}}\left(1\,\frac{\text{m}}{\text{s}} - 8\,\frac{\text{m}}{\text{s}}\right)(1 + 0.90)$$

$$v_2 = 4.8\,\frac{\text{m}}{\text{s}}$$

QUESTIONS

1. State Hooke's law as it applies to the bending of a beam. The beam is held in a horizontal position by a support on each end, and the load is applied downward at the center.

2. How does Hooke's law apply to the bending of a springboard as it is used for diving into a swimming pool.

3. Draw a graph representing Hooke's law as it applies to the stretching of a coil spring. Assume that when a 20 N weight is hung from the lower end of the spring, the spring is stretched 2.5 cm. Extend the graph to 150 N.

4. (a) Define or briefly explain the coefficient of restitution. (b) What are the smallest and largest possible values this constant can have? (c) What do these two extreme values mean when applied to the impact of two bodies, and (d) the dropping of a ball on the pavement?

PROBLEMS

1. A brass wire 2.5 m long is stretched 0.5 mm when a load of 1 N is fastened to the lower end. What force will be required to stretch the wire a total distance of 3.8 mm? [Ans. 7.60 N.]

2. A steel coil spring 0.400 m long is stretched a distance of 6.5 mm when a mass of 100 g is suspended from the lower end. Find the amount the spring is stretched when the lower end supports a mass of 1.4 Kg.

3. A coil spring made of steel is suspended by one end. When a mass of 250 g is hung from the lower end, the spring elongates by 8.0 mm. Find

(a) Hooke's law constant k, and (b) the total stretch in the spring when the lower end supports a load of 2.25 Kg. [Ans. (a) 306.2 N/m, (b) 0.072 m.]

4. A force of 50 N will compress a coiled spring by 6.2 cm. What force will stretch this same spring 4.5 cm?

5. A steel spring stretches 3.8 cm when subjected to a tension of 75 N. Calculate the force necessary to compress this same spring 2.2 cm. [Ans. 43.4 N.]

6. The free end of a diving board at one end of a swimming pool is 40 cm above the water. When a man of 55 Kg stands on the end of the board it bends to within 18 cm of the water. How heavy would the person be who bends the board to where it just touches the water?

7. A rubber ball is dropped from a height of 2.5 m onto hard concrete and bounces to a height of 1.5 m. What is the coefficient of restitution? [Ans. 0.775.]

8. A steel ball bearing is dropped from a height of 1.5 m onto an anvil, from which it bounces to a height of 1.42 m. Calculate the coefficient of restitution.

9. An ivory ball drops onto the smooth surface of the steel deck of a ship. If the coefficient of restitution is 0.680, and it bounces to a height of 1.25 m, from what height was it dropped? [Ans. 2.703 m.]

10. When a glass marble is dropped from a height of 10 cm onto the smooth horizontal surface of a thick glass plate, it bounces to nearly the same height because the coefficient of restitution is 0.97. How high will it bounce on the second bounce?

11. A super ball is dropped from a height of 1 m. If the coefficient of restitution is 0.96, how high will it bounce on each of the first five bounces? [Ans. 0.922 m, 0.849 m, 0.783 m, 0.721 m, and 0.665 m.]

12. A 2-Kg ball moving with a velocity of 6 m/s collides head-on with a 3-Kg ball at rest. If the coefficient of restitution is 0.800, find the velocity of the two balls after impact.

13. A steel ball of 3 Kg moving with a velocity of 8 m/s collides head-on with another steel ball with a mass of 5 Kg at rest. If the coefficient of restitution is 0.900, find the velocity of each ball after impact. [Ans. −1.500 m/s, and +5.70 m/s.]

14. A ball with a mass of 5 Kg, moving with a velocity of 16 m/s, makes a head-on collision with a mass of 12 Kg initially at rest. If the coefficient of restitution is 0.80, find the final velocity of each ball.

15. A wooden ball with a mass of 2 Kg and a velocity of 20 m/s collides head-on with a steel ball with a mass of 3 Kg initially at rest. If the coefficient of restitution is 0.50, find (a) the velocity of the wooden ball after impact, (b) the velocity of the steel ball after impact, (c) the total momentum before impact, (d) the total momentum after impact, (e) the total energy before impact, and (f) the total energy after impact. [Ans. (a) +2 m/s, (b) +12 m/s, (c) 40 Kg m/s, (d) 40 Kg m/s, (e) 400 J, (f) 220 J.]

16. A bronze ball with a mass of 3 Kg and a velocity of 12 m/s collides head-on with another bronze ball with a mass of 5 Kg moving in the same direction with a velocity of 2 m/s. If the coefficient of restitution is 0.60, find (a) the velocity of the 1st ball after impact, (b) the velocity of the 2nd ball after impact, (c) the total momentum before impact, (d) the total momentum after impact, (e) the total energy before impact, and (f) the total energy after impact.

12 | PROPERTIES OF LIQUIDS

Some of the properties of liquids are well known to everyone since we come constantly into contact with water in its various forms. Most liquids at room temperature are compounds, but many of the chemical elements have been made to take on the liquid state by either raising their temperature, or lowering it. The element copper, for example, is a solid at room temperature, but when heated above 1080°C will melt to become a liquid. Helium, on the other hand, is a gas at normal temperatures, but when cooled below −269°C, becomes a liquid.

Properties of solids, like bending and twisting, do not exist in liquids. Liquids, however, can be put under compression and, if placed in a thoroughly cleaned vessel or container, can be subjected to very high tensions. Although these properties are of considerable interest, they have not proved to be of much practical importance. There are physical properties of liquids, on the other hand, which are considered to be of general importance. These are pressure, buoyancy, surface tension, and capillarity.

12.1 DENSITY AND SPECIFIC GRAVITY The density of matter, whether in the solid, liquid, or gaseous state, is defined as the mass per unit of volume. Algebraically, this may be expressed as

$$\text{density} = \frac{\text{mass}}{\text{volume}} \qquad \boxed{\rho = \frac{M}{V}} \tag{12a}$$

The Greek letter ρ is commonly used to represent density. In the *cgs* system, density is given in *grams per cubic centimeter*. Since the gram is defined as the mass of 1 cm^3 of water, the density of water is 1 g/cm^3. In the *MKS* system, density is given in *kilograms per cubic meter*.

EXAMPLE 1. If 2500 cm^3 of mercury has a mass of 34.0 Kg, what is its density?

SOLUTION. Since there are 1×10^6 cm^3 in 1 cubic meter, the volume of the mercury is 0.00250 m^3, and direct substitution in Eq. (12a) gives

$$\rho = \frac{34.0 \text{ Kg}}{0.00250 \text{ m}^3} = 13{,}600 \frac{\text{Kg}}{\text{m}^3}$$

If the volume and density of a body are known, the mass is given by Eq. (12a) as

$$M = V \times \rho \tag{12b}$$

Specific gravity, another term frequently used to express the relative weights of matter, is defined as *the ratio of the weight of a given volume of substance to the weight of an equal volume of water.*

Specific gravity

$$= \frac{\text{weight of a given substance}}{\text{weight of equal volume of water}} \qquad (12c)$$

According to this definition, the specific gravity of a substance is given by the same numerical value as the density in the *cgs* system. Since it is the ratio between like quantities, specific gravity has no dimensions and is therefore the same in all systems of units. For example, the specific gravity of aluminum is 2.7, which means that any solid piece of aluminum weighs 2.7 times as much as an equal volume of water.

TABLE 12A. Densities and Weight-densities of a Few Common Substances

Material	Kg/m³	g/cm³	lb/ft³
Liquids			
alcohol. . . .	790	0.79	49.3
benzine. . .	900	0.90	56.2
blood.	1,040	1.04	65.0
kerosene . .	690	0.69	41.2
ice.	910	0.91	57.2
mercury. . .	13,600	13.6	849
olive oil. . .	918	0.918	57.3
water.	1,000	1.000	62.4
Metals			
aluminum .	2,700	2.7	168.7
brass.	8,500	8.5	530
copper. . . .	8,900	8.9	556
gold.	19,300	19.3	1205
iron.	7,900	7.9	493
lead.	11,400	11.4	712
platinum. .	21,500	21.5	1342
silver.	10,500	10.5	655
tin.	7,300	7.3	456
zinc.	7,100	7.1	446

For convenience only, engineers, as well as others, in this country usually employ the quantity *weight per unit volume* for the term *density*. For example the density of water is commonly said to be 62.4 lb per cubic foot. Since the pound is a unit of force, it is proper to distinguish such a quantity from density as defined by Eq. (12a) by calling it *weight-density*.

$$\text{weight-density} = \rho g \qquad (12d)$$

where ρ in the engineering system is the density in *slugs/ft³*, and g is the acceleration due to gravity, 32 ft/s².

12.2 PRESSURE It is essential in the following treatment of the properties of liquids to introduce the concept of *pressure* as contrasted with the meaning of *total force*. *Pressure is defined as the force per unit of area.* Written in the form of an equation,

$$\textbf{pressure} = \frac{\textbf{total force}}{\textbf{area}}$$

$$\boxed{P = \frac{F}{A}} \qquad (12e)$$

As an illustration of the distinction between pressure and total force, consider the two aluminum metal blocks in Fig. 12A. Block (a) stands on one end where the area is 200 sq in., while block (b) stands on one side where the area is 400 sq in. Weighing 1000 lb, each block separately exerts the same downward force. Standing on end as in (a), the downward pressure is given by Eq. (12e) as

$$\text{pressure} = \frac{1000 \text{ lb}}{200 \text{ in.}^2} = 5 \frac{\text{lb}}{\text{in.}^2}$$

This is read *5 pounds per square inch*. Standing on a side as in (b), on the other hand, the pressure is only one half as great.

$$\text{pressure} = \frac{1000 \text{ lb}}{400 \text{ in.}^2} = 2.500 \frac{\text{lb}}{\text{in.}^2}$$

Over each square inch in the first case there is a downward force equivalent to 5 lb, whereas in the second case there is only 2.500 lb. The pressure in (a) is, therefore, twice as great as the pressure in (b), but the total downward force is the same.

Solving Eq. (12e) for F gives:

$$\boxed{F = P \times A} \qquad (12f)$$

Total force equals pressure times area.

This is a useful relation since it enables one to calculate the total force exerted on any given

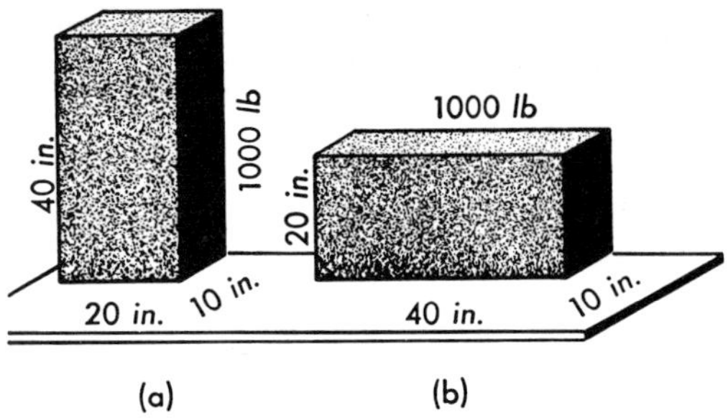

Fig. 12A. A block standing on end exerts a greater pressure than when it is lying on its side.

area. Note carefully in using it, however, that the pressure must be uniform over the area in question. If the pressure varies over the surface a suitable *average pressure* can usually be found.

12.3 LIQUID PRESSURE It is frequently necessary to determine the pressure at various depths within a liquid as well as the pressure on the bottom and sides of any containing vessel. The rule regarding pressure states that *the magnitude of the pressure at any depth is equal to the weight of a column of liquid of unit cross section reaching from that point to the top of the liquid.*

At a depth of l_1, as illustrated in Fig. 12B, the pressure p_1 is given by the weight of a column of liquid 1 sq m in cross section and l_1 meters in height. At a greater depth of l_2, the pressure p_2 is given by the weight of a column of liquid 1 sq m in cross section and l_2 meters in height.

This can be demonstrated with a glass cylinder and a thin light-weight disk as shown in Fig. 12B. With water surrounding the empty cylinder the force f, pushing up, holds the disk tight against the end. If the cylinder is gradually filled with water, an increasing downward force is exerted on the disk. Just as the water inside reaches the level of the water outside, the disk drops from the end of the cylinder, showing that the downward force and upward force at that point and at that instant become equal. We can conclude from this experiment that not only is the pressure at any given point in a liquid given by the weight of the liquid above it but that the pressure at one point is equal to the pressure at any other point at the same level.

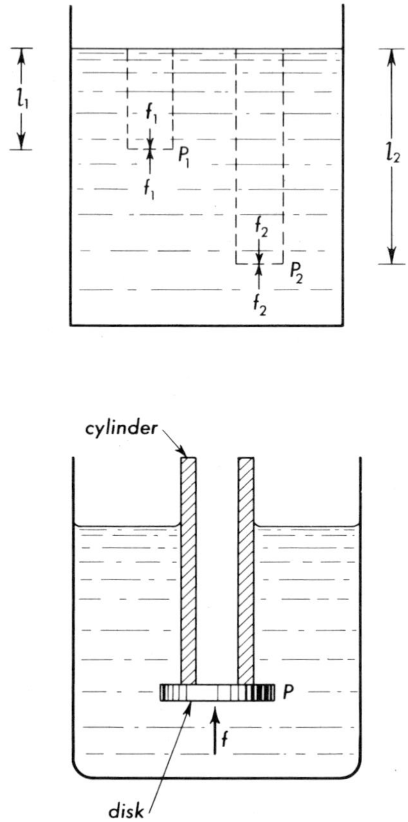

Fig. 12B. The pressure at any given depth in a liquid is equal to the weight per unit cross section of the liquid directly above.

Pressure in a liquid is defined as the normal force exerted by the liquid per unit area.

In the *MKS* system of units pressure is measured in *newtons per square meter*, and in the English system in *pounds per square inch.* Pressure is not a force—it is a *force per unit area*, i.e., force divided by area. Pressure is **not** a vector, since it acts in all directions. See Sec. 12.4.

To find a simple relation for pressure in a liquid we start with Eq. (12b). The volume of a column of liquid of height l and cross section A is given by

$$V = lA$$

Since the downward force F exerted by the liquid is given by the weight,

$$F = Mg$$

we obtain upon substitution,

$$F = \rho V g = \rho l A g$$

Dividing by A, we can write for the left side F/A, which is the pressure P, and for the right side obtain $\rho l g$.

$$P = \rho l g. \quad (12g)$$

EXAMPLE 2. Find the pressure at a depth of 240 m in fresh water.

SOLUTION. By direct substitution in Eq. (12g), and the value of the density of water from Table 12A, we obtain

$$P = 1000 \frac{\text{Kg}}{\text{m}^3} \times 240 \text{ m} \times 9.80 \frac{\text{m}}{\text{s}^2}$$

$$P = 2.352 \times 10^6 \frac{\text{Kg m}^2}{\text{m}^3 \text{ s}^2}$$

$$P = 2.352 \times 10^6 \frac{\text{N}}{\text{m}^2}$$

12.4 PRESSURE ACTS IN ALL DIRECTIONS In a liquid at rest, the force exerted by the liquid upon any surface is perpendicular to the surface. At any given point the magnitude of the force exerted on an element of surface is independent of the orientation of that surface. This can be illustrated in many ways. For example, in Fig. 12C(a) a hollow steel ball B is filled with water. By pushing down on the handle H the plunger P in the tube T forces water out of the several smaller metal tubes J leading from the sides and the bottom of the ball. Equal force in all directions is indicated by the water jets all coming to the same height as drawn.

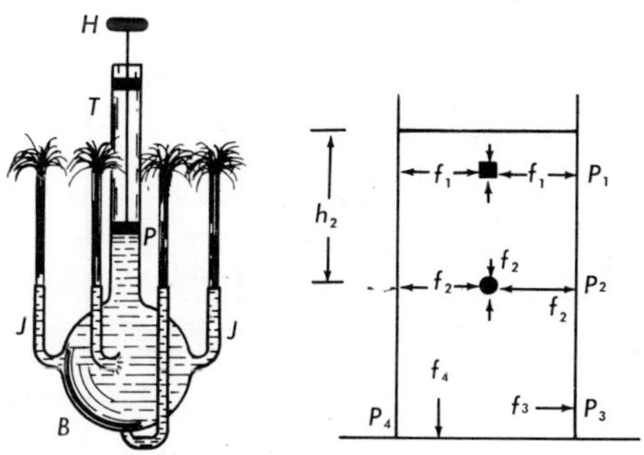

Fig. 12C. Demonstration experiments illustrating liquid pressure.

12.5 THE OCEAN DEPTHS Only recently has man descended to very great depths in the ocean to observe one of the most interesting regions of the earth's surface layers. The *bathysphere*, a hollow but thick-walled steel sphere about $4\frac{1}{2}$ ft in diameter, descended with its two American observers, Beebe and Barton, in 1931, to a depth of 1400 ft, and in 1934 to a depth of 3000 ft.

Auguste Piccard, the Swiss professor who once set a balloon record into the stratosphere (see Fig. 13B of the next chapter), set another record in 1953 by diving into the Mediterranean, just off the coast of Italy. Accompanied by his son Jacques, they descended in their *bathyscaph* to a depth of 10,300 ft.

In January of 1960, Jacques Piccard and Lt. Donald Walsh, U.S.N., descended into the Pacific, just off the coast of the island of Guam, to a depth of 37,800 ft. The pressure at this great depth of over seven miles amounts to 16,400 lb/in.^{2} (see Fig. 12D). These bathyscaphs had steel walls 3 to 5 in. thick and small portholes through which observations and photographs could be made.

12.6 ARCHIMEDES' PRINCIPLE Archimedes'* principle states that *a body floating or submerged*

* ARCHIMEDES (287–212 B.C.), Greek mathematician and inventor, was born at Syracuse in Sicily. He was the son of Pheidias, an astronomer, and was on intimate terms with—if not a relative of—Hiero, king of Syracuse. Of the many legends told of him and King Hiero, the one of the lever is perhaps the most famous. After Archimedes made the claim "Give me a place to stand on and I will move the Earth," King Hiero summoned him for an explanation. He is said to have set one end of a lever to a ship that was just ready to be launched and King Hiero himself, upon pushing lightly upon the other end, moved the ship into the water.

Another time, King Hiero, suspecting that the goldsmith had not made his crown of pure gold, as instructed, gave Archimedes the task of learning the truth without

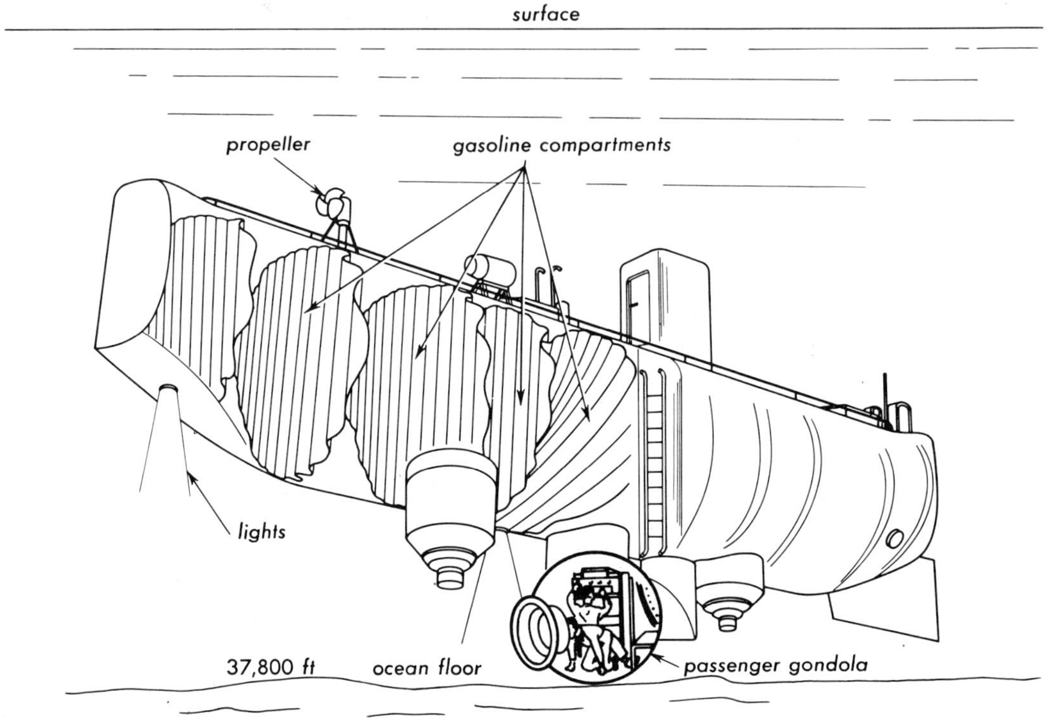

Fig. 12D. A diagram of the bathyscaph, an underwater vessel in which Jacques Piccard and Lt. Donald Walsh, U.S.N. descended to an ocean depth of 7 mi, in January 1960. The bathyscaph operates much the same as a balloon. The main part of the vessel is a thin-walled, multiple-celled tank, supporting a small, thick-walled passenger gondola at the bottom. While the buoyancy of a balloon is due to the lightweight gas it contains, the bathyscaph obtains its buoyancy by being filled with lighter-than-water gasoline. To descend, the divers release gasoline; to rise, they release ballast. Propellers on top permit the vessel to maneuver for short periods of time.

in a liquid is buoyed up by a force equal to the weight of the liquid displaced. For example, if a block of wood is floating in water, as shown in Fig. 12E, the buoyant force F holding the block up is equal to the weight of the water displaced (heavily shaded area). When the block is first placed in the water it sinks until the buoyant

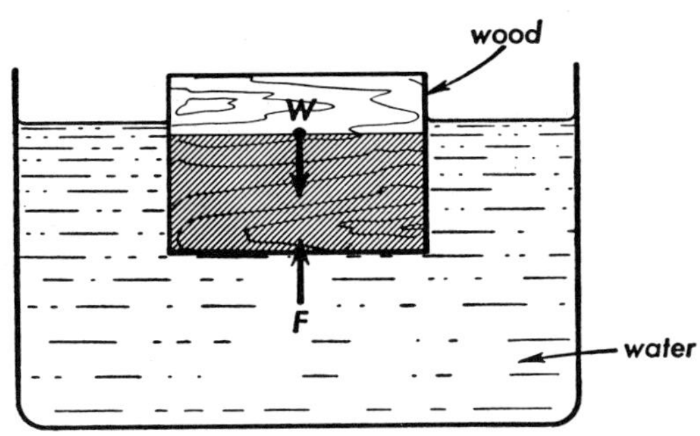

Fig. 12E. A block of wood lowered into the water sinks until the buoyant force of the water equals the weight of the block.

harming the crown. Just when he felt he would have to tell the king it could not be done, Archimedes stepped into the bath and noticed how the water ran over the edge. Springing from the bath he ran naked through the streets shouting "Eureka." ("I have found it") To find the volume of the metal had stumped him, but now he knew that by submerging the crown in a vessel filled with water, the volume of the overflow water would equal the volume of the metal. Knowing the actual weight of the crown and its volume, he calculated its density and found it to be less than the density of pure gold. A confession from the goldsmith confirmed the king's suspicions and Archimedes' experimental observations.

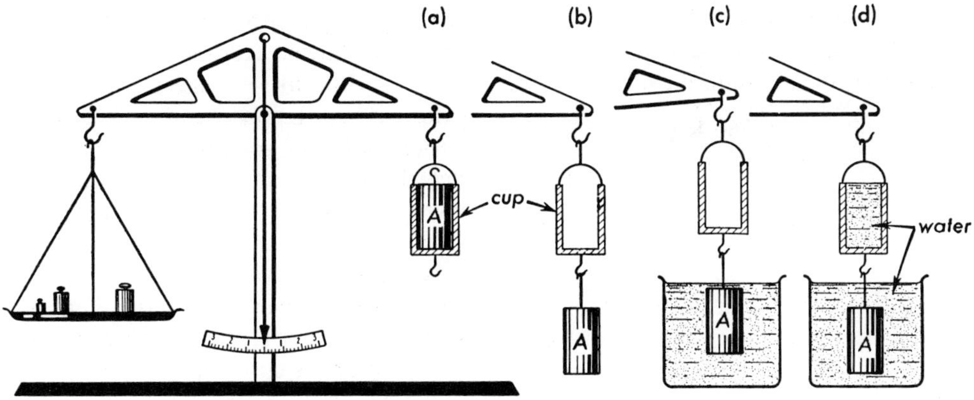

Fig. 12F. Illustration of the four steps in an experiment demonstrating Archimedes' principle.

force F becomes great enough to equalize the downward force, the weight of the block W.

An experimental demonstration of the truth of Archimedes' principle is shown in Fig. 12F. A small cylindrical cup containing a close fitting solid metal cylinder as shown is accurately balanced by a set of weights on an equal-arm balance. No change in balance occurs when, in diagram (b), the cylinder has been removed from the cup and suspended from a hook underneath. In diagram (c) a beaker of water has been raised until the cylinder is completely submerged. Balance is now destroyed because of the upward buoyant force of the water on the cylinder.

If Archimedes' principle is correct, the buoyant force is equal to the weight of a volume of water exactly equal to the volume of the cylinder. A slow addition of water to the cup in diagram (d) shows that, at the very instant the cup becomes filled, exact balance of the beam is restored. The restoring of balance confirms the principle.

12.7 MEASUREMENT OF DENSITY AND SPECIFIC GRAVITY One of the many ways of measuring the density of solids or liquids is by the use of Archimedes' principle. An experiment in which the density of a solid (aluminum) and of a liquid (kerosene) may be determined is shown in Fig. 12G. A solid aluminum cylinder is weighed with spring scales (a) in air, (b) in water, and (c) in kerosene.

Diagram (a) gives directly the mass of the cylinder as 1.400 Kg. When submerged in water the scales indicate a mass of 0.880 Kg. The difference of 1.400 − 0.880 = 0.520 Kg is the buoyant force, and therefore the mass of the displaced liquid. But since the density of water is equal to 1000 Kg/m³, a mass of 0.520 Kg of water has a volume of 0.520×10^{-3} m³. Therefore, 0.520×10^{-3} m³ is the volume of the cylinder. By Eq. (12a) the density of the aluminum cylinder is

$$\rho = \frac{M}{V} = \frac{1.400 \text{ Kg}}{0.520 \times 10^{-3} \text{ m}^3}$$

$$\rho = 2{,}700 \frac{\text{Kg}}{\text{m}^3}$$

Submerged in kerosene, the cylinder is now weighed and found to have an apparent mass of 1.040 Kg, indicating a buoyant force of 1.400 − 1.040 = 0.360 Kg of kerosene. Since the cylinder

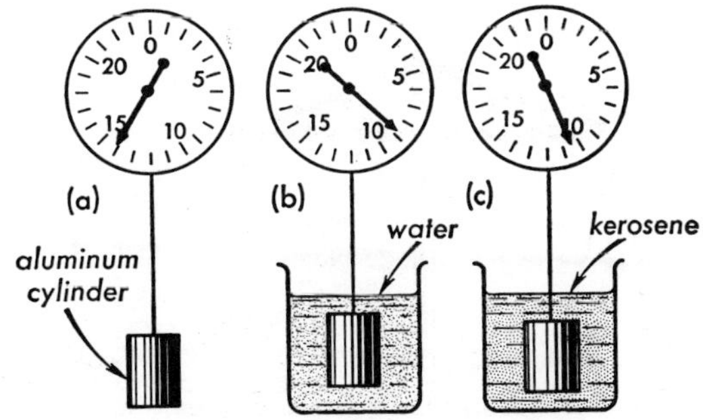

Fig. 12G. The buoyant effect of liquids may be used as a means of determining the density of solids or liquids.

displaces its entire volume of 0.520×10^{-3} m³ of liquid, and this liquid has a mass of only 0.360 Kg, the density of the aluminum cylinder is

$$\rho = \frac{M}{V} = \frac{0.360 \text{ Kg}}{0.520 \times 10^{-3} \text{ m}^3}$$

$$\rho = 690 \frac{\text{Kg}}{\text{m}^3}$$

12.8 ADHESION AND COHESION All matter is composed of atoms and molecules of one kind or another. As already stated in Chap. 10, these ultramicroscopic particles attract each other with forces which depend upon the kinds of atoms or molecules involved and upon the distance between them. The closer together two atoms or molecules are, the greater is the attractive force between them. The attractive force between different kinds of molecules is called *adhesion*, and the attractive force between two like kinds of molecules is called *cohesion*.

Although the force of attraction between two molecules is extremely small, the combined attraction of billions of molecules contained within a very small bit of matter is astonishingly great. A steel cable 1 in. in diameter, for example, will support a maximum load of 25 or more tons without breaking. This is a direct measure of the cohesive forces between hundreds of billions of atoms.

The difference between adhesion and cohesion can be demonstrated by an experiment diagrammed in Fig. 12H. A glass plate *G* is supported by one arm of a beam balance. The plate, after being balanced by weights on the left-hand scale pan, is brought into contact with the surface of water as shown. Additional weights are next added at *W* until the plate breaks free from the water surface. Upon examining the glass, water is found clinging to the under surface, showing that the break came between water molecules only. The adhesive forces between glass and water molecules therefore exceed the cohesive forces between water molecules. The weight added in the experiment is therefore a measure of the cohesive forces between water molecules.

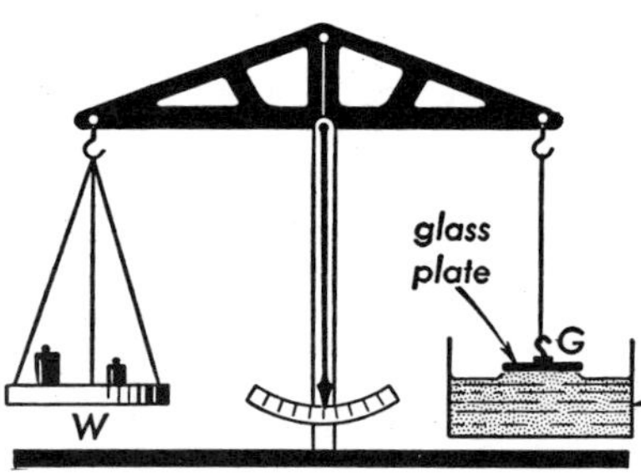

Fig. 12H. Experiment illustrating the forces of adhesion between glass and water.

If mercury is substituted for the water in the above experiment and the glass plate pulled away from the mercury surface, the added weights will measure adhesion, the force of attraction between glass and mercury molecules. This is shown by the fact that no mercury clings to the bottom of the glass plate. Thus the cohesion between mercury molecules is greater than the adhesion between mercury and glass.

12.9 SURFACE TENSION The cohesion of molecules gives rise in liquids to a phenomenon called surface tension. According to this aspect of molecular attraction, the surface of a liquid acts at all times as though it has a thin membrane stretched over it and that this membrane is under tension and trying to contract. It is for this reason that fogdrops, raindrops, soap bubbles, etc., assume a spherical shape as they fall through the air. (For any specified volume of matter a sphere has a smaller surface area than any other geometrical figure.)

An experiment designed to illustrate the spheroidal state of a liquid drop is shown at the left in Fig. 12I. Alcohol and water are poured carefully into a glass vessel where they form a separation boundary, the water with its greater density going to the bottom. Olive oil, which is not soluble in either water or alcohol, is then dropped into the liquid. The drops quickly take on a spherical form and, due to their intermediate density, settle slowly to the boundary level where they remain suspended.

A second illustration of surface tension is the floating needle experiment. A common sewing needle lowered horizontally to the surface of

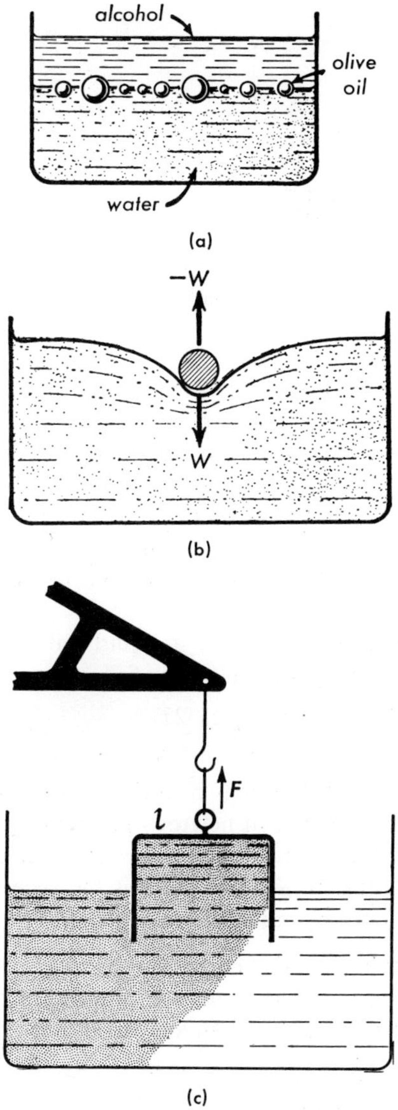

Fig. 12I. Experiments demonstrating surface tension. (a) Spheroidal state of liquid drops. (b) Steel needle floating on water. (c) Measurement of surface tension.

water in an open dish will be found to float as shown in Fig. 12I(b). (If the needle is first drawn through the fingers, a thin film of grease is deposited on the surface, making it easier to float. The adhesion between water and grease is very weak.) For the needle to break through the surface, water molecules must be pulled apart from each other. Rather than do this the surface becomes depressed until the upward force $-W$, which, by Archimedes' principle, is equal to the weight of the liquid displaced, is equal in magnitude to the weight of the needle.

An interesting experiment is performed by fastening a bit of soap to the back of a small wooden boat and then placing the boat on the surface of a pond of water as shown in Fig. 12J. As the soap goes into solution the surface tension of the water is greatly weakened behind the boat and surface tension and adhesion in front pull the boat forward. As soon as a soap film covers the whole of the water surface the boat will stop.

A similar experiment is to scrape small bits of camphor into a dish of water. Like the boat, each tiny camphor flake is propelled rapidly around on the surface of the water. The camphor, going into solution most rapidly at the pointed end of each flake, reduces surface tension more at that point than at any other.

Surface tension in the laboratory is usually measured by an arrangement illustrated in Fig. 12I(c). A small wire frame of length l is dipped into water and then pulled slowly out. As a result of both cohesion and adhesion, a thin film of water is formed in the frame. This film pulls down by a force which can be measured by suspending the frame from the arm of a beam balance. Weights are slowly added to the other scale pan, not shown (see Fig. 12F), until the film breaks. The maximum weight added without breaking the film measures directly the cohesive forces of the water film, the surface tension.

Surface tension is defined as the force of contraction across a line of unit length, the line and

TABLE 12B. Surface Tension of a Few Liquids

Liquid	T in N/m		
	0°C	20°C	50°C
Acetone.....	0.0263	0.0237	0.0199
Alcohol.....	0.0240	0.0223	0.0198
Benzine.....	0.0315	0.0289	0.0250
Mercury.....	0.5080	0.4800	0.4450
Water.......	0.0756	0.0727	0.0679

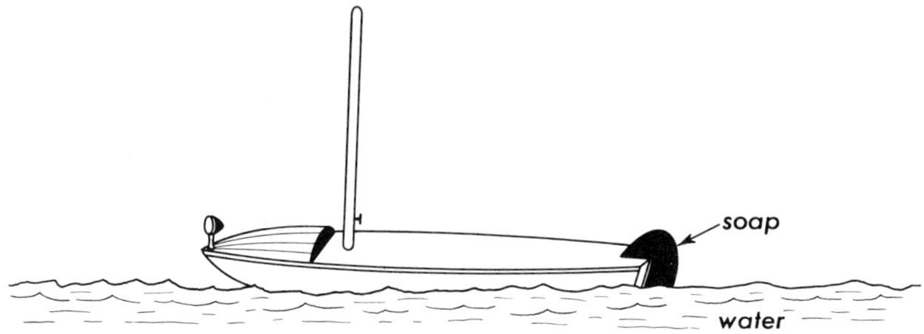

Fig. 12J. Small wooden boat propelled on water by surface tension and adhesion.

the force being perpendicular to each other, both lying in the plane of the liquid surface. If F is the maximum force applied in the above experiment and l is the length of the wire frame, the surface tension T is given by

$$T = \frac{F}{2l} \qquad (12h)$$

The factor 2 enters because there are two surfaces to a thin film, thus making the effective length of the surface equal to $2\,l$.

12.10 CAPILLARITY When a long glass tube is placed in a dish of water as illustrated in Fig. 12K(a), the water rises in the tube until it reaches a certain height and then stops. The finer the bore of the tube the higher the water rises.

Water rises in capillary tubes because the adhesive forces of glass for water are greater than the cohesive forces of water. When a fine-bore tube is first placed in water the glass walls immediately above the edge of the water attract molecules to it by adhesion. These molecules in turn attract other nearby molecules, pulling them up by cohesion. This process continues, filling up the space below as the water rises higher and higher. Surface tension, the result of cohesion, prevents any of the water from dropping back. The water continues to rise until surface tension is equalized by the weight of the liquid in the tube.

The height to which a liquid will rise in a capillary tube is given by the following formula:

$$h = \frac{2T}{r\rho g} \qquad (12i)$$

In the *MKS* system of units T is the surface tension measured in newtons per meter length, r is the radius of the bore in meters, ρ is the density of the liquid in Kg/m^3, and g is the acceleration due to gravity and equal to $9.8\ m/s^2$.

EXAMPLE 3. One end of a capillary tube is placed in a dish of water. Assuming the surface tension to be 0.070 N/m and the diameter of the

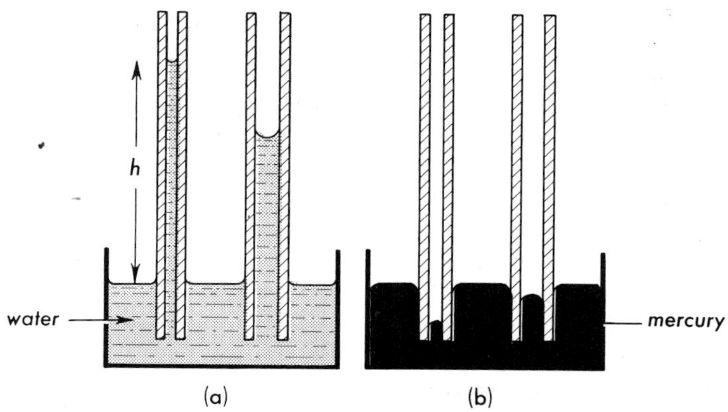

Fig. 12K. The rise of water and the fall of mercury in capillary tubes are due to adhesion, cohesion, and surface tension.

tube to be 1 millimeter, find the height of rise.

SOLUTION. By direct substitution in Eq. (12i), we obtain

$$h = \frac{0.140 \text{ N/m}}{5 \times 10^{-4} \times 1000 \text{ Kg/m}^3 \times 9.80 \text{ m/s}^2}$$

$$h = \frac{0.140 \text{ N/m}}{4.90 \text{ Kg/m s}^2}$$

$$h = 0.02857 \text{ m} = 2.857 \text{ cm}$$

With mercury in a glass tube, the strong cohesive forces form a surface which is convex downward as shown in Fig. 12K(b). Surface tension here tries to flatten out this surface thus depressing the liquid to a lower level. The same formula applied to the rise of a liquid in a capillary tube may here be applied to give the depression with mercury or any other fluid which does not "wet" the tube.

Surface tension is largely responsible for the rise of sap in trees. Carefully performed experiments show that water, by surface tension, can be raised to heights of one hundred feet or more.

QUESTIONS

1. (a) What is the difference between *density* and *weight density?* (b) What is the difference between *adhesion* and *cohesion*? (c) What is the difference between *pressure* and *force*?

2. What are the units of force in (a) the *MKS* system of units, (b) the *cgs* system, and (c) the engineering system?

3. How can the pressure at any depth in a liquid be calculated?

4. State Archimedes' principle. Why will a metal boat float when a block of metal will sink to the bottom?

5. How does the weight of a metal ball in air compare with its apparent weight in water?

6. What is the apparent weight of a boat floating on water?

7. When a fish is placed in a bowl of water does the total weight increase, even though the fish does not rest on the bottom?

8. How do the adhesive forces of water for glass molecules compare with the cohesive forces between water molecules?

9. Why do small drops of liquid take the form of spheres?

10. Why is it possible to float a steel needle and a razor blade on the free surface of water when their density is greater than water and you would expect them to sink?

11. Why does water rise in a capillary tube seemingly defying the downward force of gravity?

12. When a glass capillary tube is placed in a dish of mercury, as shown in Fig. 12K, why is the liquid depressed rather than raised?

PROBLEMS

1. Find the pressure at a depth of 50 m in a lake. [Ans. 4.90×10^5 N/m^2.]

2. The water in a lake is 180 m deep in the deepest part. Find the pressure on the bottom.

3. A gold metal award has a mass of 1200 g. Find (a) its volume, and (b) its apparent mass in water. [Ans. (a) 62.2 cm^3, (b) 1.138 Kg.]

4. A solid sphere of iron is used by a wrecking crew for demolishing old buildings. If the sphere is 1 m in diameter, find (a) its mass in Kg, and (b) its apparent mass in water.

5. When weighed in air a large stone has a mass of 1485 Kg. When weighed in water its mass appears to be 960 Kg. Find (a) the stone's volume, and (b) its density. [Ans. (a) 0.525 m^3, (b) 2830 Kg/m^3.]

6. Water at 50°C is found to rise to a height of 4.20 cm in a glass capillary tube. Find the diameter of the bore.

7. A glass capillary tube has an internal diameter of 0.25 mm. How high will the water rise in this tube if the temperature is 20°C? [Ans. 11.87 cm.]

8. A block of metal in the form of a casting has a mass of 2.85 Kg when weighed in air, and a mass of 2.54 Kg when weighed in water. Find the density of the metal.

9. An aluminum casting is suspected of being defective by having an air bubble inside. If the casting "weighs" 4.62 Kg in air and 2.67 Kg in water, find (a) the volume of the casting, (b) the volume of metal, and (c) the volume of the hole. [Ans. (a) 1950 cm^3, (b) 1711 cm^3, (c) 239.0 cm^3.]

10. The water mains in the street maintain a pressure of 45 lb/in^2. Find the water pressure in the kitchen of a house where the faucet is 25 ft higher than the mains.

13 | PROPERTIES OF GASES

The classification of familiar substances as *solids*, *liquids*, and *gases*, etc., is based upon differences in their physical properties. It is these very differences, as well as the specific properties of matter, that interest us most. The properties of gases, therefore, can be expected to be different from those of liquids and solids.

A number of chemical elements and many compounds are gases at normal room temperatures and pressures. The noble gases helium, neon, argon, krypton, xenon, and radon, as well as the more common elements hydrogen, oxygen, and nitrogen are gases under normal conditions.

13.1 THE EARTH'S ATMOSPHERE We on the earth's surface, although little conscious of the fact, are submerged in a great sea of air called the atmosphere. This air, which to the earth is the most common of all gases, is really a mixture of about 77% nitrogen, 21% oxygen, and 1% argon. The remaining 1% includes small quantities of such gases as carbon dioxide, hydrogen, neon, krypton, helium, ozone, and xenon.

Being most dense at sea level, the atmosphere extends upward to a height of from fifty to several hundred miles. The exact height of the atmosphere has no real meaning for the air gets thinner and thinner the higher one goes and eventually thins out into interstellar space. Observations show that even interstellar space, which is often referred to as the most perfect vacuum, contains a small but definite amount of matter in the gaseous state: about one molecule per cubic centimeter.

Fig. 13B is a schematic cross section of the atmosphere up to a height of twenty-five miles.

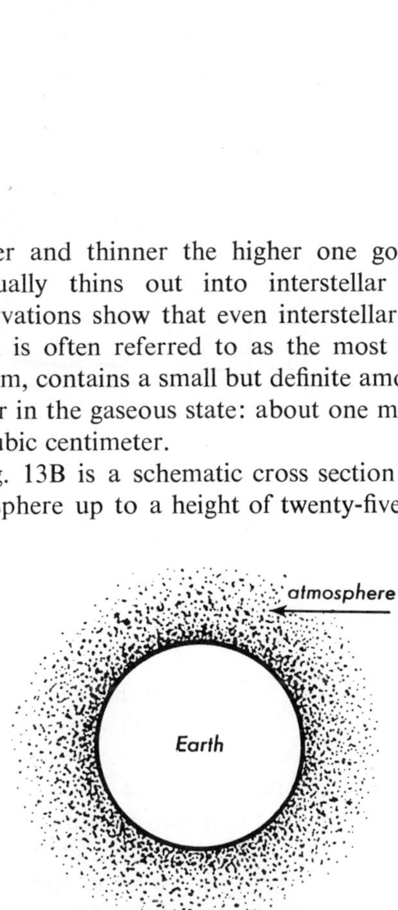

Fig. 13A. Illustration of the air surrounding the earth. The height is exaggerated to bring out the decrease in density with altitude. (If drawn to scale, the earth's atmosphere would form a layer much thinner than the line shown here representing the earth's surface.)

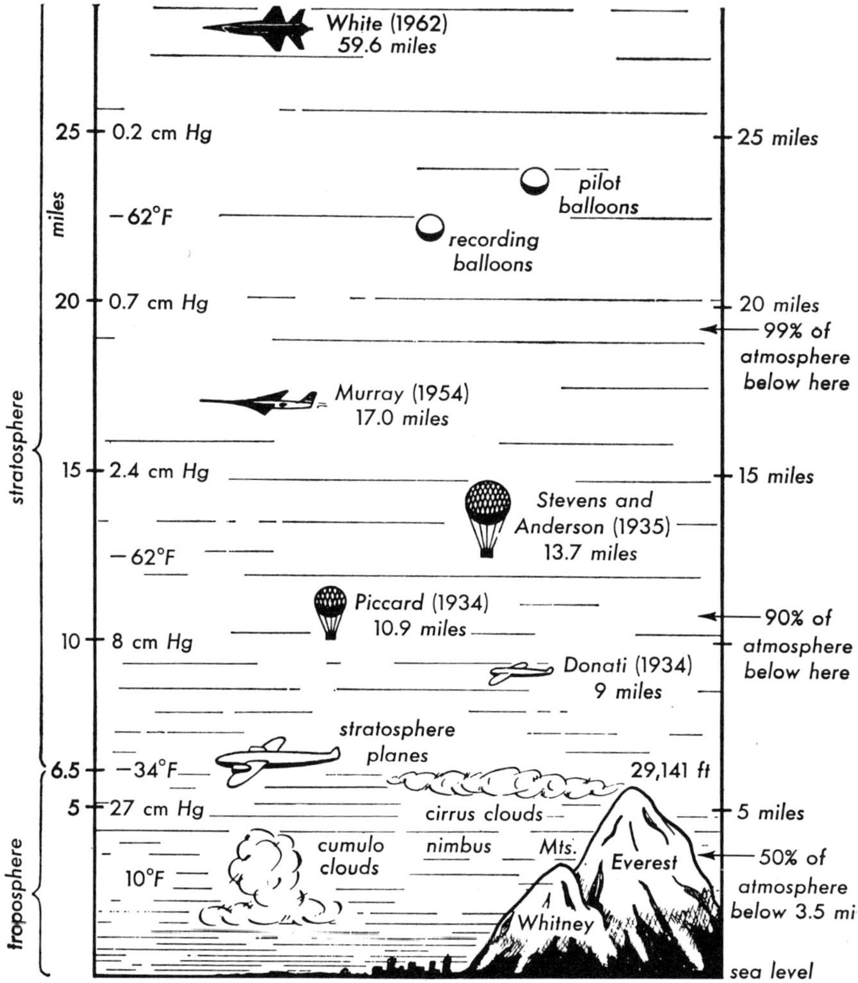

Fig. 13B. Illustration of important facts concerning the troposphere and the stratosphere and the relative heights reached by man in balloons and airplanes.

It will be noted on the right-hand side of the diagram that 50% of the earth's atmosphere lies below $3\frac{1}{2}$ mi, and that 99% lies below 19 mi. While this accounts for most of the atmosphere, experiments with radio waves show that the small amount of air existing at a height of several hundred miles is sufficient to reflect radio waves back to the earth.

Living as most of us do near sea level, we are constantly subjected to an enormous pressure due to the weight of the air above us. Unbelievable as it may seem, the air exerts a pressure of about 100,000 N for every square meter of surface. This, the atmospheric pressure, is given by the weight of a column of air 1 square meter in cross section and reaching from sea level to the top of the atmosphere.

13.2 THE DENSITY OF AIR That air has weight may be shown by one of the simplest of experiments. A hollow brass ball with a volume of 1 liter (1 liter = 1000 cm³) is first weighed when it is filled with air and again when it is evacuated. See Fig. 13C. With the air removed the vessel is found to be lighter than before by 1.29 g. If the scales are first balanced with the sphere

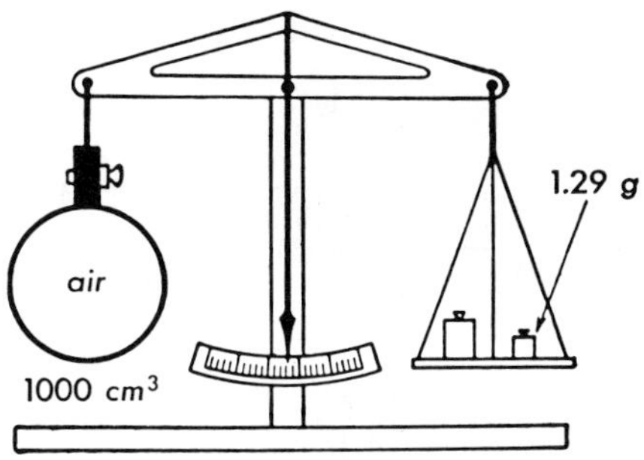

Fig. 13C. Weighing the air.

evacuated and then the air allowed to enter, a mass of 1.29 g must be added to the opposite scale pan to restore balance. Since this is the mass of 1000 cm^3 of air, the mass of 1 cm^3 will be 0.00129 g, and the mass of 1 m^3 will be 1.29 Kg.

If the above experiment is repeated at an elevation of 5 mi, the air in the brass vessel will weigh only one third as much as at sea level. The reason for this is that the lower pressure at a height of 5 mi admits only one third as much air to an evacuated vessel.

If, on the other hand, air is pumped into the hollow sphere, the weight can be made to increase considerably, thus giving a greater than normal density. *The density of a gas is* therefore standardized and *defined as the mass of 1 m^3 of the gas measured at standard pressure and temperature. Standard pressure is defined as a pressure of one atmosphere and standard temperature as zero degrees centigrade.* The densities of a few common gases are given in Table 13A. The reason for specifying temperature is that gases expand with a rise in temperature. For the numerical value of the pressure of one atmosphere see Eq. (13a).

TABLE 13A. Densities of Common Gases

Gas	Symbol	Kg/m³ or g/liter	lb/ft³
Air	N_2, O_2	1.293	0.0800
Ammonia	NH_3	0.771	0.0483
Butane	C_4H_{10}	2.519	0.157
Carbon dioxide	CO_2	1.980	0.124
Carbon monoxide	CO	1.250	0.078
Chlorine	Cl	3.214	0.201
Helium	He	0.180	0.0113
Hydrogen	H_2	0.090	0.0056
Hydrogen bromide	HBr	3.641	0.228
Krypton	Kr	3.708	0.232
Neon	Ne	0.900	0.056
Nitrogen	H_2	1.250	0.078
Nitrous oxide	N_2O	1.978	0.124
Oxygen	O_2	1.430	0.088
Sulfur dioxide	SO_2	2.927	0.183

13.3 THE MERCURY BAROMETER A barometer is a device for measuring the atmospheric pressure. There are in common use today two kinds of barometers—the mercury barometer and the aneroid barometer. The mercury barometer was invented by the Italian physicist, Evangelista Torricelli,* some three hundred years ago. Torricelli's experiment is illustrated in Fig. 13D. A long glass tube is filled with mercury and the finger placed over one end as shown in diagram (a). This tube is then inverted and with the open end in a dish of mercury the finger is removed as in diagram (b). At the instant the finger is removed the mercury level drops in the tube to a height h as shown. The mercury drops until the pressure due to its own weight inside the tube (at the level P) is equal to the atmospheric pressure outside.

At sea level the height at which the mercury column stands is about 76 cm or 30 in. This height will be the same regardless of the diameter of the tube or the length of the vacuum space at the top. Torricelli's experiment shows that a column of air 1 sq meter or 1 sq cm in cross

* EVANGELISTA TORRICELLI (1608–1647), Italian physicist and mathematician, first studied in Rome under Benedictine Benedetto Castelli, Professor of Mathematics. Inspired by Galileo, he acted as his amanuensis during the last three months of Galileo's life. After Galileo's death, Torricelli was made grand-ducal mathematician and professor of mathematics in the Florentine Academy. During his life, he wrote on fluid motion, the theory of projectiles, and on the motion of two bodies connected by a string passing over a fixed pulley. He is best known for his discovery of the principle of the barometer.

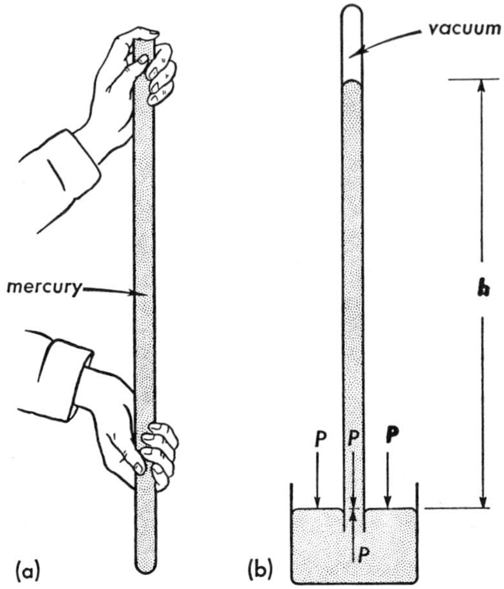

Fig. 13D. Torricelli's experiment. The making of a mercury barometer.

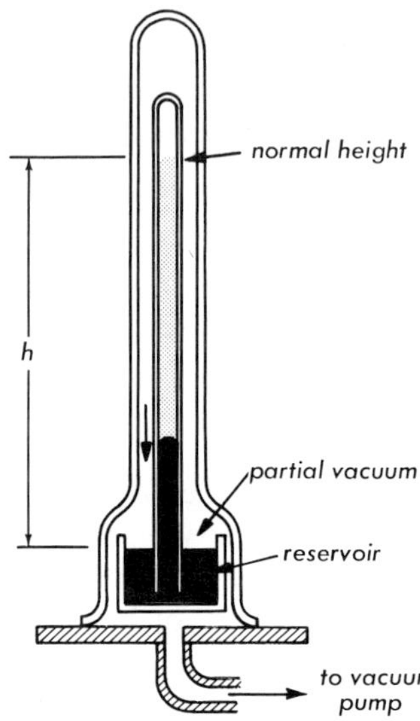

Fig. 13E. Experimental arrangement demonstrating that atmospheric pressure supports the column of mercury in a barometer tube. When the air is removed the mercury column drops from its normal height.

section and reaching to the top of the atmosphere is equal in weight to a column of mercury of the same cross section and 76 cm high.

It was the French philosopher and mathematician, Blaise Pascal,* who first showed that when a mercury barometer is taken to a high elevation like the top of a mountain, the height of the mercury column drops considerably. It drops because there is less air above that point and hence a lesser downward force on the free mercury surface.

Fig. 13E is a diagram of an experiment demonstrating that it is the atmosphere outside of a barometer pushing down on the exposed mercury surface which supports the mercury column inside and not the vacuum in the space above drawing it up. An entire barometer is placed in a tall cylinder and the air removed by means of a vacuum pump. As the air slowly leaves, the mercury column drops steadily. When the cylinder is well evacuated the level of the mercury inside the tube is the same as the level in the small reservoir outside. A return of the air forces the mercury back into the tube and up to its original height h.

The height of the mercury in a barometer measures directly the atmospheric pressure. Instead of specifying the pressure in N/m^2, $lb/in.^2$ or in $dynes/cm^2$, it is customary to give the height of the mercury column in centimeters or inches. The pressure so expressed in centimeters of Hg (the chemical symbol for mercury) is given at each 5-mi interval at the left in Fig. 13B.

If a barometer were made to employ water in place of mercury, the barometer tube would have to be at least 13.6 times as high, or 10.34 m. This is equivalent to about 34 ft. Such an instrument would be too cumbersome to be of much practical value.

* BLAISE PASCAL (1623–1662), French religious philosopher, physicist, and mathematician. Noted principally for his discoveries in pure mathematics, and for his experiments with the barometer. His experiments and his treatise on the equilibrium of fluids entitle him to rank with Galileo and Stevinus as one of the founders of the science of hydrostatics and hydrodynamics.

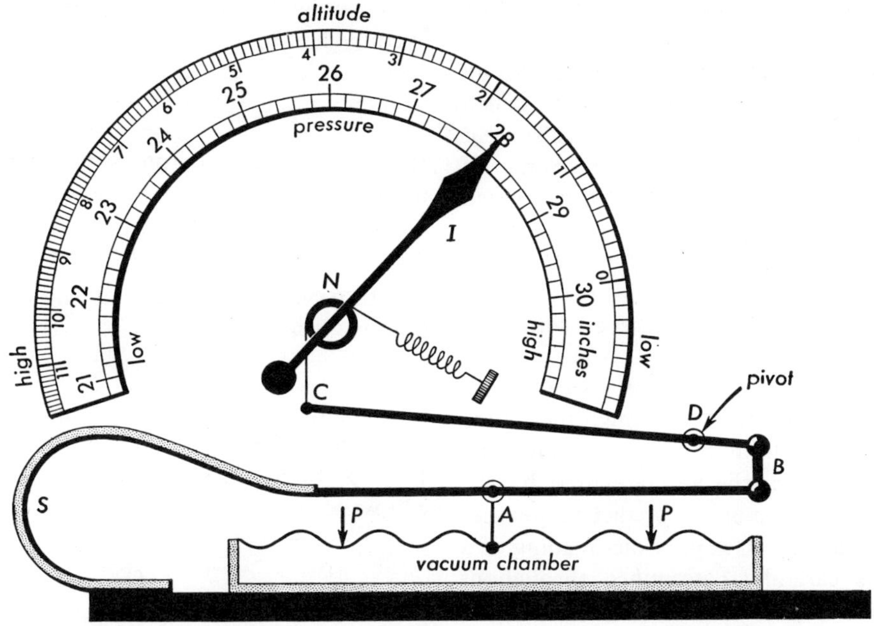

Fig. 13F. Schematic diagram and cross section of an aneroid barometer.

13.4 THE ANEROID BAROMETER The desirability of a small portable pressure measuring instrument has led to the development of the aneroid barometer. This device is frequently used as an altimeter and barometer combined. A cross-section diagram of such an instrument is shown in Fig. 13F, and photographs are reproduced in Fig. 13G(a). A small flat metal box, evacuated and with a flexible top, is attached at *A* to a multiplying system of levers. The end of the lever system is connected to a small cable C which is wrapped around a spindle *N* carrying a pointer *I*. If the atmospheric pressure *P* increases, the flexible boxtop is pushed down at *A*. This lowers the end of the lever system at *B*, and with a pivot at *D* raises the point *C*. The cable winds up on the spindle *N*, turning the pointer *I* to the right to a scale reading of higher pressure. The scale of the aneroid is calibrated by a standard mercury barometer so the pressure is always given in centimeters or inches of mercury.

Since the atmospheric pressure decreases as

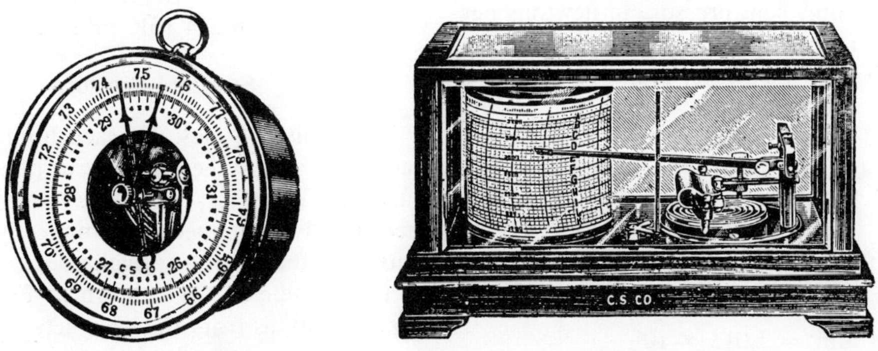

(a) (b)

Fig. 13G. (a) Aneroid barometer used for measuring atmospheric pressure. (b) Self-recording aneroid barometer, or barograph.

one goes to higher altitudes, a barometer is often used to determine elevation. As a matter of fact, aneroid barometers are frequently made with an altitude scale attached. Such instruments, called *altimeters*, are to be found on the instrument panel of every airplane and blimp. Some of these instruments are small enough to be carried in the pocket like a watch, and others are so sensitive they will indicate a change in elevation of 1 m. The altitude scale usually has its zero mark near the sea-level pressure as shown in Fig. 13F.

Atmospheric pressure not only varies with altitude but also with time. Although these time variations are small and do not follow any regular law, they can be and are used by the Weather Bureau as an aid to predicting weather conditions. If at any place the exact reading of a barometer is carefully determined throughout the day and the season, slight changes will be observed. When the barometric pressure begins to fall, it is a sign of changing weather. If the pressure continues to fall, rain usually follows. As the storm passes, the barometer rises again. So, by keeping close watch of changing barometric pressure, the weather for that locality can be forecast.

13.5 STANDARD ATMOSPHERIC PRESSURE

Standard atmospheric pressure is defined as the pressure equivalent to a column of mercury 76 cm high when the temperature is 0°C. This is equivalent to 29.92 in. of mercury at 32°F. To calculate the equivalent pressure in newtons per square meter, multiply the mercury column height 0.760 m by the density of mercury and the acceleration due to gravity.

$$0.760 \text{ m} \times 13{,}600 \frac{\text{Kg}}{\text{m}^3} \times 9.80 \frac{\text{m}}{\text{s}^2}$$

$$\boxed{\begin{aligned} 1 \text{ atm.} &= 1.013 \times 10^5 \frac{\text{Kg}}{\text{ms}^2} \\ 1 \text{ atm.} &= 1.013 \times 10^5 \frac{\text{N}}{\text{m}^2} \end{aligned}} \qquad (13a)$$

Standard Pressure

This standard pressure in the *fps* system of units is equivalent to

$$1 \text{ atm.} = 14.7 \frac{\text{lb}}{\text{in.}^2}$$

EXAMPLE 1. If a barometer were to be made using kerosene in place of mercury, how high would the liquid column be if the atmospheric pressure is equivalent to 76 cm of mercury?

SOLUTION. The pressure at any depth in any fluid is given by Eq. (12g).

$$P = \rho l g$$

Since 76 cm of mercury is equivalent to standard atmospheric pressure, we can set $P = 1.013 \times 10^5 \text{N/m}^2$. Solving the equation for l, we can write

$$l = \frac{P}{\rho g} \qquad (13b)$$

which on substitution gives

$$l = \frac{1.013 \times 10^5 \text{N/m}^2}{690 \text{ Kg/m}^3 \times 9.80 \text{ m/s}^2}$$

$$l = 14.98 \text{ m}$$

This is equivalent to 34.17 ft.

It is common practice in meteorology to measure atmospheric pressure in bars and millibars.

$$1 \text{ bar} = 1 \times 10^5 \text{ N/m}^2$$

$$1 \text{ millibar} = 100 \text{ N/m}^2$$

On this basis,

$$\boxed{\begin{aligned} 1 \text{ atm.} &= 1.013 \text{ bars} \\ 1 \text{ atm.} &= 1013 \text{ millibars} \end{aligned}} \qquad (13c)$$

Standard Pressure

The United States Weather Bureau obtains daily records of the barometric pressure from hundreds of places over the country. These records are obtained automatically at each station by means of a specially designed aneroid barometer like the one shown in Fig. 13G (b). A pen or stylus from the barometer itself moves up and down on a slowly rotating drum, thus recording the pressure at every instant. These records are then compiled by the United States

Weather Bureau, and maps are published daily of the equal-pressure areas over the North American Continent and its immediate surroundings. One such map is shown in Fig. 13H, the irregular lines representing points of equal pressure. The four digit numbers on these lines indicate the pressure in millibars [see Eq. (13b)].

Low areas, indicating stormy weather, move slowly across the map, sometimes growing in intensity and at other times breaking up and disappearing. Particularly noticeable in the diagram is the LOW pressure area in the north-central part of the United States. Like the arms of a great windmill, the cold and warm air masses, around every pronounced LOW such as this, rotate in a counterclockwise direction.

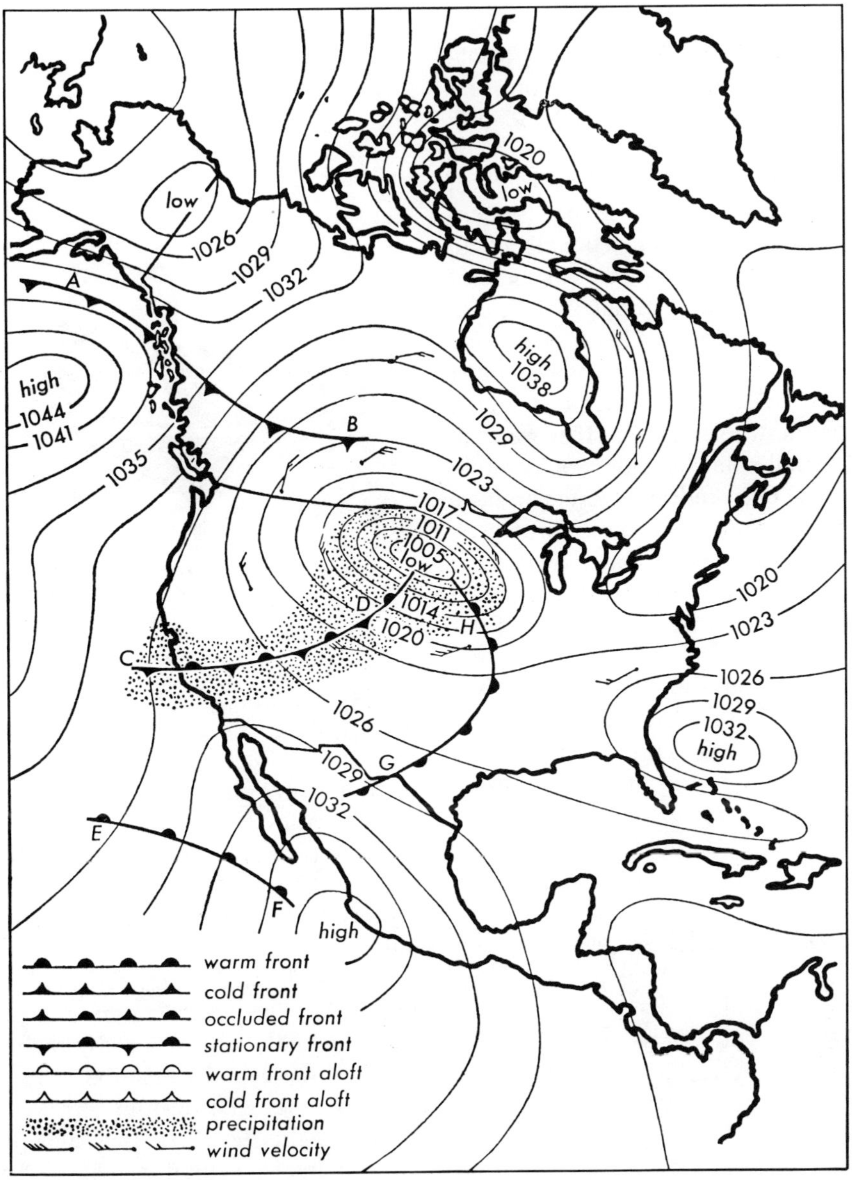

Fig. 13H. Typical weather map for the North American Continent.

As cold air fronts like AB sweep down from the polar regions and advance upon warm masses like EF moving up from the equatorial regions, they push the warm air ahead of them and up over them to form a stationary front like CD. There the warm air above the cold is cooled below its saturation temperature and precipitation of rain or snow takes place. HIGH areas are usually areas of fair weather.

13.6 EXPERIMENTS ILLUSTRATING ATMOSPHERIC PRESSURE Standard atmospheric pressure of 15 lb/in.^{2} (more accurately 14.7 lb/in.^{2}) does not ordinarily impress a person as being very great. Taken over a considerable area, however, such a pressure gives rise to a tremendous force. Consider the evacuated bulb of an incandescent lamp 3 in. in diameter. A sphere of this diameter has an area of about 28 in.^{2}, and hence a total of 28 $\times$ 15 or 420 lb is exerted inward upon its walls. The thin glass walls can withstand this because the force is distributed uniformly over the whole surface. If it were all applied at one small point the bulb would surely break. A spherical or cylindrical vessel can have thin walls and yet stand enormous pressures, whereas a vessel with flat sides may not. It is for this reason that all large vessels used for storing liquids or gases have curved walls instead of flat ones.

An experiment illustrating atmospheric pressure is diagrammed in Fig. 13I. In diagram (a) a thin sheet of rubber is first tied over the top end of a jar. When this is done the air pressure inside is the same as that outside. If now the air causing the inside force, upward on the rubber, is removed by means of a vacuum pump, the outside force pushes the rubber down inside as shown in diagram (b).

The principles of breathing in the human body are demonstrated in Fig. 13J. Muscular contraction in pulling down on the *diaphragm* creates a low pressure around the lungs and atmospheric pressure pushes air into the lungs. Retraction of the diaphragm raises the pressure and compresses the lungs, forcing air and carbon dioxide out.

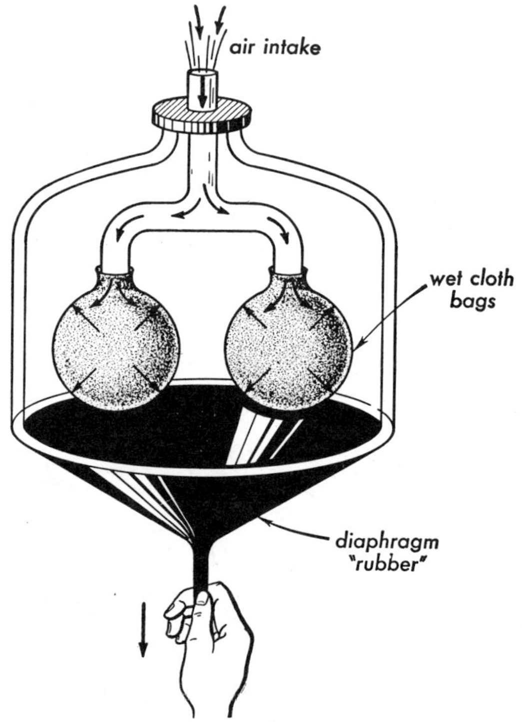

Fig. 13J. Experimental demonstration of the principles of breathing by the human body.

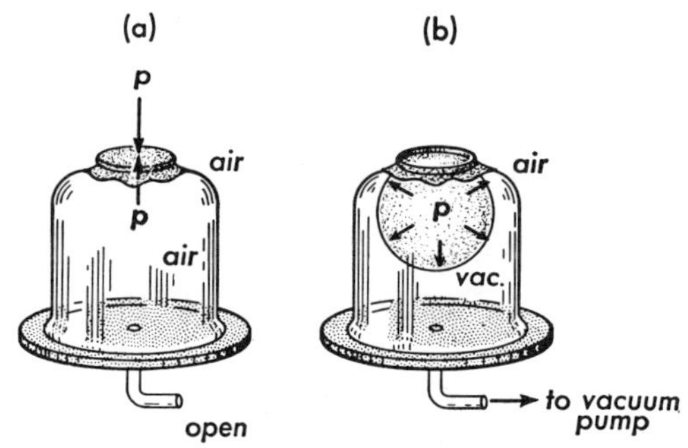

Fig. 13I. Experiments illustrating the magnitude of atmospheric pressure.

Two additional experiments are illustrated in Fig. 13K. The first diagram represents an inverted cylinder and piston. By pumping out the air from the cylinder chamber, the atmospheric pressure lifts the piston and accompanying heavy weights W. With a circular piston only 5 in. in diameter a total weight of 295 lb can be lifted. The second diagram represents the drink-

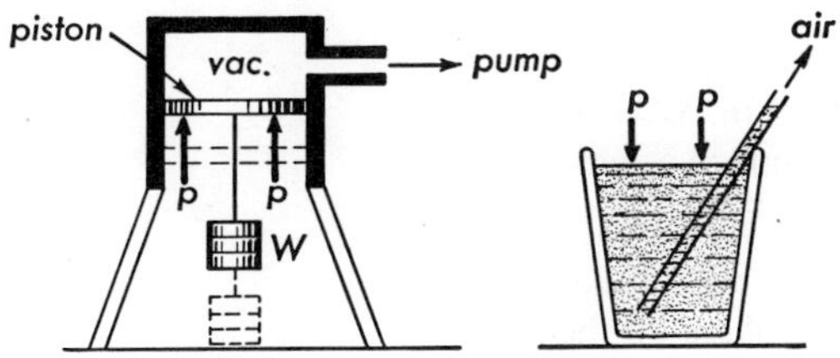

Fig. 13K. Experiments illustrating the presence of atmospheric pressure and its action in all directions.

ing of water from a glass by means of a straw. The water is not drawn up through the straw; it is pushed up from the outside. Suction at the top end of the straw removes the air, and hence the pressure at that point, and the atmospheric pressure at the liquid surface in the glass pushes the water up the straw and into the mouth. This is similar in its action to a siphon.

13.7 THE MAGDEBURG HEMISPHERES In the year 1654 Otto von Guericke* performed before the Emperor Ferdinand III, at Regensburg, the celebrated experiment of the "Magdeburg hemispheres." Two copper hemispheres, about 22 in. in diameter, were placed together to form a sphere as shown in Fig. 13L. A ring of leather soaked in oil and wax was set between them to make an airtight joint. When the sphere was evacuated, two eight-horse teams were unable to pull the hemispheres apart. The force required to pull them apart is nearly three tons.

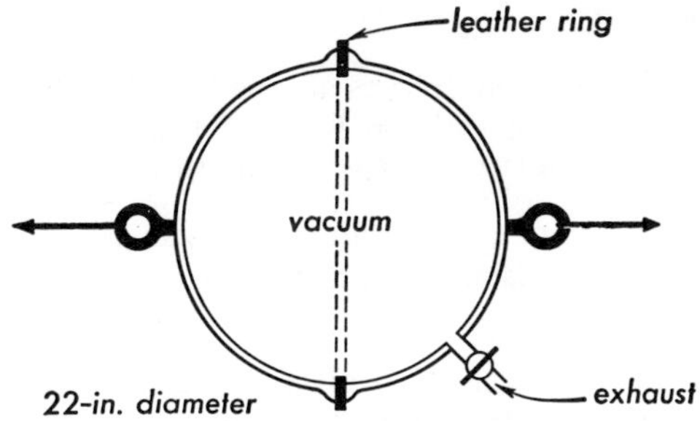

Fig. 13L. The Magdeburg hemisphere designed by Otto von Guericke.

* OTTO VON GUERICKE (1602–1686), German philosopher, lawyer, physicist, and magistrate. Incited by the discoveries of Galileo, Pascal, and Torricelli, he produced the first vacuum and made the first air pump. From researches in astronomy he predicted the periodic return of comets.

EXAMPLE 2. What minimum force in pounds is required to pull the original Magdeburg hemispheres apart if their internal diameter is 22 in.? Assume atmospheric pressure to be 14.7 lb/in.^{2}

SOLUTION. Since atmospheric pressure exerts forces that are everywhere perpendicular to the surface, the problem is simplified by imagining one hemisphere replaced by a flat plate and then evacuating. The total force on this flat plate will then be given by Eq. (12f), where the area is given by the area of a circle of radius r.

$$A = \pi r^2$$

By direct substitution in Eq. (12f), we find

$$F = PA$$

$$F = 14.7 \frac{\text{lb}}{\text{in.}^2} \times 3.142 \times (11 \text{ in.})^2$$

$$F = 5590 \text{ lb.}$$

QUESTIONS

1. Using the information given in Fig. 13B, plot a graph of the atmospheric pressure in meters of mercury against altitude in meters. Plot meters horizontally from 0 to 40 Km, and the corresponding pressure vertically.

2. Using the data given in Fig. 13B, plot a graph of the atmospheric pressure in millibars against altitude in feet. Plot altitude vertically to 1.5×10^5 ft.

3. Define or briefly explain each of the following: (a) gas density, (b) standard atmospheric pressure, (c) bar, and (d) millibar.

PROBLEMS

1. Magdeburg hemispheres are made with a diameter of 0.60 m. What is the minimum force required to pull them apart? Assume standard atmospheric pressure. [Ans. 2.864×10^4 N.]

2. The screen of a TV picture tube has an area of 320 in². Since such tubes are highly evacuated, what total force is exerted by the atmosphere on the screen?

3. A hollow metal ball with a diameter of 15 cm is highly evacuated. Calculate the total inward force on the surface if standard atmospheric pressure prevails. [Ans. 7160 N.]

4. What would be the average atmospheric pressure in millibars at an altitude of (a) 5 mi, (b) 10 mi, (c) 20 mi? See Fig. 13B.

5. A hollow glass bulb 30 cm in diameter is evacuated. What would be its mass increase in grams if it were filled with air at a pressure of (a) 1 atm, (b) 2 atm, and (c) 5 atm? [Ans. (a) 18.24 g, (b) 36.48 g, (c) 91.2 g.]

6. A jet plane maintains an internal pressure of 11.6 lb/in.² when flying at a height of 5 mi. Find the outward force acting on each window if it is a circular glass plate 14 in. in diameter. See Fig. 13B.

7. A thick-walled tin can 20 cm high and 10 cm in diameter is evacuated. Find the total force on either end. Assume standard atmospheric pressure. [Ans. 796 N.]

8. A single Magdeburg hemisphere 18 in. in diameter has a flat metal plate placed over the opening, and then evacuated. Calculate the total inward force on the metal plate. Assume atmospheric pressure of 14.7 lb/in.²

9. A cylindrical pipe 24 in. in diameter has metal plates sealed on the ends, and then evacuated. What is the inward force exerted on either end? Assume standard atmospheric pressure. [Ans. 6650 lb.]

10. A spherical balloon 30 m in diameter is filled with air. (a) Find the mass of the air entrapped. How much lighter would the balloon be if the air inside were replaced by (b) helium, and (c) hydrogen? Assume gas inside is at a pressure of 1 atmosphere.

11. The lower cylinder section of the glass vessel in Fig. 13E has a diameter of 12 cm. What is the downward force exerted by the air on this vessel, pushing it against the flat bottom plate, when the vessel is thoroughly evacuated? Give your answer in (a) newtons, and (b) pounds. [Ans. (a) 1146 N, (b) 257.7 lb.]

12. If the cylinder and piston of Fig. 13K have a diameter of 8 in., how great a load can be lifted by evacuating the cylinder?

14 | FLUIDS IN MOTION

The transportation of many fluid substances from one place to another is often done through pipes. The process of the flow of materials through pipes is called *mass transport*. The term *fluid* applies to any substance capable of flowing and includes gases, liquids, and powdered solids. Since all fluids have mass, Newton's Second Law of Motion implies that unbalanced forces are required to set them in motion. As a matter of practical interest we will consider the various means by which such forces are obtained, how they are applied to fluids, and what factors control the resultant motion. In these discussions the student would do well to keep in mind the fact that liquids are practically incompressible.

14.1 VELOCITY THROUGH AN ORIFICE Many city water supply systems store water in reservoirs on some hilltop or in a nearby water tower and from these run the water through pipes into the houses, stores and factories in and around the city. Such an arrangement is called a "gravity system."

When a hole is opened in the side of a vessel containing a liquid the velocity of flow through the orifice increases with depth. Here the unbalanced force setting the liquid in motion is gravity acting through the liquid as pressure. We have already seen how the pressure at any given depth is the same in all directions. At a depth h_1 (see Fig. 14A), the liquid exerts a pressure p_1

Fig. 14A. The velocity of efflux through a hole in the side of a vessel of water increases with depth.

against the walls and the walls exert an equal and opposite pressure against the liquid.

The instant an opening is made in the side of the vessel the wall pressure is destroyed at that point and the liquid pressure inside pushes the liquid which is directly in front of the hole, giving it an acceleration outward and normal to the plane of the opening. To find the velocity of escape, consider the potential energy of the liquid body in the vessel when the hole is first opened and then a short time later when a small amount of liquid has escaped, dropping the surface level a distance t.

As far as energy is concerned the change is the same as though the top layer of water of thickness t had been lowered a distance h and its potential energy mgh converted into kinetic energy $\frac{1}{2}mv^2$ in the emergent stream. By conservation of energy,

$$mgh = \tfrac{1}{2}mv^2 \tag{14a}$$

Since m is the same on both sides it may be dropped from the equation

$$gh = \tfrac{1}{2}v^2$$

which upon solving for v gives

$$\boxed{v = \sqrt{2gh}} \tag{14b}$$

the same as the law of falling bodies. In other words, *the velocity of efflux at any depth* h *is equivalent to the velocity acquired by free fall from the same distance* h. This relation, first discovered by Torricelli, is known as *Torricelli's theorem.*

Note that the parabolic path followed by an emerging stream of liquid is such that the greatest horizontal range R (see diagram) is obtained from an orifice midway between any base level and the top surface of the liquid and that holes equidistant above and below this point have lesser but equal ranges. The proof of this will be left as a problem for the student.

EXAMPLE 1. A water tank is located on a platform 10 m above the ground, and the water in the tank is 5 m deep. A hole develops in the side of the tank 1 m below the bottom. Find (a) the velocity of escape, (b) the time it takes for the stream to hit the ground, and (c) the distance from the platform base to the point the stream hits the ground.

SOLUTION. (a) The velocity of escape is given by Eq. (14a). By direct substitution,

$$v = \sqrt{2 \times 9.80\ \text{m/s}^2 \times 4\ \text{m}}$$
$$v = 8.86\ \text{m/s}$$

(b) Since the hole is 11 m above the ground the time taken to fall that distance can be determined by using Eq. (4b), $x = \frac{1}{2}gt^2$. Solving this equation for t, and making a direct substitution of $x = 11$ m, we obtain

$$t = \sqrt{\frac{2x}{g}} = \sqrt{\frac{22\ \text{m}}{9.80\ \text{m/s}^2}} = 1.50\ \text{s}$$

(c) With a horizontal velocity of 8.86 m/s, and in a time of 1.5 s, the stream will travel a horizontal distance of

$$x = vt = 8.86\ \text{m/s} \times 1.50\ \text{s}$$
$$x = 13.29\ \text{m}$$

14.2 MEASURING THE VELOCITY OF A STREAM

One method of measuring the velocity of water in a trough or river bed is to use an L-shaped tube called a "pitot tube." See Fig. 14B. If the water were at rest it would rise in the vertical arm to the height of the surface outside.

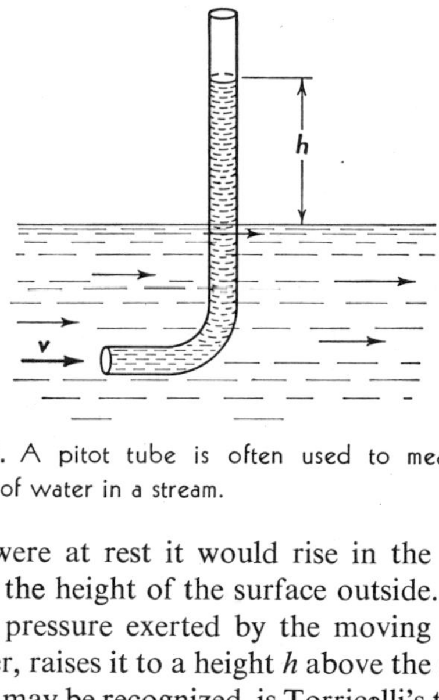

Fig. 14B. A pitot tube is often used to measure the velocity of water in a stream.

The pressure exerted by the moving stream, however, raises it to a height h above the surface. This, it may be recognized, is Torricelli's theorem in reverse, the height h measuring the velocity of the moving stream. Using Eq. (14a) and solving for h, we obtain

$$\boxed{h = \frac{v^2}{2g}} \tag{14c}$$

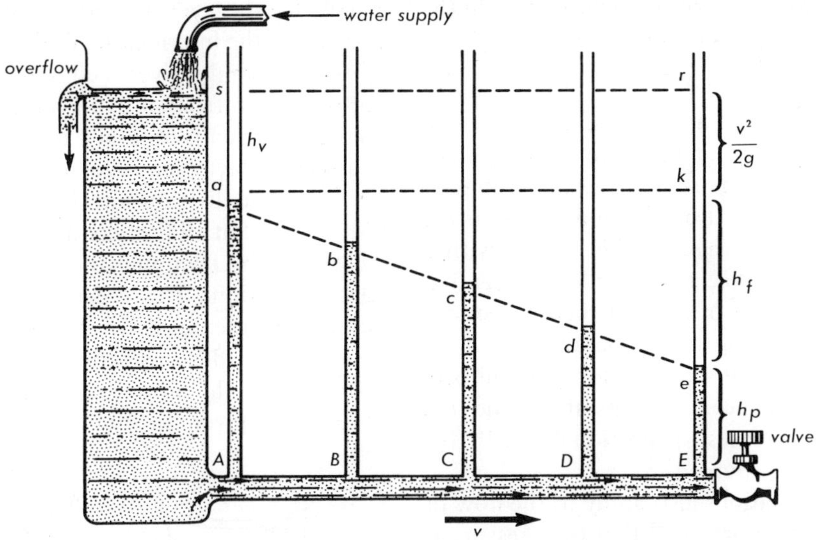

Fig. 14C. Flow of water through a pipe showing the velocity head $h_{v'}$, friction head $h_{f'}$, and the pressure head h_p.

The right-hand term $v^2/2g$ is called the *velocity head* of the moving stream.

The simplest method of measuring the velocity of a jet of water is to turn it straight upward and from the height of rise calculate the velocity from the law of falling bodies, Eq. (14b).

14.3 FLOW THROUGH A PIPE One of the factors determining the flow of water, oil, or gas through a pipe, or the flow of blood through the arteries and veins of the body, is the resistance to flow offered by the confining walls. Consider the experiment shown in Fig. 14C in which water from a vertical tank at the left is made to flow through a horizontal glass tube at the bottom. The pressure at five equally spaced points along the tube is measured by vertical standpipes, and the velocity of flow is controlled by the valve at the right. By means of a "water supply" and "over-flow" pipe the water level in the tank is maintained at a constant level, thus assuring a constant pressure and hence a steady flow.

With the valve closed the water "seeking its own level" soon brings all standpipes to the same level *sr*. Their heights thereby indicate equal pressures at all points along the pipe *A* to *E*. When the valve is partially opened and a steady flow is attained, the water in all pipes drops to different levels similar to those shown. The more the valve is opened, the more rapid is the flow and the steeper is the straight line *abcde*.

Since at all times the heights of columns *aA*, *bB*, *cC*, *dD*, and *eE* measure the pressures at the points *A*, *B*, *C*, *D*, and *E*, respectively, the straight line *ae* indicates a smooth and uniform drop in pressure all along the pipe from *A* to *E*. Such a drop in pressure, designated h_f in the figure, is due to *fluid friction* in the pipe and is called the *friction head*. By measuring the friction head for different rates of flow, a comparison of the results will show that h_f is proportional to the square of the velocity v. This may be expressed as an equation

$$h_f = Kv^2 \tag{14d}$$

where K is the proportionality constant.

The pronounced drop in level from *s* in the tank to *a* in the first standpipe measures the drop in pressure at *A* where the water is practically at rest in the tank and is then speeded up to a velocity v upon entering the pipe. Just as in Torricelli's theorem, the drop in potential energy from s to a is converted into kinetic energy in the stream, and by Eqs. (14b) or (14c),

$$h_v = v^2/2g \tag{14e}$$

Here h_v equals sa and is the *velocity-head.*

The final pressure at the point E where the water is being drawn from the pipe is directly measured by the height of the liquid column Ee, and is called the *pressure head* h_p. Thus, with a total pressure available at the tank, there is first a velocity-head drop of h_v, and then a friction-head drop of h_f, giving as the final pressure $h_p = h - h_v - h_f$. For all points along a uniform pipe the velocity head is constant while the friction head increases proportionally with the distance from the source.

The above equations also apply to the flow of gas through a pipe provided that the pressure is relatively small. If the pressure is large, the formulas must be modified to allow for the compressibility of the gas.

14.4 VISCOSITY OF LIQUIDS If a thick syrup or heavy oil is subjected to pressure and made to flow through a pipe, the rate of flow will not be as great as when gasoline or water is sent through the same pipe under the same total pressure. This difference in rate of flow is due to an internal fluid resistance called viscosity. In some ways viscosity resembles the friction between solids and in other respects it is quite different.

Consider the slow steady flow of water over the sandy bed of a river or stream. See Fig. 14D. Flow measurements show that the speed is a maximum at the top surface and decreases with depth, becoming approximately zero at the bottom.

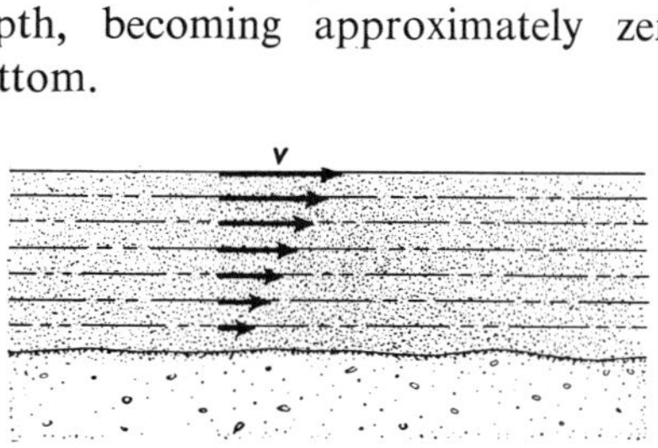

Fig. 14D. The water in a river flows fastest at the top surface.

If we imagine the water divided into thin layers as indicated, it will be seen that the motion is such that the layers are sliding over one another. Because of the interlocking forces between molecules, the sandy river bed tends to keep the bottom layer from moving at all, the bottom layer in turn tends to hold back the second layer, this second layer tends to hold back the third, etc. Since the division into layers is an arbitrary one, we see that all the way up through the liquid there must be frictional forces tending to resist relative motion. The greater the resistance to motion, the greater is the viscosity.

A simple demonstration of the effect of viscosity on the flow of a liquid around an obstacle is given in Fig. 14E. Two identical weights are dropped simultaneously, one into water and one into glycerine. In water the weight settles quickly, but in glycerine the descent is very slow.

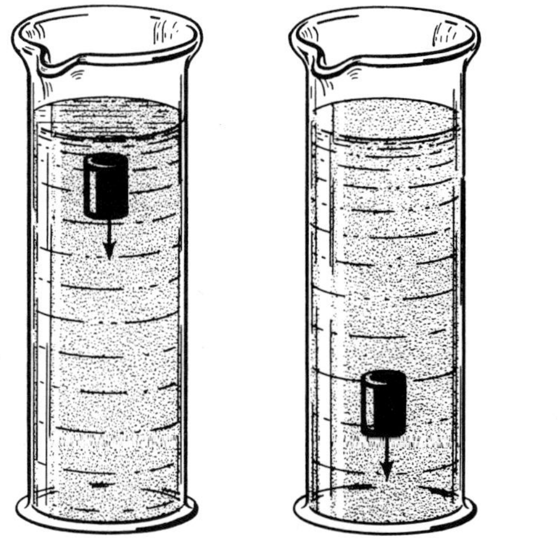

Fig. 14E. The rate at which a weight settles in a fluid is a measure of viscosity.

A demonstration of the viscosity of gases is illustrated in Fig. 14F. A cardboard disk suspended at its center by a thread is placed close to but not touching a wooden disk. When the latter is set into rapid rotation, the cardboard disk begins to rotate in the same direction and gradually picks up speed.

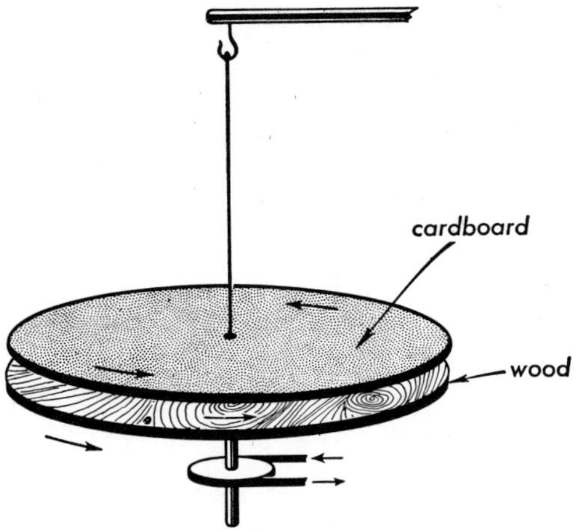

Fig. 14F. Because of the viscosity of the air the cardboard disk is carried around by the rotating disk below.

14.5 MEASUREMENT OF VISCOSITY The most common method used for measuring the coefficient of viscosity of liquids is to measure the rate of flow through a tube of relatively small bore. See Fig. 14G. Friction between the neighboring sections of the liquid and the tube walls tends to hold the liquid back.

If we imagine the liquid divided into concentric cylinders, then each cylinder will be sliding lengthwise through the other, the velocity of flow being greatest at the center and decreasing with the distance from the center to become zero at the tube walls. The motion is similar to the sliding of the tube sections of a small pocket telescope.

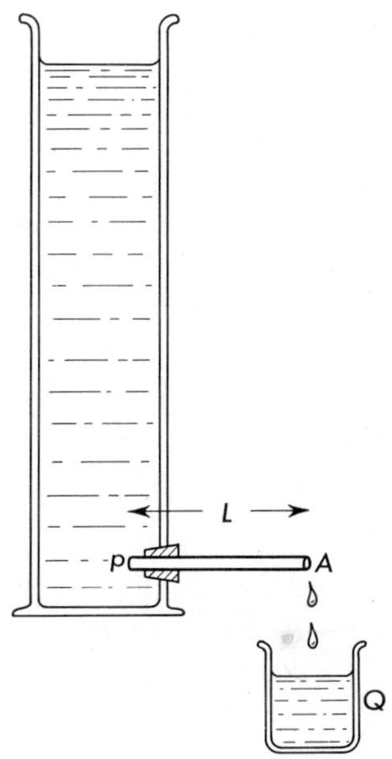

Fig. 14G. Viscosity of a liquid is measured by the rate of flow through a small tube.

14.6 BERNOULLI'S PRINCIPLE* When a river runs through broad open country, the water runs slowly; but when it comes to a narrow rocky gorge, its velocity increases manyfold. Similarly, when a gas or liquid flowing through a pipe comes to a narrow constricted section, the velocity increases as it enters the constriction and decreases again as it leaves the other end. This is illustrated by an experiment diagrammed in Fig. 14H.

The arrangement is the same as in Fig. 14C except that the horizontal flow pipe contains a short section *DE* having only half the cross-sectional area as elsewhere. When the valve is opened and a condition of steady flow exists, the water in the standpipes will have dropped from *sr* to new levels like *a*, *b*, *d*, etc., as shown. It is to be noted that, as the water enters the constriction at *C*, the velocity of flow increases and the pressure drops from *c* to *l*. Farther along where it leaves the narrow tube at *F*, the velocity decreases and the pressure rises from *f* to *k*.

This illustrates Bernoulli's principle which may be stated as follows: *where the velocity of a fluid is high the pressure is low, and where the*

* DANIEL BERNOULLI (1700–1782), a member of a famous Swiss family of scientists and mathematicians who made their home in Basle. Bernoulli spent eight years as professor of mathematics at St. Petersburg and then returned to Basle as professor of anatomy and botany. His most important contribution to science is found in his book "Hydrodynamics" (1738), in which the equilibrium between fluid velocity and pressure are considered both theoretically and practically. One of his mathematical treatments is concerned with the physical cause for the inclination for the planetary orbits. Historically, he is most famous for his "Bernoulli's Principle."

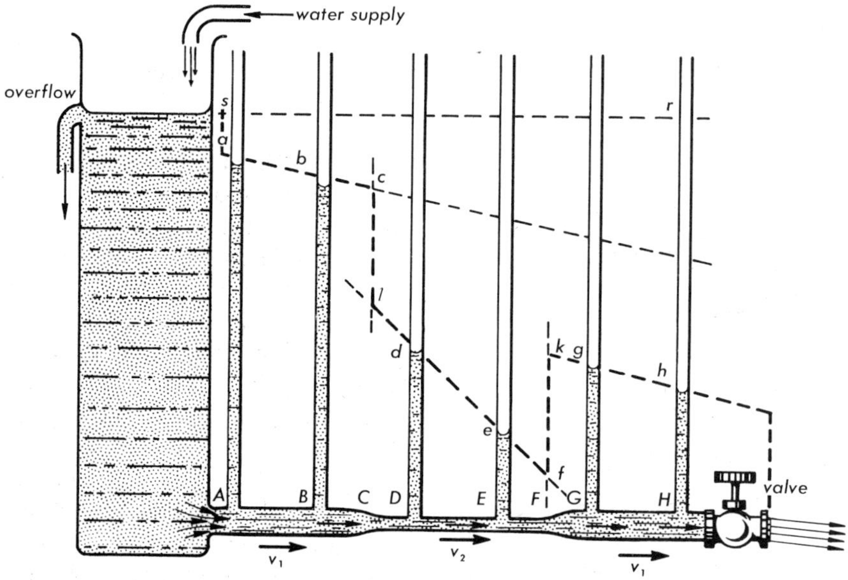

Fig. 14H. Where the velocity of a fluid increases, the pressure drops, and where the velocity of a fluid decreases, the pressure rises.

velocity of a fluid is low the pressure is high. The drops in pressure from a to b, d to e, and g to h represent friction heads for the pipe sections AB, DE, and GH, respectively. The drop from s to a gives directly the velocity head $v_1^2/2g$ for the pipe sections AB and GH, and to this must be added the drop from c to l to obtain the velocity head $v_2^2/2g$ for the narrow pipe section DE. The simplest of laboratory measurements confirm the result that narrowing the tube to half the cross section doubles the velocity and quadruples the velocity head.

14.7 EXPERIMENTS ILLUSTRATING BERNOULLI'S PRINCIPLE Bernoulli's principle is often referred to as a physical paradox and is the basis of many interesting phenomena. A number of experiments involving this principle will be described. In the first illustration (a) in Fig. 14I a blast of air from a nozzle is blown between two sheets of cardboard suspended about 3 in. apart by cords. Instead of being blown apart, as one might expect, they come together. The reason for this action is that, between the two sheets where the velocity of the air is high, the pressure p_2 is low. On the two outside surfaces where the air is not moving, the pressure p_1 (atmospheric pressure) is high and pushes the two sheets together.

In the second diagram (b) air is blown through a hole in the center of a disk AB as shown. When a piece of paper CD is placed close to the opening, it is not blown away but is drawn toward the disk. Where the velocity of air between the disk and paper is high, the pressure is low; and the higher pressure p_1 on the underside of the paper pushes it up against the disk.

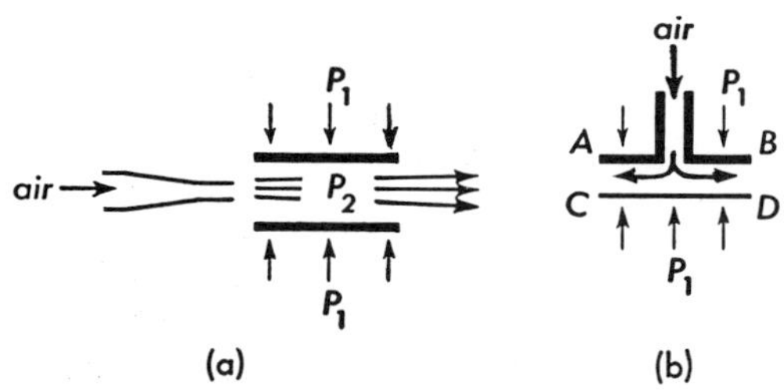

Fig. 14I. Demonstrations of Bernoulli's principle.

Sometimes, during a windstorm, tornado, or hurricane, the roofs of houses are blown off without otherwise damaging the house. This is not as freakish an accident as one might think, for there is a simple explanation. A high wind blowing over the roof creates a low pressure on top, and the atmospheric pressure inside where the wind is not blowing lifts off the roof. An excellent experiment illustrating the phenomenon is shown in Fig. 14J. A jet of compressed air

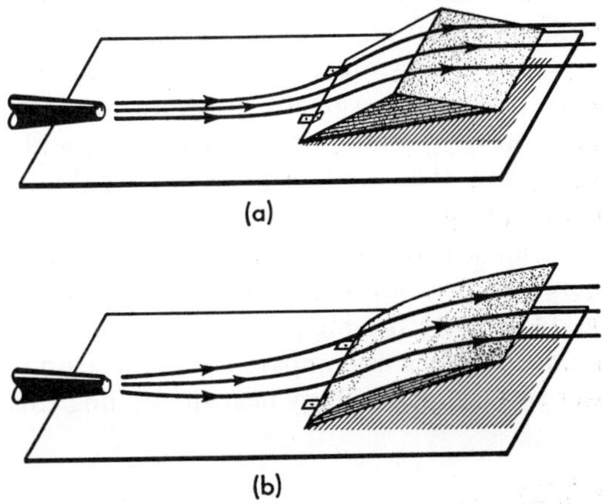

Fig. 14J. Experiments illustrating the lifting forces on (a) roof and (b) an airfoil when a fast stream of air is blown over the top surface.

is blown over the surface of a board on which is located a balsa wood model of a house roof or an airplane wing, hinged at one edge.

Diagram (a) in Fig. 14K represents a common

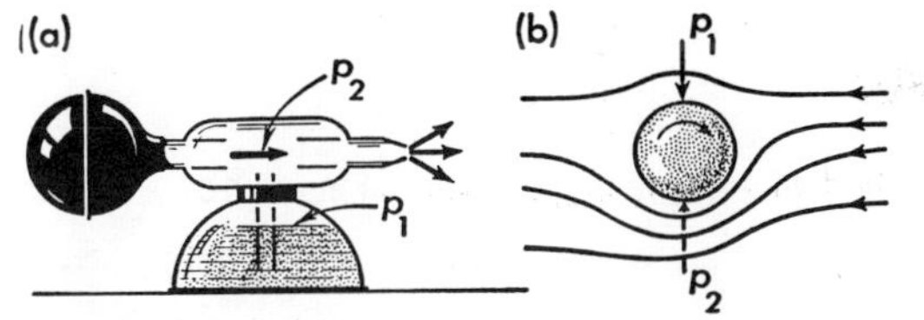

Fig. 14K. Demonstrations involving Bernoulli's principle.

form of perfume atomizer. Squeezing the bulb sends a stream of air through the central tube, creating a low pressure p_2 inside. Atmospheric pressure p_1 on the liquid surface pushes liquid up the stem to be blown out the right-hand tube with the air stream.

Most of the baseballs thrown by a pitcher are curves, some up or down and others in or out, i.e., to right or left. This is an art, accomplished by throwing the ball so that it spins rapidly about some particular axis. To produce a downward curve, i.e., a *drop ball,* the ball is given a top spin as shown in diagram (b). Here instead of having the ball moving to the right we can imagine the ball standing still, but spinning, and the air to be moving from right to left.

At the top surface where the wind and ball are moving in opposite directions, the air is slowed down by friction, giving rise to a high-pressure region. On the underside the surface moving with the wind keeps the velocity high, thus creating a low-pressure region. The resultant downward force thus causes the ball to drop faster than usual.

If a small Ping-Pong ball is placed in a vertical stream of air or water, it will rise to a given height above the nozzle and stay at that level, spinning and bobbing around without falling. If the ball goes to one side as illustrated in Fig. 14L,

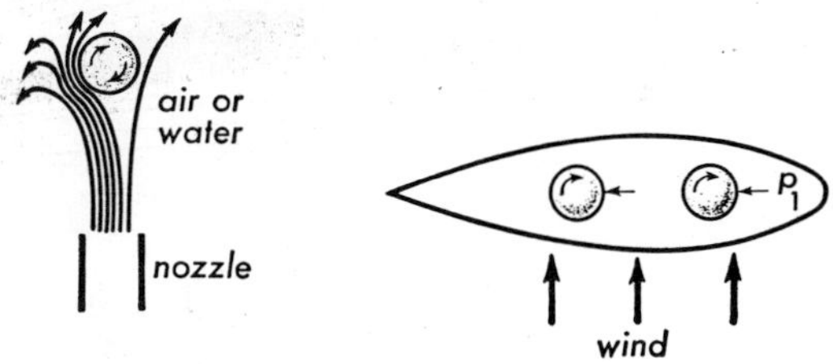

Fig. 14L. Demonstrations of Bernoulli's principle.

the fluid going by on the left side causes the ball to spin as shown. The velocity being high on the left means a low pressure. The higher pressure on the right where the velocity is low pushes the ball back into the stream.

The same principle has been applied to the historical *Flettner rotor ship* which, instead of using sails, employed two tall, motor-driven rotating cylinders. As shown by the top-view diagram, Fig. 14L, a wind broadside to the ship produces a forward force. Such a ship, carrying a cargo, crossed the Atlantic twice not many years ago. Although the trips were successful, the uncertainty of a strong wind makes shipping with such boats unreliable.

14.8 THE LIFT OF AN AIRPLANE WING The major part of the lift of an airplane wing is due to the top surface. This discovery has been made in laboratory wind tunnels by setting up sections of air foils in fast moving currents of air and measuring the pressure at various regions of the surface with pressure gauges. Fig. 14M shows how this can be done with mercury manometers connected by long tubes to small openings on the top and bottom surfaces.

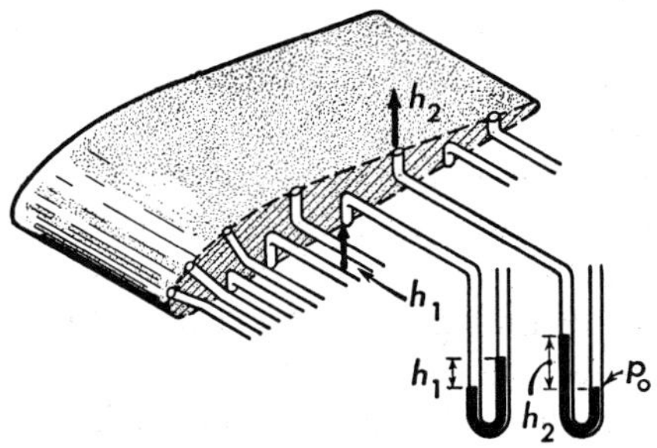

Fig. 14M. Experimental arrangement of mercury manometers showing how the air pressure can be measured at various points over the surface of an airfoil.

When the air is still, all manometer tubes show equal heights in their two arms and normal atmospheric pressure p_0 exists at all points inside the hollow wing as well as outside. When the air stream is set in motion, manometers connected to the top surface show a drop from atmospheric pressure while those connected below show a rise. Since atmospheric pressure on the outside surfaces was previously counter-balanced at all points by the atmospheric pressure inside the hollow wing, the manometer readings h_1 and h_2 give directly the resultant pressures on the two wing coverings.

A graphical representation of the manometer differential pressure readings is shown below in Fig. 14N. Note the high resultant pressures on the upper surface as compared with those on the lower surface, particularly near the leading edge.

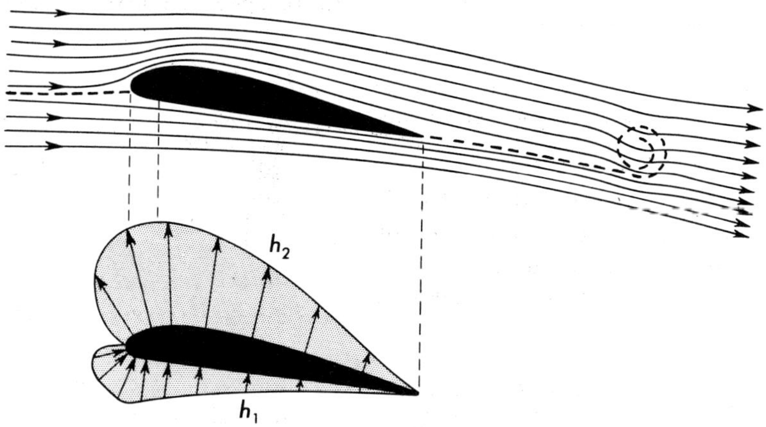

Fig. 14N. Diagram showing the air stream and lift around an airplane wing. Angle of attack 10°.

QUESTIONS

1. Define or briefly explain each of the following: (a) velocity head, (b) friction head, and (c) pressure head.

2. State Torricelli's theorem and make a diagram to illustrate the principles involved.

3. State Bernoulli's principle and give a practical example of its application.

4. Make a diagram and explain the direction a baseball must spin if the pitcher makes the ball curve toward a right-handed batter.

5. Make a diagram and explain why the shape of the top surface of an airplane wing is important to its lifting capability.

6. Why is the roof of a house so apt to be blown off when a very high velocity windstorm is active?

7. Make a diagram and explain the principles involved in a perfume atomizer.

PROBLEMS

1. Water flowing in a 3-in. pipe at the rate of 4 ft/s comes to a small 1-in. pipe, and then a 2-in. pipe. Find (a) the velocity in the last two pipes, and (b) the velocity head in all three pipes. [Ans. (a) 36 ft/s, and 9 ft/s, (b) 3 in., 20.25 ft, 1.266 ft.]

2. A water tank is located on a platform 6 m above the ground. If the water is 4 m deep at the time that a hole develops in the side of the tank 1.5 m from the bottom, how far away from the platform and tank will the escaping stream of water strike the ground?

3. A pressure gauge on a pipe leading from the bottom of a water tank indicates a pressure of 25 lb/in.^{2} When a valve in the pipe is opened to let water flow out, the gauge pressure drops to 20 lb/in.^{2} Find the speed of the water in the pipe. [Ans. 27.17 ft/s.]

4. If the internal diameter of a water pipe is 2 in., and 20 ft^3 of water flow through in 2 min, what is (a) the velocity, and (b) the velocity head?

5. A gauge on a water pipe shows a pressure of 50 lb/in.^{2} When a valve is opened, the gauge reading drops to 40 lb/in.^{2} (a) What is the water velocity? (b) How long will it take for 2 ft^3 of water to flow out if the internal pipe diameter at the gauge is 1 in.? [Ans. (a) 38.43 ft/s. (b) 9.55 s.]

6. A cylindrical tank 8 m high and full of water develops a hole in its side 2 m below the top. How far from the base of the tank will the emerging water strike the ground?

7. At what speed will alcohol emerge from a hole in the side of a tank if the hole is 7 m below the liquid surface? [Ans. 11.71 m/s.]

8. Assuming the blood in a large artery of the heart to have a velocity of 46 cm/s, what pressure drop is attributed to the velocity head? Express your answer in millimeters of mercury. Assume that the density of the blood is 1.040 g/cm^3.

9. Assuming the aorta of the heart to be a rigid pipe 1.2 cm in diameter, how much blood would be pumped through the heart with each pulse? Assume a pulse rate of 60 per min and an average velocity of flow through the aorta of 25 cm/s. [Ans. 28.27 cm^3.]

10. At what speed will the velocity head of a stream of water be equal to 26 cm of mercury? [Ans. 9.24 m/s.]

11. A pipe enters a water tank at a point 12 ft below the surface of the water. What is the water pressure in this pipe? [Ans. 5.20 lb/in.^{2}]

12. Water flowing with a velocity of 3 m/s in a 4-cm-diameter pipe enters a short section having a diameter of only 2 cm and then a section having a diameter of 8 cm. Calculate (a) the velocity in the last two pipes, and (b) the velocity head in all three.

13. A short horizontal pipe is connected to the outside of a water tank 8 ft below the surface of the water. If a hole is drilled in the top surface of the pipe, (a) how fast will the water escape, and (b) how high will the water spout rise? [Ans. (a) 22.63 ft/sec and (b) 8.0 ft.]

15 | TEMPERATURE AND EXPANSION

The physical environment in which we live on this earth is restricted to extremely narrow boundary conditions. For example, the temperature limits within which the human body can function normally without causing permanent damage, or death, are relatively narrow in comparison with conditions prevailing in most of our universe. The temperature of free space is close to absolute zero, over 500 degrees Fahrenheit below freezing, while the centers of stars like our own sun have temperatures of many millions of degrees.

Temperature is only relative and, like time, is difficult to define in terms of the simplest concepts. The word temperature means intensity of heat and may be defined as a number on a scale. To be more technical, temperature is a manifestation of the average translational kinetic energy of the molecules of a substance due to heat agitation, and is measurable by any one of many physical effects due to changes or differences in this energy.

15.1 THERMOMETERS The first authentic record of a thermometer dates back to the time of Galileo. Galileo's thermometer, as illustrated in Fig. 15A, consists of a narrow glass tube with an opening at one end and a bulb at the other. The open end of the tube is filled with colored water and inverted in a dish of water. When the temperature of the surrounding air rises, the air within the bulb expands, forcing the water down the tube. If the bulb is cooled, the air inside contracts, drawing the water up. (To be exact, atmospheric pressure outside pushes the water up.) A scale attached to the narrow tube can be calibrated to any temperature scale, low temperatures at the top and high temperatures at the bottom.

Of the many forms of temperature measuring devices, the mercury thermometer is the most common. A mercury thermometer, as shown in Fig. 15B, consists of a narrow glass tube (called a capillary), the bottom end being sealed to a small bulb. The bulb and part of the capillary are filled with mercury, and the remaining section is evacuated and sealed. When the temperature rises, the mercury and the glass bulb both expand. The mercury, however, expands more than the glass, forcing a small part of the mercury up the narrow capillary. A scale is engraved on the glass to read temperature.

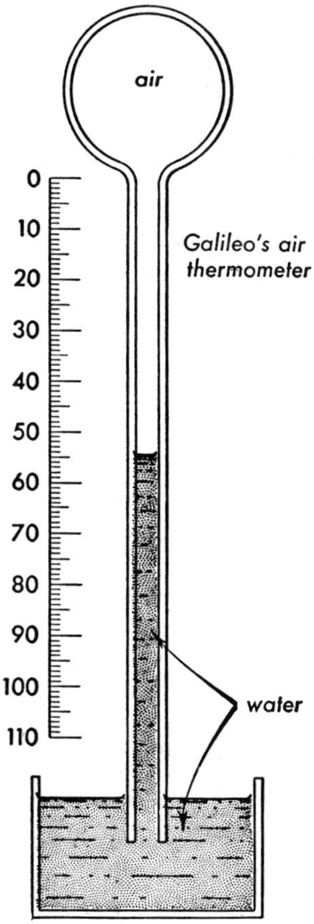

Fig. 15A. Air on heating expands and pushes the water down in the tube. On cooling, the air contracts and the water rises.

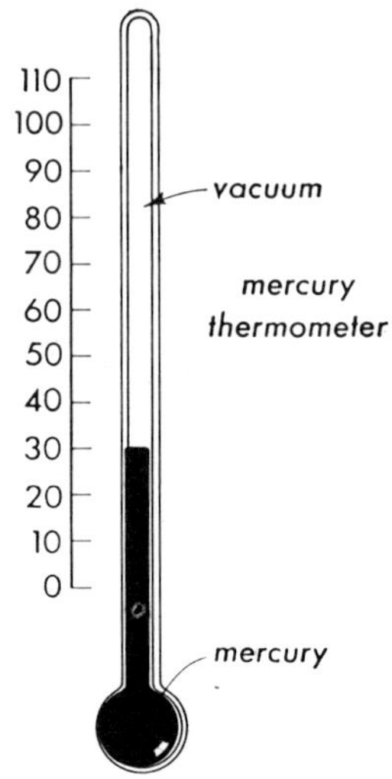

Fig. 15B. On heating a thermometer the mercury expands more than the glass tube and the mercury level rises.

15.2 TEMPERATURE SCALES There are in general use today four different temperature scales. These are the Fahrenheit, Rankine, centigrade, and Kelvin or absolute. Each scale is shown by a diagram in Fig. 15C. The thermometers are all identically made but each has a different scale. In the United States the Fahrenheit scale is commonly used in civil life, and the Rankine scale is used by engineers. The centigrade and Kelvin scales are used in all countries for scientific measurements.

All manufactured thermometers are calibrated to one of these four scales. To calibrate a thermometer, the bulb is first placed in a mixture of ice and water and the height of the mercury column marked on the side of the stem. It is next placed in steam just above boiling water and again marked. These two marks then determine the end points for whatever scale is to be used.

Between the temperatures of melting ice and boiling water there are 180° on the Fahrenheit and Rankine scales, as compared with 100° on the centigrade and Kelvin scales. The ratio of these numbers is 9:5. This comparison shows that a temperature rise of 9°F, or 9°R, is equivalent to a rise of only 5°C, or 5°K.

The lowest temperature ever reached is approximately −273.16°C, or −459.69°F. For theoretical reasons, which will be given later, this is the lowest temperature that can ever be attained. The Kelvin and Rankine scales start with the lowest possible temperature as *absolute zero*. On the basis of the centigrade and Fahrenheit scale divisions, this locates the freezing point of water, to the nearest whole number, at 273°K or 492°R, and the boiling point at 373°K or 672°R.

It is frequently necessary to change temperature readings from one temperature scale to another. Rather than develop formulas for such changes, it is more convenient to work out the simple mathematical steps by the inspection of a diagram like Fig. 15C.

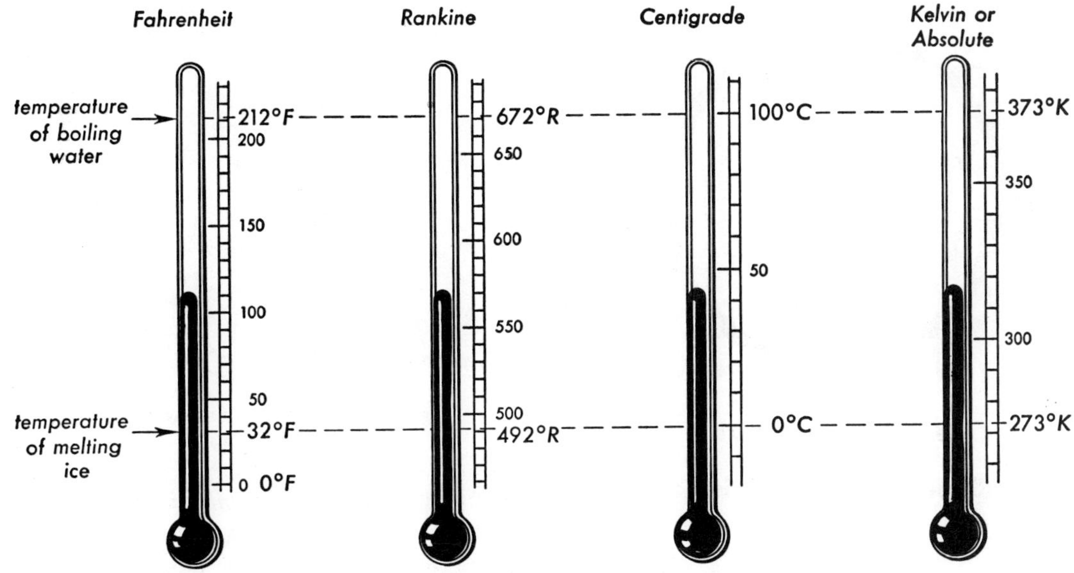

Fig. 15C. Mercury thermometers illustrating the four common temperature scales.

15.3 ELECTRICAL THERMOMETERS If very low or very high temperatures are to be measured, other than mercury thermometers must be employed. At temperatures below −39°C, mercury freezes and becomes a solid; and at high temperatures, glass melts and becomes a liquid. For both of these temperature extremes electrical thermometers are commonly used. These instruments operate upon the principle that the resistance a wire offers to a flow of electric current through it changes with temperature. The higher the temperature, the greater is the resistance.

A diagram of an electrical thermometer is shown in Fig. 15D. A fine piece of platinum wire is wound around a small spool made of silica.

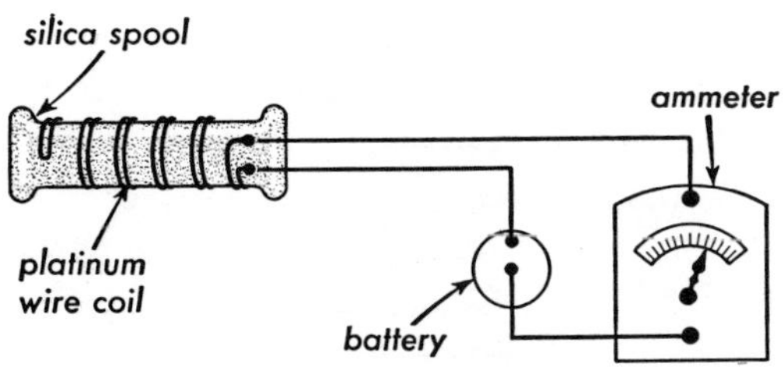

Fig. 15D. Electrical-resistance thermometer showing connections between the platinum wire coil, the battery, and the ammeter.

The ends of this wire are connected to a battery and an ammeter. The purpose of the battery is to supply the electric current and the ammeter is to determine its exact value. When the temperature of a hot body like a furnace is to be measured, the spool of platinum wire is placed inside the furnace and the battery and ammeter outside. A rise in temperature causes the resistance of the platinum wire to increase, and the current, therefore, to decrease. When the platinum wire reaches the temperature of the furnace, its resistance reaches a constant value and the ammeter pointer indicates a steady current. In many cases the ammeter scale is calibrated to give the temperature directly in degrees.

Another form of electrical thermometer, called a *thermocouple*, is illustrated in Fig. 15E. This temperature recording device is based upon a principle, discovered in 1821 by Seebeck, known as the *thermoelectric effect*. Two pieces of wire, one copper and one iron, are joined together at the ends to form a complete loop. When one junction is heated and the other kept cool, a potential difference is produced within the loop. The result is that an electric current flows around the loop in the direction indicated by the arrows. The greater the difference in temperature between

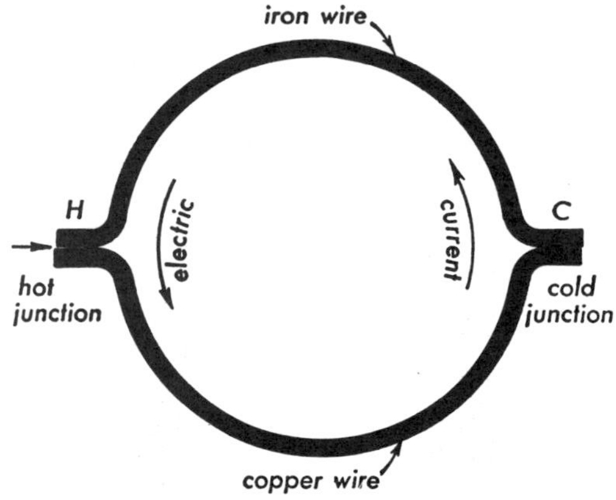

Fig. 15E. A thermocouple.

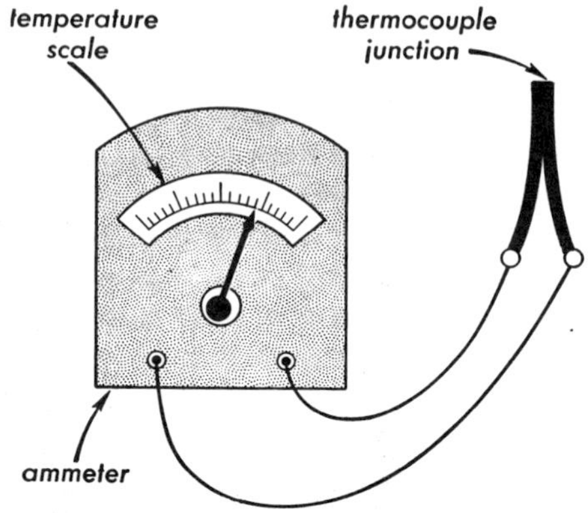

Fig. 15F. A thermocouple thermometer.

the two junctions, the greater is the electric current.

The diagram in Fig. 15F represents a thermocouple connected by wires to an ammeter. If the junction of the thermocouple is first placed in melting ice and then in boiling water, the two scale readings of the ammeter can be marked 0°C and 100°C at the appropriate points. This calibrates the instrument, making of it a direct reading thermometer.

Thermocouples are not always made of copper and iron as shown in Fig. 15E. Any two different metals when brought into contact will exhibit a thermoelectric effect. Some combinations of two metals, however, produce larger currents than others. For very high temperature measurements platinum and platinum-iridium alloys are used, owing to their very high melting point temperatures.

A set of thermocouples, when connected together as shown in Fig. 15G, form what is commonly called a *thermopile*. Small rods of two different metals are joined alternately as shown in diagram (b). One set of junctions is usually protected by placing them at the back of a small box container and the other set is exposed to heat rays through the opposite side which is left open. A funnel or horn-shaped reflector mounted over the open side will collect more heat rays from a distant hot object and thereby increase the electric current. Thermocouples containing several hundred elements can be made so sensitive that they will detect the heat of a candle flame several hundred feet away.

15.4 THERMAL EXPANSION OF SOLIDS In general, when an object is heated, whether it be a solid, liquid, or gas, it expands. There are but a few known exceptions to this. The expansion of a solid with a rise in temperature can be demonstrated by heating a long wire and measuring its

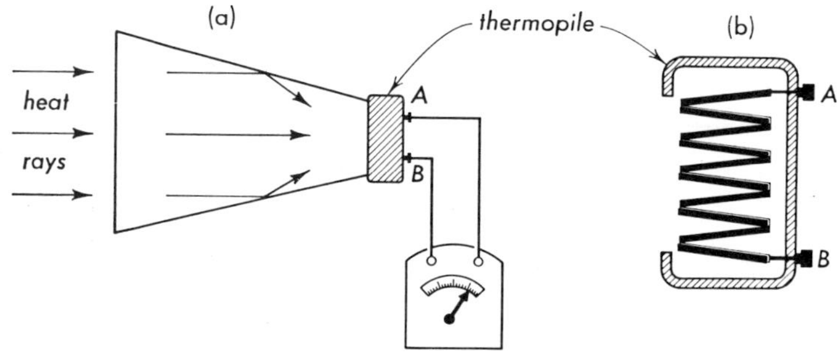

Fig. 15G. Cross section of a thermopile for measuring heat radiation from hot bodies.

overall elongation. One experimental arrangement for demonstrating this is shown in Fig. 15H. An iron wire about 2 m in length is fastened to a hook A at one end and to a weight W at the other. Between these two points the wire passes over three pulleys B, C, and D. The wire is heated by connecting it to a battery and sending an electric current through it from end to end.

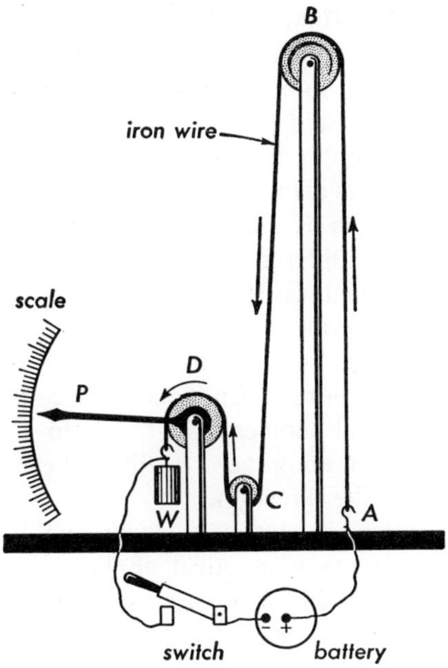

Fig. 15H. Experiment illustrating the expansion (elongation) of a wire due to a rise in temperature.

As the wire is heated and lengthens, the weight slowly falls, thus turning the pulleys as well as the pointer P. When the current is turned off by opening the switch, the wire cools and at the same time contracts to its original length. In the well-equipped laboratory accurate measurements of the rise in temperature and the lengthening of a solid rod are readily made. The graph in Fig. 15I illustrates the straight-line relation between the rise in temperature and the elongation. By the elongation is meant the increase in length and not the total length of the wire. For an iron wire 2 m long, a rise in temperature of 50°C produces an elongation of 1 mm, 100°C produces an elongation of 2 mm, 150°C

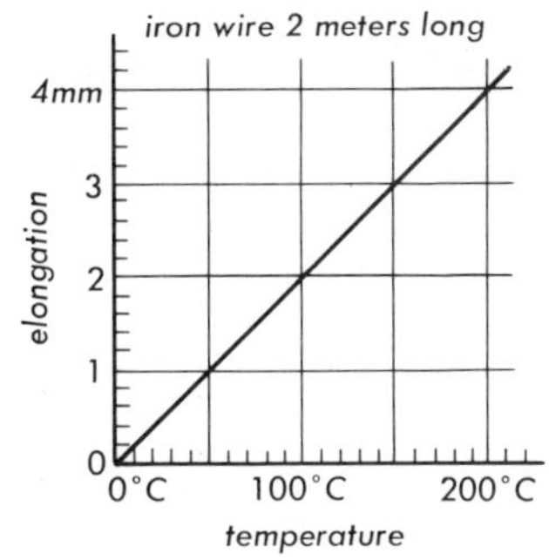

Fig. 15I. Graph of the thermal expansion of a wire.

an elongation of 3 mm, etc. The straight line means, therefore, that the elongation is directly proportional to the rise in temperature.

There are many instances in engineering where the expansion of solids is an important factor in design and construction. Particularly is this true in the construction of suspension bridges and railroads. When the steel rails of the early railroads were first put in place, small gaps were left at every union. The reason for this is that in summer, when the temperature rises, the rails expand and close these gaps. If the gaps are not large enough, the track may buckle up in summer and cause serious accidents. In winter when the rails contract and the gaps are wider, they become noisy to travel as the car wheels roll over them.

Not all substances expand by the same amount when heated through the same difference in temperature. This is illustrated by the linear coefficients of thermal expansion of a few common substances given in Table 15A.

The linear coefficient of thermal expansion α is defined as the change in length of a substance per 1° rise in temperature. Once this constant is known, the linear expansion for any sized object made of that same material can be calculated for any rise in temperature by the following formula:

$$\textbf{elongation} = \alpha \times \textbf{length} \times \textbf{rise in temp.}$$

$$\boxed{e = \alpha L(t_2 - t_1)} \qquad (15a)$$

In this equation t_1 is the original temperature

TABLE 15A. Linear Coefficients of Thermal Expansion (Centigrade)

Material	α per °C
Aluminum	25×10^{-6}
Brass	18×10^{-6}
Copper	17×10^{-6}
Glass (soda)	17×10^{-6}
Glass (pyrex)	3×10^{-6}
Gold	14×10^{-6}
Iron	11×10^{-6}
Platinum	9×10^{-6}
Steel	11×10^{-6}
Quartz	0.4×10^{-6}
Pine wood (along grain)	5×10^{-6}
Pine wood (across grain)	30×10^{-6}

of the body, t_2 the final temperature to which it is raised, and L its original length.*

EXAMPLE 1. A pyrex glass rod is ground and polished to a length of 10 cm when the room temperature is 20°C. If this same rod is heated to a temperature of 420°C, how much does it elongate?

SOLUTION. By using the above equation, we obtain

$$e = 3 \times 10^{-6} \times 10 \times (420 - 20)$$
$$= 3 \times 10^{-6} \times 10 \times 400 = 0.012 \text{ cm}$$

The rod thus lengthens by only 0.012 cm to give an over-all length of 10.012 cm. It should be noted that had the length of the rod been given as 10 in. the elongation would have been 0.012 in. In other words, e and L are always in the same units, so that the coefficients of thermal expansion are valid in the metric as well as the English system. The coefficients given in Table 15A, however, are for the centigrade temperature scale only. Should the known temperature be Fahrenheit or Rankine, the change in temperature should be converted to the centigrade scale.

* It is customary in formulas to use the letter t for temperature on either the centigrade or Fahrenheit scale. When absolute temperature must be used, however, the capital letter T is used.

15.5 DIFFERENTIAL EXPANSION In the previous section it was stated that all substances do not expand alike. Some metals, such as brass and aluminum, expand twice as much as others, like iron and platinum. See Table 15A. This difference in expansion is demonstrated by the heating of a bimetallic strip as shown in Fig. 15J(a). Two thin strips of different metals are placed side by side and welded together over their entire length. When heated, one metal expands more than the other, causing the strip to bend. The hotter it becomes the more it bends. When it cools down to its original temperature the strip becomes straight again, and if cooled still further it bends in the opposite direction.

Differential expansion as shown by this experiment finds many practical applications in industry. Bi-metallic strips are used, for example, in the making of balance wheels for fine watches and in thermostats for refrigerators, hot-water heaters, and car radiators. When, on a hot day, the spokes of the balance of a watch expand, they shift the weight of the rim farther from the center, causing the balance wheel to oscillate more slowly. By making the rim of the wheel of two bimetallic strips this can be compensated for as shown in Fig. 15J(b). With a rise in temperature

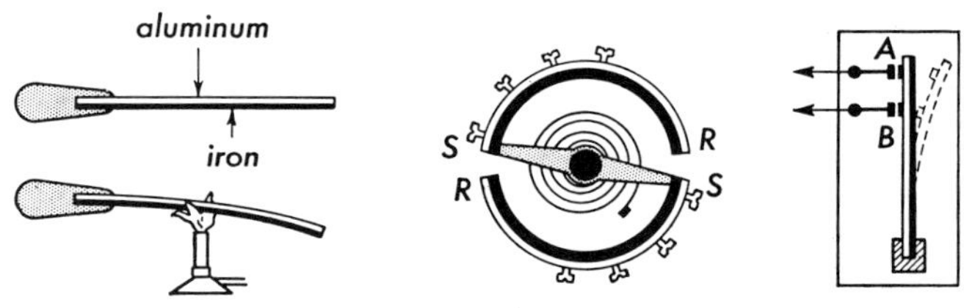

(a) (b) (c)

Fig. 15J. Different substances expand by different amounts: (a) bimetallic strip, (b) balance wheel of a watch, (c) a thermostat.

the ends of the spokes S move out and the free ends R of the bimetallic strips bend in closer to the axis of rotation. One expansion compensates the other, keeping the watch running at the same rate.

Thermostats are so commonly employed in electrical devices today that their explanation will be given here. An electrical thermostat is an automatic electric switch which closes when the temperature reaches one desired temperature and opens when it reaches another. One type of switch is shown in Fig. 15J(c). If the temperature is low, the bimetallic strip is straight and makes electrical contact between the points A and B. This operates an electrical device which, for argument's sake, might open a gas valve of a furnace in the basement of a house. When the air in the house rises to the desired temperature the bimetallic strip has bent far enough away to break the electrical contact at A and B thus turning off the furnace. As the air cools the strip straightens out and makes contact, again turning on the furnace.

15.6 AREA AND VOLUME EXPANSION When the temperature of a wire is raised, it not only lengthens but the diameter and cross-sectional area increase as well. A disk, when heated, increases in radius and area, while a sphere or cube increases in volume. In isotropic substances like copper, linear expansion takes place equally in all directions. In nonisotropic substances like wood, expansion at right angles to the grain is quite different from expansion parallel to the grain. See Table 15A. To find the increased area or volume of such materials, the linear expansion formula, Eq. (15a), must be applied to each direction separately. See Fig. 15K. The same procedure can be applied to isotropic substances, but more rapid calculations can be made by the following formulas.

For area expansion of isotropic media

$$a = 2\alpha A(t_2 - t_1) \tag{15b}$$

where A is the original area of the surface, a is the increase in area due to expansion, α is the

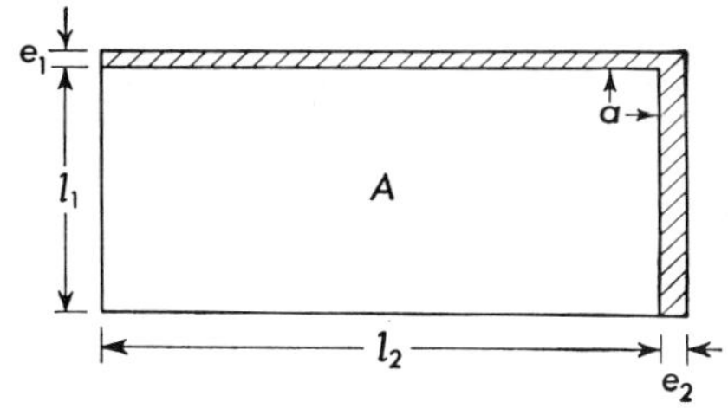

Fig. 15K. Illustration of thermal expansion of a surface area.

linear coefficient of expansion, and $t_2 - t_1$ is the rise in temperature.

For volume expansion of isotropic media,

$$v = 3\alpha V(t_2 - t_1) \tag{15c}$$

where V is the original volume of the solid, v is the increase in volume, and the remaining symbols are the same as before.

The cavity in a hollow ball or container expands as though it were a solid block of the same material. For example, a platinum cup with a volume of 1000 cm³ at 0°C will, when raised to 100°C, have the following increase in volume:

$$v = 3 \times 9 \times 10^{-6} \times 1000 \times (100 - 0)$$
$$= 2.7 \text{ cm}^3$$

Its new volume will be 1002.7 cm³.

The following question will be left as a problem for the student. If the temperature of a copper washer, i.e., a copper disk with a hole in the center, is raised, will the hole in the center become larger or smaller?

15.7 THERMAL EXPANSION OF LIQUIDS Accurate measurements of the expansion of liquids with a rise in temperature are made difficult by the simultaneous expansion of the containing vessel. This difficulty can be overcome, however, and one finds that most liquids, like solids, expand by an amount which is proportional to the rise in temperature. This is illustrated by the straight-line graphs for alcohol and mercury in Fig. 15L.

A straight-line graph here means that, for each degree rise in temperature, the increase in volume due to expansion is exactly the same. If 1 cc of

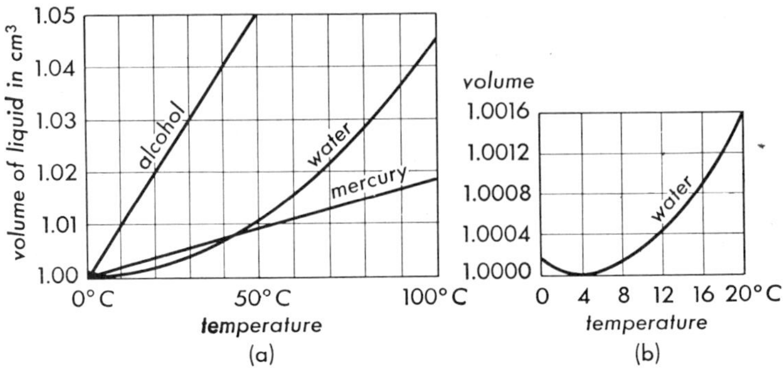

Fig. 15L. Volume-temperature graphs of the thermal expansion of liquids.

mercury at 0°C is heated to a temperature of 1°C, its volume will be 1.00018 cm³, or an increase of 0.00018 cm³. At 10° the increase will be ten times this amount, or 0.0018 cm³. Thus, for each degree rise the increase is the same and equal to 0.00018. This number is called the *volume coefficient of thermal expansion* of mercury. For alcohol, the coefficient is 0.0011 per degree centigrade.

TABLE 15B. Volume Coefficients of Expansion for Common Liquids

Liquid	per °C	per °F
Alcohol	11.0×10^{-4}	6.1×10^{-4}
Benzene	11.8×10^{-4}	6.6×10^{-4}
Carbon bisulfide	11.4×10^{-4}	6.3×10^{-4}
Glycerine	5.3×10^{-4}	2.9×10^{-4}
Mercury	1.8×10^{-4}	1.0×10^{-4}
Turpentine	10.5×10^{-4}	5.8×10^{-4}

The volume expansion of liquids like mercury or alcohol is given by a formula of exactly the same form as the one for solids. See Eq. (15a).

$$v = \beta V(t_2 - t_1) \tag{15d}$$

where v is the change in volume, β the volume coefficient of thermal expansion, V the original volume, and $t_2 - t_1$ is the difference in temperature.

15.8 ANOMALOUS EXPANSION OF WATER

Over a large range in temperature, liquids do not expand linearly. Actually their graphs curve slightly upward, indicating a more rapid rise at higher temperatures. The extent of the departure from a straight-line graph differs considerably with liquids. At 100°C, for example, alcohol expands about 20% more for each degree rise than it does at 0°C. Mercury, on the other hand, varies by less than one hundredth of one per cent and may be assumed accurately linear between these two temperatures.

Starting at its freezing temperature of 0°C, and being slowly heated, water contracts until it reaches a temperature of 4°C and then expands. At 4°C, where it reaches its minimum volume, it has its maximum density. It might be considered fortunate that water expands on being cooled from 4°C to 0°C. If this were not the case and it contracted the way most liquids do, ice would form at the bottom of lakes instead of at the top.

When a pond of water cools toward the freezing point, the surface water next to the cold air cools first. Having cooled and contracted, it has a greater density than the water below and it sinks. This process continues until all the water reaches 4°C. Now, as the surface water cools below 4°C, it expands and, becoming less dense, floats. The water on the surface therefore reaches 0°C first and freezes. In freezing, the water expands still more to become ice. Ice floats on water because its density is less than that of water. When 1 cm³ of water at 4°C is cooled down to 0°C and frozen, it expands about 9%. Its new volume as a solid at 0°C is 1.09 cm³.

QUESTIONS

1. Make a diagram of an electrical resistance thermometer. Briefly explain the principles upon which it functions.

2. Make a diagram of the four temperature scales in common use today. Label the points at which water freezes and boils.

3. Make a diagram of a thermopile and briefly explain its basic principles.

4. What is differential expansion? Name three kinds of practical devices that make use of differential expansion. Make a diagram of one application.

PROBLEMS

1. A Fahrenheit thermometer shows a temperature of 118° on a hot day. What would the temperature be on a centigrade thermometer? [Ans. 47.8°C.]

2. If a centigrade thermometer shows a temperature of 32°C, what is the corresponding temperature on a Fahrenheit thermometer?

3. What temperature on the centigrade scale is equivalent to each of the following: (a) 177°F, (b) 131°F, (c) 77°F, (d) 50°F, and (e) −31°F? [Ans. (a) 80.6°C, (b) 55°C, (c) 25°C, (d) 10°C, (e) −35°C.]

4. What temperature on the Fahrenheit scale is equivalent to each of the following: (a) 16°C, (b) 45°C, (c) 82°C, (d) 155°C, and (e) 265°C?

5. The iron rails of a railroad track have a length of 20 m when they are nailed in place on the ties. If their ends just touch each other when the temperature is 55°C what will be the width of the gaps between railends when the temperature drops to −40°C? [Ans. 2.0 cm.]

6. A radio transmitter tower is 220 m high, and made of steel. What is its change in height when the temperature rises from −20°C to 62°C.

7. An automobile is 6 m long. Find its change in length when the temperature drops from 35°C to −20°C. [Ans. 3.63 mm.]

8. A steel tape measure is made to be exactly 100 ft long when the temperature is 25°C. What will be the correction to be made if the temperature rises to 45°C?

9. A storage tank contains 600 gallons of benzene when the temperature is 30°C. What is the decrease in volume when the temperature drops to 12°? [Ans. 12.74 gal.]

10. An aluminum tank when completely filled at 25°C will hold 1000 gallons of glycerine. If the temperature rises to 35°C how many gallons will flow out? Take into account the expansion of the tank.

11. A steel tank when completely filled with turpentine at 15°C holds 800 gallons. If the temperature rises to 32°C how much turpentine will flow out? Take into account the expansion of the tank. [Ans. 13.83 gallons.]

12. A circular hole 25 cm in diameter is cut in a flat plate of aluminum, at a temperature of 25°C. Find the diameter of the hole when the temperature rises to 165°C.

13. The iron metal rim on the wheel of an old-fashioned wagon is 1.5 m in diameter when the temperature is 15°C. Find the increase in diameter when the temperature rises to 35°C. [Ans. 0.330 mm.]

16 | THERMAL CAPACITY AND HEAT TRANSFER

That heat is a form of energy and is due to the kinetic energy of molecular motion was first proposed by Count Rumford* in the late eighteenth century.

According to the kinetic theory of matter, the individual atoms of which all substances are made are in a state of rapid motion. As a body is heated to a higher temperature this atomic motion increases and the body expands. As a body cools the atomic motion decreases and the body shrinks.

All of the basic principles of heat and their related natural phenomena involve differences between temperature and quantity of heat.

The difference between the two can be illustrated by the heating of two pans of water. More gas must be burned, or electrical energy consumed, to heat a large pan of water than a small pan. Although both are started at the same temperature and both are raised to the same boiling point, 100°C, the larger pan will require more thermal energy, called heat.

16.1 THE CALORIE AND KILOCALORIE The difference between temperature and quantity of heat is well illustrated by the following experiment. See Fig. 16A. Five marbles, all of the same size but made of different materials, are heated in boiling water to a temperature of 100°C. At a given instant they are all placed on a sheet of paraffin about 0.5 cm thick and permitted to melt their way through. The iron and brass marbles are observed to drop through first, but the lead and glass marbles never do. This illustrates the fact that the heat content of the iron and brass, even though raised to the same temperature as the others, is considerably greater than the heat content of the glass and lead.

In order to determine the exact heat capacity

---

* BENJAMIN THOMPSON (COUNT RUMFORD) (1753–1814), was born in Woburn, Massachusetts, in 1753. He spent most of his adult life in Germany, where, among other things, he managed an artillery factory. From his observations about the heat developed in boring cannon, he was able to show that heat is not a pervading fluid, but a form of internal energy of the atoms or molecules forming the substance. His own expression was that heat was a mode of motion of these particles. For these and other services, the Emperor chose to appoint him to the German nobility, and Thompson chose the name of his wife's birthplace, Rumford (now Concord, New Hampshire), as his title.

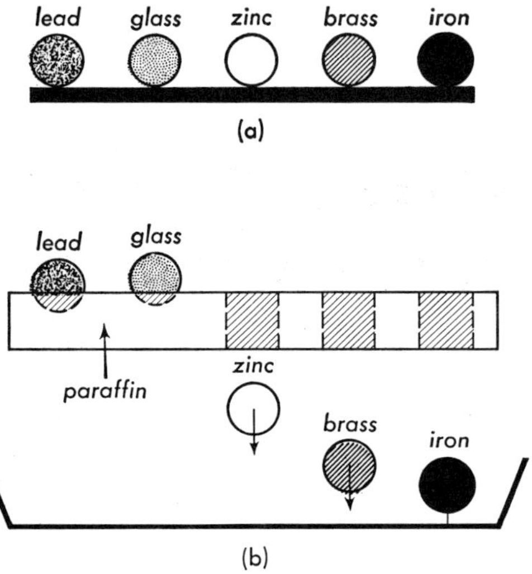

Fig. 16A. Experiment demonstrating the different heat capacities of different substances of the same size.

of a substance, we must first define the *kilocalorie*, *calorie* and the *British thermal unit.*

The amount of heat required to raise 1 Kg of water 1°C is called a kilocalorie (Kcal, or Cal).

The amount of heat required to raise the temperature of 1 g of water 1°C is called the calorie (cal).

The quantity of heat required to raise the temperature of 1 lb of water 1°F is called the British thermal unit (Btu).

The ratio between units is readily computed and found to be

$$1 \text{ Kcal} = 1000 \text{ cal}$$

$$1 \text{ Btu} = 0.252 \text{ Kcal}$$

$$1 \text{ Btu} = 252 \text{ cal} \qquad (16a)$$

Once the kilocalorie or the Btu is defined, the amount of heat required to raise any amount of water from one temperature to another may be calculated by simply multiplying the mass of water by the temperature rise. For example, to raise the temperature of 25 Kg of water from 10°C to 50°C requires $25 \times 40 = 1000$ Kcal, or to raise the temperature of 6 lb of water from 32°F to 60°F requires $6 \times 28 = 168$ Btu.

While 1 Kcal of heat will raise the temperature of 1 Kg of water 1°C, a different number of Kcal will be required to raise the temperature of 1 Kg of some other substance 1°C. For example, to raise the temperature of 1 Kg of iron 1°C requires only $\frac{1}{10}$ of a Kcal, while to raise 1 Kg of lead 1°C requires only $\frac{1}{30}$ of a Kcal. In other words, the thermal capacities of equal masses of different materials have different values.

The thermal capacity of a substance is defined as the number of kilocalories required to raise the temperature of 1 Kg of that substance through 1°C, or the number of Btu required to raise the temperature of 1 lb 1°F.

The ratio between the thermal capacity of a substance and the thermal capacity of water is called the specific heat of that substance.

Numerically, specific heat has the same value as thermal capacity; being a ratio, however, it is like *specific gravity* and has no units.

The specific heats or thermal capacities of a few common substances are given in the following table.

TABLE 16A. Specific Heats, or Thermal Capacities in Kcal/Kg

Material	c
Aluminum	0.220
Brass	0.092
Copper	0.093
Glass	0.160
Gold	0.031
Glycerine	0.60
Ice	0.50
Iron	0.105
Lead	0.031
Mercury	0.033
Silver	0.056
Zinc	0.092
Steam	0.50
Water	1.000

To illustrate the use of this table, consider the calculation of the heat content of the marbles used in the above experiment. The measured mass of each marble is given in the second column of the following tabulation and the corresponding thermal capacity or specific heat in the next row. The product of these two quantities gives the values shown in the fourth row; they represent the amount of heat in Kcal required to raise that marble 1°C. Since all marbles were raised from room temperature 20°C to the boiling point of water, 100°C, the values in the third row have been multiplied by the rise in temperature 80°C to obtain the total heat values in the last row.

TABLE 16B. Results of the Marble Experiment

Marble	m (Kg)	c	H for 1°C (Kcal)	H for 80°C (Kcal)
Lead.......	0.045	0.031	0.00139	0.111
Glass.......	0.010	0.160	0.00160	0.128
Zinc........	0.024	0.092	0.00220	0.176
Brass.......	0.030	0.092	0.00276	0.221
Iron........	0.028	0.105	0.00294	0.235

These numbers clearly indicate that, in the experiment above, iron and brass should melt through the paraffin first: they have available within them the largest amounts of stored thermal energy, 0.235 and 0.221 Kcal, respectively.

The definition of thermal capacity and the calculation of total heat content may be summarized by a generally useful formula of the following form:

$$\boxed{H = mc(t_2 - t_1)} \qquad (16b)$$

H represents the total amount of heat in Kcal, m the mass of the body in Kg, c the thermal capacity in Kcal/Kg°C, and $t_2 - t_1$ the rise in temperature in centigrade degrees.

EXAMPLE 1. Calculate the amount of heat required to raise the temperature of 25 Kg of copper from room temperature of 28°C to 316°C?

SOLUTION. Using the value of the thermal capacity of copper given in Table 16A, and substituting the known quantities in Eq. (16b), we obtain

$$H = 25\ \text{Kg} \times 0.093\,\frac{\text{Kcal}}{\text{Kg}} \times (316 - 28)$$

$$H = 670\ \text{Kcal}$$

16.2 CHANGE OF STATE The continuous addition of heat to a solid or liquid mass will eventually bring about a change of state. The general behavior of many substances can be illustrated by a detailed description of the changes that occur with the most common of all liquids, water. If a block of ice at a temperature of −50°C is placed in a closed container and put on a stove to heat, its temperature will rise slowly until it reaches 0°C.

At 0°C the temperature stops rising and the ice begins to melt. More and more ice is melted as heat is continually added and not until it has all turned to water does the temperature begin to rise. As the water becomes hotter and hotter it eventually reaches a temperature of 100°C where vigorous boiling sets in. Here again the temperature stops rising and, as heat is added, more and more water is boiled away to become steam. Finally when all has become steam at 100°C the temperature begins to rise once more.

All of these changes of temperature and changes of state are shown by a graph for a 1 Kg block of ice in Fig. 16B. The horizontal sections

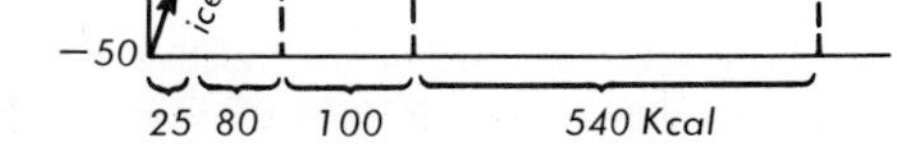

Fig. 16B. Heat-temperature graph for one gram of ice starting at −50°C. Illustrating the latent heat of fusion and vaporization.

represent changes of state without change in temperature, while the slanted sections on either side represent changes in temperature without abrupt changes in state.

16.3 HEAT TRANSFER There are numerous methods by which heat may be transmitted from one place to another. Some of these methods are slow, and others are very fast. A careful study of all known methods has led to the realization that there are but three general types of heat transfer. These are *conduction*, *convection*, and *radiation*. Conduction is a slow process by which heat is transmitted through a substance by molecular activity. Convection is a more rapid process involving the motion of heated matter itself from one place to another. Radiation of heat from one place to another takes place with the speed of light, 186,000 mi/s.

16.4 CONDUCTION Not all bodies are good conductors of heat. Metals like copper and silver are much better for this purpose than are other substances like wood, glass, paper, and water. The ability of a given substance to conduct heat is measured by what is called it *thermal conductivity*.

The relative conductivities of different substances can be illustrated by an experiment performed as follows: Similar rods of six different metals, copper, aluminum, brass, tin, German silver, and lead, are coated with a special yellow paint and arranged as shown in Fig. 16C.

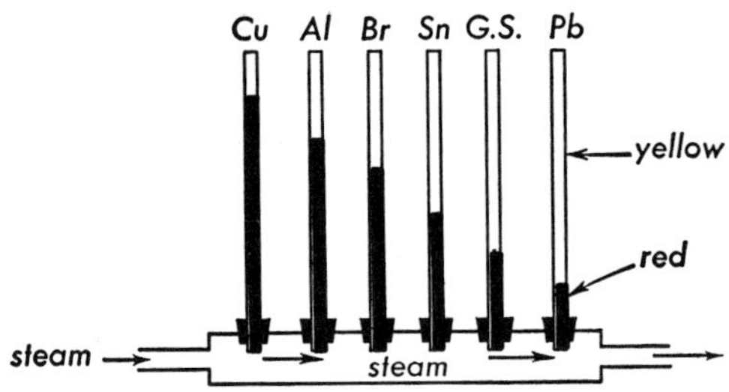

Fig. 16C. Experiment illustrating the relative heat conductivities of six different metals: copper, aluminum, brass, tin, German silver, and lead.

The ends of the rods, mounted in rubber corks, project through holes in a metal tube where their lower ends are heated to 100°C by steam passing through the tube. As the heat travels slowly up each rod, the yellow paint turns to red. After 5 or 10 min, the height to which the paint has turned color is approximately as shown by the stippled areas in the figure. Of these six metals, copper is observed to be the best conductor and lead the poorest.

In order to heat an object it is customary to bring it into contact with some other body at a higher temperature. A pan of water, for example, is generally heated by placing it over an open flame. The combustion of natural gas first sets the gas molecules into a rapid state of motion. These molecules striking the bottom of the pan set the molecules of the metal into rapid vibration. They in turn strike other metal molecules, thus transferring the motion through to the other side. This is called heat conduction. The metal molecules set the first layer of water molecules moving and they in turn set others moving. Thus molecular motion, called heat, has been given to the body of water.

Laboratory experiments show that the amount of heat flowing through a rod is proportional to the time, the cross-sectional area, and the difference in temperature between the ends, and is inversely proportional to the length.

Using appropriate symbols for each of these factors and inserting a proportionality constant, the following equation is set up:

$$\boxed{H = k\frac{A(t_2 - t_1)}{L}T} \qquad (16c)$$

H is the amount of heat flowing through a body of length L and cross-section A, k is the thermal conductivity, T is the time interval of flow, t_2 is the temperature of the hot end, and t_1 is the temperature of the cold end. It is quite clear to almost everyone that, if the temperature difference $t_2 - t_1$ or the area A is increased (see Fig. 16D), the amount of heat passing through is increased. It is not as obvious, however, that an

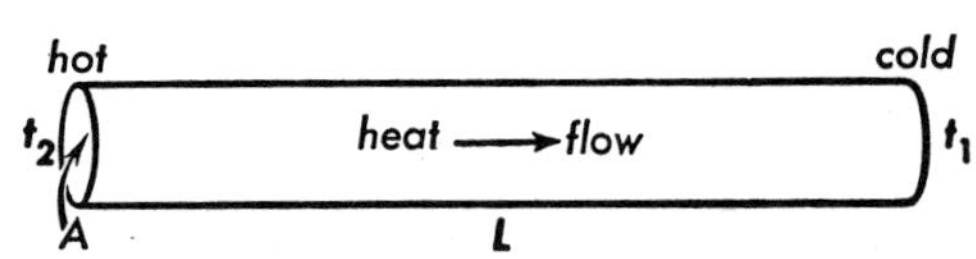

Fig. 16D. Diagram illustrating the various measurable factors involved in the flow of heat through a body by conduction.

increase in the length L causes a decrease in heat flow, or that a decrease in length produces an increase. This latter will be illustrated by two experiments.

Although paper is a poor conductor, the flow of heat through it can be made very great by increasing A, the cross-sectional area, and decreasing L, the distance it has to flow. Diagram (a) in Fig. 16E illustrates thermal conductivity

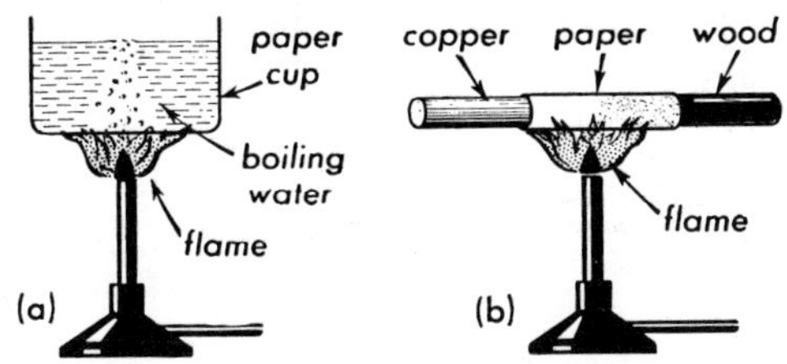

Fig. 16E. Illustration of the conduction of heat through paper.

by the boiling of water in a paper cup. Although the gas flame plays directly against the surface of the paper, the cup will not burn. The reason for this is that the heat from the lower surface of the paper is conducted through to the water fast enough to keep the temperature of the paper from rising too high. If the paper is thick the lower surface will burn. Strange as it may seem, the thinner the paper, the less is the chance of burning.

In diagram (b) of Fig. 16E, a thin piece of paper is wrapped once around a rod made half of wood and half of copper. When the flame is brought up as shown, the paper burns only where it is in contact with the wood and not at all where it is in contact with the copper. Copper, being a good conductor, carries the heat into the interior of the metal and away from the metal surface. Since wood is a poor conductor it cannot conduct the heat away from the surface fast enough and the paper heats up and soon burns.

The thermal conductivities of a few common substances are given in Table 16C.

The number k is the quantity of heat in Kcal that in 1 s will pass through a 1-m cube when two opposite faces are maintained at 1°C difference in temperature. Knowing the value of k for a

TABLE 16C. Thermal Conductivities, k, in Kcal/m²s C°

Substance	k
Silver	970×10^{-4}
Copper	920×10^{-4}
Aluminum	500×10^{-4}
Brass	260×10^{-4}
Iron	160×10^{-4}
Lead	80×10^{-4}
German Silver	100×10^{-4}
Mercury	20×10^{-4}
Tile	2×10^{-4}
Glass	2.5×10^{-4}
Water	1.4×10^{-4}
Wood	0.5×10^{-4}
Paper	0.3×10^{-4}
Felt	0.04×10^{-4}

given substance, it is possible to calculate, by means of Eq. (16c), the amount of heat flowing through any sized object made of that same material.

EXAMPLE 2. One end of an aluminum rod 40 cm long and 5 cm² in cross section is maintained at a temperature of 100°C, and the other end at 20°C. Find the amount of heat that will flow through the rod in 2 min.

SOLUTION. Substitution in Eq. (16c) gives

$$H = 500 \times 10^{-4}\,\frac{5 \times 10^{-4}(100 - 20)}{0.40}\,120$$

$$H = 0.6 \text{ Kcal}$$

16.5 CONVECTION Why is it that a poor conductor of heat like water can be heated so quickly when it is placed in a pan over a hot fire? It is due to the second method of heat transfer known as *convection*. Water on the bottom of a pan is heated first. Because of a rise in temperature it expands. Being lighter than the cold water above, it then rises to the top, permitting cold water to come to the bottom from the sides. This action sets up a flow of water called a *convection current*. See Fig. 16F. Convection currents thus keep the water stirred up as it heats.

Convection currents set up by the heating of a vessel of water are illustrated in Fig. 16G. A

Fig. 16F. Convection currents in a pan of water being heated over a stove burner.

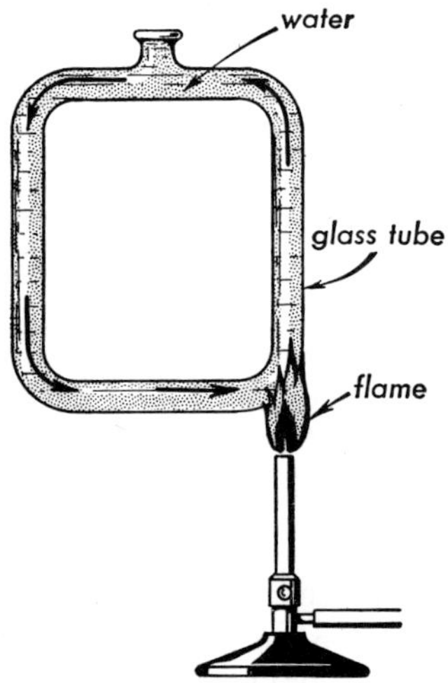

Fig. 16G. Demonstration of a convection current in water.

glass tube in the shape of a letter O is filled with water and then heated at one of the lower corners as indicated. A drop of ink admitted at the top opening will mix with the water and quickly flow around the tube in a counter-clockwise direction. This circulation is the basis of the hot water heating systems used in some houses. As illustrated in Fig. 16H, hot water from a supply tank in a lower room or basement rises and flows through several radiators only to return, somewhat cooled, to the tank again where it is reheated.

Similar to this in its action is the hot air heating system used in some houses. Air heated in a furnace in the basement rises through an outlet in or near the floor as shown in Fig. 16I. Rising up one side of the room this air travels across the ceiling, down the other side,and across the floor to return to the furnace by another opening.

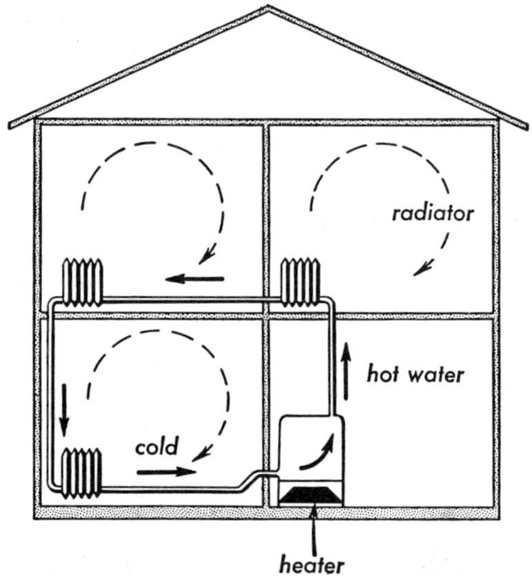

Fig. 16H. Illustrations of heat convection by the circulation of water in a pipe and air in a room.

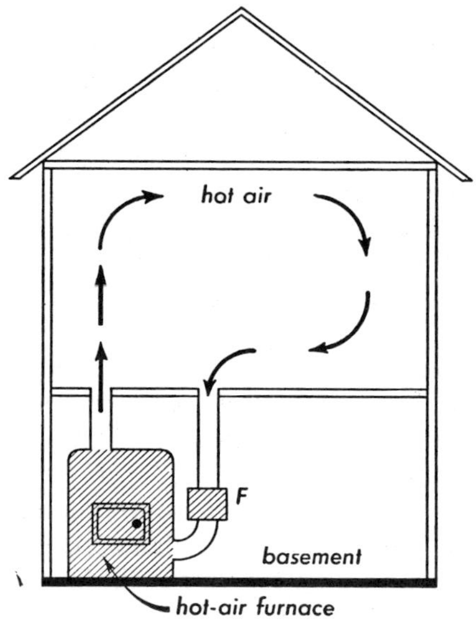

Fig. 16I. Convection hot-air currents set up by heating.

Convection currents in the atmosphere are quite noticeable and account for the wind. Along the sea coast, cool air from over the ocean comes as a sea breeze due to convection. The sun's rays are absorbed more readily by the land than they are by water and the warmed air over the land rises while cooler air from the ocean comes in to take its place. At night the land cools quickly by radiation back toward the cold sky and soon the air over the water is the warmer, and rising causes a reversal in air movement. The wind blows from the land to the sea. These air currents are readily observed by smoke from a fire built on the seashore. During the day the smoke blows inland and at night it blows seaward.

Pilots are well aware of rising air currents over certain local areas of ground. Solar radiation being absorbed more completely by a newly plowed field, for example, will warm the air sufficiently to cause convection. Flying into such an upward draft of air the plane receives a sudden lift. During certain seasons these updrafts of warm air are cooled by the air layers above and water vapor is condensed to form clouds.

16.6 RADIATION When the sun comes over the horizon in the early morning the heat can be felt as soon as the sun becomes visible. This heat, called *radiation*, travels with the speed of light, 186,000 mi/s.

Radiant heat is but one of the many forms of energy and is readily detected by means of a radiometer, thermocouple, thermister, thermometer, etc. A Crooke's radiometer, shown in Fig. 16J, will often be found in a jewelry store

Fig. 16J. A Crooke's radiometer.

window. In daylight or under a bright light the little pinwheel made of very thin mica will be found spinning around as if by perpetual motion.

Each vane of such a radiometer is shiny on one face, and blackened on the other. The black faces absorb more radiant energy than the polished surfaces and heat the adjacent air. The faster recoiling air molecules therefore exert a larger force on the blackened sides driving it around.

Should a radiometer be so highly evacuated that little air remains inside the vessel, light waves in bouncing off the polished surfaces will exert twice the force they do in being absorbed on the blackened side and the pinwheel will spin the other way.

Heat waves are electromagnetic waves having all the general properties known to visible light. The essential difference between the two is that heat rays, sometimes called infrared rays, have a slightly longer wavelength and are not visible to the human eye.

A demonstration of the reflection of infrared rays is diagrammed in Fig. 16K. A candle flame acting as a source at F emits light and heat rays in all directions. Of these rays only the ones traveling toward the concave mirror M_1 are reflected into a parallel beam. Arriving at the second concave mirror M_2 these rays are again reflected, being brought together to a focus on the exposed junctions of a thermopile T. As the junctions of the thermopile warm up an electric current is produced, causing the ammeter pointer P to move to the right. When the candle is removed the pointer returns to zero.

A practical example of heat radiation is to be found in every home where a fireplace is used as a means of heating. Contrary to most beliefs, the heat entering a room from a fireplace is practically all in the form of infrared rays originating in the flames, the coals, and the stone or brick walls. The air that is heated within the fireplace does not enter the room but is carried up the chimney as a convection current. See Fig. 16L. This rising current of air draws fresh air into the room and into the fire, thus supplying fresh oxygen to the burning wood or coal.

A Dewar flask or "thermos bottle" is an example of a practical device in which the conduction, convection, and radiation of heat are reduced as much as possible. As shown by the cross-sectional diagram in Fig. 16M, a thermos

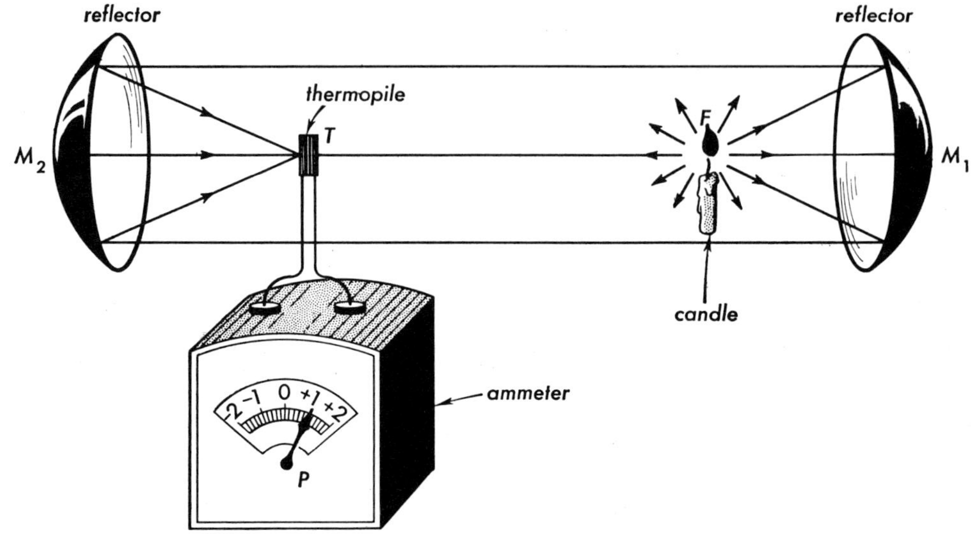

Fig. 16K. Reflection of heat rays by concave mirrors.

bottle consists of a double-walled glass vessel silvered on the inside. The purpose of the silvering is to reflect all radiant heat attempting to enter or leave the vessel. The space between the walls is highly evacuated to prevent convection, and the glass, being a poor conductor, minimizes conduction through the walls of the neck. With the exception of the vacuum space between walls, a calorimeter of the type commonly used in laboratories is similar to a Dewar flask.

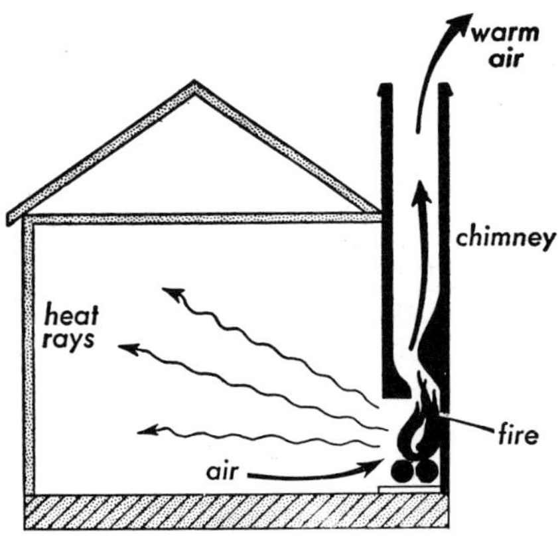

Fig. 16L. A fireplace heats a room by radiation from the flame, the coals, and the walls. Convection currents set up a draft and carry warm air and smoke out the chimney.

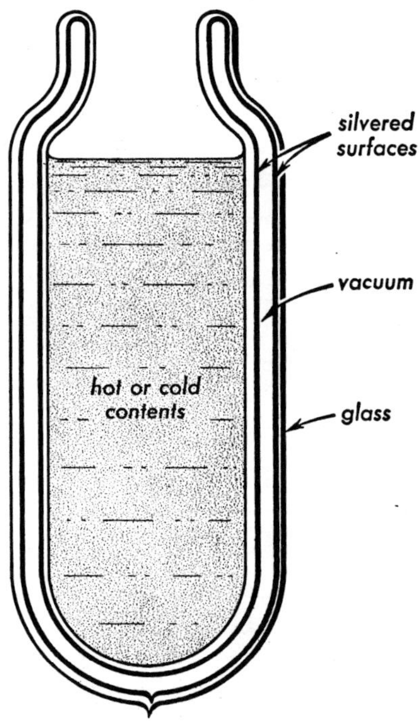

Fig. 16M. The Dewar flask, or "thermos bottle," minimizes conduction by using glass, convection by evacuating, and radiation by silvering.

QUESTIONS

1. Define or briefly explain each of the following: (a) kilocalorie, (b) calorie, (c) British thermal unit, (d) thermal capacity, and (e) specific heat.

2. Briefly describe the three methods of heat transfer.

3. What is the process by which heat is conducted through a solid? Explain the molecular process that takes place.

4. Describe a laboratory experiment that can be performed to demonstrate the transfer of heat by radiation.

5. Upon what five factors does the amount of heat conducted through a rod depend? Write down the formula.

6. Explain the process of heat transfer called convection. Explain the molecular process that takes place.

7. Make a diagram of a laboratory experiment for demonstrating convection in a gas.

8. Describe the heat transfer processes taking place in the heating of a pan of water. Make a diagram to illustrate your explanations.

9. Make a diagram of a thermos bottle and explain in what way all three methods of heat transfer are accounted for in its construction.

PROBLEMS

1. How much heat is required to raise the temperature of 25 Kg of brass from 25°C to 520°C? [Ans. 1138 Kcal.]

2. How much heat is required to raise the temperature of 50 lb of lead from 32°F to 450°F?

3. Find the quantity of heat required to raise the temperature of 35 lb of aluminum from 50°F to 370°F. [Ans. 2464 Btu.]

4. One end of an aluminum rod 40 cm long and cross-sectional area 5 cm^2 is maintained at 0°C and the other end at 100°C. How much heat will pass through in 1 hr?

5. An iron rod 28 cm long has a cross section of 12 cm^2. Find the amount of heat that flows through in 30 min if one end is maintained at 5°C and the other end at 80°C. [Ans. 9.26 Kcal.]

6. The bottom of an iron boiler is 3 mm thick and has a cross-sectional area of 0.65 m^2. If the upper surface is 100°C and the gas flames maintain the under surface at 105°C, how much heat will flow through in 1 hr?

7. The bottom of a brass boiler is 4 mm thick and has a cross-sectional area of 1.2 m^2. If the tank maintains the contained water at the boiling point of 100°C, and the gas flames maintain the lower surface at 105°C, how much heat flows through in 1 hr? [Ans. 1.404×10^5 Kcal.]

8. A cylindrical iron rod is 25 cm long and 1.2 cm in diameter. How much heat will flow through this rod in 5 min if one end is maintained at 350°C and the other end at 0°C?

17 CHANGE OF STATE AND LOW TEMPERATURES

It is well known that to change the state of most common substances we need only to raise or lower the temperature. For example, water is a liquid at room temperature and on cooling can be made to freeze to become ice, a solid. When water is heated to the boiling point it changes into steam, a gas.

The element copper is a solid at room temperature, and by raising its temperature it can be melted to become a liquid. Raised to a still higher temperature molten copper can be made to boil and become a gas.

17.1 EXPANSION AND CONTRACTION ON FUSION When molten metal of one kind or another is poured into a mold for casting, the metal may contract or expand on solidifying and then, on cooling down to room temperature, contract or expand according to its coefficient of thermal expansion. Cast iron, for example, is a substance which, on solidifying, expands slightly but then, on cooling down to room temperature, contracts about 1 per cent of its length. It is therefore well suited to casting, since slight expansion on solidifying aids in the reproduction of every detail in the mold. To allow for shrinkage in cooling, however, a pattern maker's rule is made 1% longer than the standard length. In this way the pattern itself is larger than the final casting used in making a mold.

An experiment illustrating the enormous expansion forces of freezing water is diagrammed in Fig. 17A. A small cast-iron bomb, about 2 in. in diameter and $\frac{1}{8}$ in. thick, is completely filled with water at a temperature close to 0°C. After the threaded iron plug is screwed in tightly, the bomb is packed in a freezing mixture of cracked ice and salt. After some minutes as the water freezes it explodes the bomb with a dull thud.

17.2 COOLING BY EVAPORATION When water is left in an open dish it slowly evaporates,

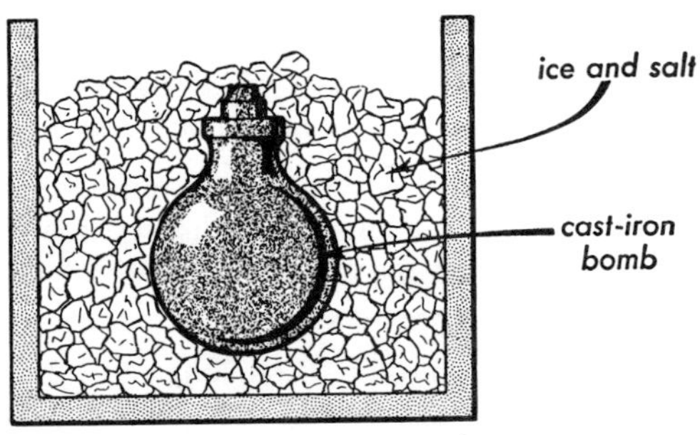

Fig. 17A. When the water inside the bomb freezes, the expansion bursts the cast-iron walls.

i.e., it goes spontaneously into the gaseous state. Evaporation therefore is a free expansion, and expansion is always accompanied by cooling. This phenomenon of cooling by evaporation, which is so important from the standpoint of its many commercial applications, is explained by the kinetic theory of matter.

Due to the random motions of the molecules of a liquid, some molecules obtain, momentarily, a very high velocity. If a molecule at the surface is given a high velocity in an upward direction, it may escape into the air above. Some of these escaped molecules soon find their way back into the liquid by chance collisions with air molecules from above the surface but many of them do not. See Fig. 17B. The sporadic escape of molecules

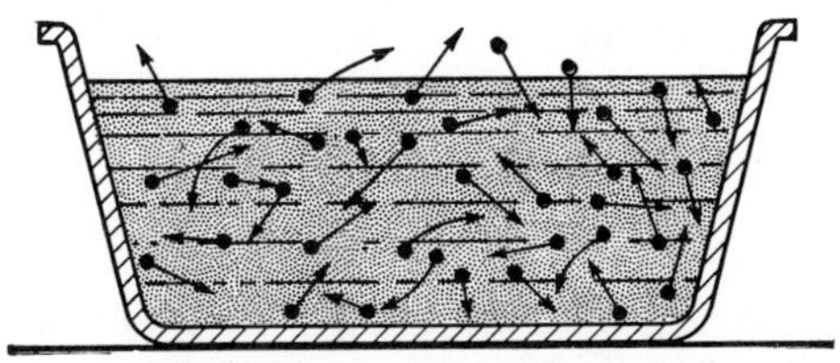

Fig. 17B. Evaporation of water from an open dish is due to the rapid motion of the water molecules and their occasional escape into the air space above.

may be speeded up by blowing air across the surface. The air carries the newly escaped molecules away before they have a chance to return to the liquid.

By virtue of the high speed of the molecules escaping from a liquid surface, considerably more than the average kinetic energy is carried away with them. A lowering of the average kinetic energy of the remaining liquid molecules means a lowering of the temperature. The more rapid the evaporation, therefore, the faster will be the cooling. This is strikingly demonstrated by pouring a small amount of ether or alcohol on the finger. Either of these liquids, and particularly ether, evaporates very rapidly, cooling the surface of the finger quickly. Ether is often used in this way by surgeons, in place of an anesthetic, to freeze local spots of the body before beginning a minor operation.

Cooling by evaporation can be demonstrated to a large group by pouring a small quantity of ether over the bulb of an air thermometer as shown in Fig. 17C. Due to the cooling of the glass bulb the air inside contracts and the external atmosphere forces water further up into the stem of the thermometer.

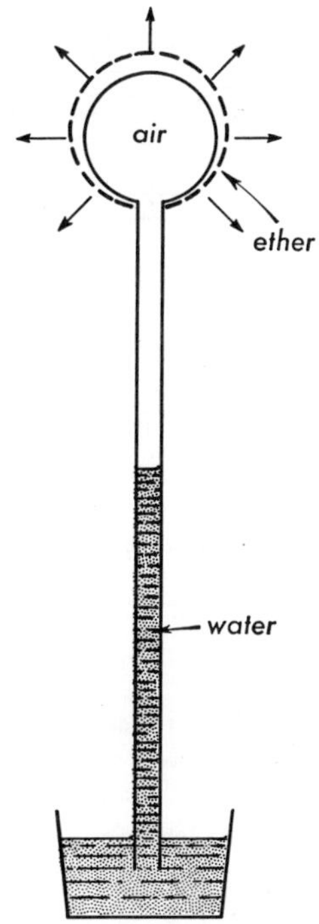

Fig. 17C. Experiment demonstrating the cooling effect produced by the evaporation of ether.

17.3 HUMIDITY When water molecules escape by evaporation from the free surface of a liquid, they mix with the air molecules above. If the space above the liquid surface is enclosed, as shown in Fig. 17D, this mixture cannot escape. Under these circumstances the water will continue to evaporate until the air above becomes saturated with water vapor, i.e., until it can hold no more. When this condition is reached, as many free water molecules will be returning to the liquid every second as are escaping.

The maximum amount of water the air can hold in the vapor state depends upon the temperature and very little upon the air pressure. This is illustrated by the values given in Table 17A.

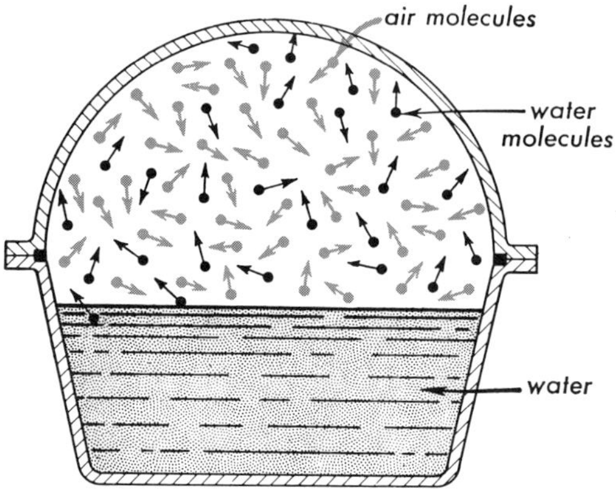

Fig. 17D. Illustration of the saturation of air with water vapor.

The temperature of the air is given in one column and the maximum amount of water that can exist in the vapor state in a cubic meter of air is given in the other.

It is clearly seen from the table that the hotter the air the greater is the amount of water it can hold in the vapor state.

TABLE 17A. Mass of Water Vapor in One Cubic Meter of Saturated Air

Temperature	Water Vapor
0°C or 32°F.	4.8 g
5°C " 41°F.	6.8 g
10°C " 50°F.	9.3 g
15°C " 59°F.	12.7 g
20°C " 68°F.	17.1 g
25°C " 77°F.	22.8 g
30°C " 86°F.	30.0 g
35°C " 95°F.	39.2 g

The atmosphere, which we might term free air, is not always saturated with water vapor. If it contains very little or no water vapor, we say the air is dry; if it contains a great deal we say it is damp.

The quantity of water vapor present in 1 m³ of air is called the *absolute humidity*. It is, therefore, a measure of the dampness of the air. It is measured by the number of grams of water vapor present in 1 cubic meter of air. For example, the absolute humidity might be said to be 14 g/m³.

It is customary in speaking of the dampness of air not to specify the absolute humidity but the *relative humidity*. The relative humidity R is defined as the ratio of the mass of water vapor actually present in any volume of air to the mass of water vapor required to saturate the same volume of air at the same temperature.

$$R = \frac{\text{water content present in air}}{\text{water content when saturated}} \qquad (17a)$$

EXAMPLE 1. Find the relative humidity inside a building where the air at 25°C is found to have 5.7 g of water vapor per cubic meter of air.

SOLUTION. If the air were saturated at this temperature it would contain, according to Table 17A, 22.8 g/m³. By direct substitution in Eq. (17a), we find

$$R = \frac{5.7 \text{ g/m}^3}{22.8 \text{ g/m}^3} = 0.25$$

It is customary to express such answers in per cent and say that the relative humidity in this case is 25%.

If air which is saturated with water vapor is cooled to a lower temperature, some of the water vapor will condense to the liquid state. These are the conditions under which rain and fogdrops are formed. The reason for this condensation is that at the lower temperature and equilibrium conditions less water can exist in the vapor state and still saturate the air. If the air cools without the formation of rain or fog the air takes on an unstable state in which it is supersaturated.

17.4 VAPOR PRESSURE The presence of water vapor in the air increases atmospheric pressure. To explain why it does so, consider again Fig. 17D. As more and more water evaporates into the space above, the pressure due to the bombardment of the walls by the water molecules becomes greater and greater, reaching a maximum at saturation. The water molecules exert pressure, each independent of the other.

Vapor pressure is expressed in centimeters of mercury and at saturation is referred to as the saturated vapor pressure. Values of saturated

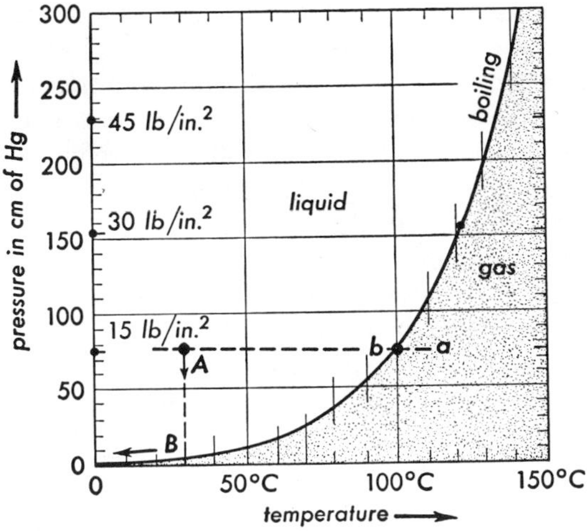

Fig. 17E. Boiling-point curve for water as it changes with pressure.

vapor pressure are given as a graph in Fig. 17E. This graph shows that at higher pressures the air becomes saturated at higher temperatures.

17.5 BOILING *The boiling of a liquid is but a state of rapid evaporation.* As the temperature of water is raised, the rate of evaporation increases until at the boiling temperature it reaches a maximum. Beyond this temperature water can exist only in the vapor state.

When water boils at normal atmospheric pressure, evaporation takes place throughout the liquid as well as at the surface. Evidence of this is seen in the bubbles of saturated vapor that form near the bottom of a vessel and increase in size as they rise to the surface. The bubbles are able to form because the vapor pressure at 100°C is 76 cm of Hg and slightly greater than the external pressure of the atmosphere. Boiling cannot occur at any point in a liquid unless the vapor pressure there is equal to or slightly greater than the pressure within the liquid.

If in the process of boiling water, while bubbles are forming within the liquid, the pressure on the liquid surface is increased, the bubbles will collapse and boiling will cease. Under this increased pressure, however, boiling can be started again by adding heat and raising the temperature to a higher value. In other words, *the higher the pressure on a liquid the higher is the boiling temperature.* Conversely, *the lower the pressure within a liquid the lower is the temperature required to make it boil.*

A graph of the boiling point of water, as it varies with pressure, is shown in Fig. 17E. Note that at standard atmospheric pressure of 76 cm of Hg, which is about 15 lb/in.², water boils at 100°C. At twice this pressure, however, a temperature of 120°C must be reached, etc.

Consider, for example, the boiling of water in a pressure cooker when the safety gauge is set for 15 lb/in.² See Fig. 17F. As the water inside

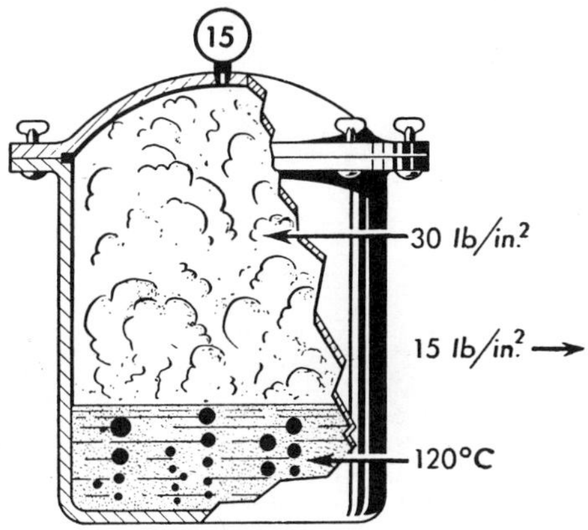

Fig. 17F. In a pressure cooker set for 15 lb/in.², water boils at 120°C.

gets hotter and hotter, some of it vaporizes and the pressure rises. This pressure continues to rise until it becomes 15 lb/in.² greater than the pressure outside. At this point the safety valve opens slightly and prevents the internal pressure from exceeding this value. Under these conditions the pressure outside the vessel is 15 lb/in.², while the pressure inside is 15 lb/in.² greater, or 30 lb/in.² From the graph it will be seen that the confined water must now reach 120°C to boil. At this higher temperature food cooks more quickly. In an open pan the water boils at 100°C, and its temperature will not exceed this value.

17.6 GEYSERS One of the great wonders of the western world is the spontaneous eruption almost hourly of the mammoth geyser "Old Faithful" in Yellowstone National Park. The following explanation of geyser activity was first given by Bunsen in 1847 and is based upon the above explanation of boiling.

Water from a nearby stream seeps into the vertical shaft or hole where, due to volcanic heat below, it is gradually heated to the boiling point. See Fig. 17G. Because the water is heated from below, and convection currents are shut off by the narrowness of the shaft, a temperature considerably higher than 100°C must be reached before the water at the bottom can boil. Since atmospheric pressure exists at the surface, the water there will boil at 100°C. Far down the shaft, however, the added pressure of nearly 70 ft of water requires a temperature of 130°C to produce boiling. Because the water is heated from below, this high temperature is reached near the bottom and boiling begins there before it does at the top. When a sufficiently high temperature is reached, the vapor pressure deep down exceeds the pressure due to the air and water column above and the rise of numerous bubbles, by pushing out the column of hot water above, starts an eruption. Nearing the surface the vapor pressure of the superheated water is so high that the remaining water is pushed out with great force.

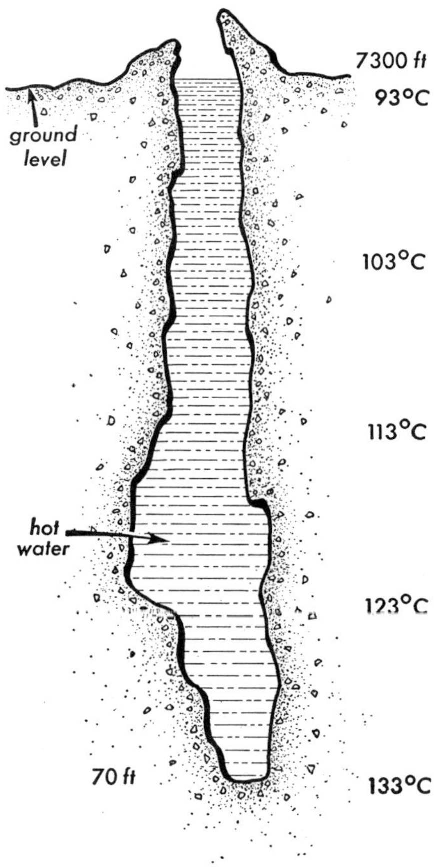

Fig. 17G. Diagram illustrating the principles of a geyser like "Old Faithful" in Yellowstone National Park.

An excellent demonstration of these principles can be performed by an experimental geyser of the form shown in Fig. 17H. Such models can be made almost any size from one foot in height to 10 ft or more. The period of their eruption depends upon size as well as the quantity of applied heat.

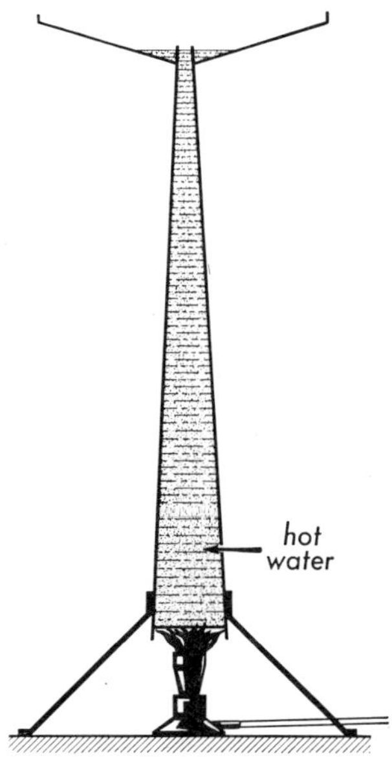

Fig. 17H. Experimental geyser that will erupt periodically

17.7 BOILING AT LOW TEMPERATURES Just as water can be made to boil at temperatures higher than 100°C by increasing the pressure, so can it be made to boil at temperatures below 100°C by reducing atmospheric pressure. Because of the practical importance of this basic fact, a detailed explanation should be given. Fig. 17E is a graph of the saturated vapor pressures of water. It also represents conditions under which water and saturated water vapor can exist together in equilibrium.

Since boiling at the surface of water takes place when the saturated vapor pressure becomes equal to the atmospheric pressure, it follows that by lowering atmospheric pressure a lower vapor pressure can bring about the conditions for boiling. The curved line in the graph is therefore a *boiling point curve;* all pressures and temperatures to the right represent the gas or vapor state, all points to the left the liquid state.

Consider, for example, water at the normal boiling point of 100°C and 76 cm of Hg. To raise its temperature without raising the pressure, heat must be added to vaporize the water and the resultant steam. To lower its temperature, heat must be given up to liquefy the steam and more heat given up to lower the temperature of all the water.

To go one step further, suppose water at room temperature, 30°C, is placed in a vacuum jar as shown in Fig. 17I and the pressure is

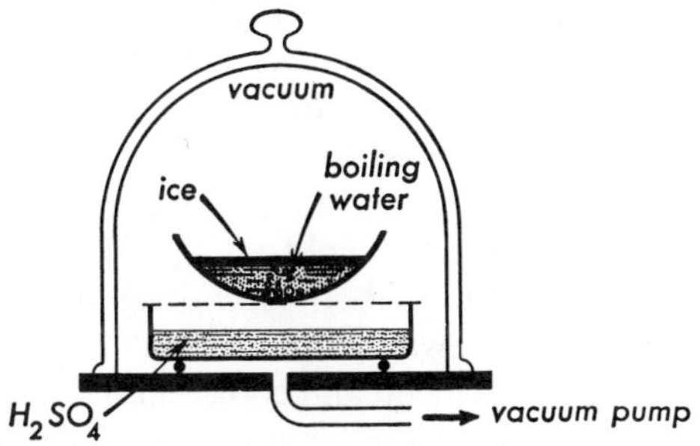

Fig. 17I. The boiling point of a liquid like water is lowered by lowering the atmospheric pressure. Water in a vacuum will boil and freeze at the same time.

slowly reduced by means of a vacuum pump. Starting at the point A in Fig. 17E the pressure decreases until it reaches B, about 3.18 cm of Hg, where bubbles form and the water begins to boil.

To vaporize water requires heat, and this is taken from the remaining water, thus cooling it to a lower temperature. Continued reduction in pressure causes continued boiling and lowering of temperature until finally the freezing point at approximately 0°C is reached. Continued evaporation cools the surface of the water until ice forms over the surface of the boiling water. Here then is a condition in which water boils and freezes at the same time at 0°C. (The small dish of sulfuric acid, H_2SO_4, placed in the vacuum chamber absorbs water vapor, thus aiding the pumps in keeping the pressure sufficiently low.)

Although water under normal atmospheric pressure at sea level boils at 100°C, water at higher altitudes boils at lower temperatures. Evidence of this fact is well known to those who like to camp in the higher mountains. There, at reduced atmospheric pressure, it takes longer than usual to cook all kinds of foods. The boiling points given in Table 17B show specific values for different elevations.

Flying at high altitudes the water in water-cooled engines boils at lower temperatures. At a height of 6 mi, gasoline boils at the normal temperature of −65°C. At 12-mi elevation, blood boils at the body temperature of 98.6°F (37°C).

17.8 LIQUEFACTION OF AIR The present method of liquefying air and other gases is based upon the principle of cooling by expansion. It was by this method that Dewar* liquefied oxygen

* SIR JAMES DEWAR (1842–1923), British chemist and physicist, noted chiefly for his experiments on the liquefaction of the so-called permanent gases. As a child he was very fond of music, and when he met with an accident which prevented him from playing the flute he worked at making violins. At this work he developed a dexterity that proved to be useful in later years. Educated at Edinburgh University, he held several professorships. In 1891 Dewar was successful in liquefying oxygen for the first time. He was awarded the famous Rumford Medal of the Royal Society in 1894, and was knighted in 1904. He built a large refrigeration plant at the Royal Institution in London, by which hydrogen gas was first liquefied at a temperature of −252°C in 1898, and solidified at a

TABLE 17B. Pressure and Boiling Point of Water at Various Distances Above and Below Sea Level

Altitude in Air	Boiling Point	Atmospheric Pressure	
ft	°C	cm of Hg	lb/in.²
100,000	9.8	0.82	0.16
50,000	48.9	8.75	1.7
40,000	58.8	14.1	2.7
30,000	69.2	22.5	4.4
20,000	79.6	35.0	6.7
15,000	84.4	42.4	8.2
10,000	89.8	52.2	10.1
8,000	92.1	56.8	11.0
6,000	94.0	61.0	11.8
4,000	96.0	65.6	12.7
2,000	98.0	70.2	13.7
Sea Level	**100.0**	**76.0**	**14.7**
−20	113.7	121.0	23.4
−40	123.3	165.0	32.0
−60	131.2	210.0	40.7
−80	138.0	255.0	49.4
−100	144.0	300.0	58.0
−120	149.0	345.0	66.7
ft	°C	cm of Hg	lb/in.²
Depth in Water	Boiling Point	Total Pressure Atmosphere + Water	

for the first time in 1891, and Linde liquefied air in 1895. Oxygen gas becomes a liquid at the extremely low temperature of −184°C and air at −191°C. On the Fahrenheit scale these correspond to −300°F and −312°F.

In the liquid air machine, see Fig. 17J, air is compressed by a pump to a pressure of about three thousand pounds per square inch. Due to compression this air is heated to a fairly high

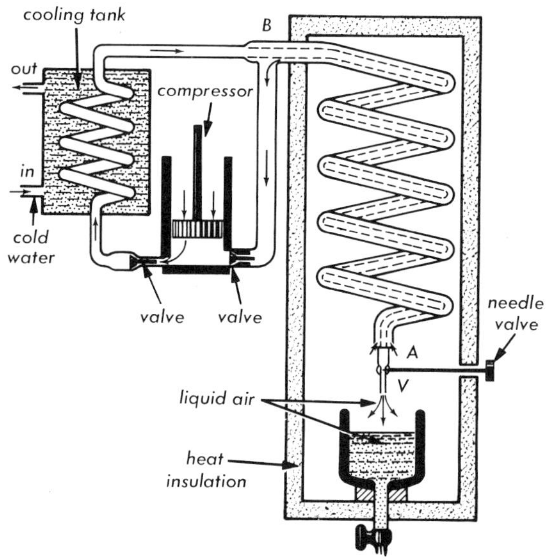

Fig. 17J. Cross section of a machine used in the liquefaction of air.

temperature. It must, therefore, be cooled by running it through a cooling tank. This cold compressed air passes through the inner tube of a double-walled coil *B* and escapes through the very narrow opening of a needle valve *A*. The escaping air expands so much that its temperature is lowered considerably below the room temperature. The continual pumping of the compressor draws this cold air up through the outer tube of the coil, thus cooling the compressed air on its way to the needle valve. This air is compressed again and cooled in the cooling tank to go around the circuit again. The cycle continues until the temperature in the region of *V* is so low that drops of liquid air form in the jet from the needle valve. These drops fall into the Dewar flask and accumulate to be drawn off later.

still lower temperature in 1899. He was known to his collaborators as an experimentalist of unusual skill, and a lecturer of remarkable eloquence.

17.9 THE LIQUEFACTION OF HYDROGEN AND HELIUM Hydrogen gas was first liquefied by Dewar at a temperature of −252°C in 1898, and helium gas was first liquefied by Onnes* at a temperature of −268°C in 1908.

The making of liquid hydrogen requires large quantities of liquid air. Hydrogen is first compressed under a very high pressure and then cooled by means of liquid air. It is then allowed to expand in an interchanger similar to that shown for liquid air in Fig. 17J. To liquefy helium also requires a considerable quantity of liquid hydrogen. The gas is compressed, cooled by liquid hydrogen, and allowed to expand.

By boiling helium in a partial vacuum, Onnes cooled liquid helium to a temperature only 0.82°C above absolute zero.

17.10 LIQUID AIR EXPERIMENTS The physical properties of matter are quite different at extremely low temperatures than they are at room temperatures. This may be illustrated by a number of experiments with a small quantity of liquid air. Liquid air has the same general appearance and density as water. When poured into an open dish or Dewar flask it boils vigorously until the container has cooled to liquid air temperature. A small coil of lead wire cooled in liquid air becomes elastic and behaves as a spring. A piece of rubber tubing or the petals of a flower become hard and break like glass under the blow of a hammer. A hollow rubber ball no longer bounces but breaks into small pieces the first time it strikes the floor. A small bunch of grapes, an egg, or a piece of beefsteak exhibits the same hardness and breaks like glass.

Mercury and gasoline will freeze when cooled in liquid air. Mercury, for example, poured into a small rectangular paper cup containing a wooden handle and then frozen can be used as a hammer to drive nails into a block of wood. See Fig. 17K.

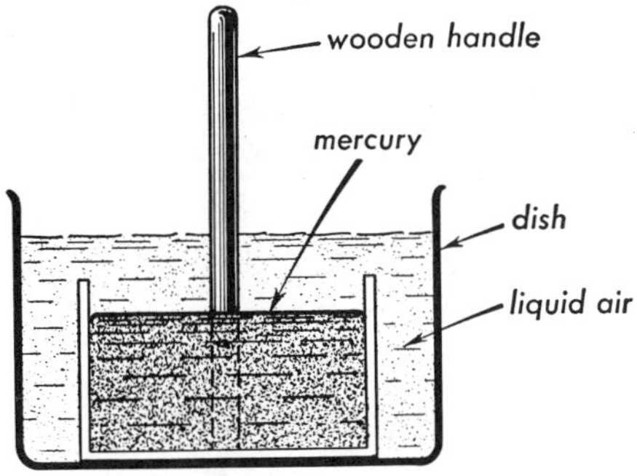

Fig. 17K. How to make a small mercury hammer.

Lead hammered into the shape of a small bell will give a dull thud when hit by a steel clapper, but when the temperature of the bell is lowered to that of liquid air it will produce a ringing sound with the same clapper. See Fig. 17L.

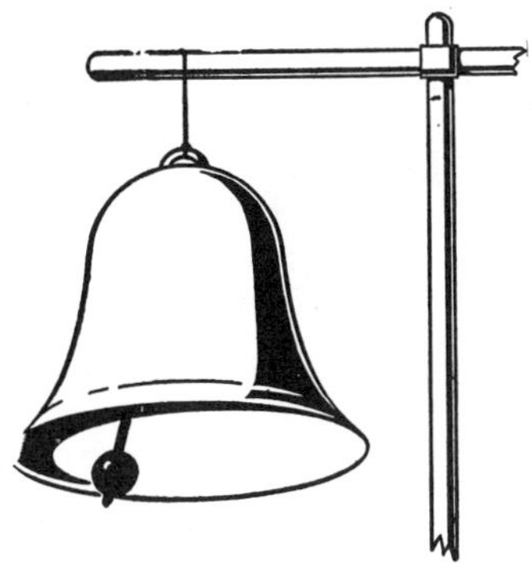

Fig. 17L. Cooled to the temperature of liquid air, a lead bell will ring.

If liquid air is left standing for some little time, its nitrogen content will have boiled off, leaving nearly pure liquid oxygen. A cigarette or cigar soaked in this liquid oxygen for a few seconds and then lighted with a match will burn rapidly with the appearance of fireworks. A small bit of wool or cotton will exhibit the same effect but will go off more explosively. When used for blasting purposes, a paper cartridge containing wood-powder or carbonized cork is soaked in liquid oxygen and then pushed into the bore-hole and fired by means of a fuse or electric spark.

* KAMMERLINGH ONNES (1853–1926), Dutch experimental physicist, noted principally for his physics experiments performed with substances at very low temperatures and for his discovery of superconductivity. In 1908 he succeeded for the first time in liquefying helium. The Nobel prize for physics was conferred upon him in 1913.

QUESTIONS

1. Why does the evaporation of a liquid produce a cooling effect?

2. What is the difference between absolute humidity and relative humidity?

3. Make a diagram of a geyser and explain why such natural phenomena are periodic in their action.

4. What is the purpose of a pressure cooker of the kind used by the modern housewife? What are the basic principles involved?

5. Can water be made to boil at room temperature?

6. How does the boiling point of water vary with altitude? How much is the boiling point of water lowered for each 1000 ft in elevation?

7. Why does food cook more slowly in the high mountains? Would a pressure cooker speed up cooking in the mountains?

8. Using the values given in Table 17B, plot a graph of altitude against the boiling point of water. Plot as far as 50,000 ft. From this graph find the boiling point of water at an altitude of 2.5 mi, and 5 mi.

PROBLEMS

1. If the relative humidity is 50% when the temperature is 25°C, what is the absolute humidity? [Ans. 11.40 g/m^3.]

2. The absolute humidity is 16.5 g/m^3 on a certain day when the temperature is 30°C. Find the relative humidity.

3. If the absolute humidity is 24.6 g/m^3 on a day when the temperature is 30°C, what is the relative humidity? [Ans. 82%.]

4. If the relative humidity is 30% when the temperature is 77°F, what is the absolute humidity?

5. When the temperature is 86°F, the relative humidity is measured and found to be 65%. What is the absolute humidity? [Ans. 19.5 g/m^3.]

6. What is the temperature on the Fahrenheit scale of (a) liquid helium, and (b) liquid hydrogen?

7. If the gauge on a pressure cooker at sea level shows an internal pressure of 17.3 $lb/in.^2$, what is the temperature of the boiling water inside? [Ans. 123.3°C.]

8. At what gauge pressure should a pressure cooker be maintained if the water inside is to be kept boiling at 138°C?

9. At what depth in a geyser would water boil at 138°C? Assume the geyser is located in the mountains at 6000 ft elevation. Use Table 17B. [Ans. 86.7 ft.]

10. At what depth in a geyser would water boil at 149°C? Assume the geyser is located in the mountains at an elevation of 8000 ft.

18 | VIBRATIONS AND WAVES

The concepts of *vibrations* and *waves* are important to all present-day theories of the physical world. They both appear in a very fundamental way in the subjects of light, sound, heat, microwaves, and in the ultramicroscopic world of atoms and nuclei and their phenomena.

Any motion, simple or complex, which repeats itself in equal intervals of time is called *periodic motion*. There are many examples in everyday life which give rise to a special kind of periodic motion called *simple periodic motion*. The swinging of the clock pendulum, the turning of the balance wheel of a watch, or the vibration of a tuning fork, are good examples of such motions. The term "simple periodic motion" applies to these because each can be described in terms of one of the simplest known types of periodic motion—namely, *uniform circular motion*.

18.1 SIMPLE HARMONIC MOTION *Simple harmonic motion is defined as the projection on any diameter of a point moving in a circle with uniform speed.*

This motion is illustrated in Fig. 18A. The point p moves around the circle of radius r with uniform speed v. If at every instant a perpendicular is drawn from p to the diameter AB, the intercept P will move with simple harmonic motion. Moving back and forth along the straight line from A to B, the velocity v_x is continually changing. At the center point C it has its greatest velocity, while at A and B it is momentarily at rest. Starting from either end of its path, the velocity increases until it reaches C; from there it slows down again, coming to rest at the opposite end of its path.

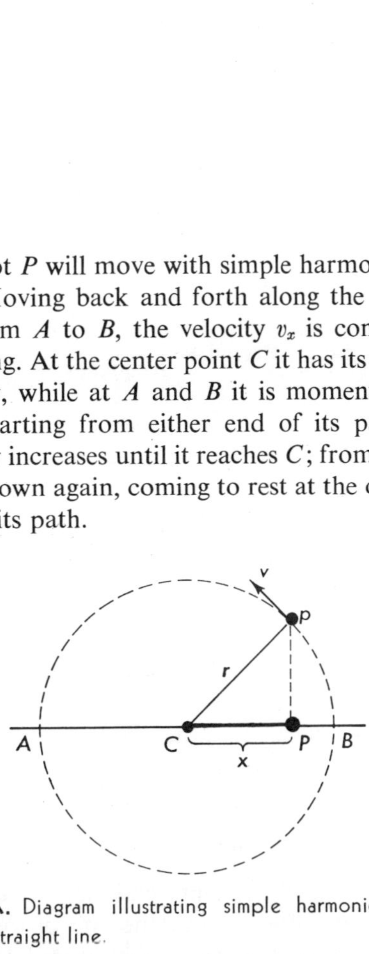

Fig. 18A. Diagram illustrating simple harmonic motion along a straight line.

The *displacement* of a simple harmonic motion is defined as the distance from the center C to the point P. As shown in Fig. 18A, the displacement x varies in magnitude from zero at C

up to r, the radius of the *circle of reference*, at A or B.

The *amplitude* r *is defined as the maximum value of the displacement* x, and the *period is defined as the time required to make one complete vibration.*

If a vibration starts at A, it is not completed until the point moves across to B and back again to A. If it starts from C and moves to B and back to C, only half a vibration has been completed. The amplitude r is usually measured in centimeters and the period T in seconds.

The frequency of a harmonic motion is defined as the number of complete vibrations per second. For example, if a particular vibrating object completes one vibration in one-half second (the period $T = \frac{1}{2}$ s), then it will make two complete vibrations in 1 s (the frequency $\nu = 2$ vib/s). If again a body completes one vibration in one-tenth of a second, $T = \frac{1}{10}$ s, it will make ten vibrations in 1 s, $\nu = 10$ vib/s. In other words, ν and T are reciprocals of each other.

$$\textbf{Period} = \frac{\textbf{1}}{\textbf{frequency}}$$

or

$$\textbf{frequency} = \frac{\textbf{1}}{\textbf{period}}$$

In algebraic symbolism,

$$\boxed{T = \frac{1}{\nu} \qquad \nu = \frac{1}{T}} \tag{18a}$$

When describing the vibration of a body in terms of motion in a circle, the vibration frequency is equivalent to the number of *revolutions per second*, or *cycles per second*.

$$1\,\frac{\text{vibration}}{\text{second}} = 1\,\frac{\text{cycle}}{\text{second}} \tag{18b}$$

As a result of international agreements reached several years ago, the unit of frequency is today called the *hertz*, abbreviated Hz. The unit is named in honor of Heinrich R. Hertz,* a German physicist who became famous the latter part of the nineteenth century for his outstanding discovery of radio waves.

$$1\,\frac{\text{cycle}}{\text{second}} = 1 \text{ hertz}$$

Abbreviated,

$$1\,\frac{\text{vib}}{\text{s}} = 1\,\frac{\text{c}}{\text{s}} = 1 \text{ Hz}$$

All three of these abbreviations are commonly used, the latter increasing in its use in newer books.

18.2 VIBRATING SPRING Any elastic body may be set into a state of vibration by first distorting it in some way and then releasing it. This is demonstrated by a weight on the end of a spring as shown in Fig. 18B. In diagram (a), the spring S and mass m hang in a state of equilibrium. At (b) a force F has been applied to stretch the

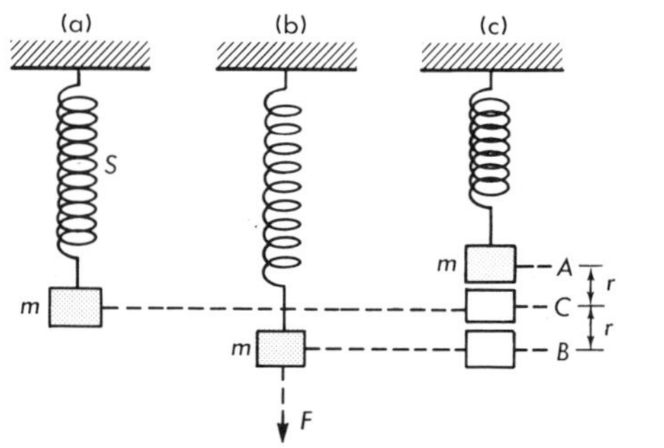

Fig. 18B. A mass on the end of a spring executes simple harmonic motion.

* HEINRICH RUDOLF HERTZ (1857–1894), German physicist born at Hamburg, February 22, 1857. He studied physics under Helmholtz in Berlin, at whose suggestion he first became interested in Maxwell's electromagnetic theory. His researches with electromagnetic waves which made his name famous were carried out at Karlsruhe Polytechnic between 1885 and 1889. As professor of physics at the University of Bonn, after 1889, he experimented with electrical discharges through gases and narrowly missed the discovery of X rays described by Röntgen a few years later. By his premature death, science lost one of its most promising disciples.

spring, displacing the mass a distance r. After being released, the mass m moves up and down with simple harmonic motion. In diagram (c) the spring is shown compressed with m at its highest point. The maximum displacement r of the vibration is determined by the original distance to which the spring is stretched, and the period T is given by the algebraic equation:

$$T = 2\pi\sqrt{m/k} \qquad (18c)$$

where m is the mass of the vibrating body in Kg and k the number expressing the stiffness of the spring. The constant k is the force in newtons required to stretch the spring 1 m and is given by Hooke's law. See Eq. (11b). This formula shows that, if m is made larger, the period is increased and that, when a stiffer spring is used (k being in the denominator), the period is decreased.

EXAMPLE 1. When a mass of 600 g is suspended from the lower end of a coil spring, as shown in Fig. 18B, it stretches the spring a distance of 12 cm. If the mass is then pulled down another 10 cm and released, the mass vibrates up and down with simple harmonic motion. Find (a) the spring constant k, and (b) the period of the vibrating mass m, and (c) the frequency v.

SOLUTION. (a) Use Eq. (11a) to find the spring constant. Solving this equation for k and substituting the known values of $x = 0.12$ m, and mg for W, we have

$$k = \frac{W}{x} = \frac{0.60 \text{ Kg} \times 9.80 \text{ Kg/s}^2}{0.12 \text{ m}}$$

$$k = 49.0 \frac{\text{Kg m}}{\text{s}^2 \text{ m}} = 49.0 \frac{\text{Kg}}{\text{s}^2}$$

(b) To find the period T, the known values are substituted in Eq. (18c).

$$T = 2\pi\sqrt{\frac{0.60 \text{ Kg}}{49.0 \text{ Kg/s}^2}}$$

$$T = 6.283 \times 0.1107 \text{ s}$$
$$T = 0.696 \text{ s}$$

(c) To find the frequency use Eq. (18a) directly, to obtain

$$\nu = \frac{1}{T} = \frac{1}{0.696} = 1.437 \text{ Hz}$$

18.3 VIBRATING STRIPS If a strip of wood or metal is clamped tightly at one end, as shown in Fig. 18C (a), it may be set into a natural state of

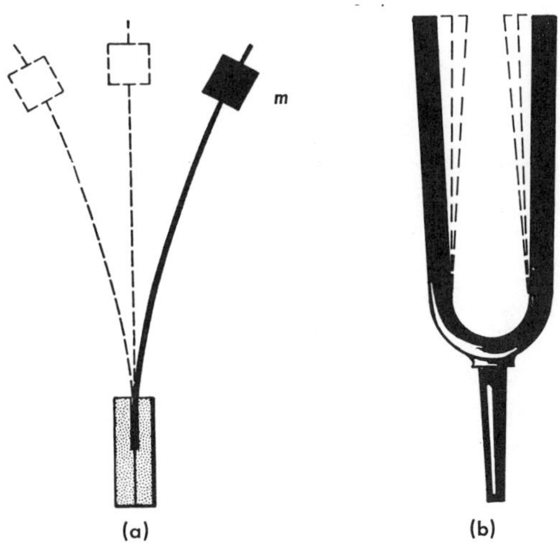

Fig. 18C. Illustration of the vibrations of (a) a strip clamped at one end and weighted near the other and (b) the prongs of a tuning fork.

vibration. Pulled to one side and then released, the free end of the strip will move back and forth with simple harmonic motion. If the mass m clamped to the free end is increased, the frequency of vibration will decrease; whereas, if the strip is made stiffer either by increasing its thickness or decreasing its length, the frequency will increase.

Shown in diagram (b) is a tuning fork as used by musicians to determine pitch. Striking one prong of a fork against some object sets both prongs vibrating simultaneously in opposite directions. The thinner the prongs of the fork the lower is the frequency of vibration; the shorter the prongs of the fork the higher is the frequency of vibration.

Besides being used as standards of pitch by musicians, tuning forks are used for scientific purposes. As scientific instruments they perform one very useful function of marking off short but equal intervals of time. A common fork used for this purpose has a frequency of one thousand, $\nu = 1000$ Hz, and is frequently kept vibrating by means of an electrical circuit similar to that of a doorbell.

18.4 THE SIMPLE PENDULUM If a simple pendulum, as shown in Fig. 18D, is set swinging in an

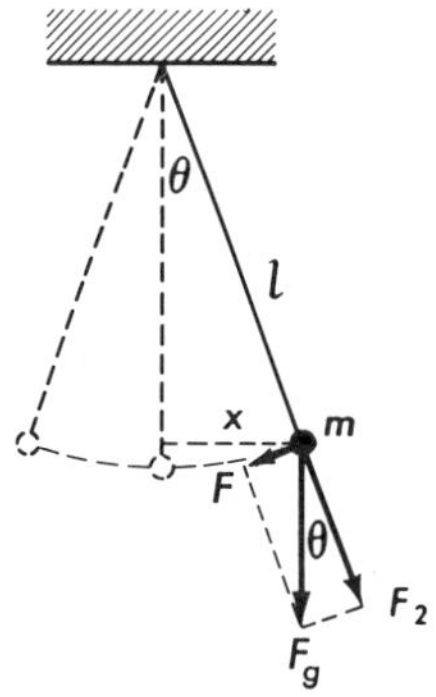

Fig. 18D. A pendulum bob executes simple harmonic motion along a curved path.

arc which is not too big, its period T is given by the following formula,

$$T = 2\pi\sqrt{\frac{l}{g}} \tag{18d}$$

T is the time of one complete vibration, l is the length of the pendulum, and g is the acceleration due to gravity. If the angle of swing is much greater than 5°, the period will be slightly increased.

The statement can now be made that the bob of a simple pendulum moves along its arc with *simple periodic motion.* At the ends of its swing the velocity is momentarily zero and at the center is a maximum. The driving force which starts the bob at one end of its path and accelerates it until it reaches the center and then decelerates it, bringing it to rest at the other end, is the component F of the force of gravity F_g. The component F_2 gives rise to the tension in the cord supporting the mass m.

The fact that the mass of a pendulum bob does not appear in Eq. (18d) signifies that pendulums of equal length but different mass should have the same period T. Started together, and with

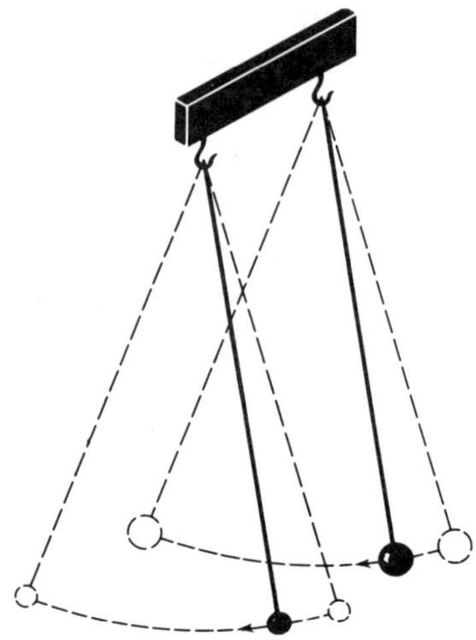

Fig. 18E. Two simple pendulums of the same length but different mass have the same period of vibration.

small amplitudes, simple pendulums of different mass will swing in synchronism (see Fig. 18E).

The effect of the acceleration of gravity g on the period is illustrated in Fig. 18F. If the pull of

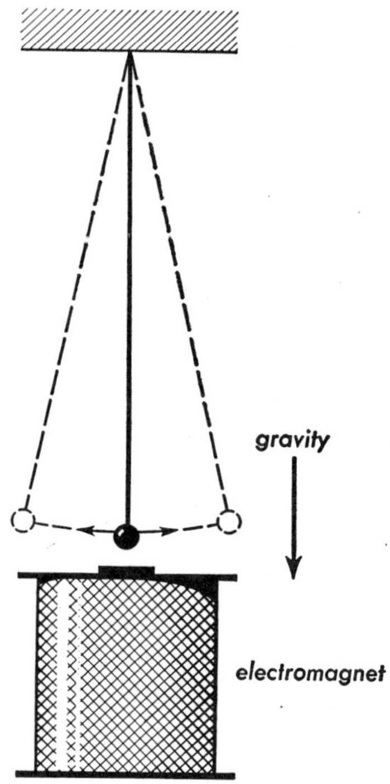

Fig. 18F. An increase in gravitational attraction, like the downward pull of the magnet, will shorten the period of a simple pendulum.

gravity could be increased, the period of all pendulums would decrease, that is, they would swing faster. If gravity could be decreased, the periods would be greater, that is, they would swing more slowly. In the diagram the pull of gravity is imitated by means of an electromagnet. When the magnet is turned on, it pulls down on the iron bob and makes it swing more quickly.

The effect of the length l on the period is shown in Fig. 18G. Two pendulums of lengths

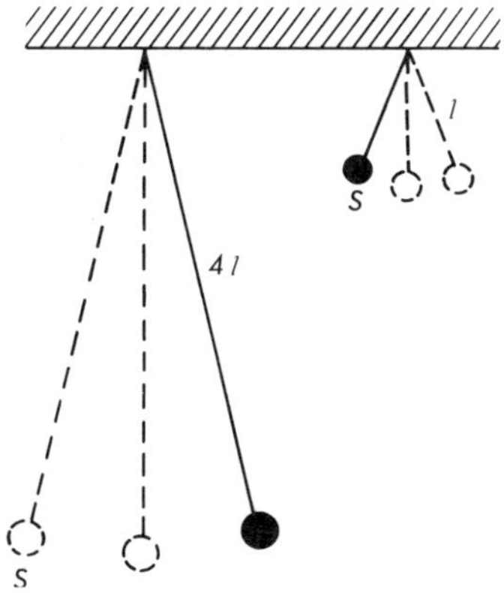

Fig. 18G. To double the period of a simple pendulum, the length must be increased fourfold.

l and $4l$ are shown at the same instant a short time after having started simultaneously at their respective origins S. While the pendulum of length l has made one complete swing, the pendulum of length $4l$ has made only half a swing. This shows that to double the time of swing the length must be made four times as large. This is in agreement with Eq. (18d), in which the length appears under the square root sign.

EXAMPLE 2. A steel ball is suspended from the lower end of a string and set swinging as a simple pendulum. If the length of the pendulum is 18 cm, find (a) the period, and (b) the frequency.

SOLUTION. The period is found by direct substitution in Eq. (18d),

$$T = 2\pi\sqrt{\frac{l}{g}}$$

$$T = 6.283\sqrt{\frac{0.18\text{ m}}{9.8\text{ m/s}^2}}$$

$$T = 6.283 \times 0.1354$$

$$T = 0.851\text{ s}$$

(b) Using Eq. (18a), we obtain

$$\nu = \frac{1}{T} = 1.175\text{ Hz}$$

18.5 SOURCES OF WAVES The motion of any material object may be considered as a source of waves. A board striking the water, the snap of a finger, or a bowed violin string are examples of this.

Suppose that the far end of a rope is fastened to a post, as shown in Fig. 18H (a), and that the

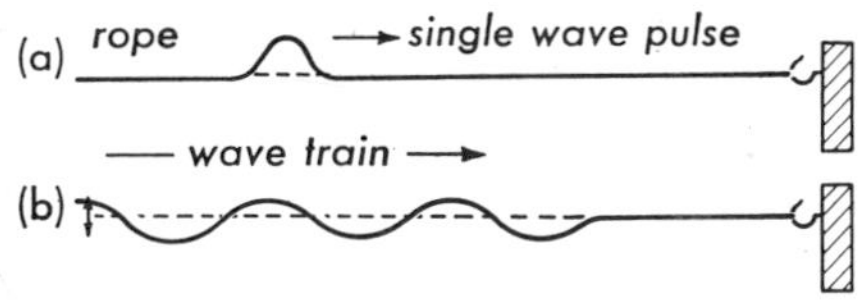

Fig. 18H. Examples of transverse waves sent out along a rope.

other end (a), which is held in the hand, is given a sudden flip up and down. The disturbance sent out along the rope travels down the rope to the post, as shown, and is then reflected back again toward the hand. This kind of wave is called a single *wave pulse.* If instead of a sudden pulse the hand is moved up and down with simple harmonic motion, a *train of waves,* like that shown in diagram (b), travels down the rope.

When the prongs of a tuning fork are made to vibrate they set the surrounding air into periodic motions. Each prong periodically strikes the air molecules next to it and these molecules in turn strike others, thus transmitting the disturbance outward. Traveling outward in all directions, such periodic disturbances constitute sound waves.

18.6 TRANSVERSE WAVES Transverse waves are those in which each particle vibrates along a line perpendicular to the direction of propagation. Along any one line of travel all particles are vibrating in one plane only. Such waves are illustrated in Fig. 18I by means of a wave machine designed for this purpose. As the handle H is turned one way or the other, the small round balls at the top move up and down with simple harmonic motion. As they move up and down, each along its own line, the wave form $ABCDEF$ will move to the right or to the left. Light is an example of transverse wave motion.

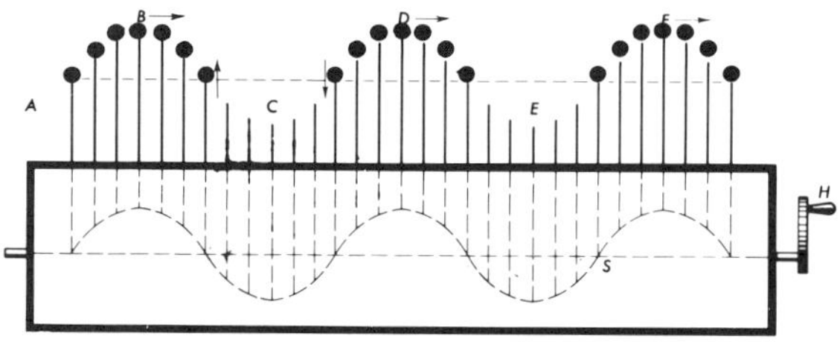

Fig. 18I. Diagram of a wave machine for demonstrating transverse waves.

18.7 LONGITUDINAL WAVES Longitudinal waves are those in which the vibrations of the particles are along straight lines parallel to the direction of propagation. This type of wave is illustrated in Fig. 18J by another wave machine.

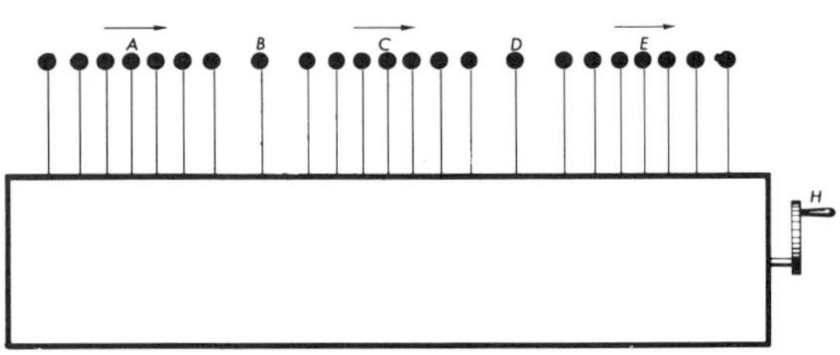

Fig. 18J. Diagram of a wave machine for demonstrating longitudinal waves.

As the handle is turned, each small ball moves horizontally and in the plane of the page with simple harmonic motion. In so doing the regions of rarefaction B and D and the regions of condensation A, C, and E move to the right, always keeping their same relative distances.

Sound waves in air are examples of longitudinal waves. Each air molecule vibrates back and forth about some equilibrium position as the wave train passes by.

18.8 WATER WAVES The motion of the surface of water as a wave passes by is a combination of both transverse and longitudinal waves. The water molecules move back and forth and at the same time up and down, i.e., in circles or ellipses. As illustrated by the diagram in Fig. 18K, each molecule moves in a circle in a vertical plane to form moving crests at A, C, and E and troughs at B and D. A small cork placed on the surface of water and observed from the side will exhibit this circular motion. When a person is swimming in a rough sea, and the crest of a wave approaches, he is first carried upward and forward in the direction the wave is traveling; after the crest has passed he falls and then is carried backward.

18.9 STANDING WAVES Nearly all sounds emanating from musical instruments are the result of standing waves. Standing waves may be produced in any substance whether it is a solid, liquid, or gas, by two wave trains of the same frequency traveling in the same medium in opposite directions. One of the ways in which this is done is illustrated in Fig. 18L. One end of a rope is fastened to a post and the other end, held taut, is moved up and down with simple harmonic motion. As the waves reach the fixed end of the rope they are reflected back to meet succeeding waves just coming up. If the waves have just the right frequency the rope will sustain both wave trains by dividing into sections as shown. The points L_1 to L_5, where the rope has a maximum up and down motion, are called

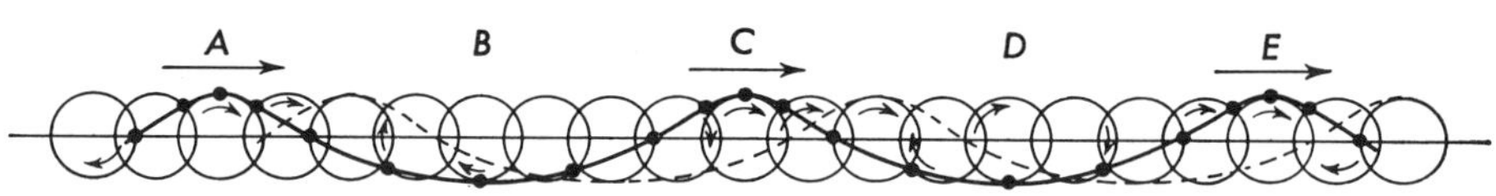

Fig. 18K. Water waves. Each water molecule moves in a circle or an ellipse as the waves pass by. This is a combination of longitudinal and transverse wave motion.

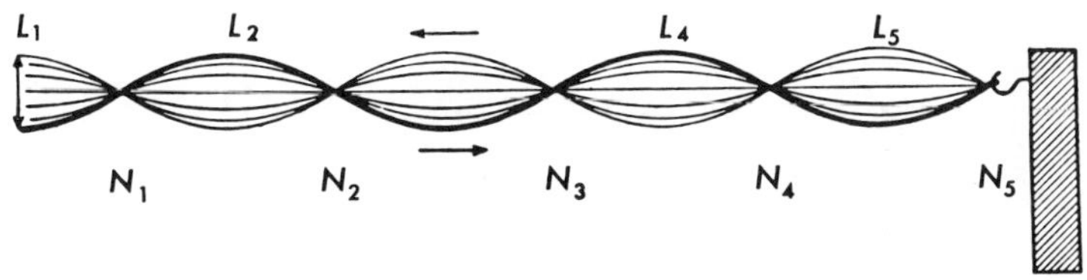

Fig. 18L. Standing waves produced by reflecting a train of transverse waves from the fixed end of a rope.

anti-nodes, and the points of no motion halfway between are called *nodes*. The heavy line represents the rope at one instant and the other lines represent it at other instants. The entire wave section between two consecutive nodes is called a *loop*.

Standing waves with longitudinal vibrations may be demonstrated with a long flexible spring as shown in Fig. 18M (a). The right-hand end of the spring is fixed and the left-hand end is moved back and forth with simple harmonic motion. If the impressed vibration has the proper frequency the waves traveling to the right in meeting the reflected waves traveling to the left set up *nodes* and *anti-nodes*. The nodes N correspond to points where there is no motion, and the anti-nodes L to points where the motion is a maximum. The dots in diagram (b) show the relative positions and motions of each individual coil of the spring at nine different times during one complete vibration. The dots at the nodes remain fixed at all times while those at the anti-nodes move back and forth as shown by the arrows in diagram (c). At one instant (3), compressions are formed at the odd-numbered nodes N_1, N_3 and N_5 and rarefactions at the even-numbered nodes N_2, N_4 and N_6. Half a vibration later (7), they change; the nodes of compression become nodes of rarefaction, and vice versa.

A direct comparison of standing transverse and standing longitudinal waves is given by diagrams (b), (c), (d) and (e) in Fig. 18M. The numbers 1, 2, 3, etc., indicate the corresponding states in the vibrations of each. Diagram (c) indicates the amplitudes of the longitudinal motions at the anti-nodes of the spring; and diagram (e), the transverse motions at the anti-nodes of the rope.

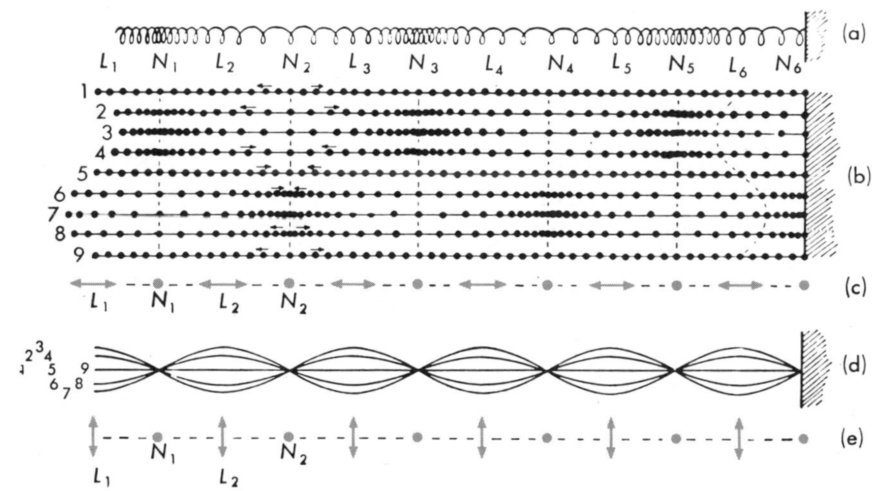

Fig. 18M. Illustration of standing waves as they are produced with (a) the longitudinal waves of a spring, (b) the longitudinal waves of sound in the air, and (d) the transverse waves of a rope. (c) and (e) indicate the direction of vibration at the anti-nodes.

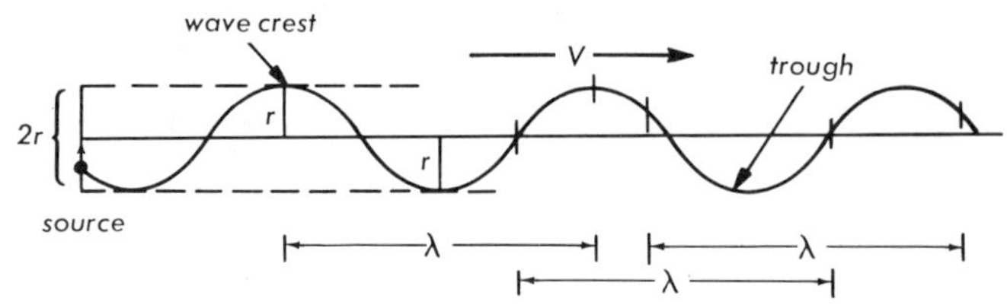

Fig. 18N. Illustration of the wavelength λ as the distance between corresponding points on two consecutive waves, and the amplitude r as the maximum displacement.

It should be explained that the dots in diagram (b) also represent the motions of air molecules when sound waves are reflected from a flat wall, back on themselves, to produce standing waves. As we shall see in the next chapter, these are like the sustained vibrations of the air in an organ pipe, a flute, or some other musical wind instrument.

While sound waves in air are longitudinal vibrations it is customary, for convenience only, to draw them as transverse waves. It is for this reason that the comparisons in Fig. 18M are made here. In the following two chapters, therefore, sound waves will usually be drawn as though they were transverse.

18.10 WAVELENGTH When a vibrating object sends out waves through a homogeneous medium the waves travel with constant velocity. If the source vibrates with simple harmonic motion and waves are transverse, they have the general appearance of the waves shown in Fig. 18N. The *wavelength* is defined as *the distance between two similar points of any two consecutive waves*, and is represented by the Greek letter lambda, λ. The distance between two consecutive wave crests, for example, is equal to one wavelength.

The amplitude of a wave is defined as the maximum value of the displacement. This is illustrated by r in Fig. 18N, the amplitude of the waves being equal to the amplitude of the source. *The frequency of a train of waves is defined as the number of waves passing any given point per second.* This is equal to the frequency of the source and is usually designated by ν. It is customary to express frequency in *vibrations per second* or in *cycles per second.*

From the definitions of velocity, frequency, and wavelength, the following very simple relation exists between them:

$$\boxed{V = \nu\lambda} \qquad (18e)$$

The length of one wave, λ, times the number of waves per second, ν, equals the total distance traveled in 1 s, V.

Since the period is defined as the time required for one wave to pass by a given point, the relation between frequency and period given for vibrating sources in Eq. (18a) also applies to waves.

EXAMPLE 3. If a train of waves moving along a rope has a velocity of 100 m/s and a wave length of 20 m, what is the frequency and period of the source?

SOLUTION. By substituting these values in the proper places of Eq. (18e), we obtain

$$100\,\frac{\text{m}}{\text{s}} = \nu \times 20\text{ m}$$

$$\nu = \frac{100}{20} = 5\,\frac{\text{vib}}{\text{s}}$$

or

$$\nu = 5\text{ Hz}$$

By Eq. (18b), we obtain

$$T = \frac{1}{5} = 0.2\text{ s}$$

The frequency is 5 Hz, and the period is 0.2 s.

QUESTIONS

1. Define or briefly explain (a) wave velocity, (b) wavelength, (c) period, and (d) frequency.

2. Define or briefly explain (a) displacement, (b) amplitude, (c) simple harmonic motion, and (d) standing waves.

3. Make a diagram illustrating transverse waves. Show the wavelength λ, and the amplitude r.

4. Write down the relation between wave velocity, wavelength, and frequency. Write down a relation between wave velocity, wavelength, and period.

PROBLEMS

1. A tuning fork with a frequency of 440 Hz sends out waves through the air. If the speed of sound waves is 1120 ft/s, what is the wavelength? [Ans. 2.545 ft.]

2. A whistle with a frequency of 880 Hz sends out waves with a wavelength of 40 cm. Find the speed of sound.

3. Transverse waves, traveling along a stretched rope with a speed of 25 m/s, have a wavelength of 0.75 m. Calculate the frequency of the source. [Ans. 33.33 Hz.]

4. One end of a coiled spring is suspended from a rigid support. When a 2-Kg mass is suspended from the lower end the spring is stretched a distance of 20 cm. The mass is then pulled down and released to set the mass vibrating up and down. Find (a) the spring constant, and (b) the frequency of vibration.

5. A coil spring suspended from the ceiling is stretched 30 cm when a 5 Kg mass is suspended from the lower end. Find (a) the spring constant, and (b) the frequency of the mass if it is set vibrating up and down. [Ans. (a) 163.3 N/m, (b) 0.910 Hz.]

6. A simple pendulum has a length of 2.48 m. Find (a) the period, and (b) the frequency.

7. A simple pendulum has a length of 8 cm. Find (a) the period, and (b) the frequency. [Ans. (a) 0.568 s, (b) 1.762 Hz.]

8. A pendulum for a grandfather clock is to have a period of exactly 1 s. What should be its exact length?

9. Standing transverse waves are produced on a long string. If the frequency of the source is 261 Hz and the nodes are 16.5 cm apart, what is the velocity of the waves? [Ans. 86.1 m/s.]

10. A 25 m long rope hangs from a bell tower. If the lower end is moved back and forth 5 times in one second, sending five complete waves up the rope, the waves reflect from the top and return to the lower end again in a total time of 4.0 s. What is (a) the wave velocity, and (b) the wavelength?

11. Longitudinal waves with a frequency of 365 Hz travel through a steel rod with a speed of 5,100 m/s. Find the wavelength. [Ans. 13.97 m.]

12. Radio waves travel through the air with a speed of 3×10^8 m/s. What is the wavelength of a station broadcasting on a frequency of 1500 K Hz?

13. A television station broadcasts on a frequency of 50 M Hz. If the speed of radio waves is 3×10^8 m/s, find the wavelength of the waves. [Ans. 6.0 m.]

14. Radar waves travel with the speed of light, 3×10^8 m/s. If a transmitter sends out waves with a length of 1.5 cm, what is the frequency of the source?

19 | SOUND

Although the subject of sound plays only a minor role in the development of atomic structure there are certain aspects of sound phenomena that definitely classify it as part of the subject we call physics. One of the most interesting parts of sound comes under the heading of the *science of music*. It is interesting to learn that *music* and *musical instruments* can be treated in a scientific manner. In fact they both involve a number of basic physical principles. Many students find sound to be the most interesting of all the subjects in physics.

19.1 SOUND TRANSMISSION The transmission of sound from one place to the other, from source to receiver, requires a material medium through which to travel. This is to be contrasted with light which travels best through a vacuum.

That sound is transmitted by air, or any other gas, may be demonstrated by placing a small bell in an evacuated jar. This is illustrated in Fig. 19A. As the air is slowly removed from the jar the ringing of the bell grows fainter and fainter until, when a good vacuum is obtained, no sound can be heard. As soon as the air is admitted, however, the ringing becomes clearly audible again. The vibrating bell strikes air molecules, knocking them away from the metal surface. These fast moving molecules strike the adjacent air molecules and they in turn strike others. Upon reaching the side of the jar, the glass walls are periodically bombarded by the molecules and set vibrating. The walls in turn set the outside air vibrating. Arriving at the observer's ear, the disturbance strikes the eardrum, setting it into motion. Without air to transmit the vibrations

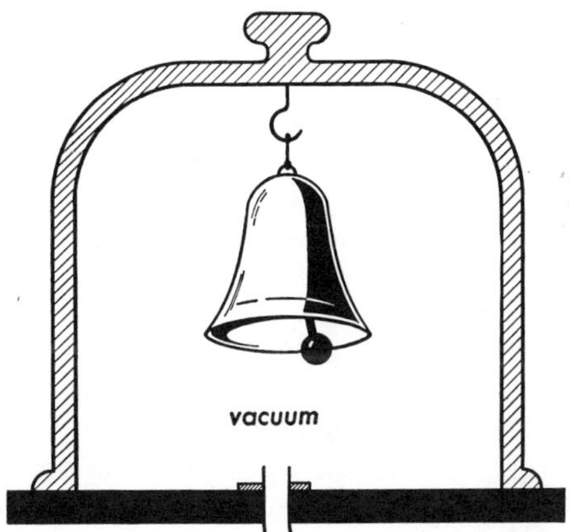

Fig. 19A. A bell ringing in a vacuum cannot be heard.

from the bell to the inside surface of the glass jar, no sound could ever leave the jar.

It should be made clear at this point that sound waves are longitudinal vibrations. The moving balls on the wave machine in Fig. 18J represent the molecular motions superposed upon the random thermal-motions of the air molecules.

The transmission of sound by liquids may be illustrated by an experiment shown in Fig. 19B(a). A tuning fork with a disk attached to its

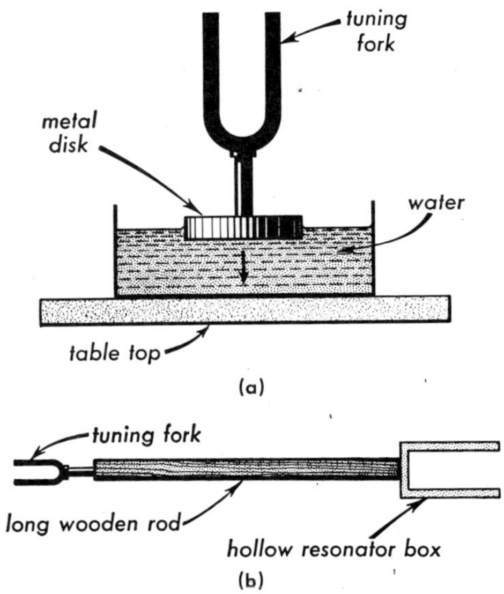

Fig. 19B. (a) Demonstration of sound waves traveling through water. (b) Demonstration of sound waves traveling through wood.

base is set vibrating and then touched to the surface of a dish of water. The vibrations of the fork and disk travel through the water to the bottom of the dish and to the table top. The table top itself is set into vibration with the same frequency as the fork, thus acting like a *sounding board* to make the sound louder.

The transmission of sound by solids is illustrated in diagram (b). A vibrating tuning fork is brought into contact with the end of a long wooden rod. The longitudinal vibrations travel down the length of the rod, setting the hollow wooden box at the other end vibrating. Sound is clearly heard coming from the box.

19.2 SPEED OF SOUND Although light and sound both travel with a finite velocity, the speed of light is so great in comparison that an instantaneous flash may be regarded as taking no time to travel many miles. When we see the light of a distant lightning flash and hear the thunder later, we know that the difference in time is due to the relatively low speed of sound. Knowing that sound requires 5 sec to travel 1 mi, the distance of a passing thunderstorm can be noted by the second hand of a watch. Similarly, when a distant train starts up and we watch for the first puff of smoke as it starts out, the arrival of the accompanying sound is not heard until an appreciable time afterward.

The earliest successful attempts to measure the speed of sound in air were made in 1640 by Marin Mersenne, a French physicist, and in 1656 by Giovanni Borelli and Vincenzo Viviani, Italian physicists. Since that time many experimenters have improved upon these earliest measurements by using various methods and devices. The most recent and probably the most accurate measurements are those made in 1934 by Miller.* With coast defense guns as a source of sound and a set of receivers located at certain distances apart, very accurate speed determinations were made. The results gave a speed of 331 m/s at a temperature of 0°C. This is equivalent to 1087 ft/s.

As a general rule, sound travels faster in solids and liquids than it does in gases. This is illustrated by the measured speed for a few common substances given in Table 19A.

---

* DAYTON C. MILLER (1866–1940), American physicist, noted for his experiments on the quality of musical sounds and on the ether drift. He collected and had in his possession the largest collection of flutes in the world. These instruments he turned over to the Smithsonian Institute in Washington, D. C., where they are now on exhibit. A member of the National Academy of Sciences, and one-time president of the American Physical Society, he has been awarded the Elliott Cresson Medal and the Cleveland Distinguished Service Medal.

TABLE 19A. Speed of Sound in Different Substances

Substance	Speed in m/s	Speed in ft/s
Air (at 0°C)......	331	1087
Hydrogen........	1269	4165
Water............	1435	4708
Alcohol..........	1213	3890
Iron..............	5130	16,820
Glass............	5000	16,410

It is well known that the temperature has a small but measurable effect upon the speed of sound. For each degree centigrade rise in temperature the speed in air increases by 61 cm/s. Written as an equation,

$$V = V_0 + 0.61t \tag{19a}$$

where V_0 is the speed in meters per second at 0°C, and t is the temperature of °C.

EXAMPLE 1. Calculate the speed of sound in air when the temperature is 34°C.

SOLUTION. Using Eq. (19a), and the value for the speed of sound in air at 0°C, we obtain

$$V = 331\,\frac{\text{m}}{\text{s}} + 0.61\,\frac{\text{m}}{\text{s °C}} \times 34\text{°C}$$

$$V = 331\,\frac{\text{m}}{\text{s}} + 21.74\,\frac{\text{m}}{\text{s}}$$

$$V = 352.7\,\frac{\text{m}}{\text{s}}$$

For each Fahrenheit degree rise in temperature the speed in air increases by 1.1 ft/s. If the speed V_0 is in ft/s at 32°F, and t is the temperature change in °F from 32°, the speed V is given by

$$V = V_0 + 1.1t \tag{19b}$$

A speed of 1087 ft/s is equivalent to 741 mi/hr. High in the stratosphere where the daytime temperature reaches 200°F, the speed of sound increases 185 ft/s. There a speed of 1272 ft/s is equivalent to 867 mi/hr.

19.3 PITCH The pitch of a musical note refers to its position on a musical scale and is determined principally by the frequency of the sound impulses sent out by the vibrating source. The dependence of pitch upon frequency can be demonstrated in many ways. Diagram (a) of Fig. 19C represents a toothed wheel (called Savart's wheel) rotating at high speed. A small card held against the teeth is set into vibration, giving out a musical note. As the wheel slows down, the vibration frequency of the card decreases and the note lowers in pitch.

Diagram (b) is a siren similar to those commonly used as factory whistles. The energy as well as the sound is obtained from compressed air which is blown through small holes in the hollow container C. The air issuing from these jets passes through similar holes in a rotating disk W. With the holes drilled at an angle as

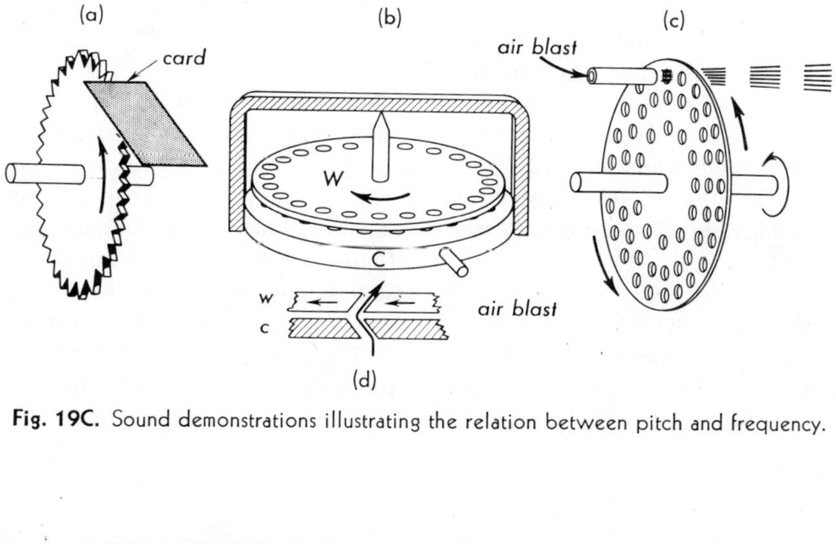

Fig. 19C. Sound demonstrations illustrating the relation between pitch and frequency.

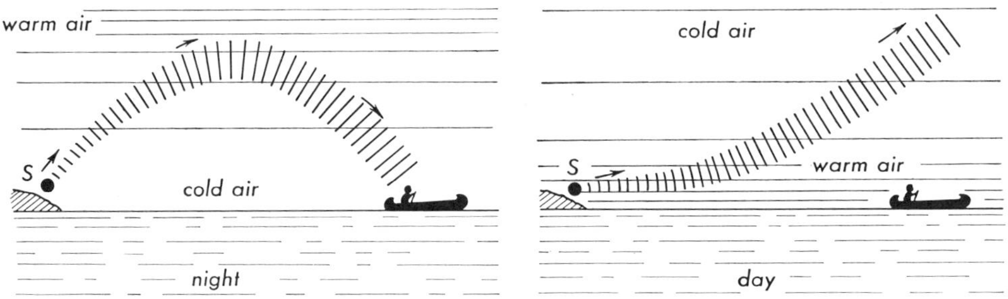

Fig. 19D. The higher speed of sound waves in warm air causes sound waves to be refracted down at night and up in the daylight.

shown in detail (d), the air blasts from the stationary holes below exert a force on one side of the holes in *W*, causing the disk to turn. As the disk turns, each blast of air is momentarily cut off until another hole arrives just above it. The intermittent pulses of air issuing from the disk give rise to a musical note.

Diagram (c) in Fig. 19C represents a siren in which a single blast of air is interrupted by a rotating disk containing several rings of holes. When the air is blown through one ring of holes, the air pulses emerging from the opposite side will produce a note whose frequency depends on the number of holes in the ring and the speed of the wheel. Since each ring may contain a different number of holes, a musical scale can be arranged and played by putting the air nozzle in front of the proper rings in succession.

Detailed experiments with many trained and untrained observers alike show that pitch and frequency are not identical. Pitch is a *subjective* measurement and is therefore a *sensory magnitude* depending on the individual, while frequency is a *physical measurement* of the number of Hz. As the intensity of a pure tone of 300 Hz is increased, for example, it appears to most observers to change in quality and at the same time to decrease slightly in pitch. Conversely, at a high frequency an increase in the intensity seems to increase the pitch.

19.4 REFRACTION OF SOUND WAVES The bending of sound waves in layers of air at different temperatures is called refraction. The phenomenon can be observed in various ways and is due to the greater speed of sound in warm air compared to the speed in cold air (see Sec. 19.2).

A good illustration is found in the frequent observation while boating on a lake or river of being able to hear the music from a quite distant radio or phonograph at night but not in the daytime. The reason for this is shown in Fig. 19D. At night the air near the water is colder than it is higher up, and the higher speed in the warmer air bends the waves back down. During the day the air close to the water is warmer and the waves bend up away from the water as shown.

Recent experiments of this kind have been performed with the very loud sounds from big guns. Sound waves refracted back from high up in the stratosphere indicate with some degree of certainty the existence of very warm layers of air at altitudes of 25 to 40 mi. Refraction in such cases as this is similar to reflection from a mirror surface since the waves travel in more or less straight lines going up and back but bend over when they enter, more or less abruptly, a warmer layer.

19.5 THE HUMAN EAR The ear is by far the most important and most universal receiver of sound. It has an enormous range of frequency and sensitivity, and can distinguish between musical tones whose frequencies differ by less than 1%. In addition to this it can analyze some sounds into their component notes and concentrate on these notes one at a time.

The process of hearing has long been a subject for discussion in many branches of science. Although most of the experts agree on the general

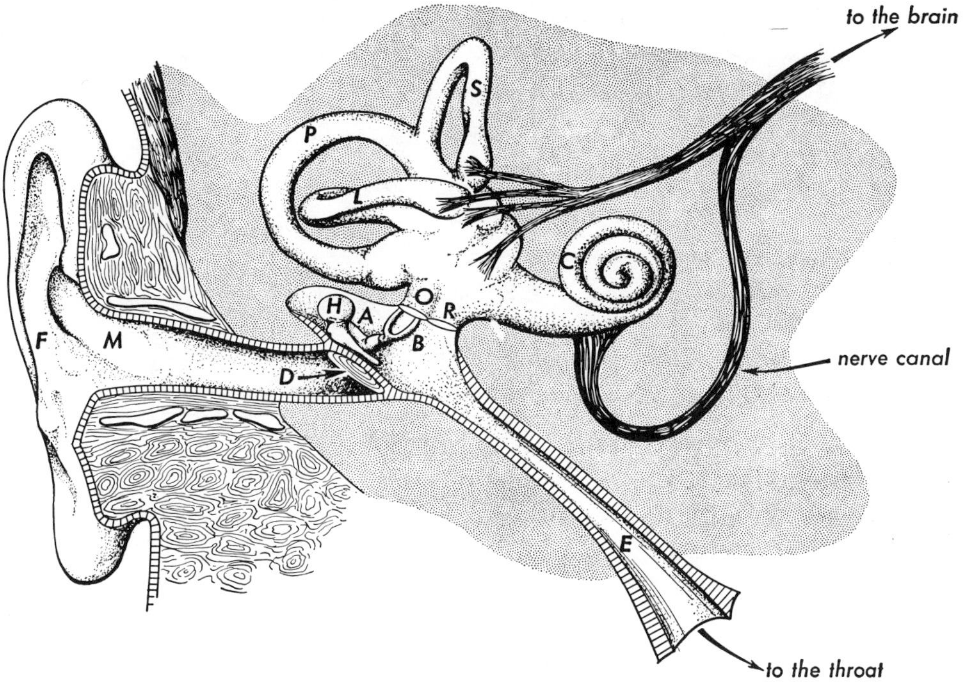

Fig. 19E. Cross section of the human ear.

structure and the mechanical motions within the ear, there still exist several controversies regarding the functions of certain parts. The entire hearing mechanism is divided into three parts: the *outer ear*, the *middle ear*, and the *inner ear* (see Fig. 19E).

The outer ear consists of the pinna *F*, used to collect the sound waves from the outside, and the ear canal *M* to carry the waves to the eardrum *D*. The middle ear contains three small bones *H*, *A*, and *B*, called the *hammer* (malleus), *anvil* (incus), and *stirrup* (stapes) respectively, and is connected to the nasopharynx and thus to the outside air by means of a small canal, the *Eustachian tube E*. The function of the three bones is to transmit the vibrations of the eardrum to the *oval window* of the inner ear. The inner ear itself consists of two essential parts: the *cochlea C*, and the semicircular canals *P*, *L*, and *S*. In the cochlea are found the nerve endings which are stimulated by sound vibrations and give rise to the sense of hearing, and in the semicircular canals are the nerve endings which give rise to a sense of balance.

The entire inner ear is contained within the cavity of a solid bony structure that is sometimes referred to as the *bony labyrinth*. This labyrinth is entirely filled with a watery liquid through which the sound vibrations from the outside are transmitted to the sensitive membranes of the cochlea. The cochlea consists of two and one-half turns of a spiral cavity shaped like a snail shell and divided lengthwise into three parts by what are called the *spiral lamina* and *Reissner's membrane*. Cross sections of the cochlea are shown in Fig. 19F. Diagram (a) represents a section directly across one turn of the spiral, and diagram (b) a lengthwise cross section as it would appear if the cochlea could be straightened out.

Throughout the total length of the basilar membrane, which is just a little over 3 cm in length, there are about 30,000 nerve endings. This amounts to 1000 nerves per millimeter length which must pass through the bony spiral

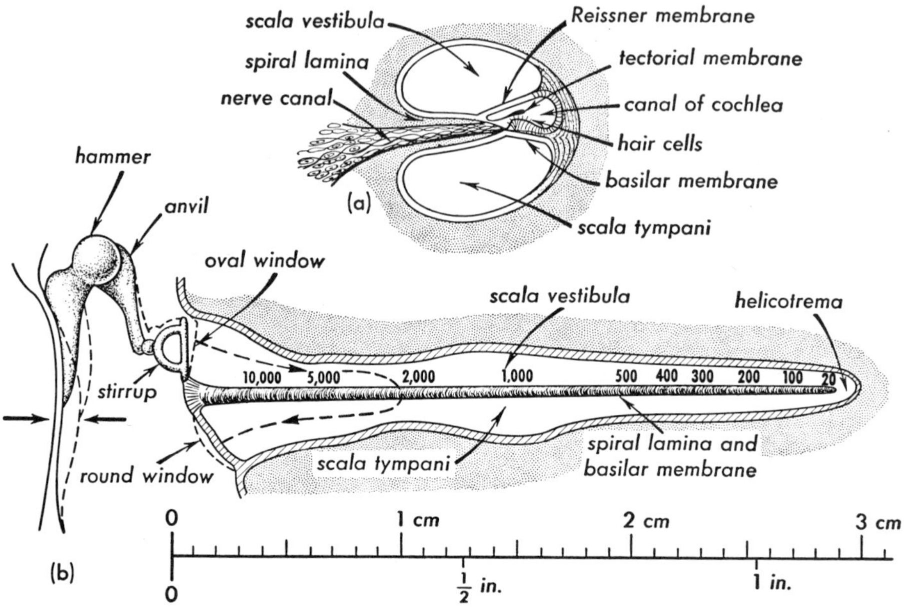

Fig. 19F. Detail of the cochlea of the human ear: (a) typical cross section and (b) straightened cochlea illustrating the various regions of the spiral lamina sensitive to sounds of various frequencies.

lamina and into the cochlear nerve canal leading to the brain. The work of many experimentalists shows that the nerve endings nearest the oval window, where the sound vibrations enter the *scala vestibula*, respond to the notes of highest pitch, whereas those at the farther end respond to those of lowest pitch. The various regions sympathetic to the entire frequency scale are shown in the diagram.

The eardrum and the bones of the middle ear act as a lever mechanism to decrease the amplitude of the vibrations from the air, a very light medium, to the liquid, a much more dense medium. This reduction in the motion gives rise to a pressure on the stirrup which is from 30 to 50 times that exerted on the eardrum. As the stirrup tips in and out with a low frequency, the entire liquid column from the oval window down the *scala vestibula* to the *helicotrema*, and back along the *scala tympani* to the round window, is set into vibration. Since liquids are practically noncompressible, the round window moves out when the oval window and stirrup move in, and vice versa.

When a higher frequency like 2000 Hz is sounded, the vibrations in the liquid, set up by the motion of the stirrup at the oval window, travel the path shown by the dotted line in Fig. 19F, diagram (b). As the waves travel through the thin Reissner membrane and across the edge of the spiral lamina, there is a relative motion set up between the *basilar membrane* and the *tectorial membrane*, which causes the local hair cells to stimulate the nerve endings at their base. Somewhere in this stimulation and motion, part of the energy is transformed into electrical impulses which travel along the cochlear nerve canal to the brain.

19.6 RESONANCE, OR SYMPATHETIC VIBRATIONS If two violin strings are tuned to the same frequency and one is set vibrating, the other stationed some distance away will soon pick up the vibrations and give out the same note. This is a case of *resonance*, a phenomenon which occurs only if two objects have the same natural frequency of vibration.

An experimental demonstration of resonance is illustrated in Fig. 19G. Two tuning forks with

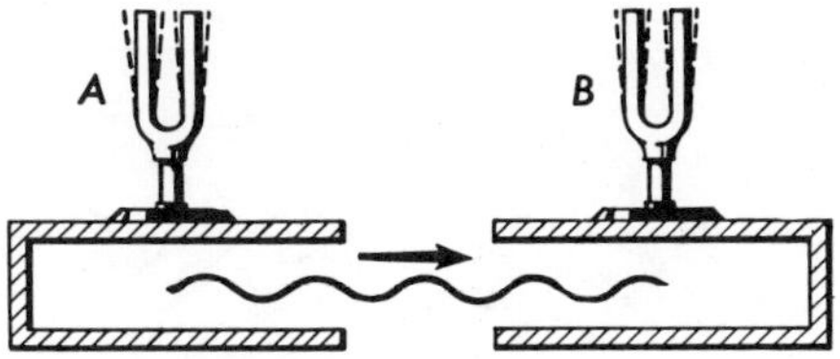

Fig. 19G. Tuning forks mounted on resonator boxes for demonstrating resonance.

exactly the same pitch are mounted on separate hollow boxes as shown. Fork A is first set vibrating for a moment and then stopped by touching the prongs with the fingers. Fork B will then be found vibrating. Taking into account the hollow boxes, whose purpose it is to act as sounding boards and intensify the sound, the explanation is quite simple. Each sound pulse emerging from the box with each vibration of fork A passes into the other box, pushing out the sides at just the right time to make the prongs of fork B move in the same direction as with the previous pulse.

By singing a sustained note directly at the strings of a piano, and then listening in all possible quietness, one of the strings can be heard faintly sounding forth with sympathetic vibrations of the same frequency.

Occasionally it has been shown in motion pictures that by singing a sustained note into a drinking glass or vase, sympathetic vibrations will cause it to break or explode. This is a hoax and has never been demonstrated in any laboratory. Sound waves contain too little energy to do this.

19.7 BEAT NOTES When two notes of slightly different pitch are sounded together, beats are heard. This phenomenon is used in organ pipes to produce the familiar *vibrato* effect. Two pipes tuned to slightly different frequencies are used for every note.

The phenomenon of beats may be demonstrated by two tuning forks mounted as shown in Fig. 19G. One fork is made slightly out of tune with the other by looping rubber bands *tightly* around the prongs. If the two forks are sounded directly, the intensity of loudness of the sound rises and falls periodically. This is illustrated by means of vibration graphs as shown in Fig. 19H. The upper curve represents the sound vibrations arriving at the ear from one fork and the second curve the vibrations from the other. Both waves arriving at the ear are first in phase, i.e., in step with each other, then out of phase, then in phase, then out of phase, etc.

The resultant action of these two waves on the eardrum is represented by the third graph. When the waves are in phase the resultant has a large amplitude equal to the sum of the amplitudes of the two. When they are out of phase the amplitude becomes zero. The number of beats per second, N, is determined by the difference between ν_2 and ν_1, the respective frequencies of the two sources producing the sound.

$$\text{Beat frequency } N = \nu_2 - \nu_1 \qquad (19c)$$

When the beat frequency lies between about 1 and 6 Hz, the ear perceives an intertone halfway between the two sounded but periodically waxing and waning in intensity. As the beat fre-

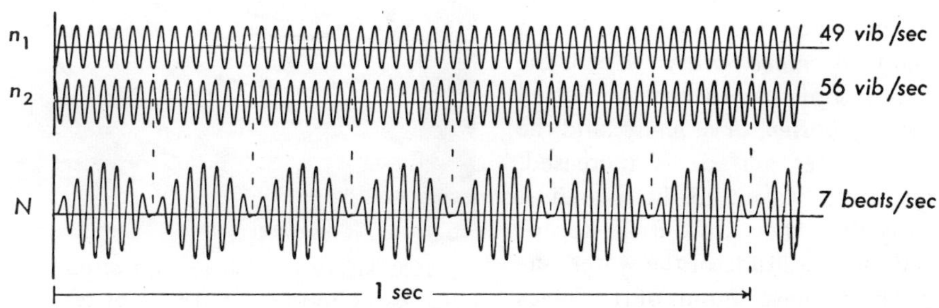

Fig. 19H. Wave graphs illustrating how beat notes are produced by two different frequencies.

quency increases, the smooth rise and fall gives way to a succession of pulses, then to a sensation of roughness, and finally to two clearly perceived tones.

19.8 INTERFERENCE OF SOUND WAVES Everyone has at one time or another dropped a stone in a pond of still water and watched the waves spread outward in ever-widening circles. Such waves are represented by concentric circles as shown in Fig. 19I. The solid lines of the circles

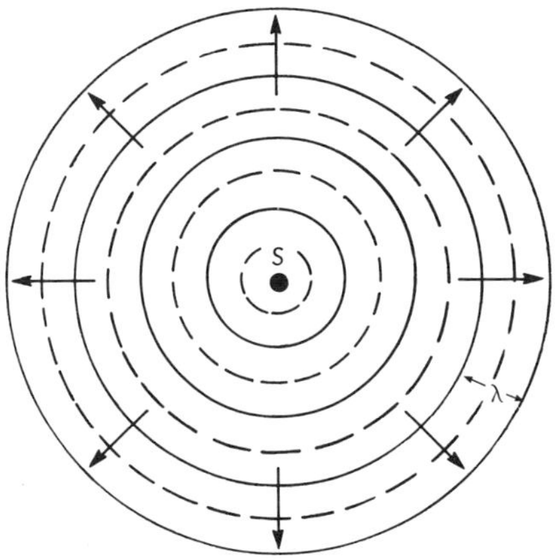

Fig. 19I. Concentric waves traveling outward from a single source *S*.

represent crests of waves and are therefore one wavelength apart, while the dotted lines represent the troughs of the waves and are also one wavelength apart.

If two stones are dropped simultaneously into the water, two sets of waves will spread outward as shown in Fig. 19J. As these waves cross each other, they act one upon the other producing what is called an *interference pattern*. Where the crests of two waves come together at the dotted intersectors, they are *in step*, or *in phase*, and the amplitude of the water surface is increased. Where the crest of one wave and the trough of another come together, they are out of step or *out of phase*, and the amplitude of the water surface is reduced. The in phase regions of the waves move outward along the dotted lines, such as *x*,

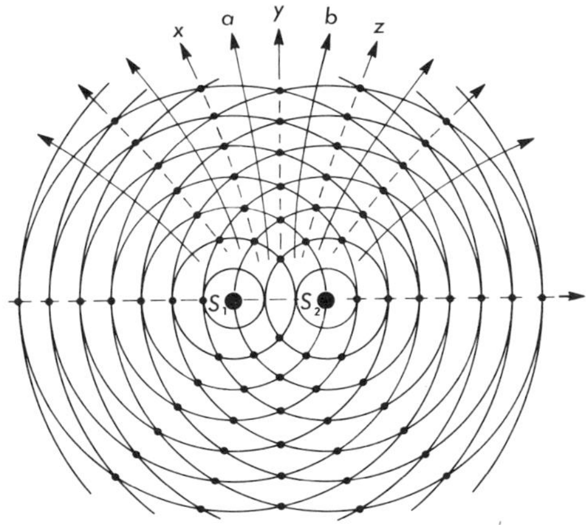

Fig. 19J. Concentric waves traveling outward from a double source, producing what is called an interference pattern.

y, and *z*, and we have what is called *constructive interference*. The out of phase regions move outward along the solid lines, such as *a* and *b*, and

Fig. 19K. Ripple tank photograph of the interference of water waves from two sources. (Courtesy of the Physical Sciences Study Committee Project.)

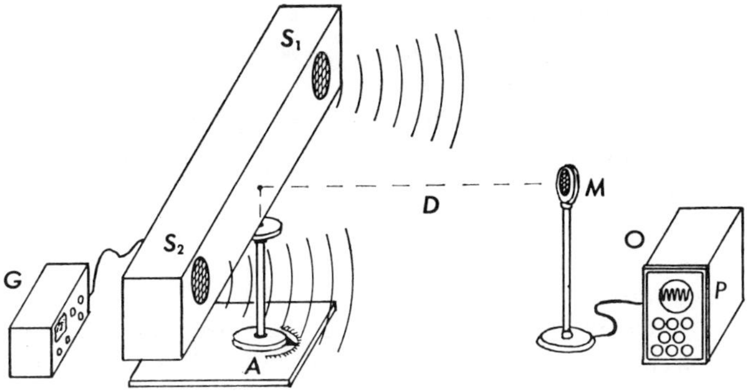

Fig. 19L. Experiment for demonstrating the interference of sound waves from two sources.

we have what is called *destructive interference.*

An instantaneous photograph of such a wave pattern is shown in Fig. 19K. Note how clearly the interference regions of the waves stand out. Photographs of this kind, as well as direct observations of such wave patterns, are readily made as follows. A glass tray for maintaining a shallow water layer can be made from a piece of window glass and a wooden frame. A thin metal strip, clamped at one end and set vibrating up and down over the water, is used as a source. A piece of wire, fastened to the vibrating end of this strip, should have one wire end dipping into the water for a single wave source, and both ends dipping into the water for a double source. Intermittent viewing of the waves through a slotted disk, or illumination by means of a stroboscopic light source, enables one to make the wave pattern stand still, or to progress in *slow motion.*

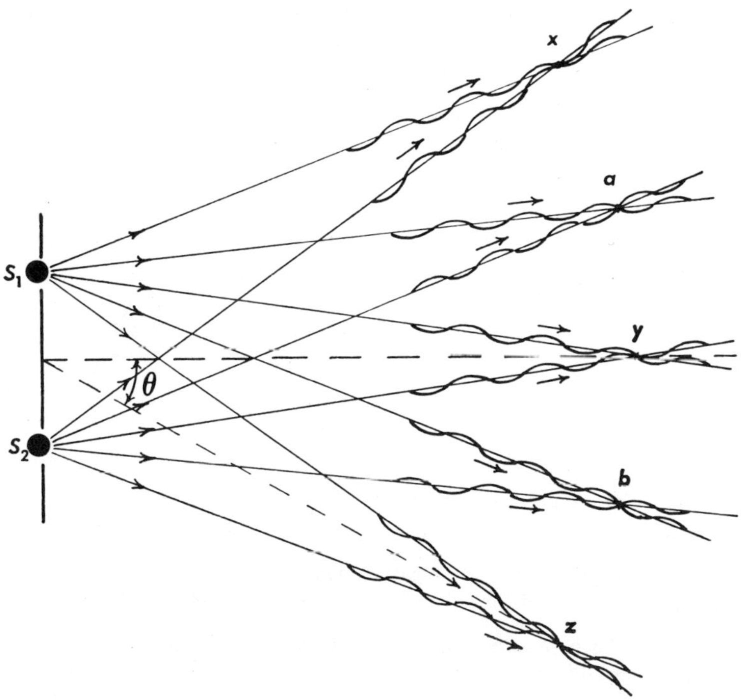

Fig. 19M. Diagram showing the in phase and out of phase of waves at certain points.

An experiment for demonstrating the interference of sound waves is shown in Fig. 19L. Two small radio loudspeakers S_1 and S_2 are mounted, about 150 cm apart, in a hollow box. The box is free to pivot about a vertical axis midway between the speaker centers. When both speakers are connected to an electronic sound generator, G, the speakers vibrate in phase and send out identical waves into the surrounding air. A microphone, M, about two meters away, is an excellent detector, and an oscilloscope connected to it will indicate the relative amplitudes of the resultant sound waves arriving at M. As the box is turned slowly in one direction, the amplitude of the oscilloscope signal P rises and falls periodically indicating constructive and destructive interference. As it is turned slowly back again, maxima and minima are again observed at the same angle positions.

An instructive diagram showing the arrival of pairs of waves at the microphone, for maximum and minimum response positions, is shown in Fig. 19M. Points x, y, and z correspond to points x, y, and z in Fig. 19J, where the waves arrive in phase, while points a and b correspond to those where they arrive out of phase.

By measuring the angles at which maximum response occurs for the experiment shown in Fig. 19L, as well as the distance between S_1 and S_2 and the distance D, the wavelength of the sound waves can be calculated. The geometry for this calculation is shown in Fig. 19N.

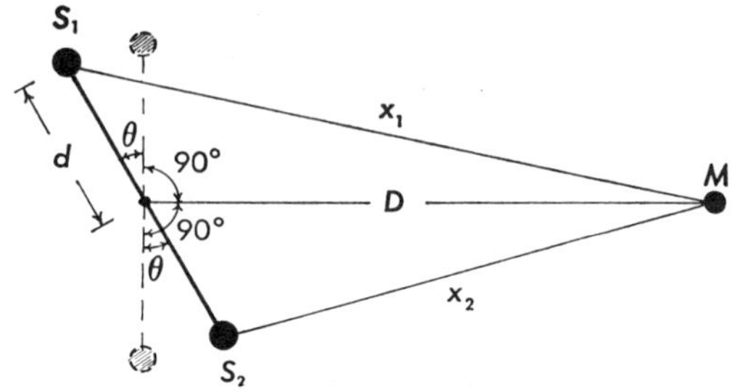

Fig. 19N. Geometrical relations for finding the wavelength of sound waves from a double source.

When the line between the two speakers makes an angle of 90° with the perpendicular bisector D, the sources are equidistant from M, the waves arrive in phase, and we have point y of Figs. 19J and 19M. When the sources are turned to position z, the line between the speakers has been turned through an angle θ, the waves arrive in phase again, but the path x_1 is exactly one wavelength longer than the path x_2. At some greater angle θ, the path x_1 can be made exactly two wavelengths longer than x_2, and at some still greater angle it can be made three wavelengths longer, etc. For all these particular angles where the waves are in phase, we can write the single equation

$$x_2 - x_1 = n\lambda \tag{19d}$$

where $n = 0, 1, 2, 3, \ldots$

By measuring distances x_2 and x_1 for the different in-phase positions, the wavelength of the sound waves can be calculated from Eq. (19d).

QUESTIONS

1. Define or briefly explain each of the following: (a) pitch, (b) beats, (c) resonance, (d) interference, (e) refraction, and (f) sympathetic vibrations.

2. Describe an experiment that might be performed with two violins to demonstrate resonance.

3. Why would you expect the temperature of a gas to have any effect on the speed of sound? Explain.

4. When listening to stereo music where should one sit with respect to the two speakers to hear all sounds in phase?

5. Do you think that by singing a sustained note directly at the strings in a piano you could get any one of the strings vibrating sufficiently to be heard at all?

6. Do you think that by singing a sustained note into a glass vessel it could be made to vibrate sufficiently to break the glass?

PROBLEMS

1. Find the speed of sound in air when the temperature is 27°C. [Ans. 347.5 m/s.]

2. What is the speed of sound in air if the temperature is 75°F?

3. If the speed of sound is measured to be 350 m/s what is the temperature? [Ans. 31.1°C.]

4. If the speed of sound in air is measured to be 1112 ft/s, what is the temperature?

5. Fireworks in a city park are observed from a hilltop 12 mi away. When a skyrocket explodes high in the air, how long after the flash of light is first seen will the sound be heard? Assume the light is seen the instant the skyrocket bursts, and the air temperature is 75°F. [Ans. 55.9 s.]

6. How long after the first puff of smoke from a locomotive, just starting up, will the sound be heard by an observer on a hilltop 2.5 mi away? Assume a temperature of 65°F.

7. If a warship 20 Km at sea fires its big guns, how long will it take for the sound to be heard ashore? Assume the temperature to be 27°F. [Ans. 60.8 s.]

8. Three strings of a piano are sounded as a chord. Find the frequencies of all the beat notes produced if their frequencies are 528, 704, and 880 Hz, respectively.

9. Four musical notes are sounded simultaneously as a chord. If their frequencies are 264, 352, 440, and 528 Hz, respectively, (a) find all the beat note frequencies produced; and (b) find the smallest whole numbers which multiplied by a common factor will give the beat frequencies and the sounded frequencies. [Ans. (a) 88, 176, and 264 Hz, (b) 1, 2, 3, 4, 5, and 6, times the factor 88 Hz.]

10. A military scout observes an enemy gun in the act of firing. Observing his watch he determines the time interval between the flash of light and the sound to be 8 s. How many yards away should he report the gun's position if the temperature is 65°F?

11. A man is observed to be driving spikes into the wooden ties as workmen are laying a railroad track. For an observer at a distance of 1 mile, how long will it take for the sound to arrive after each blow is seen to strike? Assume the temperature to be 82°F. [Ans. 4.62 s.]

12. Krakatoa, an active volcano in Sumatra, erupted violently in 1883. The sound waves from this event were detected around the world. If the average air temperature were 50°F, how long did the sound take to circle the earth?

20 | MUSICAL INSTRUMENTS

Musical instruments are classified under one of the following heads: *strings*, *winds*, and *percussions*. One who is familiar with music will realize that these terms apply to the principal vibrating elements of instruments responsible for the particular sounds they emit. It is the purpose of this chapter to consider these vibrating sources and the various factors governing the frequencies of their musical notes.

20.1 STRINGED INSTRUMENTS There are two principal reasons why stringed instruments of different kinds do not sound alike as regards to *tone quality*—first, the size and shape of the instrument, and second, the method by which the strings are set into vibration. The violin and cello are bowed with long strands of tightly stretched horsehair, the harp and guitar are plucked with the fingers or picks, and the piano is hammered with light felt mallets.

Under very special conditions a string may be made to vibrate with nodes at either end as shown in Fig. 20A. In this state of motion the string gives rise to its lowest possible note, and it is said to be vibrating with its *fundamental frequency*.

Every musician knows that a thick heavy string has a lower natural pitch than a thin one, that a short strong string has a higher pitch than a long one, and that the tighter a string is stretched the higher is its pitch. The *G* string of a violin, for example, is thicker and heavier than the high-pitched *E* string, and the *bass* strings of the piano are longer and heavier than the strings of the *treble*.

20.2 HARMONICS AND OVERTONES When a professional violinist plays "in harmonics," he touches the strings lightly at various points and sets each one vibrating in two or more segments as shown in Fig. 20B. If a string is touched at the center, a node is formed at that point and the vibration frequency becomes double that of the fundamental. If the string is touched lightly at a point one third the distance from the end, it will vibrate in three sections and have a frequency three times that of the fundamental.

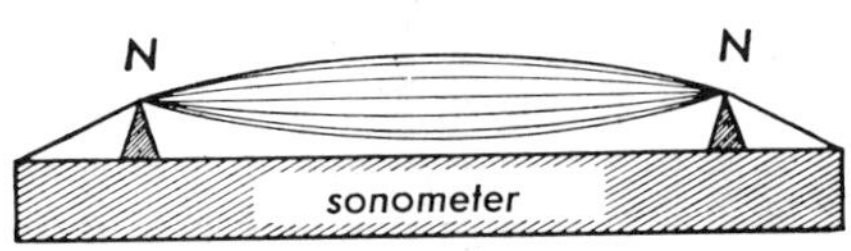

Fig. 20A. Single string vibrating with its fundamental frequency.

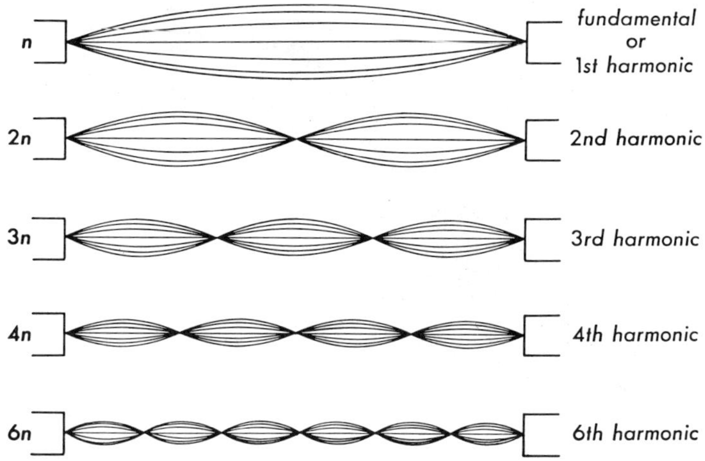

Fig. 20B. Vibration modes for strings of musical instruments.

In the elementary theory of string instruments it is assumed that each string is thin, uniform, and highly flexible, and that it vibrates with small amplitude between unyielding supports. For such an ideal string the above vibration forms, called *natural modes*, have frequencies exactly equal to whole number multiples of the fundamental frequency ν, and are called *harmonics*. To show how closely a real string meets these ideal conditions the measured frequencies of a piano string whose fundamental frequency is 32.70 Hz are given in the following table.*

TABLE 20A. Harmonic Frequencies of a Piano String

Mode Number	1	2	3	4	5	6	7	8
Meas. Freq........	32.70	65.52	98.39	131.4	164.7	198.4	232.4	266.8
Harmonic Freq.....	32.70	65.40	98.10	130.8	163.5	196.2	228.9	261.6
Ratio............	1.000	2.003	3.008	4.018	5.038	6.066	7.106	8.159

It is not difficult to set a string vibrating with its fundamental and several of its higher modes at the same time. This is accomplished by plucking or bowing the string vigorously. As an illustration, a diagram of an ideal string vibrating with two normal modes at the same time is shown in Fig. 20C. As the string vibrates in two loops with a frequency 2ν, it also moves up and down as a single loop with the fundamental frequency ν.

Fig. 20C. String vibrating with its fundamental and first overtone simultaneously.

The sound wave sent out by such a vibrating string is composed of two frequencies, the fundamental or first harmonic of frequency ν, and the second harmonic or first partial with the frequency 2ν.

An interesting experiment with a vibrating string is diagrammed in Fig. 20D. Light from an arc lamp is focused on the central section of a stretched steel string, which, except for a small vertical slot, is masked by a screen. An image of the slot and the string section seen through it is

* See R. W. Young, *American Journal of Physics*, Vol. 20, p. 177, 1952.

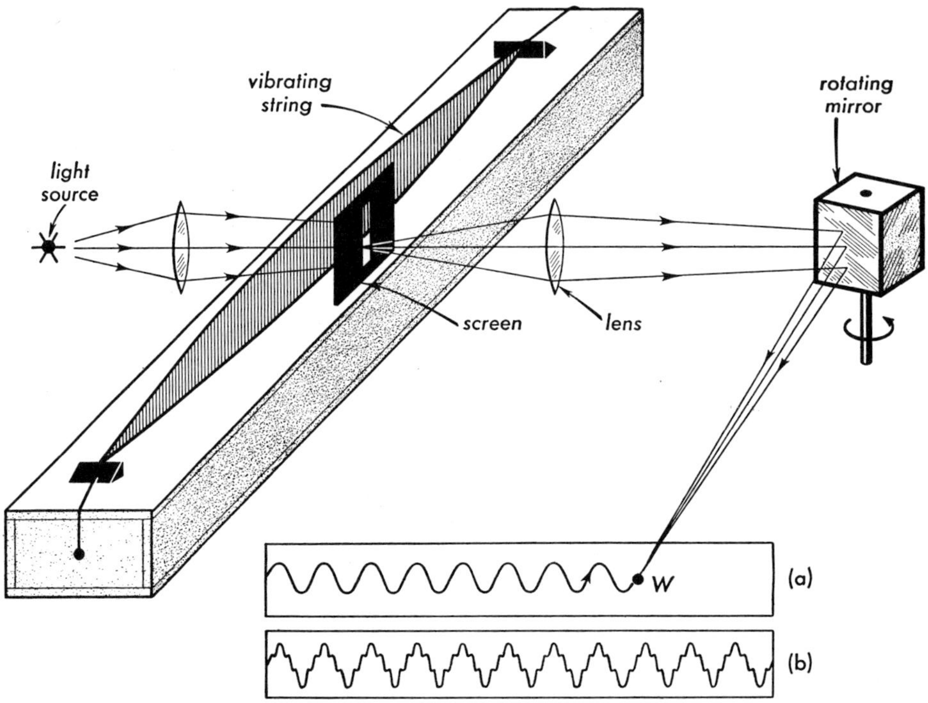

Fig. 20D. Experiment for observing the detailed vibrations of stretched string.

focused on a screen by a second lens, after reflection from a rotating mirror. As the string vibrates up and down only a blurred image of the short section of string is seen, but when the mirror is rotated the wire section draws out a clearly visible curve *W*.

If the string is plucked gently near the center, a smooth wave form (a) is drawn out on the screen; but if it is plucked hard near the end to produce a harsh sounding note, the wave form is more complex as shown in (b). In the first case the string is vibrating only with its fundamental mode, whereas in the second case various overtones, or partials, are also present.

As a string vibrates with *transverse waves* it strikes air molecules all around it, sending periodic impulses through the air as *longitudinal waves.*

20.3 THE THEORY OF VIBRATING STRINGS. A string set into vibration with nodes and loops is but an example of standing waves (see Fig. 18M). A disturbance produced at one end, as in the plucking or bowing of a string, sends a wave or train of waves along the string to be reflected back and forth from end to end. When traveling in one direction the wave is on top and, upon reflection, it flips to the under side. Because the wave is always on top when moving to the right, and on the bottom when moving to the left, a flipping up and down of the string results.

Since the fundamental frequency of vibration is equal to the number of times per second that the wave arrives at the same end, the pitch will depend upon the velocity of the waves and the distance they have to travel.

The velocity of transverse waves along a rope or string under tension is given by

$$V = \sqrt{F/m} \qquad (20a)$$

where F is the tension and m is the mass per unit length of string. When standing waves are produced, the distance L between any two consecutive nodes is just equal to half a wavelength, $\frac{1}{2}\lambda$. Accordingly,

$$\lambda = 2L \qquad (20b)$$

To obtain an equation for the fundamental frequency of a vibrating string, the general wave equation $V = \nu\lambda$ is used [see Eq. (22g)]. If we solve this equation for ν, and then substitute the above values for V and λ, we obtain

$$\boxed{\nu = \frac{1}{2L}\sqrt{F/m}} \tag{20c}$$

Accurate measurements with vibrating strings and musical instruments confirm this equation.

EXAMPLE 1. A piano string that sounds three octaves below middle C is 1.10 m long. Calculate the tension on the string when, in proper tune, its frequency is 33 Hz. Assume the mass of the string to be 160 g.

SOLUTION. The *mass per unit length* of string is

$$m = \frac{0.16 \text{ Kg}}{1.1 \text{ m}} = 0.1455 \frac{\text{Kg}}{\text{m}}$$

Solve Eq. (20c) for F, and the frequency equation becomes

$$F = 4\nu^2L^2m$$

By direct substitution,

$$F = 4 \times (33)^2 \times (1.10)^2 \times 0.1455$$

$$F = 767 \text{ N}$$

Since 1 N = 0.225 lb [see Eq. (3d)],

$$F = 172.6 \text{ lb}$$

The vibration of a string brought about by bowing is associated with friction. Because starting friction is greater than sliding friction, the resinous bow periodically engages the string, pushing or pulling it to one side. During one-half the vibration the string clings to the bow and is carried along by it. When the tension becomes too great and exceeds starting friction, the string slips back to the opposite side of its vibration. There it stops and upon reversal in direction is grabbed again by the bow.

20.4 WIND INSTRUMENTS Musical instruments often classified as "wind instruments" are usually divided into two subclasses, "wood winds" and "brasses." Under the heading of wood winds we find such instruments as the *flute*, *piccolo*, *clarinet*, *bass clarinet*, *saxophone*, *bassoon*, and *contra bassoon*, and under the brasses such instruments as the *French horn*, *cornet*, *trumpet*, *tenor trombone*, *bass trombone*, and *tuba* (or *bombardon*).

The fundamental principles involved in the vibration of an air column are demonstrated by means of an experiment shown in Fig. 20E. A

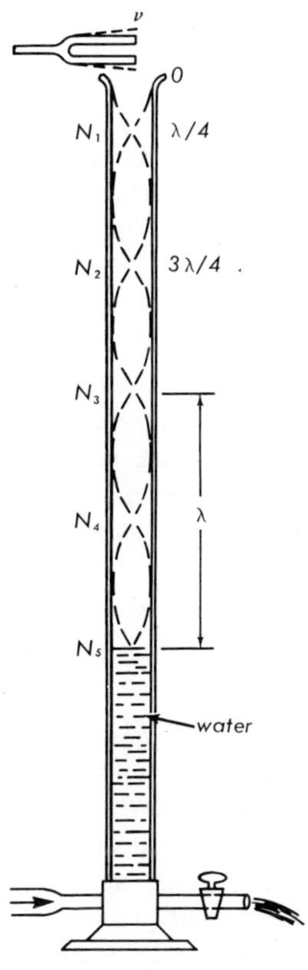

Fig. 20E. Sound waves from tuning fork set up standing waves in air column adjusted to proper length.

vibrating tuning fork acting as a source of sound waves is held over the open end of a long hollow tube containing water. Traveling down the tube with the velocity of sound in air, each train of sound waves is reflected from the water surface back toward the top. If the water is raised or lowered to the proper level, standing waves will be set up and the air column will resonate to the frequency of the tuning fork.

The first resonance occurs at N_1 when the water level is but a short distance from the top. The second resonance occurs at N_2, three times the distance of N_1 below the top, and the third N_3 at five times the distance, etc. The reason for these odd fractions is that only a *node* can form at the closed end of the pipe, i.e., at the water surface, and an *anti-node* at the open end.

Standing waves in air are longitudinal in character, and they are difficult to represent in any drawing (see Fig. 18M). For convenience only it is customary to indicate the positions of nodes and loops as if they were transverse standing waves (dotted lines in the diagram). If the frequency of the tuning fork used in the above experiment is known, the velocity of sound in air can be calculated. The distance between consecutive nodes is equal to $\lambda/2$, so that λ is equal to the length of 2 segments $N_3 - N_1$, $N_4 - N_2$, or $N_5 - N_3$ as shown in the diagram. In an actual experiment when the temperature is 27°C, and a tuning fork sounding 512 Hz, the nodes are 13.37 in. apart. Substitution of these values in the general wave equation, Eq. (18c) gives

$$V = \nu\lambda = 512 \times 26.74 = 13{,}690 \text{ in./s}$$

Dividing by 12 gives $V - 1141$ ft/s

20.5 VIBRATING AIR COLUMNS The various modes in which air columns may vibrate in open or closed pipes are shown in Fig. 20F. Starting at the top a pipe open at both ends may vibrate with (1) a single node at the middle and an anti-node at both ends, (2) two nodes and three anti-nodes, or (3) with three nodes and four anti-nodes, etc. On the other hand, a pipe closed at one end and open at the other may vibrate with (1) one node and one anti-node, (2) two nodes and two anti-nodes, or (3) three nodes and three anti-nodes, etc. *In all vibrating air columns an anti-node always forms at an open end and a node at a closed end.*

The various possible frequencies to which a pipe may resonate are definite and fixed in value and depend only upon the length of the pipe and

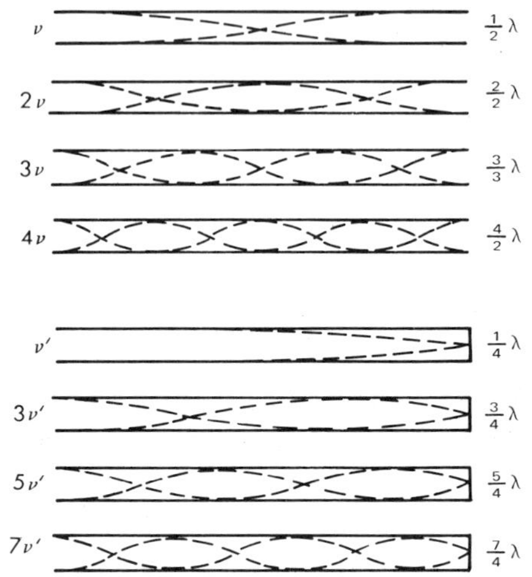

Fig. 20F. Resonating air columns, showing nodes and loops.

the velocity of sound in air. If, for example, the pipes in Fig. 20F are all 2 ft long and the velocity of sound in air is 1120 ft/s, Eq. (18e), $V = \nu\lambda$, shows that they will vibrate with the following respective frequencies,

ν	2ν	3ν	ν'	$3\nu'$	$5\nu'$
280	560	840	140	420	700

With an open pipe the lowest possible vibration frequency is called the *fundamental*, and the others with whole numbered multiples of the fundamental frequency, 2ν, 3ν, 4ν, etc., are called *harmonics*. With closed pipes the lowest frequency is again the fundamental and the others with odd integral multiples, $3\nu'$, $5\nu'$, $7\nu'$, etc., are harmonics. All these vibration modes are referred to as natural modes and their corresponding frequencies as natural frequencies. *The fundamental is also called the first harmonic.*

The existence of standing waves in a resonating air column may be demonstrated by a long hollow tube filled with illuminating gas as shown in Fig. 20G. Entering through an adjustable plunger at the left the gas escapes through tiny holes spaced at regular intervals in a row along the top. Sound waves from an organ pipe enter the gas column by setting into vibration a thin paper sheet stretched over the

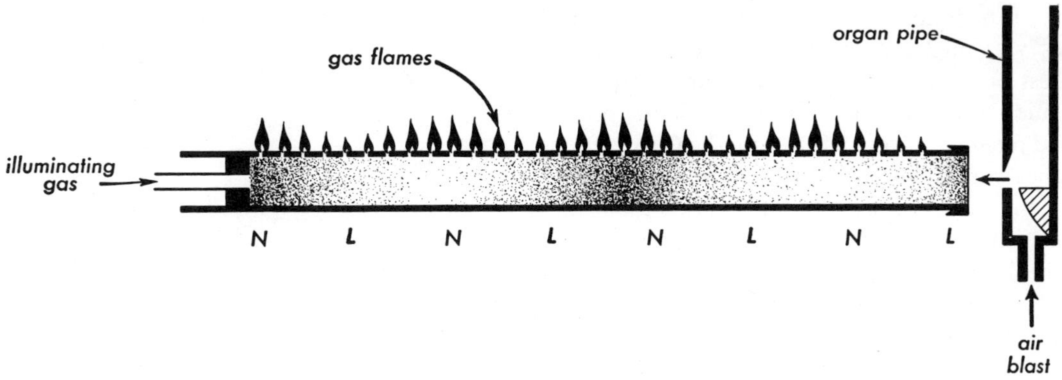

Fig. 20G. Standing waves in a long tube containing illuminating gas.

right-hand end. When resonance is attained by sliding the plunger to the correct position, the small gas flames will appear as shown. Where the nodes occur in the vibrating gas column the air molecules are not moving [see Fig. 18M(b)]; at these points the pressure is high and the flames are tallest. Halfway between are the anti-nodes —regions where the molecules vibrate back and forth with large amplitudes, and the flames are low. Bernoulli's principle is chiefly responsible for the pressure differences (see Sec. 14.6), for where the velocity of the molecules is high the pressure is low, and where the velocity is low the pressure is high.

In many of the wind instruments of the orchestra, the vibrating air columns are not entirely uniform and the far open end is considerably flared. Because of this non-uniformity the nodes are not equally spaced, and the possible frequencies are not exact harmonics of the fundamental. As an example of how nearly the various normal modes come to being harmonics of a fundamental the following measured frequencies of the *open notes* of a B♭ cornet are given in Table 20 B (ν = 116.7 Hz):

Open notes are those for which all valves are open and the entire air column is vibrating.

20.6 THEORY OF VIBRATING AIR COLUMNS

The various notes produced by most wind instruments are brought about by varying the length of the vibrating air column. This is illustrated by the organ pipes in Fig. 20H. The longer the pipe, the lower the fundamental frequency or

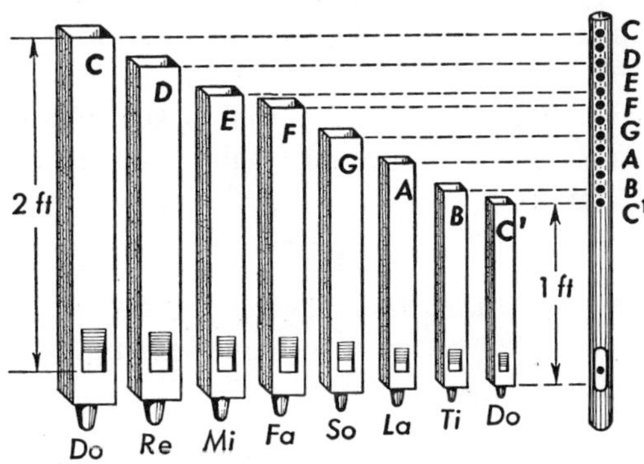

Fig. 20H. Organ pipes arranged in a musical scale. The longer the pipe, the lower is its fundamental frequency and pitch. The vibrating air column of the flute is terminated at various points by openings along the tube.

TABLE 20B. Harmonic Frequencies of a Cornet

Mode Number	2	3	4	5	6	8
Measured Freq.	233.4	349.8	467.5	587.7	706.2	984.6
Harmonic Freq.	233.4	350.1	466.8	583.5	700.2	933.6
Ratio	2.000	2.998	4.007	5.037	6.058	8.130

pitch of the note. In a regular concert organ, the pipes vary in length from about 6 in. for the highest note to almost 16 ft for the lowest. For the middle octave of the musical scale the open-ended pipes vary from 2 ft for middle C to 1 ft for C^1 one octave higher. In the wood winds, like the flute, the length of the column is varied by openings in the side of the instrument; and in many of the brasses, like the trumpet, by means of valves. A valve is a piston which on being pressed down throws in an additional length of tube.

Since a vibrating air column is a condition of standing waves, the frequency of vibration will depend upon two factors, the length of the pipe and the velocity of waves through it. The velocity of longitudinal waves in a gas is given by Newton's formula as modified by Laplace,

$$V = \sqrt{K\frac{P}{\rho}} \tag{20d}$$

where K is a number representing the compressibility of a gas, P is the gas pressure in N/m^2, and ρ its density in Kg/m^3. For monatomic gases, $K = 5/3$; for diatomic gases, $K = 7/5$; and for triatomic gases, $K = 4/3$, and V is in m/s.

For all standing waves, the distance L between any two consecutive nodes, that is, the length of one segment, is just equal to half a wavelength, $\lambda/2$. Accordingly,

$$\lambda = 2L$$

By substituting this relation and the velocity above in the general wave equation, Eq. (18e), we get the general formula

$$\boxed{\nu = \frac{1}{2L}\sqrt{K\frac{P}{\rho}}} \tag{20e}$$

Measurements with resonating pipes confirm this relation.

The effect of the density of a gas on the pitch of a note may be demonstrated by a very interesting experiment with the human voice. Voice sounds originate in the vibrations of the vocal cords in the larynx (see Fig. 20I). This source of vibration, which determines the fundamental pitch of the speaking or singing voice, is con-

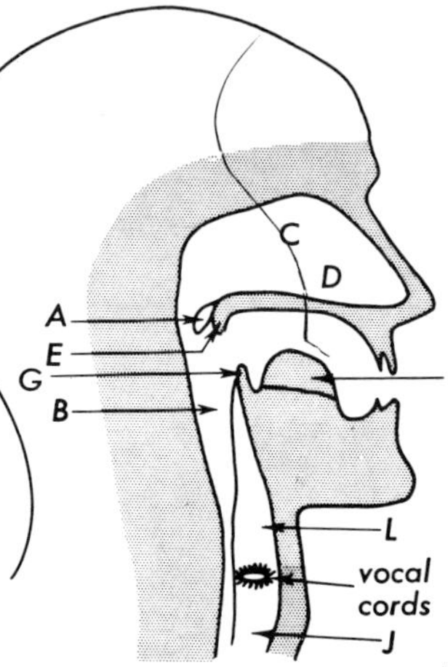

Fig. 20I. Cross-section diagram showing the vocal cords in the larynx. (*A*) Soft palate, (*B*) pharynx, (*C*) nasal cavity, (*D*) hard palate, (*F*) tongue, (*G*) epiglottis, and (*J*) windpipe.

trolled by muscular tension on the cords. The quality of the voice is determined by the size and shape of the throat, the mouth, and the nasal cavities.

If a gas lighter than air is breathed into the lungs, the above equation shows that the voice quality should change. The demonstration can be best and safely performed by exhaling completely, and then filling the lungs with helium gas (hydrogen is unsafe). Upon speaking, the experimenter will be observed to have a very peculiar high-pitched voice, which must be heard to be appreciated. The peculiarities arise from the fact that the fundamental pitch, due to the vocal-cord frequency, remains practically normal, while the harmonics from the resonating mouth, throat, and nasal cavities are raised by about $2\frac{1}{2}$ octaves.

EXAMPLE 2. A straight piece of pipe 2 m long and open at both ends is filled with carbon dioxide gas at standard temperature and pressure. Find (a) the speed of sound in carbon dioxide, and (b) the fundamental frequency of the pipe.

SOLUTION. To find the speed use Eq. (20d). Since CO_2 is a triatomic molecule, $K = 1.33$. From Table 13A, the density of CO_2 gas is $1.98\ Kg/m^3$. By direct substitution in Eq. (20d), we obtain

$$V = \sqrt{\frac{1.333 \times 1.013 \times 10^5 \text{ N/m}^2}{1.98 \text{ Kg/m}^3}}$$

$$V = \sqrt{6.80 \times 10^4 \text{ m}^2/\text{s}^2}$$

$$V = 261 \text{ m/s}$$

(b) To find the fundamental frequency, use Eq. (18e). Since an antinode forms at the open end of a pipe, the pipe length is equal to $\frac{1}{2}\lambda$, and this is equal to L, the length of one loop. By substitution in Eq. (18e), we obtain

$$\nu = \frac{1}{2 \times 2 \text{ m}} 261 \frac{\text{m}}{\text{s}}$$

$$\nu = 65.2 \text{ Hz}$$

20.7 EDGE TONES Although the pitch of the note sounded by any wind instrument is determined by the vibration of an air column according to principles of resonance, the method by which the air is set into vibration varies widely among instruments. In instruments like the saxophone, clarinet, oboe and bassoon, air is blown against a thin strip of wood called a reed, setting it into vibration. In most of the brasses the musician's lips are made to vibrate with certain required frequencies, while in certain wood winds, like the flute and piccolo and in organs and whistles, air is blown across the sharp edge of an opening near one end of the instrument, setting the air into vibration. A brief discussion of these source vibrations is therefore important here.

When wind or a blast of air encounters a small obstacle, little whirlwinds are formed in the air stream behind the obstacle. This is illustrated by the cross section of a flue organ pipe in Fig. 20J. Whether the obstacle is long, or a small round object, the whirlwinds are formed alternately on the two sides as shown. The air stream at B waves back and forth, sending a pulse of air first up one side and then the other. Although the wind blows through the opening A as a continuous stream, the separate whirlwinds going up each side of the obstacle become periodic shocks to the surrounding air. Coming at perfectly regular intervals these pulses give rise to a musical note often described as "edge tones."

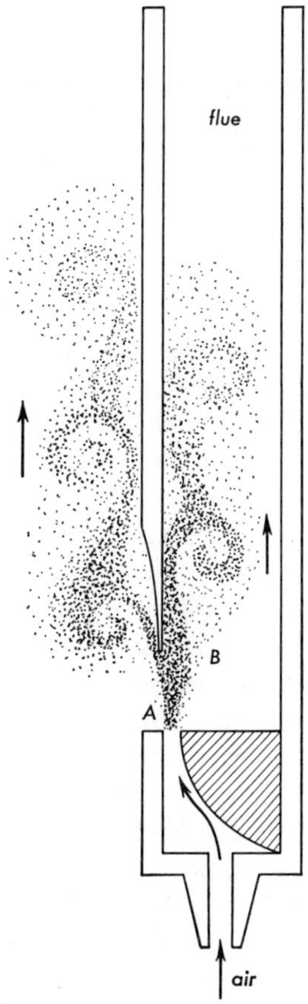

Fig. 20J. A steady stream of air blown across the lip of an organ pipe sets up whirlwinds along both sides of the partition.

The number of whirlwinds formed per second, and therefore the pitch of the edge tone, increases with the wind velocity. When the wind howls through the trees the pitch of the note rises and falls, its frequency at any time denoting the speed of the wind. For a given wind speed, smaller objects give rise to higher pitched notes than large objects. A fine stretched wire or rubber band, when placed in an open window or in the wind, will be set into vibration and give out a musical note. Each whirlwind shock to the air reacts on the obstacle (the wire or rubber band), pushing it first to one side and then the other. These are the pushes that cause the rope of a flagpole to flap periodically in the breeze, while the waving of the flag at the top of a pole shows the whirlwinds that follow each other along each side.

The air column in an organ pipe, flute, or piccolo has its own natural frequency of vibration which may or may not coincide with the frequency of an edge tone. If it does coincide, resonance will occur, the air column will vibrate with a large amplitude, and the returning pulses of air down the tube with each vibration will force the air stream out at just the right moment, thus aiding in building up the natural frequency of the pipe. If the edge tone has a frequency different from the fundamental of the string, or air column, vibrations will be set up but not as intensely as before. If the frequency of the edge tone of an organ pipe, for example, comes close to double that of the fundamental, and this can be obtained by a stronger blast of air, the pipe will resonate to double its fundamental frequency and give out a strong note one octave higher.

By blowing more sharply against the opening in a piccolo or flute, the entire scale of notes can be raised one octave above the normal playing range of the instrument. In all instruments like those mentioned above where air is blown across the sharp edge of an opening to make it sing, as in the organ pipe of Fig. 20J, an antinode is formed at that end. Whether an antinode or node forms at the other end depends upon whether it is open or closed, respectively.

20.8 PERCUSSION INSTRUMENTS, VIBRATING RODS If a number of small sticks are dropped upon the floor, the sound that is heard is described as a noise. If one stick alone is dropped one would also describe the sound as a noise, unless, of course, a set of sticks of varying lengths are arranged in order of length and each one dropped in its order. If this is done, one notices that each stick gives rise to a rather definite musical note and that the set of sticks could be cut to the proper lengths to form a musical scale. The use of vibrating rods in a musical instrument is found in the *xylophone*, the *marimba*, and the *triangle*. Standing waves in a rod, like those in a stretched string, may be any one of three different kinds—transverse, longitudinal, and torsional. Only the first two of these modes of vibration will be treated here.

Transverse waves in a rod are usually set up by supporting the rod at points near each end and striking it a blow at or near the center. As illustrated in Fig. 20K(a), the center and ends of

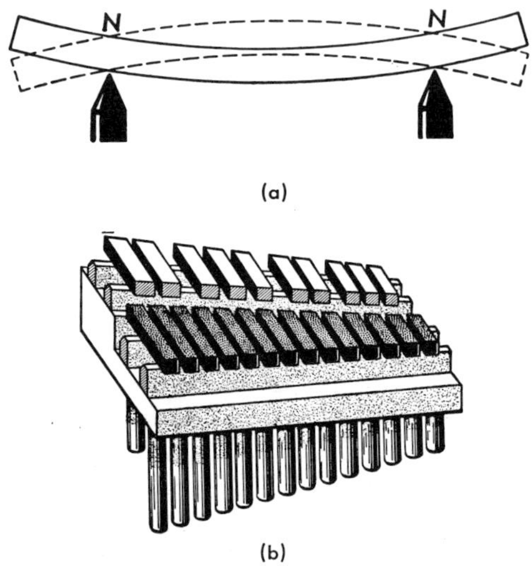

Fig. 20K. (a) Transverse vibrations of a uniform rod or bar. (b) Diagram of the bars and pipes of a marimba.

the rod move up and down, forming nodes at the two supports. Like a stretched string of a musical instrument, the shorter the rod the higher is its pitch, and the longer and heavier the rod the lower is its frequency of vibration and pitch.

The xylophone is a musical instrument based upon the *transverse vibrations* of wooden rods of different lengths. Mounted as shown in Fig. 20K(b), the longer rods produce the low notes and the shorter ones the higher notes. The marimba is essentially a xylophone with a long straight hollow tube suspended vertically under each rod. Each tube is cut to such a length that the enclosed air column will resonate to the sound waves sent out by the rod directly above. Each resonator tube, being open at both ends, forms a node at its center.

The tuning fork depends for its pitch upon the transverse vibrations of a bar. Sounding its fundamental as shown in Fig. 20L, an antinode forms at both ends. Due to the bend in the center

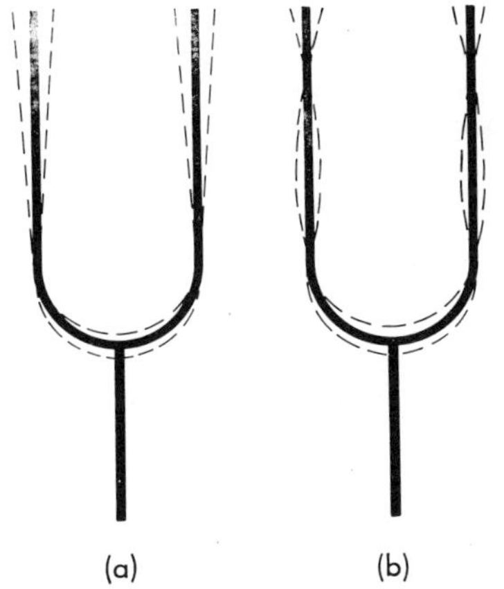

Fig. 20L. Vibration nodes of a tuning fork show (a) fundamental, (b) first overtone.

the two nodes are closer together than in a straight bar, and the antinode at the center transmits forceful vibrations of small amplitude through the shaft to any surface on which it is rested.

20.9 VIBRATING PLATES Although the drum or the cymbals should hardly be called musical instruments, they are classified as such and made use of in nearly all large orchestras and bands. The noise given out by a vibrating drumhead or cymbal plate is in general due to the high intensity of certain characteristic overtones. These overtones in turn are due to the very complicated modes of vibration of the source.

Cymbals consist of two thin metal disks with handles at the centers. Upon being struck together their edges are set into vibration with a clang. A drumhead, on the other hand, is a stretched membrane of leather held tightly at the periphery, and is set into vibration by being struck a blow at or near the center.

To illustrate the complexity of the vibrations of a circular plate, two typical sand patterns are shown in Fig. 20M. The sand pattern method of studying the motions of plates was invented in the eighteenth century by Chladni, a German physicist. A thin circular metal plate is clamped at the center C and sand is sprinkled over the top surface. Then, while touching the rim of the plate at two points N_1 and N_2, a cello bow is drawn down over the edge at a point L. Nodes are formed at the stationary points N_1 and N_2 and antinodes in the regions of L_1 and L_2. The grains of sand bounce away from the loops and into the nodes, the regions of no motion. At one instant the regions marked with a (+) sign all move up, while the regions marked with a (−) sign all move down. Half a vibration later the + regions are moving down and the − regions up. Such diagrams are called *Chladni's sand figures*.

With cymbal plates held tightly at the center by means of handles a node is always formed there, and antinodes are always formed at the periphery. With a drumhead, on the other hand, the periphery is always a node and the center is sometimes but not always an antinode.

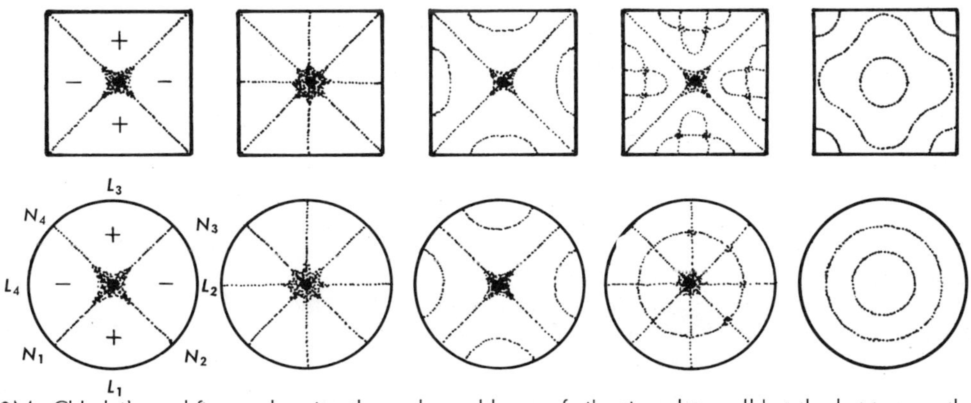

Fig. 20M. Chladni's sand figures showing the nodes and loops of vibrating plates; all but the last two on the right are clamped at the center; the last two are clamped at the periphery.

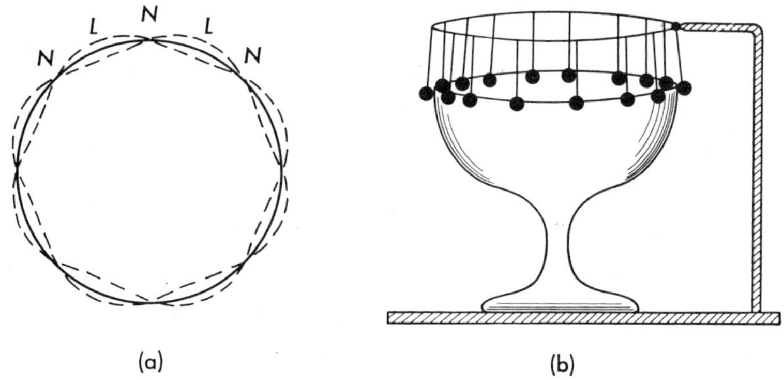

Fig. 20N. Experiment illustrating how the rim of a bell or glass vibrates with nodes and loops.

20.10 BELLS In some respects a bell is like a cymbal plate, for when it is struck a blow by the clapper, the rim in particular is set vibrating with nodes and loops distributed in a symmetrical pattern over the whole surface. The vibration of the rim is illustrated by a diagram in Fig. 20N(a) and by an experiment in diagram (b). Small cork balls are suspended by threads around and just touching the outside rim of a large glass bowl. A violin bow drawn across the edge of the bowl will set the rim into vibration with nodes at some points and loops at others. The nodes are always even in number just as they are in cymbal plates and drumheads, and alternate loops move in while the others move out.

Strictly speaking, a bell is not a very musical instrument. This is due to the very complex vibrations of the bell surface giving rise to so many loud overtones. Some of these overtones harmonize with the fundamental while others do not.

QUESTIONS

1. Upon what three factors does the frequency of a vibrating string depend?

2. Make a diagram of a string vibrating with (a) its 6th harmonic, (b) its 10th harmonic, and (c) its 1st and 5th harmonics simultaneously.

3. Make a diagram showing the vibrating air column in a pipe closed at one end only and sounding (a) its 3rd harmonic, and (b) its 6th harmonic. (c) What kind of waves are they?

4. Make a diagram of a straight pipe, open at both ends, and vibrating with (a) its fundamental, (b) its 5th harmonic, and (c) its 8th harmonic.

5. What are the differences between the treble and bass strings of a piano? Why do bass strings look like coiled springs?

6. Make a diagram showing a xylophone bar vibrating with its 4th harmonic. What kind of waves are these?

PROBLEMS

1. The fundamental frequency of a bugle is 66 Hz. All the other notes blown on this instrument are higher harmonics of this frequency. Find the frequencies of the 2nd, 3rd, 4th, 5th and 6th harmonics. [Ans. 132, 198, 264, 330, and 396 Hz.]

2. Find the mass of a piano string 65 cm long if, under a tension of 650 N, it has a frequency of 280 Hz.

3. The A string of a violin is 35 cm long and has a mass of 2 g. Find the tension in newtons when properly tuned to a frequency of 440 Hz. [Ans. 542 N.]

4. A piano string 100 cm long has a frequency of 132 Hz, and a mass of 16 g. Find the tension in newtons.

5. The G string of a violin is 35 cm long and has a frequency of 198 Hz. If the mass of this length of string is 18 g, what is the tension in the string? [Ans. 988 N.]

6. The E string of a violin has a fundamental frequency of 660 Hz. Find the wavelength of the 5th and 7th harmonics of the sound waves it produces if the speed of sound is 346 m/s.

7. A harp string with a length of 75 cm is tuned to a frequency of 440 Hz. What is the wavelength of the 4th harmonic (a) on the string, and (b) in the air? Assume the speed of sound to be 350 m/s. [Ans. (a) 37.50 cm, (b) 19.89 cm.]

8. A steel wire 1.8 m long has a mass of 62 g. If this wire is fastened at both ends and then put under a tension of 1200 N, what will be the frequency of its 4th harmonic?

9. An organ pipe with both ends open is used to produce sound waves in air in the experiment similar to that shown in Fig. 20G. If the gas tube resonates to a frequency of 1080 Hz, and the speed of sound in the gas is 348 m/s, what is the distance between consecutive points where the flames are a maximum? [Ans. 16.11 cm.]

10. A whistle has a fundamental frequency of 6000 Hz at a temperature of 30°C. If antinodes are formed at both ends, find the length of the air column.

11. An organ pipe open at both ends and filled with air at 30°C has a fundamental frequency of 132 Hz. What would be its fundamental frequency if the air were replaced by CO_2? The speed of sound in CO_2 is 272 m/s. [Ans. 102.8 Hz.]

12. An organ pipe with one end closed has a fundamental frequency of 296 Hz when the temperature is 30°C. Find its frequency when the temperature drops to −40°C.

13. Calculate the speed of sound in carbon monoxide at normal temperature and pressure. Assume the density of CO to be 1.250 Kg/m^3. [Ans. 336.8 m/s.]

14. Calculate the speed of sound in chlorine gas at normal temperature and pressure. Assume the density of the gas to be 3.210 Kg/m^3. Chlorine is diatomic.

15. Calculate the speed of sound in nitrous oxide gas at normal temperature and pressure. Assume the density of N_2O to be 1.978 Kg/m^3. [Ans. 261.3 m/s.]

16. A pipe 95 cm long, open at both ends, is filled with sulfur dioxide gas at normal temperature and pressure. (a) Find the speed of sound in sulfur dioxide, and (b) the fundamental frequency to which the pipe will sound. Assume SO_2 to have a density of 2.927 Kg/m^3.

21 REFLECTION, REFRACTION, AND DISPERSION OF LIGHT

Light and its various phenomena present some of the most interesting subjects in the whole realm of physics. They are interesting because many natural phenomena as well as laboratory experiments are revealed through the sense of vision as arrays of color.

All of the various known properties of light are conveniently described in terms of the experiments by which they were discovered and the many and varied experiments by which they are now commonly demonstrated. Numerous as they are, these experiments may be grouped together and classified under one of the three following heads: (1) *geometrical optics*, (2) *physical optics*, and (3) *quantum optics*. Each of these may be subdivided as follows:

Geometrical Optics	Physical Optics	Quantum Optics
reflection	*diffraction*	*photoelectric effect*
refraction	*interference*	*Compton effect*
dispersion	*polarization*	*pair production*

The first group, geometrical optics, is treated in this chapter.

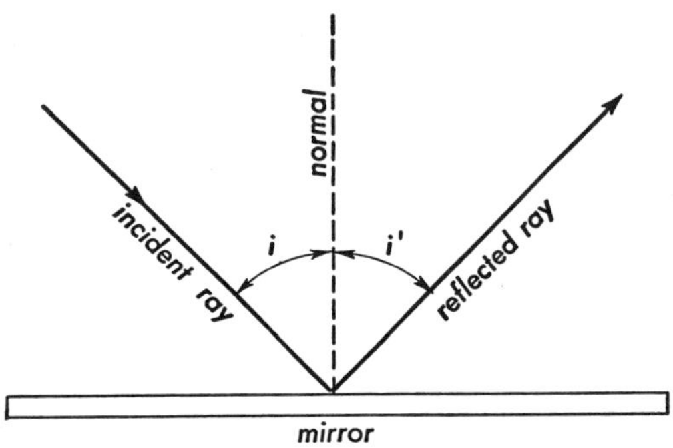

Fig. 21A. Diagram illustrating the law of reflection from a plane surface.

21.1 THE LAW OF REFLECTION Experiment shows that, whenever a ray of light is reflected from a plane surface, the nature of the reflected light can be described in terms of a number of simple and well-defined laws. The simplest of these is the one known as *the law of reflection*. According to this law, the angle at which a ray of light strikes the reflecting surface is exactly equal to the angle the reflected ray makes with the same surface. Instead of measuring the *angle of incidence* and the *angle of reflection* from the mirror surface, however, it is customary to measure both from a line perpendicular to the plane of the mirror. This line as shown in Fig. 21A is called the *normal*. As the angle i increases, the angle i' increases by exactly the same amount, so that, for all angles of incidence,

$$\text{angle } i = \text{angle } i' \qquad (21a)$$

A second part of this law stipulates that the reflected ray lies in the plane of incidence, the plane of incidence being defined as the plane containing the incident ray and the normal. In other words, *the incident ray, the normal and the reflected ray all lie in the same plane.*

In speaking of a mirror surface one does not necessarily mean a silvered plate of glass; a mirror is any surface smooth enough to produce regular reflection as it has just been described.

21.2 IMAGE IN A PLANE MIRROR The image of one's self seen in a mirror is formed by rays of light traveling in straight lines which are reflected according to the law of reflection. All objects seen in a plane mirror are images formed by reflection. This can be demonstrated by the experiment shown in Fig. 21B. A lighted candle O is placed on the table near a plate of glass MN. With the candle itself hidden in the box H, the observer at E sees only the reflected image at I. If a glass of water is placed at B, this image appears as a real candle burning under water.

As shown in the top view, all rays of light leaving the source O are reflected according to the law of reflection. To an observer anywhere between L and R on the right side of the mirror, all light appears to come from the same point I. This image point is just as far behind the mirror as the object O is in front of it; the two lie on the same perpendicular to the mirror.

The image one sees in a plane mirror is not a real image but a virtual image. A virtual image is one from which rays seem to radiate but actually do not. In the figure the rays do not come from I; they come from O and by reflection reach the observer.

This experiment illustrates a trick used to make ghost-like figures appear to move about a room or stage. Light from real persons or objects, located below or above the stage, is reflected from a large sheet of plate glass at the

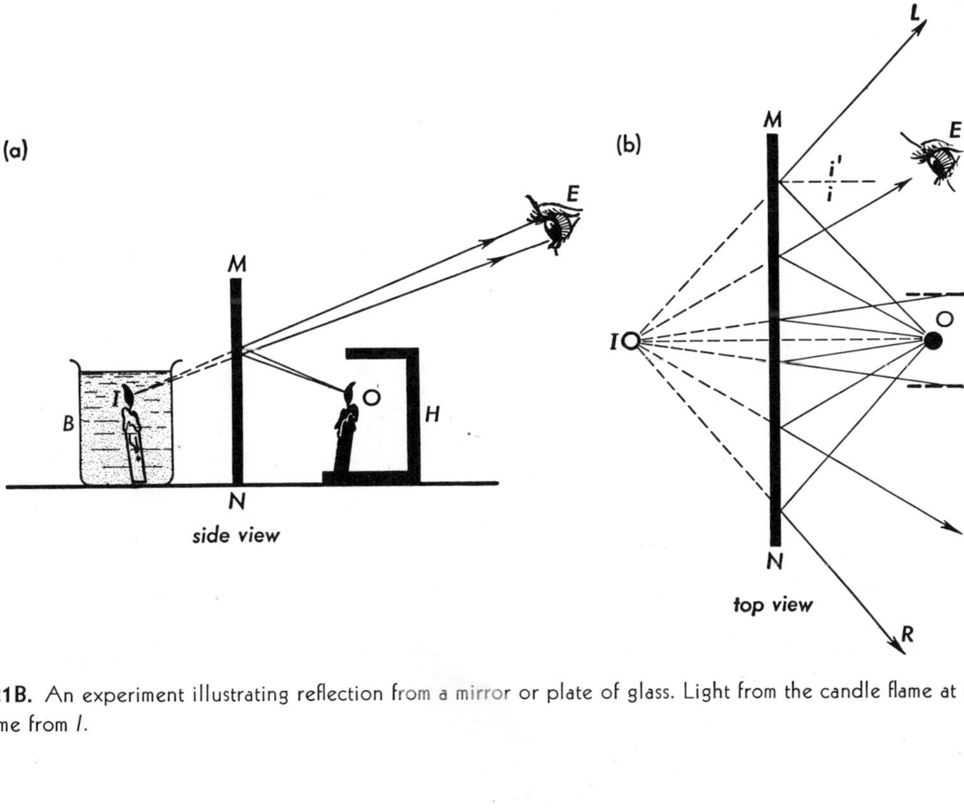

Fig. 21B. An experiment illustrating reflection from a mirror or plate of glass. Light from the candle flame at O appears to come from I.

front of the stage. With proper drapes and a darkened room, the illusion is very effective.

If one looks at his own face in a plane mirror the image observed is technically described as *perverted.* The image is the same as though the face were reproduced as a rubber mask and the mask turned inside out and viewed from the new front. The right ear of the subject becomes the left ear of the image and vice versa.

To see one's face as others see it, two front silvered mirrors should be placed 90° apart and touching each other along one edge as shown in Fig. 21C. The observer's right ear will then be

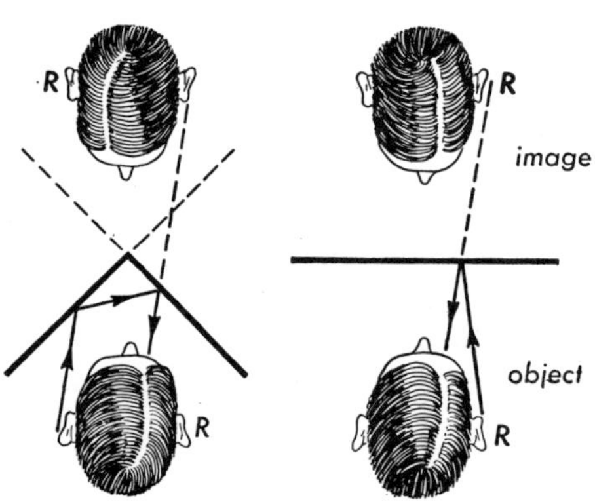

Fig. 21C. One's own image seen in 90° mirrors is normal; that seen in a plane mirror is perverted.

seen, because of two reflections, as the right ear of his image, etc. This experiment must be performed to be appreciated since many people's faces are, unknowingly, slightly unsymmetrical. Seen in 90° mirrors all such irregularities are reversed and therefore appear double in magnitude and are very noticeable.

21.3 SPEED OF LIGHT In a vacuum, or in interstellar space, light travels with a speed of 186,280 mi/s, or

$$c = 299{,}792 \text{ Km/s}$$

For most practical calculations this latter value is usually rounded off to

$$\text{speed of light} \qquad c = 300{,}000 \text{ Km/s}$$

In a transparent medium like air, glass, or water, light travels at a somewhat lower speed. In water, for example, the speed of light v is 225,000 Km/s, just three quarters of the speed in vacuum, whereas in common crown glass its speed is close to 200,000 Km/s.

21.4 THE REFRACTIVE INDEX The ratio between the velocity of light in a vacuum and the velocity in a medium is called the *refractive index*, or the *index of refraction* of the medium.

$$\frac{\textbf{velocity of light in vacuo}}{\textbf{velocity of light in a medium}} = \textbf{refractive index}$$

Symbolically,

$$\frac{c}{v} = \mu \tag{21b}$$

The Greek letter μ (mu) is frequently used to represent this ratio. Substituting the velocities given in the preceding section, the following refractive indices may be calculated:

$$\text{for water, } \mu = 1.33 \tag{21c}$$

$$\text{" glass, } \mu = 1.50 \tag{21d}$$

$$\text{" air, } \mu = 1.00 \tag{21e}$$

Very exact measurements of the refractive index of air give a value 1.00029.

21.5 REFRACTION When light falls upon the smooth surface of a transparent substance like water or glass, other than normally, part of it is reflected according to the law of reflection and the remainder is refracted into the medium as in Fig. 21D. This bending is due to the change in

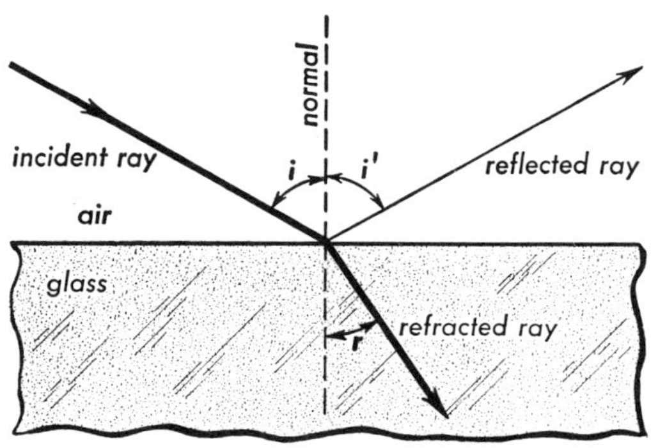

Fig. 21D. Reflection and refraction of light at the boundary of a glass surface.

the velocity of the light upon entering the second medium. The direction of the refracted ray, like the incident and reflected rays, is always measured by the angle it makes with the normal.

The angle of refraction, r, is found by experiment to depend upon two factors: (1) *the angle of incidence i*, and (2) *the index of refraction μ*. As defined in the previous section, *the index of refraction is the ratio of the speed of light in vacuo to the speed of light in the medium.* To determine the angle of refraction from these two factors we perform the following graphical construction shown in Fig. 21E.

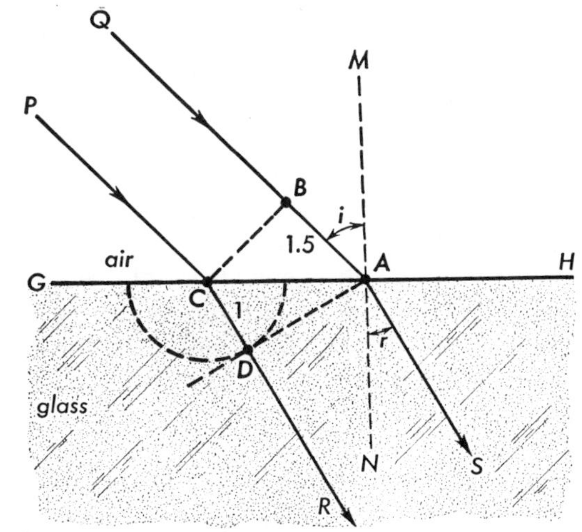

Fig. 21E. Illustration of the graphical method of determining the angle of refraction.

Assume the refracting surface is that of glass, with air above, and the refractive index $\mu = 1.5$. First a ray of light QA is selected, incident on the surface at an angle i. Using a ruler a line segment is measured back from A to a point B such that AB = 1.5 units. The unit of length to be chosen here is arbitrary and may be 1 cm, or 1 in., or any other. From B a line is next drawn perpendicular to QA, intersecting the surface GH at C. Line CP is now drawn parallel to QA to represent the other boundary of the incident beam of width CB.

Now with a compass of radius 1.0 unit, and center at C, a circular arc is drawn as shown. Through the point A a tangent is drawn, and from the point of contact D, the refracted ray CDR is drawn. The other ray AS is finally drawn parallel to CR to represent the other edge of the refracted beam of width DA.

The law of refraction may now be stated in terms of the line segments BA and CD as follows:

$$\frac{BA}{CD} = \mu \tag{21f}$$

It was the Dutch astronomer and mathematician, Willebrord Snell,* who first discovered, from experiments, that the ratio of the lengths of these two lines for a given substance is the same for all angles of incidence. In other words, μ is a constant. The relation as given by Eq. (21f) is therefore called Snell's law. The reason for the constancy of this ratio is that, while the light travels from B to A in air, it travels the corresponding lesser distance CD in water.

21.6 REFRACTION BY A PRISM When light passes through a prism it is refracted at two surfaces, once on the way in and once on the way out. If the two sides involved are parallel, as they are in Fig. 21F, diagram (a), the emergent ray is always parallel to the incident ray. If the sides are not parallel, as in diagrams (b), (c), and (d), the emergent ray has a different direction. The larger the angle A between the two refracting surfaces the larger is the angle of deviation D. Upon entering the prism at the first surface, see diagram (d), the light is bent toward the normal. Emerging into the air from the second surface, the light is bent away from the normal. Note in Fig. 21F that neither the apex nor the base of the prism has any effect on the deviation of the light.

In verifying these results by experiment, light

---

* WILLEBRORD SNELL (1591–1626), Dutch astronomer and mathematician, was born at Leyden in 1591. At the age of twenty-two he succeeded his father as professor of mathematics at the University of Leyden. In 1617 he determined the size of the earth from measurements of its curvature between Alkmaar and Bergen-op-Zoom. In 1621 he discovered the law of refraction which now carries his name.

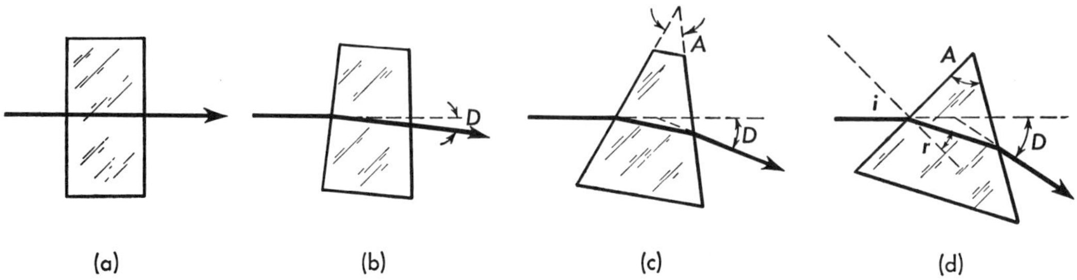

Fig. 21F. Illustration of the bending of a beam of light by prisms made of the same glass.

of only one color should be used, because white light will spread out into a spectrum of colors. Light of one color only is readily obtained by inserting a piece of red or green colored glass into a beam of white light.

One very useful and general principle concerning the behavior of light is *the reversibility of light rays.* If, in any of the experiments or illustrations already described, the light rays could be reversed in direction, they would be found to retrace their paths exactly.

A relatively simple graphical method of tracing a given ray of light through a prism is shown in Fig. 21G. Given is a ray of light VP incident at an angle i on the first face of a prism of angle A and refractive index μ. In this diagram $A = 60°$, $\mu = 1.5$, and $i = 60°$.

Starting at one side a line EF is first drawn parallel to the incident ray VP. With a center at any point O, and with radii of 1.0 unit and 1.5 units, arcs CS and BR are drawn of indefinite length. From the intersection C a line is now drawn parallel to the normal M, intersecting the second arc at R. The line OR is drawn next, and from the point P of the prism a parallel line is drawn intersecting the second face at Q. Starting at R a line is next drawn parallel to the normal N, intersecting the first arc at S. Finally the line OS is drawn in and from the point Q on the second prism face a parallel line QW is drawn as the emergent ray.

Since the above construction obeys Snell's law, the accuracy of the results depends upon the sharpness of the pencil used and the care with which the diagram is made. Furthermore the left-hand diagram as well as the prism diagram includes all of the angles involved in the prism, including the angle of deviation D.

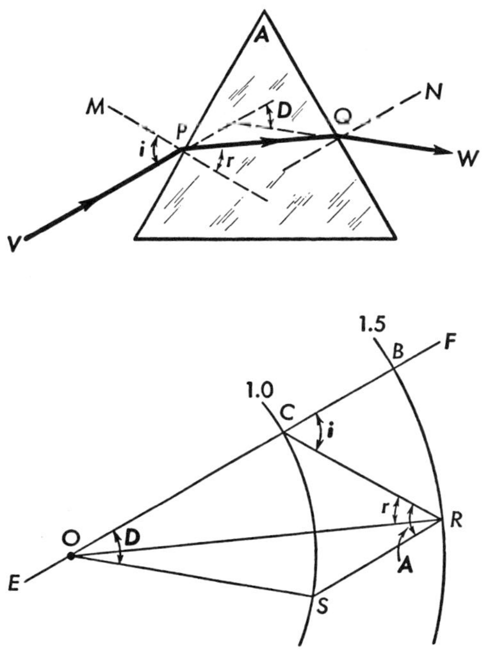

Fig. 21G. Graphical construction for refraction by a prism.

21.7 MINIMUM DEVIATION If, during the time a beam of light is refracted by a prism, the prism is rotated continuously in one direction about an axis parallel to the refracting edge A, the angle of deviation will be observed to decrease, reach a minimum, and then increase again as shown in Fig. 21H. *The smallest deviation angle produced by a prism is called the angle of minimum deviation,* and occurs at that particular angle of incidence where the refracted ray inside the prism makes equal angles with the prism faces P and Q.

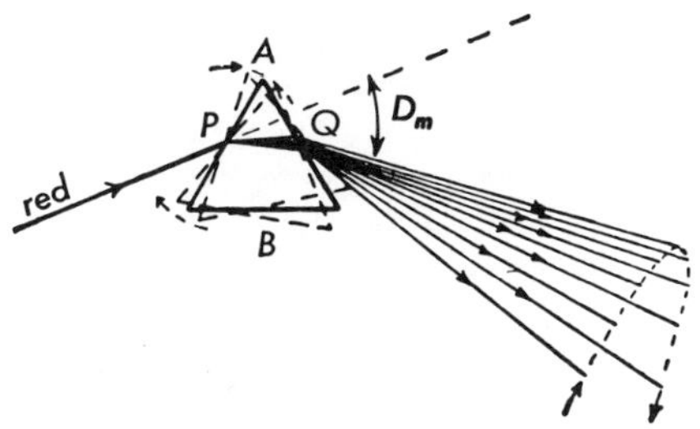

Fig. 21H. Light refracted by a prism has a minimum angle of deviation *Dm*.

21.8 DISPERSION It was known to the ancients that sunlight, on passing through transparent crystals and jewels of various kinds, would produce brilliant colors of light. The early philosophers, attempting to explain the phenomenon, attributed the origin of the colors to the crystal itself. It was Newton who first demonstrated with prisms that the colors were already present in the white sunlight and that the function of the prism was to separate the colors by refracting them in different directions. This is shown diagrammatically in Fig. 21I.

We have already seen how light of one color is refracted at the boundary of a medium like glass or water and how it is deviated by a prism. It may be seen in the diagram how, *with white light, each color is refracted by a different amount* to produce its own angle of deviation. *Red light is refracted least and violet light is refracted most.* The angular spread of all the colors produced by sending white light through a prism is called the *dispersion* and the band of color so produced is called *a spectrum.*

If white light is sent through a group of similar prisms made of different substances, each prism will be found to have a different dispersion. This can be demonstrated for solids by *flint* and *crown glass prisms,* and for liquids by *kerosene, carbon dioxide,* and *water.* It will be noted that the two glass prisms in Fig. 21J, one of flint glass and the other of crown glass, produce quite different dispersions. The liquid prisms, produced by filling thin-walled glass troughs with liquid, also disperse light by different amounts.

Since different colors are refracted by different amounts, the index of refraction is different for each color. In a vacuum all colors travel with the same speed, 300,000 Km/s, but in a transparent medium, like glass or water, they travel considerably slower and at different speeds. Among the spectrum colors, red travels the fastest and violet the slowest, with the speeds of all other colors somewhere in between. In air there is very little dispersion and in a vacuum there is absolutely none.

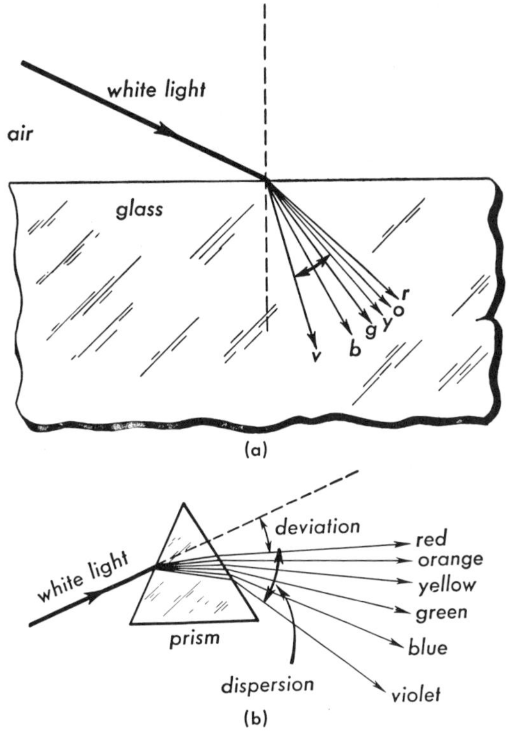

Fig. 21I. (a) Refraction of white sunlight into its spectrum colors. (b) Refraction at both surfaces of a prism produces higher dispersion.

The refractive indices for a number of transparent solids are given in Table 21A. It will be noted that, although the values for any one substance do not vary greatly between colors, the values for blue and violet are the largest and those for orange and red are the smallest. Note the relatively high values for diamond and the relatively low values for ice.

When white light is refracted at a smooth surface, the angular spread of the spectrum band, red to violet, is proportional to the difference be-

TABLE 21A. Refractive Index for Several Transparent Solids

Substance Color	Violet	Blue	Green	Yellow	Orange	Red
Crown glass	1.5380	1.5310	1.5260	1.5225	1.5261	1.5200
Light flint	1.6040	1.5960	1.5910	1.5875	1.5867	1.5850
Dense flint	1.6980	1.6836	1.6738	1.6670	1.6650	1.6620
Quartz	1.5570	1.5510	1.5468	1.5438	1.5432	1.5420
Diamond	2.4580	2.4439	2.4260	2.4172	2.4150	2.4100
Ice	1.3170	1.3136	1.3110	1.3087	1.3080	1.3060

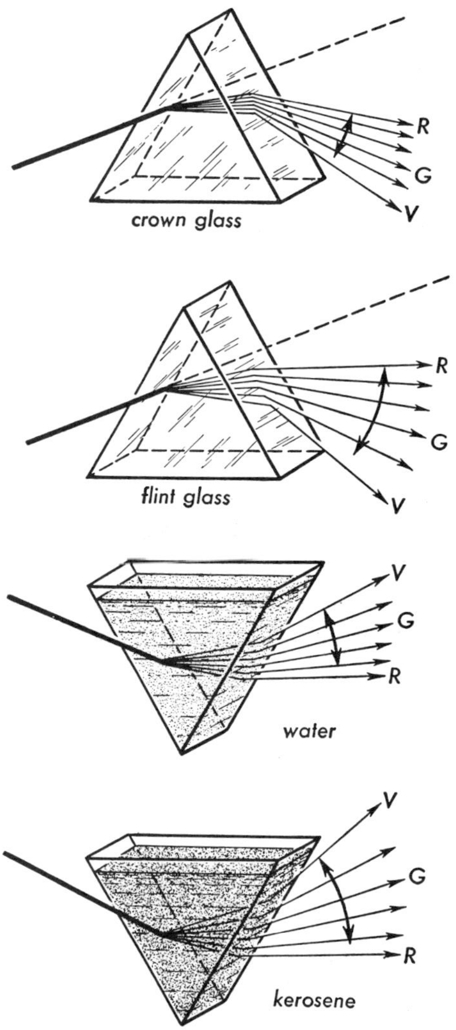

Fig. 21J. Illustration of the relative dispersions of solid as well as liquid prisms.

tween the two extreme indices of refraction. Take a specific case of a crown glass where the refractive indices for violet, yellow, and red are

$$\mu_v = 1.5380$$
$$\mu_y = 1.5225$$
$$\mu_r = 1.5200$$

Upon refraction the angular dispersion is proportional to the difference between the two extreme indices:

$$\textbf{dispersion} = \mu_v - \mu_r$$

Specifically,

$$\text{dispersion} = 1.5380 - 1.5200 = 0.0180$$

The deviation of the spectrum from the direction of the incident light is given by the deviation of the middle of the spectrum, and for this the index of the yellow light is used. Since there would be no deviation of this light if $\mu_y = 1$, the deviation of the spectrum is defined as

$$\textbf{deviation} = \mu_y - 1$$

Specifically,

$$\text{deviation} = 1.5225 - 1.0000 = 0.5225$$

The ratio of these two quantities is called the *dispersive power.*

$$\textbf{dispersive power} = \frac{\mu_v - \mu_r}{\mu_y - 1} \qquad (21g)$$

giving

$$\text{dispersive power} = \frac{0.0180}{0.5225} = 0.03445$$

21.9 THE RAINBOW The rainbow is nature's most spectacular display of the spectrum of white light. The required conditions for the appearance of the phenomenon are that the sun be shining in one part of the sky and the rain be falling in the opposite part of the sky. Turning one's back to the sun, the bright primary bow and sometimes the fainter secondary bow, with colors reversed, are seen as the arcs of circles. From a high vantage point or an airplane, these bows may form complete circles whose common center lies in the direction of the observer's shadow.

The elementary theory of the rainbow was first given by Antonius de Demini in the year 1611 and later developed more exactly by Descartes. The general characteristics of the *primary* and *secondary bows* are satisfactorily accounted for by considering only the reflection and refraction of light by spherical raindrops. To understand how the phenomenon arises, we first confine our attention to an individual raindrop as shown in Fig. 21K. A ray of sunlight is shown entering a

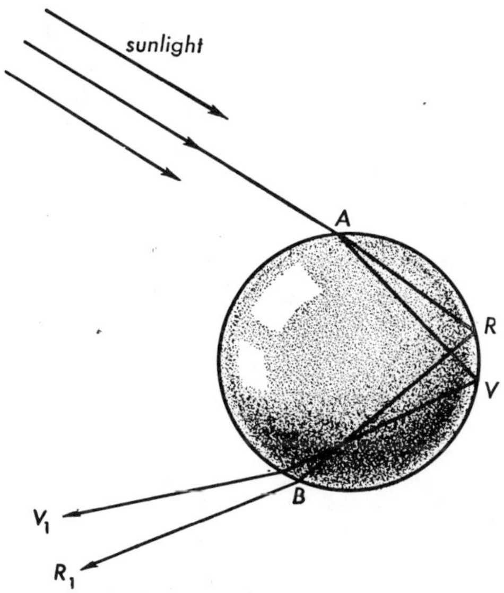

Fig. 21K. Dispersion of sunlight by a single raindrop. (Primary rainbow)

single raindrop at a point A near the top. At this point some of the light is reflected (not shown), and the remainder is refracted into the liquid sphere. At this first refraction the light is dispersed into its spectrum colors, violet being deviated the most and red the least.

Arriving at the opposite side of the drop, each color is partly refracted out into the air (not shown) and partly reflected back into the liquid. Reaching the surface at the lower boundary each color is again reflected (not shown), and refracted. This second refraction is quite similar to that of a prism (see Fig. 21I) where refraction at the second surface increases the dispersion already produced at the first. This is the path of the light in thousands of drops giving rise to the bright primary rainbow.

In Fig. 21L, a ray of sunlight, coming from

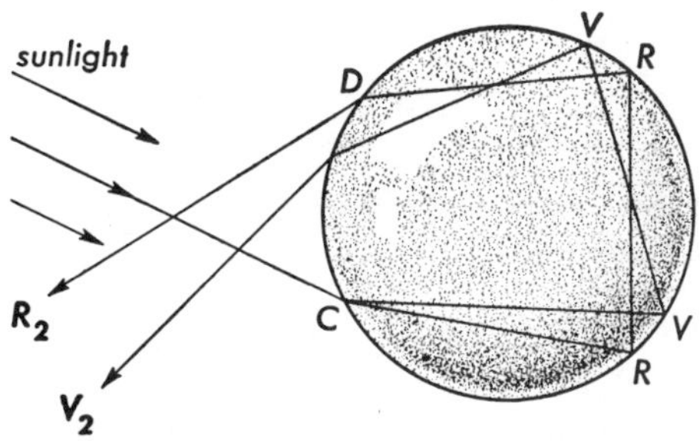

Fig. 21L. Dispersion of sunlight by a single raindrop. (Secondary rainbow)

the same direction as in Fig. 21K, is shown entering a single raindrop at a point C near the bottom. After one refraction and two internal reflections, the light is again refracted and dispersed, this time in a direction not greatly different from that in Fig. 21K. This is the path of the light in thousands of drops giving rise to the fainter secondary rainbow.

As shown in Fig. 21M, the primary bow appears inside the secondary bow and arises from sunlight entering the tops of drops properly located. Those in a region R_1 refract red light toward the observer's eye at O, and the violet and other colors over his head. Drops in the region of V_1 refract violet light to the observer's eye at O, and the red and other colors toward his feet. In other words, the light seen from any one drop is but one color, all drops giving this color lying on the arc of a circle. The reason they lie on the arc of a circle is that the angle between the incident sunlight and the refracted light of any one color is of necessity the same for each drop.

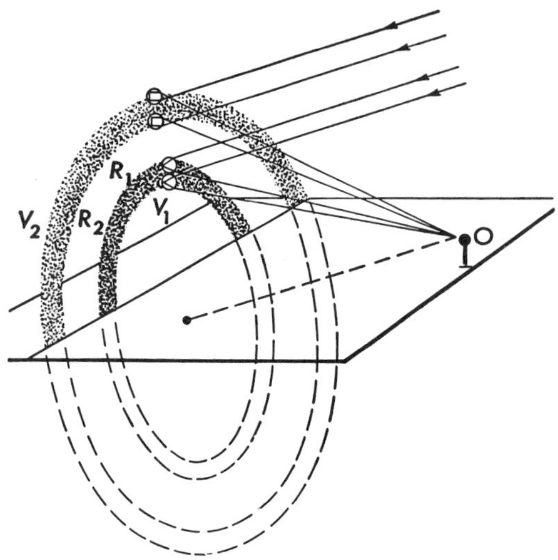

Fig. 21M. The primary and secondary rainbows as seen by an observer at O.

In the primary bow this angle is 42° for the red light and 40° for the violet.

The secondary bow is formed by similar reasoning and appears at higher angles of elevation, and with the colors reversed. The angles subtended by the secondary bow are 50° for the red and 54° for the violet.*

21.10 HALOS Halos are commonly observed as faint rainbow-like rings around the sun or moon and are due to tiny ice crystals floating in the upper stratosphere. Such crystals are hexagonal in shape and, acting like prisms, refract and disperse white light into a spectrum. Two halos are frequently observed, the brighter one making an angle of 22° with the luminary and the fainter one an angle of 46°. Both exhibit confused spectrum colors with a decided red tint on the inside.

Crystals, like prisms, have an angle of minimum deviation for each color of light and, because they are oriented at random in space, millions of them appear to each observer to concentrate the light in a circle with the luminary at the center. As shown in Fig. 21N, the angle of minimum deviation for light entering and leaving different faces of either plate-like or rod-like crystals is 22° and 46°. The 22° angle arises from rays like (3) where the two refracting surfaces are alternate faces that make an angle of 60° with each other. The 46° angle arises from rays like (4) where the two refracting surfaces make an

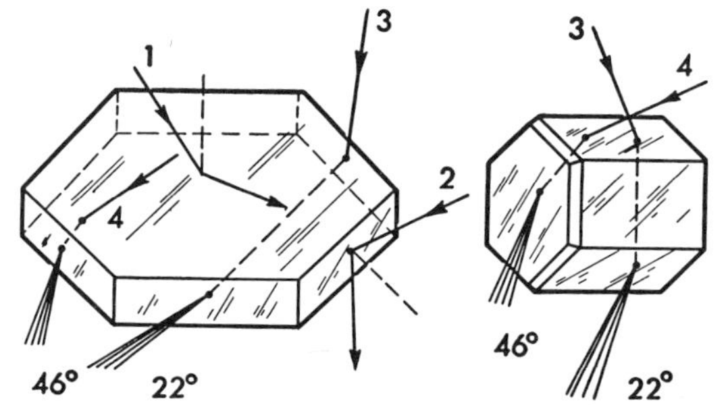

Fig. 21N. Typical forms of ice crystals showing reflected and refracted rays of sunlight giving rise to halos and "mock suns."

* For a treatment of multiple rainbows, coronas, lunar bows, sundogs, etc., see Chap. 38 of *Modern College Physics*, by H. E. White, Van Nostrand Reinhold.

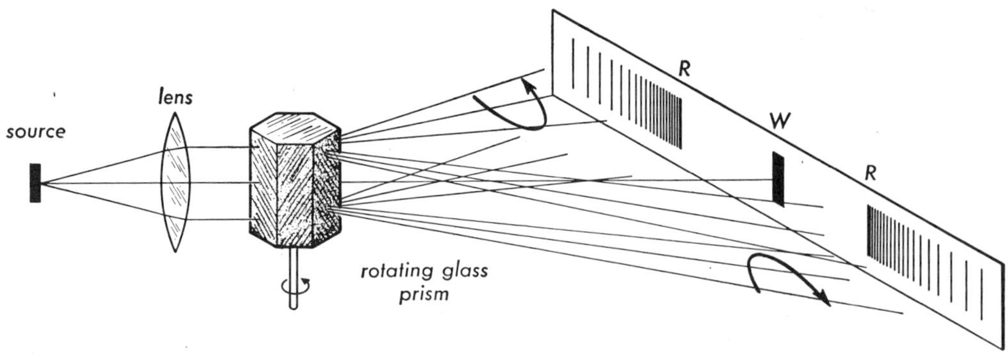

Fig. 21O. Experiment demonstrating colors produced at minimum deviation by a single rotating prism.

angle of 90°. (The refractive index of ice for red light is 1.3060.)

A demonstration of such color effects can be produced as shown in Fig. 210 by means of a hexagonal or octagonal glass prism rotated rapidly about its axis in a strong beam of white light. Dispersed light refracted by alternate faces sweeps in from either side of the screen, slows down and stops at minimum deviation, and then retreats again. The blurred patches of light on the screen are brightest at the minimum deviation angles *R*.

QUESTIONS

1. (a) How does the image in a plane mirror compare in size with the object? (b) How does the image distance from the mirror compare with the object distance?

2. (a) What is the technical name for the nature of the image seen in a plane mirror? (b) What is a virtual image? (c) What is a real image? (d) What is a perverted image?

3. When you look at your face in a plane mirror and put your hand to your left ear, which ear of the image does it appear to be?

4. What is the minimum height of a mirror that will enable a person to see his full height?

5. Make a diagram showing the refraction and dispersion of light by a single raindrop as it relates to the primary rainbow.

6. What is a halo? Make a diagram of an ice crystal and explain how, when, and where halos are formed.

7. For which color is the refractive index the greatest for a given optical material? For which color is it least?

PROBLEMS

1. Make a diagram to show how a beam of light in air is refracted at the smooth surface of glass. Assume the angle of incidence to be 60°, and the refractive index to be 1.75. [Ans. Angle of refraction 29.7°.]

2. How long will it take light to travel to the moon and back if the distance to the moon is 239,000 mi?

3. A satellite is launched into orbit 6000 mi above the earth. When the satellite is straight overhead, a laser beam is turned in its direction and the beam reflected back to earth. What is the total time it takes the light to make the round trip? [Ans. 64.4 milliseconds.]

4. How long will it take light to travel to Mars and back if the distance to the planet is 125×10^9 m?

5. How long will it take light to travel to the planet Saturn and back if the distance is 1.45×10^{12} m? [Ans. 9.67×10^3 s = 2 hr 41 min 10 s.]

6. Light is incident at an angle of 60° on one face of a 50° prism. Graphically find the total deviation of the light if the refractive index is 1.70.

7. A parallel beam of green light is incident at an angle of 72° on the surface of a light flint glass plate. Graphically determine the angle of refraction. [Ans. 36.7°.]

8. A prism with a vertex angle of 60° is made of crown glass of refractive index 1.526. Graphically find the angle of deviation D for a beam of light incident on the first surface at 50°.

9. A prism with a vertex angle of 55° is made of quartz of refractive index 1.550. Graphically determine (a) the angle of refraction at the first surface, (b) the angle of incidence on the second surface, and (c) the angle of emergence from the prism, if the angle of incidence on the first surface is 48°. [Ans. (a) 28.65°. (b) 26.35°. (c) 43.5° with the surface normal.]

10. A prism is made of clear plastic with a vertex angle of 65°. Graphically determine the angle of minimum deviation if the refractive index is 1.42. Note: Start with a ray inside the prism making equal angles with the two sides.

11. A prism is made of ice with a vertex angle of 60°. Graphically determine the angle of minimum deviation if the refractive index is 1.31. Note: Start with a ray inside the prism making equal angles with the two sides. Compare your answer with the rays shown in Fig. 21N. [Ans. 21.8°.]

12. Find the maximum angle of refraction for a ray of red light incident on the polished surface of dense flint glass. See Table 21A.

13. Find the maximum angle of refraction for a ray of violet light incident on the surface of quartz. Use graphical methods. See Table 21A. [Ans. 40.0°.]

14. If 35° prisms were made of the three materials crown glass, quartz, and diamond, respectively, which one do you think would produce the greatest dispersion?

15. Calculate (a) the dispersion, and (b) the dispersive powers for light flint glass and ice, respectively. [Ans. (a) 0.0190, 0.0110. (b) 0.03234, 0.03563.]

22 THE SCIENCE OF COLOR

The beautiful colors of precious stones have adorned the crowns of the emperors and rulers of the world for centuries. Diamonds, emeralds, rubies, and sapphires derive their colors from well known principles of optics. In the preceding chapter we have seen how prisms of diamonds, glass, and quartz, etc., disperse white light into spectral colors. In this chapter we will see how colored lights and pigments are mixed to produce many colors.

Color vision is perhaps nature's most valued gift to man. While color is for the most part a physiological phenomenon, its origin is considered to belong to the realm of physics. The science of color mixing has been made possible through the discovery that all colors can be completely analyzed by spreading them out into a prismatic spectrum.

22.1 EFFECT OF ILLUMINATION ON COLOR

To see a body in its true color, that body must be illuminated by light of the same color. If a red rose, as an illustration, is placed in the different colors of a prismatic spectrum it will appear a brilliant red in red light and gray or black in all the others.

Another experiment is illustrated in Fig. 22A

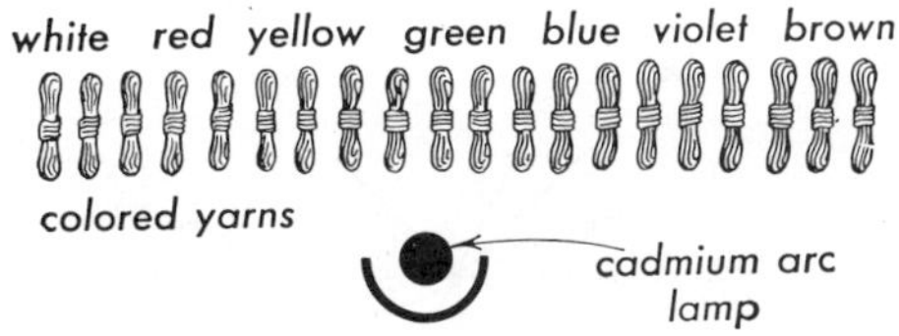

Fig. 22A. Experiment with colored skeins of yarn showing that to see an object in its true body color it must be illuminated by the proper light.

where red light from a cadmium arc lamp is shown illuminating a row of colored skeins of yarn. When the lamp is turned on, only the red yarn appears with its true color; the white yarn is red and the others are black or gray. If the same set of colored yarns is illuminated with the yellow light from a sodium arc lamp, only the yellow yarn will appear in its true color; the white yarn will now be yellow and the others will be gray or black. In other words, unless the source emits the proper colors the body cannot be seen in its true color. Sunlight will show each yarn in its true color, for sunlight contains all colors of the spectrum.

22.2 SURFACE COLOR The above experiments demonstrate what is called *surface color.* When sunlight falls on a red rose, red yarn, red paint, or red glass, all of the colors except red are absorbed and do not get through or out again. The red, as it passes through, is reflected and refracted by the fine grains of pigment and comes out in all directions as shown in Fig. 22B(a).

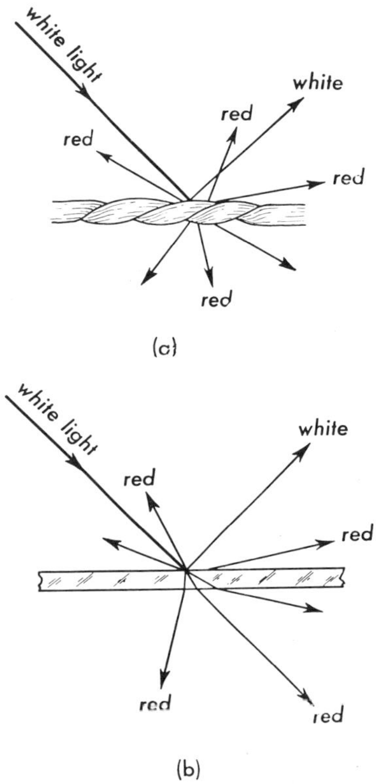

Fig. 22B. Illustration of surface color. Red yarn and red glass appear red on all sides. The other colors, orange, yellow, green, blue, and violet, present in white light, are absorbed within the body.

Not all of the other colors are completely absorbed, for a small amount of each color is reflected from the first surface the white light strikes. This may be illustrated by a polished sheet of red glass as shown in diagram (b). Although the glass appears red from both sides, a small amount of white light is reflected from the top surface, obeying the law of reflection. Some of the red, on the other hand, is reflected and refracted in the usual way at each surface as shown.

The three aspects of surface color are *hue*, lightness or *brightness*, and *saturation. Hue refers to the name of a color, brightness to the relative magnitude of the sensory response, and saturation to the color strength.* Hue is qualitative and is the most distinctive aspect of color, for without hue there is no color. Hue cannot be defined but only exemplified: red, yellow, green, blue, violet, purple, and various intermediaries between these are hues. *Brightness is a subjective intensity* and is a measure of the quantity of light given off by a body. Hue cannot exist alone, for if we have hue it has a certain brightness and saturation. Illustrations of these three concepts are shown in Fig. 22C.

Colors which do not contain any trace of white light are said to be *saturated.* The more white they contain the less saturated they become. Pink is not a saturated color since it is a mixture of red and white. This may be demonstrated by mixing a small amount of red pigment with white paint, or, what is still more striking, by pulverizing a piece of red glass. As the glass is ground finer and finer, the amount of white light reflected is increased by the increasing surface area, until the powder becomes almost white. Although the red light is still present, the white light by comparison is much more intense. A similar effect is produced by transparent substances like crystals and window glass; when powdered they become white. *The smaller the amount of white light mixed in with a color, the greater is the saturation.*

22.3 METALLIC COLOR Some substances appear to be one color by reflected light and a different color by transmitted light. This is particularly true of metals and of certain aniline dyes. Gold, for example, is always yellow-orange by reflected light but, if thin enough, is blue-green by transmitted light. This is illustrated in Fig. 22D.

White light, composed of the spectrum colors red, orange, yellow, green, blue, and violet, is incident on the thin film of gold. Although all of these colors are partially reflected and partially transmitted, the predominant colors in the trans-

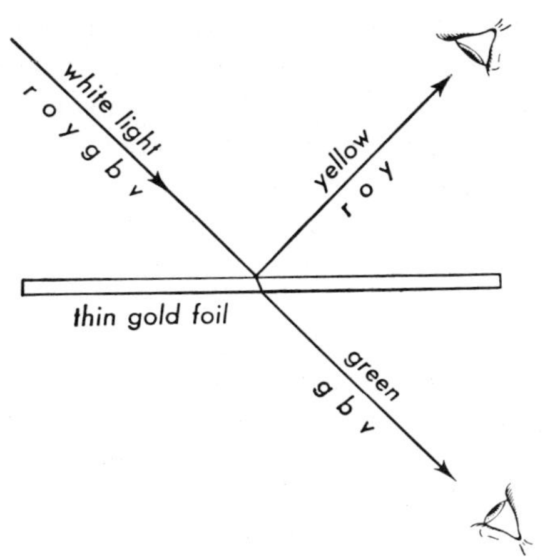

Fig. 22D. A thin gold foil appears yellow-orange by reflected light and blue-green by transmitted light.

mitted light are green, blue, and violet, while those in the reflected light are red, orange, and yellow. To the eye the mixture of reflected colors appears yellow, and the mixture of transmitted colors appears green.

22.4 MIXING SPECTRUM COLORS Over a period of many years different color charts and color theories have been proposed, some of them good and some of them bad. Because the most successful theories have, of necessity, been detailed and complicated, some simplification of their concepts and an explanation of their common principles will be given here.

As a starting point, consider the experiment shown in Fig. 22E in which a narrow beam of white light from a carbon arc and lens falls on a glass prism and is spread out into a complete spectrum. With the prism located near the center of curvature of a large concave mirror, all colors after reflection are brought to a focus on a translucent glass rod where, combined again, they produce white. A large white card is next held in front of the mirror to act as a screen to control the colors that are permitted to mix at the rod. By screening off violet, blue and green, for example, the remaining colors—red, orange and yellow—come together and the rod appears orange.

We now proceed to divide the spectrum up into three equal parts as shown at the lower left in Fig. 22F and to call these parts the *additive primaries.* When red and orange are allowed to mix, the rod appears a bright red; when yellow and green are mixed the rod appears bright green, and when blue and violet are mixed it appears blue-violet. As colors these additive primaries, red, green and blue, appear like the three large circular areas at the upper left in Fig. 22F.

The next step is to mix two *primary colors* two at a time and to observe their resultant color mixture. When primary red and primary green mix at the glass rod they produce yellow; red and blue produce magenta; and green and blue produce cyan, a light blue-green. These colors, the so-called *subtractive primaries,* are shown by the three large circles at the upper right in Fig. 22F, and the overlapping areas at the upper left. The

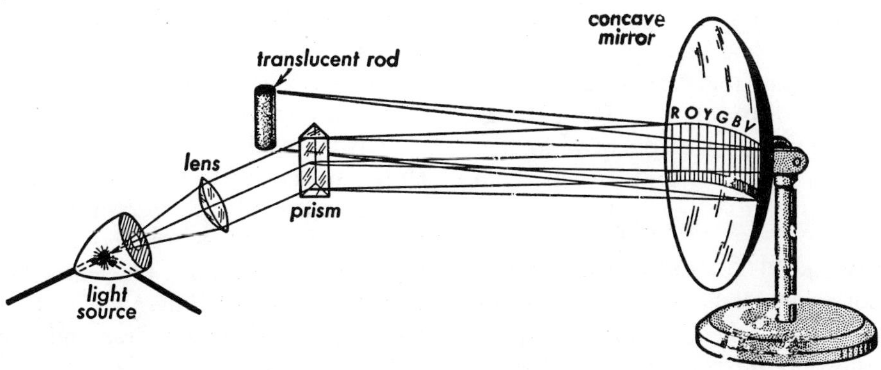

Fig. 22E. Experimental arrangement for mixing pure spectrum colors to form all primary colors.

pure spectrum colors that go to make up each subtractive primary are shown at the lower right.

22.5 THE COLOR TRIANGLE The color triangle, as illustrated in Fig. 22G, is a triangular arrangement of the additive and subtractive primaries with white at the center. Red, green, and blue are located at the corners, while magenta, yellow, and cyan are located at the sides. The order of the colors is such that the sum of any two additive primaries at the corners gives the subtractive primary between them on the sides, and the sum of all three gives white at the center.

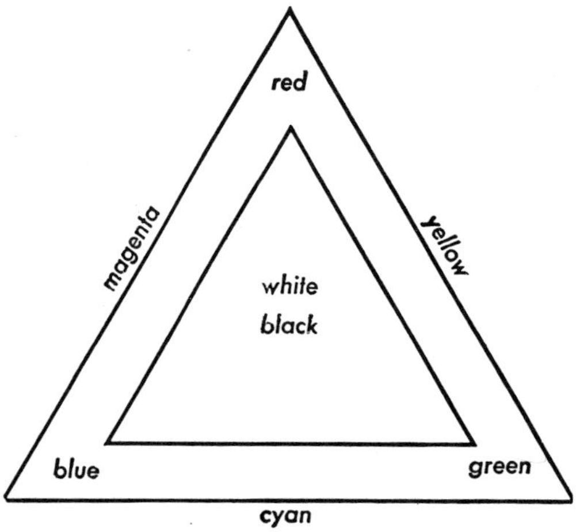

Fig. 22G. Diagram of the color triangle, with the additive primaries at the corners and the subtractive primaries at the sides.

Colors opposite each other on the color triangle are *complementary*. *Two colors are said to be complementary if when added together they produce white.* Magenta and green are complementary, for when added together, as can be seen from their spectral distributions in Fig. 22F, they contain all of the spectrum colors of white light. Similarly red and cyan, as well as yellow and blue, are complementary.

22.6 ADDITIVE METHOD OF COLOR MIXING The mixing of colored lights described in the two preceding sections is called the additive method of color mixing and differs greatly from the subtractive method to be described in the following section. An interesting experiment for demonstrating the additive method is shown in Fig. 22C. Three boxes containing white lights are arranged to illuminate separately the three sides of a white pyramid. A matched set of glass filters, one for each of the additive primary hues, red, green, and blue, respectively, are placed in front of each box opening, thereby illuminating the pyramid faces as shown in the left-hand diagram.

Upon rotating the pyramid slowly a point is reached, as shown in the right-hand diagram, where pairs of lights mix in equal amounts on each of the three faces. These mixtures are the subtractive primaries, magenta, yellow, and cyan. As the pyramid turns from position (a) to position (b), all variations of two colors are seen on the pyramid faces. Television in full color is produced by the additive method of color mixing.

22.7 SUBTRACTIVE METHOD OF COLOR MIXING This is the method most familiar to everyone, the method used in the mixing of pigments to produce various colored paints. For this purpose the subtractive primaries, *magenta*, *yellow*, and *cyan*, often referred to by artists as *red*, *yellow*, and *blue*, are the most useful. The mixing in equal amounts of any two subtractive primaries will produce the additive primary lying between them on the color triangle. When cyan and yellow paints are mixed the result is green.

At first it seems strange that yellow and cyan, neither one of which has the appearance of an additive primary, should produce green when mixed together. A spectrum analysis of these two colors, as shown at the lower right in Fig. 22F, shows that green and yellow are spectrum colors common to both.

Mixing by the subtractive method is demonstrated with prisms and filters in Fig. 22H. To see what happens to each spectral hue in each filter, the white light is first spread out into its complete spectrum. To illustrate, the yellow filter alone in diagram (h) absorbs blue and violet, and the cyan filter alone in (f) absorbs red and orange. When both are inserted as in diagram (d) only green and yellow are transmitted. To the eye this mixture appears bright green. The other two pairs of filters in diagrams (e) and (i) give the other two primaries red and blue.

To carry these experimental demonstrations to the mixing of paint, each little grain of pigment is like a piece of colored glass. See Fig. 22I.

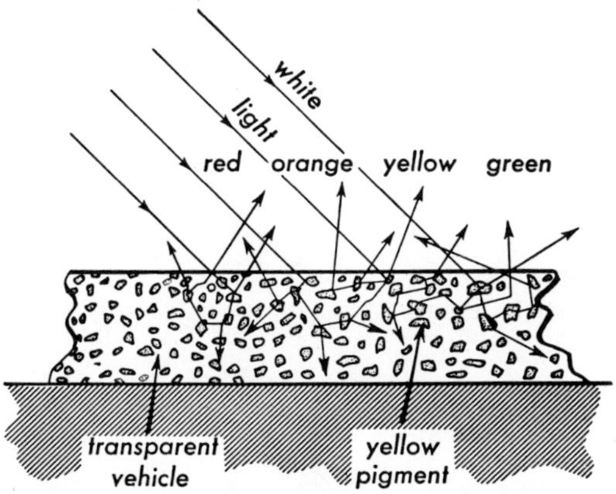

Fig. 22I. Illustration of the absorption of blue and violet light by yellow paint and the emission of red, orange, yellow, and green.

Assuming the oil in which the yellow pigment is imbedded to be transparent, white light entering the paint is reflected and refracted as shown. Wherever blue or violet rays pass through the yellow pigment grains, they are absorbed. After many reflections and refractions, the red, orange, yellow, and green can still escape. Together these four colors appear as yellow. See Fig 22F.

When yellow and cyan pigments are mixed together as illustrated by the detailed diagram in Fig. 22J, only green and yellow light is transmitted by both pigments.

The essential difference between the additive method and subtractive method of color mixing is just that suggested by the name; in the additive method the resultant color is just the *sum* of the two constituents used to produce it, and in the subtractive method it is just the *difference* between the two. Addition always produces a brighter color and subtraction produces a darker

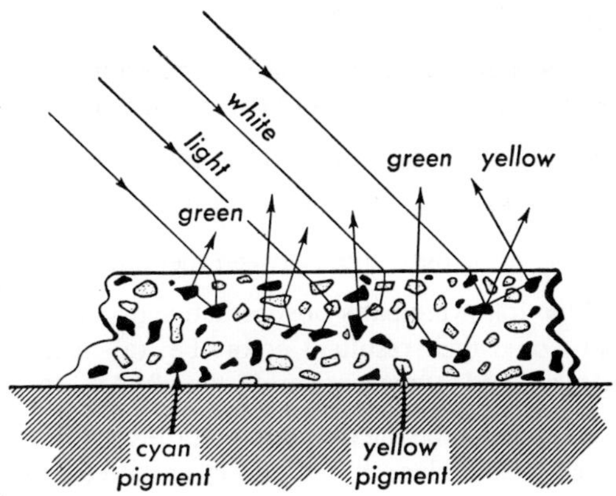

Fig. 22J. When blue and yellow paints are mixed together, green and yellow are the only pure spectral colors transmitted by both pigments.

color. Just as the additive mixing of red, green, and blue produces white, so the subtractive mixing of magenta, yellow, and cyan produces black. Similarly, two complementary colors, when mixed additively, produce white, and when mixed subtractively produce black. It is for this reason that white and black are shown at the center of the color triangle in Fig. 22G.

22.8 THE ICI CHROMATICITY SYSTEM Well-planned steps toward a quantitative measurement of color were taken by the International Commission on Illumination in 1931. At that time three additive primaries, red, green, and blue, were adopted in which the visible spectrum was divided into three overlapping spectral response regions somewhat similar to those at the lower left in Fig. 22F. Although a treatment of this standard ICI system* must be left for more advanced studies, it should be mentioned here that any given color sample can be measured with a spectroscope in terms of the three adopted primaries and the results of the measurements expressed by two numbers. These two numbers can then be plotted on a graph.

* For a treatment of the ICI color system see *Principles of Physics*, Chap. 13, by F. W. Sears, Addison-Wesley Press.

QUESTIONS

1. Make a diagram of the color triangle and include (a) the primary colors at the corners and sides, and (b) the two colors at the center.

2. Make three spectrum charts of the additive primary colors.

3. Make three spectrum charts of the subtractive primary colors.

4. Explain the difference between metallic color and body color.

5. If the following pairs of subtractive primary colors are added together, what would their resultant colors be: (a) magenta and green, (b) yellow and cyan, and (c) magenta and yellow? [Ans. (a) black, (b) green, (c) red.]

6. What would be the resultant color if the following pigments are mixed: (a) yellow and blue, (b) red and cyan, and (c) magenta and cyan?

7. What color added to green will give (a) yellow, (b) cyan, and (c) white? [Ans. (a) red, (b) blue, (c) magenta.]

8. Make diagrams and explain how magenta and yellow pigments when mixed together as paints can produce red. Do the pigment particles themselves become red?

9. Define or briefly explain complementary colors. When two complementary colors are added together name the spectrum colors present. [Ans. red, orange, yellow, green, blue, and violet.]

10. When two complementary colors are added as pigments, what spectrum colors are to be seen from the mixture?

11. What color mixed subtractively with magenta will produce (a) blue, (b) black, and (c) red? [Ans. (a) cyan, (b) green, (c) yellow.]

12. What color mixed as pigments with cyan will produce (a) blue, (b) green, and (c) black?

13. The pictures on a color television screen are produced by red, green, and blue phosphorescent dots. What two colors produce yellow? [Ans. red and green.]

14. Sunlight is allowed to pass through a piece of yellow glass. The emergent light is now incident on a piece of cyan glass. What spectrum colors emerge from the second piece of glass?

23 | LENSES AND OPTICAL INSTRUMENTS

The lens is a very old optical instrument, dating from at least 500 years B.C. A converging lens of rock crystal was found in the ruins of Nineveh, and "burning glasses" were manufactured by the Greeks in large quantities several centuries B.C.

Today lenses are used in thousands of optical devices and instruments. Some of the most common of these are: *cameras, picture projectors, microscopes, telescopes, binoculars, periscopes, range finders, transits, magnifiers, and lasers.*

23.1 LENSES The primary function of a lens is to form images of real objects. Although most lenses are made of common glass, a few special lenses are made of other transparent materials like *quartz* and *fluorite*. To understand the principles upon which a lens functions, imagine a set of several matched prisms and blocks of glass arranged in the order shown in Fig. 23A. In the first arrangement the prisms are made to refract the incoming parallel light rays in such a way that the emergent rays converge to a focus at F. In the second arrangement the parallel rays are made to diverge as if they had come from a common point F'. In each system the greatest deviation occurs at the outermost prisms, for they have the greatest angle between the two re-

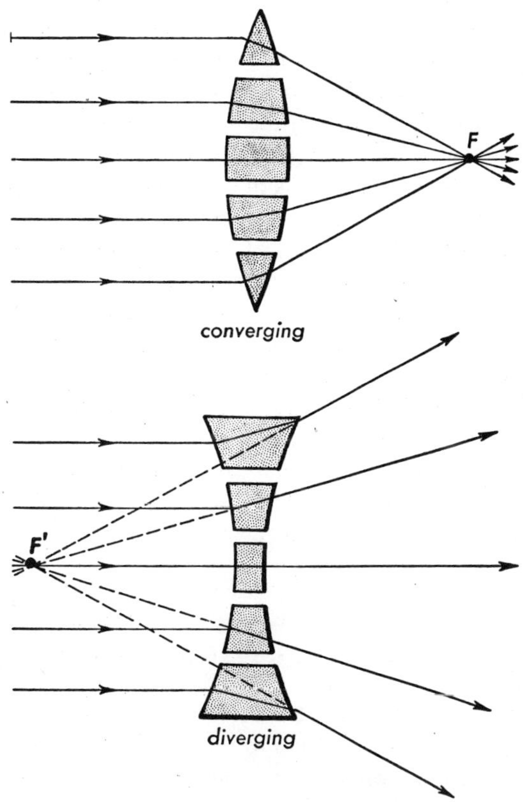

Fig. 23A. Matched sets of prisms illustrating lenslike action.

fracting surfaces. No deviation occurs for the central rays, for at that point the glass faces are parallel to each other.

A real lens is not made of prisms, as indicated in Fig. 23A, but of a solid piece of glass with surfaces ground to the form of a sphere. Cross sections of several standard forms are shown in Fig. 23B. The first three lenses, which are thicker in

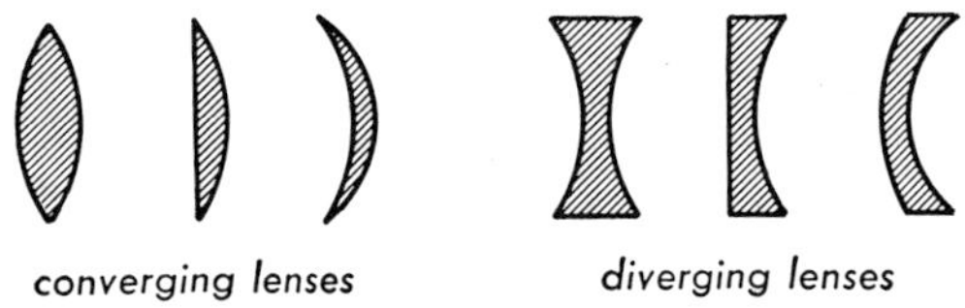

Fig. 23B. Cross sections of standard forms of common lenses.

the center, are called *converging* or *positive lenses*, while the last three, which are thinner in the center, are called *diverging* or *negative lenses*. Special names attached to each of the six lens types shown are (1) *double convex*, (2) *plano-convex*, (3) *convex miniscus*, (4) *double concave*, (5) *plano-concave*, and (6) *concave miniscus*.

Diagrams showing the refraction of light by converging and diverging lenses are given in Fig. 23C. The principal axis in each case is a straight line passing through the center of a lens perpendicular to the two faces at the points of intersection. The principal focus F lies on the principal axis and is defined for a converging lens as the point where parallel light rays are brought together, and for a negative lens as a point from which parallel light rays appear to originate. *By symmetry every lens has two principal foci, one on each side of the lens and at the same distance from the center of the lens.* The distance from the focal point to the lens is called the focal length.

$$CF = \text{focal length} = f$$

A plane perpendicular to the principal axis which passes through either principal focus is called the *focal plane*. Parallel light rays entering the lens from any other direction than shown in the diagrams will come to a focus at some point on the focal plane. This point is readily located by remembering that a ray through the lens center does not change in direction.

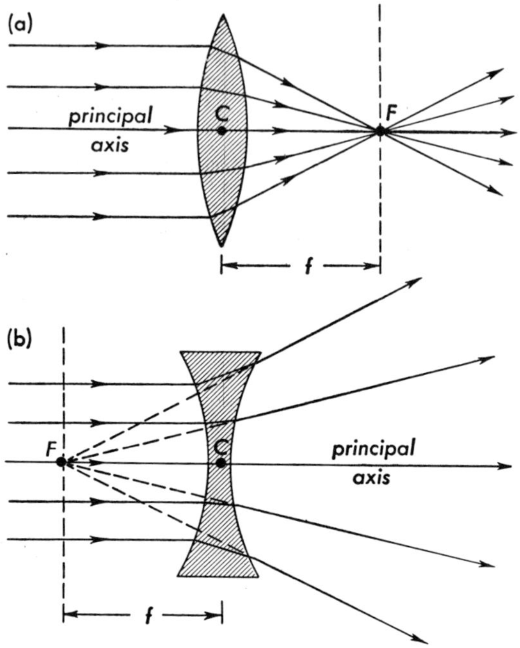

Fig. 23C. Converging and diverging lenses.

The greater the curvature of the two surfaces of a lens, the shorter is its focal length. The reason for this, as can be seen from the diagrams, is that the greater the curvature the greater is the deviation of the light rays passing through near the edges of the lens.

One important principle concerning lenses is the reversibility of light rays. If a point source of light is placed at F in Fig. 23C(a), the rays of light which strike the lens will be refracted into a parallel beam of light moving to the left. Similarly, in Fig. 23C(b), if light rays are converging toward the focal point F, they will be refracted by the lens into a parallel beam.

23.2 IMAGE FORMATION When an object is placed on one side of a converging lens beyond the principal focus, a real image will be formed on the opposite side of the lens. This is illustrated in Fig. 23D. If the object is moved closer to the focal point, the image will be formed farther away from the lens and will be bigger, i.e., magnified. As the object is moved farther away from the lens, the image is formed closer to the focal point and is smaller in size.

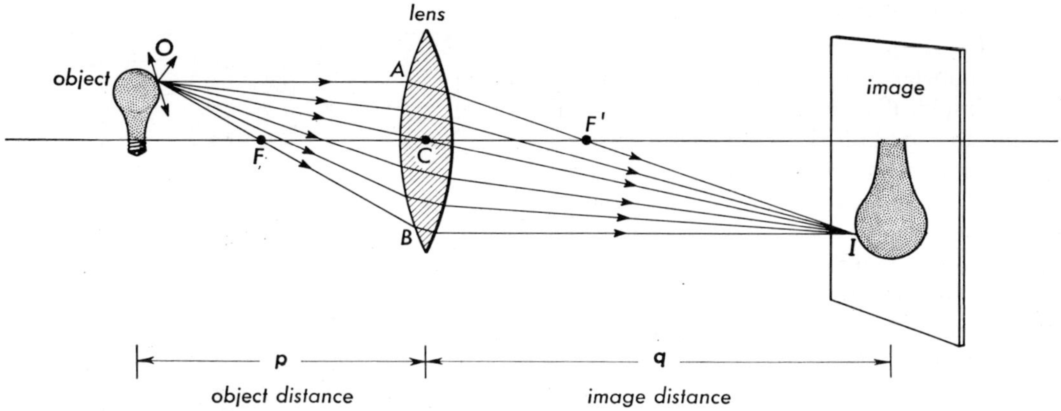

Fig. 23D. Ray diagram illustrating the formation of a real image by means of a single converging lens.

In general there are two ways of accurately determining the position of an image: one is by graphical construction and the other is by use of the following lens formula:

$$\boxed{\frac{1}{p} + \frac{1}{q} = \frac{1}{f}} \tag{23a}$$

where *p is the object distance, q the image distance*, and *f the focal length.*

The graphical method is illustrated in Fig. 23E. Consider the light emitted by some one particular point like O in the object. Of the rays going out from this point in all directions the ray OA traveling parallel to the principal axis will be refracted to pass through the focal point F. See Fig. 23C(a). The ray OC arriving at the center of the lens where the faces are parallel will pass straight through, meeting the other ray at some point I. These two rays locate the tip of the image at I. All other rays from the point O which strike the lens will be brought to a focus at this same point. To check this, note that the ray OF, which passes from O through F, by the principle of the reversibility of light rays, will be refracted parallel to principal axis, crossing the other ray at I, as shown.

The use of the lens formula can be illustrated by the following example. Let an object be placed 60 cm in front of a lens of focal length 20 cm. If we solve Eq. (23a) for q, we obtain the expression:

$$q = \frac{pf}{p - f} \tag{23b}$$

Then substituting the known quantities, we obtain

$$q = \frac{60 \times 20}{60 - 20} = 30 \text{ cm}$$

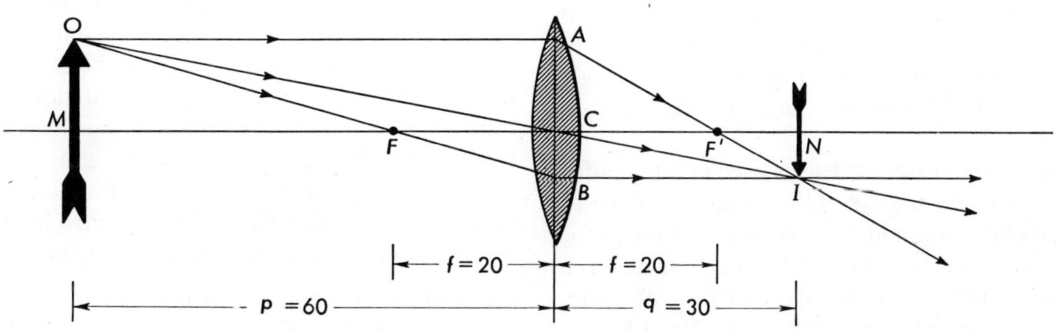

Fig. 23E. Graphical determination of image position and size.

The image is formed 30 cm from the lens or 10 cm from F.

If a centimeter rule is used to construct this problem graphically, the resultant diagram will be similar to Fig. 23E. Each line is drawn in its proper position and size and, when the image is located by all three rays its position and size are measured by the same scale. Drawn carefully, the graphical results will agree in every detail with those calculated by the above formulas.

Note in Fig. 23E that the triangle OMC is similar to triangle INC. From the fundamental theorem in geometry that corresponding sides of similar triangles are proportional, we can write

$$\frac{\text{IN}}{\text{OM}} = \frac{\text{NC}}{\text{MC}}$$

If we call OM the object height O, and IN the image height I, we can write

$$\frac{I}{O} = -\frac{q}{p} \tag{23c}$$

where the relation I/O is called the *magnifying power M*.

$$M = \frac{I}{O} \tag{23d}$$

23.3 VIRTUAL IMAGES The images formed by the lenses in Figs. 23D and 23E are *real. Real images* are defined as those which can be formed on a screen and are characterized by the fact that rays of light are actually brought together to a focus there. *Virtual images* are not real, they cannot be formed on a screen, and the rays from different points on the object do not pass through corresponding points in the image. Virtual images may be observed with a converging lens by placing an object close to a lens and inside the focal point, or by a diverging lens with the object at any point. These two examples are illustrated in Fig. 23F.

In the first case, the lens is used as a magnifier, or reading glass. Rays of light radiating from the point of the object at O are refracted in the proper direction but are not sufficiently deviated to come to a focus. To the observer's eye at E these rays appear to be coming from a point I back of the lens. This is a *virtual image, right side up and magnified*. To find this image graphically, we observe that the ray FOA must be refracted parallel to the principal axis. The ray OC through the center of the lens goes on undeviated. These two refracted rays extended backward intersect at I. If the lens formula, Eq. (23a) or (23b), is used to find the image in such a case, the image distance q will come out as a negative quantity, showing it is a virtual image on the same side of the lens as the object.

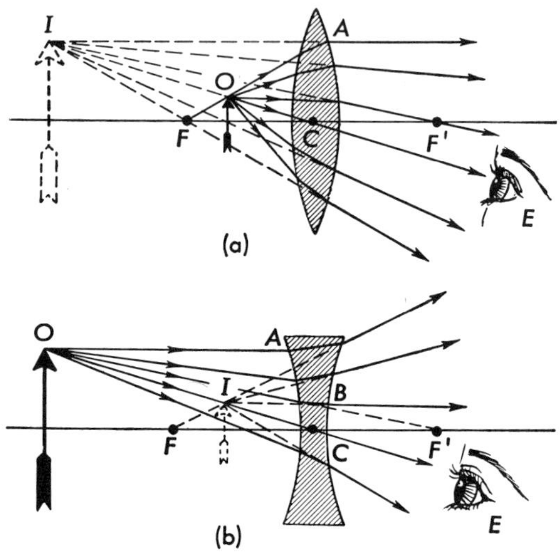

Fig. 23F. Illustration of the formation of virtual images.

In the case of a negative lens, the image is always virtual, closer to the lens, and smaller in size than the object. As shown in Fig. 23F(b), light rays diverging from the object point O are made more divergent by the lens. To the observer's eye at E, these rays appear to be coming from the point I back of but close to the lens. To find this image we observe that the ray OA parallel to the principal axis must be refracted in such a direction that it appears to come from F. The ray OC through the center goes on undeviated. Since these two directions intersect at I the image is formed there.

In applying the lens formula to a diverging lens the focal length f is always *negative* in sign. To illustrate this, consider the following example.

EXAMPLE. An object is placed 30 cm in front of a diverging lens having a focal length of -15 cm. Locate the image position.

SOLUTION. By direct substitution in Eq. (23b), we obtain

$$q = \frac{pf}{p - f} = \frac{30 \times (-15)}{30 - (-15)} = \frac{-450}{+45} = -10 \text{ cm}$$

The image is found at 10 cm from the lens, the negative sign indicating it is virtual and therefore on the same side of the lens as the object.

The sign conventions for all lens problems solved by using Eqs. (23a), or (23b) are the following:

1. A real object is always located to the left of the lens.

2. Light rays entering and passing through a lens are drawn from left to right.

3. All distances are measured from the center of the lens.

4. For any object located on the left side of the lens the object distance is positive.

5. For any image on the right side of the lens the image distance is positive.

6. For any image on the left side of a lens the image distance is negative.

When using Eq. (23a) to solve a problem one should remember always to substitute into the equation two of the quantities with the correct *sign* and *magnitude*, allowing the equation to indicate the correct sign and magnitude of the third quantity.

23.4 THE CAMERA Since the photographic camera employs but a single lens unit it may be considered as one of the simplest of all optical instruments. As illustrated by the roll-film camera in Fig. 23G, a converging lens forms a *real* and *inverted image* on the film. If the object is far away, the light rays approaching the lens are nearly parallel and the image is formed at the focal plane. If the object is close up, the image will be formed beyond the focal plane as shown in the diagram. To permit distant landscapes or "close ups" to be taken with the same camera, a bellows is used, allowing the lens distance to be varied at will. Motion of the lens to the proper image distance is called *focusing*.

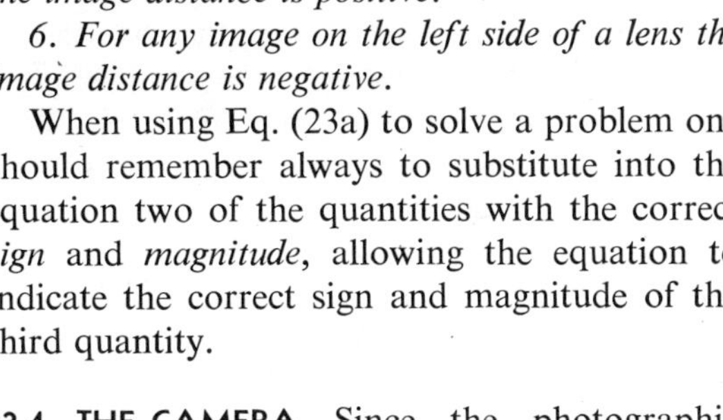

Fig. 23G. Diagram of the image formation by a camera.

Only a simple converging lens is used in the cheapest of cameras, which means that all of the common defects of images are present to give rise to a slightly blurred or diffuse image. In more expensive cameras, however, the most objectionable defects are fairly well corrected by a compound lens made of several individual lenses. As a rule, a good camera lens will contain from three to five lens elements and will partially correct for *chromatic aberration*, *spherical aberration*, *astigmatism*, and *curvature of field*.

The purpose of an iris diaphragm is to decrease the effective aperture of a lens and hence increase its *f number*. Such practice is desirable in photographing still objects because the smaller the lens opening, the sharper will be the focus of near and far as well as central and peripheral objects. *The f number of a lens is equal to its focal length divided by its diameter.*

23.5 THE EYE Some aspects of this most remarkable optical instrument, the human eye, have been presented in the Introduction. There it was pointed out that a single eye is in principle an exceptionally fine camera with an elaborate lens system on one side and a sensitive screen or photographic film called the *retina* on the other.

When light from a distant object passes through the lens system of the eye, it is refracted and brought to a focus on the retina. There a real but inverted image of the object is formed. It is a most amazing fact that, while all retinal images are inverted, as shown in Fig. 23H, they are interpreted as being erect.

Accommodation is the ability to focus the eyes on near and far objects. In a camera the focusing of a picture on the photographic film or plate is accomplished by moving the lens toward or away from the film. In the human eye, however, *focusing is brought about by changing the shape of the*

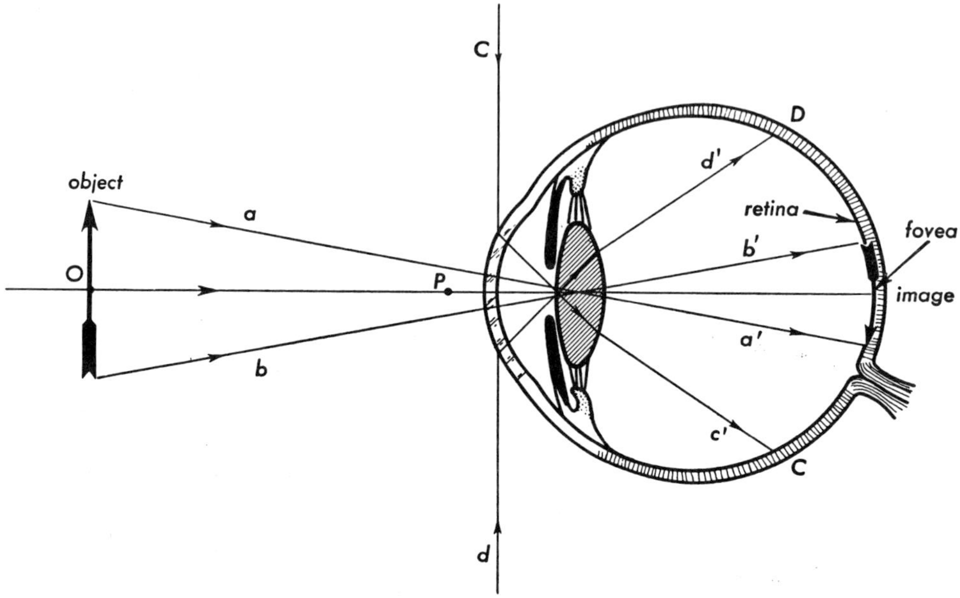

Fig. 23H. The human eye is similar to a camera. All retinal images are inverted.

crystalline lens. This is achieved by a rather complicated system of ligaments and muscles. Due to a tension which exists in the lens capsule (see Introduction, Fig. B), the crystalline lens, if completely free, would tend to become spherical in shape. The edge of the lens is surrounded by the *ciliary muscle* which, by contracting, causes the lens to bulge out. This reduces the focal length of the lens, bringing nearby objects to focus on the retina. When the ciliary muscle relaxes, the suspensory ligaments, being under tension, pull at the edges of the lens, thus tending to flatten it. Under these conditions the focal length increases, bringing distant objects to focus on the retina. This is the accommodation process.

The normal eye is most relaxed when it is focused for parallel light, i.e., for objects far away. To study the detail of an object, however, the object should be brought close to the eye. The reason for this is that the closer the object is to the eye the larger is the image formed on the retina. A distance of about 10 in. is found to be the distance of most distinct vision. Prolonged observation at distances of 10 in. or less will result in a considerable amount of fatigue and eyestrain.

It frequently happens that the cornea acquires a greater curvature in one plane than in another. Such irregularities are called *astigmatism*. Whether accommodation by the crystalline lens has been lost or not, such eyes require astigmatic spectacle lenses, i.e., lenses that have more curvature in one direction than at right angles. The surfaces of such lenses are not spherical and are a little more difficult to grind and polish.

23.6 THE TELESCOPE The first known telescope was constructed in Holland in 1608 by an obscure spectacle lens grinder, Hans Lippershey. A few months later Galileo, upon hearing that objects at a distance may be made to appear close at hand by means of two lenses, designed and made with his own hands the first authentic telescope. The elements of this telescope are still in existence and may be seen on exhibit in Florence, Italy.

Astronomical telescopes today are practically the same in principle as they were in the earliest days of their development. A diagram of a small telescope is shown in Fig. 23I. Light rays from a single point of a far distant object are shown

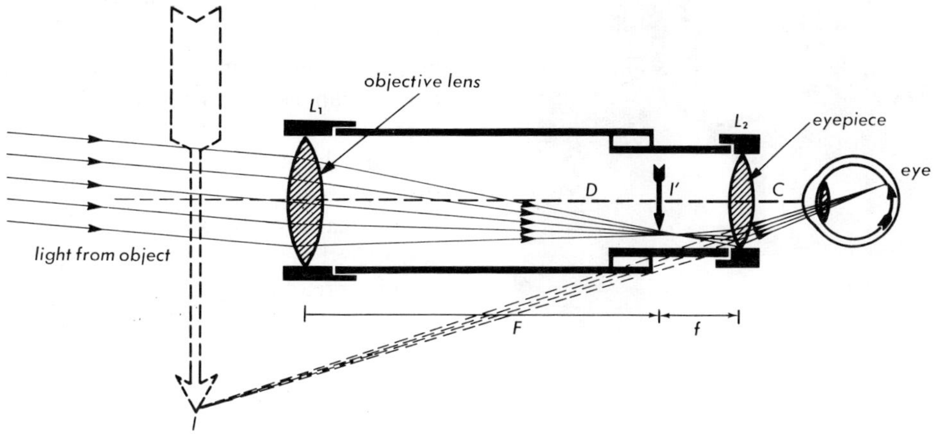

Fig. 23I. Illustration of the paths of light rays in a simple astronomical telescope.

entering the objective lens as a parallel beam. These rays are brought to a focus and form a point image at I'. In a similar manner, parallel sets of rays from other points of the same object (not shown) will form point images in the focal plane of the objective. Assuming that the distant object is an arrow pointing upward, the image, as shown in the diagram by a red arrow, is *real* and *inverted*.

The function of the second lens in a telescope is to magnify the image formed by the objective. For this purpose a converging lens of short focus, called the *eyepiece*, is usually used. If the eyepiece is moved to a position where the image I' is just inside its focal plane, a magnified image I will be seen by the eye. For normal observation, the image I' and the focal planes of both lenses are in conjunction, and the light emerging from the eyepiece is parallel. The image I under these conditions appears to be far off to the left of L_1, "at infinity," and is of approximately the same angular size as shown in the diagram. Actually, of course, the final image is the one formed on the retina by rays which appear to have come from I.

The magnifying power of a telescope is defined as the ratio between the angle subtended at the eye by the final image I′ and the angle subtended at the eye by the object itself. In other words, it is the number of times larger an object appears to be when viewed with the telescope. When plane geometry is applied to a simple light-ray diagram of a telescope, it is found that the magnifying power is equal to the ratio of the focal lengths of the two lenses.

$$\boxed{\text{Magnifying power} = \frac{F}{f}} \tag{23c}$$

where F is the focal length of the objective, and f is the focal length of the eyepiece.

As an illustration, suppose the objective lens of a small telescope has a focal length of 1 yard and that it is used with an eyepiece having a focal length of 1 in.; the magnifying power would be 36/1 or 36. Distant objects viewed through this telescope would appear to be 36 times as tall and 36 times as wide as when viewed with the unaided eye. If an eyepiece with a focal length of only half an inch is used with the same objective, the magnification would be twice as great, or 72.

23.7 PRISM BINOCULARS Prism binoculars are in reality a pair of twin telescopes mounted side by side, one for each of the two eyes. The objective lenses in front and the eyepieces at the rear are converging lenses as in the astronomical telescope, but each pair of total reflecting prisms (see Fig. 23J) inverts the rays to give erect images. The doubling back of the light rays has the ad-

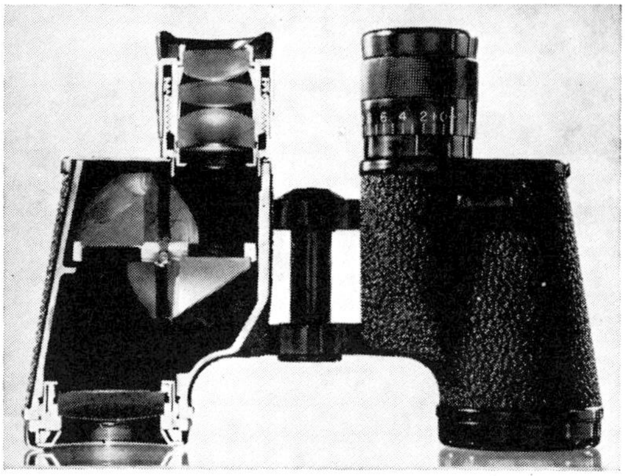

Fig. 23J. Cutaway of a prism binocular showing the lenses and reflecting prisms. (Courtesy of Bausch & Lomb Optical Company.)

vantage of enabling long focus objective lenses to be used in short tubes, thus giving higher magnification. In addition to good achromatic lenses and accurately ground prisms, three features of good binoculars are (1) *magnification*, (2) *field of view*, and (3) *light gathering power*.

For hand-held use, binoculars with a 6, 7, or 8 power magnification are most generally useful. Glasses with powers above 8 usually require a tripod mount to hold them steady. For powers less than 4, aberrations produced by the lenses usually offset the magnification and the average person can usually see better with the unaided eyes.

The field of view is determined principally by the eyepiece aperture and should be as large as is practical. For 7 power binoculars a 6° field is considered large, since in the eyepiece the same field is spread over an angle of 7 x 6 or 42°.

The diameter of each objective lens determines the light gathering power, which is important at night when there is so little light available.

Binoculars with the *specification* 6 x 30 have a magnification of 6 and objective lenses with an effective diameter of 30 mm. The specification 7 x 50 means a magnification of 7 and objective lenses 50 mm in diameter. Although glasses with the latter specifications are excellent day or night, they are considerably larger and more cumbersome than daytime glasses specified as 6 x 30's or 8 x 30's. For general civilian use these latter two types are by far the most desirable.

QUESTIONS

1. Describe an easy way to measure the focal length of a converging lens.

2. How can you distinguish between a converging and a diverging lens by feeling it?

3. Write down the lens-image distance formula from memory. Solve this equation for f as the unknown quantity.

4. What is the range of object distances that will produce a real image with a converging lens?

5. Where should an object be placed if a virtual image is to be formed with a converging lens?

6. How can you demonstrate conjugate foci? If object and image are interchanged, will the ratio of object to image size be the same?

7. Where can an object be located in front of a diverging lens in order to form a real image?

8. Define or briefly explain each of the following: (a) focal length, (b) f-number, (c) accommodation, (d) magnifying power, and (e) virtual image.

PROBLEMS

1. An object 3 cm high is located 18 cm in front of a converging lens with a focal length of +6 cm. Calculate (a) the position, and (b) the size of the image. (c) Solve this problem graphically and compare answers. [Ans. (a) +9 cm, (b) $-1\frac{1}{2}$ cm.]

2. An object 2 cm high is located 6 cm in front of a lens. A real and inverted image is formed 12 cm from the lens on the opposite side. What is the focal length of the lens?

3. An object 3 cm high is located 8 cm in front of a lens. A real and inverted image is formed 12 cm from the lens on the opposite side. Calculate the focal length of the lens. [Ans. 4.80 cm.]

4. An object 2 cm high is located 6 cm in front of a converging lens. If the focal length of the lens is +8 cm, find (a) the image distance, and (b) the image size. (c) Solve graphically and check your answers with your calculated answers.

5. An object 1.5 cm high is located 6 cm in front of a lens of focal length +10 cm. Find (a) the image distance, and (b) the image size. (c) Solve graphically and compare your answers with your calculated values. [Ans. (a) −15 cm, (b) +3.750 cm.]

6. An object 3 cm high is located 12 cm in front of a diverging lens of focal length −6 cm. Find (a) the image distance, and (b) the image size. Solve graphically and compare with your calculated values.

7. An object 4 cm high is located 12 cm in front of a diverging lens of focal length −3 cm. Find (a) the image distance, and (b) the image size. Solve graphically as well and compare your answers. [Ans. (a) −2.400 cm, (b) +0.800 cm.]

8. Where in front of a converging lens of focal length +3 cm should an object be placed if the image is to be 3 times the size of the object and real?

9. A magnifying lens with a focal length of +2.0 cm is placed 1.5 cm from a small object. Calculate the (a) image distance, and (b) magnification. (c) Solve graphically and compare your answers. [Ans. (a) −6.00 cm, (b) +4.00 cm.]

10. The objective lens of a small telescope has a focal length of 75 cm. Find the magnification if the focal length of the eyepiece is 1.5 cm.

11. A telescope objective lens has a focal length of 120 cm. Find the focal length of the eyepiece if the magnification is to be 50. [Ans. 2.400 cm.]

12. The objective lenses of a pair of binoculars have a diameter of 5.0 cm, and a focal length of 32 cm. If the eye lenses have a focal length of 4.0 cm, find the binocular specifications.

13. An object 3 cm high is located 20 cm in front of a converging lens having a focal length of 10 cm. (a) Calculate the position of the image. A second lens with a focal length of +10 cm is located 5 cm behind the first lens. (b) Find the position of the final image. (c) Solve graphically and check your answers. [Ans. (a) +20 cm, (b) +6.0 cm from the second lens.]

14. An object 3 in. high is located 24 cm in front of a converging lens with a focal length of 8 cm. Calculate (a) the image distance, and (b) the image size. (c) Solve this problem graphically and compare answers.

24 DIFFRACTION, INTERFERENCE, AND POLARIZATION

In the three preceding chapters on light, we have seen how the optical phenomena of reflection, refraction, and dispersion are treated, assuming that light travels in straight lines, and that the fundamental principles of geometry can be applied. This is the reason these various natural phenomena are referred to as *geometrical optics*.

In this chapter we are concerned with that group of optical phenomena referred to at the beginning of Chapter 21 as *physical optics*. These are the phenomena most easily explained on the assumption that all light rays are bundles of waves. In other words, light is a wave motion and different colors of the visible spectrum have different wavelengths and frequencies.

When light passes close to the edge of any object, it is bent slightly in its path and travels on in a new direction. This bending of light around corners is called *diffraction*. In the preceding chapters light has been shown to travel in straight lines and to obey the laws of reflection and refraction. Furthermore, according to the rectilinear propagation of light, it is customary to assume that an object will cast a sharp and well-defined shadow. A close examination of every shadow, however, shows that the edges are not sharp, but blurred and diffused.

If one is careful to choose a small source of light, such as the light emanating from a pinhole in a screen, the shadow of an object cast on a distant screen is bounded at the edges by narrow bands or fringes of light. To observe these effects the following simple experiment may be performed in a darkened room. A box containing a light bulb and a pinhole is placed on one side of the room and a ground glass observing screen or photographic film is placed on the other. The objects whose shadows are to be observed are then placed about halfway between the source and the screen. This is the arrangement used in obtaining the original photographs reproduced in Fig. 24A. This is a photograph of light diffracted by a pin, a small hexagonal nut, and a piece of wire screen.

24.1 HUYGENS' WAVE THEORY OF LIGHT The wave theory of light was first proposed by the English physicist, Robert Hooke, in 1665, and improved twenty years later by the Dutch scientist and mathematician, Christian Huygens.*

---

* CHRISTIAN HUYGENS (1629–1695), famous Dutch physicist and contemporary of Isaac Newton. Born at

Fig. 24A. Photographs of the shadows cast by small objects. The narrow bands are due to the diffraction of light.

Everyone has at some time or other dropped a stone in a still pond of water and watched the waves spread slowly outward in ever-widening concentric circles. See Fig. 19I. In the analogous case of a point source of light, the spreading waves form concentric spheres moving outward with the extremely high velocity of 186,300 mi/s. This is represented diagrammatically in Fig. 24B. Each circle represents the crest of a wave so that the distance between consecutive circles is one wavelength.

According to Huygens' principle every point on any wave front may be regarded as a new point source of waves. Regarding each of any number of points like *a*, *b*, *c*, etc., as point sources like *S*, secondary wavelets spread out simultaneously as shown. The envelope of these an instant later is the new wave front A, B, C, etc., and still later the wave front L, M, N, etc. Although Huygens' principle at first might seem to be a useless play with circles, it has quite general application to many optical phenomena.

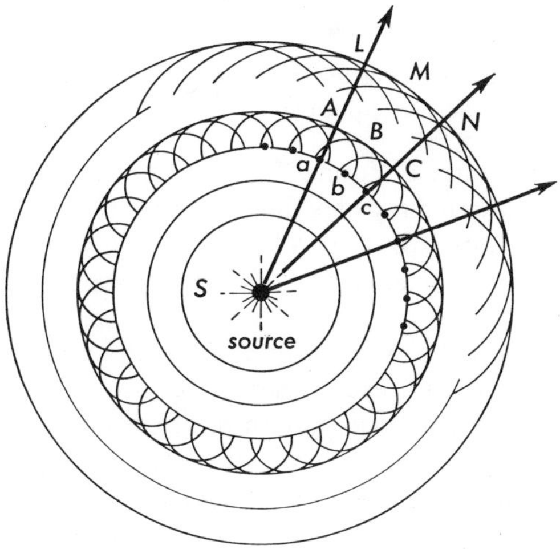

Fig. 24B. Diagram of waves spreading out from a point source. The secondary wavelets and new wave fronts illustrate Huygens' principle.

The Hague in 1629, young Christian got his first ideas about waves and their propagation by watching the ripples on the canals about his home. Although his chief title-deed to immortality is his development of the wave theory of light, he made many and valuable contributions to mathematics and astronomy. He improved upon the method of grinding telescope lenses and discovered the Orion nebula, part of which is now known by his name. He was elected to the Royal Society of London in 1663, and delivered before that august body the first clear statement of the laws governing the collision of elastic bodies. He died a confirmed bachelor at The Hague in 1695.

A direct experimental demonstration of Huygens' principle is illustrated in Fig. 24C. Plane waves approaching a barrier AB from the left are reflected or absorbed at every point except at S where they are allowed to pass on through. When the experiment is carried out with water waves, one can see the waves spreading out in all directions as if S were a point source. The lines represent crests of waves, one wavelength apart.

24.2 YOUNG'S DOUBLE-SLIT EXPERIMENT The crucial test between Newton's corpuscular theory of light and Huygens' wave theory came in 1801

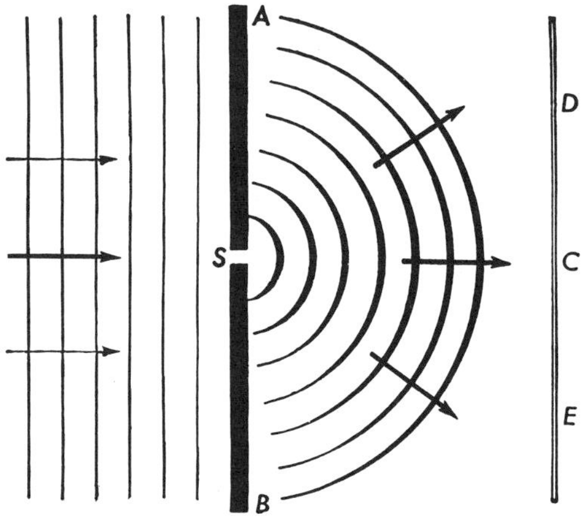

Fig. 24C. Diagram of the diffraction of waves at a small opening. Huygens' principle.

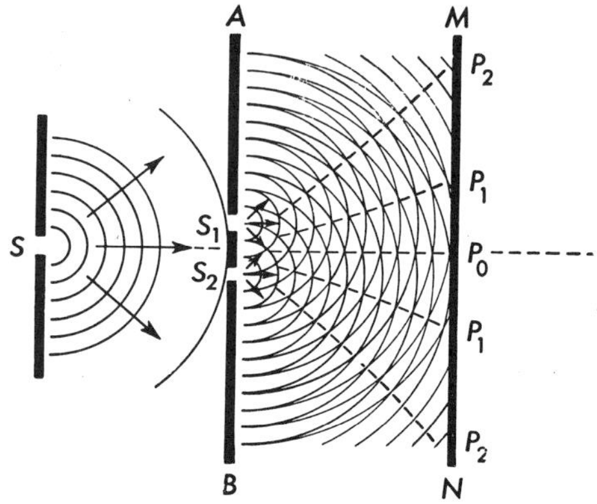

Fig. 24D. Diagram of Young's double-slit experiment illustrating the interference of light waves.

when Thomas Young* performed his now famous interference experiment. This is represented schematically in Fig. 24D. Sunlight from a pinhole S was allowed to fall on a distant screen containing two pinholes, S_1 and S_2. The two sets of spherical waves emerging from the two holes interfered with each other in such a way as to form a symmetrical pattern of bands on another screen MN. This experiment is now regarded as the first definite proof that light is a wave motion.

For convenience it is now customary to repeat Young's experiment with narrow slits in place of pinholes. If S, S_1, and S_2 in Fig. 24D represent the cross sections of three narrow slits, the light falling on the farther screen MN has the appearance of equidistant bands or fringes as shown by the photograph in Fig. 24E. The bright fringes

Fig. 24E. Interference fringes produced by a double-slit as in Young's experiment.

correspond to the points P_0, P_1, P_2, etc., and the dark fringes to the points halfway between.

As the waves travel outward from each slit S_1 and S_2, the crests cross each other only at points which lie along the dotted lines shown in the diagram. These represent the points where the crests of two waves come together and produce a maximum brightness. About halfway between these dotted lines lie other points where the crest of one wave and the trough of another cancel each other and produce darkness. This is called *interference*. Where the bright fringes are formed there is *constructive interference*, and where the dark fringes appear there is *destructive interference*.

* THOMAS YOUNG (1773–1829), English scientist, was born of a Quaker family. At the age of 14, he knew Latin, Greek, French, Italian, Hebrew, Arabic, and Persian. He studied medicine at Edinburgh and Gottingen, became a practicing physician in 1799 and was made professor of physics at the Royal Institution in 1801. Young is best known for his work in optics, particularly experiments demonstrating the wave properties of light. His work on the elastic properties of solids led to his name becoming attached to the stretch modulus of solids. In 1801, he described the vision defect known as astigmatism and put forward the hypothesis, later developed by H. von Helmholtz, that color perception depends on the presence in the retina of three kinds of nerve fibers which respond to red, green, and violet light. He should certainly be classified as one of the great men of science.

24.3 MEASURING THE WAVELENGTH OF LIGHT

A formula for the wavelength of light can be derived from the geometry of Young's double-slit experiment as shown in Fig. 24F. Let P be the

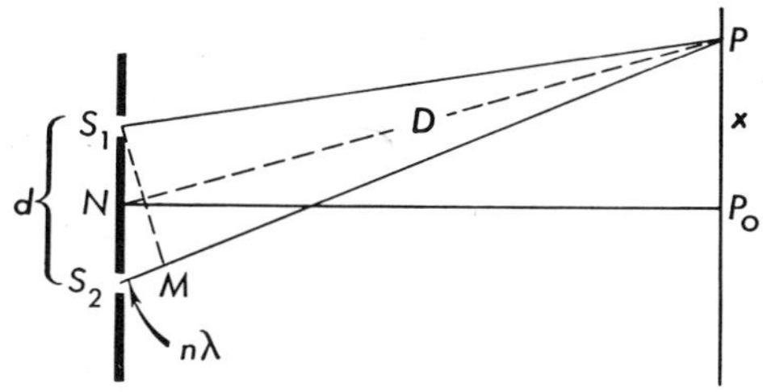

Fig. 24F. Geometrical relations for the double-slit experiment.

position of any bright fringe on the screen and x its distance from the central fringe at P_0. P_0 is located on the perpendicular bisector of the double slit S_1 and S_2. A straight line from each slit to the point P is drawn, and with a compass of radius S_1P the arc of a circle S_1M is scribed. By this construction line MP is made equal to S_1P and the short line S_2M becomes the extra distance light must travel from the lower slit. To produce a bright fringe at P the interval S_2M must be equal to one whole wavelength, two whole wavelengths, three whole wavelengths, etc., for only then will the waves from S_1 and S_2 arrive at P in phase. Therefore, S_2M must be equal to $n\lambda$, where n is a whole number, $n = 0, 1, 2, 3, 4$, etc., and λ is the wavelength of the light.

Since the distance d between slit centers is extremely small compared with the distance D to the screen, line S_1M may be considered straight and at right angles to all three lines S_1P, NP, and S_2P. With corresponding sides mutually perpendicular to each other, triangles S_1S_2M and NPP_0 are similar to each other. From the well-known theorem that corresponding sides of similar triangles are proportional,

$$\frac{n\lambda}{d} = \frac{x}{D}$$

or

$$\lambda = \frac{xd}{nD} \qquad (24a)$$

If we let x_1 be the distance from the central fringe to the first one on either side, then $n = 1$, and the equation becomes

$$\boxed{\lambda = x_1 \frac{d}{D}} \qquad (24b)$$

Because the fringes are evenly spaced, x_1 represents the spacing all along the pattern. By measuring the three distances d, x, and D, the wavelength of light can be calculated. Repeated experiments, carefully performed, give the results shown in Table 24A.

TABLE 24A. Wavelengths of Visible Light

Color	Wavelength λ	Average λ
red	0.00000070 m – 0.00000062 m	6.6×10^{-7} m
orange	0.00000062 m – 0.00000060 m	6.1×10^{-7} m
yellow	0.00000060 m – 0.00000058 m	5.9×10^{-7} m
green	0.00000058 m – 0.00000050 m	5.4×10^{-7} m
blue	0.00000050 m – 0.00000044 m	4.7×10^{-7} m
violet	0.00000044 m – 0.00000040 m	4.2×10^{-7} m

Since the complete spectrum of white light is a continuous band, each color has a range of wavelengths. Violet light, for example will vary from 0.00000040 m to 0.00000044 m, while red light will vary from 0.00000062 to about 0.00000070 m.

As shown in Fig. 24G, violet light of the visible light spectrum is composed of the shortest waves, whereas red is composed of the longest waves. Beyond the violet end of the visible spectrum, however, are found invisible light waves of even shorter wavelength called the *ultraviolet*, and beyond the red end of the visible spectrum are invisible light waves of longer wavelength called the *infrared*. See Fig. 32A.

The wave equation given in Chap. 18 applies equally well to light waves, and is usually written using different symbols as follows:

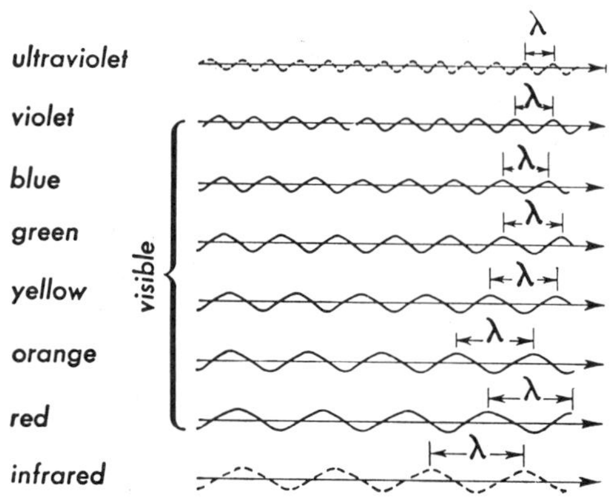

Fig. 24G. Diagram showing the relative wavelengths of light

$$c = \nu\lambda \tag{24c}$$

where c is the velocity in meters/second, ν is the frequency in vibrations/second, or hertz, and λ is the wavelength in meters. The velocity of light in a vacuum or free space is 3.0×10^8 m/s.

EXAMPLE 1. Monochromatic light (light of one wavelength only) falls normally on a double slit. If the distance between slit centers is 0.50 mm, and the distance between consecutive fringes on the screen 2.5 m away is 2.35 mm, what is (a) the wavelength of the light, and (b) its color?

SOLUTION. (a) By direct substitution in Eq. (24b), we obtain

$$\lambda = 2.35 \times 10^{-3}\text{ m}\,\frac{5 \times 10^{-4}\text{ m}}{2.5\text{ m}}$$

$$\lambda = 4.70 \times 10^{-7}\text{ m}$$

(b) Referring to Table 24A, this light is blue.

24.4 MICHELSON INTERFEROMETER The Michelson* interferometer is an optical device employing the principle of the interference of two beams of light. Its treatment here is important because historically it led to Einstein's theory of relativity.

The form of the Michelson interferometer generally found in the science laboratory is that shown in Fig. 24H. The optical parts consist of

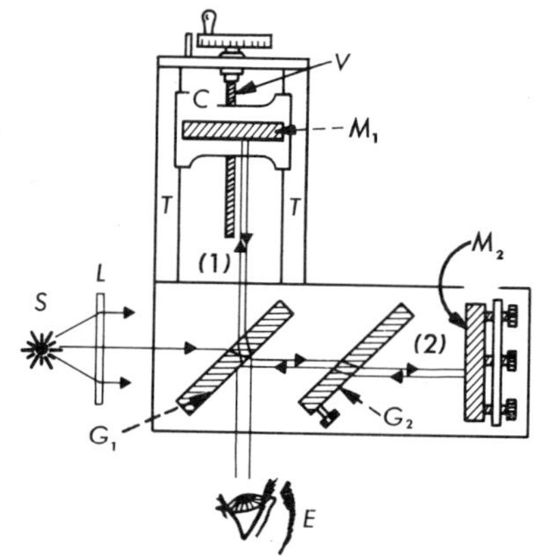

Fig. 24H. Diagram of the Michelson interferometer.

two mirrors M_1 and M_2 and two parallel plates of glass G_1 and G_2. Oftentimes the rear side of plate G_1 is lightly silvered (shown heavy in the figure) so that light coming from the source S is divided into (1) a reflected and (2) a transmitted beam of equal intensity. The light returning from M_1 passes through G_1 a third time before reaching the eye. The light returning from M_2 is reflected from G_1 and into the eye. The purpose of

* ALBERT ABRAHAM MICHELSON (1852–1931), American physicist, was born of Jewish parents in Strelno, Germany. His parents moved to San Francisco, where he studied in the public schools. He graduated from the U. S. Naval Academy in 1873 and was physics and chemistry instructor for four years thereafter. In 1883, he became professor of physics at Case School of Applied Science, and while there, developed the interferometer known by his name. In 1892, he became Chairman of the Physics Department, University of Chicago, and there directed his attention to measuring the speed of light. He received many national and international awards for his research and was the first American to receive the Nobel Prize in physics (1907). He published several books on optics and is most famous for the Michelson interferometer, as well as its application to the famous Michelson-Morley experiment on ether drift. As a college student the author had the good fortune to meet Michelson and see him making measurements on the speed of light at the Mt. Wilson Observatory during the summer of 1926.

plate G_2 is to render the path in glass equal for the two rays.

The mirror M_1 is mounted on a well-machined guide and can be moved along slowly by means of a screw V. When mirror M_2 is made exactly perpendicular to M_1 by screws on its back face, interference fringes similar to those found with a double slit may be seen, or photographed, at E. Photographs of typical fringes when mirror M_1 is at different distances are shown in Fig. 24I.

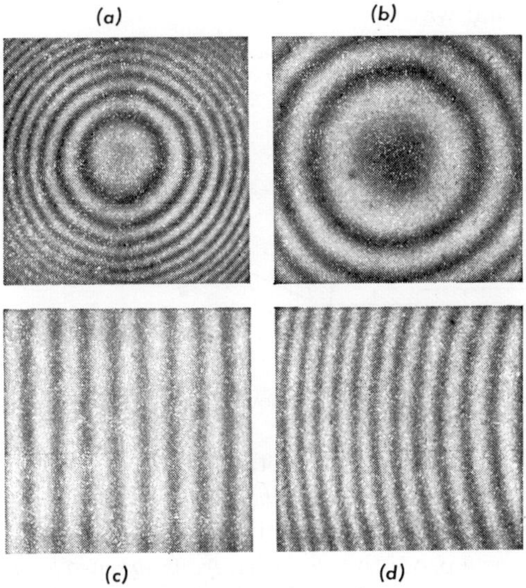

Fig. 24I. Interference fringe patterns as photographed with a Michelson interferometer.

When light of one wavelength is used as a source and the mirrors are in exact adjustment, circular fringes are observed as shown in photographs (a) and (b). If the mirrors are not exactly at right angles to each other, fringes like those in photographs (c) and (d) are obtained. If circular fringes are observed and mirror M_1 is slowly moved along by the turn of the screw V, the circular fringe pattern will expand or contract; if it expands, new fringes will appear as a dot at the center, widen, and expand into a circle; and if it contracts, fringes grow smaller, become a dot, and vanish at the center. If straight or curved fringes are observed, the motion of M_1 causes the fringes to drift across the field at right angles to the fringe lines.

The expansion, contraction, or drift in the pattern for a distance of one fringe corresponds to M_1 moving a distance of exactly one-half wavelength of light. When M_1 moves back a distance of $\frac{1}{2}\lambda$, the total light path (1) increases a whole wavelength. If M_1 moves 1λ, the pattern will move two fringes because the total light path (1) has changed by 2λ. Any bright fringe that one observes is caused by both beams coming together in phase. When the path is changed by $\frac{1}{2}\lambda$, 1λ, $\frac{3}{2}\lambda$, etc., the two beams arriving at the same field points will again be in phase.

By counting the number of fringes required to move the mirror M_1 a given distance, the wavelength of light can be calculated. This would appear to be the most accurate method for the wavelength measurements of different light sources. Knowing the accurate wavelength of any light source, one can then use the interferometer to accurately measure distances.

It is by means of the Michelson interferometer that the standard meter was determined in terms of the wavelength of orange light, $\lambda = 6057.80$ Å from a krypton lamp.*

$$1 \text{ meter} = 1{,}650{,}763.73 \text{ wavelengths}$$
$$(\textit{for orange light of krypton})$$

24.5 VELOCITY OF LIGHT IN MOVING MATTER

In 1859 the French physicist Fizeau** measured

* Adopted as the international legal standard of length on October 14, 1960, by the General Conference on Weights and Measures in Paris, France.

** ARMAND FIZEAU (1819–1896), French physicist. Born of a wealthy French family and although financially independent, he devoted his life to experimental work in science. His most important work was the measurement of the speed of light in 1849, carried on in Paris between Montmartre and Suresnes. Prior to this work, it was believed by most scientists that light had an infinite speed. He also gave the correct explanation of the Doppler principle as applied to light coming from the stars and showed how the effect could be used to measure stellar velocities. He carried out his experiments on the velocity of light in a moving medium in 1851, and showed that light is dragged along by a moving stream of water.

the velocity of light in a moving stream of water and found that the light was carried along by the stream. A schematic diagram of his apparatus is shown in Fig. 24J.

Light from a monochromatic source S is separated into two beams by means of a lens L_1. These two beams pass through tubes A and B containing water flowing rapidly in opposite directions. After reflection from M, the beams traverse opposite tubes so that upon arrival at L_1 one has traversed streams A and B in the direction of flow while the other has traversed both tubes bent always against the flow.

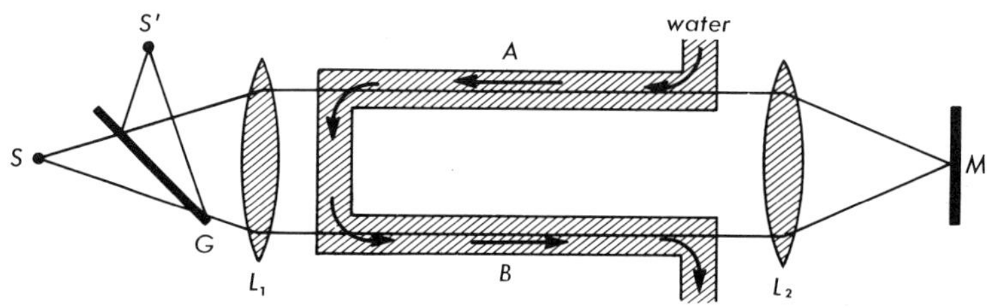

Fig. 24J. Fizeau's experiment for measuring the velocity of light in a moving medium.

If the light travels faster by one path than the other, the time will be different and the fringes formed at S′ where the beams came together will shift. Using tubes 1.5 m long and a water speed of 7.0 m/s, Fizeau found a shift of 0.46 of a fringe upon a reversal of the water stream. This shift corresponds to a decrease in the speed of light in one direction, and an increase in the other, of about half the speed of the water. In other words, the moving water has a dragging effect upon the light waves.

In 1818 the French physicist Fresnel derived a formula for this dragging effect, based upon the existence of what was then called the *ether*. His formula gives v' the increase in the velocity of light in any medium, due to motion:

$$v' = v\left(1 - \frac{1}{\mu^2}\right) \tag{24d}$$

where v is the velocity of the medium, and μ is the index of refraction. For water, with an index of 1.33, $v' = 0.43\,v$ in reasonably good agreement with Fresnel's observations.

24.6 THE POLARIZATION OF LIGHT The experiments described in the first sections of this chapter, illustrating the *diffraction* and *interference* of light, are generally regarded as proof that *light is a wave motion*. Although such experiments enable the scientist to measure accurately the wavelengths of light, they give no information of the kinds of waves involved. The reason for this is that all types of waves, under the proper conditions, will exhibit diffraction and interference. See Sec. 19.8 on the interference of sound waves.

The desired information in the case of light waves is found in another group of phenomena known as *polarized light*. The phenomena, which will be described in this chapter, are considered to be a proof that *light is a transverse wave motion* in contrast with the longitudinal wave motion in sound.

An understanding of the following experiments can best be attained by first presenting the graphical methods of representing transverse waves. We assume at the outset that each light wave is a transverse wave whose vibrations are along straight lines at right angles to the direction of propagation. Furthermore, we assume that a beam of ordinary light consists of millions of such waves, each with its own plane of vibration, and that there are waves vibrating in all planes with equal probability. Looking at such a beam end-on as in Fig. 24K, there should be just as many waves vibrating in one plane as there are vibrating in any other. This then can be referred to as perfect symmetry.

If, by some means or other, all the waves in a beam of light are made to vibrate in planes paral-

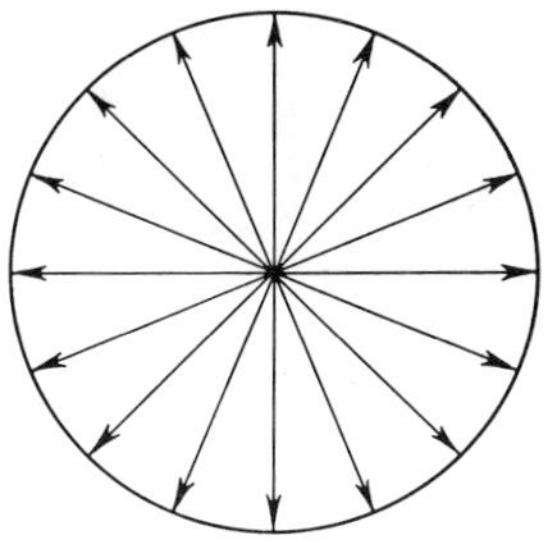

Fig. 24K. End-on view of a beam of unpolarized light illustrating schematically the equal probability of all planes of vibration.

lel to each other, the light is said to be *plane-polarized.* Diagrams illustrating such light are shown in Fig. 24L. The top diagram (a) represents plane-polarized light waves traveling to the right and vibrating in a vertical plane, while the second diagram (b) represents a ray of plane-polarized light vibrating in a horizontal plane. The dotted line indicating waves in diagram (a) is usually omitted.

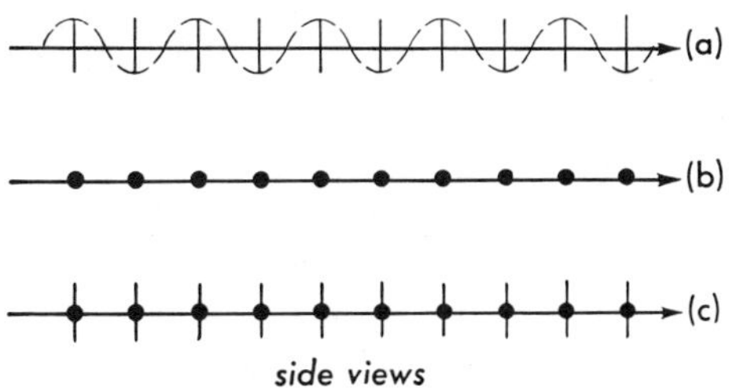

Fig. 24L. Diagrams illustrating plane-polarized rays of light.

It can be shown that a beam of ordinary unpolarized light, vibrating in all planes, may be regarded as being made up of two kinds of vibrations only, half of the waves vibrating in a vertical plane as in diagram (a) and the other half vibrating perpendicular to it as in diagram (b). Diagram (c), therefore, represents ordinary unpolarized light.

24.7 DOUBLE REFRACTION The double refraction of light by Iceland spar (calcite) was first observed by a Swedish physician, Erasmus Bartholinus, in 1669, and later studied in detail by Huygens and Newton. Nearly all crystalline substances are now known to exhibit the phenomenon. The following are but a few samples of crystals which show the effect: *calcite*, *quartz*, *mica*, *sugar*, *topaz*, *selenite*, *aragonite*, and *ice*. Calcite and quartz are of particular importance because they are used extensively in the manufacture of special optical instruments.

Calcite as found in nature always has the characteristic shape shown in Fig. 24M(a).

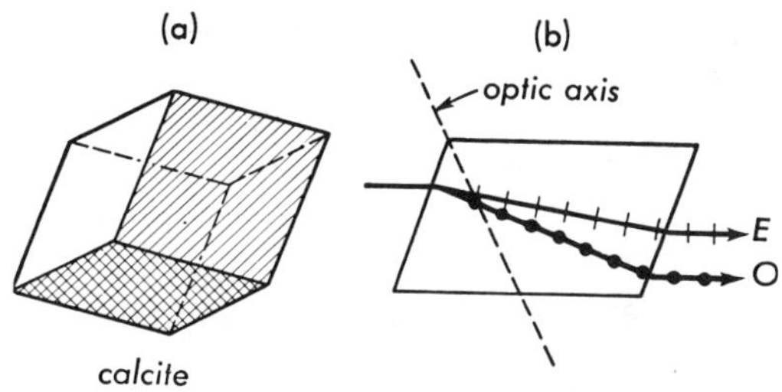

Fig. 24M. Diagrams of calcite and its behavior with light.

Not only is light doubly refracted by calcite but both rays are found to be plane-polarized. One ray, called the *ordinary ray*, is polarized with its vibrations in one plane, and the other ray, called the *extraordinary ray*, is polarized with its vibrations in a plane at right angles to the first. This polarization is illustrated by *dots* and *lines*, and can be proved by a *polarizing film*. These latter devices will be described in the next two sections.

Since the two opposite faces of a calcite crystal are always parallel to each other, the two refracted rays always emerge parallel to the incident light and are therefore parallel to each other. If the incident light falls perpendicularly upon the surface of the crystal, as in Fig. 24N,

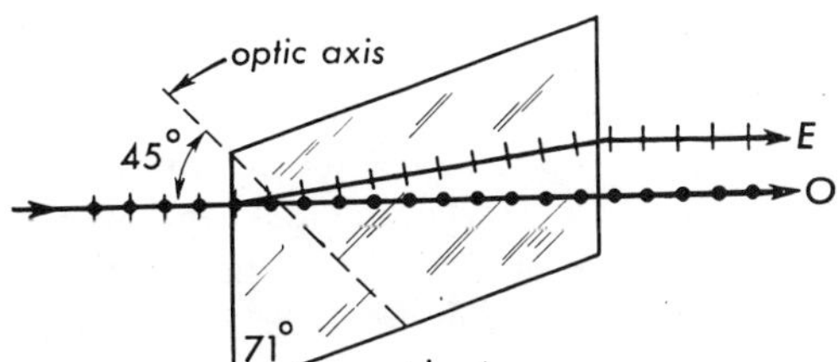

Fig. 24N. Only one ray obeys Snell's law in double refraction.

the extraordinary ray will be refracted away from the normal and come out parallel to, but displaced from, the incident beam, and the ordinary ray will pass straight through without deviation.

In general, the O ray obeys the ordinary laws of refraction, that is, it obeys Snell's law. In this way the crystal acts like glass or water. The E ray obeys no such simple law, behaves quite abnormally, and does not obey Snell's law.

In other words, the O ray travels with the same velocity regardless of its direction through the crystal, whereas the velocity of the E ray is different in different directions. This is the origin of the designations ordinary and extraordinary. The optic axis in a crystal is the one direction, shown by the dashed line in Fig. 24N, in which all vibrations travel with the same speed and there is no double refraction.

24.8 POLARIZATION BY SELECTIVE ABSORPTION When ordinary light enters a crystal of tourmaline, double refraction takes place in much the same way that it does in calcite, but with this difference: one ray, the so-called O ray, is entirely absorbed by the crystal, while the other ray, the E ray, passes on through. This phenomenon is called *selective absorption* because the crystal absorbs light waves vibrating in one plane and not those vibrating in the other.

Tourmaline crystals take in ordinary light, dispose of the O vibrations, and transmit plane-polarized light as illustrated in Fig. 24O(a). When two such crystals are lined up parallel, with one behind the other, the plane-polarized light from the first crystal passes through the second with little loss in intensity. If either crystal is turned at 90° to the other, i.e., in the *cross position*, the light is completely absorbed and none passes through.

The behavior of tourmaline and similar optical substances is due to the molecular structure of the crystal. To draw an analogy, the regularly spaced molecules of a single crystal are like the regularly spaced trees in an orchard or grove. If one tries to run between the rows of trees carrying a very long pole held at right angles to the direction of motion, the pole must be held in a vertical position. If it is held in the horizontal plane, the runner will be stopped.

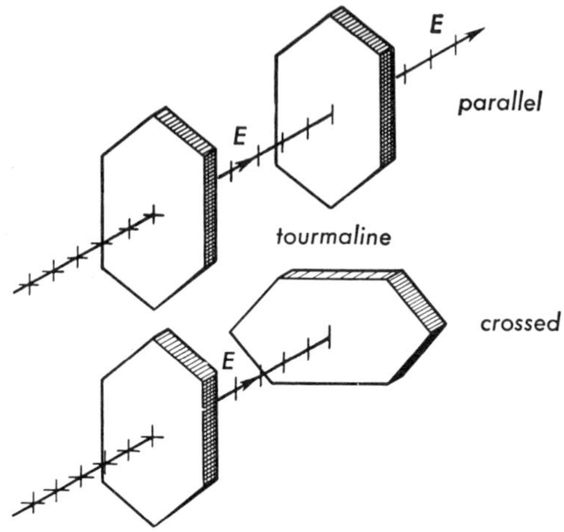

Fig. 24O. Tourmaline crystals polarize light.

A more satisfactory substance for this purpose which does transmit white light is a manufactured material made in the form of very thin films. These films have the general appearance of a plastic and are made from small needle-shaped crystals of an organic compound *iodosulfate of quinine*. Lined up parallel to each other and embedded in a *nitrocellulose mastic*, these crystals act like tourmaline by absorbing one component of polarization and transmitting the other.

Two such films mounted separately in rings between thin glass plates are shown schematically in Fig. 24P. In the crossed position no light can pass through both films, whereas in the parallel position white light vibrating in the plane indicated by the parallel lines is transmitted. Polarizing films of this kind are finding many practical applications, particularly wherever glaring light is not desired. The glaring light reflected at an angle from a table top, a book, a window pane, the water, or the road ahead when driving a car is polarized and can be partly eliminated by polarizing films.

24.9 SCATTERING AND THE BLUE SKY The blue of the sky and the red of the sunset are due to a phenomenon called scattering. When sunlight passes through the earth's atmosphere, much of

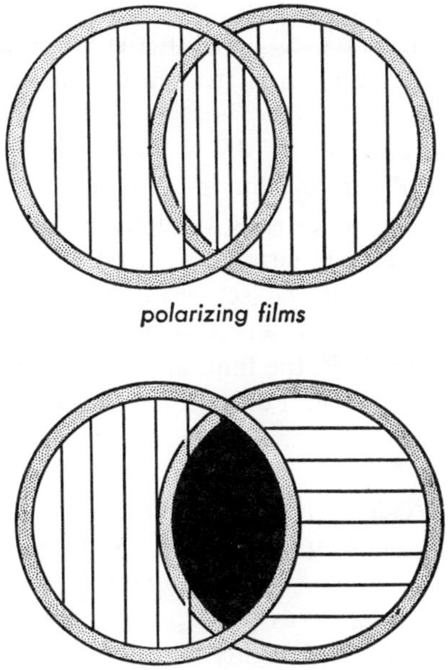

Fig. 24P. Polarizing films in the parallel and crossed positions.

the light is "picked" up by the air molecules and given out again in some other direction. The effect is quite similar to the action of water waves on floating objects. If, for example, the ripples from a stone dropped in a still pond of water encounter a small cork floating on the surface, the cork is set bobbing up and down with the frequency of the passing waves.

Light is pictured as acting in the same way on air molecules and fine dust particles. Once set into vibration by a light wave, a molecule or particle can send out the absorbed light again, sometimes in the same direction but generally in almost any other direction. This is illustrated schematically in Fig. 24Q, where waves of light are shown being scattered at random in all directions.

Experiments show, in agreement with the theory of scattering, that the shortest waves are scattered more readily than longer waves. To be more specific, the *scattering is inversely proportional to the fourth power of the wavelength.*

$$\boxed{\text{scattering} = k\,\frac{1}{\lambda^4}} \qquad (24e)$$

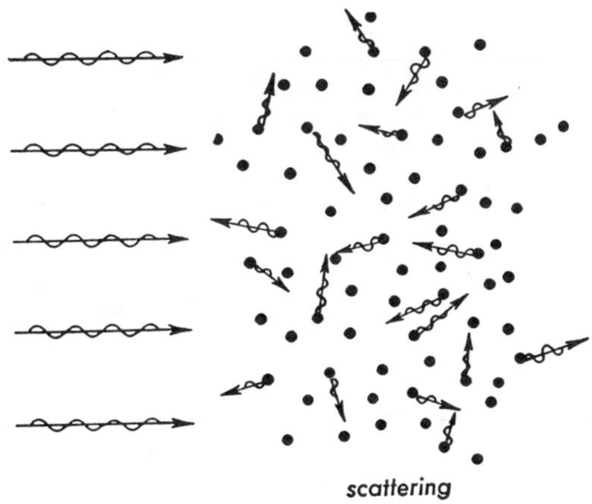

Fig. 24Q. Light waves are scattered by air molecules.

According to this law the short waves of violet light are scattered ten times as readily as the longer waves of red light. The other colors are scattered by intermediate amounts. Thus when sunlight enters the earth's atmosphere, *violet* and *blue light* are scattered the most, followed by *green*, *yellow*, *orange*, and *red*, in the order named. For every ten violet waves scattered from a beam, there is only one red wave.

red	orange	yellow	green	blue	violet
1	2	3	5	7	10

At noon on a clear day when the sun is directly overhead, the whole sky appears as *light blue*. This is the composite color of the mixture of colors scattered most effectively by the air molecules. As illustrated by the spectral colors in the lower right spectrum in Fig. 22F, light blue, like cyan of the color triangle, is obtained by the added mixture of *violet*, *blue*, *green*, and *yellow*.

24.10 THE RED SUNSET The occasional observation of an orange-red sunset is attributed to the *scattering of light* by fine dust and smoke particles near the earth's surface. This is illustrated in Fig. 24R. To an observer at A, it is noonday and the direct sunlight from overhead, seen only by looking directly at the sun itself, travels through a relatively short dust path. As a result, very little violet and blue are scattered away and the sun appears white. As sunset

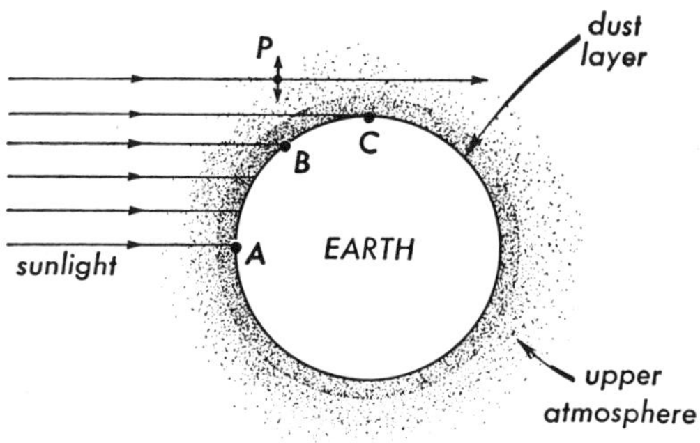

Fig. 24R. The scattering of light by a layer of dust near the earth's surface causes the sun to turn yellow, then orange, and finally red at sunset.

approaches, however, the direct sunlight has to travel through an ever-increasing dust path. The result is that an hour or so before sundown, when the observer is at B, practically all of the blue and violet have been scattered out, and owing to the remaining colors—red, orange, yellow, and a little green—the sun appears yellow. At sunset, when the observer is at C, the direct rays must travel through so many miles of dust particles that all but red are completely scattered out and the sun appears red. At this same time the sky overhead is still light blue. If the dust blanket is too dense, even the red will be scattered appreciably from the direct sunlight and the deepening red sun will become lost from view before it reaches the horizon.

An excellent demonstration of scattering by fine particles is illustrated in Fig. 24S. A parallel beam of white light from a carbon arc and lens L_1 is sent through a water trough with glass sides. After passing through an iris diaphragm at the other end, a second lens L_2 forms an image of the circular opening on the screen.*

Tiny sulfur particles develop slowly over a period of from 5 to 10 min. During this time the beam through the tank turns a sky blue, while the circular disk of light on the screen, representing the sun, turns from white to yellow to orange to red.

If the blue of the sky is observed through a piece of Polaroid film in a direction at right angles to the sun's rays, the light is found to be partially plane-polarized. This polarization can also be seen in the scattering experiment described above. Observed through a Polaroid film, the beam in the tank appears bright at one orientation of the Polaroid and darkens with a 90° rotation.

24.11 LASERS The term *laser* derives its name from the description, *Light Amplification by Stimulated Emission and Radiation.* In principle the laser is a device that produces an intense, concentrated, and highly parallel beam of light. So parallel would be the beam from a visible light laser 1 ft in diameter that at the moon the beam would be no more than a mile wide.

Historically the laser is the outgrowth of the *maser*, a device using microwaves instead of light waves. The first successful maser was built by C. H. Townes* at Columbia University in 1953. During the next seven years great strides were made in developing intense microwave beams, the principal contributions being made by the Bell Telephone Research Laboratories and the

* To produce the fine particles for scattering, about 40 g of photographic fixing powder (hyposulfite of soda) are first dissolved in about 2 gallons of water. Next, about 1 to 2 cm^3 of concentrated sulfuric acid are added and the two thoroughly mixed in the trough. The correct amount of acid to produce the best results is determined by trial. The first visible precipitate should appear after 2 or 3 min.

* CHARLES H. TOWNES (1915–), American physicist. He received his Ph.D. from the California Institute of Technology in 1939 and has received honorary degrees from a number of Universities. He has served in the physics departments at several universities, and is now Professor-at-large at the University of California. He is most noted for his outstanding work in the development of masers and lasers, for which he was awarded the Nobel Prize in Physics in 1964. He has received many national and international awards. He has served on many national science committees, such as: the President's Science Advisory Committee, Chairman of the Science and Technical Advisory Committee on Manned Space Flight, NASA, and the Science Advisory Board of the U. S. Air Force.

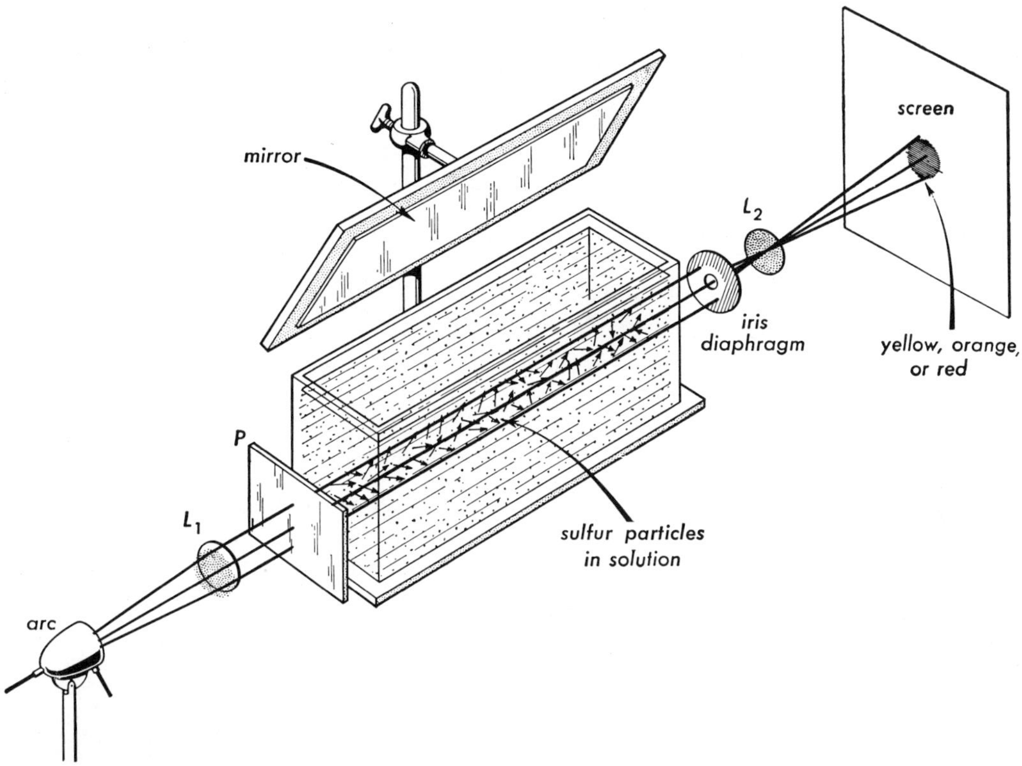

Fig. 24S. The sunset experiment. Demonstration of the scattering and polarization of light by small particles.

Lincoln Laboratories at the Massachusetts Institute of Technology.

The first successful laser, using a large synthetic ruby crystal, was built by T. H. Maiman of Hughes Aircraft Company Laboratories in 1960. Hundreds of extensive researches on laser development have been carried on since that time, and because such devices appear to have great potential in so many different fields of research and development, a brief account of their basic principles will be presented here.

Lasers use three general kinds of materials: solids, liquids, and gases. Since the first successful laser was made with a large ruby crystal, this device will be explained as representative of solid-state lasers.

Pure ruby crystals 2-10 cm long can now be grown quite readily in the laboratory. The ends of the ruby rod are ground and polished so their two surfaces are flat and parallel to one-hundredth of a wavelength of light. The ends are then coated with a thin layer of aluminum to make them highly reflecting. See Fig. 24T.

When white light enters a ruby crystal, strong absorption occurs in the blue and green part of the spectrum, and transmission of only the red region gives the ruby its red color. Absorbing light in this case means that certain electrons in the atoms store up potential energy for a period of time. Atoms with stored energy are called *excited atoms*.

By sending a beam of red light of the right frequency ν_0 into the end of the ruby rod, part of the beam enters and travels down the crystal axis. Much of this light is reflected back and forth between the ends. As a single wave passes close to an excited atom, the atom is stimulated to give up its energy by emitting a wave of the same frequency ν_0. This emitted wave will be exactly in step, that is, in phase, with the passing wave. The two waves traveling along will stimulate others to emit light, thus building up streams

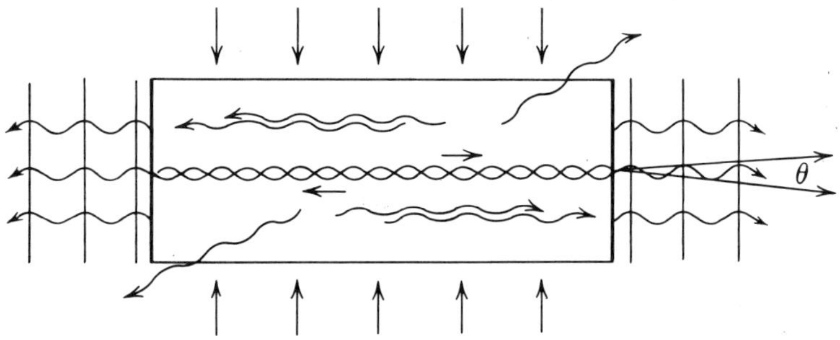

Fig. 24T. Diagram illustrating the coherent stimulation of light waves in a solid-state laser such as a ruby crystal.

of light waves bouncing back and forth between the reflecting ends. Very quickly the whole rod is filled with intense light, with some of it continually emerging from either end. See Fig. 24T.

Because the light waves emerging from the end of the laser are all in phase, the beam is said to be coherent. The angular spread of the light is given by the relation

$$\theta = \frac{2.44\lambda}{d} \qquad (24f)$$

where θ is in radians, λ is the wavelength of light, and d is the diameter of the emergent beam. This equation arises from the treatment of the diffraction of light by a circular aperture.

EXAMPLE 2. If a ruby laser were made of a crystal rod 10 cm in diameter what would be (a) the angular spread of the beam it emits, and (b) the diameter of the area it would illuminate if pointed at the moon? Assume red light of wavelength $\lambda = 7 \times 10^{-7}$ m, and the distance D to the moon 3.845 m.

SOLUTION. (a) By direct substitution in Eq. (24f), we obtain

$$\theta = \frac{2.44 \times 7 \times 10^{-7}\text{ m}}{0.10\text{ m}}$$

$$\theta = 1.708 \times 10^{-5}\text{ radians}$$

(b) Since the distance to the moon is $D = 3.845 \times 10^{8}$ m, the beam diameter at this distance is the product $D \times \theta$. See Eq. (8a).

$$\text{diameter} = 3.845 \times 10^{8}\text{ m} \times 1.708 \times 10^{-5}$$
$$\text{diameter} = 6.57 \times 10^{3}\text{ m}$$

which is approximately 4 mi.

To produce an intense laser beam, a large quantity of light must be fed into the crystal in order to store up energy as excited atoms. One of the many systems developed for doing this is shown in Fig. 24U. By placing the exciting light

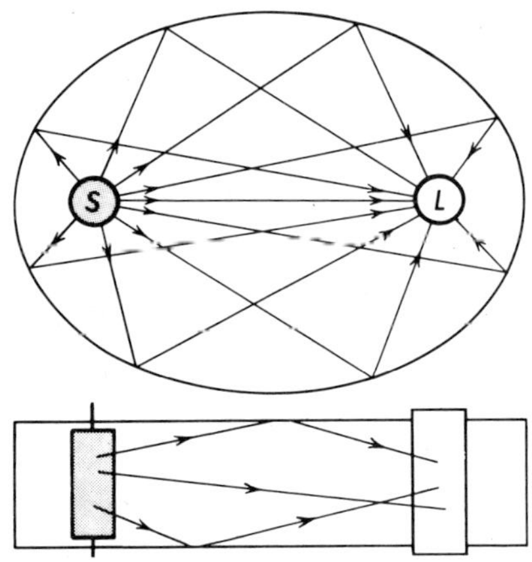

Fig. 24U. Elliptical reflector for concentrating light from a source S on a laser L.

at one focus of an elliptical reflector and the laser crystal at the other focus, high efficiency can be obtained. Such action of storing light is called optical pumping.

QUESTIONS

1. What is diffraction? What is interference?

2. What is Young's double-slit experiment? Make a diagram and explain.

3. How can one determine the wavelength of light using Young's double-slit experiment?

4. What are transverse waves? What is meant by unpolarized light?

5. What is plane-polarized light? How is unpolarized light broken down into two plane-polarized components?

6. What is double refraction? What is the nature of the light in the refracted rays?

7. Do the O and E rays obey Snell's law? Is it possible to obtain only one refracted ray in a crystal like calcite?

8. What is selective absorption? What is Polaroid? Where is Polaroid used? What is tourmaline?

9. What is scattering? How does it vary with the wavelength of light? How does scattering vary with color?

10. Why is the sky blue? If violet light is scattered more than blue, why is the sky light blue in color rather than violet?

11. Under what conditions will the setting sun turn from white to yellow, to orange, and to red? Explain.

PROBLEMS

1. Red light of wavelength 6.6×10^{-7} m is used in observing the interference fringes produced by a double slit. If the centers of the two slit openings are 0.60 mm apart, and the distance to the observing screen is 2.5 m, what is the fringe spacing? [Ans. 2.75 mm.]

2. Blue light of wavelength 4.6×10^{-7} m falls on a double slit, and 3 m away on a white screen interference fringes are formed 2.2 mm apart. Calculate the double-slit separation.

3. Light of one wavelength falls on a double slit. The distance between the slit centers is 0.70 mm, and the distance between consecutive fringes on the screen 3 m away is 2.95 mm. What is (a) the wavelength, and (b) the color of the light? [Ans. (a) 6.88×10^{-7} m, (b) red.]

4. What is the ratio of the scattering of light waves between orange light of wavelength 6×10^{-7} m and blue light of wavelength 4.6×10^{-7} m?

5. For every 500 photons of red light scattered by the air, how many photons of violet light will be scattered? Assume the wavelengths to be 7×10^{-7} m and 4×10^{-7} m, respectively. [Ans. 4690.]

6. Calculate the ratio of the numbers of light photons scattered by the air for blue light ($\lambda = 4.8 \times 10^{-7}$ m) and yellow light ($\lambda = 5.7 \times 10^{-7}$ m).

7. Find the ratio of light waves scattered by fine particles between green light ($\lambda = 5.4 \times 10^{-7}$ m) and red light ($\lambda = 7.5 \times 10^{-7}$ m). [Ans. 3.721:1]

8. A ruby laser crystal rod has a diameter of 2 cm. Find (a) the angular spread of the beam from the end, and (b) the diameter of the area it would illuminate on the moon. Assume the distance to the moon is 3.840×10^8 m, and the light has a wavelength of 7×10^{-7} m.

9. The diameter of a ruby laser rod is 3 cm. What will be the angular spread of the beam from the end, and (b) the diameter of the illuminated area on a screen 5 Km away? [Ans. (a) 5.69×10^{-5} rad. (b) 28.47 cm.]

25 | RELATIVITY

The mention of the word relativity suggests the name of Albert Einstein,* the scientist to whom we are indebted for the now famous theory. Einstein was a realist, and his theory rests upon physical facts that have been verified by repeated observations of well-planned and well-executed experiments. The first of these is the *Michelson-Morley experiment.*

* ALBERT EINSTEIN (1879–1955), German-Swiss physicist, was born of Jewish parents at Ulm, Wurttemberg, March 14, 1879. His boyhood was spent in Munich where his father, a dealer in chemicals, had settled in 1880. When the family moved to Italy in 1894, young Albert went to Switzerland to study. There he worked his way through school, finally taking his Ph.D. degree at the University of Zurich in 1902. He was appointed extraordinary professor of theoretical physics at the University of Zurich in 1909, and in 1913 he was called to Berlin as director of the Kaiser-Wilhelm Institute for Physics. While at this post he was elected a member of the Prussian Academy of Sciences, and a member of the Royal Society of London. In 1921 he received the Nobel Prize in physics and in 1925 the Copley Medal of the Royal Society. He is best known for his theory of relativity, the theory and explanation of Brownian motion, the theory of the photoelectric effect, and the theoretical derivation of the mass energy equation $E = mc^2$. Twice married, Einstein had several children. To his friends he was a quiet, sincere, and modest man who loved his pipe and violin, but disliked formality. The author had the good fortune to have met Einstein while he was still teaching at the University of Berlin in 1929.

25.1 THE MICHELSON-MORLEY EXPERIMENT

This, the most famous experiment in optics, was first performed in 1881 by the two American scientists, Michelson and Morley, in an effort to detect the motion of the earth through space. If the transmission of light through space requires an ether, that is, a medium in which it can move, then light in a laboratory on the earth should be dragged along by this ether as the earth moves along through space. In order to detect such a drift the Michelson interferometer appeared to be the most sensitive instrument to use. See Fig. 24H.

In principle the ether drift test consisted simply of observing whether there is any shift of the interference fringes of light in the Michelson interferometer when the entire instrument is turned through an angle of 90°. Let us assume that the interferometer and the earth are at rest and that the ether is moving by with a velocity v as shown in Fig. 25A. If no ether drag is effective,

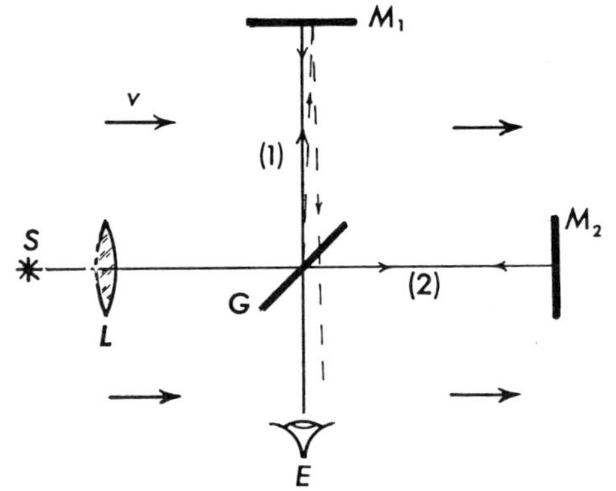

Fig. 25A. The Michelson interferometer arrangement for detecting an ether drift.

light paths (1) and (2) will be out and back as shown by the solid lines and arrows, and a set of interference bands like those shown in 24I(c) will be observed.

If we now suppose the light to be dragged along by an ether, the time it takes for light to traverse path (1) at right angles to the ether stream, and the time it takes to traverse path (2) first with and then against the stream, will both be increased by different amounts.

An ether drift should, therefore, cause a shift in the fringes observed in the interferometer. Since neither the earth's motion, nor the ether can be stopped, in an effort to observe this shift, a rotation of the interferometer through 90° should have a similar effect. By interchanging paths (1) and (2) the time difference $t_2 - t_1$ is reversed, and any fringe shift should be doubled.

Michelson and Morley made the light paths as much as 11 m long by reflecting the light back and forth between 16 mirrors, as shown in Fig. 25B. To prevent distortion by the turning of the instrument the entire apparatus was mounted on a concrete block floating in mercury, and observations of the fringes were made as it rotated slowly and continuously about a vertical axis.

If we assume the ether velocity v to be 18.6 mi/s (the speed of the earth in its orbit around the sun), and the speed of light c to be 186,280

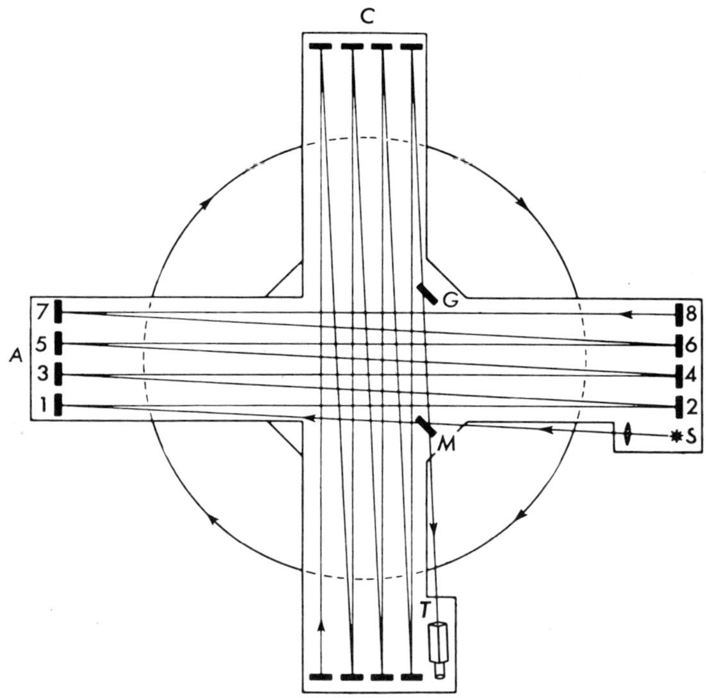

Fig. 25B. Miller's elaborate arrangement of the Michelson-Morley experiment to detect ether drift.

mi/s, a shift of one-half a fringe should have been observed. No shift as great as one-tenth of this was observed. Such a negative result was surprising and disappointing, to say the least.

The experimental fact that no fringe shift was observed led to the conclusion that either there is no such thing as an ether drift, or that the Michelson interferometer is incapable of detecting the earth's motion through space.

25.2 THE LORENTZ-FITZGERALD CONTRACTION From the time Michelson and Morley announced the negative results of their ether-drift experiment, scientists tried to explain why the experiment failed. An ingenious explanation was first advanced by Fitzgerald* in 1890. If objects moving through space have to push against the immovable ether, he suggested, they would be compressed in the direction of motion. This compression would, therefore, shorten the Michelson interferometer arms holding the mirrors and might exactly compensate for an existing ether drift.

Lorentz,** the famous Dutch physicist, studied this problem from an atomic point of view. All matter, he proposed, is made up of atoms, and atoms are made up of charged particles that produce electric and magnetic fields. These fields must exert forces on the electromagnetic ether, thus causing the atoms and molecules in moving matter to be pushed closer together. Starting with the well-known principles of electricity and magnetism, Lorentz derived the following formula for the length of any object:

$$\boxed{l = l_0\sqrt{1 - \frac{v^2}{c^2}}} \qquad (25a)$$

where v is the velocity of the object through the ether, l_0 is its length when at rest in the ether, and c is the velocity of light. Suppose that an object is at rest, so that $v = 0$. Upon substitution of $v = 0$ in Eq. (25a), we find $l = l_0$, which says the object's length l will be just equal to its rest length l_0.

EXAMPLE 1. A space ship 20 m long passes by the earth at a speed of three-fourths the speed of light. What would we on the earth observe the length to be?

SOLUTION. By direct substitution in Eq. (25a), we obtain

$$l = 20\text{ m}\sqrt{1 - \frac{(\frac{3}{4}c)^2}{c^2}}$$

$$l = 20\text{ m}\sqrt{\frac{7}{16}} = 13.22\text{ m}$$

This indicates that the moving spaceship, represented by a rod in Fig. 25C, is approximately two-thirds as long as when it is at rest.

It is interesting to see what this equation indicates if an object moves lengthwise with the speed of light, i.e., with $v = c$..The result is $l = 0$. This means that any object moving with the speed of

* GEORGE F. FITZGERALD (1851–1901), Irish physicist, was born in Dublin on August 3, 1851. His entire scientific career was spent as professor of natural philosophy at Trinity College in Dublin. His contributions to science were largely related to the electromagnetic properties of light, the magneto-optical phenomena known as the Kerr effect, the oscillatory discharge of a capacitor, the polarization of Hertzian waves, the scattering of X rays and the validity of Ohm's Law in electrolysis. While studying the results of the Michelson-Morley experiment, Fitzgerald worked out the change in length of a body due to its motion. This is known as the Lorentz-Fitzgerald contraction.

** HENDRIK A. LORENTZ (1853–1928), Dutch physicist, was born at Arnheim on July 18, 1853. He was educated at the University of Leyden, where he was appointed professor of theoretical physics at the age of 25. Of his numerous contributions to science he is best known for (1) a set of four algebraic equations, known as the Lorentz transformation, which formed the basis of Einstein's special theory of relativity, and (2) for his theoretical explanation of the Zeeman Effect. In 1922 he was awarded, jointly with Zeeman, the Nobel Prize in physics. The author had the good fortune of working with Lorentz in the preparation of six lectures he gave at Cornell University in 1927.

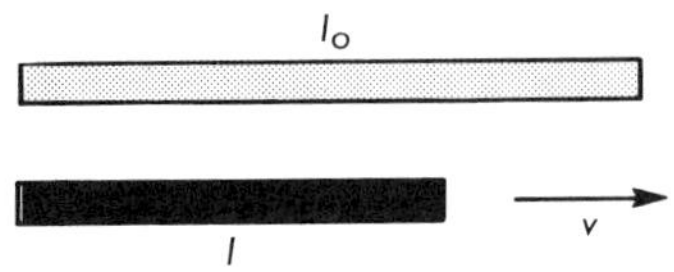

Fig. 25C. Diagram of the Lorentz-Fitzgerald contraction.

light would be compressed to zero length. The velocity of light, therefore, would appear to be an upper limit for the velocity of any moving object.

The above formula when applied to the cross arms of the interferometer used in the Michelson-Morley experiment shows that the arms are shortened by just the right amount to compensate for the expected drift. See Fig. 25B. This shortening of an object cannot be measured, for if one attempts to measure the length of a moving object, the measuring stick must move with the same velocity, and an equal length on the stick shortens by the same amount.

25.3 EINSTEIN'S SPECIAL THEORY OF RELATIVITY Relativity is divided into two parts. One is called the *special theory of relativity*, and the other is called the *general theory*. The special theory, developed by Einstein in 1905, is thoroughly backed by numerous experimental observations of high precision and deals with observers and their reference frames moving with constant velocities. The mathematics of the special theory is simple enough, and we will consider several of the relationships that are necessary for the satisfactory explanations of certain atomic phenomena.

The general theory, proposed by Einstein in 1915, deals with motions of bodies in accelerated frames of reference. The mathematics of the general theory are quite difficult, and the experimental evidence for its validity is not as well founded as for the special theory. The failure of the Michelson-Morley experiment to detect an ether drift was interpreted by Einstein to mean that we cannot find any astronomical body, nor set up any frame of reference, that will be at rest in space. Furthermore, he concluded that the velocity of light in free space will be found to have the same value regardless of the motion of the source or the motion of the observer. In other words, the velocity of light is *invariant*.

To see the meaning of this statement consider a reference frame and observer O, at rest as shown in Fig. 25D. A source of light S is set up and, by means of an experiment, the velocity of light is measured and found to be 3×10^8 m/s. Another observer O′, moving with a velocity v with respect to O, allows the light from the same source S to pass through his apparatus. Upon measuring the velocity of this same light in his frame he too finds 3×10^8 m/s.

For these two identical results to be consistent, Einstein found it necessary to modify Newton's laws of motion. To do this he assumed that distance and time are relative, i.e., that distances between two points, or the time elapsed between two events, will depend upon the relative motion of the observer. Two observers moving with different velocities will measure different distances and different times.

25.4 RELATIVISTIC VELOCITY TRANSFORMATION Suppose you are an observer O and that

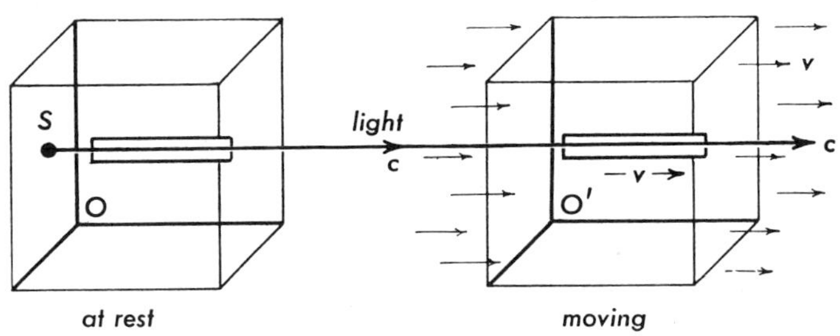

Fig. 25D. The velocity of light is the same to all observers, that is, it is invariant.

you observe an airplane B moving with a velocity u. Another observer A, moving toward or away from you with a velocity v, observes and measures the velocity of the same plane. The velocity he measures will be u', where

Special Relativity

$$u' = \frac{u - v}{1 - \frac{uv}{c^2}} \qquad (25b)$$

where c is the velocity of light.

This is the velocity transformation equation in the theory of special relativity. It shows that different observers find different velocities.

In Newtonian Mechanics, where the effects of relativity are not taken into account, the transformation equation is simply

Newtonian Mechanics

$$u' = u - v$$

EXAMPLE 2. An observer on the earth sees a space ship A receding from him at 2×10^8 m/s and overtaking a space ship B receding at 1.5×10^8 m/s. Find the relative velocity of (a) space ship B as observed by A, (b) space ship A as observed by B, and (c) space ship B relative to space ship A as observed by O.

SOLUTION. This example is shown schematically in Fig. 25E. The given quantities for (a) are $v = 2 \times 10^8$ m/s, $u = 1.5 \times 10^8$ m/s, $c = 3 \times 10^8$ m/s Upon substitution in Eq. (25b) we obtain for part (a)

$$u' = \frac{1.5 \times 10^8 \text{ m/s} - 2.0 \times 10^8 \text{ m/s}}{1 - \frac{1.5 \times 10^8 \text{ m/s} \times 2.0 \times 10^8 \text{ m/s}}{(3 \times 10^8 \text{ m/s})^2}}$$

$$u' = \frac{-0.5 \times 10^8}{1 - \frac{3 \times 10^{16}}{9 \times 10^{16}}} = -0.75 \times 10^8 \text{ m/s}$$

Observer in A observes B approaching at 0.75×10^8 m/s. For part (b) we reverse the velocity symbols: $u = 2 \times 10^8$ m/s and $v = 1.5 \times 10^8$ m/s. Upon substitution in Eq. (25b), we obtain, for part (b)

$$u' = \frac{2.0 \times 10^8 - 1.5 \times 10^8}{1 - \frac{2.0 \times 10^8 \times 1.5 \times 10^8}{(3 \times 10^8)^2}}$$

$$u' = \frac{0.5 \times 10^8}{1 - \frac{3 \times 10^{16}}{9 \times 10^{16}}} = +0.75 \times 10^8 \text{ m/s}$$

Observer in B observes A approaching at 0.75×10^8 m/s. For part (c) take just the difference between the two velocities observed by O.

$$u' = 1.5 \times 10^8 - 2.0 \times 10^8 = -0.5 \times 10^8 \text{ m/s}$$

Observer O sees A and B approaching each other at 0.5×10^8 m/s.

EXAMPLE 3. Suppose space ship B in Example 2 (Fig. 25E) is replaced by a beam of light moving from left to right which observer O measures and finds to be $c = 3 \times 10^8$ m/s. What will the velocity of this same light be as observed by space ship A?

SOLUTION. The given quantities are $u = c$, $v - 2.0 \times 10^8$ m/s, and $c = 3 \times 10^8$ m/s. Upon first replacing u by c in Eq. (25b) and solving for u', we obtain

$$u' = \frac{u - v}{1 - uv/c^2} = \frac{c - v}{1 - cv/c^2} = \frac{c - v}{1 - v/c}$$

$$= \frac{c - v}{\frac{c - v}{c}} = c$$

Hence the observer in A finds the velocity of light to be c regardless of his velocity. Hence the velocity of light is the same to all observers; it is said to be invariant.

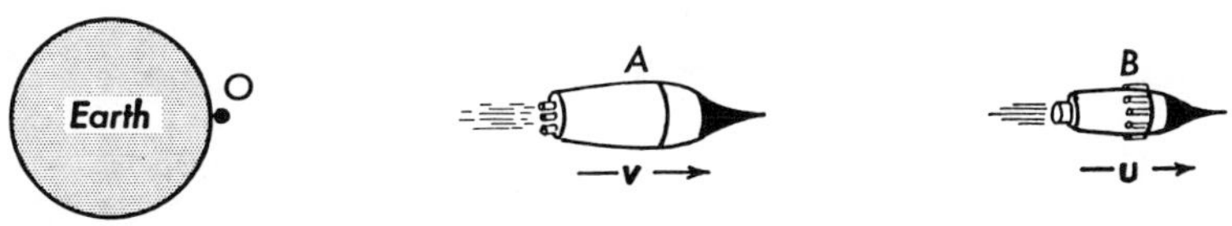

Fig. 25E. Diagram of two space ships receding from the earth with constant velocities.

Fig. 25F. Schematic diagram illustrating the relativistic increase in mass, and the Lorentz-Fitzgerald contraction, due to motion.

25.5 RELATIVISTIC MASS Einstein's special theory of relativity shows that if the mass of an object is measured by two different observers, one moving with respect to the other, the results are different. Mass, therefore, is not invariant. Although the derivation will not be given here, the special theory gives for the transformation equation, $m = \gamma m_0$, or

$$m = \frac{m_0}{\sqrt{1 - \frac{v^2}{c^2}}} \tag{25c}$$

where m_0 is the mass of an object at rest with respect to an observer O, and m is its mass when it is moving past him with a velocity v. A schematic diagram of a practical situation is shown in Fig. 25F in which the rest mass m_0 is not moving with respect to the observer O, while at the right the same mass is shown moving with a velocity v.

Table 25A gives the values of the relativistic mass of objects for a large range of velocities.

At 1% the speed of light (1860 mi/sec) the mass of a body is only one-half of one per cent greater than its rest mass. At 50% the speed of light the mass m has increased 15%, while at 99.9% the speed of light it has jumped to over 22 times its rest mass. These values are in excellent agreement with experiments on high speed atomic particles, a subject that will be considered in detail in later chapters.

It is important to note that as the speed of any given mass increases, the mass rises slowly at first, and then much more rapidly as it approaches the speed of light. No mass, however, can move with the speed of light, for by Eq. (25c) its mass would become infinite.

25.6 EINSTEIN'S MASS-ENERGY RELATION Just as sound, heat, and light are forms of energy, Einstein's special theory of relativity shows that mass is a form of energy. The expression giving the relation between mass and energy is an equation familiar to everyone. It is

$$E = mc^2 \tag{25d}$$

where m is the mass, c is the velocity of light, and E is the energy equivalence of the mass. The validity of this equation is now well established by hundreds of experiments involving atomic nuclei as well as with the general subject called atomic energy.

If an object has a rest mass of m_0, it has stored within it a total energy m_0c^2. If the same mass is moving with a velocity v, its mass has increased to m and the total stored energy is mc^2.

When a force F is applied to accelerate a given mass, the amount of work done is given in classical mechanics by Newton's second law of motion.

$$W = F \times x$$

As a result of this work done, the object, whose rest mass is m_0, is moving with a velocity v and has kinetic energy E_k.

$$F \times x = E_k$$

Applying the law of conservation of energy, we add the kinetic energy to the energy m_0c^2 it had when at rest, and obtain the total energy mc^2.

$$m_0c^2 + E_k = mc^2 \tag{25e}$$

If we solve this for E_k, we obtain for the kinetic energy of a moving mass the relation

$$E_k = mc^2 - m_0c^2 \tag{25f}$$

TABLE 25A. Relativistic Mass for Different Velocities

Velocity ratio v/c in per cent	1%	10%	50%	90%	99%	99.9%
Relative mass m/m_0....	1.000	1.005	1.15	2.3	7.1	22.3

QUESTIONS

1. What was the purpose of the Michelson-Morley experiment?

2. Describe the instrument used in the Michelson-Morley experiment. Make a diagram. What were the results?

3. What is meant by the Lorentz-Fitzgerald contraction? Write down the equation and explain its meaning from a diagram.

4. Can any material object move with the velocity of light? Explain your answer.

5. Does the mass of an object vary with its velocity? If an object is at rest and an observer is moving, will the object seem to have its rest mass? Explain.

6. Write down the mass energy equation and explain the meaning of each term.

PROBLEMS

(Use the value $c = 3.000 \times 10^8$ m/s or 186,280 mi/s for the speed of light in the following problems.)

1. Find the length of a meter stick moving lengthwise at a speed of 2.90×10^8 m/s. Assume a Lorentz-Fitzgerald contraction. [Ans. 25.60 cm.]

2. If a space ship 65 m long were to pass the earth at a speed of 2.6×10^8 m/s, what would be its apparent length, assuming a Lorentz-Fitzgerald contraction?

3. A space vehicle 100 ft long passes overhead with a velocity of 170,000 mi/s. Find its apparent length as measured by a ground observer. [Ans. 40.9 ft.]

4. An observer on the earth sees one space ship traveling at 2.6×10^8 m/s overtaking another space ship traveling at 1.6×10^8 m/s. What is the relative velocity of (a) the second ship as seen by the first, (b) the first ship as seen by the second, and (c) the relative velocity as seen from the earth?

5. An observer on the earth sees a space ship receding from the earth at 2.2×10^8 m/s launch a projectile in front of it. As seen from the earth this projectile has a speed of 2.6×10^8 m/s. What is the velocity of the projectile with respect to the space ship as seen from (a) the space ship, and (b) the earth? [Ans. (a) 1.098×10^8 m/s. (b) 0.400×10^8 m/s.]

6. An earth observer sees a space ship approaching the earth at one-quarter the speed of light. It launches an exploration vehicle, which from the earth appears to be approaching at one-half the velocity of light. What is the velocity of the vehicle with respect to the space ship as seen from (a) the space ship, and (b) the earth?

7. Two space ships are approaching each other, each with a velocity of three-fourths of the speed of light. With what velocity is each space ship approaching each other, as seen from the other ship? [Ans. 2.880×10^8 m/s, or 0.96 c.]

8. Atomic particles in the form of a beam have a velocity of 90% the speed of light. What is their relativistic mass as compared with their rest mass?

9. Atomic particles in the form of a beam have a velocity of 85% the speed of light. What is their relativistic mass compared with their rest mass? [Ans. $m/m_0 = 1.898$.]

10. An atomic particle has a rest mass of 1.82×10^{-25} Kg. Find its total mass energy when it is (a) at rest, and (b) when it has a velocity of 90% the speed of light.

11. Two atomic particles, each with a rest mass of 1.5×10^{-25} Kg, approach each other in head-on collision. If each has an initial velocity of 2.4×10^8 m/s, what is (a) the velocity of one atom as seen from the other, and (b) the relativistic mass of one as seen by the other? [Ans. (a) -2.927×10^8 m/s. (b) 6.83×10^{-25} Kg.]

12. If an atomic mass of 2.5×10^{-25} Kg were converted into energy and all of it imparted as kinetic energy to another atomic particle with a rest mass of 1.8×10^{-25} Kg, what would be the atom's velocity?

13. At what speed must a body move to have a mass ten times its rest mass? [Ans. 99.5% of c, or 2.985×10^8 m/s.]

14. A mass must move with what velocity if it is to have 100 times its rest mass?

26 | ELECTRICITY AT REST

This chapter marks the beginning of one of the most important subjects in nature, electricity. The forces of attraction and repulsion between electrically charged bodies are basic concepts that appear to hold atoms and molecules together. It is essential, therefore, that if we are to gain a clear understanding of the world around us we must obtain a clear understanding of the laws of *electrostatics*.

26.1 ELECTRIFICATION BY FRICTION It is impossible to say when electricity was first discovered. Records show that as early as 600 B.C. the attractive properties of amber were known. Thales of Miletus (640–546 B.C.), one of the "seven wise men" of ancient Greece, is credited with having observed the attraction of amber, when previously rubbed, for small fibrous materials and bits of straw. Amber (in Greek "*electron*") was used even then as it is now for ornamental purposes.

Although the electrification of amber by friction was handed down from one writer to another, nothing new about the phenomenon was discovered for more than two thousand years. It was not until the beginning of the seventeenth century that Sir William Gilbert* announced the discovery that many substances can be electrified by friction. Gilbert named this effect "electric" after the word "electron." It is now a well-established fact that all bodies when rubbed together become electrified and that amber is just one of a number of substances which show the effect most strongly.

---

* SIR WILLIAM GILBERT (1540–1603), court physician to Queen Elizabeth and a noted philosopher and experimental physicist. In 1600 he published a book on magnetism, *De Magnete*. This book was full of valuable facts and experiments on electricity and magnetism, and among other things contained many criticisms of his contemporaries, predecessors and the early philosophers. In his preface he wrote: "Why should I submit this new philosophy to the judgment of men who have taken oaths to follow the opinions of others, to the most senseless corrupters of the arts, to lettered clowns, grammatists, sophists, spouters, and the wrong-headed rabble. To you alone true philosophers, ingenious minds, who not only in books but in things themselves look for knowledge, have I dedicated these foundations of magnetic science." So strongly does he advocate here, and carry out, himself, the experimental method that he is to be classed as a scientist with his contemporary Galileo, "the father of modern physics."

26.2 ELECTROSTATIC ATTRACTION The word electrostatic means electricity at rest, and the word attraction refers to the force exerted by one body upon another at a distance. To demonstrate electrostatic attraction one often uses a rubber or amber rod and rubs it with a piece of flannel or fur. This electrifies the rod so that when it is held close over small bits of paper they jump up to the rod and hold fast.

The attraction of an electrified rubber rod for wood is illustrated in Fig. 26A. A small arrow

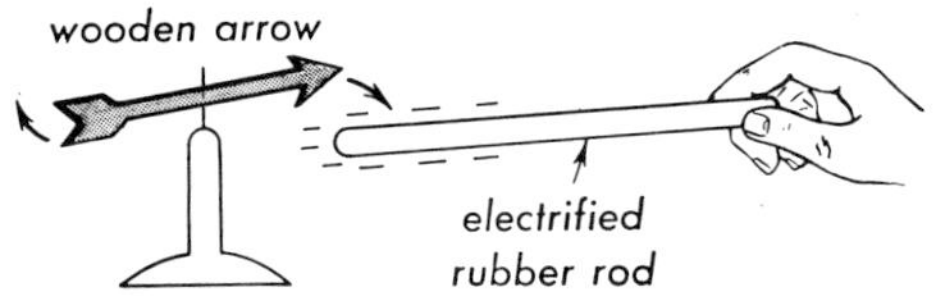

Fig. 26A. An electrified rod attracts a wooden arrow.

cut from a piece of dry wood is mounted so that it is free to turn as shown. When the electrified rubber rod is brought near the pointed end of the arrow it attracts the wood, turning the arrow until it points toward the rod. Brought near the opposite end, the wood is again attracted, turning the arrow to point away from the rubber rod.

An ordinary hard rubber comb when drawn through the hair becomes charged with electricity and will attract light objects in the same way. So great are the electrical charges sometimes produced on a comb that tiny sparks can be seen to jump between the comb and hair. This is particularly noticeable in a darkened room. These sparks are the reason for the crackling noise so often heard when hair is being combed.

A spectacular effect is produced by bringing a charged rubber rod close to one side of a very thin stream of water running smoothly from a faucet. As shown in Fig. 26B, the stream is diverted to one side and even into the horizontal before it falls again.

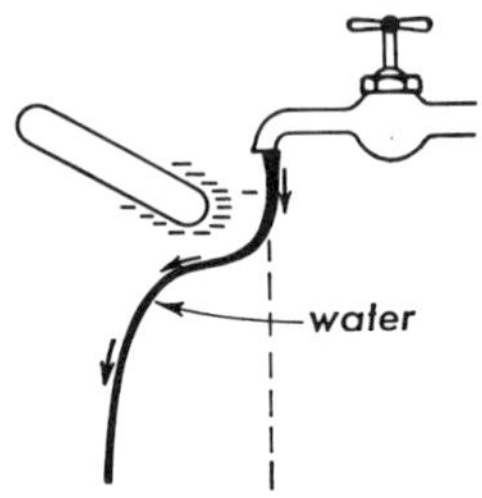

Fig. 26B. A stream of water is easily deflected by a charged rod.

An ordinary sheet of writing paper placed on the panel of a door or another similar flat surface and rubbed will hold fast and remain there for some little time without falling down.

26.3 ELECTRICITY + AND − When two different substances are rubbed together and then separated, both are found to be electrified, one with one kind of electricity and the other with another. To illustrate this, one end of a rubber rod is charged by rubbing with fur and then suspended in a small wire stirrup as shown in Fig. 26C. When the electrified end of a similarly charged rod is brought close by, as shown in diagram (a), the suspended rod turns away, showing repulsion. If the fur is brought close by in place of the rubber, the suspended rod is attracted and turns toward the fur. When a glass rod, previously rubbed with silk, is brought close by, as in diagram (b), there is attraction, and when the silk is brought up there is repulsion.

Since the fur, as well as the glass, attracts the electrified rubber rod, they each have the same kind of electrification: they are said to be *positively charged.* By similar notation the rubber and silk by their actions are said to be *negatively charged.* Positive charges are designated by a (+) sign and negative charges by a (−) sign.

Not only do the above experiments indicate the existence of two kinds of electrification but also demonstrate a rule concerning the action of one kind on another. Diagram (a), illustrating a negatively charged rubber rod repelling a similar rod, shows that two negative charges repel each other. Diagram (b) shows that positive and negative charges attract each other, and diagram (c) shows that two positive charges repel each other. The general law can therefore be stated that *like charges repel and unlike charges attract.*

26.4 THEORY OF ELECTRIFICATION Historically there have been two outstanding theories of electrification: the one-fluid theory of Benjamin

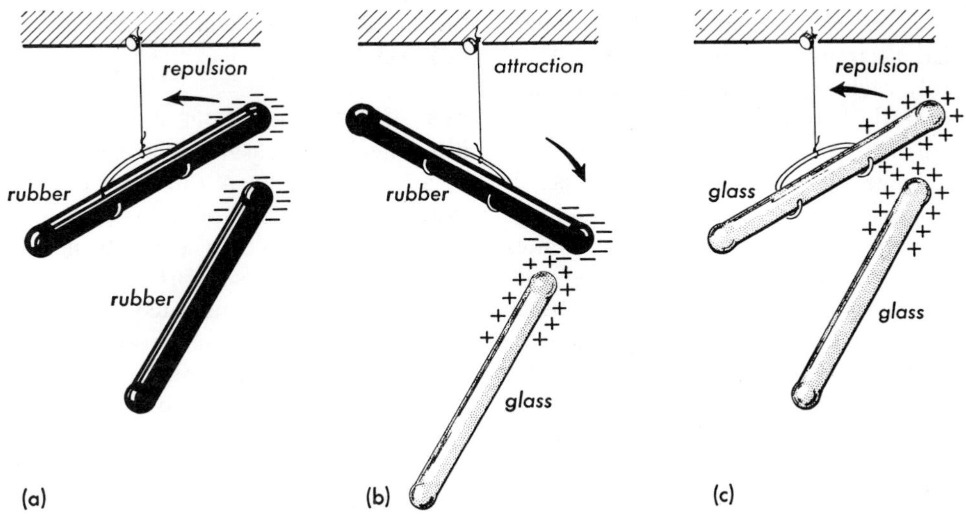

Fig. 26C. Like charges of electricity repel each other and unlike charges attract.

Franklin* and the two-fluid theory of Charles Du Fay. According to the two-fluid theory all objects contain equal amounts of two fluids. When two different substances are rubbed together, one kind of fluid (positive) is spread over one object and the other kind of fluid (negative) over the other.

According to the one-fluid theory of Franklin all bodies contain a certain specified amount of an "electric fire" or fluid to keep them in an uncharged or neutral state. When two objects are rubbed together one accumulates an excess of fluid and becomes positively charged while the other loses fluid and becomes negatively charged. To Franklin we owe the terms "plus" and "minus," "positive" and "negative" electricity.

Both of these theories are in part correct, for now we know the mechanism by which bodies become electrified by friction. The modern theory is based upon the principle already put forward—that all substances are made of atoms and molecules. Each atom contains a nucleus having a known amount of positive charge. See Fig. 26D.

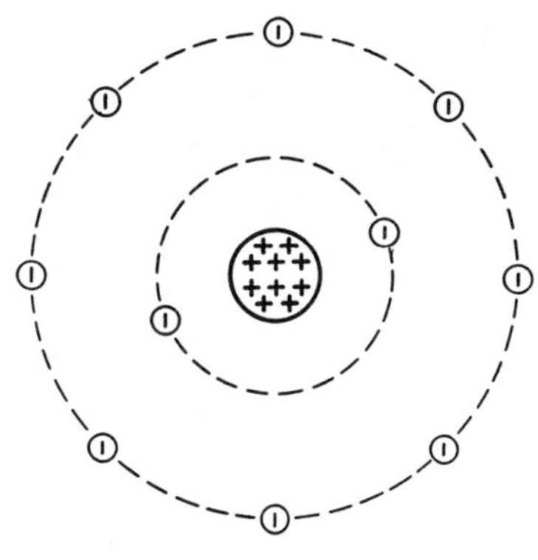

Fig. 26D. Schematic diagram of a neon atom showing its nucleus at the center with ten positive charges (called protons) surrounded on the outside by ten negative charges (called electrons).

This positive charge is due to the presence in the nucleus of a certain number of *protons*. All protons are alike and have the same mass and positive charge. Around every atomic nucleus

* BENJAMIN FRANKLIN (1706–1790). From printer's apprentice as a youth he became a man of unusual powers, not only in politics and diplomacy but also in scientific research. His most famous scientific achievement was the discovery of the electrical nature of lightning. This he did by flying a kite into the clouds on a stormy day and noting the electrical sparks at the ground end of the kite string (a copper wire). Among his many practical applications of scientific discoveries he invented the lightning rod and made the first pair of bifocal eyeglasses.

there are a number of negatively charged particles called *electrons*.

Normally each atom of a substance is electrically neutral; in other words, it has equal amounts of negative and positive charge. Since each electron has the same amount of charge as every other electron and the same amount as every proton but of opposite sign, there are just as many protons in every nucleus as there are electrons around it. While protons are much smaller than electrons in size, they contain the bulk of the mass of every atom. One proton, for example, weighs nearly two thousand times as much as an electron. The electrons therefore are light particles or objects around a small but relatively heavy nucleus.

Individual atoms or large groups of atoms and molecules have an *affinity*, or *attraction*, for additional electrons over and above the exact number which will just neutralize the positive charges of the nuclei. This attraction of the atoms for more than a sufficient number of electrons varies considerably from atom to atom and substance to substance. When, therefore, two different substances are brought into contact, the substance with greater electron affinity seizes nearby electrons from the other and thus acquires a net negative charge. Such is the case, for example, with rubber and amber when rubbed with fur. Having a strong affinity for electrons, both of these solids become strongly negative, whereas the fur becomes deficient of electrons and thereby positively charged.

26.5 THE ELECTROSCOPE An electroscope is an instrument for measuring the electrical potential of a charged body. A thin strip of gold leaf is fastened to the side of a long narrow rod of metal and mounted in a metal and glass box. See Fig. 26E. The support, which will here be called the "stem," is insulated from the box with amber. When the metal knob is touched by a charged rubber rod, some of the charge flows onto and distributes itself over the gold leaf and stem. Since like charges repel each other, the gold leaf is pushed out as shown in the diagram. When the source of charge is taken away, the electroscope retains its acquired charge, which, distributing itself more or less uniformly over the stem, causes the leaf, as shown in diagram (b), to stand out at a somewhat smaller angle. The more charge given the electroscope, the higher the gold leaf is repelled.

If an electroscope is first charged negatively as shown in Fig. 26E and then a negatively charged body is brought close to but not touching the knob, as shown in Fig. 26F, the gold leaf will rise as indicated. The reason is that the electrons are repelled away from the knob to the far end of the stem, causing the gold leaf to rise still higher. As long as the two bodies do not touch each other and allow more negatives to go

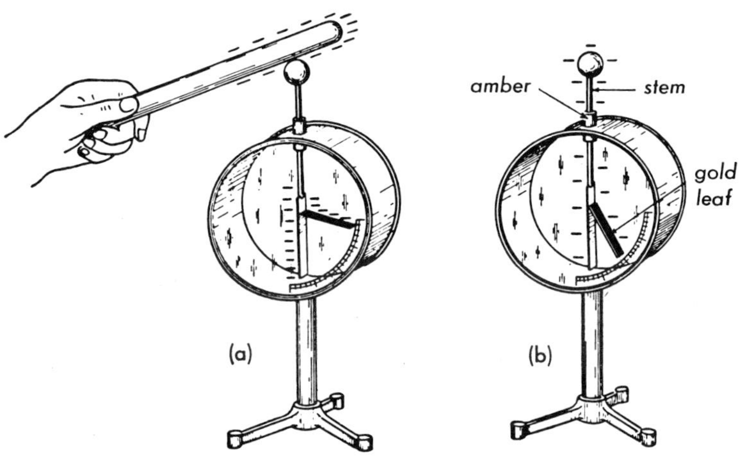

Fig. 26E. Diagram of an electroscope showing how such an instrument may be given a negative charge.

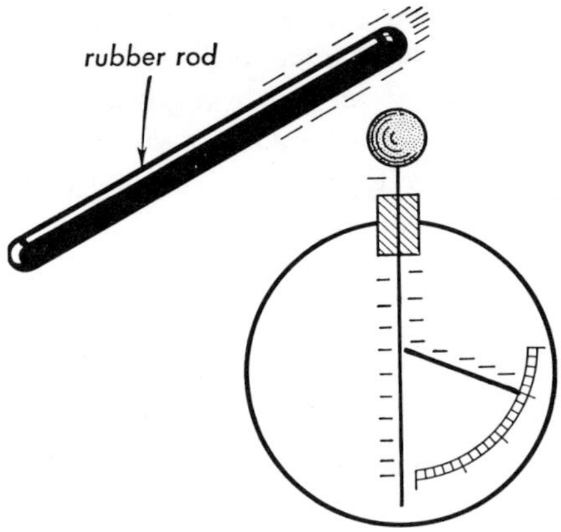

Fig. 26F. The negatively charged gold leaf rises when a negatively charged body is brought close to the knob.

to the electroscope, the gold leaf will fall back to its original angle when the negatively charged rod is taken away.

If a positively charged body is brought up as shown in Fig. 26G, negatives from the stem and

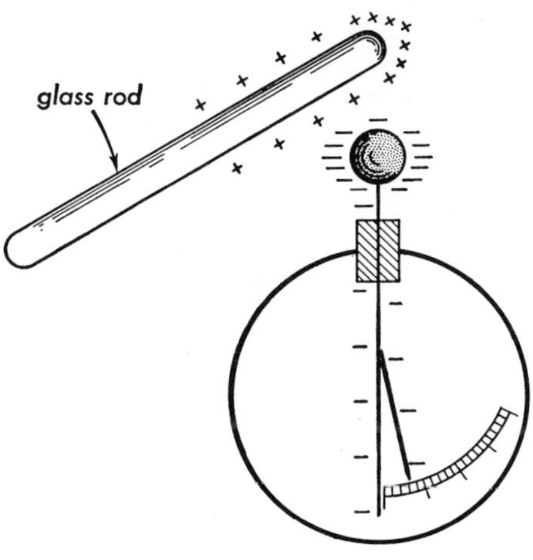

Fig. 26G. The negatively charged gold leaf falls when a positively charged body is brought close to the knob.

gold leaf are attracted to the knob, causing the gold leaf to fall. *Thus with a negatively charged electroscope a positive charge brought nearby causes the gold leaf to drop, and a negatively charged body causes it to rise.* If the electroscope is positively charged, the reverse action will take place; a positive charge causes it to rise and a negative causes it to fall.

26.6 CONDUCTORS AND INSULATORS Not all substances are good conductors of electricity. As a general rule, metals are good conductors, and nonmetals are poor conductors. The poorest of conductors are commonly called *insulators*, or *nonconductors*. Several examples of conductors and nonconductors are listed in Table 26A.

TABLE 26A. Examples of Electrical Conductors and Nonconductors

Conductors	Nonconductors
nickel	glass
platinum	amber
iron	rubber
mercury	mica
silver	sulfur
aluminum	porcelain
copper	paper
gold	silk

The property of electrical conduction is illustrated by an experiment in Fig. 26H. One end of a long thin copper wire is connected to an electroscope and the other end to a small brass knob mounted on a glass pedestal. When a charged rubber rod is touched to the knob as shown, the gold leaf of the distant electroscope

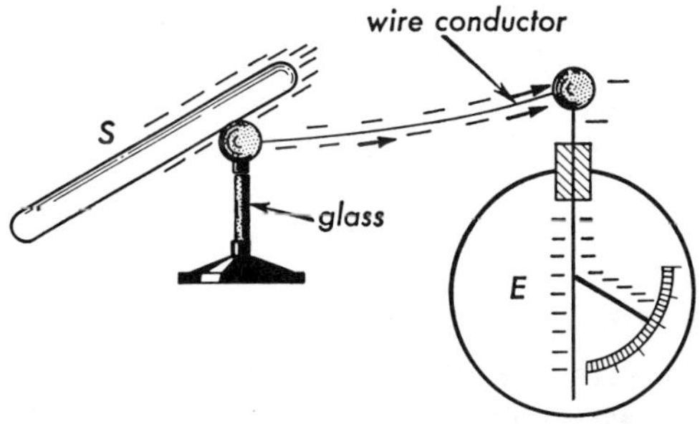

Fig. 26H. Illustration of an electron current.

rises immediately. Electrons have been conducted along the wire. If a positively charged rod contacts the knob, electrons flow away from the electroscope, leaving the gold leaf with a positive charge.

If the copper wire in the above experiment is replaced by a nonconductor, like a silk thread, the electroscope cannot be charged by the rod contacting the distant knob. Poor conductors, such as glass and amber, are used to support metal parts of electrical apparatus for the purpose of insulating them from unnecessary losses of electricity. An electroscope, for example, will retain its electric charge well if the gold leaf and stem are insulated from the electroscope case with amber as shown in Fig. 26E.

The difference between a conductor and an insulator, or dielectric, is that in a conductor there are free electrons, whereas in an insulator all the electrons are tightly bound to their respective atoms. In an uncharged body there are an equal number of positive and negative charges. In metals a few of the electrons are free to move from atom to atom; so that when a negatively charged rod is brought to the end of a conductor it repels nearby free electrons in the conductor, causing them to move. They, in turn, repel free electrons in front of them, thus giving rise to a flow of electrons all along the conductor. Hence, in Fig. 26H it is not necessarily the electrons from the charged rubber rod that actually reach the electroscope leaf, but rather the electrons from the end of the wire where it touches the electroscope knob.

There are a large number of substances that are neither good conductors of electricity nor good insulators. These substances are called *semi-conductors*. In them electrons are capable of being moved only with some difficulty, i.e., with considerable force.

26.7 THE LAW OF ELECTROSTATIC FORCE It has already been demonstrated that like charges repel and unlike charges attract. Nothing that has been said thus far, however, has indicated just how strong the repulsion or attraction might be, or how it depends on the magnitude of the charges and the distance between them.

The first quantitative measurements of the force between two charged bodies were made by Coulomb,* a French scientist and engineer, in 1780. He proved experimentally that *the force acting between two charges is directly proportional to the product of the two charges and inversely proportional to the square of the distance between them.* Symbolically this law is usually written as an algebraic equation,

$$F = k\frac{QQ'}{d^2} \qquad (26a)$$

where F is the force, Q and Q' are the charges, and d is the distance between them. See Fig. 26I.

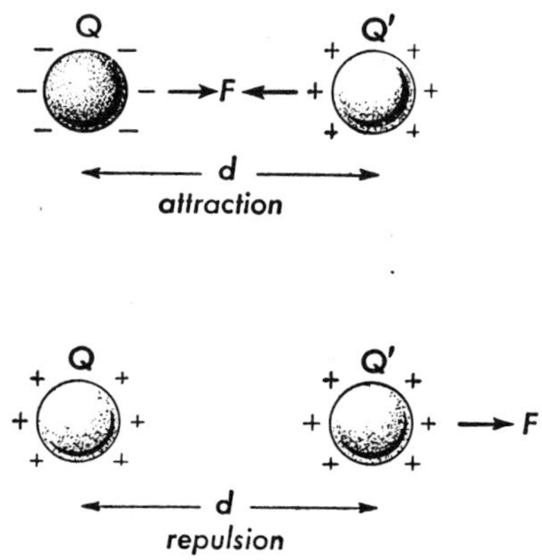

Fig. 26I. Two like charges repel each other and two unlike charges attract each other.

The constant of proportionality, k, has a value which depends upon the units of charge chosen.

In the *MKSA* system, force is given in *new-*

* CHARLES AUGUSTIN COULOMB (1736–1806), French physicist, was born in Angoulême on June 14, 1736. He was a military engineer and spent nine years in the West Indies before returning to his homeland. In poor health, he retired to a small estate at Blois and devoted himself to scientific research. He designed a very sensitive torsion balance, and with it established the inverse square law for electrically charged bodies and for magnetic poles. He also showed that an electric charge is confined to the surface of a conductor and stated that in the case of action at a distance, the intervening medium played no part. He is best known for the inverse square law known as "Coulomb's law."

tons, distance in *meters*, charge in *coulombs*, and $k = 9 \times 10^9$. The fourth fundamental unit in the *MKSA* system is the *ampere*, and is defined in Sec. 27.1. It is customary to define the coulomb in terms of electric currents. *One coulomb is that quantity of electric charge which flowing by any point in a wire in one second produces a current of one ampere.*

Experiments described in later chapters show that electrons are all alike and that each carries an amount of charge as follows:

$$\boxed{e = 1.6021 \times 10^{-19} \text{ coulomb}} \qquad (26b)$$

This means that when a body has a negative charge of one coulomb, it has an excess of $1/1.6019 \times 10^{-19}$ or 6.24×10^{18} electrons, and that a body charged positively with one coulomb has a deficiency of 6.24×10^{18} electrons.*

$$1 \text{ coulomb} = 6.24 \times 10^{18} \text{ electrons} \qquad (26c)$$

Since the unit of charge in the *MKSA* system is measured in terms of electric currents, the numerical value of k in Eq. (26a) must be determined experimentally. The best value to date is $k = 8.98755 \times 10^9$. For most practical problems the approximation $k = 9 \times 10^9$ will be used.

$$\boxed{k = 9 \times 10^9 \frac{\text{newton-meter}^2}{\text{coulomb}^2}} \qquad (26d)$$

EXAMPLE 1. A charge of $+25 \times 10^{-9}$ coulomb is located 6 cm from a charge of -72×10^{-9} coulomb. Calculate the force between them.

SOLUTION. The given quantities are $Q = 25 \times 10^{-9}$ coulomb and $Q' = -72 \times 10^{-9}$ coulomb, $d = 0.06$ m. By substitution in Eq. (26a), we obtain

$$F = 9 \times 10^9 \frac{\text{newton-m}^2}{\text{coulomb}^2} \times \frac{(25 \times 10^{-9})(-72 \times 10^{-9}) \text{ coulomb}^2}{(0.06)^2 \text{ m}^2}$$

$$F = -4.50 \times 10^{-3} \text{ newton}$$

The minus sign indicates attraction. It should be noted that the *coulomb* is an extremely large unit of charge.

* For an explanation of the power system of notation, see Appendix VII.

26.8 THE ELECTRIC FIELD In the space around a charged body is an invisible something called *an electric field.* This field is just another way of describing the action at a distance of one charge upon another. The strength of *the electric field at any point in the neighborhood of a charged body is equal to the force per unit charge exerted on any charge placed at that point.*

Since force is a vector quantity, an electric field has magnitude and direction. The field about a positive charge is therefore radially outward as shown in Fig. 26J (a). It is radially outward since a positive charge placed at any point is repelled along a line through the two charges. By a similar reasoning the field about a negative charge is radially inward as shown in diagram (b).

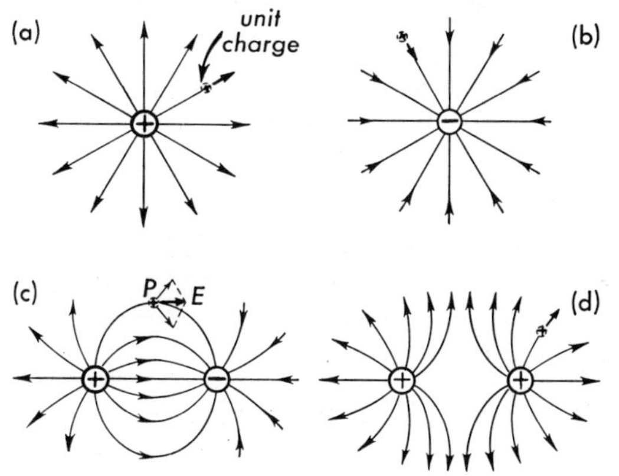

Fig. 26J. Diagrams of the electric field around charged bodies.

The electric field around two charged bodies is shown in diagrams (c) and (d). Each of these fields may be experimentally mapped by placing a positive charge at any one point and moving it always in the direction of the force F exerted on it. The lines traced out by such a charge are called *electric lines of force.* It is to be noted that as many lines as desired can be drawn and that no two lines ever cross. Furthermore, the lines themselves are imaginary and do not actually exist. They were first introduced by Michael Faraday about 1820 as an aid to the understanding of various electrical phenomena.

The direction of the lines of force are given by the arrowheads and the relative magnitude of the field at any region is given by the relative number of the lines passing through that region. In agreement with Coulomb's law *the strength of the field at any point near a single charged body is inversely proportional to the square of the distance away.*

26.9 UNIFORM ELECTRIC FIELD In many experimental studies of atomic structure a great deal of knowledge can be obtained by observing the behavior of charged atomic particles traversing a uniform electric field. To obtain such a field, i.e., a field constant in magnitude and direction over a specified volume of space, two flat metal plates are set up parallel to each other as shown in Fig. 26K.

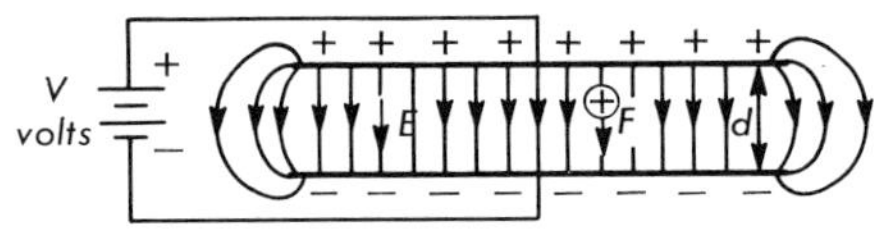

Fig. 26K. The electric field between two parallel charged plates is uniform.

When the terminals of a battery with an electromotive force of V are connected to these plates as indicated in the diagram, a uniform electric field E is produced between the plates. Outside the plates and near the ends the field is not uniform.

In mechanics *work done* is defined as *force times distance*, $W = F \times x$. The electrical equivalent of this equation follows, therefore, by direct substitution of the equivalent electrical quantities for W and F. Since work done per unit charge is called *potential difference* V, and the force per unit charge is called the *electrical field strength* E, see Eq. (6b), the work equation $W = F \times x$ becomes

$$V = E \times d \quad \text{or} \quad E = \frac{V}{d} \tag{26e}$$

V is in *volts*, d is in *meters*, and E is in *volts/meter*, or in the equivalent of *newtons/coulomb*. In other words

$$1\,\frac{\text{volt}}{\text{meter}} = 1\,\frac{\text{newton}}{\text{coulomb}} \tag{26f}$$

When a charge Q is placed in an electric field of intensity E, it will experience a force given by

$$\boxed{F = Q \times E} \tag{26g}$$

As can be seen in Fig. 26K, E is a vector quantity and is arbitrarily chosen to have the same direction as the force F acting on a positive charge Q. The force on a negative charge is opposite in direction to E.

EXAMPLE 2. Two flat metal plates are mounted parallel to each other 2.5 cm apart. The terminals of a 10,000-volt source are then connected to the plates, and a small body with a positive charge of 2.8×10^{-7} coulomb is placed between them. Find (a) the field intensity between the plates, and (b) the force on the charged body.

SOLUTION. (a) The field intensity E is obtained by using Eq. (26e).

$$E = \frac{V}{d}, \quad E = \frac{10{,}000}{0.025} = 4 \times 10^5\,\frac{\text{newtons}}{\text{coulomb}}$$

(b) The force F on the charge Q is found from Eq. (25g).

$$F = Q \times E$$

$$F = 2.8 \times 10^{-7}\,\text{coulomb} \times 4 \times 10^5\,\frac{\text{newtons}}{\text{coulomb}}$$

$$= +0.112\,\text{newton}$$

QUESTIONS

1. How many kinds of electrical charges are there? What are they called? What is meant by electrostatic?

2. What is the rule regarding repulsion and attraction? How can the rule be demonstrated?

3. What is an electroscope? How is it made?

4. What is the structure of the atom? How much negative charge does a neutral atom have?

5. What is the present-day explanation of electrification? Why are things rubbed together to charge them? What kinds of charges do not move in electrification?

6. What is meant by a nonconductor? What kinds of charges move through solid conductors?

7. When a positively charged rod is brought up and touched to the knob of an electroscope, the gold leaf rises. When the rod is taken away, the leaf drops part way back only. Why?

8. When a negatively charged rod is brought up close to the knob of a negatively charged electroscope, what happens?

9. When a negatively charged rod is brought up close to the knob of a positively charged electroscope, what happens?

10. Do you think that the flow of a liquid through a pipe might produce electrical charges? Do you know of any examples of where this occurs?

11. While combing your hair, crackling noises are sometimes heard. Why? How can you show that your comb becomes electrically charged?

12. Plastic seat covers in automobiles sometimes cause the driver to be electrically shocked. Why?

13. Why are some substances good conductors of electrons and others not? Do all substances contain electrons?

14. What is meant by an electric field? How could one find the electric field intensity at a point in space? What are the units of the electric field intensity E?

15. What is a uniform electric field? How is such a field produced? How is its magnitude determined by calculation? How is the direction of a field determined?

16. Equal positively charged bodies are located at the three corners of an equilateral triangle. Make a diagram of what you think the electric field lines should look like around these bodies.

PROBLEMS

1. A charge of -20×10^{-7} coulombs is located 10 cm from another charge of -20×10^{-7} coulomb. Calculate the force in newtons exerted by one charge upon the other. [Ans. 3.60 N.]

2. Two charges of -12×10^{-7} coulomb each are located 9 cm apart. What is the repelling force on each in newtons?

3. Two unlike charges of 75×10^{-8} coulomb each are located 25 cm apart. What is the attracting force on each in newtons? [Ans. 8.10×10^{-4} N.]

4. Two equal charges are located 25 cm apart and repel each other with a force of 5 newtons. Find the magnitude of each charge in coulombs.

5. Two small metal spheres 36 cm apart, and having equal negative charges, repel each other with a force of 4.2×10^{-3} newton. Find the total charge on the two bodies in coulombs. [Ans. 4.92×10^{-7} C.]

6. A positive charge of 5×10^{-8} coulomb is located 6 cm from a negative charge of 8×10^{-8} coulomb. Calculate the force in newtons exerted by either charge upon the other.

7. Two like charges of 15×10^{-7} coulomb are located 10 cm apart. What is the magnitude of the force on each one? [Ans. 2.025 N.]

8. A charge of -8×10^{-7} coulomb is located 4 cm from another charge of 12×10^{-7} coulomb. Calculate the force in newtons exerted by one charge upon the other.

9. Two small metal spheres are 18 cm apart. Having equal positive charges, these spheres repel each other with a force of 0.65 newtons. Find the charge on each body in coulombs. [Ans. 1.440×10^{-6} C.]

10. Two equally charged bodies are located 15 cm apart and repel each other with a force of 1.6 newtons. Find the charge on each body in coulombs.

11. What charge Q placed 18 cm from a charge of 12×10^{-8} coulomb will produce a force of 0.6 N? [Ans. 1.80×10^{-5} C.]

12. Four equal charges of $+4 \times 10^{-8}$ coulomb each are located at the corners of a square 8 cm on a side. Calculate the resultant force on each charge and show its direction on a diagram drawn to scale.

13. Two positive charges, of 6×10^{-7} coulomb each, are located diagonally opposite each other on a square 12 cm on a side. Two negative charges, of -6×10^{-7} coulomb each, are located at the other two corners respectively. Calculate the resultant force on each charge and show this resultant on a diagram drawn to scale. [Ans. 0.2056 N inward on all four.]

14. Two flat metal plates are mounted parallel to each other and 4 cm apart. If the two terminals of a 60-volt battery are applied to these plates, what is the intensity of the uniform electric field between them?

15. Two flat metal plates 3 cm apart are connected to a 6000-volt battery. A proton with its positive charge of 1.6×10^{-19} coulomb is located between these plates. Find (a) the electric field intensity between the plates, and (b) the force on the proton in newtons. [Ans. (a) 2.0×10^{5} N/C, (b) 3.204×10^{-14} N.]

16. Two flat metal plates 5 cm apart are connected to an 8000-volt source. A small charge of 2.5×10^{-9} coulomb is located in the field. Find (a) the electric field intensity, and (b) the force on the charge Q.

17. A battery of 6000 volts is applied to two parallel plates 4 mm apart. What is the force on an electron of charge 1.6×10^{-19} coulomb when it passes through the uniform electric field between the plates? [Ans. 2.403×10^{-13} N.]

18. A 4000-volt source is connected to two parallel plates 1.2 cm apart. Find the force on the nucleus of a helium atom located in the uniform electric field between the plates. The charge on a helium nucleus is $+3.2 \times 10^{-19}$ coulomb.

19. A 10,000-volt battery is connected to two parallel metal plates 1.8 cm apart. Find the force on an atomic particle having a positive charge of 1.6×10^{-19} coulomb as it passes through the uniform electric field between the plates. [Ans. 8.90×10^{-14} N.]

27 | ELECTRICITY IN MOTION

The importance of electricity in our everyday lives is never so well demonstrated as when a city electrical power house shuts down. Hospitals, hotels, office buildings, food storage plants, and research laboratories maintain standby motor generator units that automatically start the instant of a city power plant failure. For these and many other reasons everyone should have some knowledge of electricity, how it is connected to common devices, and what is proper in making simple electrical connections.

27.1 A CONTINUOUS ELECTRON CURRENT To make an electron current flow continuously along a wire, a continuous supply of electrons must be available at one end and a continuous supply of positive charges at the other. See Fig. 27A. This is like the flow of water through a pipe: to obtain a continuous flow, a continuous supply of water must be provided at one end and an opening for its escape into some receptacle at the other. The continuous supply of positive charge at the one end of a wire offers a means of escape for the electrons. If this is not provided, electrons will accumulate at the end of the wire and their repulsion back along the wire will stop the current flow.

There are two general methods by which a continuous supply of electrical charge is obtained: one is by means of *a battery* and the other is by means of *an electric generator*. The battery is a device by which chemical energy is transformed into electrical energy and the generator is a device by which mechanical energy is transformed into electrical energy.

Electric current is usually measured in a unit called the *ampere*. *The ampere*, named in honor of the French physicist Ampère,* *is the term applied to the flow of one coulomb per second.* In other words, one coulomb flowing past any given point in a wire in one second constitutes a current of one ampere (*abbr*. 1 A).

$$1 \text{ ampere} = \frac{1 \text{ coulomb}}{1 \text{ second}} \qquad (27a)$$

* ANDRÉ M. AMPÈRE (1775–1836), French physicist and mathematician. Ampère began his career as professor of physics and chemistry at Bourg at the early age of 26. He later established the relation between electricity and magnetism, and helped to develop the subject he called electrodynamics. His only son, Jan J. Ampère, also became famous as a philologist, lecturer, and historian.

If twice this quantity passes by in one second, the current is 2 A. Thus electric current is analogous to the rate of flow of water through a pipe.

$$\text{current} = \frac{\text{quantity of charge}}{\text{time}}$$

$$\boxed{I = \frac{Q}{t}} \qquad (27b)$$

Remembering that one coulomb = 6.24×10^{18} electrons, a current of one ampere means a flow of 6.24×10^{18} electrons per second past any given point. See Sec. 26.7, Eq. (26c).

This enormous number does not mean that the electrons are moving with high speed through a conductor. Actually the number of moving charges is so large that their average velocity is but a small fraction of a millimeter per second. When we picture a solid conductor as a crystal lattice, similar to that shown in Fig. 10F, the electrons are thought of as moving through the intervening spaces. This movement is not completely free, however, but is influenced by the repulsion and attraction of like and unlike charges.

27.2 BATTERIES Batteries as continuous sources of electrical energy are the result of a long series of experiments which started with the discoveries of Alessandro Volta* more than one hundred years ago. Today battery cells are manufactured in two common forms: (1) dry cells, as used in flashlights, portable radios, etc., and (2) wet cells, as used in automobiles, airplanes, boats, etc.

* ALESSANDRO VOLTA (1745–1827), Italian scientist, was for more than twenty years professor of physics at Pavia. Traveling widely throughout Europe, he became acquainted with many celebrities. In 1801 he was awarded the Copley medal of the Royal Society of London, and then was called to Paris and awarded a medal by Napoleon. In 1815 the emperor of Austria made him director of the philosophical faculty of the University of Padua. A statue now stands in his memory at Como, his birthplace.

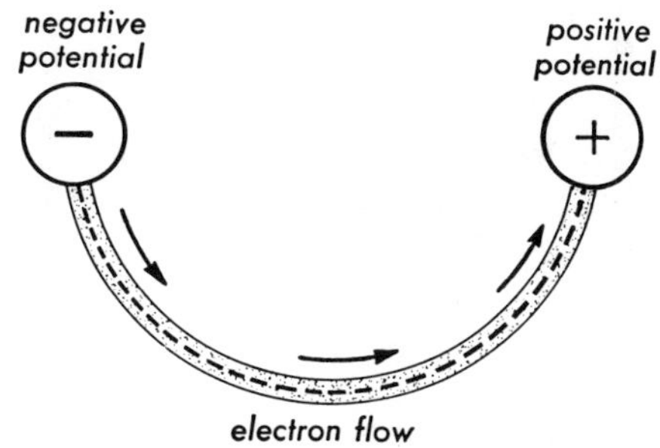

Fig. 27A. Two terminals at different potentials and connected by a conductor give rise to an electron current.

27.3 THE VOLTAIC CELL The voltaic cell, as shown in Fig. 27B, is composed of three parts,

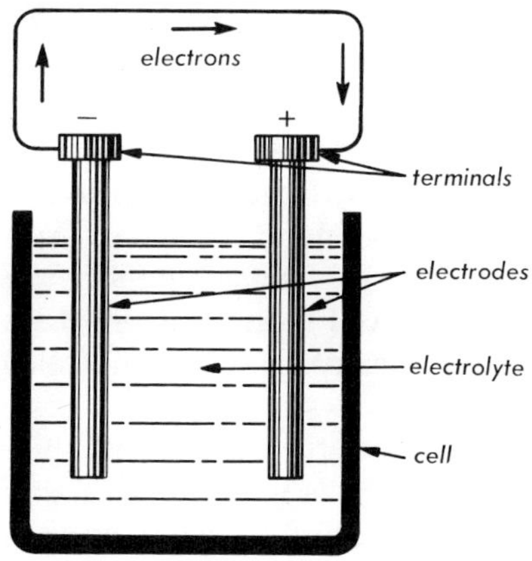

Fig. 27B. Diagram of a voltaic cell.

a pair of dissimilar metal plates called *electrodes*, a dilute acid solution called the *electrolyte*, and a nonconducting liquid container called the *cell.* While many different combinations of materials have been tried and used in such cells, zinc and carbon as electrodes, dilute sulfuric acid as the electrolyte, and a glass or hard rubber container as the cell are the most common.

Individual molecules of sulfuric acid (written H_2SO_4) are composed of seven atoms each: two hydrogens, one sulfur, and four oxygens. When concentrated acid is poured into water to form a dilute solution, a small per cent of the molecules split up, that is, they dissociate. See Fig. 27C. The two hydrogens split off from the mole-

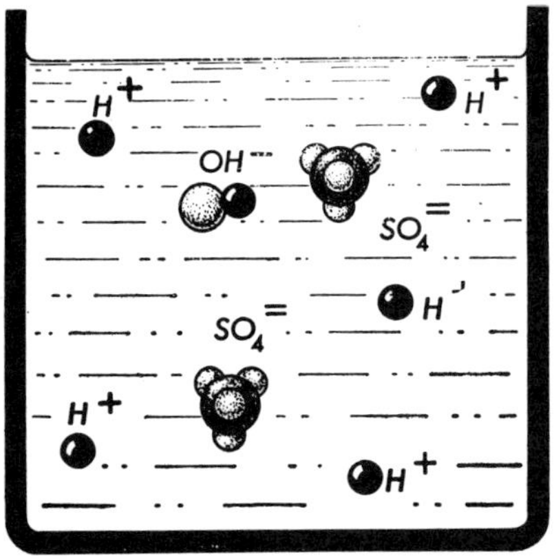

Fig. 27C. Some of the sulfuric acid molecules and water molecules in a dilute solution dissociate.

cule, each leaving an electron behind with the remaining SO_4. Similarly some of the water molecules, each composed of two hydrogens and one oxygen, dissociate by having one of the hydrogen atoms split off without an electron.

Since each hydrogen atom leaves an electron behind, it has a net positive charge and is called a *positive ion*. Each remaining SO_4 fragment from the acid and OH fragment from the water has a surplus of negative charge and is called a *negative ion*.

If we now insert a zinc rod into this electrolyte, chemical action is initiated, whereby individual zinc atoms break off from the surface leaving two electrons behind. See Fig. 27D. Going into solution as doubly charged positive ions, they strongly attract SO_4^{--} ions. With each encounter of two such ions, combination takes place, resulting in the formation of neutral $ZnSO_4$ molecules.

$$Zn^{++} + SO_4^{--} = ZnSO_4$$

As more and more zinc ions go into solution, the negative charge on the zinc electrode increases and soon reaches a maximum.

If we now insert both zinc and carbon electrodes into the electrolyte and complete the circuit by connecting the electrodes together by a conductor, or a voltmeter, as shown in Fig. 27E, each electrode acquires a charge. The carbon electrode, called the *anode*, acquires a

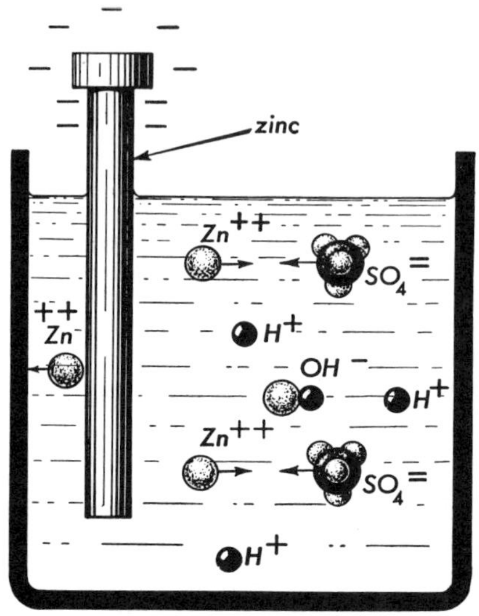

Fig. 27D. A zinc metal electrode in dilute sulfuric acid.

positive potential, and the zinc electrode, called the *cathode*, acquires a negative potential.

With the voltmeter completing the circuit as shown, the pointer indicates the difference of

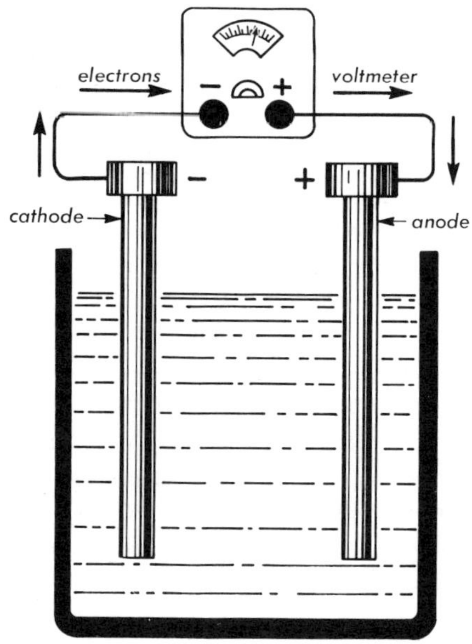

Fig. 27E. The emf of a voltaic cell is measured with a voltmeter.

potential in volts, and this we call the electromotive force (*abbr*. emf).

With the voltmeter completing the electric circuit between cathode and anode, an electron current will flow from cathode to anode through the connecting wires and the voltmeter. Electrons leaving the zinc metal surface change the cathode potential, and electrons arriving at the carbon neutralize positives, thereby changing the anode potential. With the tendency of the potential difference to fall because of electron flow, chemical action takes place immediately and maintains a continuous supply of (+) and (−) charge at the terminals.

What happens at the carbon electrode on closed circuit is shown in detail in Fig. 27F.

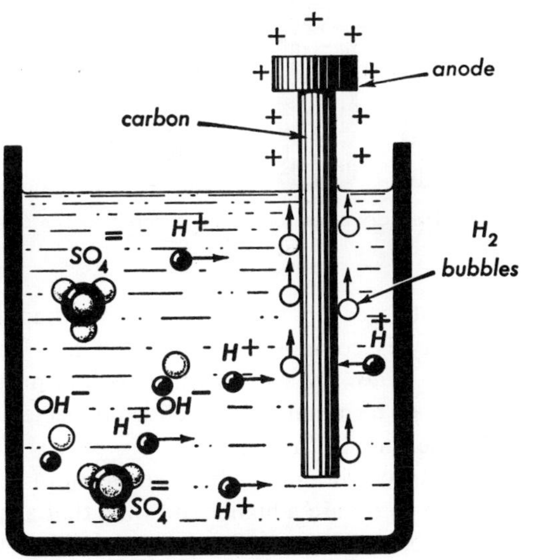

Fig. 27F. Cell activity around anode for voltaic cell when circuit is closed as shown in Fig. 27E.

Negatives flow through the voltmeter and down the carbon rod where they attract hydrogen ions to the anode. There at the surface each H^+ ion acquires an electron and is neutralized. Two such particles form a neutral hydrogen molecule H_2. As neutral hydrogen molecules accumulate at the anode surface they form bubbles of gas and can be seen rising to the surface. The action can be written

$$2(H^+) + 2(e^-) = H_2$$

It is important to note that on open circuit, that is with no wire connection from one terminal to the other, the terminals have a constant difference of potential between them, there is no current flow, and there is no chemical activity.

27.4 THE DRY CELL Probably the most common form of battery used today is composed of dry cells. While these cells are manufactured in different shapes and vary in size from $\frac{1}{8}$ in. to many inches, they all produce approximately the same emf of 1.5 volts between their two terminals (*abbr*. 1.5 V).

One common form of dry cell is shown in Fig. 27G. The negative electrode is a zinc-coated

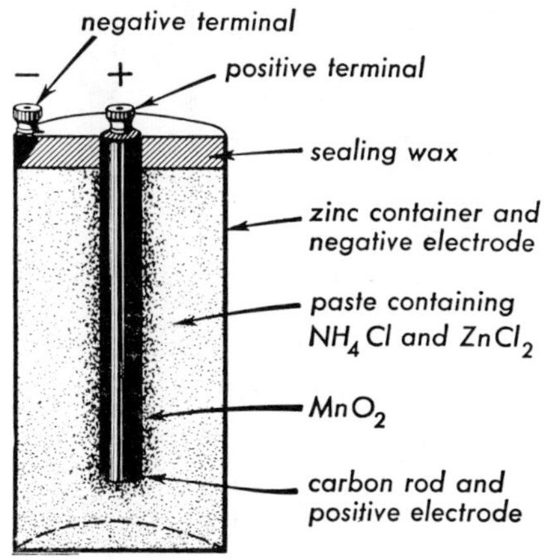

Fig. 27G. Cross-section diagram of a dry cell or "flashlight" battery, showing the essential elements.

metal container in which all chemical ingredients are sealed, and the positive electrode is a round carbon rod. In place of a liquid electrolyte we have a paste containing ammonium chloride, zinc chloride, and a little water. Surrounding the anode is a thin layer of powdered carbon and manganese dioxide.

When current is being supplied by the dry cell, zinc ions form at the cathode, while ammonium ions gain electrons at the carbon rod, forming hydrogen and ammonia gas. The hydrogen reacts chemically with the manganese dioxide, and the ammonia gas with the zinc chloride. While the emf of all dry cells is 1.5 V, the larger the cell the greater is the current and the total electrical energy that can be supplied.

27.5 THE WET CELL When the electrical energy contained in a battery composed of dry cells has been exhausted, it is thrown away. A storage battery, on the other hand, is composed of what are called *wet cells*, and when exhausted of its stored up energy, such a battery can be rejuvenated or *recharged.*

The negative electrode of a storage cell, as shown in Fig. 27H, is composed of a set of

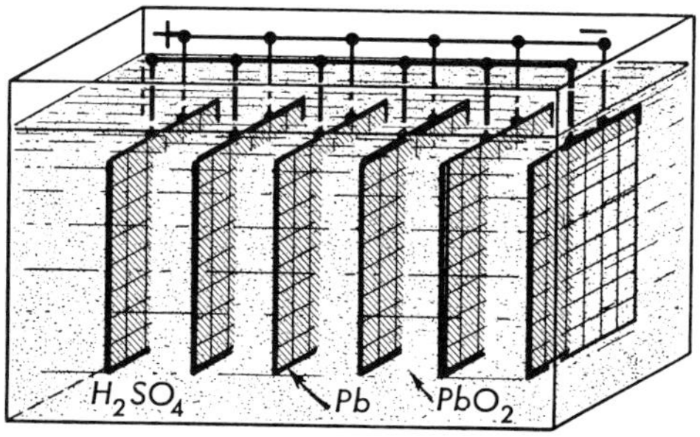

Fig. 27H. The cell of a storage battery has a number of plates.

parallel lead grills filled with *spongy lead* (Pb), while the positive electrode is another set of lead grills filled with porous *lead dioxide* (PbO_2).

When on a closed circuit a current is supplied by the cell, PbO_2 in one set of grills and Pb in the other combine with the dissociated H^+ and SO_4^{--} ions of the dielectric and form lead sulfate ($PbSO_4$) and water (H_2O). When the surfaces of all plates become coated with lead sulfate, they behave chemically alike, and no more current can be drawn from the cell.

To recharge a storage cell the two terminals are connected to a direct current source. The current flowing through the electrolyte reverses the chemical process, changing them back to their original form. Chemically, one writes

$$\begin{array}{c} PbO_2 + Pb + 2H_2SO_4 \\ \text{charge} \uparrow \; \downarrow \text{discharge} \\ 2PbSO_4 + 2H_2O \end{array} \qquad (27c)$$

Upon charging a cell too rapidly, some of the dissociated water molecules of the electrolyte are converted into hydrogen gas at the cathode and oxygen gas at the anode. These rise to the surface as bubbles and necessitate the occasional addition of water to the cell.

Regardless of the size, each such cell develops an emf of 2.2 volts. While many other kinds of storage cells have been developed, the lead battery is the most widely used in the United States.

27.6 THE BATTERY If two or more cells are connected together as shown in Fig. 27I, they

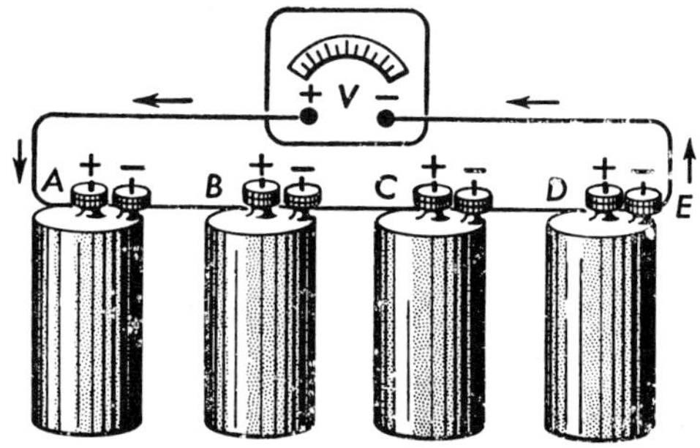

Fig. 27I. Four dry cells connected in series form a 6-volt battery.

form what is called a battery. In this diagram the battery is composed of four dry cells connected *in series.* By series connections is meant that the (+) terminal of one cell is connected to the (−) terminal of the next.

The purpose in connecting two or more cells in series is to obtain a higher emf than that available with one cell alone. The potential difference between the extreme end terminals *A* and *E* of any battery is just the sum of those for the individual cells.

Each dry cell produces an emf of 1.5 V, so that if the voltmeter is connected to the circuit, it will indicate 1.5 V between *A* and *B*, 3.0 volts between *A* and *C*, 4.5 V between *A* and *D*, and 6 V between *A* and *E*.

The common flashlight contains several dry cells connected in series as shown in Fig. 27J. When new cells are inserted, they are all turned in the same direction so that the (+) terminal at the center of each cell makes good contact

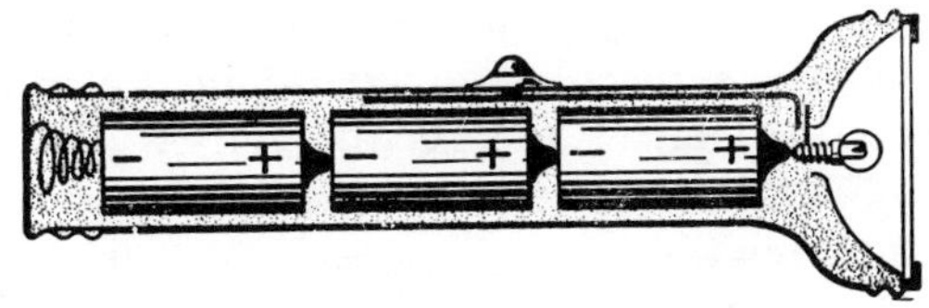

Fig. 27J. Cross section of a 3-cell flashlight.

with the (−) case of the next cell. The closing of the switch shown in the figure applies the end terminal voltage of 4.5 V to the light bulb.

The storage battery commonly used in automobiles contains six wet cells of 2 V each, connected in series as shown in Fig. 27K. Note how

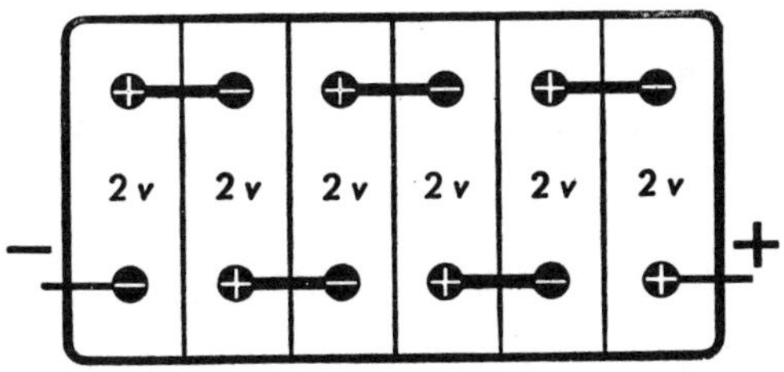

Fig. 27K. Series connections for a 12-volt storage battery.

the heavy crossbars connect the (−) terminal of any one cell with the (+) terminal of the next. With six cells in series the end terminals produce a resultant of 12 V, hence its name of *twelve-volt battery*.

It is customary in circuit diagrams to represent battery cells as shown in Fig. 27L. The (+) and

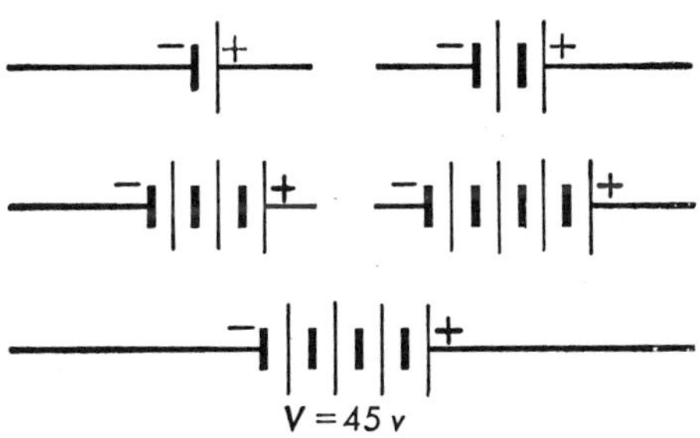

Fig. 27L. Circuit diagrams indicating the cells of a battery.

(−) terminals of each cell are indicated by long and short lines, respectively. The first four diagrams shown represent one, two, three, and four cells, respectively, connected in series. If many cells are to be represented, four or five can be drawn and the over-all terminal voltage written beneath it as shown by $V = 45$ V.

27.7 ELECTRON CURRENT If two wires are connected to the terminals of a battery and the opposite ends of the wires are immersed in a conducting liquid such as sulfuric or hydrochloric acid, positive ions in the electrolyte are attracted to the negative wire where they are neutralized, and negative ions move toward the positive wire where they are neutralized. Positives move from (+) to (−), whereas negatives move from (−) to (+). This two-way flow of charge also occurs in many gases.

Many years ago, before it was known how ions or charges moved in a solid, such as a copper wire, there seemed to be some evidence that only the positive charges moved, and not the negative. This notion became so thoroughly entrenched in the minds of those interested in electrical phenomena that in later years, when it was discovered that the negatives move in solid conductors and not the positives, it became difficult to change.

The convention that electric current flows from plus to minus is still to be found in many books and is used by some electrical engineers in designing electrical machines and appliances. The rapid growth and the importance of radio engineering and electronics, however, have brought about a change in this practice, and we shall in this text speak of current flow in a wire as one of electrons moving from (−) to (+) and call it *electron current*.

27.8 OHM'S LAW* This well-known law in electricity makes it possible to determine the

* GEORGE SIMON OHM (1787–1854), German physicist, was born at Erlavgen and educated at the university there. After teaching mathematics in Cologne for sixteen years, and in Nuremburg for sixteen more, he became professor of experimental physics at the *Technische Hochschule* in Munich. His writings were numerous and but for one exception were not of the first order. This single exception consists of a pamphlet on electric currents,

current flowing through a circuit when the resistance of the circuit and the potential difference applied to it are known. What Ohm discovered was that the ratio of the potential difference between the ends of a metallic conductor and the electron current flowing through it is a constant. The proportionality constant is called the electrical *resistance*.

$$\textbf{resistance} = \frac{\textbf{potential difference}}{\textbf{current}}$$

Symbolically, Ohm's law is often written

$$\boxed{R = \frac{V}{I}} \qquad (27d)$$

In electrical units,

$$1 \text{ ohm} = \frac{1 \text{ volt}}{1 \text{ ampere}}$$

The law is of great importance because of its very general application to so many electrical phenomena. One of its simplest applications is illustrated in Fig. 27M. A dry cell is directly connected by wires to a small light bulb. The battery maintains a potential difference of 1.5 volts across the lamp. If the electron current flowing through the circuit is 0.5 ampere, the resistance of the circuit is

$$R = \frac{1.5 \text{ volts}}{0.5 \text{ ampere}} = 3 \text{ ohms}$$

Although the resistance as found here is assumed to be the resistance of the light bulb, it really includes the resistance of the connecting wires as well. In practice one usually uses wires of such low resistance that they can be neglected in most calculations. If they are not small, they cannot be neglected and must be added in as part of the R in Ohm's law.

Consider the illustration shown in Fig. 27N where a battery of many cells maintains a potential difference of 60 V across a circuit and an electron current of 5 A through it. By Ohm's law the resistance of the circuit is given by Eq. (27d) as

$$R = \frac{V}{I} = \frac{60 \text{ V}}{5 \text{ A}} = 12 \text{ ohms}$$

Ohm's law may be written in two other useful forms. By solving Eq. (27d) for I, and then for V, we obtain

$$I = \frac{V}{R}$$

and

$$V = IR$$

Resistance in circuit diagrams is represented by a saw-toothed line as shown, and the word

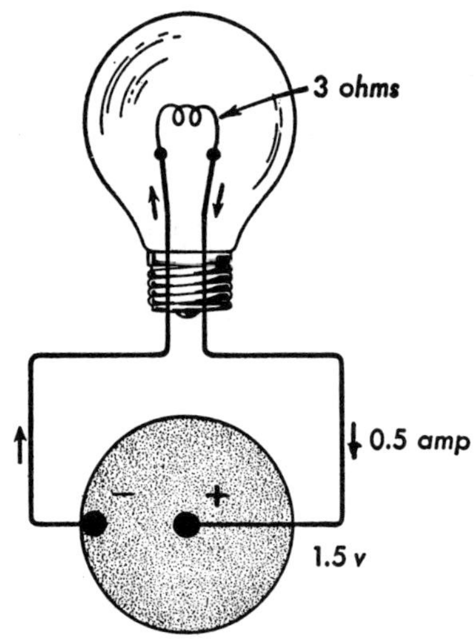

Fig. 27M. A dry cell connected to a small light bulb.

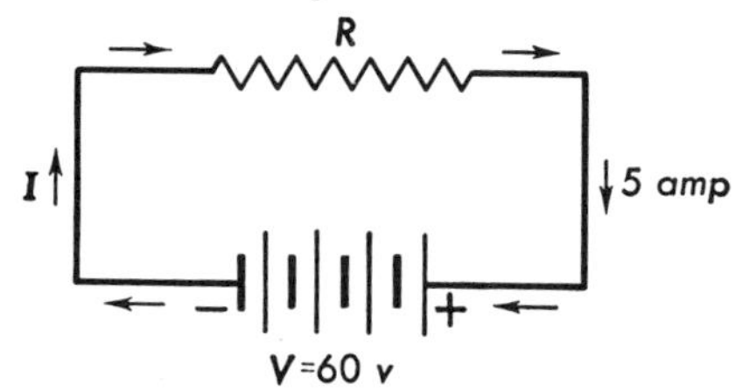

Fig. 27N. Circuit diagram of resistor *R* connected to 60-volt battery.

the most important part of which is summarized in what is now called "Ohm's law." For this work he was awarded the Copley Medal of the Royal Society of London in 1841 and made a foreign member of the society one year later.

ohm is represented by the capital Greek letter omega, Ω. In Fig. 27N the resistance

$$R = 12\,\Omega$$

Resistance may be defined as the opposition offered to a flow of current through a circuit.

To determine the resistance of any electrical circuit it is common practice to use a voltmeter and an ammeter. The voltmeter is applied across the circuit to measure the potential difference, and the ammeter is connected in series to measure the current.

Fig. 27O represents a connector board arranged for measurements of this kind. An appliance, or unknown resistor R, is connected between two terminals at the top, so the electron current supplied by the battery below must pass through it as well as the ammeter. The voltmeter is connected across R to measure the potential difference between its ends.

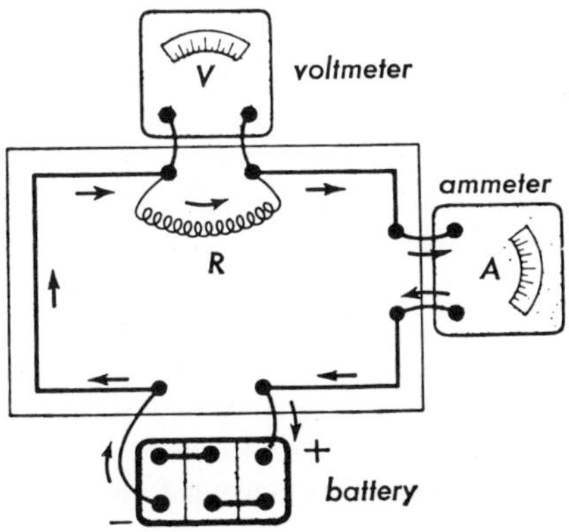

Fig. 27O. Connector board for determining electrical resistance.

Suppose that three heating elements, commonly used as replacement elements for electrical kitchen utensils, are connected in turn to the two terminals at the top of the board and a 6-volt storage battery is applied to the terminals below. The voltage V and the electron current I from a sample set of measurements give the following results.

By Ohm's law, Eq. (27d), the resistance of each element can be calculated. This will be left as an exercise for you, the reader.

TABLE 27A. Recorded Data

Element	Volts	Amperes
toaster	6.3	2.3
waffle iron	6.2	1.6
iron	6.1	1.8

27.9 RESISTORS IN SERIES

When several electrical devices are connected in series, the resistance R of the combination is equal to the sum of the resistances of the individuals. Symbolically,

$$\boxed{R = R_1 + R_2 + R_3 + R_4 + \text{etc.}} \qquad (27e)$$

This, the law of series resistances, is illustrated by an application of Ohm's law to the complete electric circuit in Fig. 27P. Three resistors $R_1 = 5\,\Omega$, $R_2 = 1\,\Omega$, and $R_3 = 3\,\Omega$ are connected in series with a battery of nine storage cells. The sum of the resistance given by Eq. (27e) is

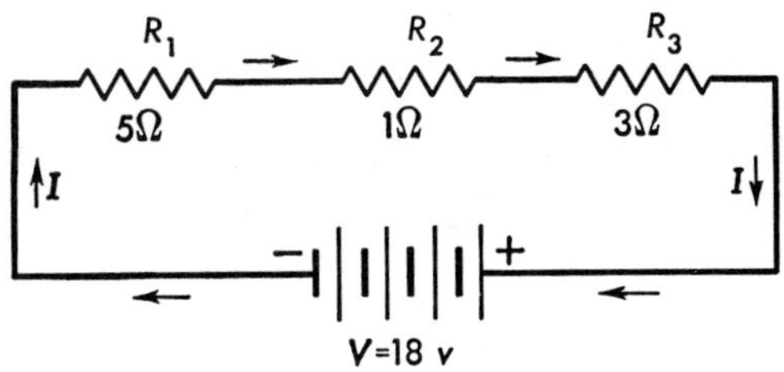

Fig. 27P. The same current flows through all resistors when connected in series.

$$R = 5 + 1 + 3 = 9\,\Omega$$

and the potential difference supplied by the battery is

$$V = 9 \times 2 = 18\text{ V}$$

Applying Ohm's law, Eq. (27d), the electron current flowing through the circuit is

$$I = \frac{18\text{ V}}{9\,\Omega} = 2\text{ A}$$

This means that the same electron current of 2 A flows through the high resistance of 5 Ω as well as through the 1-ohm and 3-ohm resistors. Like water flowing through pipes of different sizes connected one after the other, just as much water passes through one pipe per second of time as through any other, and none can accumulate at any point. The battery acts like a pump and keeps a constant and steady electron current flowing in the circuit.

If the two leads of the voltmeter are connected across the first resistance, the potential difference across R_1 is measured. Knowing the resistance of R_1 and the current I through it, one can calculate this potential difference by using Ohm's law in the form $V = IR$.

$$V_1 = I_1R_1$$
$$V_1 = 2\text{ A} \times 5\,\Omega = 10\text{ V}$$

Because the potential differs by 10 V from one side of the resistor to the other the potential difference is commonly called the *IR drop*. In a similar way the IR drop across R_2 and R_3 can be measured by connecting the voltmeter across each one in turn, or computed as just shown.

$$V_2 = 2\text{ A} \times 1\,\Omega = 2\text{ V}$$
$$V_3 = 2\text{ A} \times 3\,\Omega = 6\text{ V}$$

If we find the sum of all three IR drops around the circuit we obtain

$$10 + 2 + 6 = 18\text{ V}$$

This is known as Kirchhoff's* 1st law and is usually written in the general form

$$V = V_1 + V_2 + V_3 \qquad (27f)$$

Kirchhoff's First Law

* GUSTAV ROBERT KIRCHHOFF (1824–1887), German physicist, was born March 12, 1824, in Königsberg, Prussia. During his lifetime, he served as professor at the Universities of Breslau, Heidelberg, and Berlin. His contributions to experimental and mathematical physics were numerous and important. He extended Ohm's law to conductors in networks and generalized the relations in two laws known as Kirchhoff's Laws. He demonstrated that an electrical disturbance is propagated along a wire with the speed of light in free space, founded the science of chemical analysis by spectra with R. W. von Bunsen, and established a relationship between the emission and absorption of light known as Kirchhoff's Law of Radiation.

27.10 PARALLEL CIRCUITS A circuit diagram showing three resistors R_1, R_2, and R_3 connected in parallel is shown in Fig. 27Q. The electron

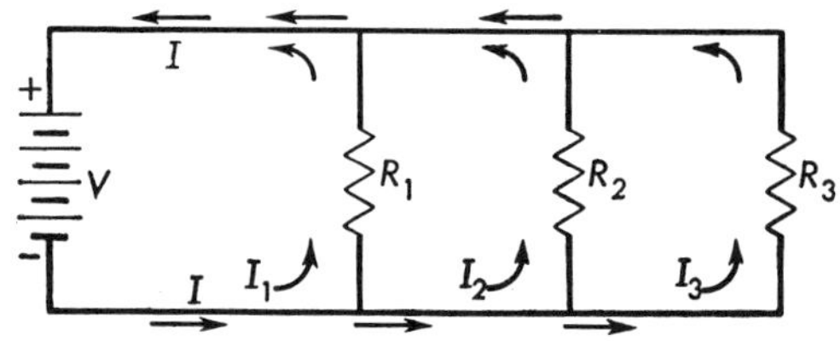

Fig. 27Q. Parallel circuit for three resistors R_1, R_2, and R_3.

current I, leaving the battery at the lower left, divides at the first junction, part I_1 going through R_1, and the remainder goes on to the next junction. Part of this electron current I_2 goes through R_2, and the remainder goes on and through R_3. These three electron currents recombine at the top junctions and form, finally, the same total electron current I returning to the battery.

It is clear from this explanation that for any number of resistors in parallel

$$I = I_1 + I_2 + I_3 + \text{etc.} \qquad (27g)$$

Kirchhoff's Second Law

where I is the toal electron current and I_1, I_2, I_3, etc., are the separate electron currents through the resistors. Eq. (27g) is Kirchhoff's 2nd law.

An excellent demonstration of Eq. (27g) is shown in Fig. 27R, where four ordinary tung-

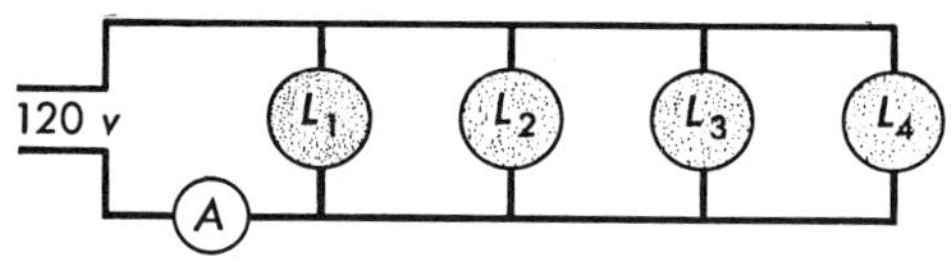

Fig. 27R. Parallel circuit of four light bulbs.

sten filament light bulbs are connected in parallel to a house lighting circuit of 120 volts. The lamps may be the same or all different.

Suppose the lamps are rated as 25, 50, 75, and 100 watts, respectively, and we begin with all lamps sufficiently loose in their screw bases to be disconnected. Each lamp in turn should now be tightened in its socket, the ammeter current read and recorded, and then that lamp loosened again.

Suppose these recorded electron currents are those shown at the left in Table 27B.

Now tighten each lamp in turn and leave it, and as each new lamp is added, record the current. These current totals could well appear like those shown below in Table 27B.

TABLE 27B. Recorded Data

$I_1 = .24$ A	$I_1 = .24$ A
$I_2 = .44$ A	$I_1 + I_2 = .68$ A
$I_3 = .65$ A	$I_1 + I_2 + I_3 = 1.33$ A
$I_4 = .85$ A	$I_1 + I_2 + I_3 + I_4 = 2.18$ A

Such current sums represent an experimental confirmation of Kirchhoff's law, as given by Eq. (27g).

27.11 THE LAW OF PARALLEL RESISTANCES

Another type of diagram frequently drawn for parallel circuits is shown in Fig. 27S. Three

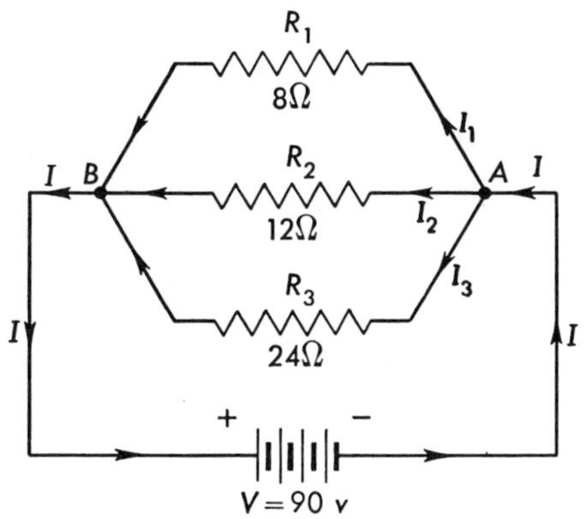

Fig. 27S. Three resistors in parallel, illustrating Kirchhoff's law.

resistors R_1, R_2, and R_3 are connected in parallel to a 90-volt battery. To find the total electron current through such a circuit we proceed to find a single resistance R which when substituted for the parallel combination of R_1, R_2, and R_3 will result in the same electron current. This equivalent resistance is given by *the law of parallel resistances*, as

$$\frac{1}{R} = \frac{1}{R_1} + \frac{1}{R_2} + \frac{1}{R_3} + \text{etc.} \qquad (27h)$$

To illustrate its use, let $R_1 = 8\,\Omega$, $R_2 = 12\,\Omega$, and $R_3 = 24\,\Omega$, as shown in Fig. 27S. By direct substitution in Eq. (27h)

$$\frac{1}{R} = \frac{1}{8} + \frac{1}{12} + \frac{1}{24}$$

Since the common denominator is 24,

$$\frac{1}{R} = \frac{3}{24} + \frac{2}{24} + \frac{1}{24} = \frac{6}{24}$$

from which

$$R = \frac{24}{6} = 4\,\Omega$$

If we now imagine the parallel combination of three resistors replaced by a single resistor $R = 4\,\Omega$, the circuit will have the general form as Fig. 27N, and the current I will be given by Ohm's law as follows:

$$I = \frac{V}{R} = \frac{90\text{ V}}{4\,\Omega} = 22.5\text{ A}$$

This is the total current I supplied by the battery to the parallel circuit of Fig. 27S.

To find how this current divides at A into three parts I_1, I_2, and I_3, we note that the full 90 V are directly applied to the opposite ends of each resistor. Therefore Ohm's law can be applied to each resistor separately as follows:

$$I_1 = \frac{90\text{ V}}{8\,\Omega} = 11.25\text{ A}$$

$$I_2 = \frac{90\text{ V}}{12\,\Omega} = 7.50\text{ A}$$

$$I_3 = \frac{90\text{ V}}{24\,\Omega} = 3.75\text{ A}$$

If we now apply Kirchhoff's law, Eq. (27g) we find

$$I = 11.25 + 7.50 + 3.75 = 22.5\text{ A}$$

and this is a check upon the previous total current. Note that the largest of the three currents, $I_1 = 11.25$ A, flows through the smallest resistance, and the smallest current I_3 flows through the highest resistance.

Kirchhoff's Law of Currents is frequently stated as follows:

The sum of all the currents flowing into any junction point is equal to the sum of all the currents flowing out.

An inspection of junction A or junction B in Fig. 27S will show how this definition gives us Eq. (27g).

QUESTIONS

1. How is current defined? What is the ampere? How many electrons pass by any point in a wire when a current of 1 amp flows?

2. What is a voltaic cell? What is a battery? What are the three principal constituents of a battery cell?

3. What are ions? Where are they produced?

4. What is electromotive force? In what units is it measured?

5. What is the source of electrical energy in a cell? What is the emf of a dry cell? Of a storage cell?

6. Of what is the storage cell electrolyte composed? Of what elements are the electrodes composed?

7. What is meant by dissociation? What is a terminal? What is a cathode? What is an anode?

8. How are cells usually connected to form a battery? What determines the emf of a battery?

9. What is the equivalent resistance of two or more resistors connected in series?

10. What is the abbreviated notation for battery cells in series? How is a resistor shown in a circuit diagram?

11. Does the same current flow through each resistor in series connections?

12. If the resistance of a circuit, and the applied voltage were known, how could you determine the current?

13. What is Ohm's law? In what units is each quantity measured?

14. What common electrical instruments are used to determine the factors in Ohm's law?

15. State the law of series resistances. Give a simple numerical example. Make a circuit diagram of your example and properly label each part.

16. State the law of parallel resistances. Give a simple numerical example. Make a circuit diagram of your example and properly label each part.

17. How does the current divide in a parallel circuit? Which resistor carries the largest current? Which carries the smallest current?

18. If several resistors are in parallel and connected to a battery, how is the current through each resistor calculated? What is the same for all resistors?

PROBLEMS

1. Find the voltage across the following batteries composed of the following cells connected in series: (a) 15 wet cells, (b) 24 wet cells, (c) 8 dry cells,

(d) 14 dry cells, and (e) 6 wet cells. [Ans. (a) 33 V, (b) 52.8 V, (c) 12 V, (d) 21 V, (e) 13.2 V.]

2. A car battery of 12 V is connected to a resistance of 4.5 Ω. Calculate the current.

3. The bulb of a 5-cell flashlight has a resistance of 20 Ω. Find the current supplied by the battery when the flashlight is turned on. [Ans. 0.3750 A.]

4. A 3-cell flashlight has a resistance of 14 Ω. Find the current flowing when it is turned on.

5. A electric toaster connected to a 115 V circuit draws a current of 3.2 A. What is the resistance? [Ans. 35.94 Ω.]

6. An electric iron connected to a 120 V source of power draws a current of 8 A. Calculate the resistance.

7. A battery of 6 dry cells is used in series in a spotlight having a bulb with a resistance of 18 Ω. What electron current flows when the light is turned on? [Ans. 0.50 A.]

8. A 12-volt storage battery supplies a current of 60 A when the starter in a car is turned on. What is the resistance of the starter motor?

9. Three resistors of 6 Ω, 8 Ω, and 24 Ω, respectively, are connected in series to a battery composed of 25 dry cells. Calculate the current through the circuit. [Ans. 0.987 A.]

10. Six resistors of 4 Ω, 8 Ω, 12 Ω, 16 Ω, 20 Ω, and 24 Ω, respectively, are connected in series to a lighting circuit of 120 V. Find the current through the circuit.

11. Three resistors of 5 Ω, 8 Ω, and 12 Ω, respectively, are to be taken at a time and connected in series to a 110 V battery. Find the current for each of the three cases. [Ans. 5.50 A, 6.47 A, and 8.46 A.]

12. A string of 8 Christmas tree lights, each with a resistance of 15 Ω, is connected in series to a house lighting circuit of 120 V. Find the current drawn when the lights are turned on.

13. Three resistors of 3 Ω, 6 Ω, and 11 Ω, respectively, are connected in series with a battery of 12 dry cells. What is the current? [Ans. 0.90 A.]

14. Two resistors of 12 Ω and 24 Ω, respectively, are connected in parallel to a battery supplying an emf of 24 V. Find (a) the equivalent resistance of the parallel circuit, and (b) the current through each resistor.

15. Two resistors of 8 Ω and 24 Ω, respectively, are connected in parallel to a battery 120 V. Find (a) the equivalent resistance of the parallel circuit, (b) the total current, and (c) the current through each resistor. [Ans. (a) 6 Ω, (b) 20 A, (c) 5 A, 15 A.]

16. Three appliances of 5 Ω, 10 Ω, and 40 Ω, respectively, are connected in parallel to a 110 V line. Find (a) the equivalent resistance of the parallel circuit, (b) the total current supplied by the line, and (c) the current through each appliance.

17. Three hot plates 4 Ω, 5 Ω, and 20 Ω, respectively, are connected in parallel to a 96 V battery. Find (a) the circuit resistance, (b) the total current, and (c) the current through each hot plate. [Ans. (a) 2.0 Ω, (b) 48.0 A, (c) 4.80 A, 19.20 A, 24.0 A.]

18. Four appliances of 4 Ω, 5 Ω, 10 Ω, and 20 Ω, respectively, are connected in parallel to a house lighting circuit of 110 V. Find (a) the equivalent resistance of the parallel circuit, (b) the total current supplied by the house circuit, and (c) the current through each appliance.

19. Two resistors of 8 Ω and 24 Ω, respectively, are connected in parallel. This parallel circuit is connected in series with a 6 Ω resistor, and the entire circuit is connected with a 120 V supply line. Find (a) the resistance of the parallel circuit, (b) the resistance of the entire circuit, and (c) the current through each resistor. [Ans. (a) 6 Ω, (b) 12 Ω, (c) 2.5 A, 7.5 A, 10 A.]

20. Two resistors of 3 Ω and 6 Ω, respectively, are connected in parallel. Two other resistors of 8 Ω and 12 Ω, respectively, are connected in parallel. These two circuits are connected in series with each other and with a fifth resistor of 7.2 Ω, and a voltage supply of 112 V. Find (a) the resistance of the entire circuit, and (b) the current through the 7.2 Ω resistor.

28 / MAGNETISM

From the time we were very young we played with small permanent magnets, and discovered their magic powers of attraction for iron. From a historical standpoint the attractive powers of a natural magnetic material called *lodestone* were known to the early Greek philosophers. Lodestone, sometimes called *magnetite*, is a solid natural ore composed of nearly one-hundred per cent pure iron, found in many parts of the world. A piece of lodestone attracts iron in much the same way as a permanent magnet.

That lodestone can be used as a compass is a very old idea. A Chinese author writing as early as the beginning of the twelfth century explains that a needle, when rubbed with lodestone and suspended free to turn, will point toward the south. This appears to be the first evidence that a piece of iron could be magnetized and used as a compass. The action of a lodestone or a bar magnet when suspended free to turn about a vertical axis is illustrated in Fig. 28A, diagrams (a) and (b).

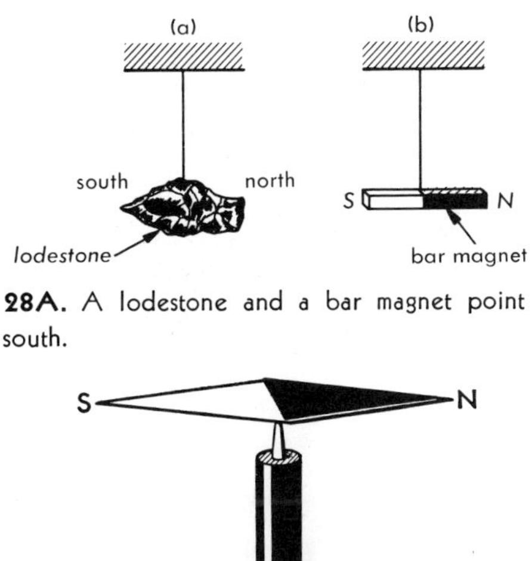

Fig. 28A. A lodestone and a bar magnet point north and south.

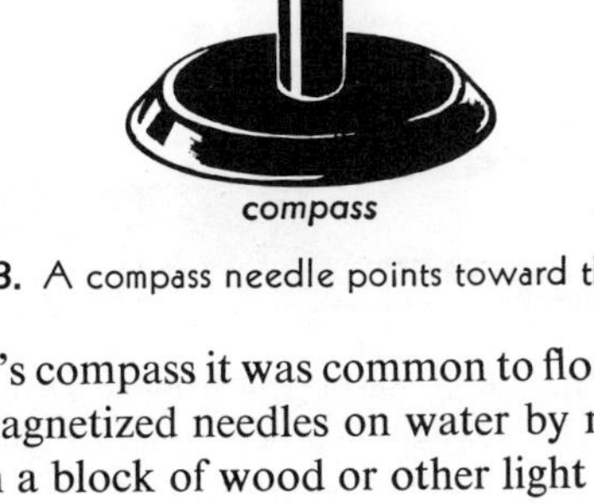

Fig. 28B. A compass needle points toward the north.

A compass as it is often made for demonstration purposes consists usually of a straight steel needle which has been magnetized and mounted free to turn on a sharp pointed rod as shown in Fig. 28B. In the early days of the mariner's compass it was common to float several small magnetized needles on water by mounting them on a block of wood or other light material.

In more recent designs a compass needle with a jewel in its center is set upon the sharp point of a hard metal rod, much shorter than that shown in Fig. 28B, and placed in a small brass box with a glass top. Such compasses are familiar to everyone.

28.1 MAGNETS Until recent years magnets have been made of hardened steel and molded or rolled into many shapes. Perhaps the most common of these is the horseshoe magnet shown in Fig. 28C, or the straight bar magnet shown in Fig. 28D. The strongest magnets are now made of an alloy containing aluminum, cobalt, nickel, iron and some of the rare earth elements. Small magnets of this alloy are strong enough to lift hundreds of times their own weight.

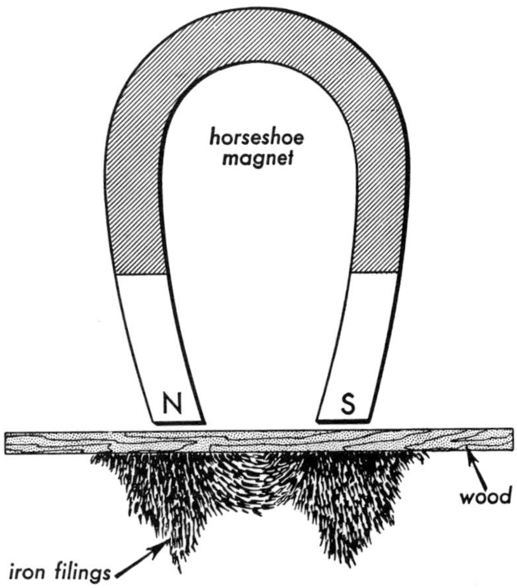

Fig. 28C. The attraction of a magnet for iron acts through all substances.

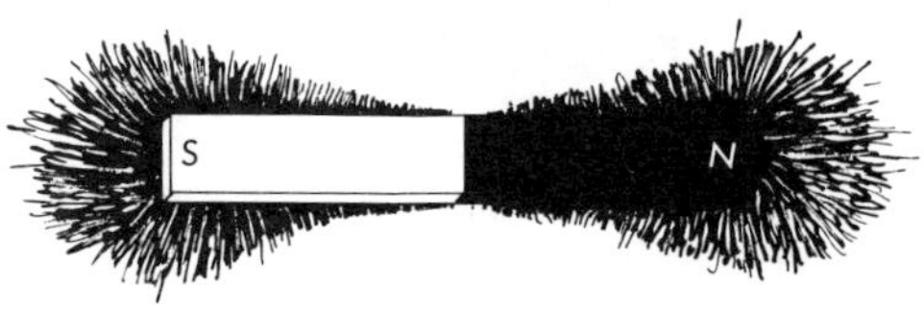

Fig. 28D. The attraction of iron filings by a straight bar magnet shows greater attraction near the ends. These regions of greatest attraction are called magnetic poles.

Pure iron, sometimes called soft iron, when magnetized will not retain its magnetism and is therefore useless in making what are called permanent magnets. Soft iron is used, however, in the construction of electromagnets. These devices will be discussed and demonstrated in Chap. 29. Of the many practical applications of permanent magnets, the compass, the telephone receiver, and the radio loud-speaker are perhaps the most common.

28.2 THE POWER OF ATTRACTION Nearly everyone has at sometime or other played with a small horseshoe magnet and discovered for himself that it attracted only things containing iron. On drawing the same magnet through the dry sand or dirt you probably discovered that it picks up small grains of iron ore.

If more extensive experiments are carried out, a magnet can be shown to attract magnetic substances at a distance even though matter lies in the intervening space. In other words, magnetic attraction acts right through matter of all kinds. This can be demonstrated as shown in Fig. 28C by picking up iron filings on one side of a thin wooden board when holding a magnet close to the other side. If a sheet of copper, or brass, or, as a matter of fact, any substance, is placed over the magnet, the power of attraction is not destroyed. A small region of space can be partially shielded from magnetic fields if it is entirely surrounded by layers of soft iron.

While a few metals are known to be feebly attracted by a magnet, most substances like aluminum, copper, silver, gold, wood, glass, paper, etc., do not exhibit any noticeable effect. Of those weakly affected, nickel and cobalt are the most important. These two metals, as mentioned above, alloyed with other metals in the proper proportions, are found to exhibit stronger magnetic susceptibility than the best grades of iron or steel. As a pure element, however, iron is by far the most strongly magnetic.

28.3 MAGNETIC POLES When an ordinary straight bar magnet is dipped into a box of iron filings the tiny bits of iron are observed to cling to the ends as shown in Fig. 28D. These preferred regions of attraction are called *magnetic poles*. If this same magnet is suspended by a thread as shown in Fig. 28A (b), it will come to rest in a position close to the north-south direction. The end toward the north is therefore called the N or *north-seeking pole*, and the other end the S or *south-seeking pole*.

That the N and S poles of a magnet are different may be shown by bringing the magnet close to a compass needle. Such an experiment is illustrated in Fig. 28E. When the S pole of the

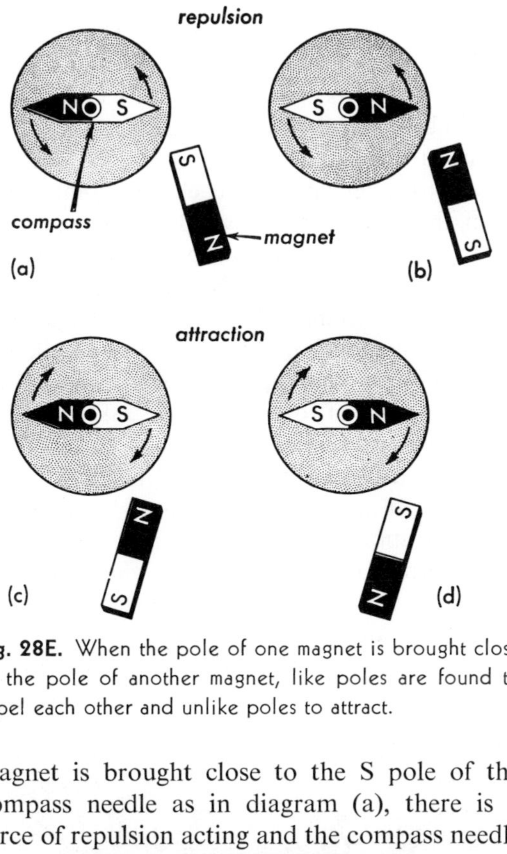

Fig. 28E. When the pole of one magnet is brought close to the pole of another magnet, like poles are found to repel each other and unlike poles to attract.

magnet is brought close to the S pole of the compass needle as in diagram (a), there is a force of repulsion acting and the compass needle turns away as shown. A similar repulsion occurs between the two N poles as shown in diagram (b). If the N and S poles are brought near to each other, however, a very strong attraction arises and the compass needle turns toward the other as shown in diagrams (c) and (d). These experiments show, therefore, that *two kinds of magnetic poles exist* and that *like poles repel and unlike poles attract*.

Permanent magnets can now be made so strong that one magnet can be lifted by the repulsion of another. This is illustrated in Fig. 28F. Unless guide rods of glass or some other

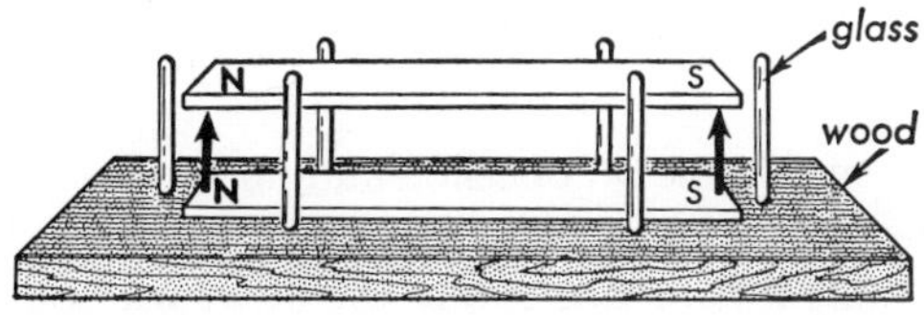

Fig. 28F. One magnet may be suspended in mid-air by the strong repulsion of like poles from another magnet.

substance are used, however, the floating bar will move to one side and then fall. In other words, the forces of repulsion are such that the upper bar is not in stable equilibrium. If the floating magnet is turned end for end, opposite poles become adjacent and the two magnets attract each other. Although the forces may be great enough, no one has ever succeeded in floating a permanent magnet in mid-air, against the pull of gravity and without guide rods, by means of permanent magnets.

It should be pointed out that each magnetic pole in a magnetized body is not confined to a single point but extends over a finite region. From a distance, however, each polar region acts as though it were concentrated at a point, similar to that of the center of mass in mechanics.

28.4 POLES EXIST IN PAIRS If a magnet is broken in the middle in an attempt to separate the poles, one finds new poles formed at the broken ends. If one of these pieces is again broken, each piece is again found to contain two poles of opposite kind. As long as this process is repeated the same result is obtained—a magnetic pole of one kind is always accompanied by a pole of opposite polarity.

This is conveniently illustrated by magnetizing a hack-saw blade and breaking it successively into smaller and smaller pieces as shown in Fig. 28G. Each time a piece is broken each fragment, upon being tested with a compass, is found to

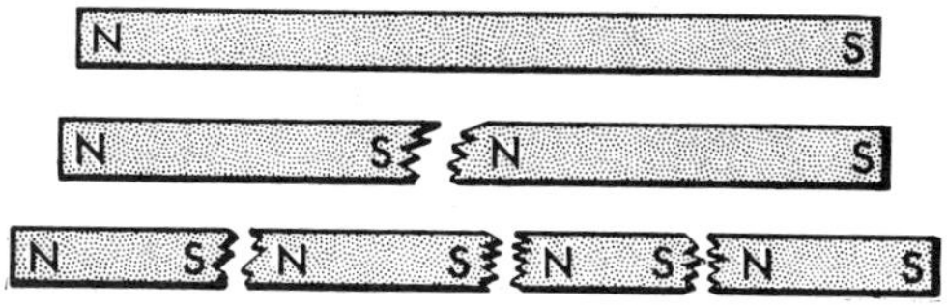

Fig. 28G. The poles of a magnet cannot exist alone. When a bar magnet is broken, poles appear on either side of the break, such that each piece has two opposite poles.

have an N pole on one side and an S pole on the other A hack-saw blade is readily magnetized by stroking it from one end to the other with one of the poles of a magnet.

It is possible to magnetize a bar of steel so that it has three or more polar regions. This is illustrated in Fig. 28H where a hack-saw blade

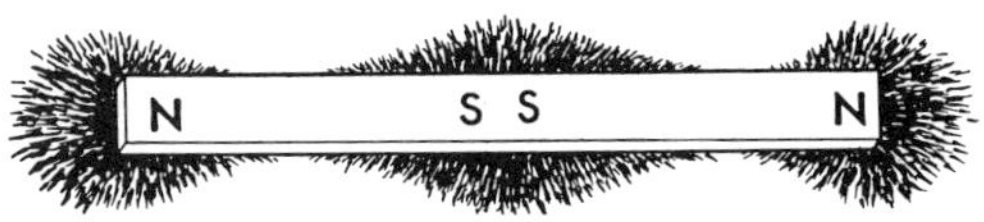

Fig. 28H. Diagram of a bar magnet with iron filings showing three polar regions.

has been magnetized with an N pole at each end and an S polar region in the center. The combined strength of the N poles is seen by the quantity of iron filings to be equal to the S pole strength in the center. We might say, therefore, that the magnet has four poles: an N pole at either end and two S poles at the center.

28.5 THE MAGNETIC FIELD In the space surrounding every magnet there exists what is called a magnetic field. Although this field cannot be seen, it can be demonstrated and mapped out in the following way.

If a very small compass is placed at some point near the N pole of a straight bar magnet and then moved always in the direction the compass is pointing, the center of the compass will trace out a smooth line called a *magnetic line of force*. Starting at various points many such lines may be drawn as shown in Fig. 28I. Each line starts at some point near the N pole and ends at a corresponding point near the S pole.

These *magnetic lines of force* do not really

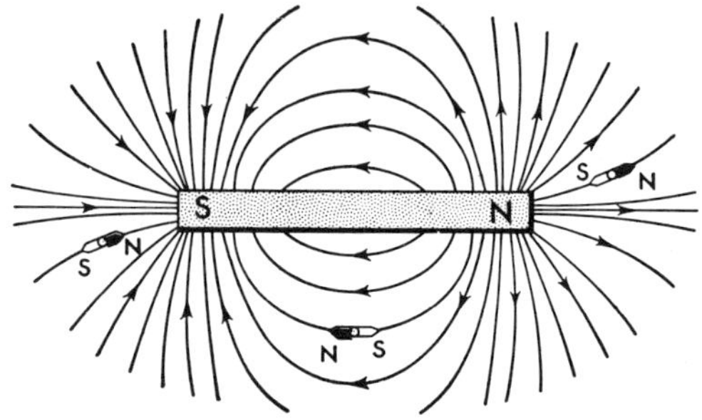

Fig. 28I. Diagram of the magnetic field and magnetic lines of force about a straight bar magnet, as obtained with a small compass needle.

exist; they are but useful devices that may be used in describing the many different magnetic phenomena to be taken up in later chapters. It should be noted that, where the magnet exerts its strongest attraction near the poles, the lines are closest together and that each line points away from the N pole and toward the S pole. This latter is an arbitrary assignment, being the direction indicated by the N pole of the compass.

A close examination of the iron filings clinging to a magnet shows that each tiny needle-like piece of iron lines up in the direction of the magnetic lines of force. See Figs. 28D and 28H. The reason for this is that each filing has become magnetized by the magnet, and having its own N and S poles, acts like a compass. An excellent demonstration of the field and its direction can be performed by laying a plate of glass or a sheet of paper over a magnet and then sprinkling iron filings over the top. By gently tapping the glass or paper the filings turn and line up as shown in the photographic reproduction given in Fig. 28J.

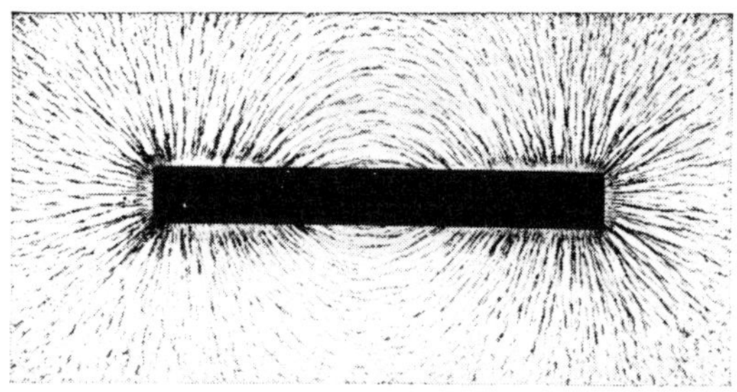

Fig. 28J. Photograph of the iron filings lined up by the magnetic field of a permanent straight bar magnet.

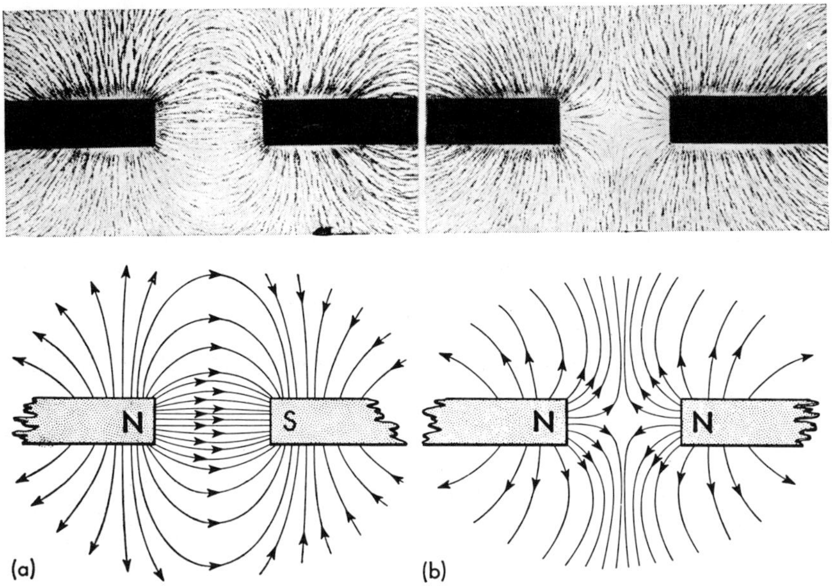

Fig. 28K. Diagrams and photographs illustrating the magnetic fields around pairs of magnetic poles.

28.6 THE FIELD ABOUT SEPARATE POLES When two magnetic poles are brought close together, the mutual action of the two is such as to produce a complicated magnetic field. This is illustrated by compass-made drawings at the top in Fig. 28K and by photographic reproductions of the iron-filing method of observation at the bottom.

A simplified explanation of many electric and magnetic phenomena can be given by assuming that these imaginary *lines of force* are endowed with certain properties. Lengthwise along the lines they act as though they were stretched rubber bands under constant tension, whereas sideways they act as if they repelled each other. Both of these properties are illustrated in Fig. 28K. When the two poles are of different polarity as in diagram (a), the lines of force acting like stretched rubber bands tend to pull the poles together. In diagram (b) where the poles are alike, the lines repel each other, pushing the poles apart.

28.7 THE EARTH'S MAGNETIC FIELD To Sir William Gilbert we owe the view that the earth is a great magnet. To prove his theory Gilbert shaped a lodestone into a sphere and demonstrated that a small compass placed at any spot of the globe always pointed, as it does on the earth, toward the North Pole.

The earth, therefore, has been schematically pictured in Fig. 28L as a large magnetized sphere of iron or as though it contained a huge per-

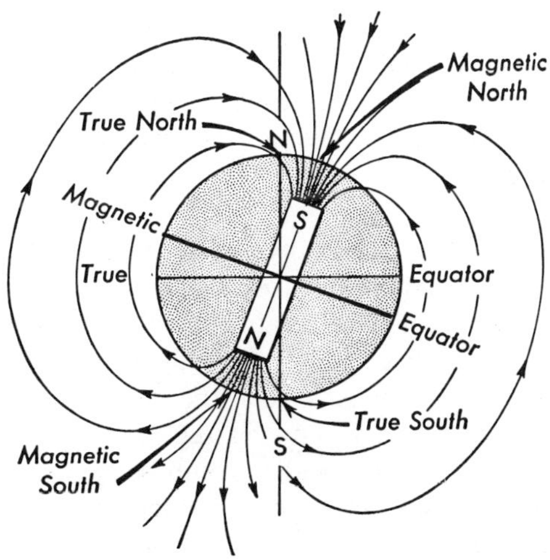

Fig. 28L, Schematic diagram illustrating the earth as a huge magnet surrounded by a magnetic field extending far out into space.

manent straight bar magnet. Since the magnetic axis is at an angle with the polar axis, the earth's magnetic poles are not at the *true North* and *true South Poles.* The true North and true South Poles are points located on the earth's rotational axis.

The *North Magnetic Pole* is located in far northern Canada, while the *South Magnetic Pole* is located almost diametrically opposite in the Southern Hemisphere. As for polarity, the North Magnetic Pole is an S pole and the South Magnetic Pole is an N pole. This becomes apparent from the magnetic lines of force which always start from an N pole and are directed toward, and end, at an S pole.

Although the reason for the earth's magnetism is not completely understood, several reasonable theories have been proposed. The earth is known to contain large iron ore deposits, some of these deposits being almost pure iron. One theory proposes that during the ages past, all of these iron deposits gradually became magnetized, in very nearly the same direction, and that together they act like one huge permanent magnet. Another theory, and a very plausible one, is that the magnetism is due to large electric currents which are known to be flowing around the earth, not only in the earth's crust but also in the air above. These earth currents seem to be connected in some direct way with the earth's rotation. Recent researches show that the earth's magnetic polarity has changed periodically and even reversed its direction completely since its creation.

28.8 MAGNETIC DECLINATION Since the earth's magnetic and polar axes do not coincide, a compass needle does not in general point toward true north. Because of the influence of the irregular iron deposits near the earth's surface, the magnetic field is not as regular as it is pictured in Fig. 28L and a compass needle may deviate considerably from magnetic north. *The angle that a compass needle deviates from true north is called the angle of declination.*

A map showing the angle of declination for the United States during the year 1954 is shown in Fig. 28M. The more or less vertical, irregular lines are lines of equal declination and are called *isogonic lines.* At every point along the line marked 20°E, for example, a compass needle actually points 20° east of true north. In the region of San Francisco the declination is seen to be about 18°E, while in the region of New York it is about 11°W. *The line through points where a compass points true north, 0°, is named the agonic line.*

28.9 MAGNETIC DIP If a compass needle is mounted free to turn about a horizontal axis as shown in Fig. 28N, it will not come to rest in a

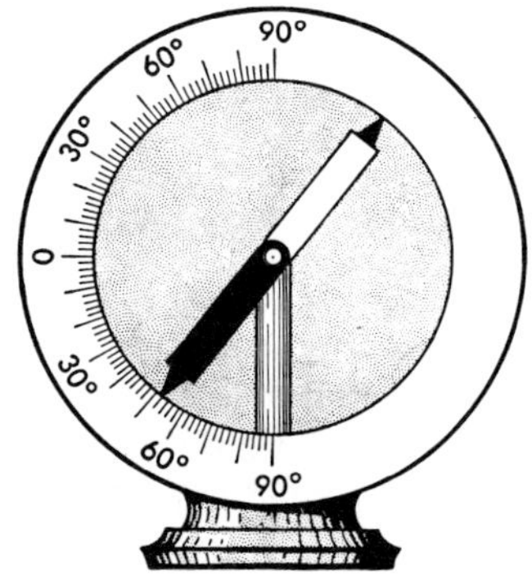

Fig. 28N. A magnetic dip needle

horizontal position but will dip down at some angle with the horizontal as shown. This direction, called the *dip*, is the angle the earth's field makes with the earth's surface at the point in question. Referring to Fig. 28L, it is seen that in the far north and south the angle of dip is quite large, whereas near the equator it is small. Fig. 28O indicates the approximate dip at different latitudes for one cross section of the entire globe. At a region on the Boothia Peninsula just north of Hudson Bay and 20° from true north, a dip needle points straight down, perpendicular to the earth's surface, and locates the North Magnetic Pole. At a region about 18° from the true South Pole a dip needle points straight up, at 90° from the horizontal, and locates the South Magnetic Pole.

On maps of terrestrial magnetism all points that have equal dip angles are connected by a line

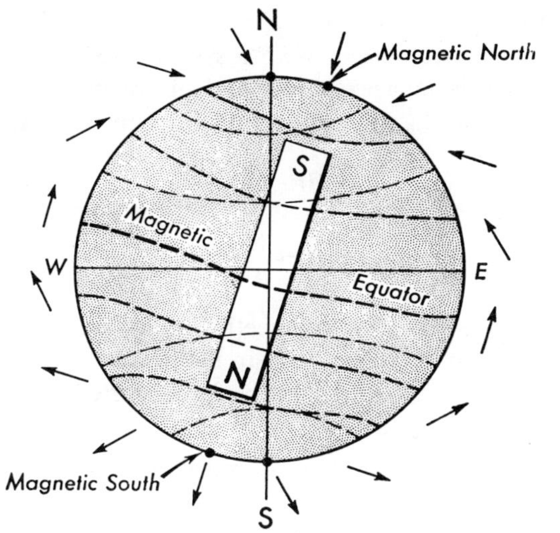

Fig. 28O. Arrows show the direction a dip needle takes at different parts of the earth's surface.

called an isoclinic line. Such lines for different angles form a set of nearly parallel lines as shown on the map of the United States in Fig. 28M.*

Careful and accurate measurements of the *declination* and *dip* show that over a period of hours, days, and months the earth's magnetic field is continually changing. Although these changes are extremely small they are somewhat periodic and at times quite erratic.

In Sec. 28.5 the magnetic field over a region of space is graphically represented by what are called *magnetic lines of force*. In order to specify the strength or intensity of a magnetic field at any point in space a vector quantity called the *magnetic induction B* is used. The direction of the *magnetic induction B* at any point is tangent to the line of force passing through that point and its magnitude is proportional to the force that would act on an N pole if placed there. These lines of force are also called *lines of induction*.

*B specifies the strength of a magnetic field. In the *MKSA* system of units B is expressed in webers/meter^2, whereas in the *cgs* system it is given in gauss. In San Francisco the earth's magnetic induction B is about 5.4×10^{-5} webers/meter^2, or 0.54 gauss and the angle of dip about 62°. This gives a horizontal component of 0.25×10^{-4} webers/meter^2 or 0.25 gauss.

28.10 MAGNETIZATION When a strong magnet is brought close to a piece of soft iron, the iron takes on all the properties of a new but somewhat weaker magnet. This phenomenon, called *magnetization*, is illustrated in Fig. 28P (a). As

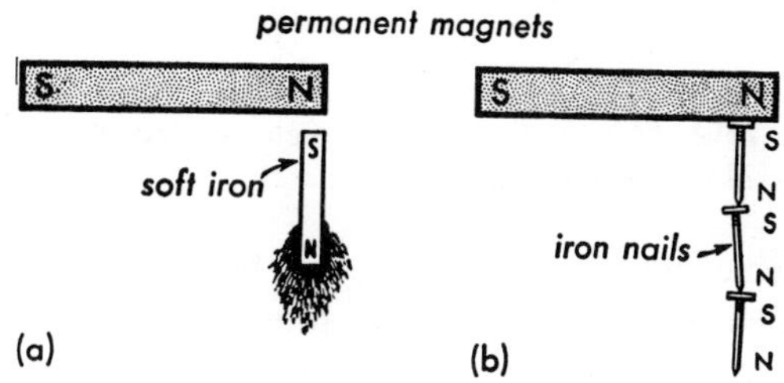

Fig. 28P. Soft iron may be magnetized by induction at a distance or by contact.

long as the permanent magnet is held close to the soft iron bar the iron filings cling to the end as shown. When the permanent magnet is removed, however, the soft iron immediately loses its magnetism and the iron filings drop off.

When a piece of iron is magnetized, a pole of opposite sign is created at the points of closest approach as shown in the diagrams. If a common iron nail is brought up to the N pole of a permanent magnet, it will become magnetized with an S pole at the point of contact and an N pole at the other end as shown in diagram (b). Having two poles, the nail is thus magnetized and will attract another nail and magnetize it in the same way. With a good strong magnet this process can be repeated by adding one nail after the other.

It is now clear how iron filings line up with the lines of force of a magnet. Each filing becomes magnetized and, like a small compass, turns parallel to the field in which it is located.

28.11 COULOMB'S LAW FOR MAGNETIC POLES For a study of the law of force between magnetic poles, specially designed magnets have been used. The necessity for this is realized when it is remembered that single magnetic poles cannot be isolated by breaking a magnet in two. The special magnets consist of thin steel rods about 18 in. long, with a small steel ball on either end. When

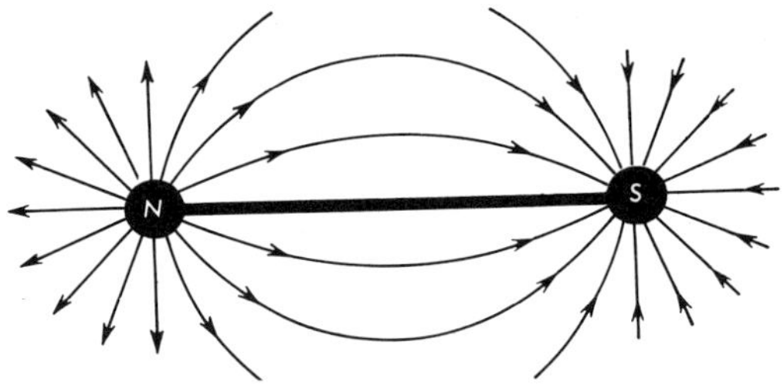

Fig. 28Q. Map of the magnetic field around a special type of permanent magnet.

magnetized the N and S poles become concentrated in the steel balls as shown in Fig. 28Q.

As used in the Hibbert balance shown in Fig. 28R, one magnet is balanced on a special

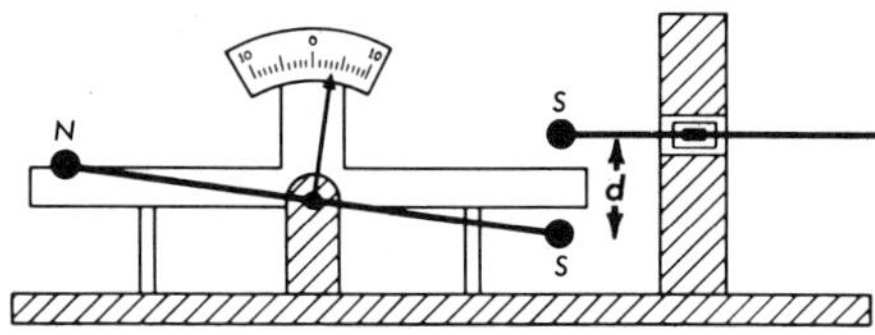

Fig. 28R. Diagram of the Hibbert magnetic balance for measuring the force between two magnetic poles.

set of scales and another is held tightly in an adjustable clamp stand. If the two adjacent poles, when they are brought together, are alike, the repulsion will throw the one magnet off balance as shown. The weight that must then be added to the left-hand side to restore balance again is a direct measure of the force of repulsion. Carrying out experiments of a similar nature, Coulomb was the first to find that *the force acting between two magnetic poles is inversely proportional to the square of the distance between them.* Having discovered this relation, he compared the pole strengths of different magnets with each other and found that *the force between two poles is proportional to the product of the pole strengths.* Combining these two relations, Coulomb proposed as a general law for magnetic poles,

$$F = \kappa \frac{Mm}{d^2} \qquad (28a)$$

for magnetic poles

where F is the force M and m are the strengths of the poles, and d is the distance between them. In the rationalized *MKSA* system used in this text, F is in *newtons*, M and m are in *ampere meters*, and κ, the proportionality constant, is

$$\kappa = 10^{-7} \frac{\text{newton}}{\text{amperes}^2}$$

$$\kappa = 10^{-7} \frac{\text{N}}{\text{A}^2} \qquad (28b)$$

EXAMPLE 1. Two magnetic S poles are located 12 cm apart. If the pole strengths are 6 Am and 8 Am, respectively, find the force exerted by one pole on the other.

SOLUTION. By direct substitution in Eq. (28a), we obtain

$$F = 10^{-7} \frac{\text{N}}{\text{A}^2} \frac{6\ \text{Am} \times 8\ \text{Am}}{(0.12\ \text{m})^2}$$

$$F = 3.333 \times 10^{-4}\ \text{N}$$

When two magnets are arranged in the configuration shown in Fig. 28S, the two poles near-

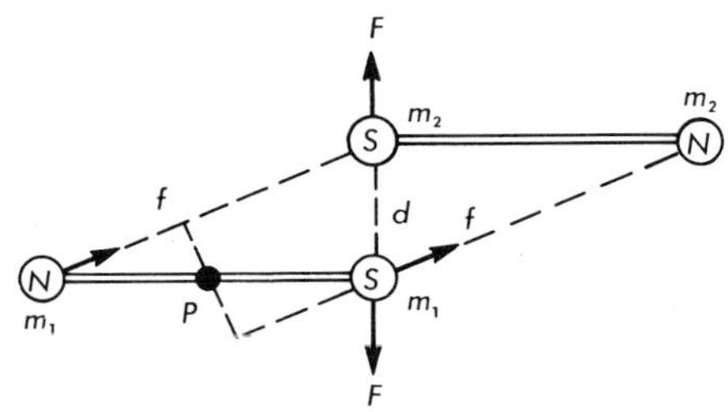

Fig. 28S. Geometry of the forces and torques involved in the Hibbert balance.

est each other will exert by far the strongest forces on each other. These forces, represented by F in the diagram, are an *action* and *reaction pair.*

To measure the force F exerted on the lower S pole the lower magnet is initially balanced carefully on a pivot P at its center of mass. Under these conditions the two attractive forces f, due to the poles of the upper magnet, will cancel out since they exert equal and opposite torques.

If the magnets are long and the distance d relatively small, the force of repulsion between the extreme end poles will be exceedingly small. To a first approximation, this force and the torque it exerts can be neglected.

28.12 MAGNETIC INDUCTION B It is quite common practice to refer to the strength of a magnetic field as the *magnetic induction.*

The magnetic induction B at any point in space may be defined as the force per unit N pole acting on any pole placed at that point.

Algebraically,

$$\boxed{B = \frac{F}{m}} \tag{28c}$$

Suppose, for example, that when a unit pole of *1 ampere meter* is placed at a given point in space it experiences a force of *5 newtons.* The magnetic induction B at that point is then said to have a magnitude of *5 newtons per ampere meter* (5N/Am). If a pole with a strength of m_1 ampere meters is placed at this same point where the magnetic induction is known to be B newtons/ampere meter, the force acting on the pole in newtons is

$$F = m_1 B \tag{28d}$$

From the above definition that $B = F/m$, and Coulomb's law, a relation for the magnetic induction at any point near a single pole of strength M may be obtained. When Eq. (28a) is solved for F/m, we obtain

$$\frac{F}{m} = \kappa \frac{M}{d^2}$$

or

$$B = \kappa \frac{M}{d^2} \tag{28e}$$

where d is the distance from the field producing pole m to the point in question.

Since magnetic induction B specifies the *strength* or *intensity* of the magnetic field at any point in space, it is informative to express the constant k of Coulomb's law in fundamental *MKSA* units. Starting with Eq. (28b), we change newtons to Kg m/s², to obtain

$$k = 10^{-7}\frac{\text{N}}{\text{A}^2} = 10^{-7}\frac{\text{Kg m}}{\text{A}^2\text{s}^2} \tag{28f}$$

The *derived unit* of magnetic induction, N/Am, has recently been named the *tesla* in honor of Nikola Tesla.* Following this adopted we write

$$1\,\frac{\text{newton}}{\text{ampere meter}} = 1 \text{ tesla}$$

abbreviated

$$1\,\frac{\text{N}}{\text{Am}} = 1\,\text{T} \tag{28g}$$

Since magnetic induction B has magnitude and direction it is a vector quantity. The direction of B at any point in space is the direction of the force acting on an N pole placed there.

EXAMPLE 2. If an N pole of a permanent magnet has a strength of 5 Am, and is brought up close to a strong magnet where it is repelled by a force of 0.28 N, what is the magnitude of the strength of the field at that point?

SOLUTION. Apply Eq. (28c), and substitute the known quantities as follows,

$$B = \frac{F}{m} = \frac{0.28\text{ N}}{5\text{ Am}} = 0.056\,\frac{\text{N}}{\text{Am}}$$

$$B = 5.6 \times 10^{-2}\,\frac{\text{N}}{\text{Am}}$$

$$B = 5.6 \times 10^{-2}\,\text{T}$$

28.13 MOLECULAR THEORY OF FERROMAGNETISM The modern theory of magnetism which is now quite firmly established as being correct is that a piece of iron consists of myriads of tiny elementary magnets. These tiny ultramicroscopic

---

* NIKOLA TESLA (1857–1943), American electrician and inventor, was born at Smiljan, Yugoslavia. He studied at the Realschule in Karlstadt in 1873, the Polytechnic School in Graz in 1876, and the University of Prague in 1879. After coming to the United States in 1884, he first worked for the Edison Co. at Orange, N. J., and then left to found his own research laboratory. There he did a great deal of research on alternating current electricity. In 1888 he invented and patented the induction motor. He invented the principle of a rotating magnetic field, and did significant research on capacitors. He is best known for his development of a high voltage, high frequency impulse transformer known as a "Tesla coil."

magnets may consist of individual atoms and molecules themselves or of groups of atoms aligned to form small elementary iron crystals. How single atoms can act as magnets will be explained later. Before a piece of iron or steel has been magnetized, these elementary magnets may be thought of as being oriented more or less at random throughout the metal as shown in Fig. 28T.

During the time a piece of iron is being magnetized, the elementary magnets are turned around and lined up parallel to each other and to the magnetizing field. This is shown by the schematic representation in diagram (b). Lined up in this way, the small N and S poles are adjacent to each other and cancel each other's effect on external objects. At one end there are many free N poles and at the opposite end an equal number of free S poles.

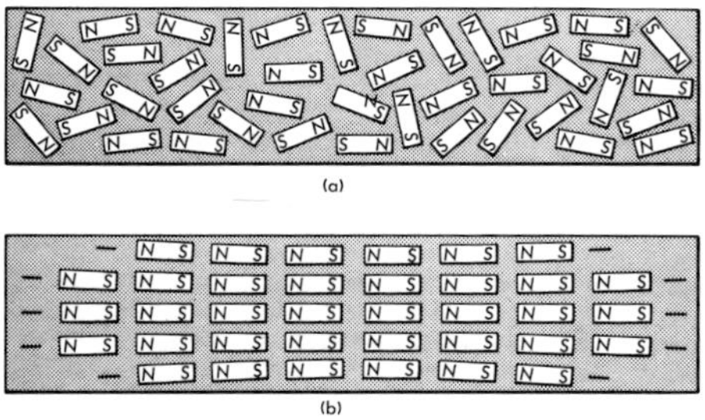

Fig. 28T. Schematic diagrams of the elementary magnets within a piece of iron, (a) unmagnetized and (b) magnetized.

When a magnet is broken at any point *free* S *poles* are exposed at one side of the break and *free* N *poles* at the other. It is therefore clear why poles always exist in pairs and that no matter how many times a magnet is broken each piece will contain an N pole at one end and an S pole at the other.

When soft iron is magnetized by induction and the permanent magnet taken away, the elementary magnets return to their original random orientations, but, when hardened steel becomes magnetized, they remain lined up after the magnetizing field is taken away.

QUESTIONS

1. What is a bar magnet? What is a horseshoe magnet?

2. What are magnetic poles? How many kinds of poles are there? What are their names? What is the origin of these names?

3. Can magnets be made strong enough to lift one another by repulsion?

4. Is a magnetic pole a point or a distributed area on the surface of a magnet? Can a magnet have one pole only? Can the two poles of a magnet be broken apart?

5. Can a magnet have more than two poles? Can a magnet have 1000 poles?

6. What is magnetic induction? How can it be demonstrated?

7. What is the molecular theory of magnetism? How does soft iron differ in behavior from steel?

8. Does a compass point true north at all times? Does it ever point true north?

9. What is magnetic declination? Why does it arise?

10. What is meant by magnetic dip? How is it determined?

11. What is the agonic line? What is an isogonic line? What is an isoclinic line?

12. Does the earth's magnetic field remain always the same?

13. Why are the isogonic lines so irregular from point to point?

14. What is a magnetic field? Can it be seen? Can it be detected? Can a magnetic field be mapped out?

15. Can you draw from memory the magnetic field around a straight bar magnet with two poles?

16. Can you draw from memory the magnetic field between two magnets with two like poles close together?

17. Can you draw from memory the magnetic field between two magnets with two unlike poles close together?

18. Can you draw from memory the earth's magnetic field? Can you label the four poles and two equators?

PROBLEMS

1. Two S poles of separate magnets are located 6 cm apart as shown in Fig. 28S. If the poles of each magnet have a strength of 10 Am and are 8 cm apart, find (a) the force exerted by one S pole upon the other, (b) the force of one N pole on the S pole of the other magnet, and (c) the force of one N pole on the other. [Ans. (a) 2.778×10^{-3} N, (b) 1.0×10^{-3} N, (c) 3.425×10^{-5} N.]

2. Two magnetic N poles are located 8 cm apart as shown in Fig. 28S. If the poles of each magnet have a strength of 12 Am, and are 10 cm apart, find (a) the force exerted by one S pole upon the other, (b) the force of one N pole on the S pole of the other magnet, and (c) the force of one N pole on the other.

3. Two magnets of the type shown in Fig. 28Q are arranged with their pole centers at the four corners of a square 20 cm on each side. If each pole has a strength of 5 Am, and like poles are diagonally opposite each other, find the force on each pole. [Ans. 5.712×10^{-5} N each.]

4. Two magnets of the type shown in Fig. 28Q have poles 12 cm apart and each with a strength of 16 Am. If the poles are located at the four corners of a tetrahedron, find the direction and magnitude of the force on each pole. Neglect the force between the two poles of the same magnet.

5. The two poles of a bar magnet have a strength of 15 Am each and are 5 cm apart. Find the magnitude of the magnetic induction at a point in line with the two poles, (a) 5 cm beyond the S pole, and (b) 10 cm beyond the N pole. [Ans. (a) -4.50×10^{-4} T, (b) 8.33×10^{-5} T.]

6. A straight bar magnet has two poles of 15 Am each, spaced 5 cm apart. Find the magnetic induction at a point in line with the poles, (a) 5 cm from the N pole, (b) 10 cm from the N pole, and (c) 20 cm from the N pole.

7. A magnet of the type shown in Fig. 28Q has poles with a strength of 10 Am, and located 10 cm apart. Find the magnetic induction at a point in line with the poles and at a distance beyond the N pole of (a) 5 cm, (b) 10 cm, (c) 15 cm, (d) 20 cm, and (e) 25 cm. Plot a graph of B vs x, where x is the distance from the N pole of the magnet. [Ans. (a) -3.556×10^{-3} T,

(b) -7.50×10^{-4} T, (c) -2.844×10^{-4} T, (d) -1.389×10^{-4} T, (e) -0.784×10^{-4} T.]

8. A magnet of the type shown in Fig. 28Q has poles of strength 5 Am and located 20 cm apart. Find the magnetic induction along a line through the center of the magnet perpendicular to the magnetic axis and at distances from the center of (a) 5 cm, (b) 10 cm, (c) 15 cm, (d) 20 cm, and (e) 25 cm. (f) Plot a graph of B vs x, where x is the distance from the center of the magnet.

9. A bar magnet with two poles, each with a strength of 12 Am, is located in a magnetic field of 1.50 T. Find the magnitude of the force acting on each pole. [Ans. 18.00 N.]

10. A small magnet of the kind shown in Fig. 28Q has two poles 10 cm apart and each of 16 Am. If this magnet is located in a uniform magnetic field of 1.5 T, with its axis at right angles with B, find (a) the force on each pole, and (b) the motion of the magnet if it is free to move.

29 EFFECTS OF ELECTRIC CURRENTS

When an electric current is sent through a wire, a number of different recognizable effects take place. While these effects are all closely related, we separate various aspects of the action to reveal the different effects in terms of familiar and accepted concepts.

Everyone is more or less familiar with the electrical appliances of the modern household—the electric lights, electric toaster, iron, refrigerator, vacuum cleaner, washing machine, etc. All of these devices depend for their operation upon one or more of five general effects produced by electric currents; these are

1. heating effect
2. magnetic effect
3. mechanical effect
4. lighting effect
5. chemical effect

It is the purpose of this chapter to consider the first three of these different phenomena and to take up in some detail the important principles involved.

29.1 THE HEATING EFFECT OF AN ELECTRIC CURRENT When an electron current is sent through a wire, heat is generated and the temperature of the wire rises. If the current is increased, the rate at which heat is generated increases rapidly until the wire itself glows a deep red. A still further increase in current will heat the wire to a yellow or white heat. Beyond this point, if it has not already done so, the wire will reach a temperature where it will melt.

Whether a wire is only warmed by an electron current or heated to incandescence depends upon a number of factors, the two principal ones of which are the current and the resistance. Experiment shows that the quantity of heat developed in a wire is directly proportional (1) *to the resistance R*, (2) *to the square of the electron current I*, and (3) *to the time t*.

Expressed as an algebraic equation,

$$\boxed{\text{Heat} = kI^2Rt} \qquad (29a)$$

where k is the proportionality factor. To be able to calculate exactly the amount of heat produced in kilocalories (*abbr*. Kcal), the value of k must be known. Experimentally this was first determined by James Joule, an English physicist, in 1843, and later by many other experimenters, and found to have the value 2.390×10^{-4} Kcal/joule. If the electron current is measured in *amperes*, the resistance in *ohms*, and the time in *seconds*, then

$$\text{heat in kilocalories} = 2.390 \times 10^{-4}\, I^2Rt \qquad (29b)$$

Eq. 29a is known as Joule's law.

Since heat is a form of energy, Joule's law may be considered as another aspect of the law of conservation of energy. When a given amount of electrical energy I^2Rt is expended or transformed into heat, it always produces the same number of kilocalories.

To calculate the heat generated in an electric circuit, consider the following simple problem.

EXAMPLE 1. If the heating element of an electric toaster has a resistance of 22 Ω and is connected to an ordinary house lighting circuit of 110 V, how much heat will be generated in one minute?

SOLUTION. From Ohm's law, Eq. (27d), the electron current is first calculated

$$I = 110/22 = 5 \text{ A}$$

Making use of Joule's law by substituting in Eq. (29b),

$$\text{heat} = 2.390 \times 10^{-4} \times (5)^2 \times 22 \times 60 = 7.89 \text{ Kcal}$$

Since by Eq. (27d) the resistance $R = V/I$, the substitution of V/I for R in Eq. (29b) provides a somewhat more useful equation for Joule's law—namely,

$$\text{heat in kilocalories} = 2.390 \times 10^{-4}\, IVt \qquad (29c)$$

where I is in amperes, V is in volts, and t is in seconds.

For some electric appliances, heating is a desired effect while in others it is a source of trouble and even danger. In an electric iron, hot plate, or toaster, for example, heat is the main objective of the device. In such appliances a relatively large electron current of several amperes is sent through a coil or element of special wire having a resistance of several ohms. As a rule the wire is of some alloy, such as nichrome, and of such a size that the heat developed will not raise the temperature higher than red hot. Diagrams of typical heating elements used in three different household appliances are shown in Fig. 29A.

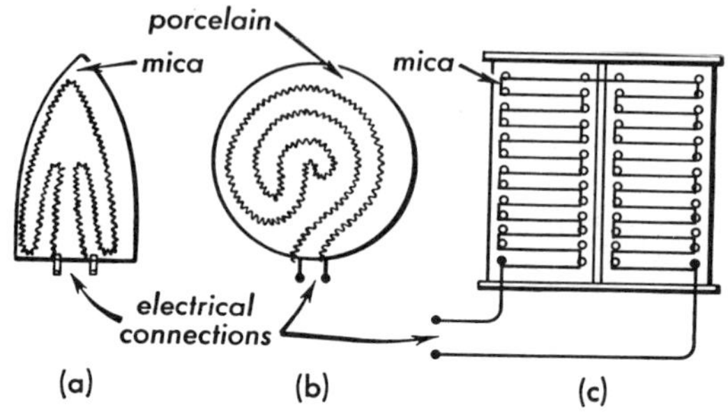

Fig. 29A. Diagrams of the heating elements of various electrical appliances found in many modern homes: (a) electric iron, (b) electric stove, and (c) electric toaster.

29.2 MAGNETIC EFFECT, OERSTED'S EXPERIMENT

The first discovery of any connection between electricity and magnetism was made by Oersted* in 1820. Often during his lectures, at the University of Copenhagen, Oersted had demonstrated the nonexistence of a connection between electricity and magnetism. His usual custom was to place a current-carrying wire at right angles to and directly over a compass needle and show that there was no effect of one on the other. On one occasion at the end of his lecture, when several of the audience came up to meet him at the lecture room desk, he placed the wire parallel to the compass needle and, not the least expecting it, saw the needle move to one

* HANS CHRISTIAN OERSTED (1777–1851), Danish scientist. Born the son of an apothecary, Oersted spent part of his boyhood teaching himself arithmetic. At the age of twelve he assisted his father in his shop and there became interested in chemistry. Passing the entrance examinations at the University of Copenhagen at the age of seventeen he entered the medical school to graduate six years later with his doctorate in medicine. At twenty-nine he came back to the university but this time as professor of physics. It was at one of his demonstration lectures on chemistry and metaphysics that he discovered the magnetic effect bearing his name. The discovery not only brought him many endowments and prizes but made him one of the most eminent personalities in his own country.

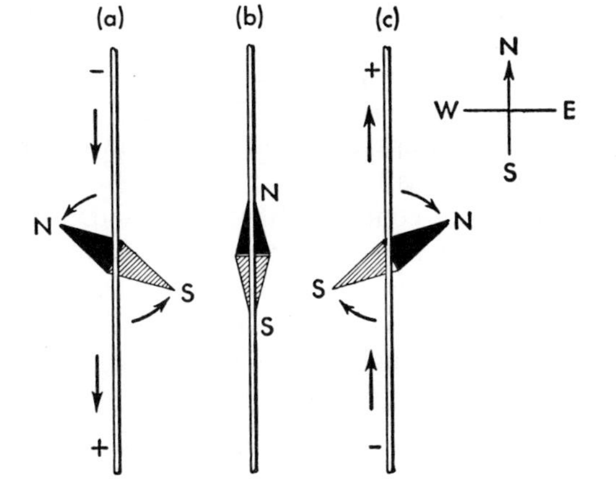

Fig. 29B. Diagram of Oersted's experiment illustrating the effect of an electric current upon a compass needle.

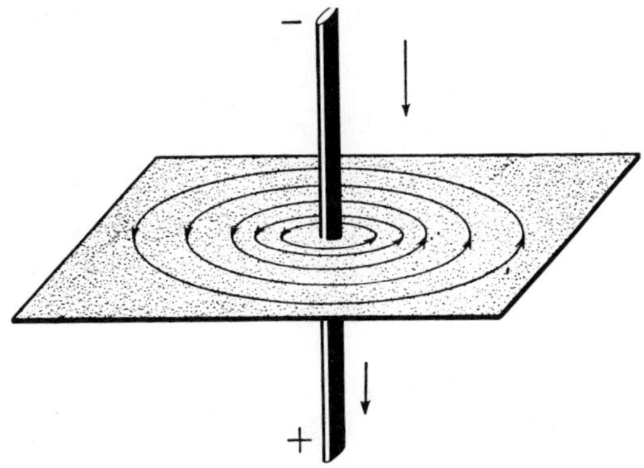

Fig. 29C. Experiment demonstrating the magnetic field about a straight wire carrying an electric current.

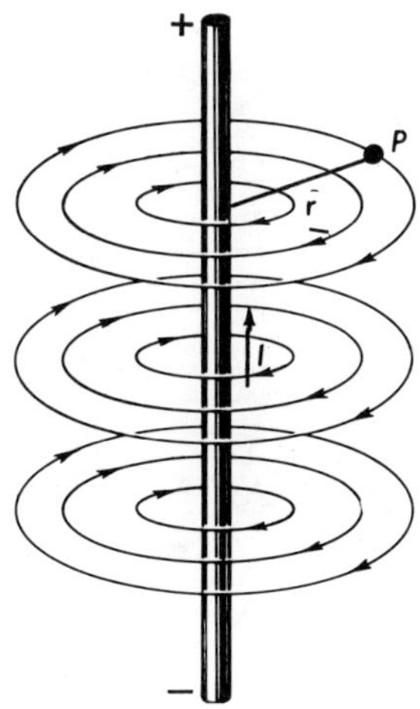

Fig. 29D. The magnetic field around a long, straight conductor.

side. See Fig. 29B. Upon reversing the current in the wire, the needle, to his amazement and perplexity, deviated in the opposite direction. Thus this great discovery was made quite by accident, but, as Lagrange once said of Newton on a similar occasion, "such accidents come only to those who deserve them."

29.3 THE LEFT-HAND RULE Oersted's experiment is interpreted as demonstrating that *around every wire carrying an electron current there is a magnetic field.* The direction of this field at every point, like that around a bar magnet, can be mapped by means of a small compass or by iron filings. If a wire is mounted vertically through a hole in a plate of glass or other suitable nonconductor and then iron filings sprinkled on the plate, there will be a lining up of the filings parallel to the magnetic induction. The result shows that the magnetic lines of force or *lines of induction* are concentric circles whose planes are at right angles to the electron current. This is illustrated by the circles in Figs. 29C and 29D.

The left-hand rule used in electromagnetism can always be relied upon to give the direction of the magnetic induction due to an electron current in a wire. Derived from experiment, the rule states that *if the current-carrying wire were to be grasped with the left hand, the thumb pointing in the direction of the electron current,* (−) *to* (+), *the fingers will point in the direction of the magnetic induction.*

The magnitude of the induction at any point P, close to a long, straight wire, is given by

$$B = \frac{\mu_0}{4\pi} \cdot \frac{2I}{r} \qquad (29d)$$

for straight wire

where r is the perpendicular distance to the point P in meters, I is the electron current in amperes, and μ_0 is a constant.

$$\mu_0 = 4\pi k \frac{\text{N}}{\text{A}^2}$$

where k is the Coulomb's law constant given by Eq. (28b). This gives

$$\mu_0 = 12.57 \times 10^7 \frac{\mathrm{N}}{\mathrm{A}^2}$$

29.4 MAGNETIC PROPERTIES OF A SOLENOID

Not long after the announcement of Oersted's discovery of the magnetic effect of a current carrying wire, Ampère found that a loop or coil of wire acted as a magnet. This is illustrated by a single loop of wire in Fig. 29E. In this case, the

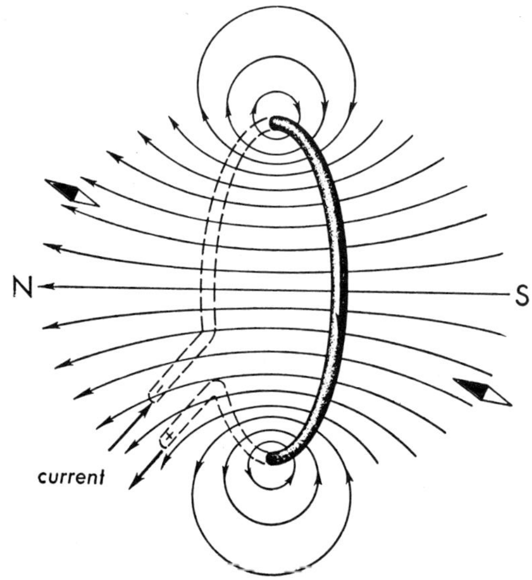

Fig. 29E. Diagram of the magnetic field through and around a single loop of wire carrying an electric current.

lines of induction are such that one side or end of the coil acts like an N magnetic pole and the other side or end like an S magnetic pole.

At all points in the region around a coil of wire carrying an electron current the direction of the magnetic induction, as shown by a compass, can be predicted by the left-hand rule. Inside each loop or turn of wire the lines point in one direction, whereas outside, and in the plane of the loop, they are oppositely directed.

The strength of the magnetic field in and around a single loop of wire is difficult to calculate and requires very complicated formulas. The magnetic induction at the very center, however, is given by a simple equation,

$$B = \mu_0 \frac{I}{2r} \tag{29e}$$

where I is in amperes, r is the radius in meters, and B is in teslas.

For a wire conductor that is wound in the form of a flat circular coil having a number of turns N, and a radius r, the magnetic field in and around the coil is greatly increased over that produced by a single turn (see Fig. 29F).

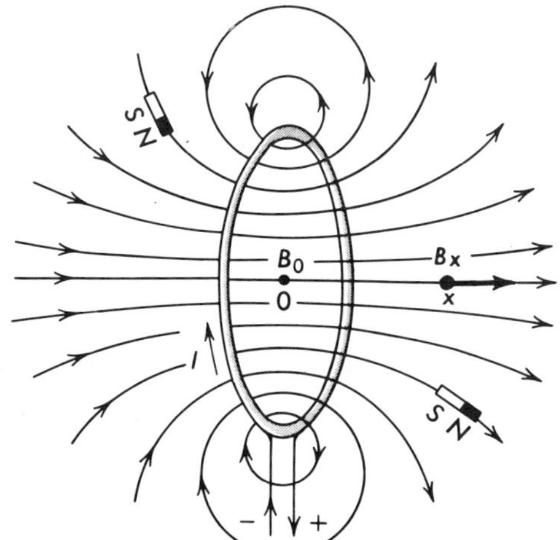

Fig. 29F. Diagram of the magnetic field through and around a flat circular coil carrying an electron current I.

Along the axis of the coil the magnetic induction B is perpendicular to the plane of the coil and its magnitude is given by

$$\boxed{B = \mu_0 \frac{NI}{2r}} \tag{29f}$$

center of a flat coil

Outside the coil the lines go from N to S in quite the same way they do about a permanent bar magnet, whereas inside they go from S to N.

Not only does one coil of wire act like a magnet, but two coils may be used to demonstrate the repulsion and attraction of like and unlike poles.

Another left-hand rule which must not be confused with the one in the preceding section, but which follows directly from it, is the following: *if the coil were to be grasped with the fingers pointing in the direction of the electron current, around the coil from (−) to (+), the thumb would point in the direction of the internal lines of induction as well as the N pole.*

EXAMPLE 2. A flat circular coil of 50 turns carries a current of 5 A. If the coil radius is 8 cm, what is the magnetic field strength at the very center?

SOLUTION. By direct substitution in Eq. (29b), we find

$$B = 12.57 \times 10^{-7} \frac{\text{N}}{\text{A}^2} \frac{5\text{ A}}{2 \times 0.08\text{m}}$$

$$B = 3.93 \times 10^{-5} \frac{\text{N}}{\text{Am}}$$

or, by Eq. (28f),

$$B = 3.93 \times 10^{-5}\text{ T}$$

29.5 UNIFORM MAGNETIC FIELDS While there are numerous ways of producing uniform magnetic fields, the methods commonly employed in the research laboratory are few in number. One of these, shown in Fig. 29G, consists of sending a

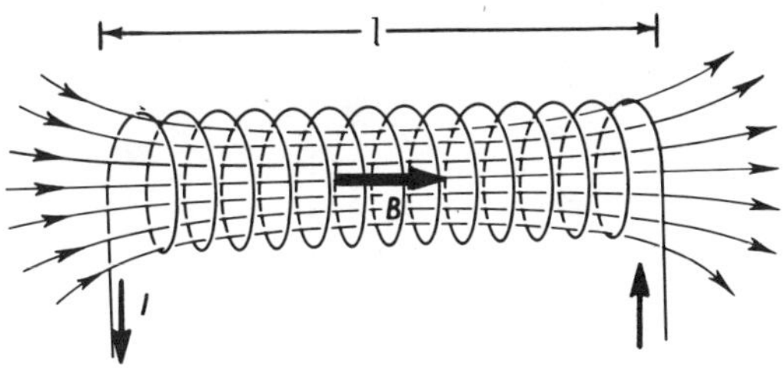

Fig. 29G. A solenoid of N turns of wire.

current through a long straight solenoid. As indicated by the lines of induction drawn in the figure, the number of lines per square meter of cross section is fairly constant over a considerable volume of space at the center.

The magnetic induction at the center of such a solenoid is given by

$$\boxed{B = \mu_0 \frac{NI}{l}} \qquad (29g)$$

long solenoid

where N is the number of turns, I is the current in amperes, and l is the length of the coil in meters.

EXAMPLE 3. A uniformly wound solenoid 80 cm long and 10 cm in diameter contains 650 turns of copper wire. If a current of 0.5 A is sent through this coil, what will be the strength of the uniform field at the center?

SOLUTION. By direct substitution in Eq. (29d), we obtain

$$B = 12.57 \times 10^{-7} \frac{650 \times 0.5\text{ A}}{0.8\text{ m}}$$

$$B = 5.11 \times 10^{-4} \frac{\text{N}}{\text{Am}}$$

$$B = 5.11 \times 10^{-4}\text{ T}$$

29.6 THE ELECTROMAGNET Five years after Oersted's discovery and Ampère's demonstration of the magnetic properties of a solenoid, William Sturgeon filled the center of a coil of wire with soft iron and thereby produced a powerful magnet. This is illustrated in Fig. 29H. As long as the

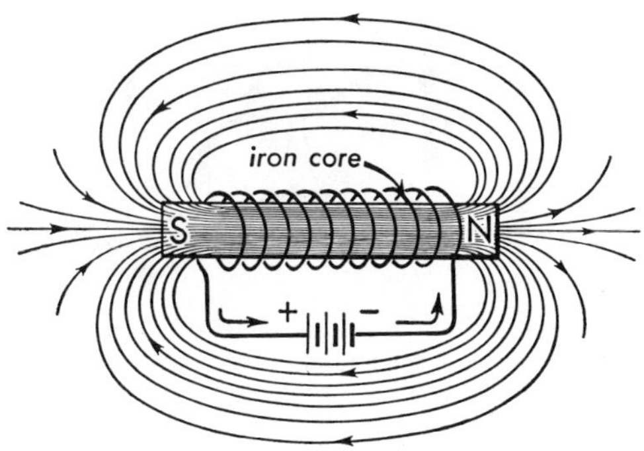

Fig. 29H. Diagram of an electromagnet and the field surrounding it when a current flows through the coil.

electron current continues to flow, the addition of the iron core produces a magnet hundreds of times stronger than does the solenoid alone. A nearby compass needle if set oscillating will demonstrate this by vibrating quickly with the iron core in place and slowly with it removed.

Again, with the iron core in the solenoid, a nearby rod of soft iron, magnetized by induction as shown in Fig. 28P, will attract iron filings much more strongly than when the iron core is removed.

29.7 MECHANICAL EFFECTS OF ELECTRIC CURRENTS It was on Christmas day, 1821, that Michael Faraday discovered that when a wire carrying a current is placed in the field of a magnet a mechanical force is exerted on the wire. This is the principle upon which the modern electric motor is based. A demonstration of Faraday's discovery is shown in Fig. 29I, where

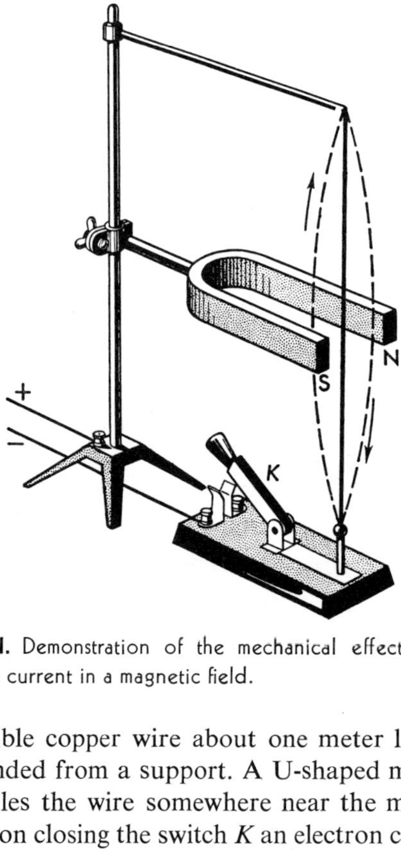

Fig. 29I. Demonstration of the mechanical effect of an electric current in a magnetic field.

a flexible copper wire about one meter long is suspended from a support. A U-shaped magnet straddles the wire somewhere near the middle.

Upon closing the switch K an electron current flows up through the wire and the wire moves to the left. If the battery connections are reversed, thereby reversing the electron current, the deflection of the wire will be to the right.

If the magnet is turned over, thereby interchanging N and S poles, the deflection of the wire will again reverse. In other words, the reversing of either the magnetic field or the direction of the electron current will reverse the direction of the force acting on the wire. The reversal of both will make it the same.

29.8 FORCE ON A MOVING CHARGE To find the force on a current carrying wire in a magnetic field we should consider its known origin. A current in a wire consists of a flow of electrons from *minus* to *plus*, and it is on these electrons that the magnetic field exerts the force. Since the electrons are confined to the wire, the total force on that wire is the sum of all the forces on the moving electrons within.

A single charge q moving with a velocity v, through a magnetic field at right angles to B, experiences a force given by

$$F = Bqv \tag{29h}$$

See Fig. 29J.

In the *MKSA* system, F is in newtons, B is in teslas, v is in meters/second, and q is in coulombs. As shown in Fig. 29J, the vectors B, v, and F are all mutually perpendicular to each other.

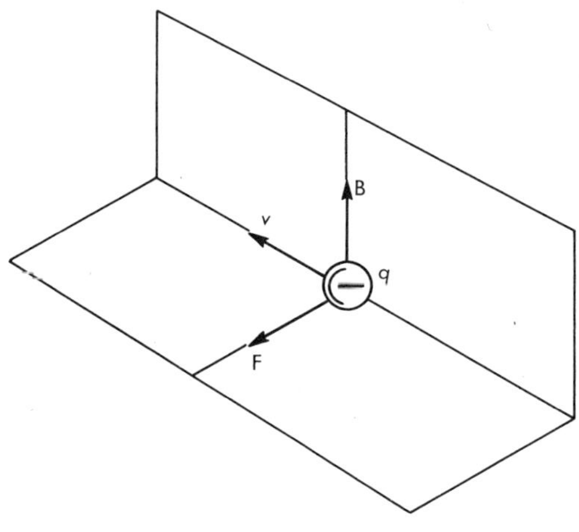

Fig. 29J. Force on a negative charge moving across a magnetic field.

When a stream of electrons is injected into a uniform magnetic field the force F causes each electron to move in the arc of a circle. Since F is always at right angles to v, it constitutes the centripetal force given by Eq. (8d).

$$F = m\frac{v^2}{r}$$

Equating these two different equations for the same force, we obtain

$$Bqv = \frac{mv^2}{r} \tag{29i}$$

which upon solving for r becomes

$$r = \frac{mv}{Bq} \tag{29j}$$

EXAMPLE 4. A beam of electrons with a speed of 3×10^6 m/s enters a uniform magnetic field B of 2×10^{-4} T at right angles to the field lines. If the charge and mass of each electron are 1.6×10^{-19} C and 9.0×10^{-31} Kg, respectively, find the radius of their path.

SOLUTION. By direct substitution of the given quantities in Eq. (29j), we find

$$r = \frac{9 \times 10^{-31}\text{ Kg} \times 3 \times 10^6\text{ m/s}}{2 \times 10^{-4}\text{ T} \times 1.6 \times 10^{-19}\text{ C}}$$

$$r = 8.44 \times 10^{-2}\text{ m}$$

$$r = 8.44\text{ cm}$$

29.9 FORCE ON A CURRENT CARRYING WIRE

To find the force on a current carrying wire, as shown in Fig. 29K, we make use of the force on

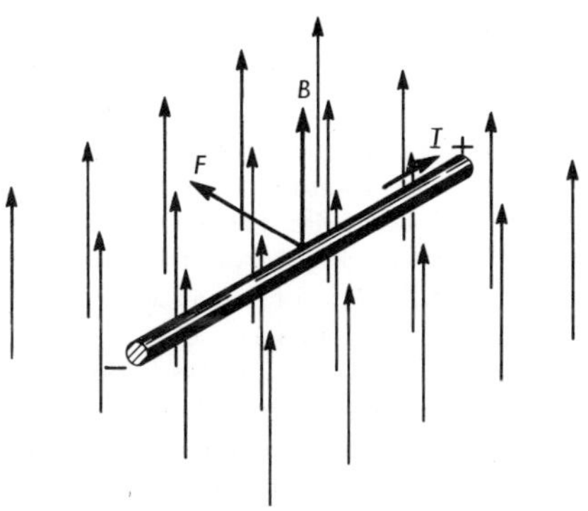

Fig. 29K. The force on a current-carrying wire in a magnetic field.

each individual electron. See Eq. (29h). Since current is defined as the charge flowing per second,

$$I = \frac{q}{t}$$

we can write

$$q = It$$

Moving with a velocity v, an electron will in a time t travel a distance $l = vt$. This gives $v = l/t$. Substituting It for q and l/t for v, in Eq. (29h), we obtain

$$F = B\,It\,\frac{l}{t}$$

or

$$\boxed{F = BIl} \tag{29k}$$

EXAMPLE 5. A straight wire 40 cm long and carrying an electron current of 12 A is placed in a field where the magnetic induction is 0.25 T. If the wire and field are perpendicular to each other, find the force on the wire.

SOLUTION. By direct substitution in Eq. (29k), we find

$$F = 0.25\text{ T} \times 12\text{ A} \times 0.40\text{ m}$$

which, by Eq. (28g), becomes

$$F = 1.20\text{ N}$$

29.10 THE ELECTRIC MOTOR

An electric motor is a device by which electrical energy in the form of an electron current is transformed into mechanical energy. The principle of the motor is illustrated in Fig. 29L. A wire carrying an electron current is bent into a loop and placed between two magnetic poles as shown. In this horizontal position the resultant magnetic field

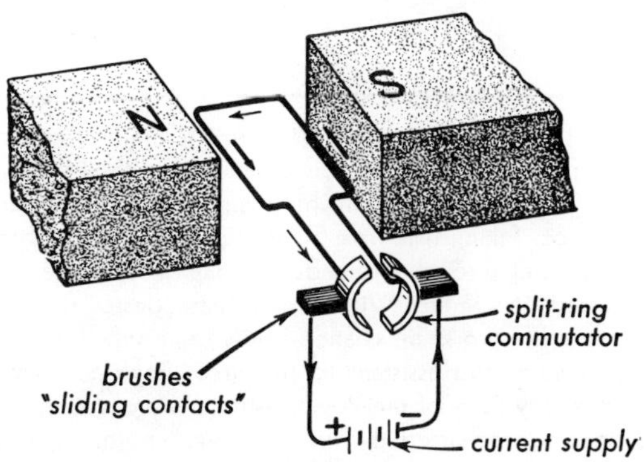

Fig. 29L. Principal elements for demonstrating the principles of an electric motor.

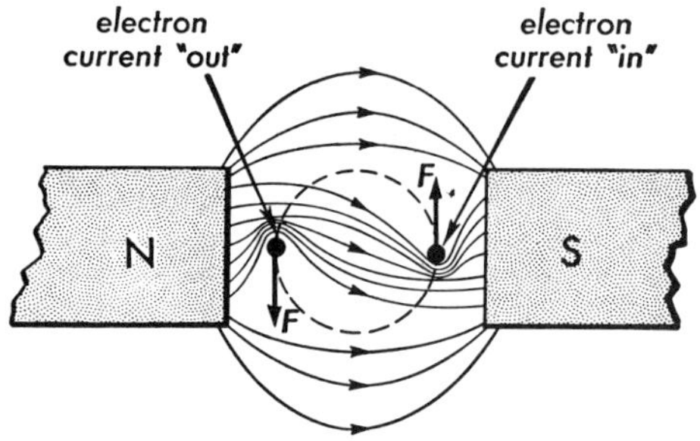

Fig. 29M. Diagram of the magnetic field around a current-carrying loop as found in the electric motor.

is warped as shown in Fig. 29M, forcing one wire down and the other up.

Mounted free to turn about an axis, the loop rotates until it is in a vertical plane. At this point the electron current in the loop is reversed in direction by means of *sliding contacts* and a *split-ring commutator*. The reversal of the electron current reverses the forces so that the side of the loop which was previously pushed up is now pushed down and the side previously pushed down is now pushed up. The loop therefore rotates through half a turn more and at that instant the current again reverses.

A repetition of this reversing process at each half turn gives rise to a continuous rotation, the left side of the coil or loop always moving down and the right side always moving up.

29.11 INDUCED ELECTRIC CURRENTS The discovery of induced electric currents goes back more than one hundred years to 1831 and the well-planned experiments of Michael Faraday.* A

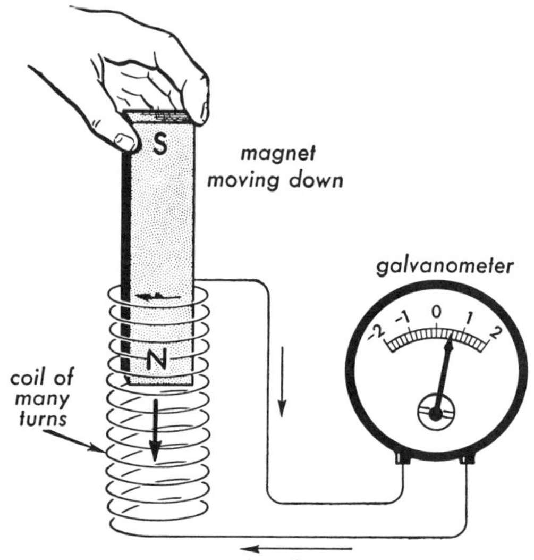

Fig. 29N. Diagram of Faraday's experimental discovery of induced electric currents.

straight-bar magnet plunged into a coil of wire was found to produce an electric current. The experiment is illustrated in Fig. 29N. As the N pole of the magnet is plunged into the coil, a galvanometer pointer deflects to the *right;* when it is withdrawn the pointer deflects to the *left*, indicating a current in the opposite direction. If the S pole is moved down into the coil the needle

* MICHAEL FARADAY (1791–1867), English experimental physicist. Born the son of a blacksmith, Faraday's early life was spent earning his living as a bookbinder's apprentice. Taking time from his work to read some of the books passing through his hands, Faraday became intensely interested in science. With a passionate desire to make science a life work, his chance finally came when he was made a valet and assistant to the great English scientist Sir Humphry Davy of the Royal Institute. As a young man he openly proclaimed that women were nothing in his life, and even wrote and published a poem in criticism of falling in love. At the age of twenty-nine he saw, fell desperately in love with, and married Sarah Barnhard who became a devoted and inspiring companion for the nearly fifty remaining years of his life. Four months after his marriage he made the famous discovery of the motion of a wire carrying a current in the field of a magnet. Since a current-carrying wire would move in a magnetic field, should not the reverse be true and a magnet be made to produce current in a wire? For days he experimented with magnets and coils of wire until in desperation he plunged a magnet down into a coil and observed that a current was generated in the coil. Why had he not discovered this before? The motion was the thing; it was the connecting link he had failed to realize. For this discovery the whole scientific world sought to honor him. So many universities gave him honorary degrees that he soon had to turn down such honors. He refused the presidency of both the Royal Institute and the Royal Society of London, and also refused to be knighted. Like many great scientists he loved his work more than these honors.

deflects to the *left* and as it is withdrawn the deflection is to the *right*.

The relative motion of the coil and magnet is what produces the electron current and it makes no difference whether the coil alone moves, whether the magnet alone moves, or they both move. In any case, when the relative motion ceases, the electron current stops. A "somewhat old-fashioned" way of describing the action is to say that only when a wire is cutting the lines of force is there an induced emf. A somewhat more acceptable statement at the present time is, in effect, that only when the total magnetic flux linking a closed electrical circuit is changing is there an induced emf. To demonstrate this concept, a simple experiment like that shown in Fig. 29O may be performed.

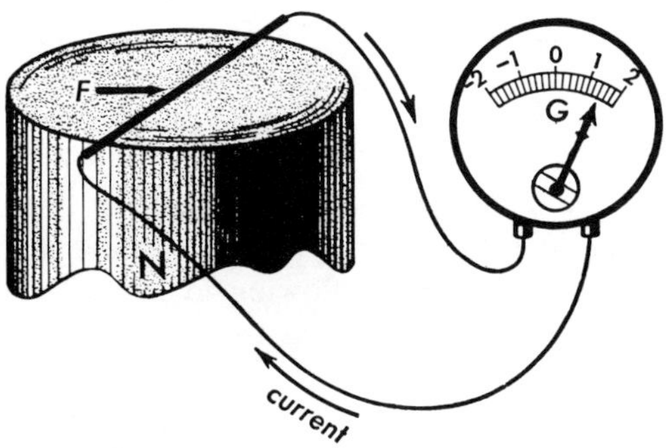

Fig. 29O. Experimental arrangement for demonstrating induced electron currents.

A flexible wire connected to an ammeter, and held in the hands, is moved in various ways across the pole of a magnet. When a straight section of the wire is held over the N pole and moved to the right, an electron current flows in the direction shown by the arrows. If the wire is moved in the opposite direction, the induced emf and current reverse direction. If the wire is moved vertically upward or downward, parallel to the magnetic induction, no current flows. In other words, *there is an induced emf only when the total number of lines of induction through the closed circuit is changing.*

The *left-hand rule* may be used to predict the direction of the induced emf in any section of wire. If we imagine grasping the wire in the left hand, as it moves through the magnetic field, the fingers pointing in the direction of the magnetic induction immediately in front of it, the thumb will point in the direction of the induced emf.

The fact that an electron current is produced means electrical energy has been created; it has been created at the expense of mechanical work, for in moving the wire across the field, a force F had to be exerted for a distance s. The faster the wire moves and the stronger the field of the magnet the greater will be the required force and the greater will be the induced electron current and electromotive force. If the wire stops moving in mid-field, the electron current stops. These are the essential principles of the electric generator.

The left-hand rule may be used to predict the direction of an induced electron current. As the wire moves across the pole of the magnet as shown in Fig. 29O, the imaginary magnetic lines of induction, acting as though they were stretched rubber bands, are warped under the action of the force F and bend away. If we now imagine the wire to be grasped, with the fingers pointing in the direction of the lines of induction on the "strong field" side of the wire, i.e., where the lines are pushed together, the thumb will point in the direction of the induced electron current.

29.12 THE ELECTRIC GENERATOR An electric generator is constructed the same as an electric motor, with a rotating armature containing coils of wire, pole pieces, field windings, brushes, and a commutator. Instead of supplying an electron current to obtain mechanical rotation, mechanical work is done to turn the armature, thus producing an electron current.

If, in the construction of a generator, two solid rings are used as a commutator, as shown in Fig. 29P, the electron current delivered to the brushes flows first in one direction, then in the other. The reversal of the electron current with each half turn of the armature is due to the fact that each wire moves up across the field at one instant and down at the next. At one instant the one terminal is positive and the other negative; at the next instant the first terminal is negative and the second positive. This periodically

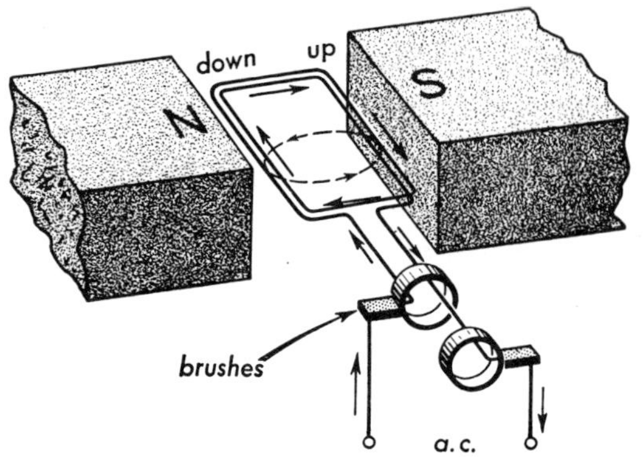

Fig. 29P. Illustration of the principles of the electric generator.

reversing emf produces what is called *an alternating emf*.

If a direct current is desired, the comutator of the generator must be of the split ring type illustrated in Fig. 29Q. It can be seen with this

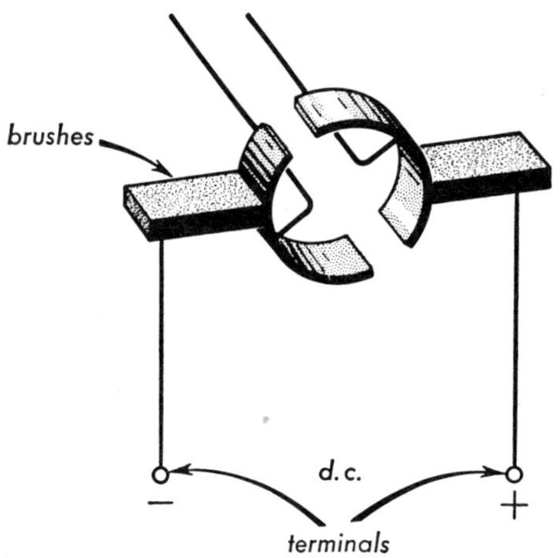

Fig. 29Q. Split-ring commutator for a direct-current generator.

arrangement that one brush is at all times in contact with wires moving up across the field while the other is in contact with wires moving down across the field. This produces a unidirectional electron current in the external circuit, and the whole machine is called a *direct-current* (d.c.) *generator*.

It is important to note that a generator does not make electricity. The electricity, or electric charge, is always in the wire, and a generator sets it into motion. A generator produces an electron current.

29.13 SUPERCONDUCTIVITY As the temperature of a wire increases, its electrical resistance also increases, and as the temperature is lowered, the resistance decreases. Although the reverse is true for some substances the above rule holds for most of the common metals we call conductors. This is the principle upon which the electrical resistance thermometer, described in Sec. 15.3, depends for its operation.

As the temperature of some metals is lowered to near absolute zero, they exhibit a sudden drop in resistance to almost zero. In other words, they become almost perfect conductors offering practically no resistance to an electric current. This change occurs at 8.2° K for lead, 4.1° K for mercury, 3.7° K for tin, and 2.4° K for thallium.

Superconductivity was first demonstrated by the Dutch physicist Kammerlingh Onnes, who immersed a ring of lead wire in liquid helium at a temperature of 4° K. By means of a magnet he then started a current flowing in the wire and observed that it continued to flow for several days. On one occasion he started a current flowing in a super-conducting metal ring in his laboratory in Holland, and then flew with it by airplane to England where he demonstrated before the Royal Society that the current was still flowing. In contrast to this, the same current started in the same ring maintained at room temperature would stop in a very small fraction of a second.

At normal temperatures the so-called *free electrons* in a wire are in a rapid state of random vibration, much the same as the molecules in a gas. Connecting the ends of a wire to a battery simply causes a slow drift of these electrons in the direction of the + terminal. Among the random motions of the electrons there are moments when a large number of electrons, by pure chance, are moving in the same direction. Such chance motions result in minute random currents first in one direction then in another. These random currents can be heard by amplifying them with a suitable radio circuit. It is well known that radio amplification is limited by the noise from the random motions of electrons in some of the connecting wires.

As the temperature of a wire is lowered the vibrations of the electrons, as well as those of the atoms forming the lattice of the conductor, slow down, until near absolute zero they are almost at rest. It is under these conditions that the electrons in some metals seem to be able to move freely through the lattice and give rise to what has been called *superconductivity*.

QUESTIONS

1. What are the five effects of electric currents? What, specifically, is the heating effect?

2. Does the same amount of electrical energy always produce the same amount of heat? What is Joule's law?

3. What three factors determine the electrical energy converted into heat? What are the units of heat energy? What are the units of electrical energy?

4. If V in Eq. (29c) is replaced by the Ohm's law value of IR, what would the equation become? If I is replaced by V/R, what would the equation become?

5. Who discovered the magnetic effect of an electron current? What did he do? What happens when the current is reversed?

6. To what kind of current does the left-hand rule apply? How does the rule apply to a straight wire?

7. How does the left-hand rule apply to a coil of wire?

8. What effect does a soft iron core have upon the strength of the magnetic field around a solenoid? Does a current have to flow to produce the field?

9. Who discovered the mechanical effects of electric currents? What are the necessary conditions for producing the mechanical effect of an electric current?

10. How is the mechanical effect used in the electric motor? What is the split ring used for? Why are sliding contacts needed?

11. Who discovered induced electron currents? How was the phenomenon produced?

12. Can a current be induced in a wire that is stationary? Can a current be induced in a wire by a stationary magnetic field?

13. If a wire moves through a magnetic field parallel to the lines of induction, is there an emf developed within the wire?

14. How can an alternating current be induced in a wire?

15. How are alternating currents generally produced?

PROBLEMS

1. An electric toaster draws a current of 5 A when connected to a house lighting circuit of 115 V. How many kilocalories of heat are produced in one minute? [Ans. 8.25 Kcal.]

2. An electric iron having a resistance of 18 Ω is connected to a 120 V supply. Find the heat developed in 10 min.

3. Calculate the heat developed by an electric soldering iron in 5 min if it draws 2.8 A from a 110 V line. [Ans. 22.08 Kcal.]

4. An electric toaster with a resistance of 20 Ω draws a current of 4.5 A when connected to a house lighting circuit. If it takes 50 s to make dark toast, how many Kilocalories are required?

5. The heating element of an electric stove connected to a 208-V line draws a current of 4.5 A. Find the amount of heat produced in 5 min. [Ans. 67.1 Kcal.]

6. An electric hot plate draws a current of 8 A when connected to a 120-V house lighting circuit. Find the heat produced in 5 min.

7. A waffle iron draws a current of 6 A when connected to a house circuit of 115 V. How much heat will be produced in 10 min. [Ans. 98.9 Kcal.]

8. An electric coffee maker draws a current of 6 A when connected to a 120-V house lighting circuit. How long will it take to raise the temperature of 2 Kg of water from 0°C to the boiling point? Assume no heat is lost.

9. A teakettle containing 1 liter (1 Kg) of water at a temperature of 15°C is heated on an electric stove. If the heating element draws an electron current of 6 A from a 208-V line, one-half of the heat generated going to heat the water, how long will it take the water to reach the boiling point? [Ans. 9 min 30 s.]

10. An electric coffee pot containing 2 liters (2000 cm^3) of water at 15°C is connected to a 120-V line. If the current drawn is 3.8 A and 60% of the heat developed goes into the water, how hot will the water be in 10 min?

11. Three resistors, 4 Ω, 6 Ω, and 10 Ω, respectively, are connected in series to a battery. If these resistors are immersed in a tumbler containing 400 cm^3 of water and a current of 3.6 A sent through them, how long will it take to raise the temperature of the water from 10°C to the boiling point of 100°C? [Ans. 9 min 41 s.]

12. A long straight wire carries a current of 12 A. Find the magnetic induction at a distance of (a) 1 mm, (b) 1 cm, and (c) 5 cm.

13. A long straight wire carries a current of 20 A. Find the magnetic induction at a distance of (a) 5 cm, (b) 10 cm, and (c) 20 cm from the wire. [Ans. (a) 8×10^{-5} T, (b) 4×10^{-5} T, (c) 2×10^{-5} T.]

14. A flat circular coil of 150 turns, and diameter of 10 cm, carries a current of 2.5 A. What is the magnetic induction at the very center?

15. A flat circular coil of 40 turns, and diameter of 12 cm, carries a current of 4.2 A. Find the strength of the magnetic field at the very center. [Ans. 3.520×10^{-3} T.]

16. A solenoid 25 cm long contains 120 turns of copper wire. What will be the magnetic induction at the center when a current of 5 A is sent through it?

17. A solenoid 60 cm long contains 400 turns of copper wire uniformly wound along its length. Find the magnetic induction at the center when a current of 2.6 A is sent through it. [Ans. 2.160×10^{-3} T.]

18. A beam of protons with a speed of 5×10^6 m/s is sent through a uniform magnetic field perpendicular to the magnetic induction. If the charge and mass of each proton are 1.6×10^{-19} C and 1.67×10^{-27} Kg, respectively, and the field strength is 0.2 T, find (a) the force on the proton, and (b) the radius of the circular path the beam traverses.

19. A beam of electrons with a speed of 6×10^6 m/s is sent through a uniform magnetic field where the magnetic induction is 2×10^{-3} T. Find (a) the force on each electron, and (b) the radius of the circular path the beam traverses. [Ans. (a) 1.920×10^{-15} N, (b) 1.690 cm.]

20. A straight wire 30 cm long is located in a uniform magnetic field of 0.18 T. Find the force on the wire if it carries a current of 4 A.

21. A straight wire 2 m long and carrying an electron current of 8 A is located in a uniform magnetic field where the magnetic induction is 0.40 T. Calculate the force on the wire. [Ans. 6.40 N.]

30 THE DISCOVERY OF ELECTRONS

Historically, a study of electrical discharges through gases and the discovery of atoms and electrons mark the beginning of a new branch of physical science called "*modern physics*." Modern physics, dealing principally with atoms, molecules, and the structure of matter, has developed at such a tremendous rate within the past four score years that it now occupies the center of attention of many leading scientists the world over. It is not exaggeration to state that recent discoveries in atomic physics have had and will continue to have a tremendous influence on the development of civilization. Because the subject of atomic physics is relatively new, it is logical to treat the subject matter associated with each major discovery in a more or less chronological order.

30.1 ELECTRICAL DISCHARGE THROUGH A GAS In 1853, an obscure French scientist by the name of Masson sent the first electric spark from a high-voltage induction coil through a partially evacuated glass vessel and discovered that, instead of the typical spark observed in air, the tube was filled with a bright glow. Several years later, Heinrich Geissler,* a German glass blower in Tübingen, developed and began the manufacture of gaseous discharge tubes. These tubes, made in diverse sizes, shapes, and colors of glass, and resembling the modern neon and argon signs used in advertising, attracted the attention of physicists in the leading scientific institutions and universities of the world, who purchased many of these "Geissler tubes" and used them for study and lecture demonstrations.

In 1869 W. Hittorf† of Münster, with im-

* HEINRICH GEISSLER (1814–1879), German physicist, was born in Igelshieb, Germany. He was educated as a glass blower, received an honorary Ph.D. from the University of Bonn in 1868, discovered the density of water, and invented a mercury pump for producing a high vacuum. He invented glass tubes of elaborate shapes in which electrical discharges in rarefied gases produced different colors. These were the forerunners of today's advertising neon signs and were called "Geissler" tubes.

† JOHANN WILHELM HITTORF (1824–1914), German physicist, was born at Bonn on March 27, 1824. He served most of his long and fruitful life as Professor of Physics and Chemistry at the University of Münster. His most important work was on the motion of ions during electrolysis and the study of the spectra of gases and vapors. He studied electrical discharges in gases and discovered a number of properties of cathode rays.

proved vacuum pumps, observed a dark region near one electrode of the electrical discharge which grew in size as the exhaustion was continued. This is but one of a number of phases of the study of electrical discharge through gases observed and studied a few years later by Sir William Crookes.*

In Fig. 30A a long glass tube about 4 cm in diameter and 150 cm long is shown connected to a mercury diffusion pump and a mechanical vacuum pump. The purpose of the pumps is to enable one to observe continuously the changes in the electrical discharge as the air is slowly removed from the tube. The purpose of the *trap* is to freeze out any mercury vapor and prevent it from reaching the discharge. High voltage from an induction coil is shown connected to the two electrodes, one at either end of the tube.

* SIR WILLIAM CROOKES (1832–1919), English physicist and chemist. At twenty-two he became an assistant at the Radcliff Observatory in Oxford. He was knighted in 1897, received the Order of Merit in 1910, and was president of the Royal Society from 1913 to 1915. He invented and made the first focusing type of X-ray tube. His experiments with electrical discharges through rarefied gases led to his discovery of the dark space which now bears his name.

Although an induction coil does not deliver current, its characteristics are such that the potentials are higher on half of the alternations than they are on the other, and the two electrodes act nearly the same as if a high-voltage direct current were used. The negative electrode under these circumstances is called the *cathode;* and the positive electrode, the *anode.*

As the long tube is slowly pumped out, an emf of 10,000 to 15,000 volts will produce the first discharge when the pressure has dropped to about one hundredth of an atmosphere, i.e., at a barometric pressure of about 10 mm of mercury. This first discharge, as illustrated in diagram (a) of Fig. 30B, consists of long, thin, bluish-colored streamers. As the gas pressure drops to about 5 mm of mercury, sometimes called a Geissler-tube vacuum, the discharge changes to pink and at the same time widens until it fills the whole tube as shown in diagram (b). At a still lower pressure of about 2 mm, diagram (c), a dark region called the *Faraday dark space* appears in the region of the cathode, dividing the bright discharge into two parts, a long pinkish section called the *positive column* and a short bluish section called the *negative glow.* As the pressure drops still further, diagram (d), the Faraday dark space grows in size and the negative glow moves away from the cathode, producing

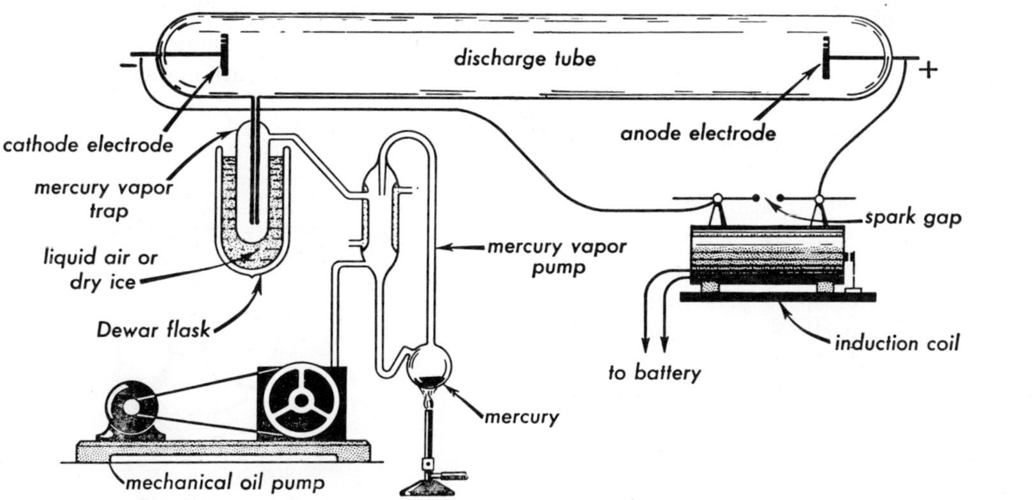

Fig. 30A. Diagram of a gaseous discharge tube showing the electrical connections as well as the vacuum pumps and accessories.

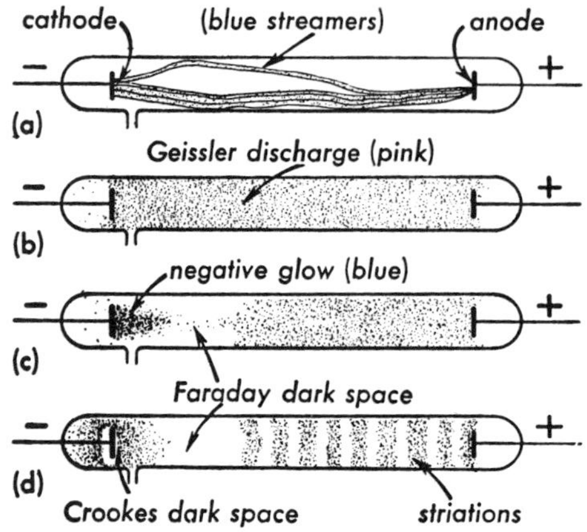

Fig. 30B. Sketches of the general appearance of a high-voltage electric discharge through rarefied air at various stages of evacuation.

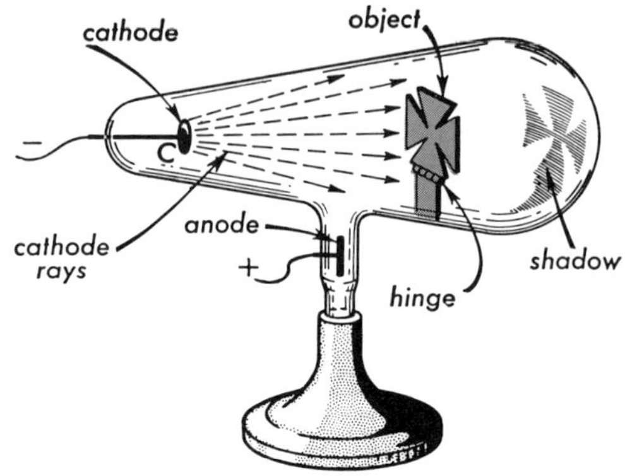

Fig. 30C. A Crookes' discharge tube for demonstrating that cathode rays travel in straight lines.

another dark space between it and the cathode. With the appearance of this second dark region, called the *Crookes dark space*, the positive column divides into a number of equally spaced layers called striations.

As the pumping proceeds, the striations and the negative glow grow fainter, and the Crookes dark space widens until finally at a pressure of about 0.01 mm it fills the whole tube. At this point a new feature appears; the whole glass tube itself glows with a faint greenish light.

30.2 CATHODE RAYS The green glow in the final stage of the gaseous discharge just described was soon found to be a *fluorescence of the glass produced by invisible rays emanating from the cathode itself*. These *cathode rays*, as they are called, believed by Sir William Crookes to be an "ultra gaseous state" and by Johann W. Hittorf to be a "fourth state" of matter, turn out to be tiny corpuscles which we now call *electrons*. In the relatively free space of a highly evacuated tube, cathode particles, torn loose from the atoms of the cathode, stream down the length of the tube, seldom colliding with a gas molecule, until they hit the glass walls.

The first important discovery concerning the nature of cathode rays was that they travel in straight lines. This was first observed by Hittorf in 1869 by his experiment of casting shadows of objects placed inside the discharge tube. This is usually demonstrated by a tube of special design as shown in Fig. 30C.

Where the rays strike the walls of the tube the glass fluoresces green, while in the shadow it remains dark. Under continuous bombardment of the walls by cathode rays, the fluorescence grows fainter because of a fatigue effect of the glass. This is demonstrated by tipping the object down on its hinge, permitting the rays to strike the fresh glass surface. Where the shadow previously appeared, a bright green image of the object is clearly visible.

That *cathode rays have momentum and energy* was first demonstrated by Crookes in 1870 using a tube of special design, as illustrated in Fig. 30D. Leaving the cathode and acquiring a high

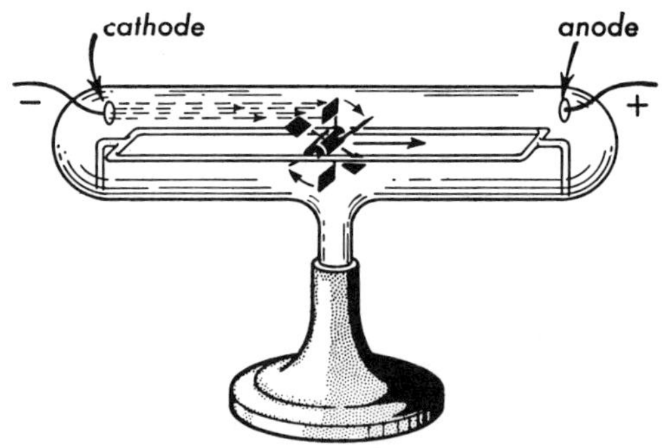

Fig. 30D. Demonstration of experiment showing that cathode rays have momentum and energy. Cathode rays striking the vanes of a small pinwheel cause it to roll from one end of the tube to the other.

speed on their way toward the anode, the rays strike the mica vanes of a small pinwheel and exert a force, causing it to turn and thus roll along a double track toward the anode. When it reaches the end of the track, a reversal of the potential, making the right-hand electrode the cathode, will send it rolling back toward the anode, now at the left. From this experiment Crookes concluded that cathode particles have *momentum* and that, therefore, they have *mass velocity*, and *kinetic energy* $\frac{1}{2}mv^2$.

That *cathode rays are negatively charged particles* was first discovered in Paris in 1895 by Jean Perrin. A discharge tube of special design usually used to demonstrate their charge is illustrated in Fig. 30E. A beam of cathode rays is

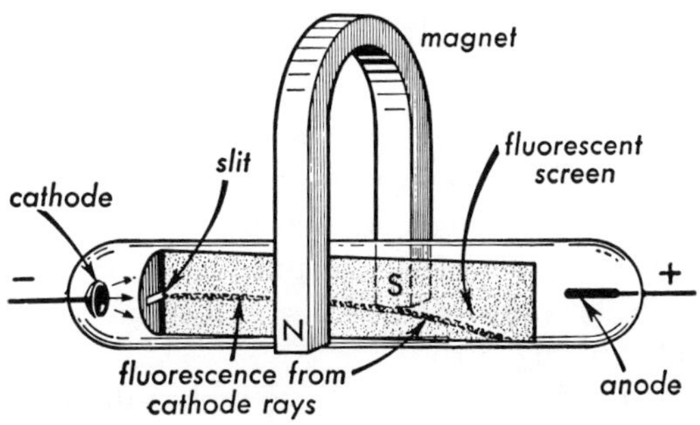

Fig. 30E. The bending of a beam of cathode rays in the field of a magnet demonstrates that cathode rays are negatively charged particles.

narrowed down to a thin pencil or ribbon of rays by a narrow slit near the cathode. The path of the rays is made visible by allowing them to strike a long strip of metal painted with zinc sulfide, a fluorescent paint. By placing a horseshoe magnet over the outside of the tube as illustrated, the path of the cathode rays is bent down. If the polarity of the magnet is reversed, the path is bent up. The bending shows they are charged, and the direction of bending shows the kind of charge. Being charged, a stream of particles is like an electron current. From the direction of the magnetic induction, the direction of the current, and applying the left-hand rule (see Sec. 29.3), the charge is found to be negative. [Remember that the left-hand rule applies to a current from (−) to (+).]

The penetrating power of cathode rays was first demonstrated by Heinrich Hertz and his assistant, P. Lenard, by passing cathode rays through thin aluminum foils. Out in the air the rays were found to retain sufficient power to cause fluorescence and phosphorescence of nearby uranium compounds.

30.3 MILLIKAN'S OIL-DROP EXPERIMENT Millikan* began his experiments on the electronic charge e in 1906. His apparatus is illustrated by a simple diagram in Fig. 30F. Minute oil drops from an atomizer are sprayed into the region just above the circular metal plates E^+ and E^-. Shown in cross section, the upper plate is pierced with a tiny pinhole P through which an occasional oil drop from the cloud will fall. Once between the plates, such a drop, illuminated by an arc light from the side, is observed by means of a low-powered microscope. See Fig 30G.

With the switch S in the "up" position, the condenser plates are grounded so that they are not charged. Under these conditions the oil drop falling under the pull of gravity has a constant velocity. This *terminal velocity*, as it is called, is reached by the drop before it enters the field of view and is of such a value that the downward pull of gravity, F_G, in Fig. 30H (a), is exactly equalized by the upward resisting force of the air. By measuring this velocity of

* ROBERT ANDREWS MILLIKAN (1868–1953), American physicist, educated at Oberlin College and Columbia University, for twenty-five years professor of physics at the University of Chicago and for thirty years president of the Norman Bridge laboratory at the California Institute of Technology in Pasadena. He served during World War I in the research division of the Signal Corps with the rank of lieutenant colonel. His principal contributions to science have been his measurement of the charge on the electron, his photoelectric determination of the energy in a light quantum, and his precision study of cosmic rays. He was the second American to be awarded the Nobel Prize in physics (1923). He has also been awarded the Edison Medal, the Hughes Medal of the Royal Society, the Faraday Medal, and the Mattenci Medal.

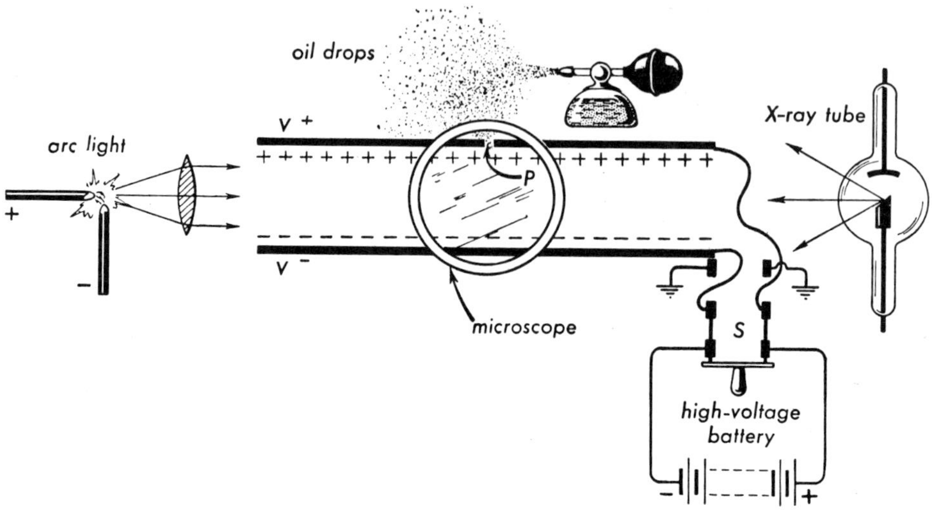

Fig. 30F. Schematic diagram of Millikan's oil-drop experiment. From this experiment the charge of the electron was determined.

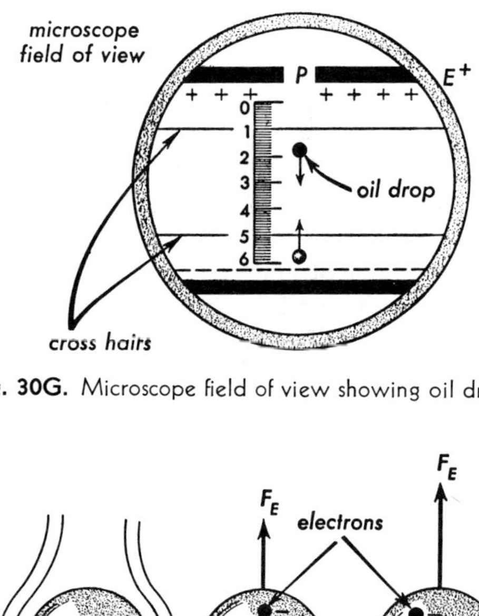

Fig. 30G. Microscope field of view showing oil drop.

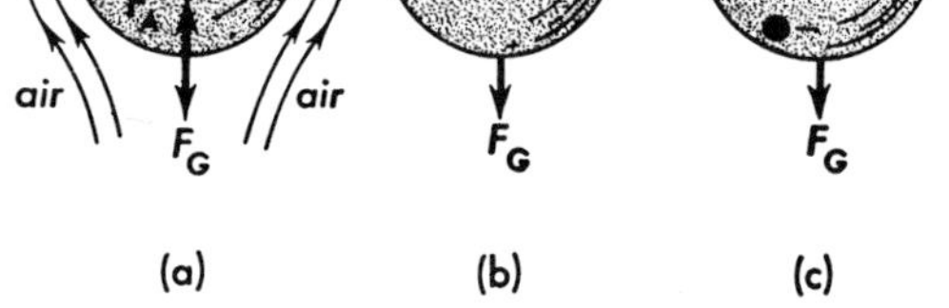

Fig. 30H. Diagrams of oil drop with extra electronic charges.

fall, the force F_G can be calculated and from it the mass of the oil drop determined. The velocity of the drop can be determined by using a stop watch and measuring the time required to fall the distance between the two cross hairs illustrated in Fig. 30G.

As the drop nears the bottom plate the switch S is thrown "down," charging the two parallel plates positive and negative. If now the drop has a negative charge, as illustrated in Fig. 30H (b), there will be an upward electrostatic force F_E acting to propel the drop up across the field of view. The drop will move upward with a constant velocity if F_E is greater than the gravitational force F_G. Again using the stop watch, this time to measure the velocity of rise, the upward force F_E can be calculated. Knowing the force, and the voltage on the condenser plates, the charge on the drop can be computed.

As the drop nears the top plate the switch S is thrown "up" and the plates are again grounded. Under these conditions the drop falls again under the pull of gravity alone. Upon nearing the bottom plate the switch is again thrown "down" and the drop rises once more. When this process is repeated, a single drop may be made to move up and down many times across the field of view. *Each time it falls, the velocity is measured and the mass computed; whereas each time it rises, the velocity is measured and the charge computed.*

Millikan found that if X rays were allowed to pass through the apparatus while an oil drop was being observed, the charge on the drop could be increased or decreased almost at will. One time on rising, the velocity would be low due to a small charge [see diagram (b) in Fig. 30H], while the next time the velocity would be high due to a larger charge, as in diagram (c). Regardless of the amount of charge, the rate of fall for a given drop is always the same. The reason for this is that the total mass of a number of electrons is so small compared with the mass of the oil drop that their added mass is not perceptible.

Millikan, and numerous other experimenters who have repeated these experiments, found that the charge on a drop is never less than a certain minimum value and is always some integral multiple of this value. In other words, any one electron is like every other electron, each carrying this minimum charge called e.

$$e = -1.60210 \times 10^{-19} \text{ coulomb} \tag{30a}$$

This is the most recent and probable value of the electronic charge.

By measuring exactly how much an electron beam bends as it passes through a magnetic field (see Fig. 30E) and using the value of the electronic charge given above, the mass of an electron is found to be

$$m = 9.10918 \times 10^{-31} \text{ Kg} \tag{30b}$$

This mass is unbelievably small; yet its value has been determined many times and by many experimenters, and the masses of all electrons are the same.

30.4 AN ELECTRON GUN A schematic diagram of an electron gun is shown in Fig. 30I. The source of electrons is an oxide-coated cathode K. Such cathodes consist of nothing more than a thin metal cup, coated on the outside surface with a thin layer of chemical element or com-

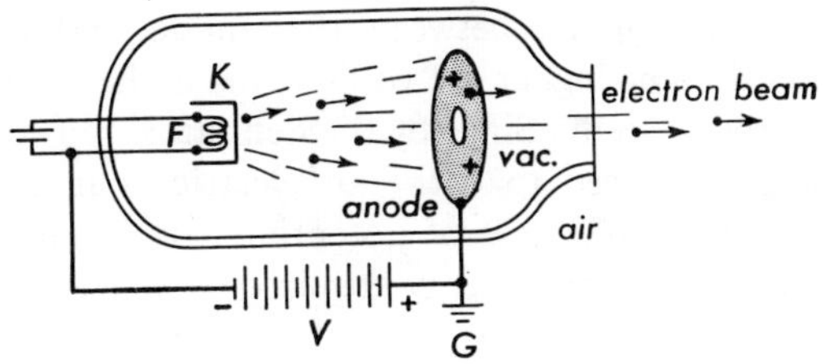

Fig. 30I. Electrons are accelerated by an applied potential V.

pound, such as thorium, strontium, or caesium oxide. These particular oxides are copious emitters of electrons when heated to a dull red heat by the filament F. The positive electrode, or anode, is a disk of thin metal with a hole in the center.

The cathode and filament are connected to the negative terminal, and the circular disk at the center, to the positive terminal of a high-voltage battery V. Starting from rest at the cathode, the electrons are accelerated along the electric lines of force acquiring at the anode disk a velocity v.

By connecting the (+) terminal to the ground, the anode is brought to the potential of the surrounding walls of the room, and the electrons are not attracted back toward the anode disk but continue on with constant velocity. With a thin aluminum foil at the end and a high applied voltage V, electrons may be projected into the air beyond.

One of the results of J. J. Thomson's* experiments with cathode rays was the discovery that the velocity of electrons depends upon the

* SIR JOSEPH JOHN THOMSON (1856–1940), English physicist, educated at Owens College, Manchester, and at Trinity College, Cambridge. He was appointed Cavendish professor at Cambridge in 1884, and professor of physics at the Royal Institution, London, in 1905. He was awarded the Nobel Prize in physics in 1906, was knighted in 1908, and elected to the presidency of the Royal Society in 1915. He became master of Trinity College in 1918 and helped to develop at Cambridge a great research laboratory attracting scientific workers from all over the world.

potential applied between the anode and the cathode. The higher the voltage, the higher is the electron velocity. Since by the definition of V the energy required to carry an electric charge Q through a difference of potential V is given by $V \times Q$ the kinetic energy acquired by an electron of charge e falling through a difference of potential V will be $V \times e$. Equating this to the kinetic energy $\frac{1}{2}mv^2$,

$$\boxed{Ve = \tfrac{1}{2}mv^2} \qquad (30c)$$

where V is the applied accelerating potential in volts, m is the mass in Kg, v the velocity in meters per second, and e the charge on the particle in coulombs.

EXAMPLE 1. Calculate the velocity of electrons accelerated by a potential of 10,000 volts. (The electronic charge $e = 1.6021 \times 10^{-19}$ coulomb, and $m = 9.1091 \times 10^{-31}$ Kg.)

SOLUTION. Direct substitution in Eq. (30c) gives

$$10{,}000 \times 1.6021 \times 10^{-19} = \tfrac{1}{2}(9.1091 \times 10^{-31}) \times v^2$$

from which

$$v = \sqrt{\frac{10{,}000 \times 1.6021 \times 10^{-19} \times 2}{9.1091 \times 10^{-31}}} = 5.93 \times 10^7 \text{ m/s}$$

This is just one fifth the velocity of light. (Velocity of light $c = 3 \times 10^8$ m/s.)

Instead of calculating the velocity of electrons in meters per second it is customary to refer to their kinetic energy in terms of the applied voltage. For example, in the problem above the energy gained by the electrons is said to be 10,000 *electron volts* (*abbr.* 10,000 *ev*). They are also sometimes referred to as 10,000 volt electrons.

QUESTIONS

1. What is the general appearance of the first vacuum discharge in a tube when the pressure is lowered?

2. What is the negative glow? What is the positive column?

3. What are striations? What is the Faraday dark space? What is the Crookes dark space?

4. What is a cathode? What is an anode?

5. How can it be shown that cathode rays travel in straight lines? Where do cathode rays originate?

6. How can it be shown that cathode rays have momentum and energy? Do cathode rays have mass?

7. How can it be shown that cathode rays (a) are charged particles and (b) that the charge is negative?

8. Do cathode rays have penetrating power?

9. What evidence is there in the experiments described here that cathode rays probably are all alike and have the same mass and charge?

10. Upon what does the color of the light from the positive column of an electrical discharge tube depend? How is this related to present-day "neon signs?"

11. What is Millikan's oil-drop experiment? What was its purpose?

12. What determines the speed of fall of an oil drop? Is the velocity of fall constant?

13. What determines the speed of rise of an oil drop? Is the velocity the same each time the same drop rises?

14. What is determined by the rate of fall of an oil drop? What is determined by the rate of rise?

15. What was the purpose of the X rays in the oil-drop experiment?

16. What conclusions can be drawn from J. J. Thomson's experiment? What conclusions can be drawn from the Millikan oil-drop experiment?

17. Make a diagram of an electron gun. Briefly explain where the electrons originate, and what determines their maximum speed.

PROBLEMS

1. If a 3000-volt battery is applied to an electron gun, what will be the maximum speed of the electrons accelerated? [Ans. 3.249×10^7 m/s.]

2. A 2200-V battery is connected to the electrodes of an electron gun. Find the speed of the electrons.

3. What voltage applied to an electron gun will give electrons one-fiftieth the speed of light? Assume the speed of light to be 3×10^8 m/s. [Ans. 102.3 V.]

4. If the electrons from an electron gun have a speed of one-twelfth the speed of light, what voltage is applied? The speed of light is 3×10^8 m/s.

5. The electron beam in an oscilloscope is accelerated by an applied potential difference of 2500 V. Find the speed of the electrons. [Ans. 2.966×10^7 m/s.]

6. The electron beam sweeping across the picture screen of a television receiver is accelerated by a potential difference of 12,000 V applied to an electron gun. Find the speed of the electrons.

7. What voltage applied to an electron gun of a television receiver will give the electrons in the beam, which scans the picture on the screen, a speed of one-tenth the speed of light? [Ans. 2559 V.]

8. How many electrons are equivalent to a mass of one kilogram?

31 THE DISCOVERY OF ATOMS

Although no one has ever seen individual atoms, there is no doubt that such particles really exist. To the physicists and chemists who have built up and established the present-day theories of the structure of matter, atoms are as real as any material objects large enough to be seen with the eyes or to be felt with the hands. Their reality is evidenced by hundreds of experiments that can be planned and executed in the research laboratory.

As the subject of atomic physics is developed in this and the following chapters, it will become more and more apparent that, although a physicist requires an extremely imaginative mind, the accumulated knowledge of atoms, their structure, and their behavior under a multitude of conditions is based upon exact results of experiments performed with the greatest of accuracy and precision.

31.1 THE DISCOVERY OF POSITIVE RAYS During the latter part of the nineteenth century, when many physicists were investigating the

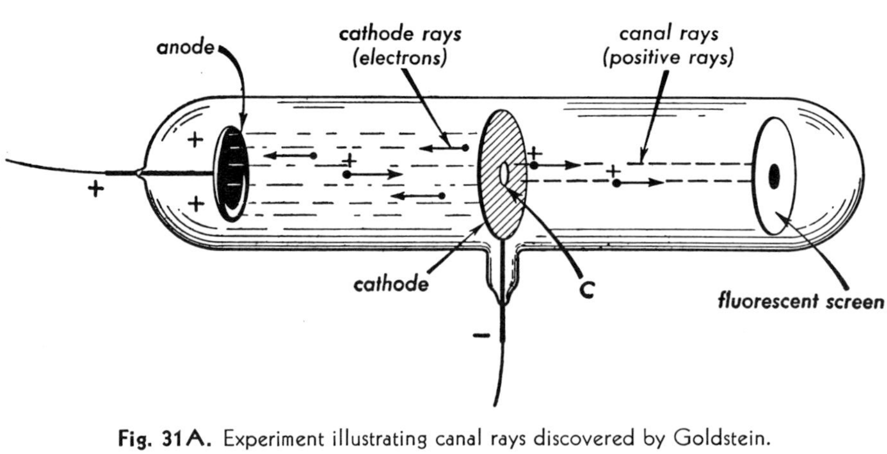

Fig. 31A. Experiment illustrating canal rays discovered by Goldstein.

various properties of cathode rays, Goldstein* designed a special discharge tube and with it discovered new rays called *canal rays*. The name "canal rays" is derived from the German "kanal-strahlen" and the fact that the rays, traveling in straight lines through a vacuum tube in the opposite direction to cathode rays, pass through and emerge from a hole in the cathode. A tube designed to illustrate this is shown in Fig. 31A.

In 1896, W. Wien† deflected a beam of canal rays in a magnetic field and came to the conclusion that the rays consisted of positively charged particles. Due to this and other experiments, canal rays have become more commonly known as *positive rays*.

* EUGEN GOLDSTEIN (1850–1930), German physicist, was born in Gleiwitz, Germany, and educated at the University of Berlin. He remained at that Institution as a research assistant until 1878 when he was transferred to the Berlin Observatory. From 1890–96, he did research at the Physikalische Technische Institut, and from 1896–1927, did independent research at Berlin-Schöneberg. His principal contributions to knowledge were in the field of electrical discharges in rarefied gases, and through these studies discovered rays, the electrical deflection of cathode rays, the spectra of a number of ionized atoms, and the band spectrum of the helium molecule.

† WILHELM WIEN (1864–1928), German physicist, was born in Goffken, East Prussia. He studied at the Universities of Göttingen, Heidelberg, and Berlin, and then served as assistant to Helmholtz. He wrote on optics and especially on black-body radiation for which in 1911, he was awarded the Nobel Prize. He also wrote on discharges through gases, cathode rays, X rays, and positive rays. Wien's most important contributions to black-body radiation are contained in two laws named after him. He developed a formula for the energy density at different wavelengths radiated by a hot body, and the wavelength displacement of the maximum with a change in temperature. He showed that positive rays undergo electric and magnetic deflection and showed a keen interest in modern theoretical physics. In 1913, he lectured at Columbia University in New York.

Since the time of Goldstein's discovery, positive rays have been found to be charged atoms of different weights. The origin of the charge carried by such atoms is explained briefly as follows. As the electrons from the cathode stream down the tube toward the anode, they occasionally collide with the atoms and molecules of the small quantity of remaining gas, knocking electrons from them. This process, called *ionization*, is illustrated by a schematic diagram of a single oxygen atom in Fig. 31B. Before the

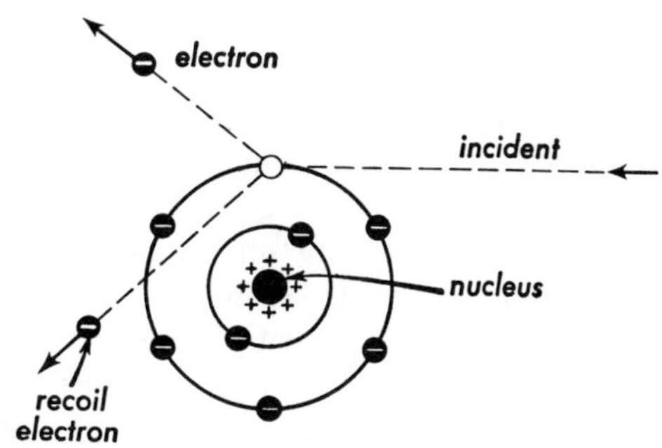

Fig. 31B. Schematic diagram of an oxygen atom in the process of becoming ionized by a collision with a high-speed electron.

collision, the atom as a whole, with its eight electrons and eight equal positive charges on the nucleus, has no net charge. After one of the electrons is removed by collision it has but seven electrons and therefore a net positive charge equivalent in amount to the negative charge of one electron.

Since the atom is now positively charged the anode repels and the cathode attracts such atoms, accelerating them toward the cathode. There exists, therefore, between the cathode and the anode, two streams of particles: electrons moving toward the anode, and positively charged atoms or molecules moving toward the cathode.

Of the many particles striking the cathode, in Fig. 31A, the ones moving toward the small opening C, constituting the observed canal rays, pass straight through to the fluorescent screen. As each atom or molecule strikes the screen, a tiny flash of light is produced. These tiny flashes, which can be seen individually in the field of view of a microscope, are called *scintillations.*

Any process by which an electron is removed from an atom or molecule is called *ionization*, and the resulting charged particle is called a *positive ion*. The amount of charge carried by a singly charged positive ion is equal to *the electronic charge* but opposite in sign.

$$e = +1.60210 \times 10^{-19} \text{ coulomb}$$

31.2 THE THOMSON MASS SPECTROGRAPH

Ever since the time canal rays were shown to be positively charged atoms or molecules of the gas contained within the discharge tube, physicists have tried to determine with ever-increasing accuracy the mass and charge of the individual ray particles. Although the charge and mass of every electron were known from Thomson's and Millikan's experiments to be the same as that for every other electron, it could be postulated that the mass of the positive rays should be different for the atoms of different chemical elements. The further postulation could be made that, if each positive ion is produced by the removal of one electron from a neutral atom, all positive ions should have the same amount of charge. This in part was anticipating what is now known.

In 1911, J. J. Thomson developed a method of measuring the relative masses of different atoms and molecules by deflecting positive rays in a magnetic and an electric field. The apparatus he developed for doing this is shown schematically in Fig. 31C and is called *Thomson's mass spectrograph.*

The entire spectrograph, enclosed in an airtight glass chamber, is first thoroughly evacuated, and then a small quantity of the gas, the masses of whose atoms are to be measured, is admitted to the bulb at the left. When a high voltage is applied to this chamber, electrons from the cathode ionize atoms and molecules in the region between the anode A and the cathode C. Traveling to the right, many of these positively charged particles pass through the narrow hole in the cathode, thus forming a very narrow pencil of rays. Leaving the cathode with a constant velocity they then pass between the poles of an electromagnet and two parallel plates and thence to a fluorescent screen at the far end of the chamber.

The two parallel plates, when charged, create a uniform electric field which exerts an upward force on the particles, deflecting them from the point O toward E. The uniform magnetic field, on the other hand, with its magnetic lines vertically downward and in the plane of the page, exerts a force at right angles to this, deflecting the particles "into" the page from the point O toward H.

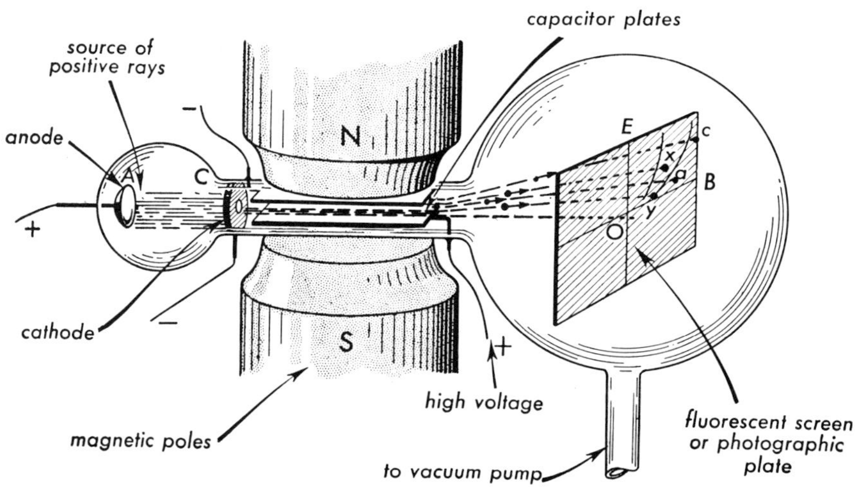

Fig. 31C. Diagram of J. J. Thomson's mass spectrograph.

Suppose now that the apparatus contains a pure gas like helium, all of the atoms of which have exactly the same mass. Of these atoms, the ones that are ionized in a region near the cathode C cannot attain a very high velocity before reaching the cathode. Remaining longer in the deflecting fields, their paths are bent considerably up and back to a point such as *c* on the screen. Particles ionized near the anode A, on the other hand, attain a high velocity upon reaching the cathode and, being under the influence of the deflecting fields for a shorter time, have their paths bent only a little, to a point like *a* on the screen. Since the velocities of the particles vary considerably, a bright streak or line of fluorescence will appear on the screen. From a calculation of the forces exerted by both fields it is found that the line on the screen should have the shape of a parabola.

If the gas in the apparatus is not pure but contains two kinds of atoms, the positive ions passing through the cathode will have two different masses. Although each ion will contain the same positive charge and will therefore experience the same electric and magnetic forces when passing through the fields, the heavier particles will not be deflected as much as the lighter ones. The net result is that the heavier particles form one parabolic curve like *xy*, and the lighter particles another curve like *ac*.

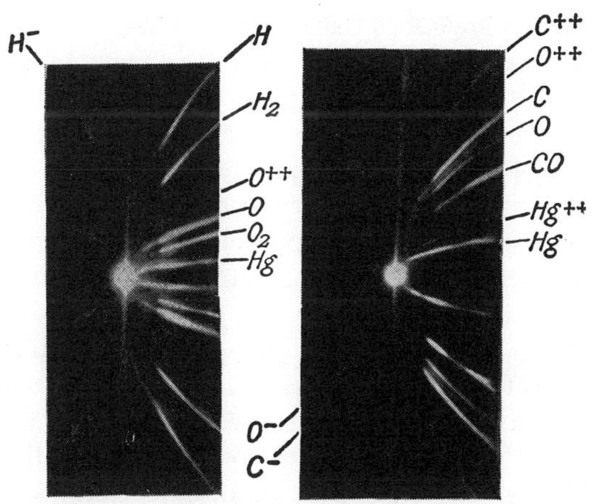

Fig. 31D. Reproduction of the photographs of parabolas made with Thomson's mass spectrograph.

By substituting a photographic plate for the fluorescent screen and exposing it to the rays for several minutes, photographs like those reproduced in Fig. 31D are obtained. The continual bombardment of the photographic plate by atoms and molecules has the same effect as does light, and images are produced upon development. The upper half of each picture is taken with the connections as shown in Fig. 31C and the lower half by reversing the polarity of the electromagnet, and exposing for an equal length of time.

When photograph (a) was taken the spectrograph contained *hydrogen*, *oxygen*, and *mercury*, and the magnetic field was relatively weak. From the known strengths of both the electric and magnetic fields and the assumption that each atom carries a unit positive charge, the mass of the atoms producing each parabola can be calculated. The results of these calculations show that the two largest parabolas are due to ionized hydrogen atoms (H^+) of mass 1, and ionized hydrogen molecules (H_2^+) of mass 2. The next three are due to ionized atoms (O^+) of mass 16, ionized oxygen molecules of mass 32, and ionized mercury atoms (Hg^+) with a mass of approximately 200.

When photograph (b) was taken, the mass spectrograph contained carbon monoxide gas and mercury vapor; and the magnetic field was relatively strong. Upon calculating the masses of the particles producing the different parabolas, the four intense lines were identified as due to ionized carbon atoms (C^+) of mass 12, ionized oxygen atoms (O^+) of mass 16, ionized carbon monoxide molecles (CO^+) of mass 28, and ionized mercury atoms (Hg^+) of mass about 200. The three faint parabolas in the photograph are due to doubly ionized atoms of *carbon*, *oxygen*, and *mercury*.

A doubly ionized atom or molecule is one which has lost two electrons in place of one and, having a net positive charge of two units, is designated by two ($^+$) signs as superscripts. Since the particles have double charges, the electric and magnetic forces exerted on them are double those for singly ionized atoms and they, undergoing greater deflections, produce larger parabolas.

The principal conclusions to be drawn from Thomson's experiments are (1) *that positive rays or canal rays are charged atoms or molecules of whatever gas is present in the apparatus*, (2) *that hydrogen and oxygen molecules have twice the weight of their single atoms and are therefore diatomic*, (3) *that a chemical compound like carbon monoxide is composed of diatomic molecules CO, each with a weight equal to the sum of the weights of the separate atoms.*

It is significant to point out that, while Thomson found that many atoms could be doubly and some even triply ionized, hydrogen could never be found more than singly ionized and helium more than doubly ionized. The reason for this, as will be seen later, is that neutral hydrogen atoms have but one electron and neutral helium atoms but two. All other elements have more than two electrons.

31.3 THE PERIODIC TABLE OF ELEMENTS From present-day knowledge of physics, chemistry, and astronomy, it seems quite certain that the entire universe is made up of approximately 90 elements. By an element is meant a substance composed of atoms having identical chemical properties. About 90 of these elements have been found in the earth's crust, some of them in much greater abundance than others. Silicon and iron are examples of abundant elements, whereas platinum is an example of a rare element.

Long before the Thomson mass spectrograph had been devised and used to measure the relative masses of atoms, the chemist had arranged all of the elements in a table according to their atomic weights. The most common form of this arrangement is given in Appendix II. Divided as they are into eight separate groups, all elements in the same column have similar chemical properties. In Group I, for example, the elements Li, Na, K, Rb, Cs, and Fr known as the *alkali metals*, have one set of chemical properties, whereas the elements Be, Mg, Ca, Sr, Ba and Ra in Group II, known as the *alkaline earths*, have another set of chemical properties. The largest group of elements having similar chemical properties is that of the fourteen rare earth elements listed by themselves at the bottom of the table.

The names of the elements are all indicated by one-letter and two-letter symbols. The full names are given in the second column of Appendix V. The number preceding each name is the order number of that element and is called *the atomic number*. The average weight of atoms of that element, called *the atomic weight*, is given in the last column.

The atomic weights of all elements are based upon the weight of carbon-12. This is purely an arbitrary selection of a unit of weight but one which has considerable significance when it is noted that the weights of the first 25 elements, with the exception of Mg and Cl, atomic numbers 12 and 17, are very close to whole numbers. Early in the history of atomic structure this suggested the possibility that the weights of all atoms are really whole number units of the unit of weight, the hydrogen atom, and that those weights of an element which differ considerably from whole numbers are incorrectly determined values. On the strength of this, Prout was the first to propose the hypothesis that all elements are made of hydrogen atoms as building stones. These suppositions, as will be seen later, are only partly true.

31.4 THOMSON'S DISCOVERY OF ISOTOPES In 1912 Thomson, in comparing the mass of the neon atom with the known masses of other elements, discovered two parabolas for neon in place of one. Upon computing the masses of the particles involved, the stronger of two parabolas was found to be due to particles of mass 20 and the other, a fainter parabola, to particles of mass 22.

Since the atomic weight of neon was then known to be 20.2, Thomson expressed the belief that neon is composed of two kinds of atoms, 90% of which have a mass of 20 and the other 10% a mass of 22. Because these two kinds of atoms exist as a mixture and cannot be separated chemically, their atomic weight, when measured by chemical methods, is found to be their average value 20.2.

The discovery of two kinds of neon atoms, identical chemically but differing in atomic weight, suggested the possibility that all other elements whose atomic weights were not whole numbers might also be mixtures of atoms which do have whole number weights. Not only has this been confirmed by experiment but a large majority of the elements have been found to be mixtures of from two to ten different kinds of atoms.

To all atoms of different mass belonging to the same element, Soddy gave the name *isotopes*. The external structure of all isotopes of a given element is identical. The two atoms Ne-20 and Ne-22, shown in Fig. 31E, are neon isotopes. Each of these neutral atoms, before it is ionized to become a positive ray, has ten external electrons and ten positive charges on the nucleus. They differ only in the mass of the nucleus.

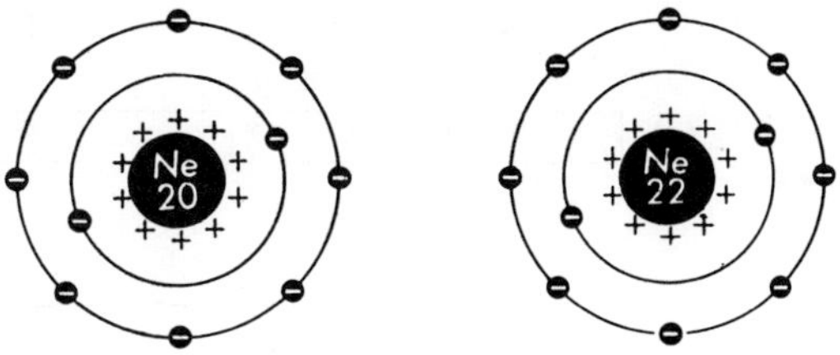

Fig. 31E. Schematic diagrams of the two different kinds of neon atoms, one of mass 20 and the other of mass 22. The external electron structures of two such isotopes are identical.

31.5 ASTON'S MASS MEASUREMENTS Immediately following World War I in 1919, F. W. Aston* developed a new and improved type of mass spectrograph. Employing both electric and magnetic fields, the device involved the chief improvement over Thomson's mass spectrograph of "focusing" the rays of different velocities to the same point on the screen or photographic plate. This had two important effects: (1) it made it possible to observe rare isotopes which might otherwise escape detection, and (2) it produced sharper images of the different masses on the photographic plate, so that their masses could be more accurately measured.

An Aston mass spectrogram is reproduced at the top of Fig. 31F. In taking this particular photograph, Aston had introduced into his apparatus, among other things, a little *hydrochloric acid* (HCl), *carbon monoxide* (CO), and *sulfur dioxide* (SO_2). Being close together in the periodic table, these elements furnish an excellent demonstration of the linear shift of atoms and molecules, differing in mass by one unit. It is found from this and other photographs that sulfur has three isotopes with masses 32, 33, and 34, and that chlorine has two isotopes of mass 35 and 37.

Since the atomic weight of chlorine is 35.46, then for every atom of mass 37 in a given quantity of chlorine gas there are four of mass 35. Mixed together in these proportions they give an average mass of 35.4.

The photographic lines corresponding to masses 28, 36, and 38 are due to diatomic molecules CO and HCl, each molecule having the combined mass of its constituent atoms. Since there are two relatively abundant chlorine isotopes there are two kinds of HCl molecules. One type, H^1Cl^{35}, has a mass of 36, and the other type, H^1Cl^{37}, a mass of 38.

A CO molecule of the type producing the strong line at mass 28 in Fig. 31F (a) is shown schematically in Fig. 31G. As a neutral molecule there are just as many electrons surrounding the two bound atoms as there are positive charges on the nuclei (six on the carbon nucleus and eight on the oxygen nucleus). When the molecule be-

* FRANCIS WM. ASTON (1877–1945), British scientist, was born in Birmingham and educated at Malvern College and Cambridge University. He became assistant lecturer in physics at the Birmingham University in 1909, and received the Mackenzie Davidson Medal of the Röntgen Society in 1920. In 1922 he was awarded the Hughes Medal of the Royal Society and the coveted Nobel Prize in chemistry for his work on atomic mass measurements. He has written an authoritative book entitled "Isotopes," in which a full account of his work is given.

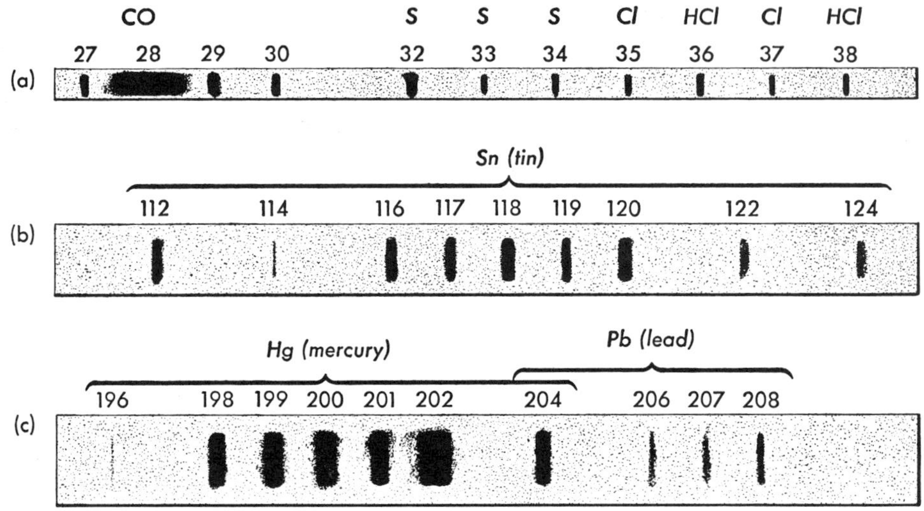

Fig. 31F. Drawings based on photographs taken with a mass spectrograph illustrating the linear shift of atoms differing by one unit of mass. (a) Carbon monoxide, sulfur, chlorine, and argon lines. (b) and (c) isotopes of mercury, tin, and lead.

comes ionized and is moving through the apparatus as a positive ray, it contains one less electron than the number shown. Since the mass of the electrons is negligibly small, the mass of the molecule is 12 + 16 or 28 mass units.

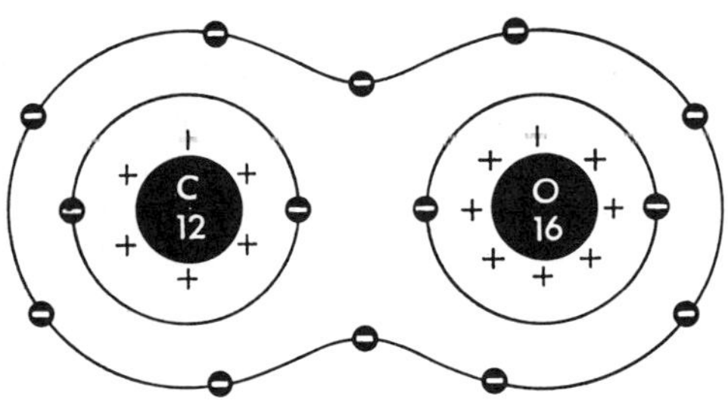

Fig. 31G. Schematic diagram of a diatomic molecule, carbon monoxide, CO.

So successful was Aston with his mass measurements and his determination of isotopes of different elements that he attempted an investigation of the entire periodic table. All of the known elements are listed in Appendix V, with all of their observed isotopes. In each case the most abundant isotope is given in heavy type, while the very rare isotopes, i.e., those present to less than 1%, are given in parentheses. Where more than one isotope is given in heavy type the isotopes occur with almost equal abundance. The isotopes Rb 87 and Tc 99, shown in italics, represent unstable atoms, a condition giving rise to *radioactivity*, the subject of Chap. 34. Appendix VI is composed almost entirely of radioactive isotopes.

Recent developments in mass spectroscopy have made it possible to detect exceptionally rare isotopes. In neon, for example, an isotope of mass number 21 has been found, making three in all, with relative abundances as follows:

Isotope	Ne 20	Ne 21	Ne 22
Abundance, per cent	90.4	0.6	9.0

In a pure carbon monoxide gas all of the molecules are diatomic and alike in every respect except for mass. Since there are two carbon isotopes, 12 and 13, and three oxygen isotopes, 16, 17, and 18 (see Appendix V), there are six different combinations of atoms to form molecules. These are $C^{12}O^{16}$, $C^{12}O^{17}$, $C^{12}O^{18}$, $C^{13}O^{16}$, $C^{13}O^{17}$, and $C^{13}O^{18}$. The relative abundances of all but the C^{12} and O^{16} isotopes are so small, however, that more than 90% of the molecules in a given quantity of gas are of the type $C^{12}O^{16}$, with a mass of 28.

31.6 ISOBARS Another mass spectrograph of remarkably high precision was devised in 1933 by an American physicist, K. T. Bainbridge.

Two photographs taken with this instrument are shown in Fig. 31F. The middle picture (b) shows the many isotopes of tin, and the lower plate (c) the isotopes of mercury and lead. The rare lead isotope 204 falls on top of the strong mercury isotope 204. Such occurrences are called *isobars.*

Atoms having the same mass but belonging to different chemical elements are called isobars. The first pair of isobars (see Appendix V) occurs in argon and calcium. The principal isotope of argon, atomic number 18, has a mass of 40, as does also the principal isotope of calcium, atomic number 20. Other examples are Cr 54 and Fe 54, Ge 76 and Se 76, Rb 87 and Sr 87, Zn 92 and Mo 92. The isobars Hg 204 and Pb 204 are illustrated in Fig. 31F (c).

All isobars and isotopes of all elements are called *nuclides.* In other words, every different nucleus is called a nuclide.

31.7 UNIT MASS AND THE HYDROGEN ATOM Until 1927 all oxygen atoms were thought to have the same mass and were arbitrarily chosen to be the standard by which all atomic masses were measured. At this time Giauque and Johnson discovered the existence of two rare oxygen isotopes with masses 17 and 18. So rare are these heavier particles that in every ten thousand oxygen atoms only sixteen of them have a mass of 18, and only three a mass of 17.

Retaining natural oxygen with its mixture of three isotopes as a standard, the chemists continued to use $\frac{1}{16}$ the weight of this mixture as *unit atomic weight.* The physicists, on the other hand, found it most convenient for their purposes to adopt a mass scale based upon the mass of the most abundant oxygen atom as exactly 16, and $\frac{1}{16}$ of this as *unit atomic mass.*

As isotope mass determinations (measured largely by research physicists) became more and more accurate, and mass differences became more important, the slight differences between the two scales became more and more of a problem.

In 1960 the International Union of Pure and Applied Physics (IUPAP) met in Ottawa and adopted the atomic mass scale based upon the carbon-12 isotope as having a mass of exactly 12. The same scale was adopted by the International Union of Pure and Applied Chemistry (IUPAC), at their Montreal meeting in 1961. While this new atomic weight scale differs ever so slightly from the older O = 16 scale, the accurate masses of isotopes show greater differences, as can be seen in Appendix IV.

Standard unit atomic mass is now generally accepted to be $\frac{1}{12}$ of the mass of the carbon-12 isotope. On this basis very accurate measurements give for the mass of the hydrogen atom 1.0081456 atomic mass units (amu), a value nearly 1% higher than unity.

For many practical purposes it is convenient to know the masses of atoms in kilograms. For easy calculations the following value may be used:

$$\boxed{M = 1.6604 \times 10^{-27}\ \text{Kg}} \qquad (31a)$$

This number multiplied by the "atomic weight" of any atom will give its mass in kilograms.

Compared with the mass of the electron, namely,

$$m = 9.10908 \times 10^{-31}\ \text{Kg} \qquad (31b)$$

an atom of unit mass would be 1824 times as heavy. The hydrogen atom is slightly heavier than one unit mass and is about 1840 times heavier than the electron. This latter number is convenient to remember for it is often quoted to illustrate the enormous difference between the mass of the nucleus of a hydrogen atom and the mass of its one and only electron.

Hydrogen has two isotopes, masses 1 and 2. The heavier isotope, as compared with the more common hydrogen-1, has a natural abundance of 1 part in 4000. Gas containing atoms and molecules of hydrogen-2 only is called *deuterium.* The nucleus of a deuterium atom is called a *deuteron.*

Atomic number Z is defined as that number ascribed to an element specifying its position in the periodic table of elements. See column 1, Appendix V.

Mass number A is defined as that whole number nearest the actual mass of an isotope measured in atomic mass units. See column 4, Appendix V.

Atomic weight is defined as the average weight of all the isotopes of an element weighted according to relative abundance and expressed in atomic mass units.

Unit atomic mass, or atomic mass unit, is defined as one twelfth the mass of a carbon-12 atom.

QUESTIONS

1. What are canal rays? Who discovered them? How are they detected?

2. What is a mass spectrograph? Make a diagram of one and explain how it operates.

3. Make a diagram of Thomson's mass spectrograph. What is it used for, and how are the results interpreted?

4. Who discovered isotopes? What are isotopes?

5. Define or briefly explain each of the following: (a) positive rays, (b) atomic number, (c) isotopes, (d) molecular weight, and (e) ionization.

6. Define or briefly explain each of the following: (a) canal rays, (b) isobars, (c) atomic weight, (d) mass number, and (e) positive ion.

7. Name the first ten elements of the periodic table in the order of their atomic number.

8. Name the last ten elements of the periodic table in the order of their atomic number.

9. What is meant by relative abundance?

10. What are isobars? Give an example.

11. How much greater is the mass of a hydrogen atom than the mass of an electron?

12. What do the more or less equal spacings of the mass spectrograms shown in Fig. 31F suggest regarding the relative masses of atoms?

13. Name five members of each of the following classifications: (a) alkali metals and (b) alkaline earths. See Appendix II.

14. What chemical element has the greatest number of isotopes? See Appendix III.

15. Make a list of elements having (a) atoms of one mass only and (b) only two isotopes. See Appendix III.

PROBLEMS

1. If the atomic weight of copper is 63.54 how many atoms are there in one kilogram of copper metal? [Ans. 9.48×10^{24} atoms.]

2. The atomic weight of zinc is 65.37. How many atoms are there in one kilogram of zinc metal?

3. If the atomic weights of silicon and oxygen are 28.09 and 16.00, respectively, find the mass in kilograms of a silicon dioxide molecule. [Ans. 9.97×10^{-26} Kg.]

4. The atomic weights of hydrogen, carbon, and oxygen are 1.01, 12.00, and 16.00, respectively. How many methyl alcohol molecules are there in one kilogram of the liquid? See Fig. 10A.

5. Calculate the mass, in Kg, of a benzine molecule. The atomic weights of the individual atoms are carbon 12.01, and hydrogen 1.0078. [Ans. 1.297×10^{-25} Kg.]

6. Find the mass of a single atom of (a) mercury, (b) radium, and (c) plutonium.

7. Determine the mass of a single atom of (a) lithium, (b) beryllium, and (c) aluminum. See Appendix IV. [Ans. (a) 1.152×10^{-26} Kg, (b) 1.496×10^{-26} Kg, (c) 4.48×10^{-26} Kg.]

8. Find the number of molecules in a mass of one kilogram of water.

9. Determine the mass of a hydrochloric acid molecule. The atomic masses are hydrogen 1.0078, and chlorine 34.969. [Ans. 5.97×10^{-26} Kg.]

10. Calculate the mass of an acetylene molecule made up of H^1 and C^{12} molecules. See Fig. 10A. Assume the masses of single atoms are 1.0078 and 12.0000, respectively.

11. Find the mass of an ethyl alcohol molecule if it is made up of C^{12}, H^1, and O^{16} atoms. See Fig. 10A. For atomic masses, see Appendix IV. [Ans. 7.64×10^{-26} Kg.]

12. The atomic weight of barium is 137.34. Find the number of barium atoms in one kilogram of the metal.

32 ATOMIC SPECTRA AND ATOMIC STRUCTURE

When light from any source is sent through a prism, the different colors of which the light is composed are spread out into a spectrum as shown in Figs. 22E and 32C. Such bands of colors constitute wavelength displays, each with the shortest of the violet waves on the left end and the longest of the red waves on the right.

The wavelengths of light emitted by hot solids, such as the tungsten wire in an incandescent lamp, or by hot gases, such as the mercury atoms in a fluorescent lamp, depend upon the nature of the emitting substance, as well as its temperature. While the distribution of wavelengths from a hot gas differs widely between chemical elements, the radiation from different hot solids like the metals, raised to the same high temperature, is nearly the same.

Through detailed studies of the wavelength spectrum of different light sources, physicists have been able to learn a great deal about the detailed structure of most of the atoms of the periodic table of elements.

32.1 CLASSIFICATION OF SPECTRA *A spectrum may be defined as an array of the different colors, or wavelengths, of radiant energy emitted by a light source.*

All spectra may be grouped into four classes:

(*a*) *Continuous emission spectra*
(*b*) *Line emission spectra*
(*c*) *Continuous absorption spectra*
(*d*) *Line absorption spectra*

32.2 CONTINUOUS EMISSION SPECTRA When a block of metal like iron or copper is heated slowly to incandescence, the first noticeable change in its appearance occurs at a temperature of about 1000°K. At this temperature the metal has a dull red glow. As the temperature continues to rise, the color changes slowly to orange, then to yellow, and finally to white.

If the slowly heating metal is observed through a prism, the first appearance of visible light will be found at the extreme red end of the spectrum. As the temperature rises the light spreads slowly out across the spectrum until, at white heat, the entire band of visible colors from red to violet is seen. At the orange stage where the temperature is about 1500°K the pure spectrum colors contain red, orange, and yellow, and when the yellow stage is reached where the temperature is about 2000°K the spectral green is included. When the white stage is reached at about 3000°K, and the

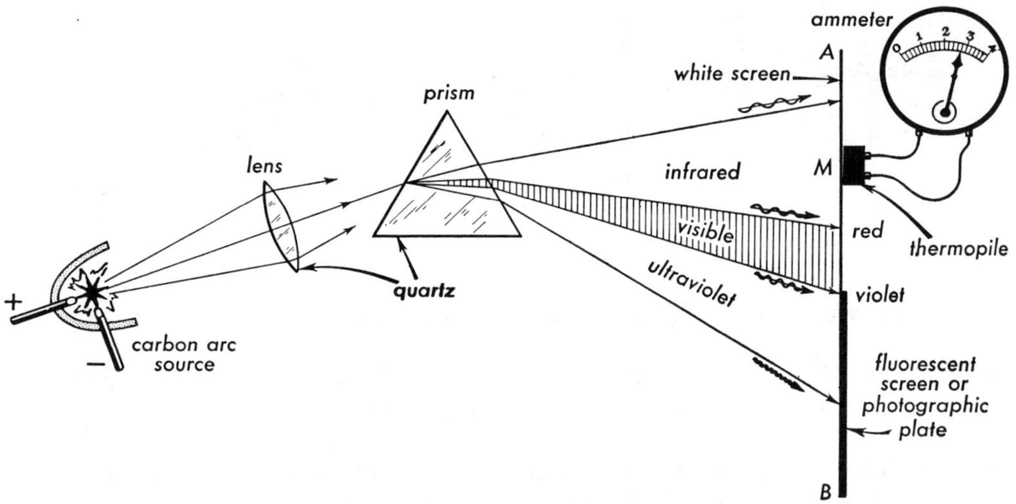

Fig. 32A. Experiment demonstrating the existence of the ultraviolet and infrared rays beyond the visible spectrum.

spectrum is complete, a further rise in temperature continues to increase the intensity of each color without a noticeable change in color.

What the prism has done in such an experiment is to separate all of the light waves according to their wavelengths, the longest waves of red light at the one side, the shortest waves of violet light at the other, and the intermediate waves at their proper places in between. See Fig. 24G. The fact that the color is continuous from red through violet is characteristic of the spectrum of all solids and liquids and means that there is a continuous set of different wave lengths present.

To demonstrate the existence of an ultraviolet and infrared spectrum, an experiment of the type illustrated in Fig. 32A may be performed. The visible light from a carbon arc lamp is made to pass through a quartz lens and prism to be focused on a nearby screen.

If at the violet end of the spectrum the screen is painted with luminous paint, a bright fluorescence will be observed for some little distance beyond the visible violet. When the screen is replaced by a photographic plate, the exposed and developed picture will again show the extension of the spectrum into the ultraviolet.

To detect the presence of the infrared radiations, a thermopile is conveniently used as shown at the top of the screen. Connected to an ammeter a thermopile measures the amount of light energy falling upon its front surface. If the thermopile is first placed to receive violet light and slowly moved across the visible spectrum and out into the infrared region beyond, the ammeter will show a steady rise in current. The current will continue to rise until a maximum is reached at a point in the region of M and then it will drop off slowly as the thermopile approaches the end of the screen at A.

If a graph of the radiated energy is plotted against the wavelength of the light, a curve similar to that shown in Fig. 32B is obtained. If the

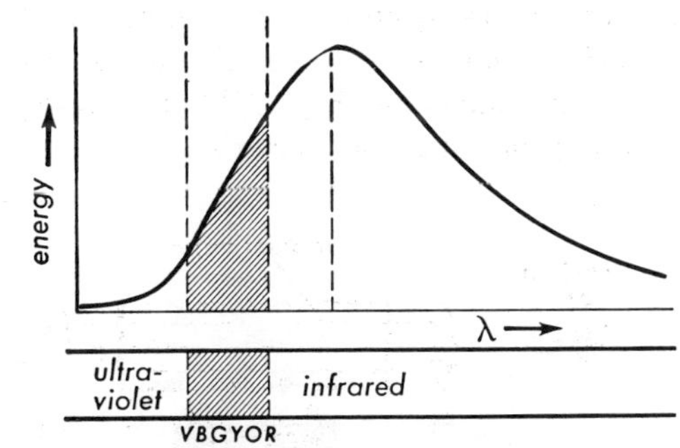

Fig. 32B. Graph of the energy emitted by a hot solid.

source temperature is increased this curve will rise at all points but the peak will shift to shorter wavelengths. When the sun's surface temperature

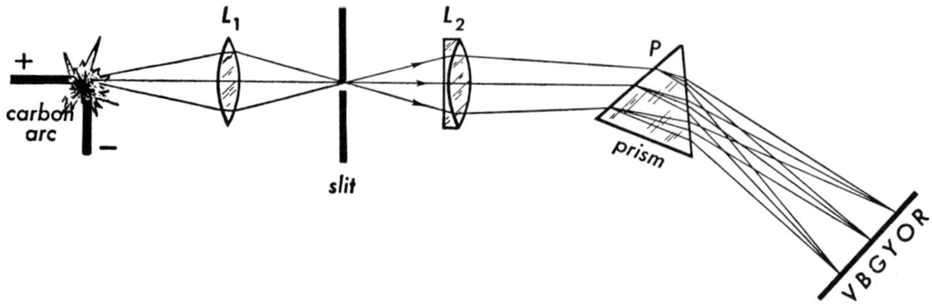

Fig. 32D. Experimental arrangement used in demonstrating spectrum lines in emission.

of 6000°K is reached the peak is located in the yellow green part of the visible spectrum.

32.3 LINE EMISSION SPECTRA When the slit of a prism spectrograph (see Fig. 32D) is illuminated by the light from a mercury arc, a sodium lamp, or a neon discharge tube, a number of bright lines appear on the screen or photographic plate, in place of a continuous spectrum. See Fig. 32C.

It is important to realize that *line spectra* derive their name from the fact that, before entering the prism, the light passes through a narrow slit, and it is the different colored images of this slit, formed by the lens, that constitute the lines. If a small circular opening were used in place of a slit a disk image would appear in the place of each line in Fig. 32C.

The most intense sources of spectrum lines are obtained from metallic arcs and sparks. The flame of a carbon arc may be used for demonstration purposes by previously soaking the *positive carbon rod* in various chemicals. (The arrangement shown in Fig. 32D may also be used for projecting the spectrum on a large screen.) Common salt water (sodium chloride in solution) gives a brilliant yellow line characteristic of sodium. Solutions of strontium or calcium chloride will show other strong spectrum lines in the red, green, and blue.

While a continuous emission spectrum arises from hot solids, *a line spectrum always arises from a gas at high temperatures*. It is the gas flame of the carbon arc that gives rise to the line emission spectrum, in the above experiment.

32.4 CONTINUOUS ABSORPTION SPECTRA Continuous absorption spectra are usually produced by passing a continuous emission spectrum through matter in the solid or liquid state. Good demonstrations can be performed by allowing white light to pass through colored glass. When the light is later dispersed by a prism, the missing colors will in general cover a wide band of wavelengths. See spectrum (b) in Fig. 32C. The same experiment was described in Chap. 22, and is illustrated for this specific example in diagram (g) of Fig. 22H.

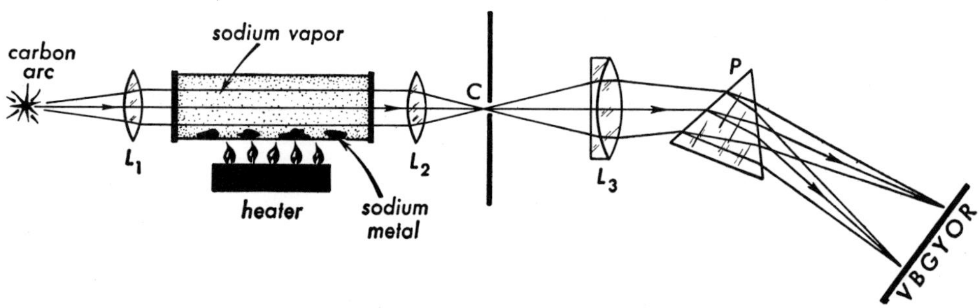

32E. Experimental arrangement for demonstrating the line absorption spectrum of sodium vapor.

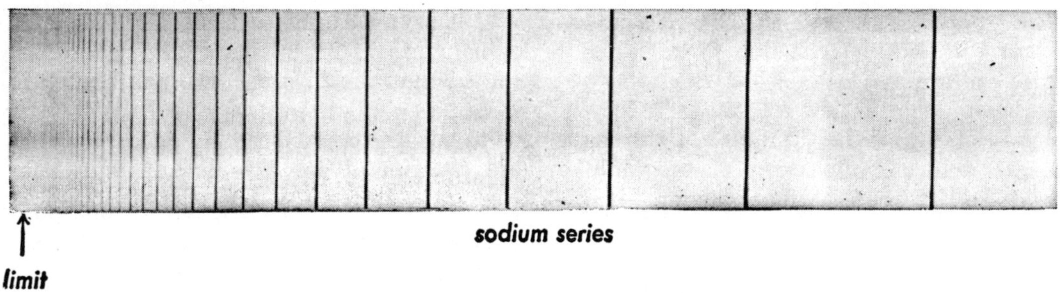

Fig. 32F. Absorption spectrum of sodium vapor. The principal series of sodium. (After Jenkins.)

32.5 LINE ABSORPTION SPECTRA Line spectra in absorption are produced by sending continuous white light through a gas. Experimentally the gas or vapor is inserted in the path of the light as shown in Fig. 32E. Light from a carbon arc, after passing as a parallel beam through a glass tube containing sodium vapor, is brought to a focus at the slit *C*. From there the light passes through a lens L_3 and a prism *P* to form a spectrum on the observing screen.

Sodium is chosen as an example for demonstration purposes because of its convenience. The vapor is produced by inserting a small amount of metallic sodium in a partially evacuated glass tube and heating it with a small gas burner. As the metal vaporizes, filling the tube with sodium vapor, a dark line will appear in the yellow region of the spectrum. See color plate (d) in Fig. 32C.

If a photograph is taken of this absorption, and the photographic plate is long enough to extend into the ultraviolet, many absorption lines as shown in Fig. 32F are detected. A systematic array of absorption lines like this occurs only with a few elements, principally with the alkali metals, lithium, sodium, potassium, rubidium, and caesium. All elements in the gaseous state, however, give rise to a number of absorption lines, usually in the ultraviolet region of the spectrum.

32.6 THE THOMSON ATOM Early in the twentieth century, while Rutherford, Geiger, and Marsden were doing their experiments on the scattering of α rays by thin metallic films, J. J. Thomson proposed a type of electron shell structure for all atoms. His model structures were worked out by mathematics from Coulomb's law for charged particles and soon became known as the *plum-pudding atom.*

Thomson visualized all the positive charge of an atom as being spread out uniformly throughout a sphere about 10^{-8} cm in diameter, with the electrons as smaller particles distributed in shells somewhat as shown in Fig. 32G. While the net

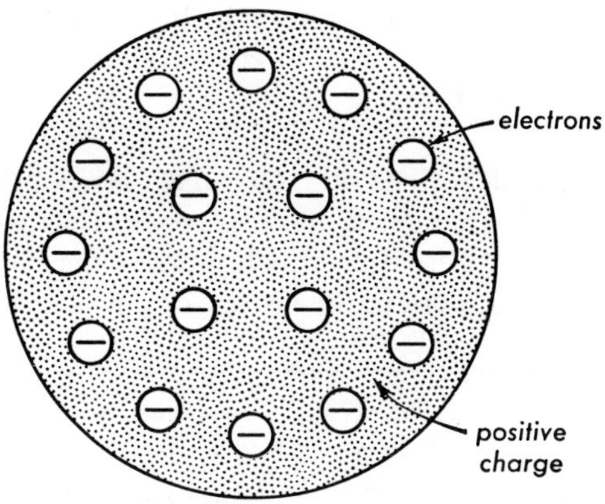

Fig. 32G. Diagram of the Thomson atom model.

force exerted by the positively charged sphere on each electron is toward the center of the sphere, the electrons mutually repel each other and form rings.

An excellent demonstration of the tendency to form rings can be obtained as shown in Fig. 32H. A glass dish 15 to 20 cm in diameter is wound with about 30 turns of No. 14, insulated copper wire. The most common steel sewing needles are then mounted in small corks, 8 mm diam. and 8 mm long, as shown at the right, and the needles magnetized by stroking from top to bottom with the N pole of a strong Alnico magnet.

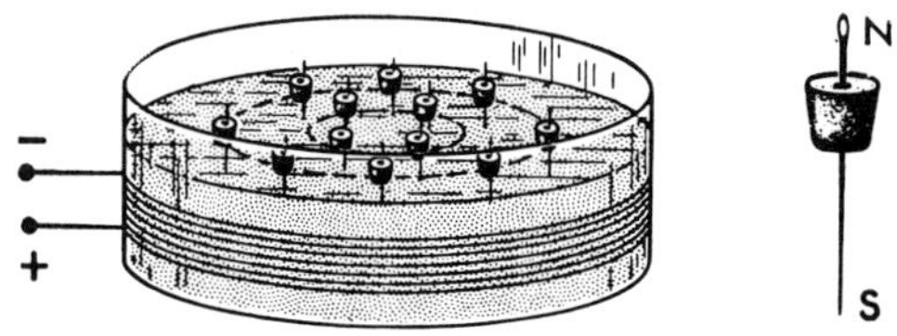

Fig. 32H. Floating needles in magnetic field demonstrating electron shell structure of Thomson atom.

With water in the dish, and a current of 1 to 2 amps through the coil, a single needle is placed upright in the water. Released, it will migrate to the center where the magnetic field is strongest. The addition of needles, one after another, near the edge of the dish will result in the formation of geometrically symmetrical patterns and rings.

An increase or decrease in current will cause any given pattern to shrink or expand, corresponding to a greater or lesser positive charge. The stability of such ring patterns undoubtedly influenced the later extension by Bohr and Stoner of the quantized orbit model of the hydrogen atom to all atoms.

32.7 BOHR'S THEORY OF THE HYDROGEN ATOM In 1913 Niels Bohr* proposed a theory of the hydrogen atom which marked the beginning of a new era in the history of physics. With his theory Bohr not only gave a satisfactory explanation of the Balmer series of hydrogen but a model for the structure of all other atoms as well. See the bottom color plate in Fig. 32C.

Starting with what should be the simplest of all atoms, Bohr assumed that a hydrogen atom, $Z = 1$, consists of a nucleus with one positive charge $+e$, and a single electron of charge $-e$ revolving around it in a circular orbit of radius r. See Fig. 32I. Because it is 1840 times heavier than the electron, the nucleus could be assumed at rest.

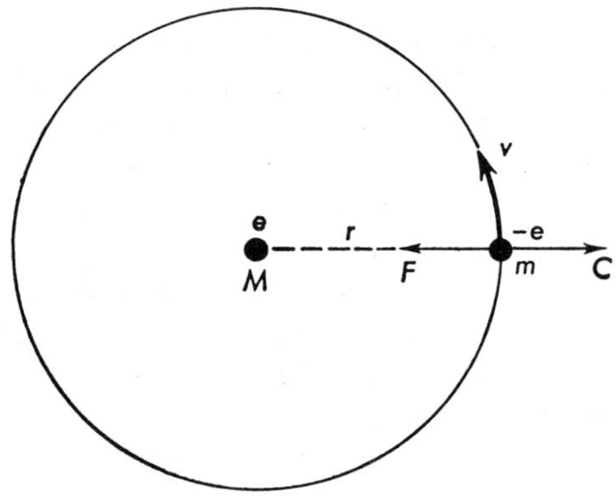

Fig. 32I. Orbital diagram of the hydrogen atom according to the Bohr theory.

To keep the electron in its orbit and prevent it from spiraling in toward the nucleus, or away from it to escape, Bohr next assumed that the inward centripetal force E is due to and, therefor, equal to, the inward electrostatic force E. From Eq. (8d) the centripetal force is mv^2/r, and from Eq. (26a), Coulomb's law, the electrostatic force is kee/r^2. Equating these two,

$$\boxed{m\frac{v^2}{r} = k\frac{ee}{r^2}} \qquad \text{(32a)}$$

At this point Bohr introduced his second assumption, *the quantum hypothesis*. The electron, he assumed, cannot move in any sized orbit, stable under the conditions of the equation above, but in just certain *definite and discrete orbits*. The sizes of these orbits, he said, are governed by Eq. (32a) and the rule that *the*

* NIELS BOHR (1885–1962), Danish physicist, was born at Copenhagen, the son of Christian Bohr, professor of physiology at the University of Copenhagen. After taking his Ph.D. degree at Copenhagen in 1911 he studied for one year under J. J. Thomson at Cambridge, and one year under Ernest Rutherford at Manchester. Returning to Copenhagen in 1913, with the results of the Rutherford scattering experiments fresh in his mind, he worked out and published his now famous theory of the hydrogen atom. In 1920 Bohr was appointed head of the institute for theoretical physics at the University of Copenhagen. In 1921 he was awarded the Hughes Medal of the Royal Society and in 1922 the Nobel Prize in physics. In 1937 he proposed the waterdrop theory of the atomic nucleus. The author had the good fortune of meeting with Bohr on several occasions. In his lifetime he was the most honored scientist in Denmark.

angular momentum mvr is equal to an integer n times a constant h divided by 2π.

$$mvr = n \frac{h}{2\pi} \qquad (32b)$$

The constant h is called Planck's constant, and its value is

$$h = 6.62559 \times 10^{-34} \text{ joule seconds}$$

In this equation n is called the *principal quantum number* and, because it can take only whole number values 1, 2, 3, 4, etc., it fixes the sizes of the allowed orbits. Since n is always a whole number $h/2\pi$ represents a unit of angular momentum. This quantity is important since its value is basic to all knowledge of atomic structure.

To find the radius of these Bohr circular orbits, Eq. (32b) is solved for v, then squared and substituted in Eq. (32a) to give

$$r = n^2 \frac{h^2}{4\pi^2 me^2 k} \qquad (32c)$$

Putting into this equation the known values of the constants e, m, h, and k,

$$e = -1.60210 \times 10^{-19} \text{ coulomb}$$
$$m = 9.10908 \times 10^{-31} \text{ Kg}$$
$$h = 6.62559 \times 10^{-34} \text{ joule seconds}$$
$$k = 8.98755 \times 10^{9} \text{ newton meters}^2/\text{coulomb}^2$$

we obtain for the orbits in Fig. 32J,

$$r_n = n^2 \times 0.529 \times 10^{-10} \text{ m} \qquad (32d)$$

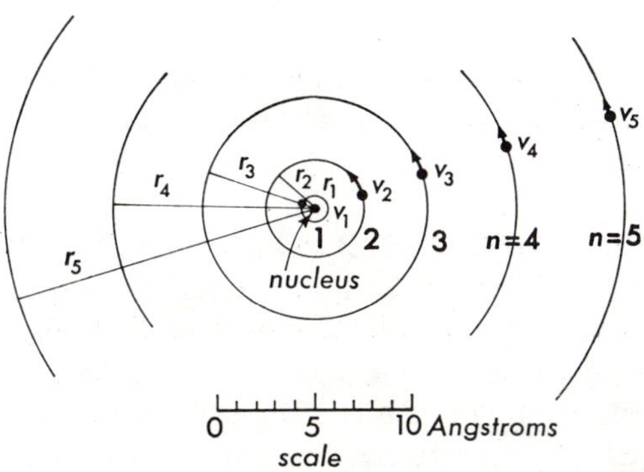

Fig. 32J. Scale diagram of the Bohr circular orbits of hydrogen.

The innermost orbit, with $n = 1$, has a radius

$$r_1 = 0.529 \times 10^{-10} \text{ m}$$

which is equal to

$$r_1 = 0.529 \text{ Å}$$

where

$$1 \text{ A} = 10^{-10} \text{ m}$$

The second orbit is 4 times larger, the third orbit 9 times larger, etc.

The speed of the electron in any orbit can be obtained as follows. By substituting the value of r from Eq. (32c) in Eq. (32b), and solving for the speed v, we obtain

$$v_n = \frac{1}{n} \cdot \frac{2\pi e^2}{h} k \qquad (32e)$$

Substituting the values of the atomic constants e, h, and k, we obtain

$$v_n = \frac{1}{n} \times 2.18764 \times 10^6 \frac{\text{m}}{\text{s}} \qquad (32f)$$

In the innermost orbit, $n = 1$, the speed becomes

$$v_1 = 2.188 \times 10^6 \frac{\text{m}}{\text{s}}$$

which is equal to $\frac{1}{137}$ the speed of light.

With such small orbits and high speeds, the number of revolutions per second becomes very high. Since the circumference of any circular orbit is $2\pi r$, the frequency with which the electron goes around each orbit is given by

$$f = \frac{v}{2\pi r} \qquad (32g)$$

EXAMPLE. 1. Calculate (a) the radius, (b) the speed, and (c) the frequency of the electron in the second Bohr circular orbit.

SOLUTION. (a) The radius of orbit $n = 2$ is obtained by direct substitution in Eq. (32d).

$$r_2 = 4 \times 0.529 \times 10^{-10} \text{ m}$$
$$r_2 = 2.117 \text{ Å}$$

(b) The speed is obtained by substituting in Eq. (32f).

$$v_2 = \frac{1}{2} \times 2.188 \times 10^6 \frac{\text{m}}{\text{s}}$$

$$v_2 = 1.094 \times 10^6 \frac{\text{m}}{\text{s}}$$

(c) The frequency is found by using Eq. (32g).

$$f = \frac{1.094 \times 10^6}{6.283 \times 2.117 \times 10^{-10}}$$

$$f = 8.22 \times 10^{14} \text{ Hz}$$

This value of the orbital frequency is very nearly the same as the frequency of visible light. Violet light for example, has a frequency of 7.5×10^{14} Hz.

It should be noted that the one and only electron in each hydrogen atom can occupy only one orbit at any one time. If the electron changes its orbit, it must move to one of the allowed orbits and never stop in between.

32.8 BOHR'S ELECTRON-JUMPS Bohr's third and final assumption regarding the hydrogen atom concerns the emission of light. Bohr postulated that light is not emitted by an electron when it is moving in one of its fixed orbits, but only when the electron jumps from one orbit to another, as illustrated in Fig. 32K. The frequency of this

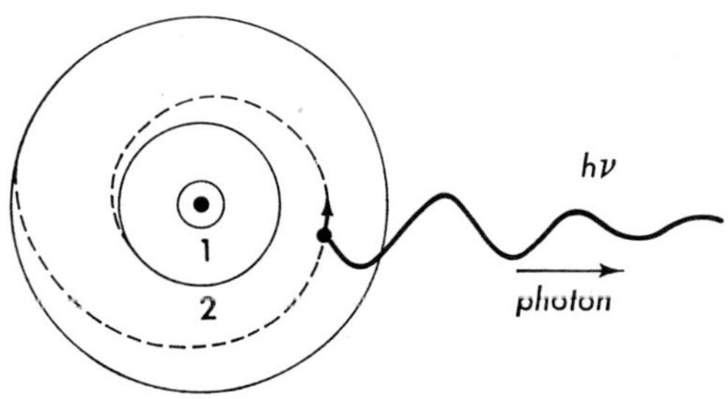

Fig. 32K. Schematic diagram of Bohr's quantum hypothesis of the radiation of light from an hydrogen atom.

light, he said, is not determined by the frequency of revolution but by the difference in energy between the initial and final orbit:

$$\boxed{E' - E'' = h\nu} \qquad (32h)$$

where E' is the energy of the *initial orbit*, E'' the energy of the *final orbit*, h is Planck's constant, and ν is the frequency of the light.

To illustrate this, let E_1, E_2, E_3, E_4, etc., represent the total energy of the electron when it is in the orbits $n = 1, 2, 3, 4$, etc., respectively. When, for example, the electron is in orbit $n = 3$ where its energy is E_3, and it jumps to orbit $n = 2$ where the energy is E_2, the energy difference $E_3 - E_2$ is ejected from the atom in the form of a light wave of energy $h\nu$ called a *photon*. Here then is the origin of light waves from within the atom.

The success of Bohr's theory is not to be attributed as much to the mechanical picture or model of the atom just proposed as to the development of an equation which agrees exactly with experimental observations.

32.9 NORMAL AND EXCITED ATOMS When the single electron of an hydrogen atom is in the innermost orbit, $n = 1$, the atom is said to be in its *normal state*. As the name implies, this is the condition of most free hydrogen atoms in a gas under normal room temperature and pressure. If an electrical discharge is sent through a vessel containing hydrogen gas, cathode rays (electrons) moving at high speed make frequent collisions with electrons, knocking some of them out of the atom completely and some of them into one of the outer allowed orbits, $n = 2, 3, 4$, etc.

When the electron is completely removed from the atom, the atom is said to be ionized; whereas, when it is forced into an outer orbit, the atom is said to be excited. Once in an excited state an atom will not remain that way long, for the electron, under the attraction of the nucleus, will jump to an inner orbit. By jumping to an inner orbit the electron loses all or part of the energy it had gained.

When an electron is excited it does not necessarily return to the innermost orbit by a single jump but may return by several different jumps, thereby emitting several different light waves, or quanta.

32.10 BOHR-STONER SCHEME OF THE BUILDING UP OF ATOMS Bohr and Stoner proposed an extension of the orbital model of hydrogen to include all of the chemical elements. As shown by the examples in Fig. 32L, each atom is composed of a positively charged nucleus with a number of electrons around it.

Although the nucleus is a relatively small

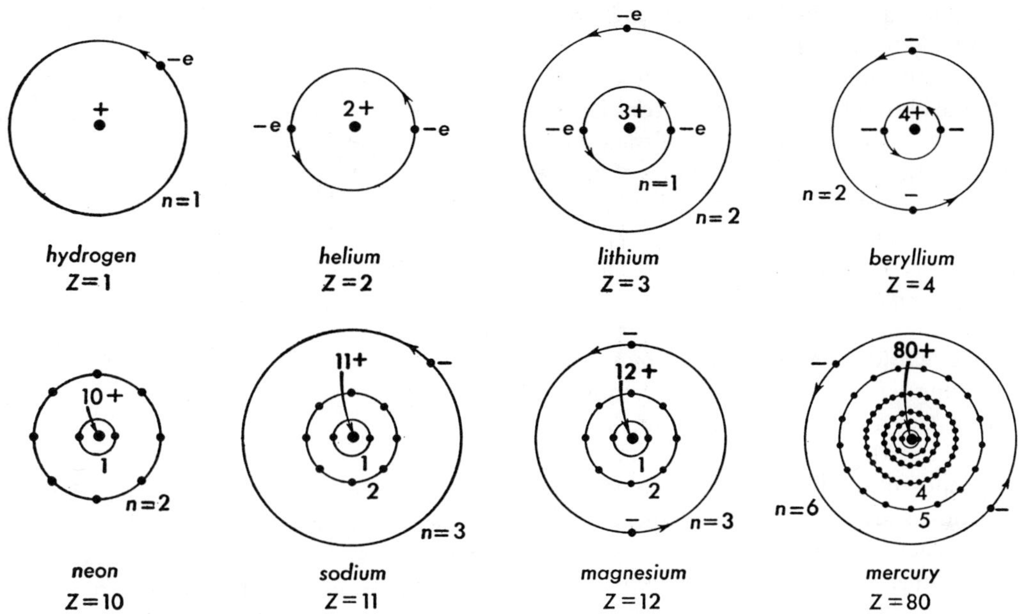

Fig. 32L. Bohr-Stoner orbital models for the light and heavy atoms of the periodic table.

particle less than 10^{-12} cm in diameter it contains almost the entire mass of the atom, a mass equal in *atomic mass units* to the *atomic weight. The positive charge carried by the nucleus is equal numerically to the atomic number, and it determines the number of electrons located in orbits outside*. A helium atom, atomic number $Z = 2$, has two positive charges on the nucleus and two electrons outside. A lithium atom, atomic number $Z = 3$, contains three positive charges on the nucleus and three electrons outside. A mercury atom, atomic number 80, contains 80 positive charges on the nucleus and 80 electrons outside.

The orbits to which the electrons are confined are the Bohr orbits of hydrogen with $n = 1, 2, 3$, etc., and are called electron shells. Going from element to element in the atomic table, starting with hydrogen, electrons are added one after the other, filling one shell and then another. A shell is filled only when it contains a number of electrons given by $2n^2$. To illustrate this, the first shell $n = 1$ is filled when it has 2 electrons, the second shell $n = 2$ when it has 8 electrons, the third shell $n = 3$ when it has 18 electrons, etc. $2 \times 1^2 = 2$, $2 \times 2^2 = 8$, $2 \times 3^2 = 18$, etc.

Quantum number	$n = 1$	$n = 2$
No. of electrons	**2**	**8**

$n = 3$	$n = 4$	$n = 5$	$n = 6$
18	**32**	**50**	**72**

Among the heavier elements there are several departures from the order in which the shells are filled. Although these departures are not important from the present standpoint, their nature is illustrated by the mercury atom, Fig. 32L. The four inner shells $n = 1, 2, 3$, and 4 are entirely filled with 2, 8, 18, and 32 electrons, respectively, while the 5th shell contains only 18 electrons and the 6th shell 2 electrons. The reasons for such departures are now well understood and are indicative of the chemical behavior of the heavy elements.

It is important to note that, as the nuclear charge increases and additional electrons are added in outer shells, the inner shells, under the stronger attraction by the nucleus, shrink in size. The net result of this shrinkage is that the heaviest elements in the periodic table are not much larger in diameter than the lighter elements.

The schematic diagrams in Fig. 32L are drawn approximately to the same scale.

The experimental evidence of these upper limits to the allowed number of electrons in each shell is now considered direct confirmation of one of the most fundamental principles of nature. A sound theoretical explanation of this principle of atomic structure was first given by W. Pauli† in 1925 and is commonly referred to as the *Pauli exclusion principle*.*

32.11 THE SODIUM ATOM Guided by the simple appearing series of spectrum lines known to arise from the alkali metals, lithium, sodium, potassium, rubidium, and caesium, Bohr and Stoner proposed that the filling of each new shell of electrons starts with these elements. This means that each alkali atom has one electron outside all the others and in an orbit by itself. This is illustrated for sodium, atomic number $Z = 11$, in Fig. 32L, and again in Fig. 32M.

In all normal sodium atoms the outer electron, called the *valence electron*, is located in orbit $n = 3$. When an electric discharge is sent through sodium vapor many of the atoms become excited and subsequently emit light waves in a manner similar to the hydrogen atom illustrated in Fig. 32K. In each excited atom the valence electron is in one of the outer virtual orbits $n = 4, 5, 6$, etc. Jumping back to the normal state $n = 3$ a photon of frequency ν is emitted.

* For a more detailed treatment of atomic structure see, "Introduction to Atomic Spectra," by H. E. White, McGraw-Hill Book Co., Inc.

† WOLFGANG PAULI (1900–1958), Austrian physicist, was awarded the Nobel Prize in Physics for the discovery of the exclusion principle, which carries his name. He received his early education in Vienna, took his Ph.D. degree under Arnold Sommerfeld at Munich in 1921, and served as assistant to Max Born at the University of Göttingen. He served as professor of theoretical physics at the Federal Institute of Technology in Zurich, Switzerland, and in 1931, visiting professor at the Institute for Advanced Study at Princeton, N. J., University of Michigan in 1931 and 1941, and Purdue University in 1942. In 1946, he became a naturalized U. S. citizen. Pauli was the first to propose the existence of the neutrino, a small weightless, chargeless atomic particle that travels with the speed of light.

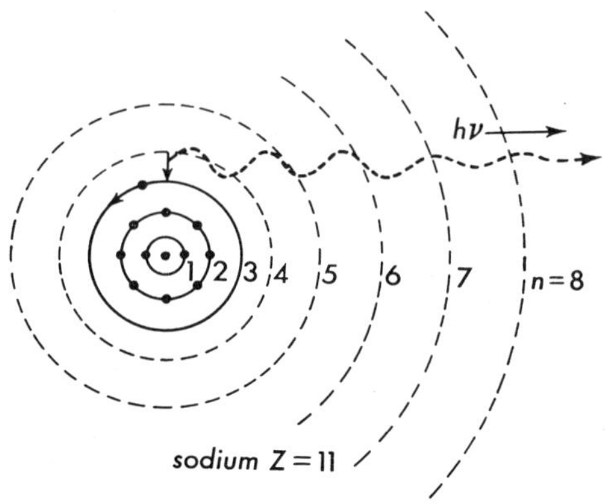

Fig. 32M. Schematic diagram of a sodium atom showing the virtual orbits.

The energy content of each photon is governed by the quantum theory expressed by Eq. (32d). A series of jumps from outer orbits all ending on the normal state $n = 3$ give rise to a series of spectrum lines as shown in Fig. 32F. A jump from $n = 4$ to $n = 3$ gives rise to the first line of the series; a jump from $n = 5$ to $n = 3$, the second line of the series, etc.

The absorption of light by sodium atoms is the reverse process of emission. When a light wave or photon of just the correct frequency approaches a normal sodium atom it will be absorbed by the atom and the energy used to raise the outer electron from its normal orbit $n = 3$ to some outer orbit. This is a process of excitation of atoms by the absorption of light. In returning to the normal state the electron may emit the light again.

32.12 SPINNING ATOMIC PARTICLES Careful studies of the spectrum lines arising from various kinds of atoms led to the proposal in 1925, by the two Dutch-American physicists Goudsmit‡

‡ SAMUAL A. GOUDSMIT (1902–), American physicist, was born in The Hague, Holland. He obtained

and Uhlenbeck,* that all electrons are spinning like a top. Now it is known that both electrons and protons spin about an axis, much the same as the earth spins about its polar axis, and that each one has exactly one-half a unit of angular momentum, $h/2\pi$.

$$\text{angular momentum} = \frac{1}{2}\frac{h}{2\pi} \tag{32i}$$

A schematic diagram of the spinning proton and electron of the hydrogen atom is shown in Fig. 32N.

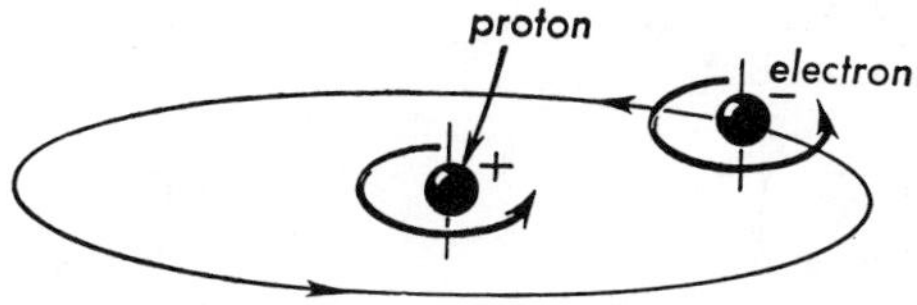

Fig. 32N. The electron and proton both spin around an axis through their center.

32.13 DE BROGLIE'S ELECTRON-WAVES In 1924 de Broglie,† a French theoretical physicist, derived an equation predicting that all atomic particles have associated with them waves of a definite wavelength. In other words, a beam of electrons or atoms should, under the proper experimental conditions, act like a train of light waves or a beam of photons. The wavelength of these waves, as predicted by de Broglie, depends upon the mass velocity of the particles according to the following relation:

$$\lambda = \frac{h}{mv} \tag{32j}$$

This is known as *de Broglie's wave equation.* For an electron moving at high speed the denominator mv is large and the wavelength is small. In other words, the faster an electron moves, the shorter is the wavelength associated with it.

To acquire some notion of the relative wavelengths of electrons moving with different velocities, several values have been computed from de Broglie's equation and given in Table 32A. The velocities are listed in miles per second in column 2 and in per cent of the velocity of light in column 3. The potentials listed in column 1 are the voltages required by Eq. (30c) to give an electron any one of the velocities listed in columns 2 and 3 (see Fig. 30I). It will be noted that the wavelengths at the bottom correspond closely to those for X rays and γ rays, to be taken up in succeeding chapters.

32.14 ELECTRON WAVES WITHIN THE ATOM The most recent development in the theory of atomic structure has shown that the Bohr picture

---

his Ph.D. in physics at Leiden in 1927, and was married the same year. He became an Instructor in Physics at the University of Michigan in 1927, and a United States citizen not long afterward. He served as professor from 1932–1946, and joined the Brookhaven National Laboratory in 1948. He has received the Max Planck Award and is best known for his prediction of electron spin and his work on atomic spectra. The author is happy to have known Goudsmit for many years.

* GEORGE E. UHLENBECK (1900–), American physicist, was born in Batavia, Java. He received his Ph.D. at Leiden, Holland, in 1927. He became Professor of Physics at the University of Michigan in 1930, and assumed the Henry S. Carhart Professorship from 1939 to 1960. He has received many honorary awards and is best known for his work with S. A. Goudsmit on the spinning electron.

† LOUIS V. P. R. DUC DE BROGLIE (1892–), French physicist, was born in Dieppe, France. After being educated at the Sorbonne, Paris, he served as Instructor and later as Professor at the same institution from 1926–62. He received many national and international awards and was elected to membership in many science societies. He has written a number of books, particularly on microphysics, and is best known for his theory that all moving objects have wave properties associated with them. For this now proven theory, which laid the ground work for wave mechanics, he was awarded the Nobel Prize in physics for 1929.

TABLE 32A. Wavelengths Associated with Electrons Moving with Different Velocities According to de Broglie's Wave Equation*

V (Applied Voltage)	v (velocity in mi/sec)	$\frac{v}{c}$ (velocity in per cent)	λ (wave length in angstroms)
1	370	0.20	12.23
10	1100	0.62	3.87
100	3700	1.98	1.22
1000	18,000	6.26	0.38
10,000	36,000	19.50*	0.12
100,000	100,000	54.80*	0.03
1,000,000	175,000	94.10*	0.01

* These values take into account the increase in mass of the electron due to the theory of relativity.

of the atom with sharply defined electron orbits is not correct. The new theory does not discard the Bohr theory entirely but only modifies it to the extent that the electron does not behave as though it were a particle. The electron behaves as if it were made up of waves (sometimes called de Broglie waves) of the type described in the previous sections.

The new theory of the hydrogen atom was worked out independently by the two German theoretical physicists, W. Heisenberg* and E. Schrödinger,† in 1925 and was later modified and improved by the English theoretical physicist, P. Dirac,‡ in 1928. Schrödinger, making use of de Broglie's idea of electron waves, pictures the single electron in the hydrogen atom as moving around the nucleus as a sort of *wave packet*. This wave packet, as it is called, is formed in

* WERNER KARL HEISENBERG (1901–), German physicist, was awarded the Nobel Prize in Physics in 1932, for "the creation of quantum mechanics," and is best known for his Uncertainty Principle. He was born at Würtzburg, where his father was a professor. He studied under Arnold Sommerfeld at Munich, where he obtained his doctorate in 1923. He studied under Max Born at Göttingen in 1924, and from 1927 to 1941 he was Professor of Theoretical Physics at Leipzig. He became Director of the Max Planck Institute for Physics at Berlin, and later in Göttingen. Heisenberg's work on quantum theory profoundly influenced the development of atomic physics.

† ERWIN SCHRÖDINGER (1887–1961), Austrian physicist, was awarded the Nobel Prize in physics in 1933 for his work on wave mechanics and its application to atomic structure. He was educated at the University of Vienna and later served as Professor of Physics in Stuttgart, Breslau, and Zurich. He succeeded Max Planck as Professor of Physics at the University of Berlin in 1927. It was there that the author attended his lectures on wave mechanics in 1929. His famous wave equation, which carries his name, applied the de Broglie matter waves to the electron of the hydrogren atom by assuming that the electron waves are an exact multiple of the orbital length. These then led to a form of spherical harmonies with nodes and antinodes for the electron cloud around the nucleus.

‡ PAUL ADRIEN MAURICE DIRAC (1902–), British mathematical physicist, received the Nobel Prize in Physics in 1933 at the age of 31 for his pioneer work in the quantum mechanics of the atom. He developed the transformation theory of quantum mechanics and was the co-discoverer of the Fermi-Dirac statistics. He replaced the conventional single second-order Schrödinger wave equation by four first-order equations, the solutions of which solved various properties of the electron spin and anomalous magnetic moment. These equations suggested the possible existence of the positron, a particle later discovered by C. D. Anderson.

somewhat the same way that standing waves are set up and maintained in sound waves.

We have seen in Sec. 18.8 and Fig. 18L how a stretched string may be set vibrating with a number of different natural modes. The frequencies ν of these different modes are called *harmonics* and are given by

$$\nu = n\nu_1 \tag{32k}$$

where ν_1 is the frequency of the *first harmonic*, or *fundamental*, and

$$n = 1, 2, 3, 4, 5, \ldots \tag{32l}$$

To set up similar conditions in a Bohr circular orbit of hydrogen, we visualize the electron as a wave with a de Broglie wavelength long enough to extend more than once around the atom. With a complete overlap of waves the system should stabilize into one of standing waves. A little reasoning applied to such a system will show that each wave pattern should contain an even number of nodes and loops.

Since the circumference of a circle is $2\pi r$, and the de Broglie wavelength $\lambda = h/mv$, the conditions for standing waves may be written: circumference $= n\lambda$,

$$2\pi r = n\lambda$$

from which

$$2\pi r = n\frac{h}{mv} \tag{32m}$$

where

$$n = 1, 2, 3, 4, 5, \ldots \tag{32n}$$

Multiplying both sides of this equation by $mv/2\pi$, we obtain,

$$mvr = n\frac{h}{2\pi} \tag{32o}$$

which is exactly the equation first given by Bohr in his theory [see Eq. (32b)].

A schematic diagram of this standing-wave model for the first three orbits of hydrogen is shown in Fig. 32O. Note that the first orbit has two nodes and a circumference of $1\lambda_1$, the second has four nodes and a circumference of $2\lambda_2$, and the third has six nodes and a circumference of $3\lambda_3$.

For increasing n, from 1 to 2 to 3, the wavelengths increase from $1\lambda_1$ to $2\lambda_1$ to $3\lambda_1$, the radii increase from $1r_1$ to $4r_1$ to $9r_1$, while the wave velocity decreases from $1v_1$ to $\frac{1}{2}v_1$ to $\frac{1}{3}v_1$.

In this representation the electron is not thought of as a particle located at some point within the atom but as though its mass and charge were spread out symmetrically throughout the space immediately surrounding the nucleus of the atom. It is interesting to point out that, while the Bohr circular orbits were confined to a plane, the wave model allows the electron distribution to be three-dimensional.

Even though the new theory of the hydrogen atom is an improvement upon the older Bohr orbit theory and gives a more satisfactory explanation of all known phenomena, it is more

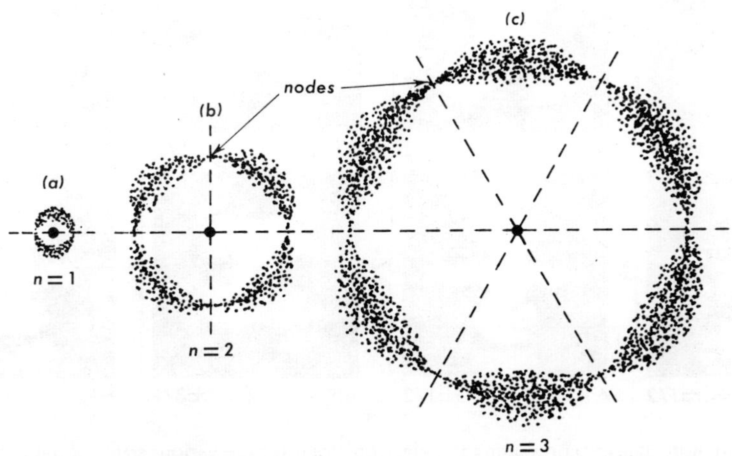

Fig. 32O. Schematic diagrams of the waves of the orbital electron in the first three quantum states of the hydrogen atom.

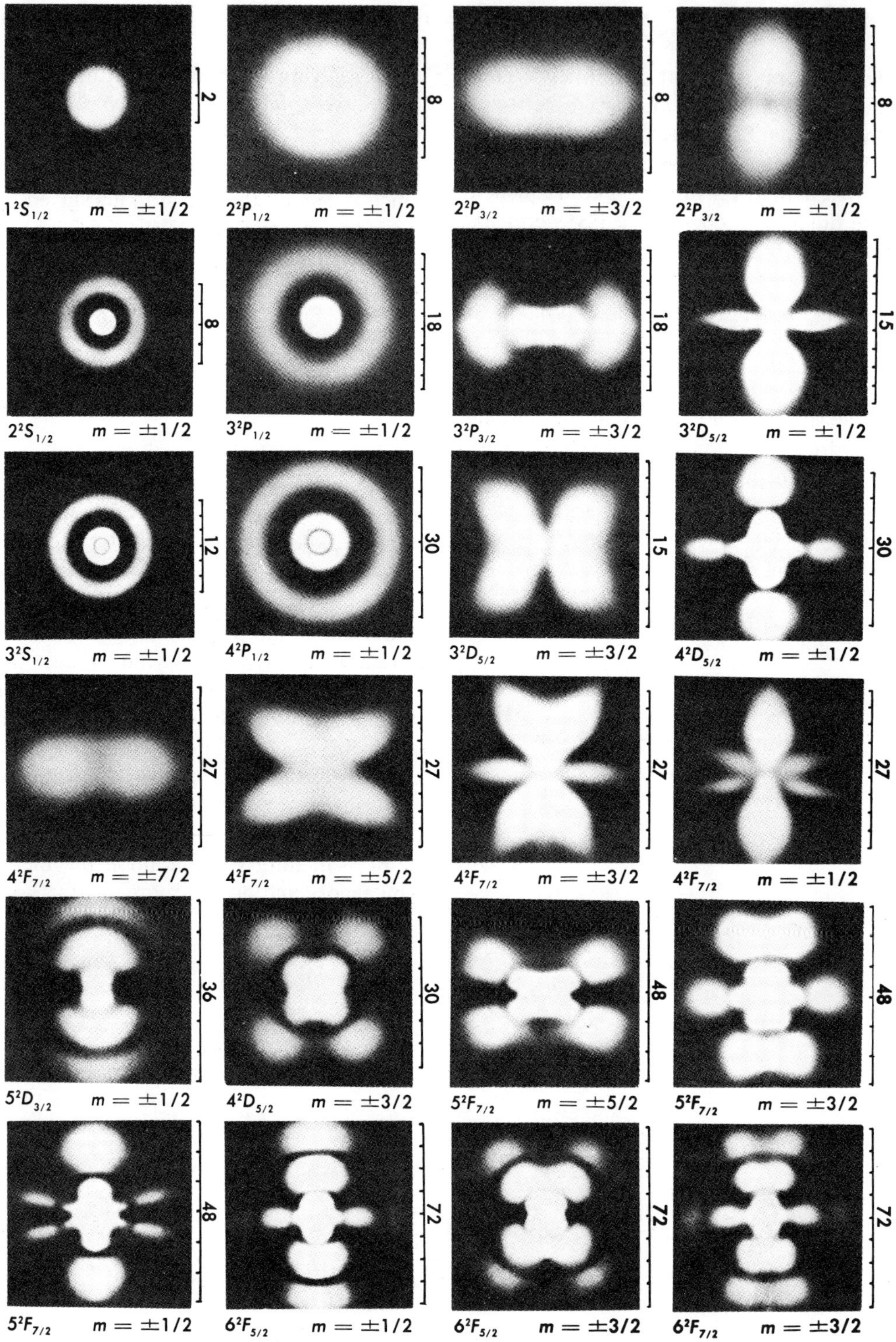

Fig. 32P. Probability density figures representing the electron cloud for the various states of the hydrogen atom according to the Dirac Wave Equations. (Photographs made by the author.)

difficult to form a mental picture of what an atom might look like. Indeed, the modern theoretical physicist goes so far as to say that the question, "what does an atom look like?" has no meaning, much less an answer. There are others, however, who still maintain that only those things that can be pictured are the things that are understood and that all mental thought processes are made in terms of things we detect by sight or touch. For this reason an interpretation is often given to the theory and its resultant equations that the amplitude of the electron waves within an atom represents the distribution of the electronic charge and mass. At the nodes where the motion is practically zero there is assumed to be little or no charge, while at the loops there is a maximum amount of charge.

Photographs representing a few of the possible states of the single electron in hydrogen are shown in Fig. 32P. These are not pictures of real atoms but are made to represent them. They are made by photographing a specially designed mechanical top. Where the electronic charge density is large the figure is white, and where it is practically zero at the nodes it is dark. The three-dimensional distributions can be visualized by imagining each figure as rotating about a vertical axis through the center. The third figure, for example, in three dimensions would have a shape similar to a smoke ring.

32.15 THE NEW ATOMIC PICTURE Although the Bohr atom has been replaced by the more satisfactory model of *a nucleus surrounded by electron waves*, it is still customary, for convenience only, to talk about electron *shells* and *orbits*. The reason for this is that there is a close analogy between the old and new models. When the Bohr-Stoner scheme of the building-up of the elements is extended to the new theory of electron waves, the electrons are found to distribute their charge in such a way that something analogous to shells is formed.

This is illustrated by the graph for a rubidium atom, atomic number 37, in Fig. 32Q. The

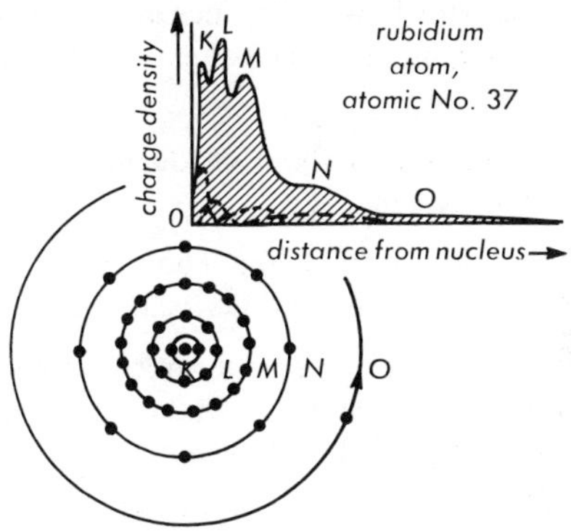

Fig. 32Q. Diagrams comparing the new and the old theories of atomic structure, according to the wave mechanics and the Bohr-Stoner model.

shaded area above represents the distribution of the charge of 37 electrons on the new theory, and the lower orbital model represents the electron shells on the old theory. The new model is represented by a graph because it is spherically symmetrical in space, while the old model is represented by orbits because it is confined to one plane. Proceeding out from the nucleus it is seen that the charge rises to several maxima at distances corresponding closely to the discrete *K*, *L*, *M*, *N*, and *O* shells of the orbital model. In other words, the new atom also has a shell-like structure

QUESTIONS

1. When a piece of metal is slowly heated until it attains a white heat, what sequence of colors is seen (a) when looking at the metal directly, and (b) when looking through a prism?

2. What is a spectrum? What are the four classes of spectra called?

3. What is a *continuous emission spectrum?* How is it produced, and how is it detected?

4. From what kind of light source does one obtain a line emission spectrum?

5. What are spectrum lines? Why are they in the form of lines?

6. Briefly describe *ultraviolet light* and *infrared light*. How are they produced, and how are they detected? What are the properties of such radiations?

7. Briefly describe how to produce a *continuous absorption spectrum*. Make a diagram showing the principal elements of the apparatus needed, and the resultant spectrum.

8. How would you go about producing a *line absorption spectrum?* Make diagrams and explain.

9. What was the form of the Thomson atom? Where did he locate the positive charges? Where did he locate the negative charges?

10. State clearly the three assumptions made by Bohr in formulating his theory of the hydrogen atom. Make a separate diagram for each assumption and properly label each essential element.

11. Briefly explain what is meant by (a) a normal atom, (b) an excited atom, and (c) an ionized atom.

12. What is the Bohr-Stoner scheme of the building up of atoms? How does it differ from the Thomson atom?

13. How many electrons can each shell have? What is the formula?

14. Make a diagram of a zinc atom (atomic number 30) according to the Bohr-Stoner scheme of the structure of atoms. Show all electrons in their proper orbits.

15. Make a diagram of an arsenic atom (atomic number 33) according to the Bohr-Stoner scheme of atomic structure. Show all electrons in their proper orbits.

16. What is the principal quantum number N? What is the smallest value this quantum number can have? What is its largest possible value?

PROBLEMS

1. If the innermost orbit of a hydrogen atom has a diameter of 1.058 Å, what is the diameter of the eighth orbit? [Ans. 67.7 Å.]

2. What is the diameter of the twentieth orbit of hydrogen atoms if the diameter of the innermost orbit is 1.058 Å.

3. Calculate the speed of the electron of a hydrogen atom if it is in the tenth circular orbit. [Ans. 2.188×10^5 m/s.]

4. Compute the speed of the electron of a hydrogen atom if it is in the twelfth orbit.

5. If the diameter of the sixth circular orbit of hydrogen atoms is 38.1 Å, what is the frequency of revolution? [Ans. 3.057×10^{13} m/s.]

6. If the one and only electron of a hydrogen atom is in the sixteenth circular orbit of diameter 270.9 Å, what is the frequency of revolution?

7. Find (a) the radius, and (b) the velocity of the electron in the fiftieth orbit of hydrogen. (c) What is the frequency of revolution of the electron? [Ans. (a) 1.323×10^{-7} m. (b) 4.38×10^{4} m/s. (c) 5.26×10^{10} Hz.]

8. If the electron of a hydrogen atom is in the orbit $n = 30$, find (a) the orbital radius, (b) the orbital speed, and (c) the orbital frequency.

9. Starting with Eqs. (32a) and (32b), derive Eq. (32c) for the radius of the Bohr circular orbits. [Ans. Solve Eq. (32a) for v^2, square Eq. (32b), and substitute the value of v^2 from Eq. (32a) for v^2 in the squared Eq. (32b).]

10. Starting with Eqs. (32a) and (32b) derive Eq. (32c) for the speed of the electron in the different Bohr circular orbits.

33 X RAYS AND THE PHOTOELECTRIC EFFECT

One of the most interesting episodes in the history of modern science began with the accidental discovery of X rays by Wilhelm Röntgen* in 1895. While studying the green fluorescent stage of an electrical discharge in a Crookes tube, Röntgen observed the bright fluorescence of some nearby crystals of barium platino-cyanide. Even though the discharge tube was in a darkened room and entirely surrounded with black paper to prevent the escape of visible light, a distant screen covered with crystals would fluoresce brightly, when the discharge was turned on. Röntgen reasoned, therefore, that some kind of invisible yet penetrating rays of an unknown kind were being given out by the discharge tube. These rays he called X rays, the letter X meaning, as it so often does in algebra, an unknown.

In the short series of experiments that followed his discovery, Röntgen found that the unknown rays were coming from the glass walls of the tube itself and, in particular, from the region where the most intense part of the cathode-ray beam was striking the glass. So great was the importance of this discovery that, within but a few weeks of Röntgen's announcement, X rays were being used as an aid to surgical operations in Vienna. This, along with other practical applications and uses to be made of a single scientific discovery, is a good example of the role played by modern science in the rapid advancement of civilization.

33.1 X-RAY TUBES The Crookes tube with which Röntgen made his discovery bears very little resemblance to the modern X-ray tube. In form it had somewhat the appearance of the tube

* WILHELM KONRAD VON RÖNTGEN (1845–1923). Born at Lennep on March 27, 1845, Röntgen received his education in Holland and Switzerland. His scientific career began at the age of twenty-five when he became an assistant in the physics laboratory at Würzburg, Germany. After a teaching career extending over a period of twenty-five years, which carried him to the University of Strasbourg, then to Hohenheim, back to Strasbourg, then to Giessen and finally again to Würzburg, he discovered X rays in his laboratory there in 1895. For this discovery he received the Rumford Medal of the Royal Society in 1896 and the first Nobel Prize in physics in 1901. Röntgen also conducted researches in light, heat, and elasticity, but none of these works compare in importance with his discovery of X rays.

shown in Fig. 30C. Within a short period of time after Röntgen's discovery quite a number of noteworthy improvements upon tube design were made. The first important contribution in this direction came immediately following the discovery that it is the sudden stopping of electrons that gives rise to X rays.

In X-ray tubes of early design, the electrons from the cathode were not allowed to strike the glass walls but were directed toward the anode as a target, as shown in Fig. 33A. By curving the

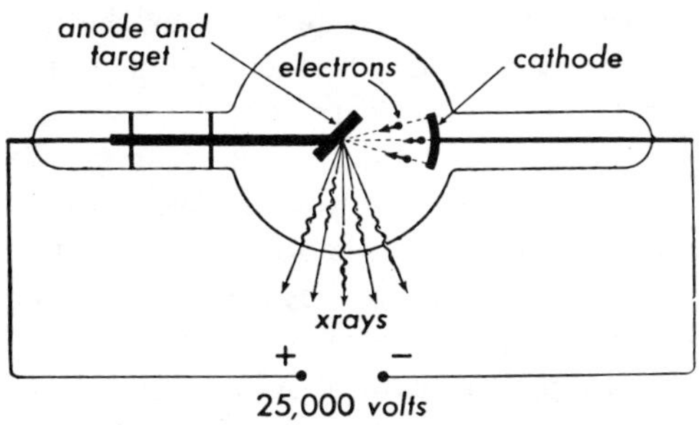

Fig. 33A. Diagram of early form of X-ray tube.

cathode like a concave mirror it was found possible to focus the electrons on one spot on the target, thus making of that spot a localized source of X rays. Radiating outward in all possible directions these Röntgen rays, as they are sometimes called, have no difficulty in passing through the glass walls of the tube.

The biggest improvement in X-ray tube design was made by Coolidge, an American physicist, in 1913. In the Coolidge tube, now a commercial product (see Fig. 33B), a tungsten wire filament

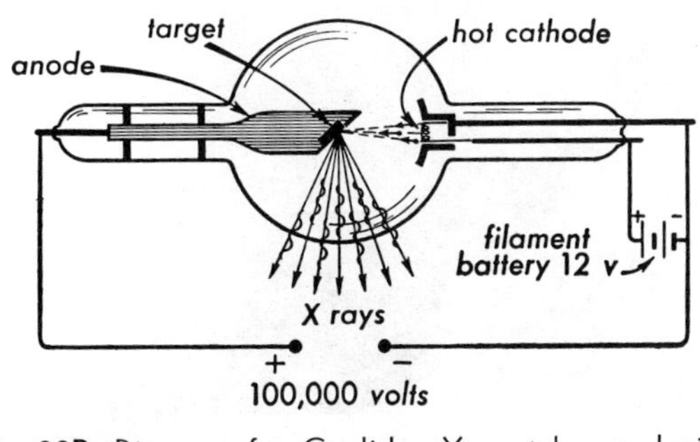

Fig. 33B. Diagram of a Coolidge X-ray tube employing a hot cathode.

is placed at the center of the cathode and heated to incandescence by a storage battery or low-voltage transformer. This filament, being a copious source of electrons, gives rise at the target to a far more intense source of X rays than was previously possible with a cold cathode. Under the terrific bombardment of the target by so many electrons, most metals will melt. To overcome this difficulty, a metal with a high melting point, like tungsten or molybdenum, is imbedded in the face of a solid copper anode to become the target. Copper, being a good heat conductor, helps to dissipate the heat.

The early source of high voltage applied to the anode and cathode of X-ray tubes was induction coils of various descriptions. Although some of these sources are still in use, they have been almost entirely supplanted by a more efficient high-voltage transformer. The emf generated by these transformers varies between 50,000 and 2,000,000 volts. The normal emf used for surgical work is about 100,000 volts, whereas for the treatment of diseases the higher emf's are employed. The high-voltage alternating emf supplied by a transformer is not applied directly to the X-ray tube but is first changed into direct current by means of rectifier tubes.

33.2 PENETRATION OF X RAYS Four useful and important properties of X rays are their ability (1) to penetrate solid matter, (2) to cause certain chemical compounds to fluoresce, (3) to ionize atoms, and (4) to affect a photographic plate. The penetration of X rays depends upon two things: first, the voltage applied between the anode and cathode of the X-ray tube; and second, the density of the substance through which the rays must travel. The higher the voltage applied to the tube, the greater is the penetration. *X rays of great penetrating power are called hard X rays, whereas those having little penetrating power are called soft X rays.*

The relation between density and penetration may be illustrated in several ways. When X rays are sent through a block of wood containing nails, or a closed leather purse containing coins, a clear and well-defined image of the nails, or coins, can be formed and observed on a fluorescent screen. The experimental arrangement is the same as that shown in Fig. 33C. When X rays

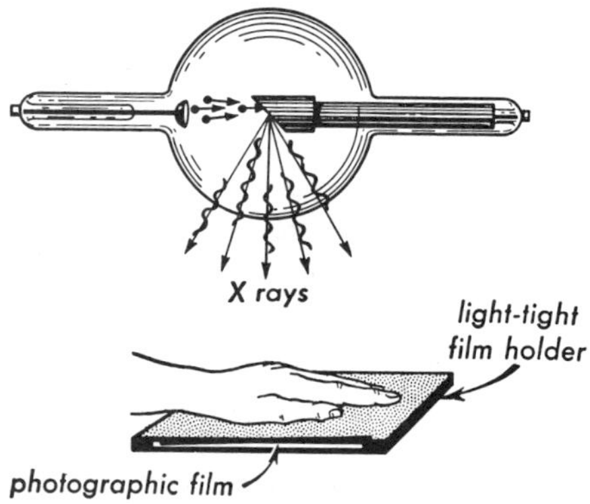

Fig. 33C. Arrangement for taking X-ray photographs of the bones of the hand.

are sent through the hand or any part of the body to obtain photographs of the bones, it is the difference in penetration between the flesh and the bones that permits a picture to be made. Materials like paper, wood, flesh, etc., composed principally of light chemical elements like those at the beginning of the periodic table, are readily penetrated by X rays. In other words they are poor absorbers of X rays. For materials like brass, steel, bone, gold, etc., composed partly of heavy elements, like those further along and near the end of the periodic table, the penetration of X rays is very poor. Hence heavy elements, or dense substances, are good absorbers.

The bones of the body, containing large amounts of calcium, are relatively good absorbers of X rays, whereas the flesh, composed principally of much lighter elements—hydrogen, oxygen, carbon and nitrogen—is a poor absorber. This explains the general appearance of X-ray photographs. X-ray pictures like the ones in Fig. 33D are similar to shadows cast by the objects being photographed. The focus point on the X-ray target, being bombarded by high-speed electrons, acts as a point source of rays. These spread out in straight lines as shown in Fig. 33C. On passing through the hand to the photographic film, more X rays are absorbed by the bones than by the flesh. Where only flesh is traversed, the X rays penetrate through to the photographic film, causing it to develop out black. The X rays penetrating the bones are very weak and the photographic film develops out clear. The bones therefore appear white against a darker background. If this "negative film," as it is called, is printed on paper as in Fig. 33D, it becomes a "positive" with the bones appearing black.

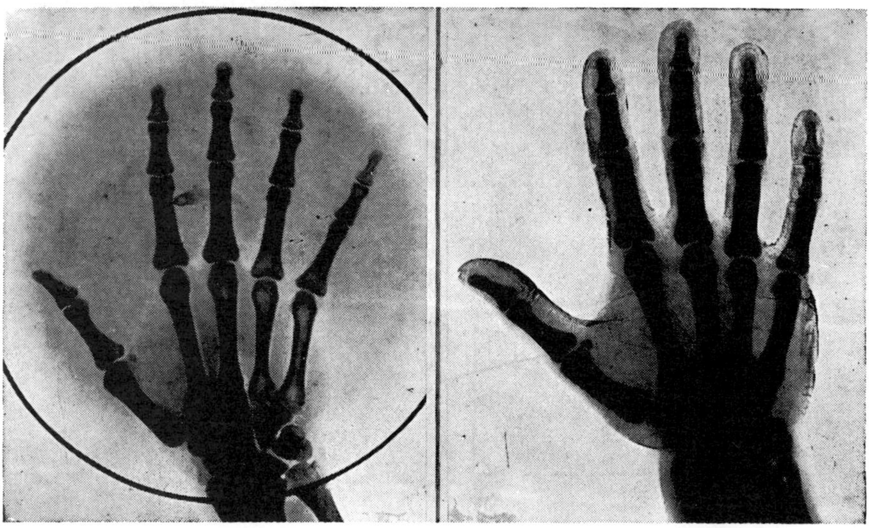

Fig. 33D. X-ray photographs of the wrist bones of the hand. *Left:* with hand in water. *Right:* with lead oxide ointment spread on hand. (Courtesy of the Stamford Research Laboratories, American Cyanamid Co.)

If the photographic film is placed farther away from the hand than shown in the diagram, the shadow picture will be larger and less distinct. The best pictures are obtained by placing the film as close in contact with the object to be photographed as is physically possible. Whenever a film is being exposed for an X-ray picture, it is mounted in a black paper envelope or thin aluminum box. This prevents visible light from reaching the film but allows the X rays to pass through.

33.3 IONIZING POWER As X rays pass through matter in the solid, liquid, or gaseous state, they are found to *ionize* atoms and molecules. This can be shown by charging a gold-leaf electroscope positively or negatively and placing it some 10 to 15 ft away from an X-ray tube. When the X-ray tube is turned on, the gold leaf falls, showing discharge. See Fig. 33E.

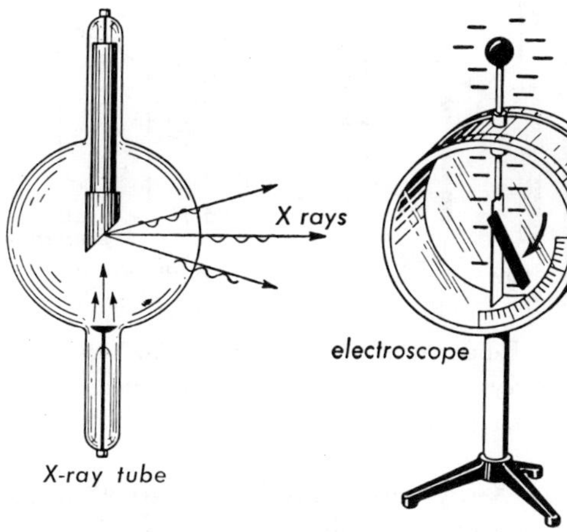

Fig. 33E. X rays discharge on electroscope.

The explanation of this experiment is as follows: X rays pass through the electroscope and ionize the air by removing electrons from many of the oxygen and nitrogen molecules. Leaving these particular molecules with a net positive charge, the freed electrons move about until they are picked up by other neutral molecules, thus giving them a net negative charge. The result is that the passage of X rays through matter produces both *positively charged* and *negatively charged ions*. If the electroscope is negatively charged, it attracts the positively charged ions to the gold leaf, neutralizing the charge and repelling the negatively charged ions to the "grounded" walls where they, too, become neutralized. If the electroscope is positively charged, it attracts the negative ions to it, again neutralizing the charge. The positive ions in this case are repelled to the walls. In either case, whether the electroscope is positively or negatively charged, the gold leaf falls, showing discharge.

It is the ionization of atoms and molecules in a substance that limits the penetrating power of X rays. Heavy elements contain more electrons than light elements, thus placing more electrons in the path of the X rays to stop them. The stopping power of a thin sheet of lead, for example, is equivalent to the stopping power of a sheet of aluminum several times thicker. Lead atoms each contain 82 electrons, whereas aluminum atoms each contain only 13.

33.4 PRACTICAL APPLICATIONS During the first few weeks following Röntgen's discovery of X rays, reports from all over the world were received by the editors of scientific journals telling of how the new rays could be put to practical use. A few examples of the first applications were (1) the location of a bullet in a patient's leg, (2) the observation and photography of the healing of a broken bone, (3) the detection of contraband in baggage, (4) the distinction between artificial and real gems, (5) the detection of pearls in oysters, and (6) the examination of the contents in parcel post. In 1897 Dr. Morton in New York exhibited an X-ray picture of the entire skeleton of a living and fully-clothed adult.

The biological effects became important when it was found that X rays killed off some forms of animal tissue more rapidly than others. This makes them a possible means of cure for certain skin diseases. The application in particular to the treatment of well-known forms of cancerous growths in animals and human beings has yielded amazing results, and oftentimes a cure. When an internal cancer is treated by sending a beam of X rays directly through the body, the cancerous tissue as well as the normal tissue is slowly destroyed. It is principally because the normal tissue grows in again more rapidly than the cancerous tissue that it is possible to bring about a cure. Periodic radiation allows the normal tissue to build up in the intervals.

Although only certain diseases can be treated successfully by X rays, a great deal of research work is still being carried on with extremely high-voltage X rays in the hope of discovering new and more effective medical aids. It is generally believed that the destruction of cell tissue by X rays is due primarily to the ionization of the molecules within the individual cells.

The importance of X rays in some phases of the field of engineering cannot be overestimated. This can be appreciated when it is realized that metal castings or welded joints sometimes contain internal flaws or blowholes that otherwise escape detection. Because of the disastrous results which might occur by the insertion of defective castings or welded joints into a bridge or building, many such metal parts are examined by X rays before being used.

33.5 X RAYS ARE WAVES Not long after Röntgen's discovery of X rays, there arose in scientific circles two schools of thought concerning the nature of these penetrating rays. The one school held to the belief that X rays are high-speed particles like cathode rays but more penetrating, and the other school held to the idea that they are electromagnetic waves of extremely high frequency. Although many experiments were performed to test these two hypotheses, several years passed before the wave theory was proved to be correct.

The crucial experiment came in 1912 when von Laue* suggested to his associates W. Friedrich and P. Knipping that they try diffracting X rays by sending them through a thin crystal. Believing that the ultramicroscopic structure of a crystal is a three-dimensional array of regularly spaced atoms, von Laue reasoned that the equally spaced layers of the atoms would act like mirrors and reflect the rays in different directions.

The experiment, as it was performed, is shown diagrammatically in Fig. 33F. X rays from a cold

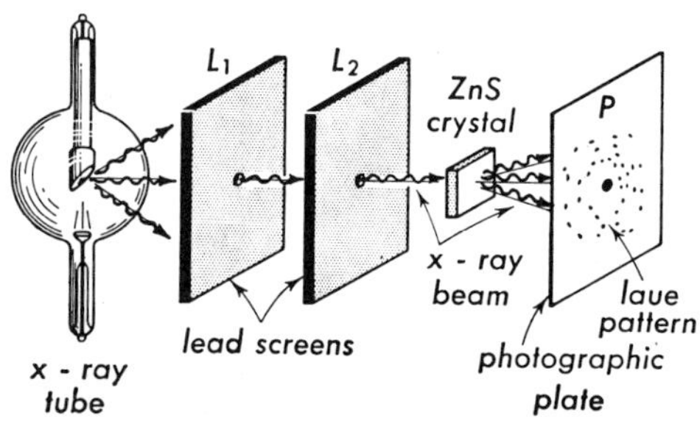

Fig. 33F. Experimental demonstration of the wave property of X rays. Diffracted by the atoms in a crystal, a Laue pattern is photographed.

cathode X-ray tube, and limited to a narrow pencil of rays by a pinhole in each of the two lead screens L_1 and L_2, are shown passing through a thin crystal to a photographic film or plate at *P*. In addition to the central beam, the major part of which goes straight through to produce a blackened spot at the center of the film, there are many other weaker beams emerging in different directions to produce other spots on the same film. The pattern of spots obtained in this way is always quite symmetrical and is referred to as a *Laue pattern*.

Photographs of two Laue patterns obtained with single crystals are reproduced in Fig. 33G.

* MAX VON LAUE (1879–1960). Born near Coblenz, Germany, in 1879, he was educated in the German Universities of Strasbourg, Göttingen, and Munich. Following this his teaching and research work carried him to the University at Munich, Zurich, Frankfurt on the Main, and finally Berlin. Being interested in theoretical physics, his early attentions were confined to various phases of Einstein's theory of relativity and to Bohr's quantum theory of atomic structure. His chief contribution to physics, however, was the instigation and supervision of experiments leading to the diffraction of X rays by crystals. For this work, which proved the wave-nature of X rays, he was granted the Nobel Prize in 1914. The author had the good fortune of knowing von Laue through working in the same laboratory as an International Research Fellow 1929–30.

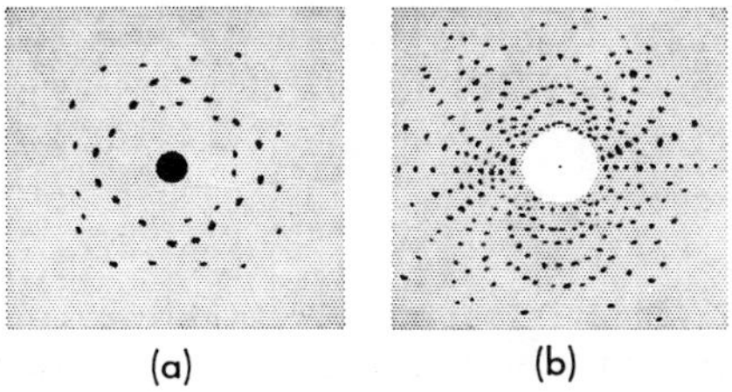

Fig. 33G. X-ray diffraction patterns from crystals: (a) Zinc sulfide crystal (face centered cubic crystal), (b) Sugar crystal (a complex crystal structure).

The small number of spots in (a) is indicative of a relatively simple crystal structure for zinc sulfide, ZnS, and the large number of spots in (b) signifies a relatively complex crystal structure for sugar, $C_{12}H_{22}O_{11}$. While the picture of sugar was being taken, the central beam was masked off by a small lead disk placed just in front of the film to prevent excessive blackening. Simple Laue patterns in general arise from simple crystal structures. Common salt is an example of a simple crystal, containing, as shown in Fig. 10F(a), sodium atoms (Na) and chlorine atoms (Cl) in equal numbers arranged in a three-dimensional cubic lattice. Fig. 33H is a cross

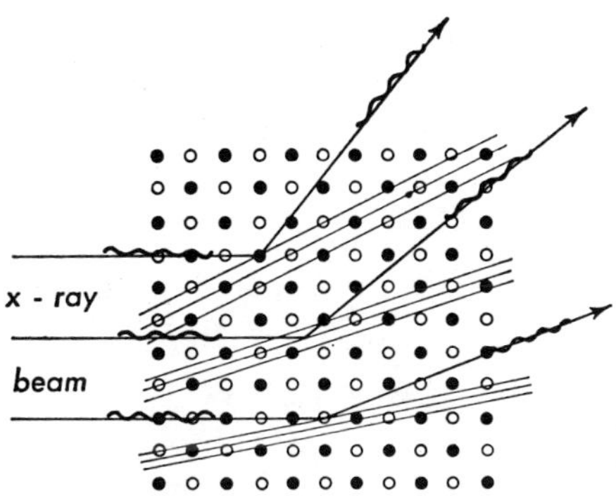

Fig. 33H. Illustration of the reflection of X rays from the various atomic planes in a cubic crystal lattice.

section through such a crystal showing the alternation of atoms in two of the three directions. Here in this two-dimensional array, the origin of the different spots on a Laue pattern is illustrated.

Each spot arises from the reflection of some of the incident X rays from one of the various sets of parallel crystal planes, three of which are shown by the sets of parallel lines. The rays always obey the law of reflection that the angle of incidence equals the angle of reflection. While the reflection planes shown in the diagram are all perpendicular to the plane of the page, there are many other planes in a three-dimensional lattice to reflect the rays off in other directions.

The success of the Laue experiment proves the correctness of two postulates: (1) that X rays are light rays of very short wavelength, and (2) that the atoms of a crystal are arranged in a regular three-dimensional lattice. These are the results for which von Laue was granted the Nobel Prize in physics in 1914. As a direct result of the Laue experiment, two new and important fields of experimental physics were opened up: (1) the study and measurement of X-ray wavelengths, and (2) the study of crystal structures by their action on X rays.

The shorter the X-ray wavelength, we say "the harder they are," and the greater is the penetrating power.

33.6 THE ORIGIN OF X RAYS X rays, like visible light, originate from the jumping of an electron from one orbit to another within an atom. When high-speed electrons from the cathode of an X-ray tube strike the target, they ionize many of the atoms comprising the surface layers of the metal.

Due to their very high speeds (about one tenth the velocity of light) the electrons penetrate the atoms and remove an electron from the inner shells by collision. This is illustrated in Fig. 33I, where an electron is knocked out of the *K* shell. The designations *K*, *L*, *M*, *N*, *O*, *P*, etc., for the

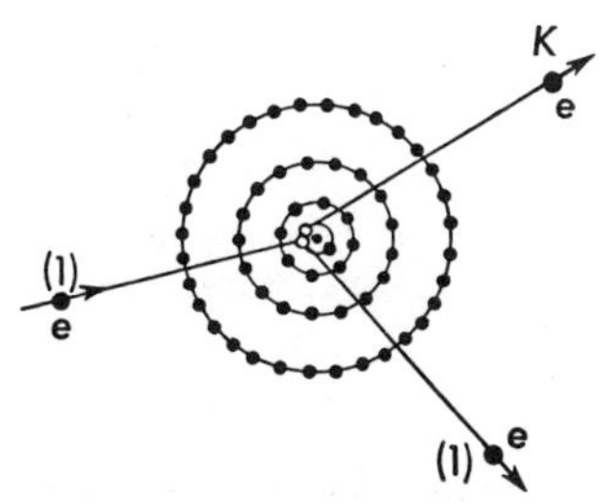

Fig. 33I. Schematic diagram illustrating the ionization of an atom by a high-speed electron.

various electron shells originated with the X-ray spectroscopist and are identical with the quantum numbers $n = 1, 2, 3, 4, 5, 6$, etc. When an electron is missing in the innermost K shell, a nearby electron from the next shell beyond jumps into the vacant space, simultaneously emitting a photon of energy $h\nu$. We see therefore that while visible light arises from an electron jumping between certain outer orbits of an atom (see Fig. 32M) X rays arise from an electron jumping between inner orbits (see Fig. 33J). Such X rays

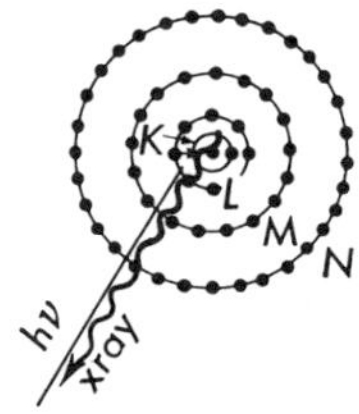

Fig. 33J. Diagram illustrating the jumping of an inner electron with the simultaneous emission of an X ray.

arising from millions of atoms produce the effect described above.

Since the L shell now has one less electron, an M electron can jump into the L-shell vacancy, with the consequent emission of another but different X-ray frequency. The jumping process continues until the outermost shell is reached, where an electron jumping in gives rise to visible light. Thus we see it is possible for a single atom to emit X rays of different wavelengths.

Another process that gives rise to the emission of an X-ray photon is shown in Fig. 33K. These radiations, called Bremsstrahlung, are due to the deflections and slowing down of high-speed electrons as they pass close to the nuclei of atoms within the target of the X-ray tube. In passing through the atom, an electron is attracted by the positive charge of the nucleus and hence deflected in its path.

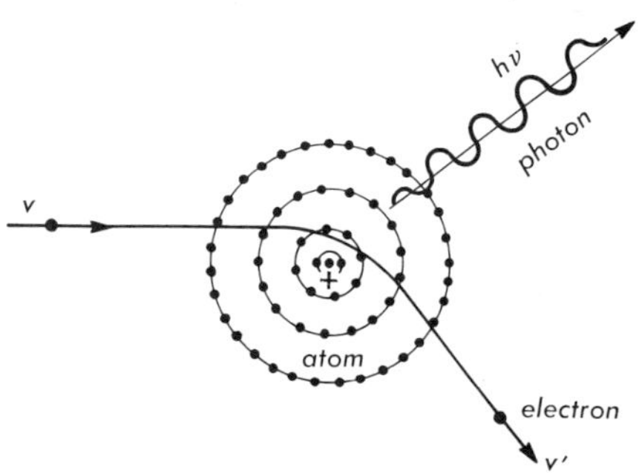

Fig. 33K. Diagram illustrating the production of a photon by a high-speed electron as it passes through an atom close to the nucleus. ("Bremsstrahlung.")

During the deflection of the electron in the strong electric field of the nucleus, a light wave of energy $h\nu$ is emitted. Since the law of conservation of momentum must hold for such a collision, the electron is deflected off to one side of the atom and the photon is deflected off to the other.

A very simple relation has been found to exist between the high voltage V applied to accelerate the electrons in an X-ray tube, and ν_{max} the highest frequency X rays emitted as Bremsstrahlung. This relation is

$$\boxed{Ve = h\nu_{max}} \qquad (33a)$$

EXAMPLE 1. A 20,000-volt power source is applied to an X-ray tube. Find the maximum frequency of the X rays emitted, and the corresponding wavelength.

SOLUTION. Here we apply Eq. (33a), and by solving for ν_{max}, and substituting the value of e from Eq. (30a), and h from Eq. (32b), we obtain

$$\nu_{max} = \frac{Ve}{h}$$

$$\nu_{max} = \frac{20{,}000 \text{ V} \times 1.6021 \times 10^{-19} \text{ C}}{6.6256 \times 10^{-34} \text{ Js}}$$

$$\nu_{max} = \frac{3.2042 \times 10^{-15} \text{ VC}}{6.6256 \times 10^{-34} \text{ Js}}$$

and since VC = J, and $\frac{1}{s}$ = Hz,

$$\nu_{max} = 4.8361 \times 10^{18} \text{ Hz}$$

For the wavelength of these X rays, use Eq. (24c), $c = \nu\lambda$, and solve for λ.

$$\lambda = \frac{c}{\nu} = \frac{2.997925 \times 10^8}{4.8361 \times 10^{18} \text{ Hz}}$$

$$\lambda = 0.619905 \times 10^{-10} \text{ m}$$

$$\lambda = 0.620 \text{ A}$$

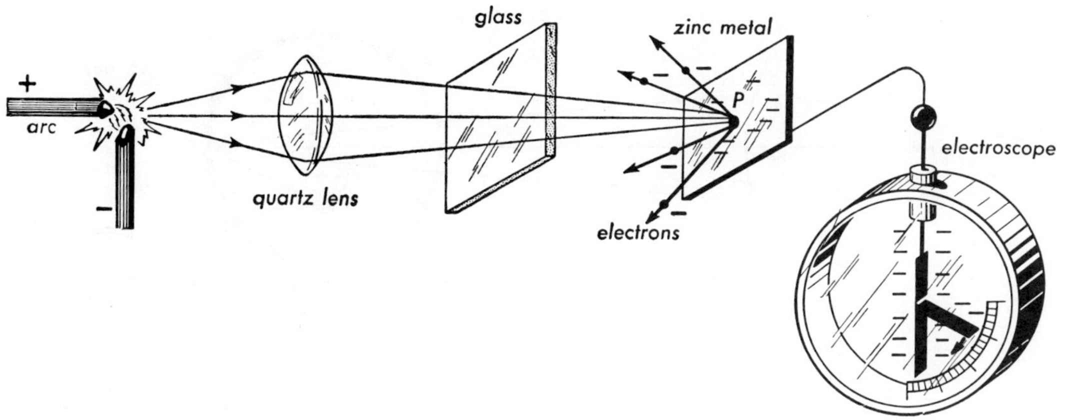

Fig. 33L. Experimental arrangement for demonstrating the photoelectric effect. When the glass plate is inserted, the effect stops.

Note that these waves are ten thousand times shorter than visible red light.

33.7 THE PHOTOELECTRIC EFFECT The photoelectric effect in its simplest form is demonstrated in Fig. 33L. Light from a carbon arc is focused by means of a quartz lens onto a freshly polished plate of zinc metal. When the plate is charged negatively and the light is turned on, the gold leaf of the attached electroscope slowly falls. It falls because the electrons, under the action of the light, leave the zinc plate at the illuminated spot *P*. When the plate is positively charged, the gold leaf does not fall, showing that the plate retains its charge. The same result of no discharge is observed if the zinc plate is negatively charged and a sheet of glass is inserted as shown in the figure. When the glass is removed the gold leaf again falls. Since common glass transmits visible and infrared light but not ultraviolet, we conclude that electrons are liberated only by ultraviolet light. This is also generally true for nearly all of the known metals.

Although these electrons are the same negatively charged particles that constitute cathode rays, they are called *photoelectrons* because they are liberated by means of light.

A few elements—namely the alkali metals *lithium*, *sodium*, *potassium*, *rubidium*, and *cesium* —are exceptions to this, for they will eject photoelectrons when visible light falls on them. For this reason the *alkali metals* are often used in the manufacture of photoelectric cells.

33.8 THE PHOTOELECTRIC CELL Photoelectric cells are usually made by depositing a thin layer of an alkali metal on the inner surface of a small vacuum tube. See Fig. 33M. If the cell is to operate in ultraviolet light it is made of quartz, whereas if it is to be used in visible light it is made of common glass. The cell must be thoroughly evacuated as the oxygen content of the air will combine chemically with the active metal layer, contaminating its surface and making it insensitive to visible light. A small section of the cell is always left clear to serve as a window for the incoming light. Photoelectrons, upon leaving the metal surface, are attracted and collected by the positively charged electrode *C*. The negative charge on the metal film and the positive charge on the central collector electrode are maintained at a constant potential by the battery *B*.

A beam of light shining through the window of a photoelectric cell acts like a switch which completes an electric circuit. When the light strikes the metal *P*, there is a flow of electrons to the collector *C*, thus causing a current to flow around the circuit. This current can be measured by means of an ammeter at *A*. If the intensity of the light increases, the number of photoelectrons increases and the current therefore rises. When the light is shut off, the photoelectric action

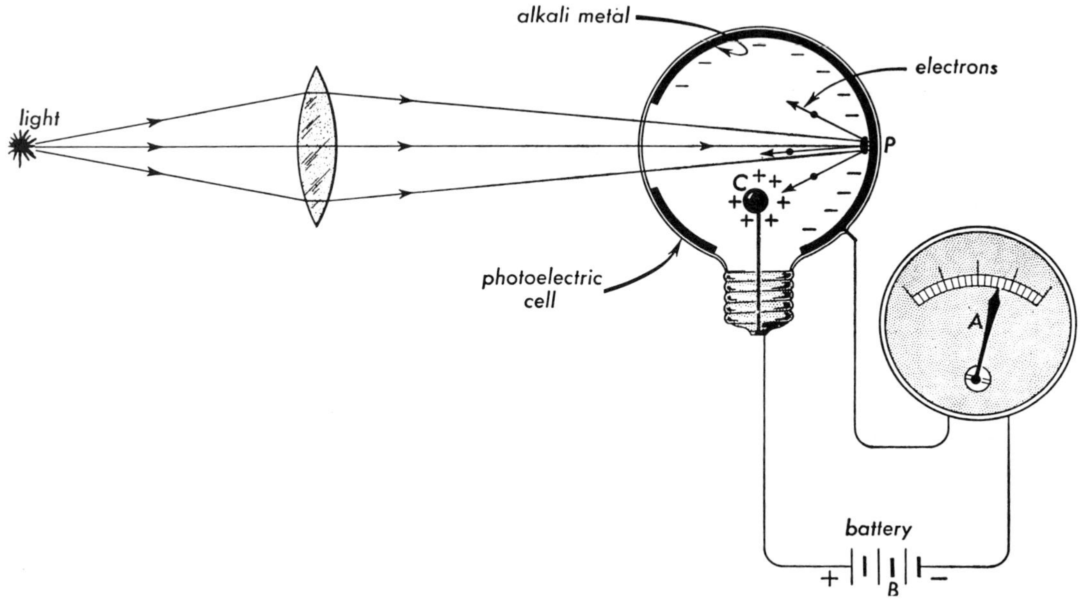

Fig. 33M. Diagram of a photoelectric cell showing the light beam and electrical connections necessary for its operation.

ceases and the current stops. If the metal film is positively charged, the cell becomes inactive to light since electrons attempting to leave the plate are held back by electrostatic attraction. All of these factors are readily demonstrated by a simple electrical circuit arranged as shown in the figure.

33.9 EINSTEIN'S PHOTOELECTRIC EQUATION

Following an earlier idea of Planck's that light waves consist of tiny bundles of energy called *photons* or *quanta*, Einstein proposed an explanation of the photoelectric effect as early as 1905. His ideas are expressed by one simple relation, an algebraic equation, destined to become famous in the annals of physics. Two Nobel Prizes, one to Einstein in 1921 and one to Millikan in 1923, have been granted on this, the photoelectric equation,

$$h\nu = W + \tfrac{1}{2}mv^2$$

The first term $h\nu$ represents the total energy content of a single quantum of light incident on a metal surface, as shown in Fig. 33N. The letter h is *Planck's constant of action*, which as we have seen in Bohr's theory, Eq. (32b), has the same value for all light waves regardless of the frequency ν. At or beneath the surface of the metal this *light quantum*, better known as a *photon*, is completely absorbed and, in disappearing, imparts its total energy to a single electron. Part of this energy W is consumed in getting the electron free from the atoms and away from the metal surface, and the remainder is used in giving the electron kinetic energy $\frac{1}{2}mv^2$, and therefore a velocity. For some metals like platinum the energy required to pull an electron away from the surface is large, whereas for other metals like

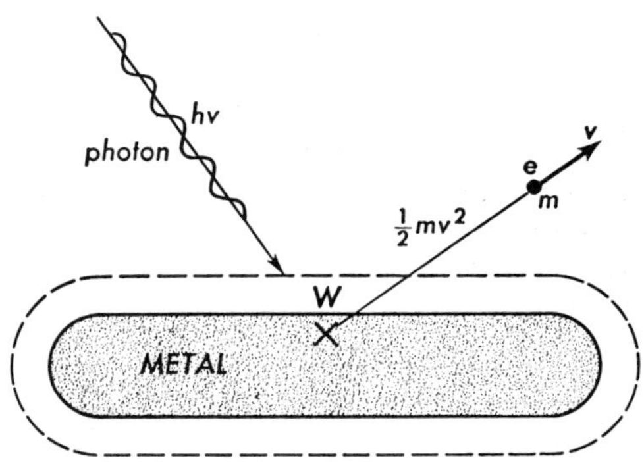

Fig. 33N. A light quantum (photon) of energy $h\nu$, incident on a metal surface, ejects an electron with a velocity v given by Einstein's equation.

the alkalies it is quite small. W is called the *work function* of the metal.

The work function W is found experimentally to be determined by the frequency that will just liberate electrons from the surface of a metal. In other words it is the threshhold frequency ν_0 which multiplied by Planck's constant h gives the threshhold energy $h\nu_0$ that will free an electron. Einstein's photoelectric equation becomes, therefore,

$$\boxed{h\nu = h\nu_0 + \tfrac{1}{2}mv^2} \qquad (33b)$$

33.10 SECONDARY ELECTRONS When electrons strike the surface of a metal plate, they knock additional electrons free from the surface. These are called *secondary electrons*, and the process is called *secondary emission*. See Fig. 33O. As the

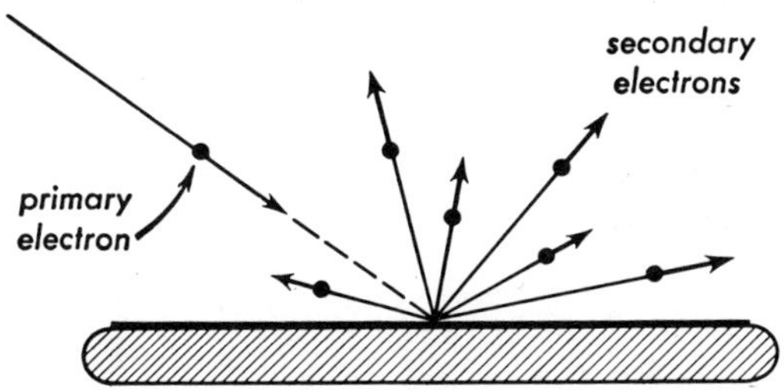

Fig. 33O. The impact of a single electron liberates additional electrons from a metal surface.

speed of a primary or incident electron increases from zero to a few hundred volts, the number of secondaries increases toward a definite maximum. For most metal surfaces this maximum is in the neighborhood of two, while for certain alkali metal films it may be as great as eight or ten. In general it is greatest for surfaces having a *low work function*.

33.11 PHOTOMULTIPLIER TUBES The process of secondary electron emission is widely used in a special type of photoelectric cell used most effectively in detecting faint light. A cross-section diagram of such a photomultiplier tube is given in Fig. 33P.

The number of photoelectrons from the photocathode A is proportional to the intensity of the incident light. These are attracted toward the next dynode B, more positive by 100 volts, where

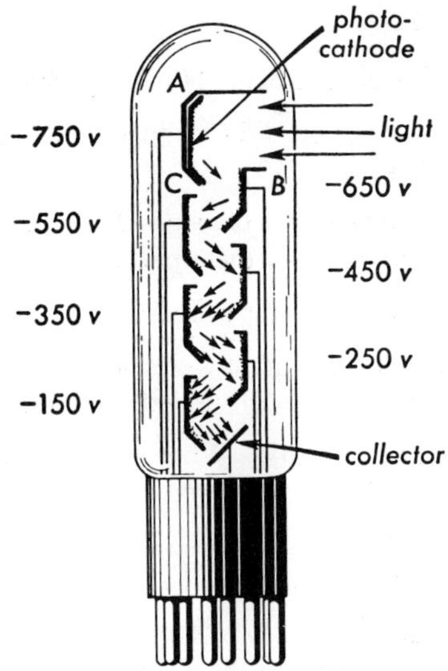

Fig. 33P. Photomultiplier tube with six stages. Cesium-oxide silver-coated photo cathode.

upon impact additional electrons are liberated. Attracted to the next more positive dynode C, still more electrons are liberated. By the time the collector plate has been reached, a small avalanche of electrons has developed and a correspondingly large charge and current are led off through that electrode to a suitable recording device.

The whole device therefore is capable of receiving a light beam of varying intensity, called a light signal, and by the photoelectric effect and secondary electron emission producing a very strong electrical current rising and falling in intensity the same as the incident light.

If each electron on impact releases 4 electrons, then in a tube with six stages, the number arriving at the collector would be

$$4 \times 4 \times 4 \times 4 \times 4 \times 4 = 4096$$

In general if each electron on impact releases n secondaries, then in a tube with k stages, the number arriving at the collector would be n^k. For example, if $n = 6$, and if $k = 5$, then $n^k = 7776$ electrons. This is an enormous gain over the signal obtained from a standard photo tube. Photomultiplier tubes have been used most successfully with faint light, not only with visible light but with infrared and ultraviolet as well.

QUESTIONS

1. Who discovered X rays? How were they discovered?

2. What contribution did Coolidge make to the design and construction of X-ray tubes?

3. Make a diagram of a Coolidge X-ray tube. Label each part and briefly describe the function of these parts.

4. Name three important properties of X rays.

5. Briefly explain why X rays enable one to see the bones of the body, even though the bones are surrounded by flesh.

6. What are hard X rays? What are soft X rays? How is hardness or softness related to X-ray wavelengths?

7. If a small child swallowed a safety pin, why would an X-ray photograph clearly show the location of the pin?

8. Upon studying the two X-ray pictures in Fig. 33D, explain why the skin shows so clearly in the right-hand photo and not in the left. (Note: Water is composed of H and O atoms, while lead-oxide is composed of Pb and O atoms.)

9. How is X-ray absorption related to the periodic table of elements?

10. Which of the following materials is the best absorber of X rays: (a) beryllium, (b) magnesium, (c) calcium, (d) copper, (e) gold, (f) lead, or (g) uranium?

11. What is the process of ionization by X rays? What happens to the liberated free electrons?

12. Make a list of specific places in which X rays would prove to be of considerable use.

13. How is the penetrating power of X rays related to the voltage applied to the tube?

14. Make a diagram and explain how a Laue pattern is produced. What is the significance of such photographs?

15. Make diagrams and explain the atomic origin of X rays.

16. What are "Bremsstrahlung"? Make a diagram and briefly explain.

17. Make a diagram of an experiment that may be performed to demonstrate that X rays produce ions as they pass through the air. Briefly explain.

18. What is the photoelectric effect? What are photoelectrons? How do they differ from cathode rays?

19. What is a photoelectric cell? What can it be used for?

20. What determines the velocity of photoelectrons? What is the Einstein photoelectric equation?

21. What is a photon? What is a light quantum?

22. What is Planck's constant? What is its significance?

23. What are secondary electrons? How is the principle employed in a photomultiplier tube?

24. Make a diagram of a photomultiplier tube with a cathode, eight dynodes, and a collector plate.

25. What is the work function of a metal? Is the work function for alkali metals larger than for most other metals or smaller?

26. How would you go about the problem of sending sound waves, like those from your voice, over a beam of light? How would you make the light source vary in intensity similar to the sound waves of your voice? How would you convert the varying light beam back into sound waves?

27. Make a diagram showing how a photoelectric cell could be used to count the number of cars passing a given point on a highway in a single day.

PROBLEMS

1. Find the energy equivalent to a light wave of wavelength 6×10^{-7} m. Such light is in the red region of the visible spectrum. [Ans. 3.311×10^{-19} J.]

2. Calculate the energy in joules of ultraviolet light of wavelength 2.5×10^{-7} m.

3. A 15,000-V power source is applied to an X-ray tube. What is (a) the frequency, and (b) the wavelength of the maximum frequency of the X rays emitted? [Ans. (a) 3.628×10^{18} Hz. (b) 0.827 Å.]

4. If 50,000 volts are applied to an X-ray tube, find (a) the frequency, and (b) the wavelength of the maximum frequency of the X rays emitted.

5. If X rays with a wavelength of 6 Å fall on a metal plate, what would be the maximum velocity of the photoelectrons emitted? Assume the work function to be negligibly small, $W = 0$. [Ans. 2.696×10^{7} m/s.]

6. If green light of wavelength 5000 Å falls on a metal surface and liberates electrons with a velocity of 3×10^{5} m/s, what is the wavelength of the photoelectric threshhold?

7. Find the over-all gain of a nine-stage photomultiplier tube if the average number of secondary electrons produced by each primary electron is five. [Ans. 1.953×10^{6}.]

8. An eight-stage photomultiplier tube has an overall gain of 100,000. Find the average number of secondary electrons produced by each primary electron.

34 | RADIOACTIVITY

During the latter part of the nineteenth century many scientists were experimenting with the fluorescence and phosphorescence of the chemical compounds of uranium. With the presence of such compounds in the laboratories, it was only natural that some keen observer should discover the radioactivity of the chemical element uranium.

Radioactivity may be defined as a spontaneous disintegration of the nucleus of one or more atoms. The phenomenon was discovered originally by Becquerel* in 1896 and is confined almost entirely to the heaviest elements in the periodic table, elements 83 to 104. What Becquerel discovered was that uranium, element 92, gave out some kind of rays that would penetrate through several thicknesses of thick black paper and affect a photographic plate on the other side. When the same phenomenon was confirmed several months later by Pierre and Marie Curie,† these rays became known as Becquerel rays.

* ANTOINE HENRI BECQUEREL (1852–1908), French physicist. Born in Paris on December 15, 1852, Antoine succeeded to his father's chair at the Museum of Natural History in 1892. In 1896 he discovered radioactivity, the phenomenon for which he is most famous. The invisible but penetrating rays emitted by uranium and other radioactive elements are now called Becquerel rays. For these researches he was granted the Nobel Prize in physics in 1903.

† PIERRE CURIE (1859–1906), French physicist, and MARIE CURIE (1867–1936), Polish physical chemist. Pierre Curie was educated at the Sorbonne where he later became professor of physics. Although he experimented on piezoelectricity and other subjects, he is chiefly noted for his work on radioactivity performed jointly with his wife, Marie Sklodowska, whom he married in 1895. Marie was born in Poland on November 7, 1867, where she received her early scientific training from her father. Becoming involved in a students' revolutionary organization, she left Poland for Paris where she took a degree at the university. Two years after the discovery of radioactivity by Becquerel, Pierre and Madame Curie isolated polonium and radium from pitchblende by a long and laborious physical-chemical process. In 1903 they were awarded the Davy Medal of the Royal Society, and (jointly with Becquerel) the Nobel Prize in physics. Professor Curie, who was elected to the Academy of Sciences in 1905, was run over and killed by a carriage in 1906.

34.1 DISCOVERY OF RADIUM Unlike the discovery of many new phenomena, the discovery of radium by Pierre and Madame Curie in 1898 was brought about intentionally by a set of carefully planned experiments. Having found that pitchblende was active in emitting Becquerel rays, the Curies treated chemically a ton of this ore in the hope of isolating from it the substance or element responsible for the activity. The first concentrated radioactive substance isolated was called *polonium* by Madame Curie, a name chosen in honor of her native country, Poland. Five months later came the isolation of a minute quantity of *radium*, a substance which was a powerful source of Becquerel rays. Continued experiments by the Curies, and others, soon led to the isolation of many other substances now recognized as radioactive elements. Some of the more common of these are *ionium*, *radon*, and *thorium*.

34.2 THE PROPERTIES OF BECQUEREL RAYS It is to the experimental genius of Rutherford* that we owe the complete unraveling of the mystery surrounding the nature of Becquerel rays. As the result of an extensive series of experiments, Rutherford and his co-workers discovered that these penetrating rays are of three quite different kinds. A simplified experiment demonstrating this is illustrated in Fig. 34A. A small sample of

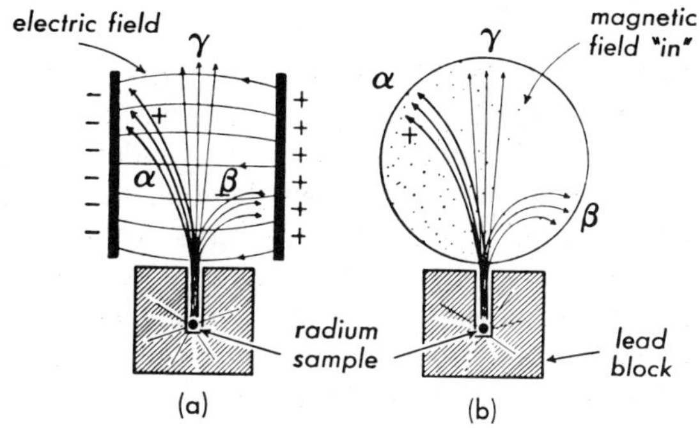

Fig. 34A. The bending of Becquerel rays in (a) an electric field and (b) a magnetic field.

radium is dropped to the bottom of a small drill hole made in a block of lead. This produces a narrow beam of rays emerging from the top of the block since those rays entering the walls of the lead are absorbed before reaching the surface. When electrically charged plates are placed at the side of this beam as shown in diagram (a), the paths of some rays are bent to the left, some to the right, and some not at all. A magnetic field as shown in diagram (b) exhibits the same effect. Those paths bending to the left indicate positively charged particles called *α rays* or *α particles*, those bending to the right indicate negatively charged particles called *β rays* or *β particles*, and those going straight ahead indicate no charge and are called *γ rays* or *photons*.

Rutherford, by a series of experiments, was able to show that each *α ray* is in reality a *doubly ionized helium atom*, i.e., a helium atom with both of its electrons gone. Such a particle is nothing more than a bare helium nucleus with double the positive charge of a hydrogen nucleus or proton, and a mass number or atomic weight four times as great. The *β rays* he found are ordinary electrons with a mass of 1/1840th the mass of a *proton* or 1/7360th the mass of an *α particle*, while *γ rays* are electromagnetic waves of about the same frequency as X rays. Although

---

Succeeding him as professor at the university, Madame Curie in 1911 was awarded the Nobel Prize in chemistry. She has the unique distinction of having had a share in the awards of two Nobel Prizes.

---

* LORD RUTHERFORD (1871–1937), British physicist, was born in New Zealand where he attended the university. In 1898 he became Macdonald professor of physics at McGill University, Montreal, Canada, and in 1907 professor of physics at Manchester University. In 1919 he became professor and director of experimental physics at the University of Cambridge, and in addition held a professorship at the Royal Institution in London. He is most famous for his brilliant researches establishing the existence and nature of radioactive transformations and the electrical structure of the atom. For this work, and until the time of his death in 1937, he was acclaimed by many as the greatest living experimental physicist. He was awarded the Nobel Prize in chemistry in 1908, and was knighted in 1914.

γ rays all travel with exactly the velocity of X rays and visible light, α rays are ejected with a speed of from one tenth to one hundredth the velocity of light. β particles move faster than α particles, some of them traveling with 99% the velocity of light.

34.3 IONIZING POWER When Becquerel rays penetrate matter in the gaseous, liquid, or solid state, they do not continue to move indefinitely, but are brought to rest slowly by ionizing atoms all along their path. Being ejected from their radioactive source with tremendously high speeds, all three types of rays collide with electrons and knock them free from atoms. They are, therefore, *ionizing agents*. The relative number of ionized atoms created along the path of an α particle, however, is much greater than the number created by a β particle or γ ray. If, in traveling the same distance in a given material, a γ ray produces one ionized atom, a β particle will, on the average, produce approximately one hundred, and an α particle will produce about ten thousand. Thus α particles are powerful ionizing agents, while γ rays are not.

As stated above, an α particle, ejected from a radioactive atom, is but the nucleus of a helium atom and lacks the two electrons necessary to make of it a neutral atom. As this particle speeds through matter it picks up and loses electrons at a rapid rate. No sooner does an electron become attached than it is swept off again by other atoms. Finally upon coming to rest, however, each α particle collects and retains two electrons, becoming a *normal helium atom*.

34.4 PENETRATING POWER At each collision with an atom, Becquerel rays lose on the average only a small part of their initial energy. Usually an α particle or β particle will make several thousand collisions before being brought to rest. At each collision some of the kinetic energy is expended in ionizing the atom encountered while giving that same atom a certain amount of kinetic energy. Since α particles produce the greatest number of ions in a given path, they penetrate the shortest distance and therefore have the poorest penetrating power. The penetrating powers of the three kinds of rays are roughly inversely proportional to their ionizing power.

	Relative ionizing power	*Relative penetrating power*
α	10,000	1
β	100	100
γ	1	10,000

34.5 METHODS OF DETECTING BECQUEREL RAYS There are several well-known methods for detecting and measuring radioactivity; the most common of these are the gold-leaf electroscope, the Wilson cloud chamber, the Geiger-Mueller tube counter, the scintillation counter, and the ionization chamber. We have already seen in Sec. 33.3 how X rays passing through an electroscope cause the charge to disappear and the gold leaf to fall. This same action may be demonstrated with α, β, and γ rays. The stronger the source of rays or the nearer the sample is brought to the electroscope, the more rapid is the discharge. Experiments show that if the walls of the electroscope are too thick, only the γ rays get through to produce ionization on the inside. For this reason, specially designed electroscopes made with thin windows of light material like aluminum are used for measuring α and β rays.

34.6 THE WILSON CLOUD CHAMBER In 1912 C. T. R. Wilson* devised a method by which one

* CHARLES THOMSON REES WILSON (1869–1959), Scottish physicist, was awarded the Nobel Prize in physics in 1927 for his development of the so-called Wilson Cloud Chamber. This device causes the formation of fog drops on ionized atomic particles, thereby making visible the paths of electrically charged particles as well as their collisions with other particles. The cloud chamber led to the discovery of recoil electrons from X rays, the so-called Compton recoils, the tracks of alpha particles, and the discovery by C. D. Anderson of positrons and mesons. C. T. R. Wilson was educated at Owens College, Manchester, and Sidney Sussex College, Cambridge. He was elected to the Royal Society and served for years as lecturer in physics and Professor of Natural History at Cambridge.

may actually observe the paths of α and β particles. As will be seen in the following chapters this method is used extensively in modern atomic physics as a means of studying many different atomic processes. The device by which this is accomplished consists of an expansion chamber in which water vapor is made to condense upon ions produced by the high-speed particles that have previously passed through it.

To begin with, the conditions under which water in the vapor state will condense into fogdrops are critical. These conditions are: *first*, there must be water vapor present; *second*, there must be dust particles or ions on which the drops can form; and *third*, the temperature and pressure must be brought to a definite value. That water drops will condense only upon ions or dust particles can be demonstrated with an ordinary glass jar containing a little water as shown in Fig. 34B. If allowed to stand for a short period of time, some of the water will evaporate and fill the bottle with vapor. Ions are next formed in the bottle by momentarily inserting a small gas flame as shown in diagram (a). Compressed air is then injected into the bottle through a tube so that when the stopper is quickly removed the sudden expansion will produce a dense fog as shown in diagram (b). If the flame is not first inserted to produce ions, no appreciable fog can be formed.

The purpose of the compressed air and subsequent expansion of the chamber is to lower the temperature, thus causing the air to become supersaturated with water vapor. Under these conditions the vapor will condense on all ionized molecules present.

When an α or β particle "shoots" through the air, positive and negative ions are formed all along its path. The removal by collision of each electron from a neutral atom or molecule leaves a positively charged ion. The electron attaching almost immediately to another neutral atom or molecule makes it a negatively charged ion. If immediately after an α particle has gone through a cloud chamber, an expansion takes place, fogdrops will form on the newly created ions, revealing clearly the path the particle has taken. As illustrated by the photographs in Fig. 34C, such α-ray tracks are straight and quite dense, whereas the β-ray tracks are crooked and sparsely lined with drops. The β rays, being very light particles, are easily deflected by collision, while the relatively heavy α particles "plow" right through thousands of atoms with only an occasional deflection.

A diagram of a laboratory type of Wilson cloud chamber is shown in Fig. 34D. The arrangement is made from an ordinary flat-bottomed flask with a rubber bulb attached to

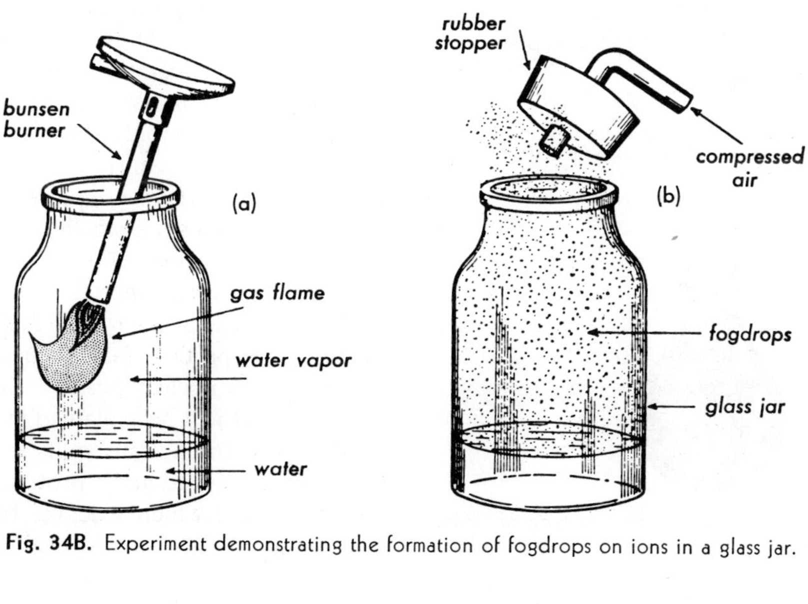

Fig. 34B. Experiment demonstrating the formation of fogdrops on ions in a glass jar.

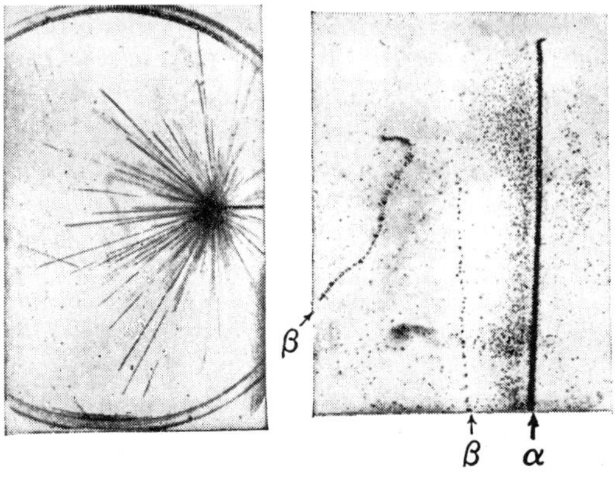

Fig. 34C. (a) α-ray tracks from radium as seen in a Wilson cloud chamber. (b) One α-ray and two β-ray tracks. (After C. T. R. Wilson, Proceedings of the Royal Society of London, Vol. 87, 1912, p. 292.)

the neck. A tiny deposit of radium or polonium is inserted in the end of a thin-walled glass tube as indicated. When the rubber bulb is squeezed to compress the air in the top, and then released to cause an expansion, fogdrops will form on the ions created by the α particles. See Fig. 34C. The battery and the wires leading to the wire ring in the top of the chamber and the water below are for the purpose of quickly removing ions previously formed in the chamber. This clears the field of view for newly formed tracks.

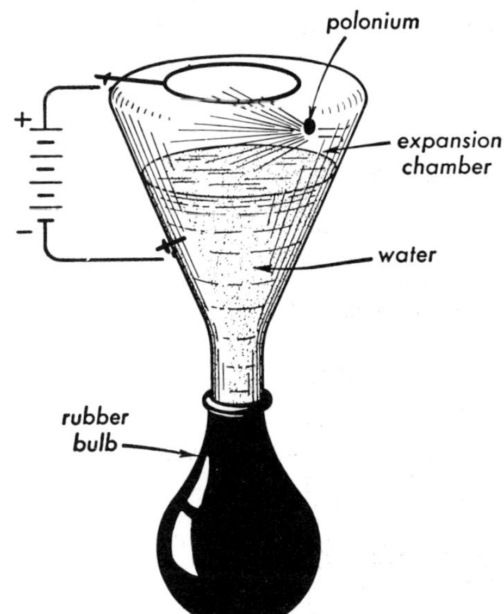

Fig. 34D. Diagram of a small laboratory type of Wilson cloud chamber.

Gamma rays are never observed in a cloud chamber since they produce so few ions. In passing through several feet of air a single γ ray will, on the average, produce only one or two ions. This is not enough to produce a recognizable cloud track. If a very strong source of γ rays is available, however, their presence can be observed in a cloud chamber by the chance collisions some of them have made with electrons. These recoiling electrons are called "Compton electrons."

Another type of instrument used for the detection of Becquerel rays is known as the Geiger-Mueller tube counter, or more briefly as the Geiger counter. This device will be described in detail in Chap. 36, dealing with cosmic rays.

34.7 TRANSMUTATION BY SPONTANEOUS DISINTEGRATION A careful study of radioactivity indicates that α, β, and γ rays originate from within the nucleus of the atom. When a radium atom disintegrates by ejecting an α particle, the nucleus loses a net positive charge of two. Since the number of positive charges on the nucleus determines the exact number of electrons outside of the atom and this in turn determines the chemical nature of an atom, the loss of an α particle, with two positive charges, leaves a new chemical element. Thus a *radium atom*, for example, in disintegrating, changes into a new atom called *radon*. We say there has been a *transmutation*. Not only does a nucleus lose a double charge by emitting an α particle and thereby *dropping down two places in atomic number*, but it also loses a weight of four units and thus *drops down four units in atomic weight*, or *four units of mass number*.

When a nucleus like *radium B* disintegrates by ejecting a β particle (an electron) to become *radium C*, the nuclear positive charge *increases by one unit*. It will be explained in Chap. 38 that a neutron in the nucleus splits up into an electron, which it ejects, and a proton, with an equal positive charge which it leaves behind. Such a transmutation yields a new element one atomic number higher in the chemical table. Since an electron weighs only 1/1840th part of a hydrogen atom or proton, the change in mass due to a β particle leaving a nucleus is too small to change the mass number. Although the loss in weight is measurable, it changes the atomic weight so slightly that for most purposes of discussion it can be and is neglected. A γ ray, like the β ray particle, changes the weight of a nucleus by a negligible amount and, since it has no charge, it does not alter either the atomic number or the mass number. These changes, if not clearly understood now, will be explained in more detail in Chap. 38.

34.8 HALF-LIFE *The half-life of a radioactive element is the time required for half of a given quantity of that element to disintegrate into a new element.* For example, it takes 1600 years for half of a given quantity of radium to change into radon. In another 1600 years half of the remainder will have disintegrated, leaving one quarter of the original amount. The half-life of radium is therefore said to be 1600 years.

The rate at which a given quantity of a radioactive element disintegrates, that is, *decays*, is found by observing the activity of a given sample over a period of time and plotting a graph of the type shown in Fig. 34E. For *polonium* the activity drops to half of its original value in 140 days. In another 140 days it again drops to half value, etc. The term activity may be defined as the number of rays given off per second of time, or as the number of ionized atoms produced each second by the rays.

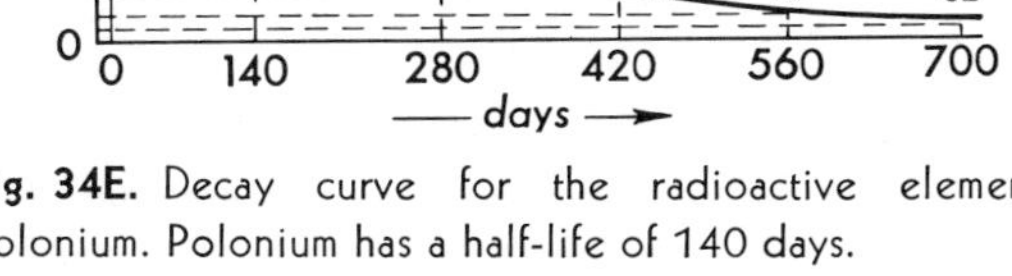

Fig. 34E. Decay curve for the radioactive element polonium. Polonium has a half-life of 140 days.

The only difference between the decay curve of one element and that of another is the horizontal time scale to which they are plotted. To turn Fig. 34E into a decay curve for radium the times 140, 280, 420 days, etc., need only to be changed to read 1600, 3200, 4800 years, etc., respectively. Since, therefore, all radioactive decay curves follow the same law, one does not have to wait for half of a given sample to disintegrate to be able to calculate how long it will be before half will have changed. This would require too many years of waiting for some elements.

34.9 RANGE The *range* of an α particle is defined as the distance such a particle will travel through dry air at normal atmospheric pressure. In a partial vacuum where there are fewer air molecules per centimeter to bump into, the distance traveled before coming to rest will be greater, whereas in air under higher than normal atmospheric pressure there are more molecules per centimeter and the distance will be diminished. Experiments show that some radioactive elements eject α particles with a higher speed than others. The higher the initial speed the greater is the range. The range of the α particles from *radium* is 3.39 cm, whereas the range of those from *thorium C′* is 8.62 cm.

The ranges of α particles in general have been determined in three different ways: *first*, by the Wilson cloud chamber, *second*, by the number of ions produced along the path, and *third*, by scintillations produced on a fluorescent screen.

In the Wilson cloud chamber photograph of Fig. 34F, α particles of two different ranges are observed. The radioactive sample used to obtain this picture was a mixture of *thorium C* and *thorium C′*. The shorter tracks with a 4.79 cm range are due to the α particles from *thorium C* disintegrating to become *thorium C″*, and the longer tracks of 8.62 cm range are due to the α particles from *thorium C′* disintegrating to become lead (see the last table in Appendix VI).

34.10 RADIOACTIVE SERIES It was Rutherford and his colleagues who discovered that when one radioactive atom disintegrates by ejecting an α or

Fig. 34F. Wilson cloud-chamber tracks from thorium C and C′. (After Rutherford, Chadwick, and Ellis; courtesy of Cambridge University Press.)

β particle, the remaining atom is still radioactive and may sooner or later eject another particle to become a still different atom. This process they found to continue through a series of elements, ending up finally with a type of atom that is stable and not radioactive. It is now known that nearly all natural disintegration processes, occurring among the heaviest elements of the periodic table, finally end up with *stable lead atoms*, atomic number 82.

There are at least four known radioactive series or chains of elements, one starting with uranium-238, a second with thorium-232, a third with uranium-235, and a fourth with plutonium-241. The first of these series is given in Table 34A. All four series are given in a graphical tabulation in Appendix VI.

When a uranium atom of mass number 238 and atomic number 92 disintegrates by ejecting an α particle, the remainder is a new atom, *uranium* X_1, of mass number 234 and atomic number 90. When a uranium X_1 atom disintegrates by ejecting a β particle to become *uranium* X_2, the mass number remains unchanged at 234, while the atomic number increases by one to become 91. This increase of one positive charge is attributed to the loss of one negative charge. These processes of successive disintegration continue until *lead*, a *stable* atom, is all that is left.

It is common practice among physicists to designate all atomic nuclei in an abbreviated form. The nucleus of radium, for example, is

TABLE 34A. Radioactive Series of Elements. The Uranium-238 Series

Element	Old Symbol	New Symbol	Atomic Number	Mass Number	Particle Ejected	Range in Air	Half-life
Uranium I.....	UI	$_{92}U^{238}$	92	238	α	2.70 cm	2×10^9 yr
Uranium X₁....	UX_1	$_{90}Th^{234}$	90	234	β	....	24.5 days
Uranium X₂....	UX_2	$_{91}Pa^{234}$	91	234	β	....	1.14 min
Uranium II....	UII	$_{92}U^{234}$	92	234	α	3.28	3×10^5 yr
Ionium........	Io	$_{90}Th^{230}$	90	230	α	3.19	83,000 yr
Radium.......	Ra	$_{88}Ra^{226}$	88	226	α	3.39	1600 yr
Radon........	Rn	$_{86}Rn^{222}$	86	222	α	4.12	3.82 days
Radium A.....	RaA	$_{84}Po^{218}$	84	218	α	4.72	3.05 min
Radium B.....	RaB	$_{82}Pb^{214}$	82	214	β	....	26.8 min
Radium C.....	RaC	$_{83}Bi^{214}$	83	214	α, β	....	19.7 min
Radium C′.....	RaC′	$_{84}Po^{214}$	84	214	α	6.97	10^{-6} sec
Radium C″....	RaC″	$_{81}Tl^{210}$	81	210	β	....	1.32 min
Radium D.....	RaD	$_{82}Pb^{210}$	82	210	β	....	22 yr
Radium E.....	RaE	$_{83}Bi^{210}$	83	210	β	....	5 days
Polonium......	Po	$_{84}Po^{210}$	84	210	α	3.92	140 days
Lead..........	Pb	$_{82}Pb^{206}$	82	206	stable	....	infinite

written $_{88}Ra^{226}$. The subscript to the left of the chemical symbol gives the *atomic number*, i.e., the number of positive charges on the nucleus, and the superscript on the right gives the *mass number*, or weight.

As explained in Sec. 31.4, all atoms with the same atomic number but different mass number are called isotopes of the same element. For example, $_{82}RaB^{214}$, $_{82}ThB^{212}$, $_{82}AcB^{211}$, $_{82}RaD^{210}$, $_{82}Pb^{209}$, $_{82}Pb^{208}$, $_{82}Pb^{207}$, and $_{82}Pb^{206}$ are isotopes of the same chemical element, lead (see Appendix VI). Even though the first five of these are radioactive, i.e., unstable, the other three are stable. Chemically they behave exactly alike and are separated only with difficulty. The isotopes 214, 210, and 206 belong to the uranium-238 series, 212 and 208 to the thorium series, 211 and 207 to the uranium-235 series, and 209 to the neptunium series.

The disintegration of radioactive nuclei may be written in the form of simple equations called *nuclear reactions*.

For radium,

$$_{88}Ra^{226} \rightarrow {}_{86}Rn^{222} + {}_{2}He^{4} \tag{34a}$$

for polonium,

$$_{84}Po^{210} \rightarrow {}_{82}Pb^{206} + {}_{2}He^{4} \tag{34b}$$

and for radium B,

$$_{82}RaB^{214} \rightarrow {}_{83}RaC^{214} + {}_{-1}e^{0} + \gamma \text{ ray} \tag{34c}$$

In each reaction the sum of the subscripts on the right side of the arrow is equal to the subscript on the left. The same is true for the superscripts. The designation $_{2}He^{4}$ represents the α particle, and $_{-1}e^{0}$ represents the β particle. In nearly all radioactive disintegrations where a β particle is emitted, one finds a γ ray also. In such cases, as shown by the example in Eq. (34c), *radium B* ejects a β particle and a γ ray to become *radium C*, a nucleus higher in atomic number by unity but with the same mass number.

QUESTIONS

1. What is a Wilson cloud chamber? How does it work?

2. What kinds of particles do not leave tracks in a Wilson cloud chamber? Why not?

3. What is spontaneous disintegration?

4. What is transmutation? What change takes place in the nucleus of an atom when an α particle is emitted? What change takes place when a β particle is emitted?

5. Who discovered radioactivity? What were the circumstances?

6. Who discovered radium, polonium, and thorium?

7. What names are given to the different radioactive rays? Who unraveled this mystery?

8. What are α rays, β rays, and γ rays?

9. What are the relative ionizing powers and penetrating powers of the different rays?

10. Define or briefly explain each of the following: (a) radioactivity, (b) half-life, (c) range, (d) stable atom, and (e) cloud chamber track.

11. Name the first and last element of a radioactive series as well as three other elements belonging to the same series. List them in the order of decreasing mass number.

12. What is meant by the half-life of a radioactive isotope?

13. What is meant by the range of radioactive rays? Which of the three kinds of rays should have the greatest range?

14. What is the abbreviated designation for atomic nuclei? Give three examples.

PROBLEMS

1. What nuclides are isotopes of polonium? See Appendix VI. Give new symbols, atomic numbers as subscripts, and mass numbers as superscripts.

2. What nuclides are isotopes of thorium? See Appendix VI. Give new symbols, atomic numbers as subscripts, and mass numbers as superscripts.

3. What nuclides are isobars of lead-210, lead-211, and radium 228? See Appendix VI. Use the new symbols.

4. Make a table of the radioactive series starting with uranium-235, and ending with lead-207. See Appendix VI. Use the first six columns of Table 34A as a guide.

5. Make a table of the radioactive series starting with thorium-232, and ending with lead-208. See Appendix VI. Use the first six columns of Table 34A as a guide.

6. Make a table of the radioactive series starting with neptunium-237, and ending with bismuth-209. See Appendix VI. Use the first six columns of Table 34A as a guide.

7. Write down the nuclear reaction for the decay of (a) thorium-234, (b) uranium-234, and (c) radium-223. See Appendix VI.

8. Write down the nuclear reaction for the decay of (a) radium-228, (b) thorium-227, and (c) protoactinium-233. See Appendix VI.

9. Write down the nuclear reactions for the decay of (a) plutonium-238, (b) protoactinium-234, and (c) astatine-217.

10. Write down the nuclear reaction for the emission of an α particle by a polonium-211 nucleus. See Appendix VI.

11. Write down the nuclear reaction for the emission of an α particle by bismuth-211. See Appendix VI.

12. Write down the nuclear reaction for α particle emission by uranium-235. See Appendix VI.

13. Write the nuclear reaction for the emission of a β particle by bismuth-211. See Appendix VI.

14. If the radioactivity of a sample drops to one-sixteenth of its initial value in 2 hr and 20 min, what is its half-life?

15. If the activity of a radioactive sample drops to one-thirty-second of its initial value in 2 days, 12 hr, and 40 min, what is its half-life? [Ans. 12 hr, 8 min.]

16. How long will it take a sample of radium to decrease to 10% if its half-life is 1600 years?

17. How long will it take a sample of bismuth-210 to decrease to 10% of its original amount if its half-life is 5 days. [Ans. 16.61 days.]

35 ATOMIC COLLISIONS AND NUCLEAR DISINTEGRATIONS

In centuries past it was the dream of the alchemists to discover a process for changing base metals into gold. At long last it has been accomplished by the physicists, but not on the kind of paying basis dreamed of by so many.

The discovery of the transmutation of stable elements by controlled experiments is attributed to the great experimental genius of Lord Rutherford. Outsiders might say that the discovery was an accident, but to those who knew him well it was the result of a long series of well-planned experiments. True, he did not predict the phenomenon and then discover it, but his long experience with radioactivity and his keen insight enabled him to recognize the meaning and importance of the phenomenon when it was first observed. Due credit must also be given to the admirable work of his collaborators and to experimenters in other laboratories who have since carried the work much further.

35.1 ELASTIC COLLISIONS BETWEEN ATOMS

Collisions between free atomic particles were first studied by Rutherford with apparatus as shown in Fig. 35A. A long glass tube, containing a small sample of radioactive material *R*, was first thoroughly evacuated by means of a vacuum pump and then filled with a gas of known constitution. Alpha particles from the radioactive source were then permitted to travel through the gas to the other end of the tube where, upon passing through a thin aluminum foil to a fluorescent screen *S*, they could be observed as scintillations in the field of view of a microscope *M*.

With air in the tube *T* and *radium C′* as a source of α particles, scintillations could be observed with the screen as far back as 7 cm. With hydrogen in the tube, it was found that the distance could be greatly increased. Inserting thicker and thicker aluminum foils between *F* and *S* in front of the fluorescent screen, the range of the particles was calculated to be equivalent to 28 cm of air. The conclusion Rutherford drew from this result was that an α particle occasionally collides with a hydrogen atom, much the same as a large ball collides with a lighter one, imparting to it a greater velocity and hence a greater penetrating power. Of the many recoil hydrogen atoms from collisions with α particles, some of them undergo "head-on" collisions and go shooting off in a forward direction again with over half (more accurately 1.6 times) the speed of the incident α particle.

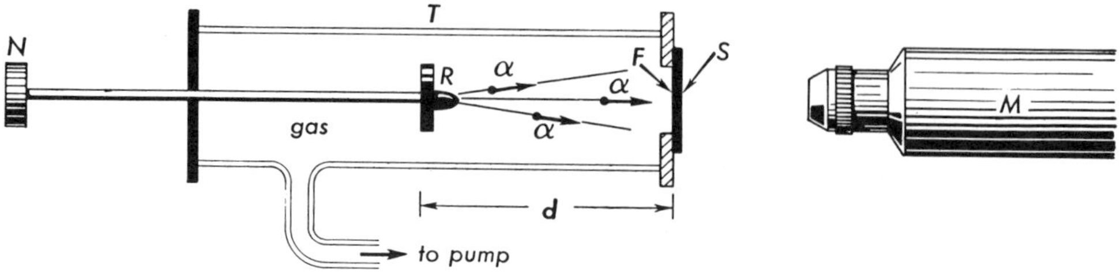

Fig. 35A. Rutherford's apparatus used in observing atomic collisions between α particles from radium and the atoms of a gas like hydrogen, helium, nitrogen, oxygen, etc.

The enormous increase in range is due principally to a higher velocity. Each hydrogen nucleus, called a *proton*, is stripped of its orbital electron, and, with but a single positive charge, produces fewer ions per centimeter path than an α particle with its double positive charge. Curiously enough, protons or α particles having the same velocity would have about the same range. The reason for this is that, whereas an α particle has double the charge of a proton and produces more ions per centimeter of its path, tending to slow it down more rapidly, it also has four times the mass and therefore four times the energy. Later, in Fig. 35H, we will see that, for *protons* and *α particles* with the same kinetic energy, the protons have about ten times the greater range.

A more convincing study of such atomic collisions can be made with a Wilson cloud chamber. Out of thousands of cloud chamber photographs of the ion tracks made by α particles from radioactive elements, one occasionally observes forked tracks of the type reproduced in Fig. 35B. When each of these pictures was taken, the cloud chamber contained a different gas. For photograph (a) the cloud chamber contained hydrogen, for (b) it contained helium, and for (c) it contained oxygen. Schematic diagrams of these same collisions are illustrated in Fig. 35C. Note in (b) that the angle between the two recoiling particles is 90°, a right angle.

Most of the elastic collisions between atoms are not "head-on" collisions but ones in which the incident particle strikes the other a glancing blow. When an α particle having an atomic mass of 4 collides with a hydrogen atom of atomic mass 1, the α particle is deviated only a little from its path, whereas the hydrogen atom nearly always recoils off at quite a large angle. This is in

(a) (b) (c)

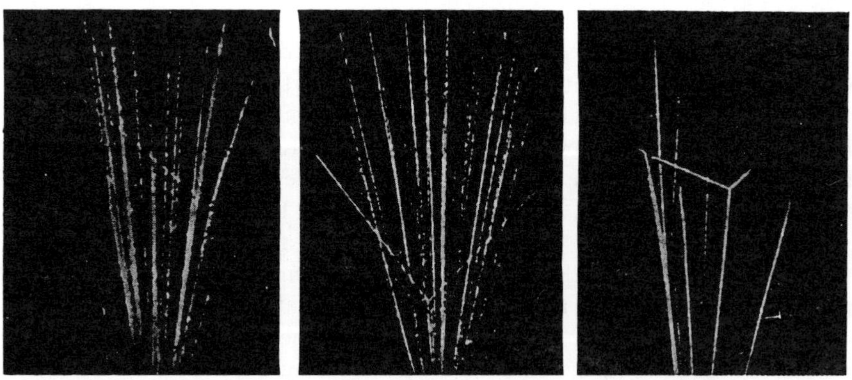

Fig. 35B. Wilson cloud-chamber photographs of collisions between α particles and (a) a hydrogen atom, (b) a helium atom, and (c) an oxygen atom. (After Rutherford, Chadwick, and Ellis.)

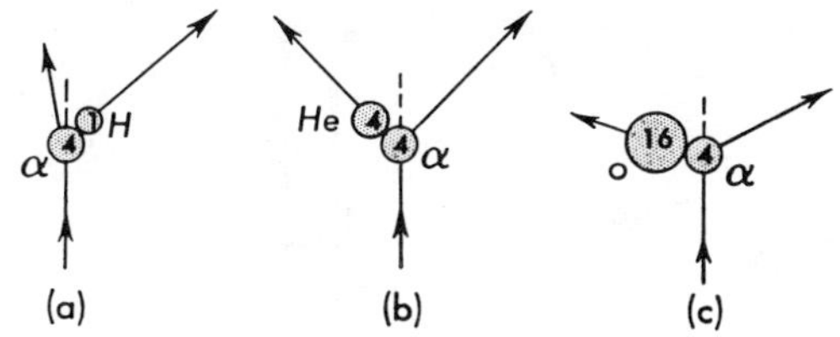

Fig. 35C. Diagrams of collisions between α particles and atoms of different mass.

agreement with the laws of conservation of energy and momentum applied to two perfectly elastic spheres.

When an α particle collides with a helium atom, both particles have the same atomic mass of 4, and the two almost always glance off at right angles to each other. The laws of mechanics show that for a "head-on" collision between perfectly elastic spheres of equal mass, one moving and one at rest, the incident particle is stopped by the collision and the second body goes on in the forward direction with all of the velocity.

When an α particle collides with an oxygen atom having an atomic mass of 16, the oxygen atom recoils to one side with a relatively low velocity, and the α particle glances off to the other side with a high or low velocity depending upon the angle of recoil. The oxygen atom with its greater mass and charge ionizes more particles per centimeter path and therefore leaves a shorter and heavier track.

It should be pointed out that atomic collisions involving such high velocities take place between the heavy nuclei of the atoms and are little affected by the light orbital electrons. If a nucleus is hit hard by a collision it will be partially or wholly denuded of its electrons. When it comes to rest it will soon pick up enough electrons to make it a neutral atom again.

35.2 THE DISCOVERY OF NUCLEAR DISINTEGRATION Upon repeating the range experiments illustrated in Fig. 35A, with a heavy gas in the tube *T*, Rutherford in 1919 made a new and startling discovery. When nitrogen gas (atomic weight, 14) was admitted to the tube, scintillations could be observed at a distance of 40 cm or more from the source. No such long-range particles had ever been observed before. What were these long-range particles? They could not be electrons or γ rays, for these are not capable of producing visible scintillations. Rutherford allowed the new rays to pass through a magnetic field and from their deflection discovered that they had the mass and charge of protons. In other words, the long-range particles were hydrogen nuclei.

Rutherford was not long in coming forward with the correct explanation of the phenomenon. An α particle, near the beginning of its range where its velocity is high, may make a "head-on" collision with a nitrogen nucleus and be captured. This capture is then followed immediately by a disintegration in which a proton is ejected with high speed. The process is illustrated in Fig. 35D, and the transformation can be represented by the following simple reaction.

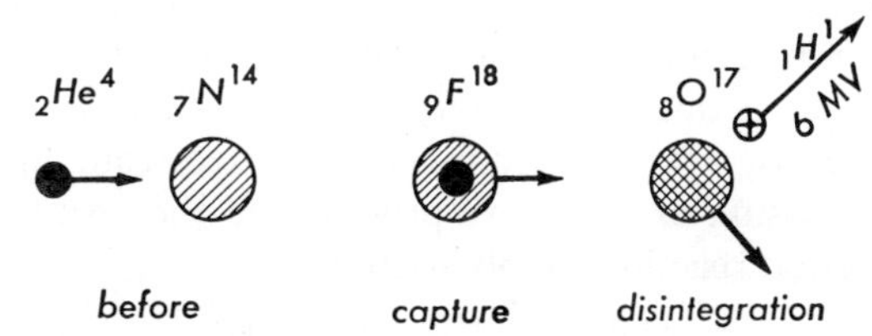

Fig. 35D. The disintegration of a nitrogen nucleus by a high-speed α particle.

$$_2He^4 + {}_7N^{14} = ({}_9F^{18}) = {}_8O^{17} + {}_1H^1 \quad (35a)$$

When the α particle with a charge of +2, and mass 4, collides with the nitrogen nucleus with a charge of +7 and mass 14, they form a single particle with a charge of +9 and mass 18. Since an atom with a nuclear charge of +9 would be expected to have all the chemical properties of *fluorine*, atomic number 9, the newly formed nucleus is labeled $_9F^{18}$.

An examination of the table of isotopes, however (see Appendix III), shows that no such isotope exists in nature. This reason becomes apparent when it is realized that such a combination of particles is not stable. A fluorine nucleus of mass 18 is unstable and disintegrates by discharging a

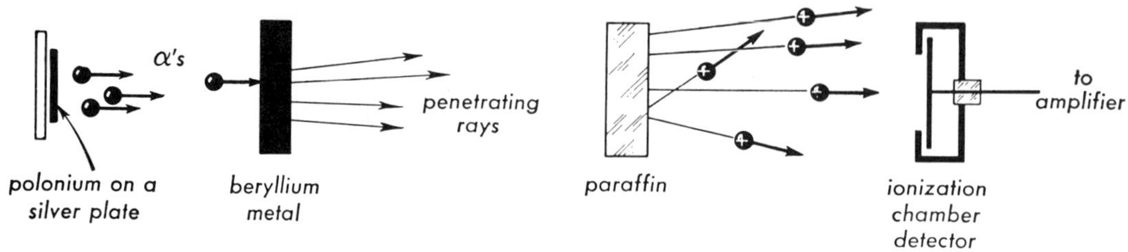

Fig. 35E. The experiment by which Chadwick discovered the neutron. The penetrating rays are neutrons and α rays.

proton, a particle with a charge of +1 and a mass of 1. This leaves behind a residual nucleus with a charge of +8, and a mass of 17. Under atomic number 8 in Appendix III, an oxygen isotope of mass 17 is seen to be stable, and one that is found in nature.

Thus the above disintegration process started with two stable nuclei, *helium* and *nitrogen*, and out of them were created two new stable nuclei, *oxygen* and *hydrogen*. This is called a *transmutation* of elements. Because the intermediate step indicates but a momentary existence of a fluorine nucleus, ${}_9F^{18}$, this step is often omitted from any discussion of the above process and the disintegration reaction simply written

$$ {}_2He^4 + {}_7N^{14} = {}_8O^{17} + {}_1H^1 \qquad (35b)$$

The equal sign is often replaced by an arrow.

Such transformation reactions are like equations and must balance: first, the total amount of charge must remain the same, and second, the mass numbers must balance. The first of these is accomplished by having the sum of the subscripts on one side of the reaction equal to the sum of the subscripts on the other side, and the second by having the sum of the superscripts the same on both sides. In every known atom the subscript, representing the nuclear charge, is the sole factor determining the chemical element to which the atom belongs.

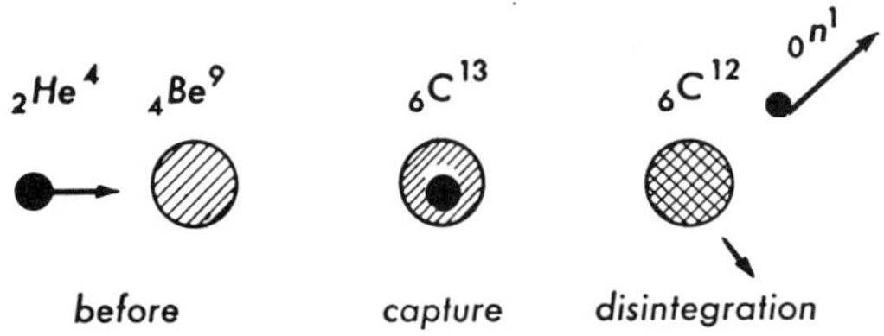

Fig. 35F. Discovery of the neutron.

35.3 CHADWICK'S IDENTIFICATION OF THE NEUTRON In 1932 Chadwick,* in England, performed an experiment for which he was later awarded the Nobel Prize in physics in 1935. As diagramed in Fig. 35E, his experiment consisted of bombarding a beryllium target with α particles. Penetrating particles emerging from the beryllium were permitted to impinge upon a block of paraffin from which protons were found to emerge with high speed. From energy calculations he was able to show that the penetrating rays were uncharged particles with the mass of protons, and these he called *neutrons*. The disintegration taking place in the metal target is the following (see Fig. 35F):

$$ {}_2He^4 + {}_4Be^9 = {}_6C^{12} + {}_0n^1 \qquad (35c)$$

---

* SIR JAMES CHADWICK (1891–), British physicist, received the Nobel Prize in physics in 1935 for the "discovery of the neutron." He was born and educated in Manchester, where he came under the influence of Ernest Rutherford at Victoria University. In 1913, he studied at the Charlottenburg Institution, Berlin, under Hans Geiger, a former associate of Rutherford's, and at the outbreak of World War I, was detained in a concentration camp in Ruhleben, Germany. After the war, he returned to England and worked under Rutherford at the Cavendish Laboratory in Cambridge. Chadwick became Professor of Physics at the University of Liverpool in 1935, was knighted in 1945, awarded the Copley Medal in 1950, and the Franklin Medal in 1951.

The α particle, ${}_2He^4$, makes a collision and unites with a beryllium nucleus, ${}_4Be^9$, causing a disintegration; whereupon a neutron, ${}_0n^1$, is expelled with high velocity. The residual particle with a charge of +6 and mass of 12 units is a stable carbon nucleus such as found in nature.

The penetrating rays from the beryllium block in Fig. 35E are mostly neutrons which, in bombarding the paraffin block, collide elastically with hydrogen atoms, knocking them out on the other side. An elastic "head-on" collision between two particles of the same weight, like a neutron and proton, finds the entire velocity of one transferred to the other; the neutron is stopped and the proton goes on. The protons, having a positive charge, can be observed by their tracks in a Wilson cloud chamber, whereas neutrons cannot.

The reason fast neutrons have such a high penetrating power is that they are not slowed down by ionizing atoms as they pass close by them. A proton, electron, or α particle has a charge and can ionize atoms by attracting or repelling electrons from a distance, but a neutron without a charge cannot do this. It must make a direct collision with another particle to be slowed down or stopped.

Neutrons can now be produced in such intense beams that they are often used in place of X rays wherever radiations of high penetrating power are desired.

35.4 THE NUCLEUS CONTAINS NEUTRONS AND PROTONS Since the time of Chadwick's identification of the neutron as an elementary particle, our ideas concerning the nucleus of the atom have had to be modified. We now believe that the nucleus contains but two kinds of particles, neutrons and protons. Each neutron has a mass of one unit and no charge, whereas each proton has a mass of one unit and a positive charge of one unit. This differs from the older idea that the nucleus contains protons equal in number to the atomic weight, and enough electrons to neutralize the surplus charge in excess of the amount specified by the atomic number.

Since only the proton has a charge, any given nucleus of atomic number Z and mass number A is now believed to have Z protons and A-Z neutrons, and that in a neutral atom the number of protons is equal to the number of orbital electrons. The number of neutrons A-Z is called the neutron number N. The nuclear particles of a few of the elements of the periodic table are given in Table 35A, as examples.

TABLE 35A. Number of Neutrons and Protons in the Nuclei of a Few Elements

Atom	Protons Z	Neutrons N	Nucleons A
${}_1H^1$	1	0	1
${}_1H^2$	1	1	2
${}_2He^4$	2	2	4
${}_3Li^6$	3	3	6
${}_3Li^7$	3	4	7
${}_4Be^9$	4	5	9
${}_4Be^{10}$	4	6	10
${}_5B^{11}$	5	6	11
${}_7N^{13}$	7	6	13
${}_8O^{16}$	8	8	16
${}_{11}Na^{23}$	11	12	23
${}_{29}Cu^{65}$	29	36	65
${}_{80}Hg^{200}$	80	120	200
${}_{92}U^{238}$	92	146	238

Schematic diagrams of the nucleus of five different atoms are given in Fig. 35G.

The nuclei of all elements, whether they are stable or radioactive, are called *nuclides*. Since all nuclides are composed of protons and neutrons, these nuclear "building blocks" are called *nucleons*.

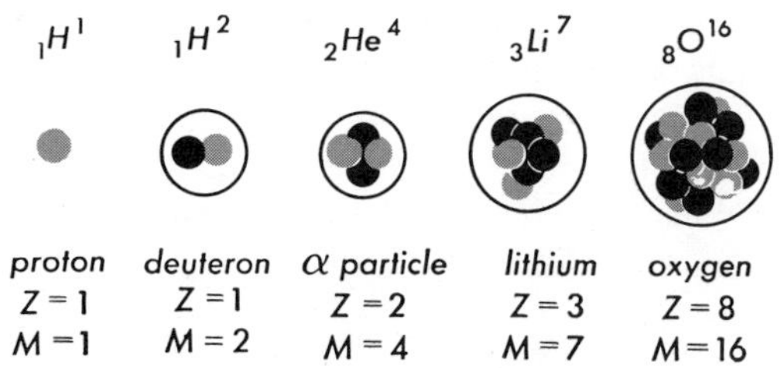

Fig. 35G. Diagrams showing the number of protons and neutrons in the nuclei of hydrogen, deuterium, helium, lithium, and oxygen.

35.5 MASS IS A FORM OF ENERGY Einstein, in working out the theory of relativity, arrived at a number of simple equations concerning the nature of the physical world. It is important at this point to consider one of these equations, since its proof can be demonstrated by, and is needed to explain, atomic disintegration experiments. The relation referred to is known as *Einstein's mass-energy equation*. See Sec. 25.6. In its simplest form this equation is written

$$\boxed{E = mc^2} \tag{35d}$$

where m represents mass, c the velocity of light, and E the energy equivalence of mass.

This relation would predict that mass can be turned into energy, or energy into mass. In other words, mass is a form of energy, for if a quantity of mass m could be annihilated, a definite amount of energy E would appear in some other form. To illustrate this, suppose that a 1-Kg mass could be completely annihilated and the liberated energy given to another body in the form of kinetic energy.

$$\begin{aligned} E &= 1 \text{ Kg} \times (3 \times 10^8 \text{ m/s})^2 \\ &= 9 \times 10^{16} \text{ joules} \end{aligned} \tag{35e}$$

In foot-pounds of energy this is equivalent to 7×10^{16} ft-lb, or enough energy to propel the largest battleship around the world.

The annihilation of mass then is a source of undreamed of energy. Disintegration is one means, however, whereby mass can be annihilated or created more or less at the will of the experimenter.

If an atom, a part of an atom, or an electron is annihilated, the energy may either be transformed into kinetic energy and given to another atomic particle in the form of a velocity or appear as a γ ray of specified frequency ν and energy $h\nu$. To find the equivalence between *mass energy*, *γ-ray energy*, and *kinetic energy*, all of the following quantities are equated to each other.

$$\boxed{E = mc^2 = h\nu = \tfrac{1}{2}mv^2 = Ve} \tag{35f}$$

It is customary among physicists to express each of these energies in terms of V in volts. Thus one speaks of a million volt γ ray, a three million volt electron, or a 12.5 million volt proton, etc. This terminology is used for convenience only, and denotes the value of V in the above equation which, with the electronic charge substituted for e, gives the energy of the γ-ray photon, or of the moving atomic particle in electron volts.

To give several examples, suppose that an electron were to be annihilated and we wished to express the liberated energy in volts. To calculate V we make use of the equality between the second and last terms of Eq. (35f); that is,

$$mc^2 = Ve \tag{35g}$$

Substituting the known electronic mass $m = 9.1 \times 10^{-31}$ Kg, $e = 1.60 \times 10^{-19}$ coulomb, and $c = 3 \times 10^8$ m/s, we obtain $V = 511{,}000$ volts.

$$\text{Annihilation energy of 1 electron} = 0.511 \text{ MeV} \tag{35h}$$

This means that, if the energy liberated by the annihilation of an electron could be given to another electron in the form of kinetic energy, that electron would have a velocity equivalent to half a million volts. In other words, it would have the same velocity and energy as an electron which has been accelerated by a potential of 511,000 volts in a tube of the kind shown in Fig. 30I.

If the velocity v in the kinetic energy term, $\frac{1}{2}mv^2$, is greater than 10% the speed of light, the kinetic energy E_k in Eq. (35f) must be replaced by the relativity term.

$$E_k = m_0c^2\left[\frac{1}{\sqrt{1 - v^2/c^2}} - 1\right]$$

In certain processes to be described in the next chapter, atomic mass is sometimes converted into a γ ray. To calculate the frequency of a single γ ray, made by the annihilation of an electron, we make use of the second and third terms in Eq. (35f), that

$$mc^2 = h\nu \tag{35i}$$

and find $\nu = 1.2 \times 10^{20}$ Hz. Such a frequency is 200,000 times that of visible green light and the wavelength about two hundredths of an angstrom unit. As a proof of Eq. (35i), experiments show that γ rays of twice this frequency, or higher, will produce electron pairs by collision. See Fig. 36H.

As a third and last example, we may calculate the energy liberated by the annihilation of one *atomic mass unit*, i.e., the energy equivalent to one twelfth the mass of an ordinary carbon atom. Substituting the known values of m, c, and e in Eq. (35g), the following is obtained:

Annihilation energy of 1 atomic mass unit

$$= 931.478 \text{ MeV} \qquad (35j)$$

This is a useful number to memorize, for it is frequently used in the following chapters. Atomic mass unit is abbreviated *amu*.

35.6 ATOMIC MASSES ARE NOT WHOLE NUMBERS When an α particle collides with the nucleus of an atom and produces a disintegration as shown in Fig. 35D, the total energy before collision must be equal to the total energy after collision. To verify this, all forms of energy involved in the process must be included; namely, (1) *the kinetic energy of all particles*, (2) *the energy of a γ ray if one is involved*, and (3) *the mass-energy*. The latter is necessary since disintegration experiments show that the total mass of the two colliding particles is not in general equal to the total mass after disintegration. To test this change, it is necessary that we know the exact masses of all atoms individually.

Mass spectrographic measurements by Aston, Bainbridge, and others show that the masses of atoms are not exactly whole number values as given in Appendix III. A list of the most accurate mass determinations of some of the lighter elements of the periodic table is given in Appendices III and IV. These values are all based upon the carbon isotope 12 as having a mass of exactly 12.0000.

35.7 CONSERVATION OF ENERGY IN NUCLEAR DISINTEGRATIONS To illustrate the law of conservation of energy as it applies to nuclear disintegrations, consider Rutherford's first experiment, shown in Figs. 35A and 35D, where α particles from ${}_{84}Po^{214}$ passing through nitrogen gas make collisions with, and disintegrate, nitrogen nuclei. The total energy of any two particles before impact will be the sum of the masses of the two nuclei, ${}_2He^4 + {}_7N^{14}$, plus their kinetic energy E_{k1}. The total energy after impact will be the sum of the masses of the two nuclei, ${}_8O^{17}$ and ${}_1H^1$, plus their kinetic energy E_{k2}. If we insert accurately known masses, the reaction becomes

$${}_2He^{4.002604} + {}_7N^{14.003074} + E_{k1} = {}_8O^{16.999133} + {}_1H^{1.007825} + E_{k2} \qquad (35k)$$

It is customary to express the energies E_{k1} and E_{k2} in *mass units* or in million electron volts, MeV. The kinetic energy before impact is confined to the α particle from ${}_{84}Po^{214}$, which has been measured and found to be equivalent to 7.7 MeV. Dividing this value by 931, from Eq. (35j), gives the equivalent of 0.00827 mass unit. Adding mass-energy for both sides of Eq. (35a), we must obtain the same total.

Using mass values given in Appendix III we obtain

$$\begin{array}{rlrl} {}_2He^4 = & 4.002604 & {}_8O^{17} = & 16.999133 \\ {}_7N^{14} = & 14.003074 & {}_1H^1 = & 1.007825 \\ E_{k1} = & 0.008270 & E_{k2} = & ? \\ \hline \text{Total} = & 18.013948 & & 18.013948 \end{array} \qquad (35l)$$

Simple addition and subtraction show that, to yield the proper sum for the right-hand column, E_{k2} must be equal to 0.006990 mass unit. Multiplying by 931 gives, this time, 6.51 MeV as the energy liberated. This is the liberated energy utilized in the "explosion" which drives the proton and oxygen nuclei apart.

In general, when an atomic nucleus disintegrates by splitting up into two particles, the annihilation energy is divided between them. Experiments show that this division takes place according to the ordinary laws of mechanics. When one particle is much more massive than the other, *the kinetic energies of the two particles are approximately inversely proportional to their respective masses.*

When, in the above example, the available energy is divided between an oxygen nucleus of mass 17 and a proton of mass 1, the ${}_8O^{17}$ nucleus acquires an energy of 0.36 MeV, and the proton an energy of 6.15 MeV. A proton with this kinetic energy and velocity has a range of 49 cm in air.

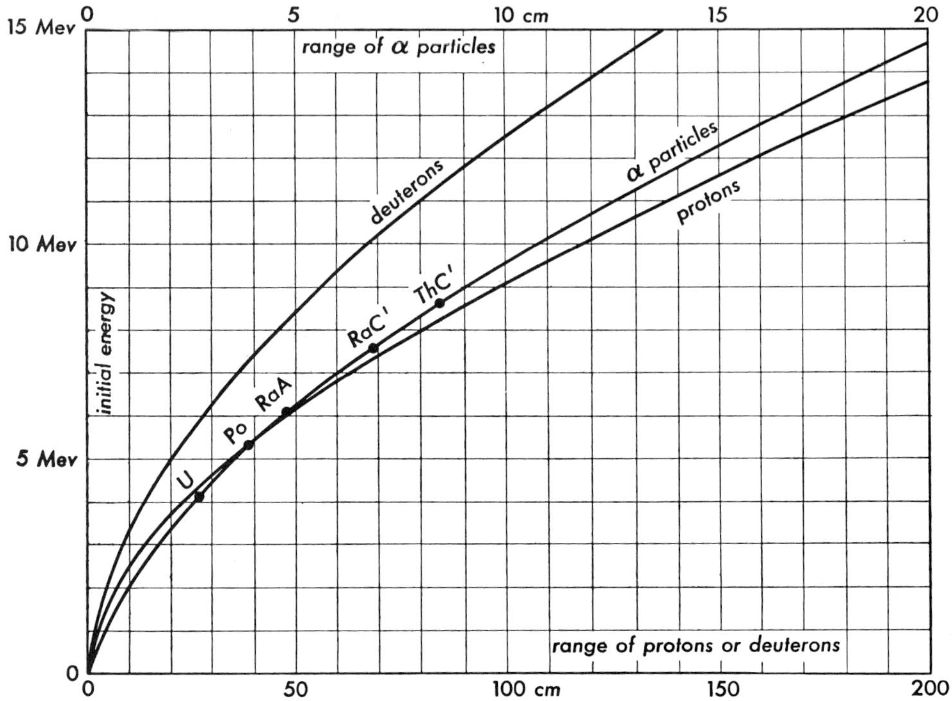

Fig. 35H. Graphs giving the range of protons, deuterons, and α particles for different initial energies in million electron volts.

Recent repetitions of Rutherford's experiment give measured ranges of 48 cm and an energy of 6 MeV, in good agreement with the calculation.

Graphs showing the ranges of protons, deuterons and α particles for different energies are drawn in Fig. 35H. These three curves are extremely useful in disintegration work, for knowing the range of a particle, its initial energy becomes known, or knowing the initial energy, the range becomes known. It is interesting to note that for the same energy a proton has ten times the range of an α particle.* Deuterons will be considered in detail in succeeding chapters.

* *Radioactivity and Nuclear Physics*, by J. M. Cork, D. Van Nostrand Co., Inc.

QUESTIONS

1. What is meant by a perfectly elastic collision? Do atomic nuclei ever collide perfectly elastically?

2. Briefly explain Rutherford's experiment for measuring the range of α particles in different gases. Make a diagram. What unusual results did he observe?

3. Who discovered nuclear disintegration by α-particle collisions? What was the experiment? Can you write down the nuclear reaction?

4. Who discovered the neutron? What was the experiment? Why don't neutrons produce cloud chamber tracks?

5. How many protons are to be found in a given nucleus? How many neutrons are there?

6. Make a table of the following atoms, giving the number of protons and neutrons in each nucleus: (a) Na^{23}, (b) Co^{59}, (c) Sn^{120}, (d) Bi^{209}, and (e) Cf^{244}.

7. What is the Einstein mass energy equation? What does it mean?

8. How can energy be expressed in volts? What is the basic relation between the various forms of atomic energy and the equivalent energy in volts?

PROBLEMS

1. Calculate the mass equivalent of an energy of (a) 15 MeV, (b) 25 MeV, and (c) 100 MeV. [Ans. (a) 0.01611 amu. (b) 0.02685 amu. (c) 0.1074 amu.]

2. Find the atomic mass equivalent of an energy of (a) 6 MeV, (b) 9.6 MeV, and (c) 12.4 MeV.

3. From the graphs plotted in Fig. 35H write down the ranges of the following particles: (a) 5 MeV protons, (b) 7 MeV protons, and (c) 9 MeV alpha particles. [Ans. (a) 33 cm. (b) 63 cm. (c) 9.0 cm.]

4. Using the graphs of Fig. 35H, find the ranges of the following particles: 4.5 MeV protons, (b) 9.8 MeV protons, and (c) 14 MeV alpha particles.

5. Protons from a radioactive element are found to have a range of 125 cm in air. From the graphs in Fig. 35H, find their energy in MeV. [Ans. 10.3 MeV.]

6. Complete the following disintegration reactions:

$$_1H^1 + {_4Be^9} \rightarrow {_3Li^6} + ?$$
$$_2He^4 + {_{13}Al^{27}} \rightarrow {_{15}P^{30}} + ?$$
$$_1H^2 + {_{15}P^{31}} \rightarrow {_{15}P^{32}} + ?$$
$$_1H^2 + {_6C^{12}} \rightarrow {_7N^{13}} + ?$$

7. Complete the following nuclear reactions:

$$\text{(a)}\ {_2He^4} + {_5B^{11}} = {_7N^{14}} + ?$$
$$\text{(b)}\ {_2He^4} + {_7N^{14}} = ? + {_1H^1}$$
$$\text{(c)}\ {_2He^4} + {_8O^{16}} = {_9F^{19}} + ?$$
$$\text{(d)}\ {_2He^4} + {_{11}Na^{23}} = {_{12}Mg^{26}} + ?$$

8. Complete the following disintegration reactions:

$$_1H^2 + {_8O^{16}} \rightarrow {_7N^{14}} + ?$$
$$_2He^4 + {_{13}Al^{27}} \rightarrow {_{14}Sl^{30}} + ?$$
$$_1H^2 + {_5B^{10}} \rightarrow {_6C^{11}} + ?$$
$$_1H^1 + {_3Li^6} \rightarrow {_2He^4} + ?$$

9. Alpha particles from a radioactive element are found to have range of 22 cm in air. Extrapolating the graph plotted in Fig. 35H, find their energy in MeV. [Ans. 15.8 MeV.]

10. Deuterons from a small cyclotron are found to have a range of 160 cm in air. Extrapolating the graph in Fig. 35H, find their energy in MeV.

11. If a deuteron could be completely annihilated, how much energy would be created, in MeV? See Appendix III. [Ans. 1875 MeV.]

12. If an entire carbon atom could be annihilated, ${}_6C^{12}$, how much energy, in MeV, would be created?

13. If a neutron were to be completely annihilated, how much energy in MeV would be produced? [Ans. 940 MeV.]

14. If an entire helium atom could be annihilated, how much energy in joules would be produced?

15. If one gram of mass could be completely annihilated, how many joules of energy would be produced? [Ans. $8.9^9 \times 10^{13}$ J.]

16. If the following atomic masses are annihilated, how much energy in MeV would be liberated? (a) 0.01735 amu, (b) 0.13680 amu, and (c) 0.00254 amu.

17. If the following masses are annihilated, how much energy in MeV is liberated? (a) 0.01460 amu, (b) 0.00571 amu, and (c) 0.13295 amu. [Ans. (a) 13.59 MeV, (b) 5.32 MeV, (c) 123.8 MeV.]

18. What is the equivalent energy in electron volts of a gamma ray with a wavelength of 0.0250 Å?

19. What is the energy content of a gamma ray of wavelength $\lambda = 0.00472$ Å? [Ans. 2.626 MeV.]

20. If 1% of a kilogram of uranium could be converted into energy, how many joules would be produced?

36 | COSMIC RAYS

At or near sea level an average of twenty-eight cosmic rays pass through the average human being per second. Whether such low intensity radiation has cumulative beneficial or detrimental effects is not yet known.

Karl K. Darrow has described the subject of cosmic rays "as unique in modern physics for the minuteness of the phenomena, the delicacy of the observations, the adventurous excursions of the observers, the subtlety of the analysis, and the grandeur of the inferences." * It is impossible for anyone to say when and by whom cosmic rays were first studied. From the time of the discovery of radioactivity by Becquerel (in 1896) and the discovery of radium by the Curies but a few months later, the radioactive rays from the ground, air, and outer space have been investigated by many scientists. Extending over a period of some fifty years these investigations have led to some of the most interesting and important discoveries in the structure of atomic nuclei.

36.1 EARLY EXPERIMENTS It has long been known that a charged electroscope, if left standing for some little time, will discharge regardless of how well the gold leaf is insulated. Realizing that the rays from radioactive materials can be stopped by a sufficient thickness of heavy matter, Rutherford and Cooke (in Canada, 1903) surrounded an electroscope with a thick wall of brick and found very little decrease in the rate of discharge. McLennan and his co-workers (also in Canada) lowered an electroscope into a lake hoping that the thick layer of water would screen off the rays. This experiment, like the other, failed. In 1910 Glockel, with an electroscope, rose nearly 3 mi in a balloon in order to get away from the ground radiation, but to his astonishment he found that the rate of discharge did not decrease but increased the higher he went. The same effect was observed by Hess* (in Austria,

* This quotation is from a monograph on cosmic rays by Karl K. Darrow, American physicist and for many years a writer on modern physics at the Bell Telephone Laboratories.

* VICTOR FRANCIS HESS (1883–1964), Austrian physicist, shared the 1936 Nobel Prize with C. D. Anderson for his discovery of cosmic rays. Born at Schloss, Waldstein, he was educated at Graz, Vienna. He performed experiments on radioactivity at the Institute for Radium Research of the Vienna Academy of Sciences, Vienna, and lectured in physics at the Vienna Veterinary College. He became Director of the U. S. Radium Corporation in 1921, and consultant to the U. S. Department of the In-

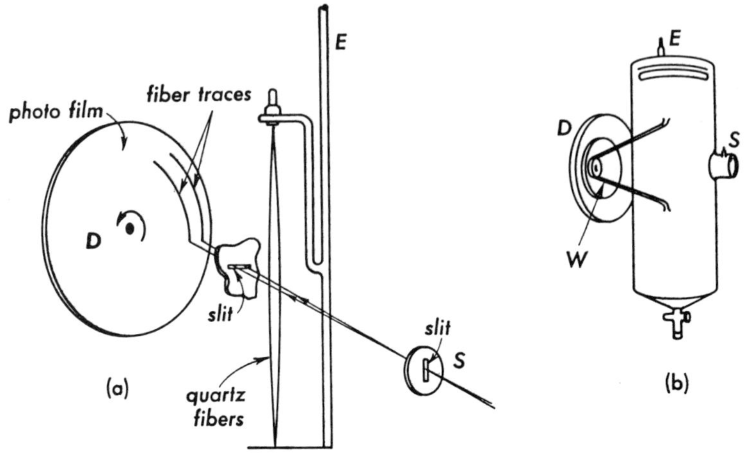

Fig. 36A. Diagrams of one of the sensitive electroscopes sent up into the stratosphere by Millikan and Bowen to measure cosmic rays. (a) Schematic diagram, (b) scale drawing of entire instrument 6 in. high.

1911) and Kolhörster (in Germany, 1914). Rising to heights as great as $5\frac{1}{2}$ mi, both of these observers independently found that the intensity of these unknown rays became more intense the higher they went. Because in one of his scientific publications concerning these results Hess suggested the possibility that some kind of penetrating rays were entering the earth's atmosphere from outer space, he is usually credited with the discovery of cosmic rays. For this reason he was granted the Nobel Prize in physics for the year 1936.

36.2 MILLIKAN AND BOWEN'S DISCOVERY Soon after World War I (1922), R. A. Millikan, with the help of I. S. Bowen, constructed several small self-recording string electroscopes. Making use of their wartime experiences with sounding balloons, they sent these electroscopes high into the stratosphere by fastening each one to two sounding balloons. As shown in Fig. 36A, the string electroscope E consisted of two gold-covered quartz fibers insulated and mounted with their ends together. When they were charged the fibers spread apart due to mutual repulsion, and as they discharged they slowly came together.

Daylight, passing through a narrow vertical slit S in the instrument case, cast a shadow of the center section of the fibers on a rotating disk D containing a photographic film. As the film turned slowly and the fibers came together, they left a double trace as indicated in the diagram. On the same film a small oil manometer recorded the height of ascent and a small thermometer recorded the temperature. The film was driven by a watch W, the whole apparatus weighing only 7 oz. On one of the best record flights only one of the balloons burst at a height of 10 mi and the other brought the instruments safely to earth. Like the earlier results obtained by other experimenters, Millikan and Bowen found the ionization to increase with increasing altitude. After extending the observations of previous workers to higher altitudes Millikan and Bowen became convinced, and announced their belief, that the rays were coming from interstellar space.

36.3 THE PENETRATION OF COSMIC RAYS In order to determine the nature of the new rays, Millikan and his co-workers, Otis, Cameron, and Bowen, in the fall of 1922, began an extensive study of the penetrating power of cosmic rays. Since cosmic rays penetrate our atmosphere of

terior, Bureau of Mines. In 1938, he became Professor of Physics at Fordham University, New York, and became a naturalized citizen of the U. S. A. in 1944.

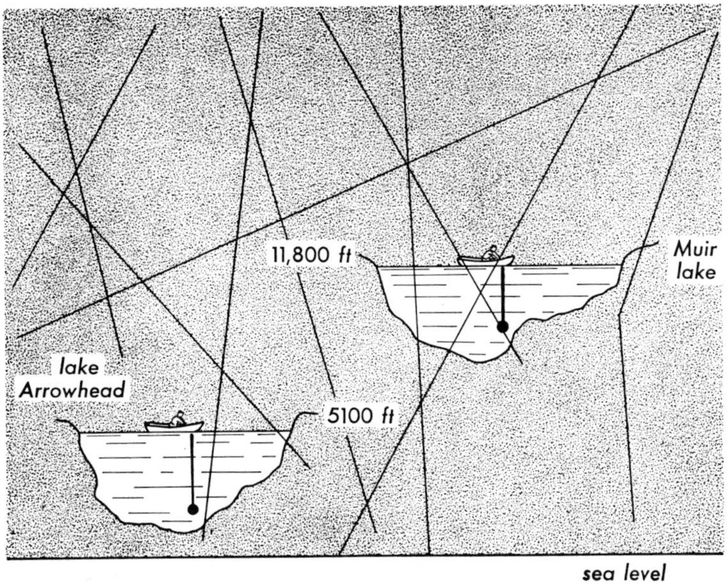

Fig. 36B. Self-recording electroscopes lowered into deep snow-fed lakes are used to measure the absorption of cosmic rays.

many miles of air, how far might they penetrate beyond?

Self-recording electroscopes were lowered to various depths in snow-fed lakes as illustrated schematically in Fig. 36B. Measurements taken at Arrowhead Lake in Southern California (at an elevation of 5100 ft) agreed approximately with those taken at Muir Lake near Mt. Whitney (at an elevation of 11,800 ft), provided one took into account the increased air path for the lower elevation. The extra mile and a quarter of air is equivalent in weight to 6 ft of water. As cosmic rays penetrate deeper and deeper below the surface of water their number decreases, until at a depth of 100 ft the intensity is reduced, to about one ten-thousandth of that at the surface. With very sensitive electroscopes, cosmic radiation capable of penetrating 2000 ft of water has more recently been detected. This is a far greater penetrating power than that possessed by any known X rays or γ rays.

36.4 THE GEIGER-MUELLER TUBE COUNTER

There are a number of devices in common use for the observation and measurement of cosmic rays. These are:

- electroscopes
- cloud chambers
- bubble chambers
- Geiger-Mueller counters
- scintillation counters
- ionization gauges
- photographic emulsions
- semi-conductors
- spark chambers

The Geiger-Mueller tube counter, named after its inventors, is one of the simplest electrical instruments ever designed. See Fig. 36C. It consists of an open-ended copper cylinder from 1 to 25 in. long fitted inside a thin-walled glass cylinder with a fine tungsten wire stretched along the middle. After the tube has been partially evacuated (a pressure of from 5 to 10 cm of mercury is convenient), a potential of about 1000 volts is applied, the positive to the center wire and the negative to the cylinder.

When a single cosmic ray or high-speed parti-

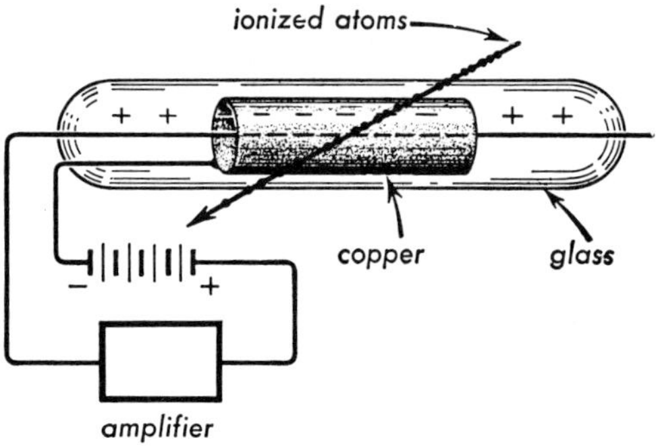

Fig. 36C. Diagram of a Geiger-Mueller tube.

cle from a radioactive source goes through a Geiger-Mueller tube, ions are created by the freeing of electrons from air molecules. These freed electrons are attracted by the positively charged wire and move toward it, acquiring within a very short distance a high velocity of their own. Because of this velocity they, too, can ionize other atoms, thus freeing more electrons. This multiplication of charges repeats itself in rapid succession, producing within a very short interval of time an *avalanche of electrons toward the central wire*. This sudden surge of charge is equivalent to the flow of a small current impulse along the electrical circuit. When this current has been intensified by an amplifier it may be made to operate an electric switch, a radio loudspeaker, or any kind of an electrical device.

Quite frequently the impulses of a Geiger-Mueller tube are made to operate a small counting device. Each cosmic ray particle passing through the tube is therefore counted automatically. The number of counts received per second depends upon the size of the counter tube. An average-sized tube 1 in. in diameter and several inches long, at sea level, gives from 50 to 100 counts per minute.

To observe the direction of the greatest cosmic ray intensity, a cosmic ray telescope is used. Such a telescope is made by connecting two or more Geiger-Mueller tubes *in coincidence*, and mounting them on a common support some distance apart. Tubes in coincidence are so connected electrically that a current will flow in the accompanying electric circuit only when both tubes discharge at the same time.

When the tubes are set one above the other, as shown in Fig. 36D (a), a single cosmic ray on

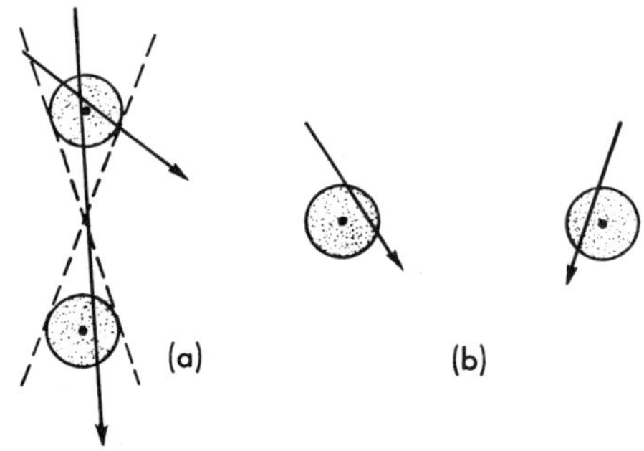

Fig. 36D. Diagram of two Geiger counters. Connected in coincidence, they form a cosmic-ray telescope.

going through both cylinders will cause a current to flow and a count to be made. If, however, a particle goes through one and not the other, no count is recorded. Experiments at sea level show that when the telescope is mounted in the horizontal position (b), few counts are made; whereas when it is mounted in a vertical direction, many more counts are recorded. The interpretation to be made, therefore, is that cosmic rays come principally from overhead.

36.5 THE WILSON CLOUD CHAMBER In one important application of the Geiger counter development a Wilson cloud chamber (see Sec. 34.6) is inserted between two Geiger counter tubes as shown in Fig. 36E and a photograph of each cosmic ray is taken. Thousands of such photographs are made automatically by having a single cosmic ray take its own picture. This is accomplished by allowing the sudden electric current from the two counter tubes, produced by a single ray in transit, to open and close a camera shutter, to cause the cloud chamber to expand, and to flash a light, illuminating the fog track that forms.

In the reproduction of Fig. 36E, either one of the two cosmic rays would have tripped the electrical devices and taken the picture. It should be noted that both rays passed right through a

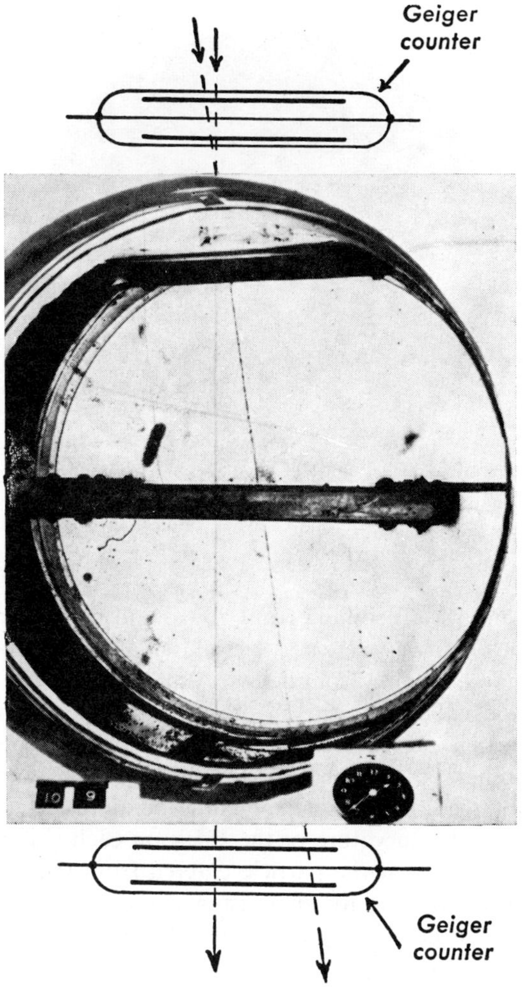

Fig. 36E. With a Wilson cloud chamber mounted between two Geiger tubes connected in coincidence the cosmic rays are made to take their own picture. (After Brode.)

0.5 in. lead plate at the center of the cloud chamber without being deviated. Cloud chamber pictures are not photographs of cosmic rays but of the path traversed by the rays.

36.6 PRIMARIES AND SECONDARIES Experimental observations show that the cosmic rays entering our atmosphere are almost entirely composed of positively charged atomic nuclei. About 89% of these so-called *primary cosmic rays* are protons, and the other 11% (by mass) are about 90% α particles and 10% heavier nuclei like carbon, nitrogen, oxygen, iron, etc. See Table 36A.

Upon entering the atmosphere, a high-energy primary particle soon collides with another atomic nucleus, splitting one or both particles into a number of smaller nuclear fragments, each one of which carries away some of the primary's energy. These high-speed particles in turn collide with other nuclei, further dividing their energy to produce other high-speed particles. All of these with the exception of the primary particle are called *secondary cosmic rays*.

One of the results of cosmic-ray collision processes is the creation of very high frequency and highly penetrating gamma rays. These photons, too, are included in the classification *Secondary Cosmic Rays*.

At a height of some 15 mi, about ten to fifteen times as many secondary cosmic rays exist as have entered the atmosphere as primaries. At this level, where more than nine tenths of the earth's atmosphere still lies below, as many rays are observed moving in a horizontal direction as in the vertical. From the diagram presented in Fig. 36F it becomes quite clear why the primaries are difficult to distinguish from the far greater number of secondaries.

At lower altitudes the total intensity decreases since many of the secondaries produced above are stopped by collision. In other words, so much energy is lost by successive collisions that the energy is gradually absorbed as heat motion by the air molecules. By the time sea level is reached, the remaining rays consist principally of a few high-speed secondaries and primaries. Even at sea level some of these have enough energy left to penetrate several hundred, and even several thousand, feet of earth and water.

It should be borne in mind that between collisions of the type indicated in Fig. 36F, each cosmic ray particle is continually being slowed down as it "plows through" thousands of air molecules, knocking electrons free to produce ions. These are the ions on which fogdrops form, revealing the path in a cloud chamber.

Although γ rays also lose energy by collisions with atoms to produce Compton electrons, the

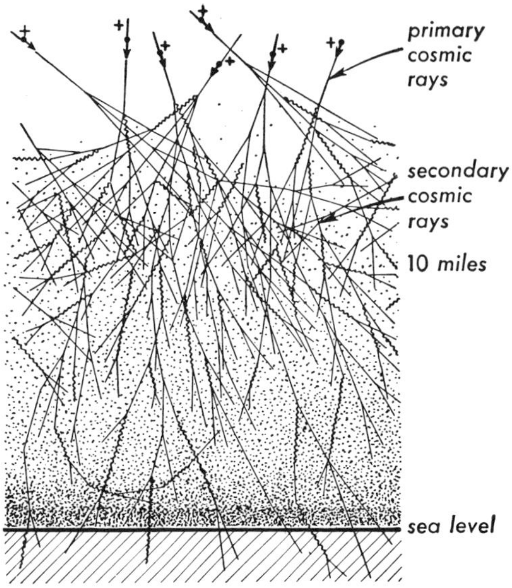

Fig. 36F. Schematic illustration of secondary cosmic rays produced from primaries entering the earth's atmosphere.

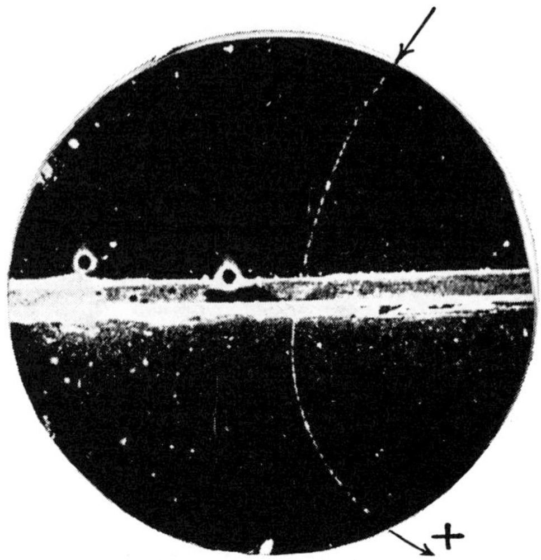

Fig. 36G. Wilson cloud-chamber photograph of a positron. (After Anderson.)

γ rays themselves do not leave visible tracks in a Wilson cloud chamber. The reason for this is that collisions are few and far between, and the resulting fogdrops are too far apart.

36.7 DISCOVERY OF THE POSITRON The positron, or positive electron, was discovered by Anderson* in 1932 by photographing the tracks of cosmic rays in a Wilson cloud chamber. Under the influence of a strong magnetic field applied perpendicular to the face of the cloud chamber, positively charged particles should bend to the right and negatively charged particles should bend to the left.

In order to be certain that those bent one way were not all coming from above and those bent the other way were particles of the same kind and charge coming from below, Anderson inserted a block of lead in the chamber to slow down the particles. Under these conditions photographs similar to the one shown in Fig. 36G were obtained. Here Anderson could be quite certain, from the curvature of the track on each side of the lead, that the particle entered from the side shown above; for in passing through the lead plate it could only have been slowed down and not speeded up.

Knowing the direction of motion, the direction of the field, and the direction of bending, Anderson concluded that such a particle had a positive charge. Comparing the track with well-known electron tracks and α-particle tracks, he concluded that the new particle had about the same mass as the electron. Later experiments continued to give more positive proof of the existence of a positive electron. Now very strong beams of positrons can be produced in the laboratory. For his discovery of the positron, Anderson was awarded the Nobel Prize in physics in 1936.

36.8 CREATION OF ELECTRON PAIRS Soon after Anderson's discovery of the positron, J. R.

* CARL DAVID ANDERSON (1905–), American physicist. He obtained his Ph.D. in physics in 1930 at the California Institute of Technology, where he later became Professor of Physics in 1939 and Chairman of Division of Physics, Mathematics, and Astronomy in 1962. He has received many medals and honorary degrees, including the Nobel Prize in physics in 1936. He is best known for his discovery of mesons as well as positively charged electrons in cosmic rays.

Oppenheimer attempted to calculate the conditions under which a positron might exist in nature. An extension of the quantum theory of the electron, proposed earlier by P. Dirac, led him to the prediction that if a high-energy photon, i.e., a high-frequency γ ray, were to come close enough to the nucleus of an atom, the electric field of the nucleus would be strong enough to annihilate the γ ray and create in its place a *pair of particles—an electron and a positron.* These two particles, the theory predicts, should have the same mass, and equal but opposite charges. A schematic diagram of pair production is given in Fig. 36H.

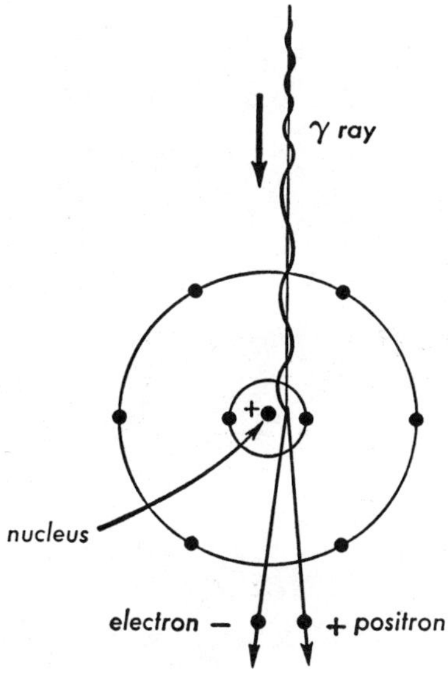

Fig. 36H. Schematic diagram of pair production.

Blackett, Anderson, and others looking for such pairs in a cloud chamber soon found them exactly as predicted. Gamma rays, from a radioactive element like *thorium C″*, in passing through matter were observed to produce pairs of electrons. One photograph of such an incident is shown at the top in Fig. 36I.

When an electron pair is created, *conservation of energy and momentum* require the two particles to move almost straight forward. Without a magnetic field applied to the cloud chamber, the particles travel side by side in almost parallel paths;

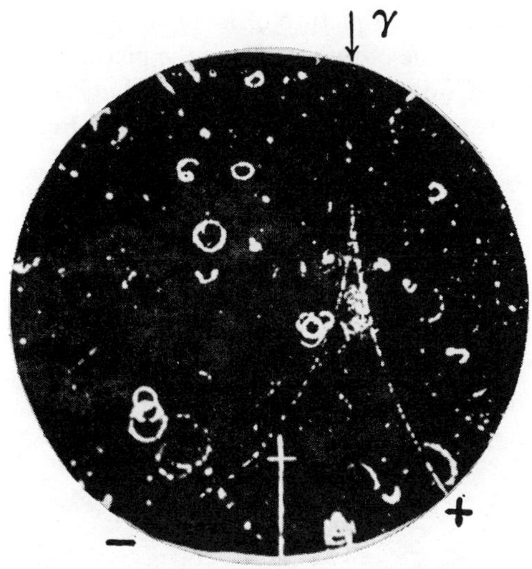

Fig. 36I. A γ ray, entering the cloud chamber from above, creates a pair of electrons, one (+), the other (−). (After Lauritson and Fowler.)

but with a magnetic field, the path of the positron bends to one side and that of the electron to the other.

The reason positrons were not discovered earlier in the history of physics is that they do not exist long in the free state. As soon as a positron meets with an electron, the two are annihilated.

Experiments indicate that all electrons and positrons spin around an axis through their center of mass. There is good evidence that, when a positron and an electron come close together, they frequently combine by revolving around each other like a double-star, with their spin axes parallel to one another. As such a pair they are called *positronium.*

Positronium is very short lived, for soon the two particles disintegrate completely and in their place γ rays are created. If the particles were spinning in the same direction, they would disintegrate into three γ rays of different energies; whereas, if they were spinning in opposite directions, they would produce two γ rays.

36.9 COSMIC RAY SHOWERS Out of hundreds and hundreds of cloud chamber photographs of cosmic rays, the experimenter is occasionally rewarded with a picture of a cosmic-ray shower. Instead of one or two tracks in the picture, in this instance, one finds anywhere from half a dozen or more to several hundred, as shown in Fig. 36J.

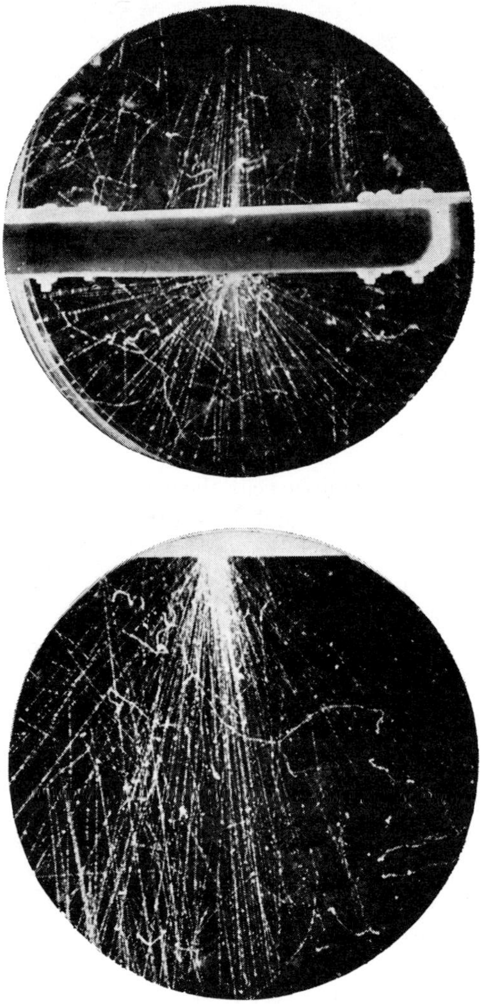

Fig. 36J. Cloud-chamber photographs of cosmic ray showers. (Courtesy of R. B. Brode, and C. D. Anderson and the Physical Review.)

Direct evidence that such showers originate with a single high-energy particle is shown in Fig. 36K. Here, in a cloud chamber with five equally spaced lead plates, a relatively large shower is seen to have grown from but one, or possibly two, particles at the top. Not only do the particles of this avalanche grow in number with each traversal of a lead plate, but the relatively small spread of the tracks indicates how nearly each new particle recoils along with the others in the forward direction. In this picture one observes in the small space of several inches the process that, in Fig. 36F, requires several miles of air.

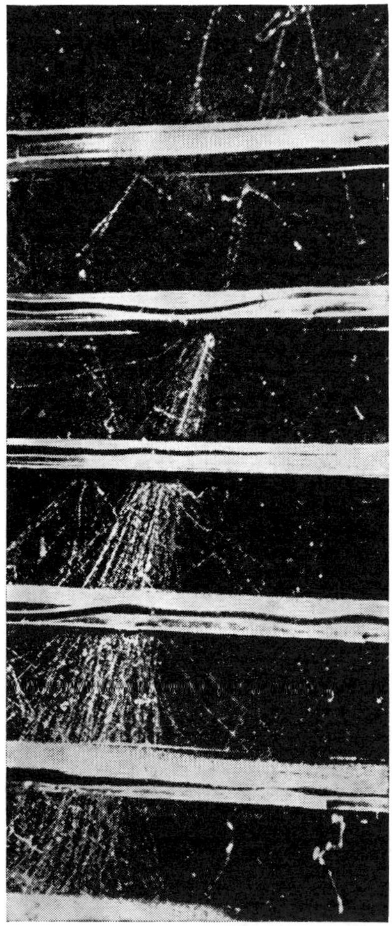

Fig. 36K. Photograph showing cascade shower growth of cosmic rays in five layers of lead. (Taken on Mt. Evans, Colo., by W. Powell.)

36.10 MESONS The presence in cosmic rays of charged particles having a mass several hundred times that of an electron, yet considerably lighter than a proton, was discovered by Anderson and Nedermeyer in 1938. These particles, now called *mesons*, are of several kinds, and experimental data taken in balloons and airplanes show that most of them are produced high in the atmosphere by the collisions of primary cosmic rays with air nuclei.

In these collisions, positively and negatively charged pions (π^+, π^-) are produced along with neutral pions (π^0), protons (p), and neutrons (n), as shown in Fig. 36L. The pions, each with a mass of about 273.3 m_e, along with other nuclear particles, recoil forward with speeds close to that of light. (m_e = mass of an electron.) The term nuclear particles is here applied to only those particles believed to exist in atomic nuclei, and they consist of protons, neutrons, pions, and α particles.

The possible existence of, and the spontaneous disintegration of, pions was first predicted by Yukawa* in 1935, and first photographed by Williams and Roberts in 1940. All charged pions seem to have a half-life of 2×10^{-8} s, and each one decays into a charged muon and a lightweight neutral particle called a neutrino. The charged μ^- mesons in turn decay, with a half-life of 2×10^{-6} s, into an electron and two neutrinos as shown in Fig. 36M. The μ^+ particle decays into a positron e^+ and two neutrinos. See the cloud chamber photograph in Fig. 36N.

The neutrino ν, an uncharged particle, was postulated originally by physicists in order to explain nuclear phenomena in keeping with the fundamental laws of the conservation of energy and of momentum.

* HIDEKI YUKAWA (1907–), Japanese physicist, won the Nobel Prize for Physics in 1949 for "investigations in theoretical physics, especially the theory of elementary particles." He was born in Tokyo, graduated from Kyoto University in 1929, and while lecturer at Osaka University in 1935, proposed a new theory of the nuclear forces that bind atomic nuclei together. Through this work, he predicted the existence of mesons, later discovered in cosmic rays by C. D. Anderson. He served as Professor of Physics at Kyoto from 1939 to 1950; Director of the Research Institute for Fundamental Physics in 1953; Visiting Professor at the Institute for Advanced Studies, Princeton, N. J., in 1948; and Lecturer at Columbia University, 1949–1953.

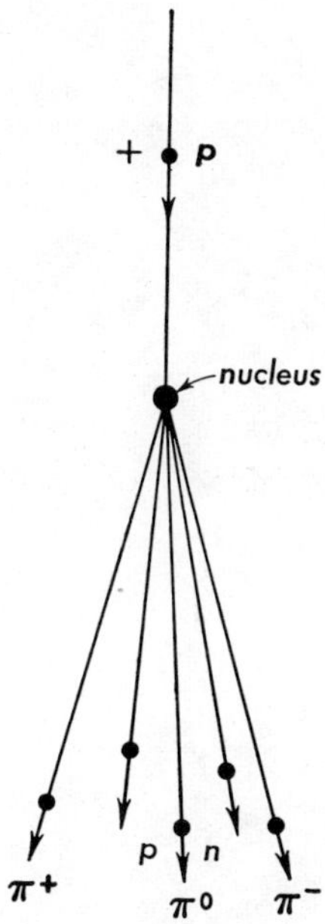

Fig. 36L. Primary cosmic ray produces π mesons by nuclear collision.

The decay of pions may be written as interactions as follows:

$$\pi^- \longrightarrow \mu^- + \nu \qquad (36a)$$

$$\pi^+ \longrightarrow \mu^+ + \nu \qquad (36b)$$

The uncharged pions are very unstable. With a half-life of less than 10^{-14} s, they decay into two

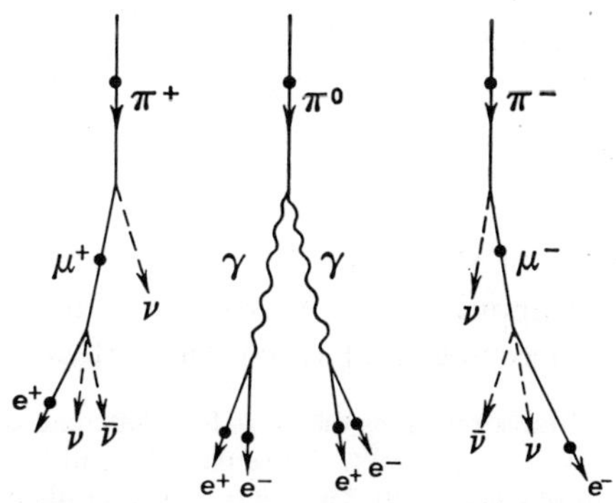

Fig. 36M. Pions disintegrate into μ mesons, γ rays, electrons, and neutrinos.

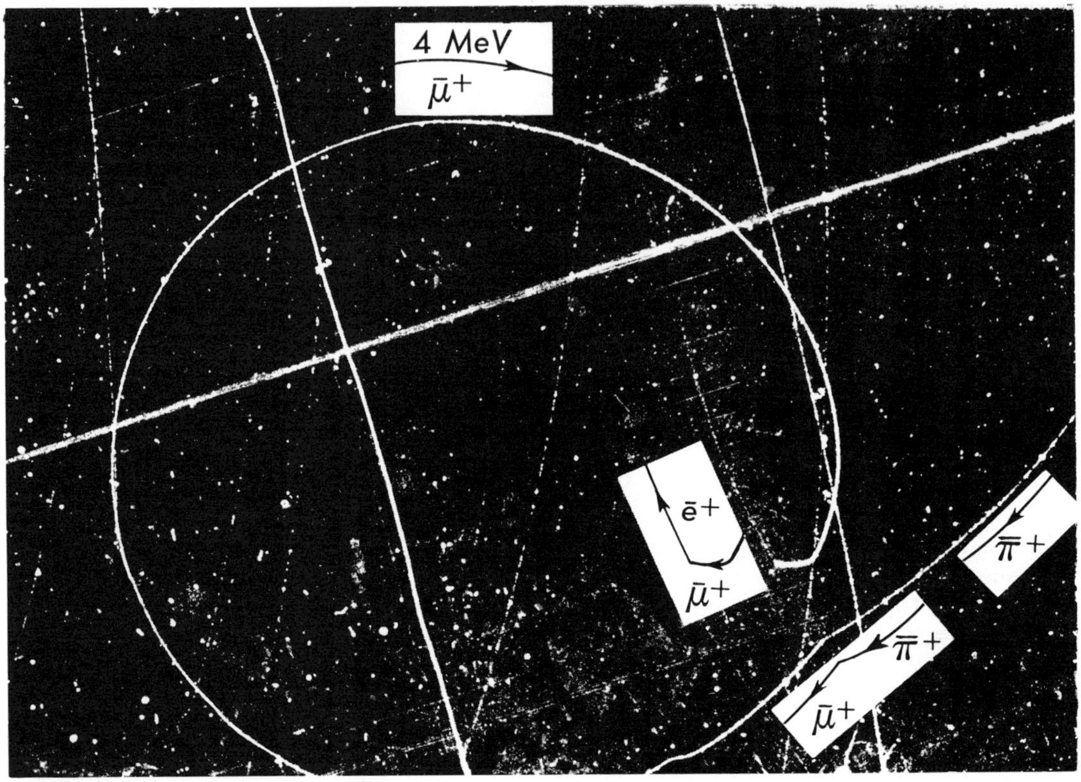

Fig. 36N. Wilson cloud chamber photographed in a magnetic field of 0.8 tesla showing decay of a pion plus into a muon and the muon into a positron. (Courtesy W. Powell.)

γ rays. In the upper atmosphere these γ rays create cascade showers of electrons by electron pair-production and bremsstrahlung (see Fig. 36K).

The decay of a neutral pion can be written

$$\pi^0 \longrightarrow \gamma + \gamma \tag{36c}$$

The decay of the muons may be written as follows:

$$\mu^- \longrightarrow e^- + \nu + \nu \qquad \text{(36d)*}$$

$$\mu^+ \longrightarrow e^+ + \nu + \nu \qquad \text{(36e)*}$$

Many of the charged μ mesons, with their mass of about 207 m_e, traverse the atmosphere before decaying, and reach the surface of the earth. At sea level the charged cosmic rays are nearly 70% μ mesons and 29% electrons and positrons, with about 1% heavier particles like protons, deuterons, α particles, etc.

In traversing solid matter, negatively charged pions frequently slow down to such a speed

* In the above reactions there are four different kinds of neutrinos, ν_μ, $\bar{\nu}_\mu$, ν_e, and $\bar{\nu}_e$. These particles will be treated in detail in Chapter 40. The neutrino subscripts indicate the electrons and muons with which they were associated, while the bar over a particle signifies it is an *antiparticle*.

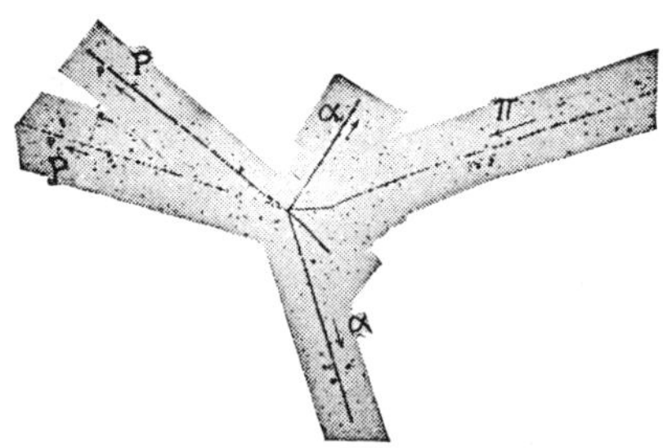

Fig. 36O. "Star" in photographic emulsion showing the explosion of a nucleus resulting from the capture of a slow pion. (Courtesy, C. F. Powell's laboratory, University of Bristol, England.)

that, upon an encounter with a nucleus, they are attracted by the + charge and captured. In this process the meson mass is transformed into energy exciting the nucleus to such a state that it literally explodes by shooting out a number of heavier particles like protons, deuterons, α particles, etc. Figure 36O shows such a "star" event in a photographic emulsion.

Recent studies of cosmic-ray tracks made in photographic emulsions and cloud chambers indicate the presence of particles with various other masses and charges (to be discussed in Chapter 40).

Most recent cosmic-ray observations at high altitudes and at sea level show the following:

TABLE 36A. Composition of Primary and Secondary Cosmic Rays

Primaries		Sea Level	
H	89%	μ Mesons	70%
He	9%	e^+ and e^-	29%
Li, Be, B	0.5%	Heavier particles	1%
C, N, O	0.5%		
Ne, Mg, Si	0.1%		
Fe	0.03%		

QUESTIONS

1. Who discovered cosmic rays? How does the cosmic ray intensity vary as one goes up into the air above the earth's surface?

2. What kinds of detectors were used in the early cosmic ray experiments? Name six other detection and measuring devices.

3. How are cosmic rays made to take their own photographs?

4. What are primary cosmic rays? Of what are they composed?

5. What are secondary cosmic rays? Where do they come from? Of what are they composed?

6. What are positrons? Who discovered them? In what ways are positrons and electrons alike?

7. Make a diagram of a Geiger-Mueller tube counter and briefly explain how it functions as a detector of high-speed atomic particles.

8. Draw a graph of how you think the cosmic ray intensity varies as one ascends high into the atmosphere? Plot altitude horizontally and intensity vertically.

9. How did Anderson, in his discovery of the positron, know that the track was produced by a positively charged particle from above, and not from an ordinary electron from below?

10. What is meant by "pair production"? Briefly describe the process. What finally becomes of positrons?

11. What is positronium? Explain what eventually becomes of such atomic particles.

12. What is a cosmic ray shower? What kinds of particles are believed to initiate and create cascade electron showers?

13. Make a list of the different kinds of mesons. Give (a) their charge, and (b) their relative mass.

14. Make a list of all atomic particles you know of having a mass equal to or less than the mass of an α particle.

PROBLEMS

1. Assuming all the energy of a γ ray could be used in creating a pair of electrons (a positron and an electron), what must be (a) its frequency, and (b) its wavelength? [Ans. (a) 2.471×10^{20} Hz. (b) 0.01213 Å.]

2. What is the minimum energy in volts that a γ ray must have to produce an electron pair? See Eq. (35f).

3. When a positron combines with an electron the two are annihilated. What is (a) the frequency of the two γ rays produced, and (b) their wavelength? [Ans. (a) 1.235×10^{20} Hz. (b) 0.02427 Å.]

4. If a γ ray in coming close to a nucleus could create a pair of π mesons (π^+ and π^-), what is the minimum energy it must have in volts?

5. If a γ ray in coming close to a nucleus could create a pair of protons, one + and the other −, what is the minimum energy it must have in volts? [Ans. 1.877 GeV.]

37 | ATOMIC ACCELERATORS

Believing that the disintegration of atomic nuclei might be accomplished by using other than α particles as projectiles, Rutherford instigated in 1930 the construction of a high-voltage, direct-current generator at the Cavendish laboratory in England. The purpose of this *million volt* source of potential was to accelerate hydrogen nuclei, *protons*, to high speeds and then cause them to strike known substances. In this way he hoped to produce new and various kinds of disintegrations.

At the same time that Rutherford and his assistants were struggling with their high-voltage system, Ernest Lawrence in the United States conceived of accelerating atomic particles to high speeds by repeatedly applying high voltages to the same particles. This idea led to the development of the first cyclotron, an accelerator only 10 cm in diameter. This in turn led to a long line of accelerators of larger and larger diameters, until today instruments many hundreds of feet in diameter are to be found in many countries.

37.1 THE COCKCROFT-WALTON EXPERIMENT

Becoming impatient with the relatively slow progress of their million-volt source for accelerating atomic particles, Rutherford suggested to Cockcroft and Walton that lower voltages be tried in the meantime to see if, by chance, disintegrations might occur. In 1932 Cockcroft* and Walton† announced that they had successfully

* SIR JOHN D. COCKCROFT (1897–1967), British physicist. Educated at Manchester University and St. John's College, Cambridge, Sir Cockcroft won the Nobel Prize in Physics, jointly with E. T. S. Walton, in 1951. These two men, working in the Cavendish Laboratory with Rutherford, were the first to transmute atomic nuclei by artifically accelerated atomic particles. Cockcroft served his country in several capacities: as Chief Superintendent, Air Defense Research and Development, Director of the Atomic Energy Ministry of Supply, and several others. He was elected a fellow of the Royal Society in 1936 and was knighted in 1948. The author had the good fortune to meet Sir Cockcroft in the Cavendish Laboratory and to see their disintegration equipment in 1930.

† ERNEST T. S. WALTON (1903– Irish physicist. He was educated at Bainbridge, County Down, at Cookstown, County Tyrone, and in the Methodist College, Belfast. While working with Sir John D. Cockcroft, under Rutherford, at the Cavendish Laboratory, the first transmutation experiments with laboratory accelerated atomic

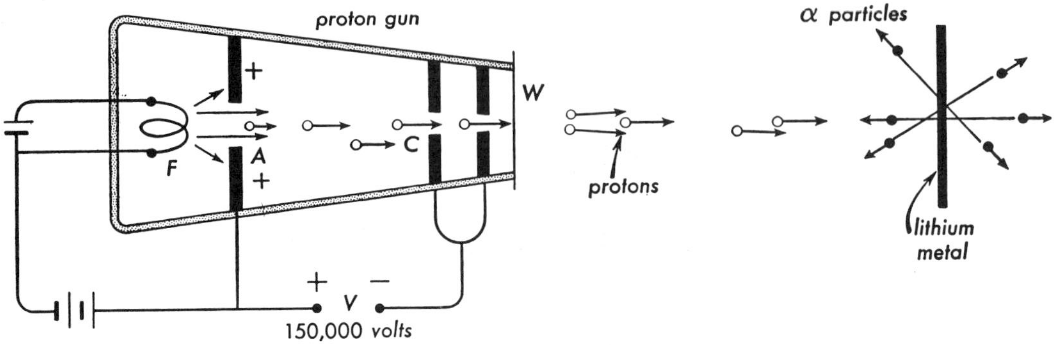

Fig. 37A. Schematic diagram of the Cockcroft-Walton experiment. Lithium is disintegrated by 150,000-volt protons.

disintegrated lithium atoms with protons accelerated by relatively low voltages. Their apparatus is schematically represented in Fig. 37A.

Electrons from a hot filament F, passing through hydrogen gas in the region of A, ionize many hydrogen atoms. These protons with their positive charge are then accelerated toward the other end of the tube by a potential V of 150,000 volts. Upon passing through the opening C and a window W, they emerge from the acceleration chamber as a narrow beam of protons.

This tube, acting as a "proton gun," is aimed at a target consisting of lithium metal. Cockcroft and Walton observed α particles emanating from the metal with a range of 8 cm, an energy equivalent to 8.5 MeV. Considering the relatively low energy of the bombarding protons of only 0.15 MeV, this is a tremendous release in atomic energy. The transmutation taking place here is written as follows:

$$\boxed{{}_1H^1 + {}_3Li^7 + E_1 = {}_2He^4 + {}_2He^4 + E_2} \tag{37a}$$

This reaction, illustrated in Fig. 37B, shows a proton, ${}_1H^1$, of energy $E_1 = 0.15$ MeV, entering

---

particles were performed. He was awarded the Hughes Medal of the Royal Society in 1938, and elected Erasmus Smith Professor of Natural Philosophy at the University of Dublin in 1946. He was awarded, jointly with Sir Cockcroft, the Nobel Prize in Physics for 1951.

---

a lithium nucleus, ${}_3Li^7$, to form a new but unstable beryllium nucleus, ${}_4Be^8$. Being unstable, this compact structure of eight particles splits up into two α particles which are driven apart with great violence. Since the measured energy of each α particle is equivalent to 8.5 MeV, each disintegration involves the liberation of 17.0-MeV energy. The source of energy is to be found in the annihilation of a part of the total atomic mass.

The loss in mass can be calculated from the table of atomic weights, given in Appendix III. Listing the involved masses in two columns and adding, we obtain

$$\begin{array}{rlrl} {}_1H^1 = & 1.007825 & & \\ {}_3Li^7 = & 7.016005 & {}_2He^4 = & 4.002604 \\ E_k = & \underline{0.000161} & {}_2He^4 = & \underline{4.002604} \\ & 8.023991 & & 8.005208 \end{array}$$

E_k is the mass equivalent to the energy of the incident proton and is obtained by dividing 0.15 MeV by 931 [see Eq. (35j)]. The difference between the two sums, $8.023991 - 8.005208 = 0.018783$ mass unit, represents the loss in mass by the disintegration. When multiplied by 931, this gives 17.48 MeV as the liberated energy, a value in good agreement with the experimentally determined value of 17.0 MeV.

It might be thought that such a disintegration as the one described above could be used as a source of energy but as yet it has not been feasible to do so. While each nuclear collision and disintegration liberates at least one hundred times as much energy as that supplied to the proton, it takes many particles to make a few

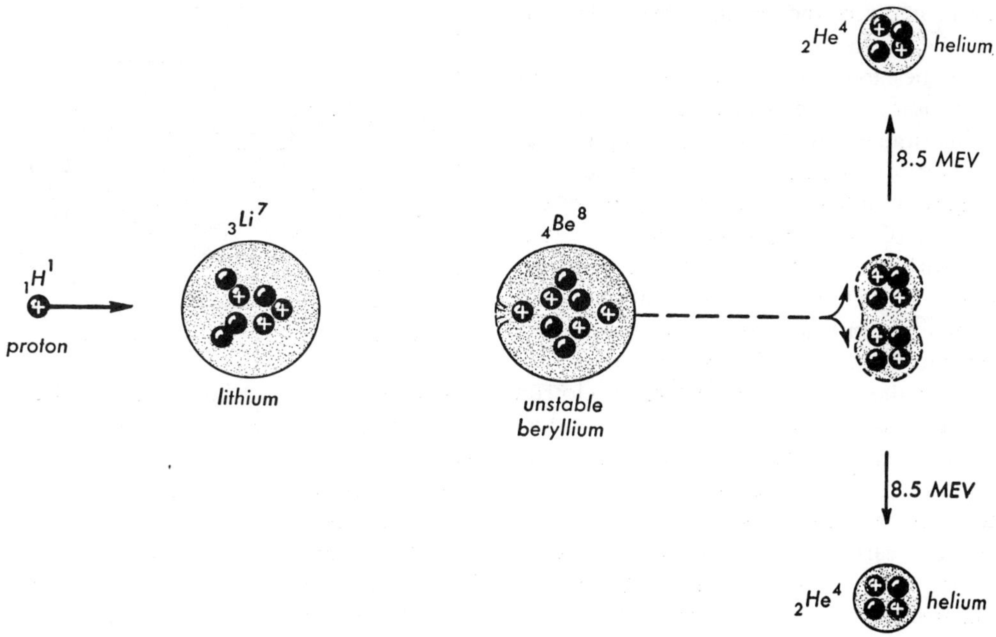

Fig. 37B. Disintegration of a lithium nucleus by a proton of 0.15 MeV energy. The Cockcroft-Walton experiment.

collisions. In other words, only a small percentage of the proton bullets hit the tiny nuclear targets as they pass through matter. Most of them are slowed down by electron collisions and the ionization of atoms. To a proton bullet the lithium nuclei, as targets in the lithium metal, present an area millions of times smaller than the space between them.

37.2 THE LAWRENCE CYCLOTRON At the time Cockcroft and Walton were performing their first disintegration experiments, E. O. Lawrence,* an American physicist, was developing a new type of atomic accelerator called a cyclotron, which soon attracted the attention of the leading physicists the world over. So successful was this "atomic machine gun" in producing high-speed atomic projectiles for disintegration experiments that a new and larger cyclotron was soon constructed and put into operation. Now a cyclotron of considerable size occupies a most prominent

* ERNEST O. LAWRENCE (1901–1958), American experimental physicist. Deriving his early education in South Dakota, Lawrence obtained the A.B. degree at the University of South Dakota in 1922, the master's degree at Minnesota in 1923, and the Ph.D. at Yale University in 1925. After two years as a National Research Fellow he became, at the early age of twenty-six, Assistant Professor of Physics at Yale University. The following year he was appointed Associate Professor of Physics at the University of California, and in 1930 was made full Professor. Having built up the Radiation Laboratory at the same institution, he became its Director in 1936. In 1937 he was awarded the Comstock Prize of the National Academy of Sciences, the Cresson Medal of the Franklin Institute, and the Hughes Medal of the Royal Society of London. Lawrence was a member of the National Academy of Sciences. He was noted principally for his invention and development of the cyclotron and its application to the production of induced radioactivity. It is for these discoveries that he was awarded the Nobel Prize in 1939. During World War II he directed one of the main research projects leading to the isolation of uranium-235 used in atomic bombs.

position in many of the leading physics laboratories of the world.

One of the large cyclotrons built at the University of California, and called the "sixty-inch," was an instrument capable of producing intense beams of protons, deuterons, or α particles having energies of 10, 20, and 40 MeV, respectively. The purpose of these high-speed particles, as is the case with the beams from all such instruments, is to subject various known substances to bombardment and thus produce disintegrations and transmutations of all kinds.

Although the operation of a large cyclotron requires an elaborate outlay of apparatus and equipment, the principles upon which it operates are quite simple. Cross-section diagrams of a cyclotron are shown in Figs. 37C and 37D.

The very heart of the instrument consists of two short, hollow, half cylinders D_1 and D_2, mounted inside of a vacuum chamber V, between the poles of a powerful electromagnet, and connected on the outside to the two terminals of a high-frequency alternating current generator. It is interesting to point out that this generator is really a short-wave radio transmitter supplying energy to the dees (D_1 and D_2) instead of to the antenna. When a trace of hydrogen gas is admitted to the evacuated chamber, the hot wire filament F ionizes some of the hydrogen atoms, thereby producing the protons to be used as atomic bullets. At the particular instant when D_1 is charged positively and D_2 is charged negatively, a proton in the neighborhood of F will be accelerated toward D_2. Moving through the strong magnetic field of the huge magnet, this positively charged particle traverses a circular path as shown in the diagram. If, after making a half turn, the potential is reversed so that D_1 becomes negatively charged and D_2 positively charged, the proton will be attracted by D_1 and repelled by the D_2, causing it to increase its speed. With added speed it therefore moves in the arc of a larger circle as shown. After this second half turn, the potential again reverses, making D_1 positive and D_2 negative, and again the proton speeds up. Thus, as the potential reverses periodically, the proton travels faster and faster,

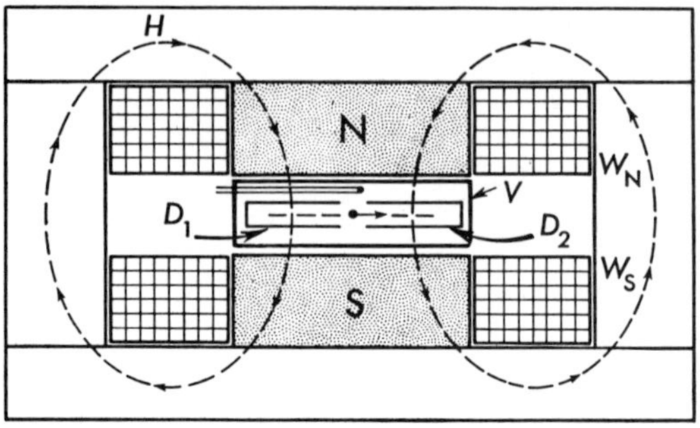

Fig. 37C. Cross-section diagram of a cyclotron.

moving in ever-expanding circles, until, reaching the outer edge, it passes through a narrow open window W.

Upon leaving W, all protons must pass close to a negatively charged plate P where, by attraction, their paths are straightened out and they become a separated beam of projectiles. Whatever substance is to be bombarded is then placed in this beam, and the disintegrated fragments are studied by means of various detecting devices.

The fundamental principle that makes the cyclotron work at all is the fact that *the time required for a charged particle to make one complete turn within the dees is the same for all speeds.* The faster a particle travels, the larger is the circle

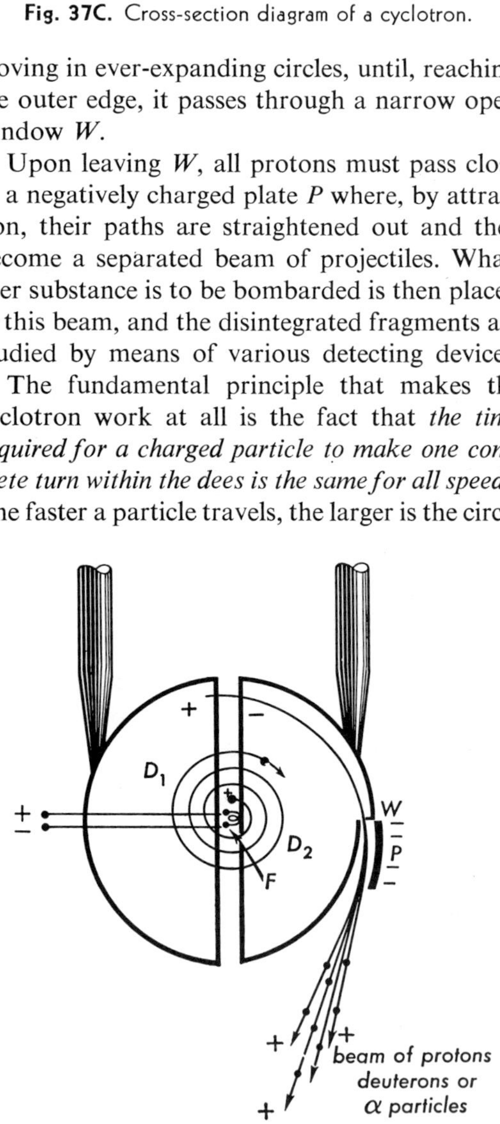

Fig. 37D. Detailed diagram of the D's of a cyclotron.

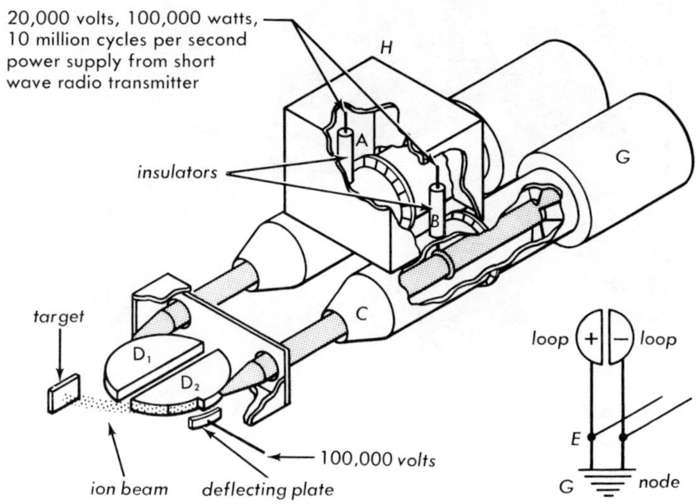

Fig. 37E. Assembly diagram showing the *D*'s and their mounting supports, the deflecting plate, and the electrical supply connections.

it must traverse, thus keeping the time constant. Hence, with a constant frequency of the alternating current supply, some particles may be just starting their acceleration near the center while others farther out have already acquired higher speeds. The result is a more or less continuous stream of protons emerging from the window *W*.

If the alternating current voltage applied between the dees of the cyclotron is 200,000 volts, with each half turn a particle obtains an added velocity equivalent to 200,000 volts. If a proton makes twenty-five complete revolutions before leaving the chamber at *W*, it will have acquired a velocity equivalent to 200,000 times 25 times 2, or 10,000,000 volts. Here, then, is a beam of 10-MeV protons acquired by the application of a potential only one fiftieth as great.

When hydrogen in the evacuated chamber of the cyclotron is replaced by *deuterium* and the magnetic field strength is doubled, a beam of high-energy deuterons is obtained. Having twice the mass but the same charge as protons, these particles acquire twice as much energy. If helium gas is used in place of deuterium, many of the atoms become doubly ionized at the source and, after acceleration, emerge from the cyclotron window with an energy of about 40 MeV. By increasing or decreasing the frequency of the potential applied to the dees and properly adjusting the magnetic field, protons, deuterons, or α particles of 10, 20, and 40 MeV, respectively, can be produced.

Some of the details of the cyclotron shown in Figs. 37C, 37D, and 37E are as follows: The dimension, 60 in., refers to the diameter of the poles of the cyclotron magnet, and this in turn limits the size of the dees and therefore the maximum energy available in the form of atomic projectiles. Most of the instrument's total weight of 200 tons lies in the solid core, and the pole pieces located inside the field windings. The latter, consisting of many turns of thick copper wire, are encased in tanks W_S and W_N through which cooling fluid is continually circulated.

The cut-away diagram in Fig. 37E shows the dee supports as long hollow tubes rigidly clamped and grounded at the one end *G*. The high voltage a.c., applied through vacuum tight joints *AB*, sets each of the freely suspended dees into electrical oscillation.

Fig. 37F. Photograph by Paul Donaldson of an 11-MeV deuteron beam from the Harvard cyclotron. (Courtesy of the Harvard University Press and A. K. Solomon.)

These two long rods oscillate electrically with a *voltage node* (zero potential) at the clamped end *G* and a *voltage loop* or peak at each of the dees. With the connections made at *E*, about one fifth of the distance from *G* the peak potential at the dees reaches approximately five times that of the peak voltage applied. The electrical oscillations are analogous to the vibration of the prongs of a tuning fork. The applied voltage is approximately 20,000 volts, whereas each dee reaches approximately 100,000 volts. In some respects this step-up in voltage is similar to a step-up transformer.

A photograph of an 11-MeV deuteron beam from the Harvard University cyclotron is shown in Fig. 37F. From the point where they emerge from the cyclotron window at the left center to where they come to rest in mid-air at the lower right, the high-energy particles ionize the air molecules and atoms, causing them to emit visible light.

37.3 THE VAN DE GRAAFF GENERATOR This machine developed in 1931 by R. Van de Graaff at Princeton University employs the principle of the electrostatic generator discovered many years ago. A typical installation, as shown in Fig. 37G, consists of a large hollow sphere, supported on insulating columns and charged by a belt conveying electrical charges from a battery at ground

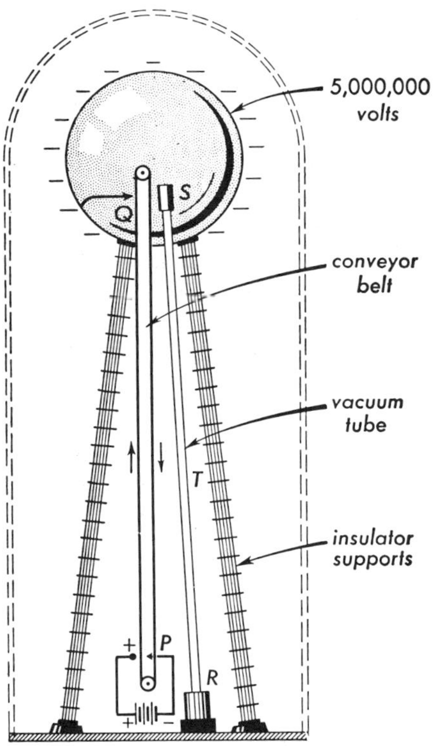

Fig. 37G. Diagram of a Van de Graaff generator of high voltage.

potential and depositing it inside the sphere. The fabric conveyor belt, a foot or more in width, and running over well aligned rollers, travels about 3 m/s.

As the belt passes between the metallic surface and row of needle points at *P*, electrons from the points jump toward the positive electrode and are caught by the belt. Upon entering the sphere at the top the electrons jump to the needle points *Q* where they go quickly to the outside surface of the sphere. The "spraying" of electrons *to* and *from* the points is assured by keeping the battery potential high (about 50,000 volts) to maintain a "brush discharge." As more and more electrons arrive at the sphere, its negative potential rises higher and higher until leakage into the surrounding air and through the insulators becomes equally fast.

Atomic particles to be accelerated are generated inside a vacuum tube source *S* inside the sphere. Starting at the top of a long straight vacuum tube *T*, electrons are accelerated downward toward ground potential, where, acquiring the full energy of the available voltage, they are allowed to bombard whatever target is being studied. Where installations are designed for accelerating protons, deuterons, or α particles, the battery potential is reversed and the sphere acquires a high positive potential.

37.4 THE BETATRON The *betatron*, invented in 1941 by D. W. Kerst* at the University of Illinois, is an electron accelerator capable of producing electron beams of high energy as well as X rays of extremely high penetrating power. This ingenious device differs from the cyclotron in at least two fundamental respects; first, the electrons are accelerated by a rapidly changing magnetic field, and second the circular orbit of the particles has a constant radius.

A cross-section diagram of a 20-MeV betatron is shown in Fig. 37H. A glass vacuum tube in the shape of a *doughnut*, and containing an *electron gun*, is mounted between the poles of an electromagnet. An alternating current (180 Hz) applied to the coils causes some of the magnetic lines of force to pass through the vacuum tube at the electron orbit and the remainder through the orbit center in Fig. 37H. Electrons are injected only at the beginning of each quarter cycle when the field begins to increase in the "up" direction. The increasing field through the center of the orbit gives rise to an electromotive force, tangent to the orbit, speeding up the electrons, whereas the increasing field at the orbit is just sufficient to increase the centripetal force and keep the electrons from spiraling outward. The

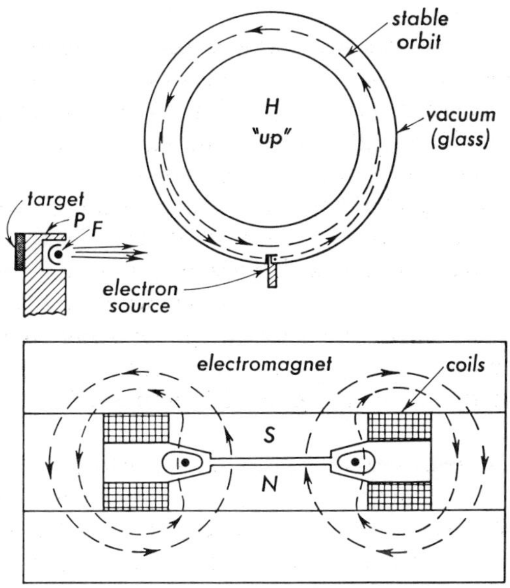

Fig. 37H. Cross-section diagrams of an electron accelerator, called the betatron.

* DONALD WILLIAM KERST (1911–), American physicist. Educated at the University of Wisconsin, he began his career as an X-ray tube developer with the General Electric X ray Corporation in 1937, and became an Instructor in Physics at the University of Illinois in 1938, and served as Professor in 1942–57. He then spent five years with the Atomic Division of the General Dynamics Corporation, and, in 1962, became Professor of Physics at the University of Wisconsin. He is a member of the National Academy of Science and has been awarded their Comstock Prize and the Scott Award of the Franklin Institute. He is most noted for his invention of the betatron while at the University of Illinois.

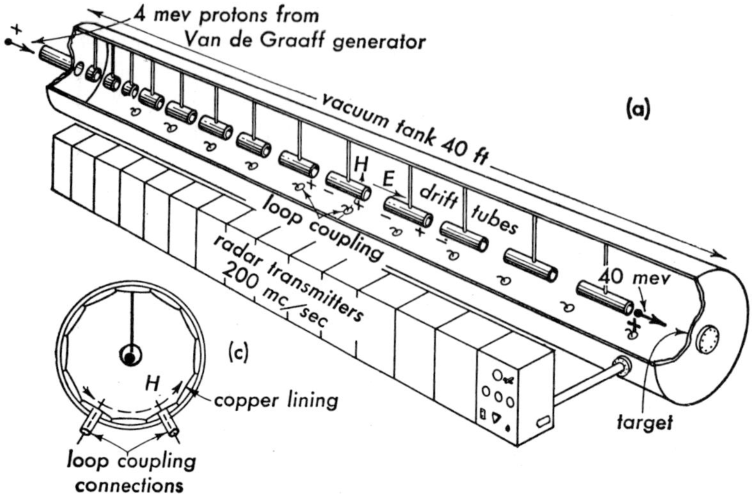

Fig. 37I. Diagrams showing the main section of a linear accelerator.

stability of such an orbit is brought about by properly shaping the pole faces of the magnet, adjusting the frequency and strength of the magnetic field, and injecting the electrons at the proper voltage at the appropriate time.

During World War II a 350-ton betatron was constructed by the General Electric Company and put into use as a source of extremely penetrating X rays. In this instrument electrons accelerated to 100-MeV energy, and impinging upon a target, give rise to X rays capable of penetrating many feet of solid iron and lead.

37.5 THE LINEAR ACCELERATOR Although linear accelerators were proposed as early as 1929, and several were constructed, they had never proved satisfactory until years later. Applying the principles of tubular wave guides and resonant cavities, L. Alvarez* and his collaborators immediately after World War II constructed the first successful linear accelerator. Since a giant machine 2 miles long employing the same principles is now in operation at Stanford University the principles involved are of some importance.

A cut-away diagram of part of the Alvarez apparatus is given in Fig. 37I. Protons are initially produced and accelerated to 4 MeV by a Van de Graaff Generator (see Fig. 37G) and then injected at that energy into one end of a 60 m tank as shown at the upper left. Once inside they are further accelerated as they pass through a series of "drift tubes," and arrive at the other end with an energy of about 40 MeV.

* LUIS W. ALVAREZ (1911–), American physicist, is the son of a famous American physician and author, Dr. Walter C. Alvarez. He obtained his Ph.D. degree in physics at the University of Chicago in 1936, under the direction of A. H. Compton. He joined the physics staff of the University of California in 1936 and began his post-graduate research career in nuclear physics with Ernest O. Lawrence of the Radiation Laboratory. During World War II, he developed the Ground Control Approach Radar System widely used at major airports for landing planes in dense fog cover. For this he received the National Collier Trophy in 1946. Since then, he has won the Medal of Merit, the Einstein Medal and the Michelson Award. His greatest contribution to physics is the development of the liquid hydrogen bubble chamber, and with his colleagues, his use of it to identify a large number of elementary atomic particles. For this and other work, he was awarded the Nobel Prize in physics for 1968.

The tank cavity within the copper lining, fed by 30 radar transmitting oscillators, is set resonating at its *dominant mode* at a frequency of 200 MHz. The "standing-wave" conditions set up are such that the electric field E is parallel to the tube axis and is everywhere rising and falling together. The lengths of the drift tubes gradually increase so that the protons cross each gap when the field E is to the right, and are inside the tubes, in a field free space, when the field is to the left. It now seems feasible, but not very practical, that additional tank sections could be added end to end to

TABLE 37A. GeV Accelerators

Location	Name	Particle	GeV Energy	Completion
United States				
LRL Berkeley	Bevatron	deuterons	6.2	1954
Princeton	PPA	deuterons	3.0	1963
ANL	ZGS	deuterons	12.5	1963
Brookhaven	AGS	deuterons	33.0	1960
Stanford	Mark III	electrons	1.2	1950
Stanford	SLAC I	deuterons	20.0	1966
Cal. Tech.	Synchrotron	deuterons	1.5	1952
Cornell	DO	electrons	2.2	1955
Cornell	Synchrotron	deuterons	10.0	1968
Cambridge	CEA	deuterons	6.0	1962
Planned Construction	GeV	protons	200.0	(1972)
		deuterons	800.0	(1979)
France				
Saclay	Saturne	deuterons	3.0	1958
Orsay	Linac	electrons	1.3	1961
Germany, Hamburg	DESY	deuterons	6.0	1964
Italy, Rome	Synchrotron	deuterons	1.1	1959
Japan, Tokyo	DO	deuterons	1.3	1960
Sweden, Lund	DO	deuterons	1.2	1962
Switzerland, CERN	PS	protons	28.0	1959
United Kingdom				
Birmingham	Synchrotron	deuterons	1.1	1953
Daresburg	NINA	electrons	5.0	1967
Harwell	NIMROD	deuterons	8.0	1963
USSR				
Moscow, ITER	Synchrotron	deuterons	7.0	1961
Dubna, JINR	DO	deuterons	10.0	1957
Kharkov	Linac	electrons	2.0	1965
Serpukov	70	protons	70.0	1966
Yerevan	Synchrotron	electrons	6.0	1965

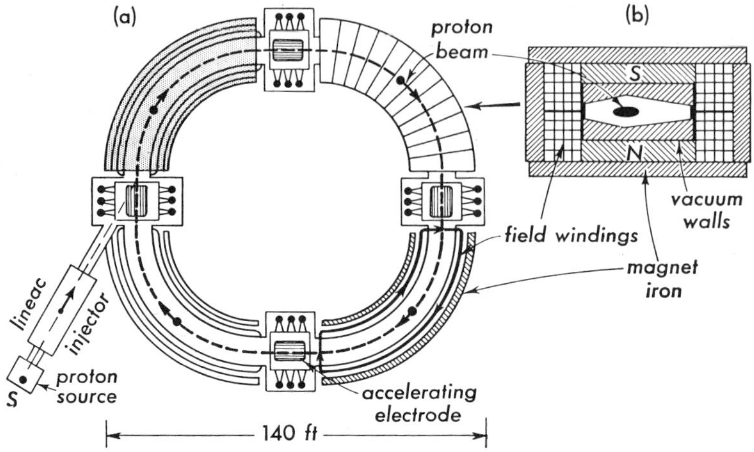

Fig. 37J. Berkeley Bevatron designed to produce 6.2-GeV protons.

Fig. 37K. Aerial photograph of the area housing the giant 31-GeV proton accelerator at Brookhaven, Long Island, New York. The circular mound of earth covers the accelerator ring proper, the ground fill serving to keep the temperature constant and as an absorber of unwanted stray radiation. Note the parked cars as an indication of size. (Courtesy Brookhaven National Laboratories.)

this system to obtain almost any desired energy. Calculations indicate about 3 MeV per meter can be expected from such a system.

37.6 BILLION ELECTRON VOLT ACCELERATORS

The design and construction of an instrument capable of accelerating particles to energies of billions of electron volts (BeV or GeV) involve many problems. Not the least of these is the enormous cost of such an instrument as well as its subsequent maintenance.

A number of accelerators producing particles of one GeV or more are now in operation and others of greater and greater energy are being planned.

The basic design of one of the large instruments, the *bevatron*, stems from proposals of W. M. Brobeck. The magnet arrangement was first proposed by H. R. Crane and consists of four quadrant segments spaced so that the particle orbits are quarter-circles connected by straight sections. See Fig. 37J. The electrical power supplied to the 10,000-ton magnet is provided by a motor generator with a large flywheel. During build-up of the magnetic field, a peak power of 100,000 kilowatts is drawn from the flywheel and stored in the magnet. As the field is reduced between beam pulses, the generator acts as a motor and returns energy to the flywheel.

Protons from the source at the lower left in Fig. 37J are first accelerated by a "Cockcroft-Walton" transformer cascade accelerator to 500,000 electron volts. After deflection by a magnetic field they are speeded up to 10 MeV by a linear accelerator. At this energy they are inflected into the 130 m race track proper where they attain a speed equivalent to 6.2 GeV (over six billion electron volts).

A photograph showing the ground area that houses one of the worlds largest accelerators in operation is reproduced in Fig. 37K.

QUESTIONS

1. What was the Cockcroft-Walton experiment? What was the interaction?

2. Where does the energy come from that causes lithium to split up with a hundredfold increase in kinetic energy?

3. What is a cyclotron? What is its purpose? What does it accelerate?

4. Make a diagram of the essential elements of a cyclotron employing two dees. Label the essential parts. Show where the ionized atomic particles start, and where they finally emerge as a beam.

5. What is deuterium? Is deuterium the same as hydrogen?

6. What is a Van de Graaff generator? What is it frequently used for? Can it be used to accelerate protons and electrons?

7. Make a diagram and briefly explain the Van de Graaff generator. Show the battery connections and discharge points as they would be arranged to charge the metal sphere positively.

8. What is a betatron? From what do you think it gets its name? What is it used for?

9. What is a linear accelerator? What is its purpose? From what do you think it gets its name?

10. Make a diagram and briefly explain the operation of the bevatron. Explain its purpose and the origin of its name.

11. Approximately what is the highest energy to which atomic particles have been accelerated in the laboratory?

38 | TRANSMUTATION OF THE ELEMENTS

The use of high-speed atomic particles, whether they come from radioactive materials, from cyclotrons, or from cosmic rays, has led in recent years to the discovery of many secrets that lie hidden within the atomic nucleus. While the thousands of experiments performed have led to the discovery of many new elementary particles, several hundred radioactive isotopes of the known elements have been produced by the process of nuclear bombardment.

38.1 PROTON AND DEUTERON DISINTEGRATIONS When high-energy protons or deuterons are used to bombard different known elements, various disintegration products are formed. An experimental arrangement in which the cyclotron acts as the source of high-speed particles is shown in Fig. 38A. To determine the nature of the disintegration taking place within the substance under bombardment, it is common practice to identify the penetrating rays emerging from the other side by the use of suitable detectors.

Numerous experiments have shown that the disintegration products to be looked for may be *protons*, *α particles*, *neutrons*, *γ rays*, or even *electrons*, *positrons*, and *mesons*. For some of these penetrating rays one kind of detector may be more suitable than another. The *Geiger-Mueller tube counter*, for example (see Fig. 36C), is particularly useful in detecting γ rays and electrons, whereas the *Wilson cloud chamber* and *ionization chamber* are useful in detecting protons, α particles, or neutrons.

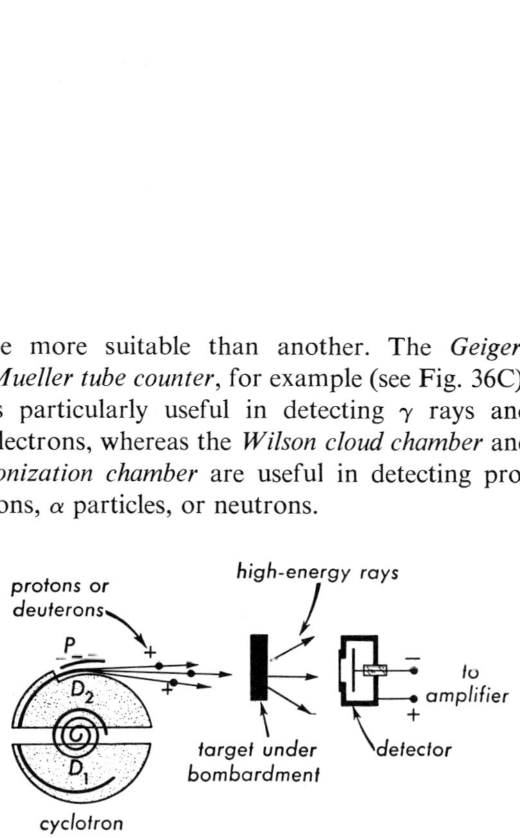

Fig. 38A. Experimental arrangement generally used for bombarding known substances with high-speed deuterons from the cyclotron and for detecting the disintegration products with an ionization chamber as a detector.

The detector shown at the right in Fig. 38A represents a so-called "shallow ionization chamber." This device consists of a small copper box filled with air or some other suitable gas. When a proton or α particle passes through the thin metal foil window at the front of the box, it produces numerous ions inside. The positive ions are attracted toward a negatively charged plate at the center, where they give rise to minute electric currents along a wire to a radio tube amplifier. The amplifier in turn can be made to operate a high-speed counting device, or to activate a loud-speaker from which each impulse can be heard and counted. By inserting thin foils one after the other between the target and detector, the nature and energy of the incoming particle can be determined.

When a Wilson cloud chamber is used to identify disintegration products, charged particles can be identified by the density of their fog tracks, and their energy can be determined by the curvature of the tracks when a magnetic field is applied. This is illustrated in Fig. 38G for positrons. Once the nature of the emerging rays from a bombarded target is known, the recoil product of the disintegration also becomes known by writing down a reaction equation. Six examples of such reaction equations are given by the following:

Reactions	Q values
${}_1H^1 + {}_9F^{19} = {}_8O^{16} + {}_2He^4$	8.1 MeV
${}_1H^1 + {}_5B^{11} = {}_6C^{12} + \gamma$ ray	15.8 MeV
${}_1H^2 + {}_7N^{14} = {}_6C^{12} + {}_2He^4$	13.6 MeV
${}_1H^2 + {}_8O^{16} = {}_7N^{14} + {}_2He^4$	3.1 MeV
${}_1H^2 + {}_3Li^6 = {}_3Li^7 + {}_1H^1$	5.0 MeV
${}_1H^2 + {}_4Be^9 = {}_5B^{10} + {}_0n^1$	4.4 MeV

It is customary to omit the *mass energy* of the bombarding particle from the left-hand side of all reaction equations and to designate the total energy liberated by the disintegration as shown at the right. The values of Q given therefore represent the experimentally determined values of the energy over and above that supplied by the incident projectile.

EXAMPLE 1. Find the Q value for the nuclear reaction shown by the first reaction above.

$${}_1H^1 + {}_9F^{19} = {}_8O^{16} + {}_2He^4$$

SOLUTION. Instead of adding masses of atoms on the two sides of the equation, and taking the difference to find the total mass converted into energy, it is easier to use column three in Appendix III, as follows: The mass excess over or under whole number values is already converted to MeV equivalents in the table. These values for the four atoms involved are

$$\begin{array}{ll} {}_1H^1 = +7.288 \text{ MeV} & {}_8O^{16} = -4.736 \text{ MeV} \\ {}_9F^{19} = -1.486 \text{ MeV} & {}_2He^4 = +2.425 \text{ MeV} \\ \hline \quad +5.802 \text{ MeV} & \quad -2.311 \text{ MeV} \end{array}$$

The difference between the two sums gives a Q value of 8.113 MeV.

To see how Q values may be used, consider the fifth reaction above in which the Q value is 5.0 MeV.

$${}_1H^2 + {}_3Li^6 = {}_3Li^7 + {}_1H^1$$

Let us assume that a beam of 3.0 MeV deuterons from a cyclotron bombards a target of lithium metal. From the other side of the target a stream of high-energy protons is found to be coming off with an energy of 7.0 MeV. Since the total available energy from the reaction is 5.0 MeV from mass conversion, plus 3.0 MeV from the incident beam energy, there is but 1.0 MeV left for the recoiling lithium-7 nuclei within the target. In other words, the incident energy is added to the Q value to obtain the total energy available to be divided between the recoiling particles.

As a second example, consider the sixth reaction in which deuterons, bombarding beryllium metal, produce high-speed neutrons and recoiling boron nuclei. This particular disintegration is important experimentally because it is used as a means of obtaining intense beams of neutrons for use as projectiles in other disintegrations. The nuclear changes are illustrated schematically in Fig. 38B. The available energy from the loss in mass alone is the Q value 4.4 MeV; so that, if deuterons with an energy of 7 MeV are used to bombard the beryllium target, the available energy becomes about 11.4 MeV, 1 MeV going to the recoil boron nucleus and approximately 10.4 MeV to the neutron.

It can be shown from the laws of conservation of momentum and energy that, when a nucleus disintegrates into two fragments, the total energy is divided between them in approximately the

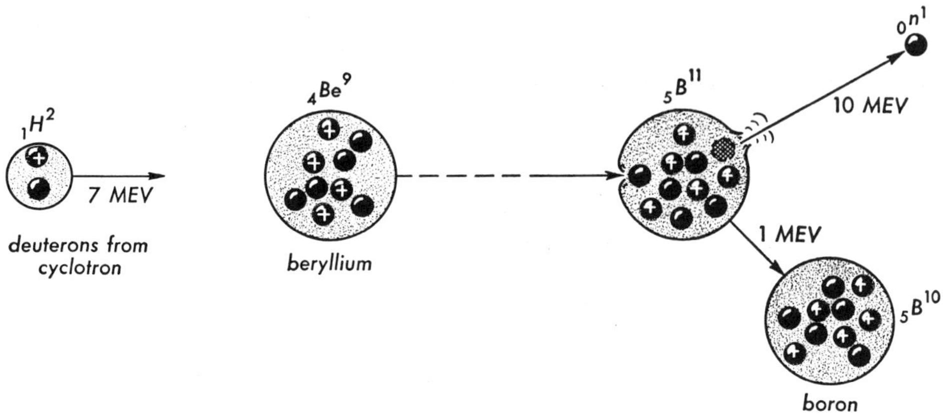

Fig. 38B. Deuteron disintegration of a beryllium nucleus to produce high-speed neutrons.

inverse ratio of their masses. If for example the total energy liberated is 15 MeV, and the fragments are a proton-1 and a beryllium-9 nucleus, 90% of the 15 MeV, or 13.5 MeV is given to the proton, and 10% of 15 MeV, or 1.5 MeV to the beryllium nucleus.

38.2 MULTIPLE DISINTEGRATIONS A study of certain disintegration experiments shows that some of the unstable nuclei created by the capture of a proton or deuteron by a stable nucleus split up into more than two stable nuclei. Examples of this arise when boron is bombarded by protons and when nitrogen is bombarded by deuterons. In the case of boron (see Fig. 38C), the proton is first captured by a $_5B^{11}$ nucleus to form an unstable carbon nucleus, $_6C^{12}$. This composite structure disintegrates by the expulsion of an α particle with several million volts energy, leaving behind a beryllium nucleus, $_4Be^8$.

$$\begin{aligned} {}_1H^1 + {}_5B^{11} &= {}_4Be^8 + {}_2H^4 \\ &= {}_2He^4 + {}_2He^4 + {}_2He^4 \end{aligned} \qquad (38a)$$

This nuclear combination is still unstable and splits apart into two more α particles. When the phenomenon was first observed it was thought that all three α particles came apart simultaneously, but further observations showed that first one and then two were ejected. The total energy liberated has been measured to be about 11 MeV and checks almost exactly with the value obtained from the loss in mass.

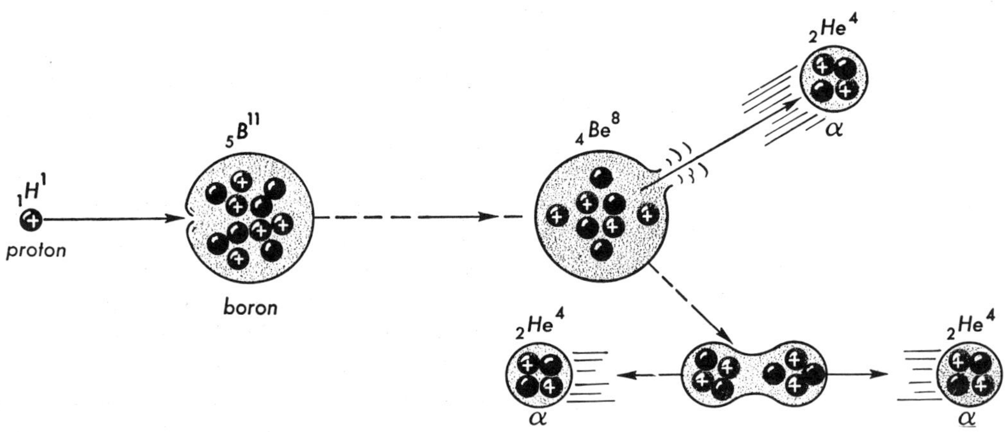

Fig. 38C. Diagram of the disintegration of a boron nucleus of mass 11 by a proton to produce three α particles.

38.3 DISCOVERY OF INDUCED RADIOACTIVITY

The discovery of induced radioactivity was made in 1934 by F. Joliot† and I. Curie Joliot.*

For years the Joliot-Curies, as they are now often called, had been exposing various substances to the α rays from naturally radioactive elements and had been studying the various disintegrations that took place. In the specific instance referred to above, they bombarded aluminum with *α particles from polonium* and measured the energies of the ejected neutrons by the recoiling of protons from paraffin. See Fig. 38D.They observed that even after the polonium source was taken away the detector continued to respond to some kind of penetrating radiation. Upon investigating the nature of these rays they found positively charged electrons coming from the aluminum.

Repeating the experiments to make certain of the results, they came to the conclusion that, under the bombardment of α particles, the aluminum had become radioactive in its own right. What was happening has since been verified: α particles striking aluminum nuclei are captured, and the resulting nuclei disintegrate with the violent ejection of neutrons.

$$_2He^4 + {}_{13}Al^{27} = {}_{15}P^{30} + {}_0n^1 \qquad (38b)$$

The newly created recoil particles, with a charge of +15 and mass 30, have been identified as phosphorus nuclei which are not stable but radioactive. Spontaneously disintegrating, these radioactive phosphorus nuclei $_{15}P^{30}$ shoot out positrons, leaving behind them stable silicon atoms of charge +14 and mass 30.

$$_{15}P^{30} = {}_{14}Si^{30} + {}_1e^0 \qquad (38c)$$

The *half-life* of this activity, which measures the rate of decay of the phosphorus into silicon, is only 2.5 min. For the meaning of half-life see Sec. 34.8.

Although the mass of the electron is not zero, it is so small compared with unit mass (the mass of one electron, it will be remembered, is 1/1840th of one atomic mass unit) that e is written with a zero superscript. According to this notation, a positron is written $_1e^0$ and an electron $_{-1}e^0$.

Because the phosphorus does not all disintegrate immediately it has been possible to identify the activity as coming from the newly created phosphorus atoms in the following way. A piece of aluminum metal, immediately after being bombarded, is dissolved in hydrochloric acid

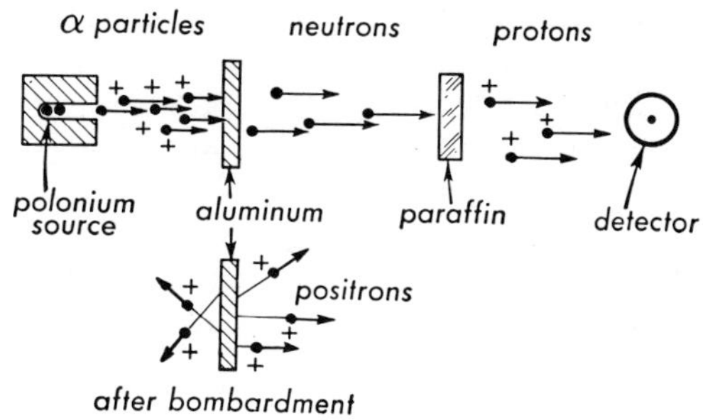

Fig. 38D. Experimental arrangement used by the Joliot-Curies when they discovered induced radioactivity.

---

* IRENE CURIE (1897–1956), daughter of the most famous woman physicist, was the wife of Frederic Joliot. Because of the now famous name of Curie, the physicists of the world hyphenate the name and call them the Joliot-Curies. It is not so strange a coincidence that induced radioactivity was discovered by these observers since for years they had worked with radioactive substances in the famous laboratory of the late Mme. Marie Curie at the Radium Institute in Paris.

† FREDERIC J. JOLIOT (1900–1958), French physicist, also known as Joliot-Curie after his marriage in 1926 to Irene Curie. In 1925 he was appointed to the staff of the Paris Radium Institute where he later became Professor. In bombarding boron, aluminum, and magnesium with alpha particles, he produced radioactive isotopes of nitrogen, phosphorus, and aluminum. In 1935 he was awarded, jointly with his wife, the Nobel Prize in Chemistry for the discovery of new radioactive elements prepared artificially. In 1944, he was appointed Director of the French Centre Nationale des Recherches Scientifiques, Paris; and in 1946 was made High Commissioner for Atomic Energy, but was removed by the government in 1950 because of his strong Communist views.

together with some ordinary inactive phosphorus and a standard chemical separation made. Testing each part separately, the radioactivity is found to be present with the phosphorus residue and not with the aluminum.

38.4 THE DISCOVERY OF RADIOACTIVE SODIUM Immediately after the discovery of induced radioactivity by the Joliot-Curies, Lawrence bombarded sodium with 2-MeV deuterons from the cyclotron and found that it too, like aluminum, became radioactive. See Fig. 38E. Upon

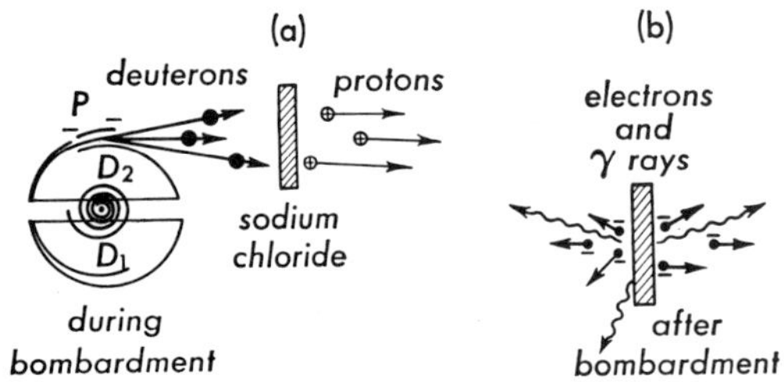

Fig. 38E. Experimental arrangement used by Lawrence in discovering radioactive sodium.

testing for the nature of the rays given off during bombardment, Lawrence found protons with an energy of about 7 MeV. When the sodium target was removed from the deuteron beam, as shown in diagram (b), and then tested for activity it was found to be emitting both electrons and γ rays. The bombarding reaction, therefore, is

$$_1H^2 + {}_{11}Na^{23} = {}_{11}Na^{24} + {}_1H^1 \qquad (38d)$$

followed by the radioactive decay of the unstable sodium nuclei,

$$_{11}Na^{24} = {}_{12}Mg^{24} + {}_{-1}e^0 + \gamma \text{ ray} \qquad (38e)$$

The first stage of the disintegration process is shown at the left in Fig. 38F, and the radioactive decay is shown at the right.

The residual nucleus $_{11}Na^{24}$ of the first disintegration is called *radio-sodium.* Having a charge of +11 and mass of 24, it must be an isotope of sodium not found in nature. Since measurements of the activity of radio-sodium give a *half-life* of only 15 hr, it is clear why such atoms are not found in nature.

Up to the present time more than two hun-

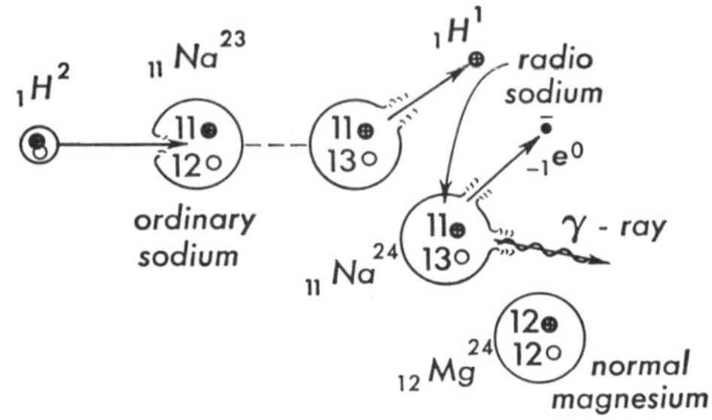

Fig. 38F. Diagrams illustrating the production and disintegration of radio sodium.

dred different kinds of radioactive atoms have been produced in the laboratory. Two examples in addition to those already given are illustrated by the following reactions:

$$_1H^2 + {}_{15}P^{31} = {}_{15}P^{32} + {}_1H^1$$

$$_{15}P^{32} = {}_{16}S^{32} + {}_{-1}e^0$$

$$_1H^2 + {}_6C^{12} = {}_7N^{13} + {}_0n^1$$

$$_7N^{13} = {}_6C^{13} + {}_1e^0$$

The first of these reactions forms *radioactive phosphorus* $_{15}P^{32}$ which is *electron active* with a half-life of 15 days.

A cloud chamber photograph of the positrons emitted by radioactive nitrogen $_7N^{13}$ is reproduced in Fig. 38G. The magnetic field bends all

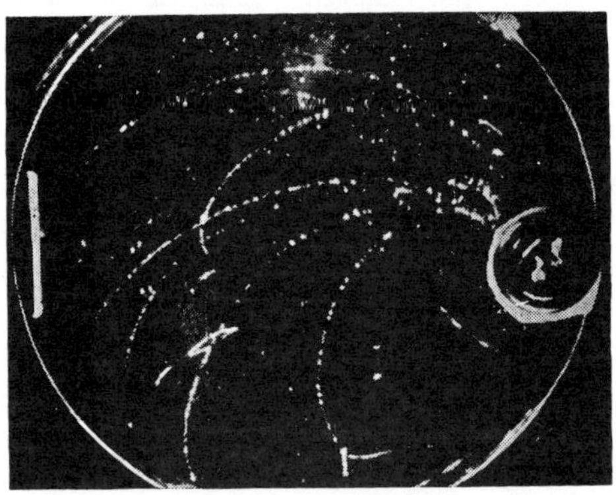

Fig. 38G. Photograph of the Wilson cloud-chamber tracks of positrons ejected by radio nitrogen, $_7N^{13}$.

the rays in the same direction, indicating all are positive charges. The low-density fogdrops forming the tracks indicate particles with the mass of an electron.

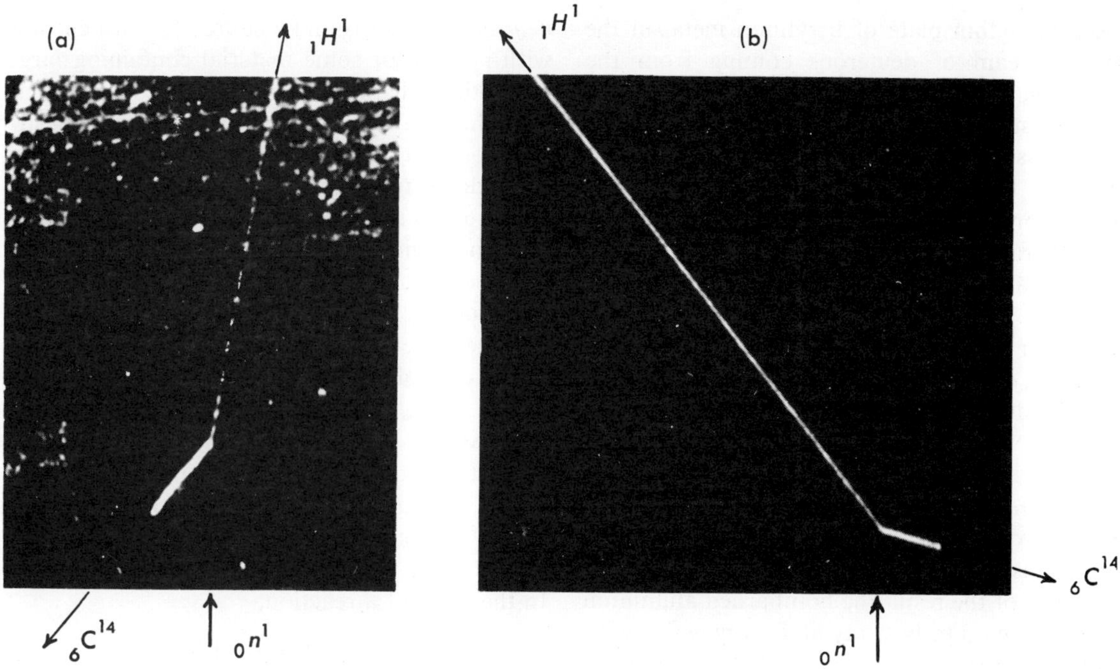

Fig. 38H. Cloud-track photographs of neutron disintegrations of nitrogen. (After Feather and Rasetti.)

38.5 NEUTRON PROJECTILES PRODUCE DISINTEGRATIONS The first disintegrations produced by high-speed neutrons as atomic projectiles were announced in 1932 by the English physicist Feather. Immediately following Chadwick's discovery of these neutral particles, Feather allowed neutrons from beryllium to enter a Wilson cloud chamber containing pure nitrogen gas. Numerous expansions of the chamber and the simultaneous clicks of a camera shutter gave many photographs of the ion tracks left by recoiling nitrogen atoms.

Although most of the photographs indicated elastic collisions between nitrogen atoms and neutrons, an occasional photograph showed a forked track, indicating a disintegration of a nitrogen nucleus.

$$_0n^1 + {}_7N^{14} = {}_6C^{14} + {}_1H^1 \tag{38f}$$

Two photographs of several such disintegrations are reproduced in Fig. 38H. Although hundreds of neutrons enter the cloud chamber every second, they do not ionize atoms as charged particles do, and hence leave no tracks. When a head-on nuclear collision occurs, however, the disintegrated nuclei, possessing as they do high speeds and positive charges, leave a trail of ions behind them. The fork in (a) shows a proton track of considerable length originating at the same point as the more dense, short-ranged track of the recoiling carbon nucleus.

The second photograph, like the first, shows a long thin track of the proton and a short heavy track of a heavier carbon recoil nucleus.

Strong sources of neutrons are produced by

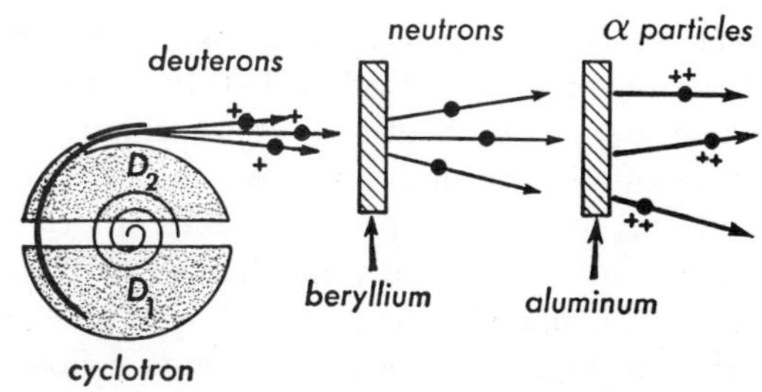

Fig. 38I. Experimental arrangement for producing intense beams of neutrons by bombarding beryllium with deuterons. The neutrons are then used as projectiles for further disintegrations as illustrated here for aluminum.

inserting a thin plate of beryllium metal in the intense beam of deuterons coming from the cyclotron as shown in Fig. 38I.

Into such a beam of chargeless particles numerous substances of known chemical constitution have been inserted and the disintegration products studied with suitable detectors. To illustrate by an example, suppose that a thin sheet of aluminum is inserted into the beam of neutrons as shown in the diagram. In this particular instance α particles are observed emerging from the aluminum, thus enabling one to write down the following neutron reaction,

$$_0n^1 + {}_{13}Al^{27} = {}_{11}Na^{24} + {}_2He^4 \qquad (38g)$$

Thus *radioactive sodium*, produced originally by the deuteron bombardment of ordinary sodium, is here produced by a different reaction. As proof of the result the bombarded aluminum target is found to be β ray and γ ray active with a half-life of 15 hr [see Eq. (38d)]. There are at least two other known disintegration processes by which radio-sodium is produced: one by the neutron bombardment of silicon, and the other by the α particle bombardment of magnesium,

This is but one example of the many known radioactive elements that can be manufactured in four different ways. As a matter of fact, with sufficiently energetic atomic bullets it is now possible to produce hundreds of new atomic nuclei not found in nature.

38.6 SLOW NEUTRON DISINTEGRATIONS The fact that neutrons, slowed down to very low speeds, have the ability to disintegrate certain atoms was first discovered and investigated by the Italian physicist Enrico Fermi* and his collaborators. A neutron approaching the nucleus of an atom does not experience a repulsive force, as does a proton, deuteron, or α particle, and consequently its chances of penetration into, and of being captured by, a nucleus is relatively large. It is for this reason that slowly moving neutrons are able to bring about disintegrations that slowly moving charged particles cannot.

The customary method of producing *slow neutrons* is to surround a source of fast neutrons with paraffin or some material containing large quantities of hydrogen or deuterium. As neutrons pass through this "moderator" material, elastic collisions with hydrogen nuclei continually slow them down until, at a distance of several centimeters from the source, they have lost nearly all of their original energy. What little energy they do have is picked up by regular thermal collisions with other atoms. Their resultant motions become quite the same as the random motions of atoms or molecules in a gas.

38.7 MEAN LIFE OF A NEUTRON Numerous laboratory experiments show that a free neutron has an average life of about 11.7 min, or 700 s before it spontaneously decays. The neutron decays into a proton, an electron, and a neutrino. In the form of a reaction,

$$n \rightarrow p + e^- + \nu \qquad (38h)$$

In relation to the time scale in which atomic particles interact with each other during a collision (about 10^{-8} to 10^{-15} s), the mean life of a neutron is extremely long; even a thermal neutron will, on the average, travel more than 1000 mi before decaying. A 2 MeV neutron will travel 10 million miles.

By comparing the decay of a neutron with the radioactive decay of unstable nuclei, we can speak of the above equation as one of beta decay.

* ENRICO FERMI (1901–1954), American physicist, born in Italy. He studied at Pisa, Göttingen, and Leiden, and later taught physics at the universities of Florence and Rome. He was Professor of Physics at Columbia University (1939–1945) and the University of Chicago (1945–1954). For his experimental discoveries of disintegration, induced by slow neutrons, Fermi was awarded the 1938 Nobel Prize in physics. His creation of the first self-sustaining uranium chain reaction at Chicago in 1942 led to the development of the atomic bomb. He later worked on the hydrogen bomb and served on the General Advisory Committee of the Atomic Energy Commission. The element fermium (At. No. 100) was named after him.

38.8 THE SCINTILLATION COUNTER This is a sensitive device used principally for the detection of high-speed atomic particles. In principle it is based upon the earliest discoveries in radioactivity that α particles upon striking a fluorescent material, like zinc sulfide, produce a tiny flash of light. These flashes, called scintillations, can be seen by the dark-adapted eye, or they can be detected by a photomultiplier tube.

It is now well known that when charged particles pass through certain transparent materials, fluorescent light is produced all along the path. For many crystals and plastics, this fluorescent light is blue or violet in color.

A typical scintillation counter tube is shown in Fig. 38J. A block of fluorescent material is

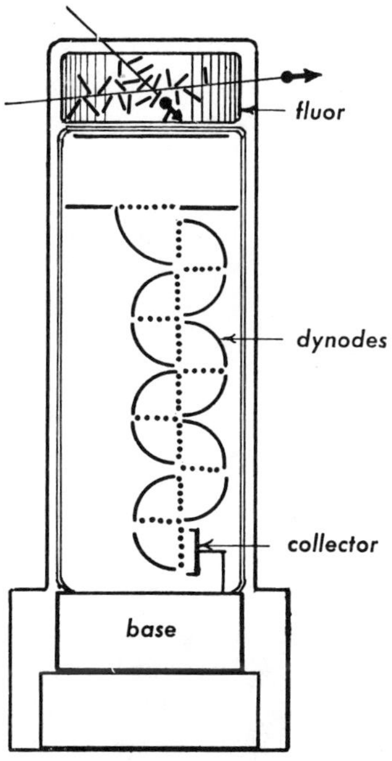

Fig. 38J. Scintillation detector tube using a photomultiplier tube with a fluorescent block.

mounted on the flat end of a photomultiplier tube, and then encased in a thin-walled, light tight, aluminum shield. When a particle traverses the fluor, the light ejects electrons from the photocathode by the photoelectric effect. The electron build-up by the eight or more dynodes makes a sizable voltage pulse that activates an electrical counting circuit.

When γ rays are to be detected, the fluorescent materials frequently used are crystals of sodium iodide, NaI, and cesium iodide, CsI. For high energy β rays such plastics as polystyrene, impregnated with anthracene, are used. For α particles, with their relatively low penetrating power, a thin layer of zinc sulfide deposited on the face of the photomultiplier is used.

The principal advantages of scintillation counters over other detectors of nuclear radiation are first, they operate in air or in a vacuum; second, they deliver an electrical impulse that is roughly proportional to the energy lost by the traversing particle; and third, they can count at amazingly high speeds because of their extremely short dead time of 10^{-5} to 10^{-9} s.

If atomic particles enter a crystal with almost the speed of light, 3×10^8 m/s, their speed in the crystal will be greater than the speed of light in the crystal.

The light waves produced by such high-speed particles are shown in Fig. 38K and are called

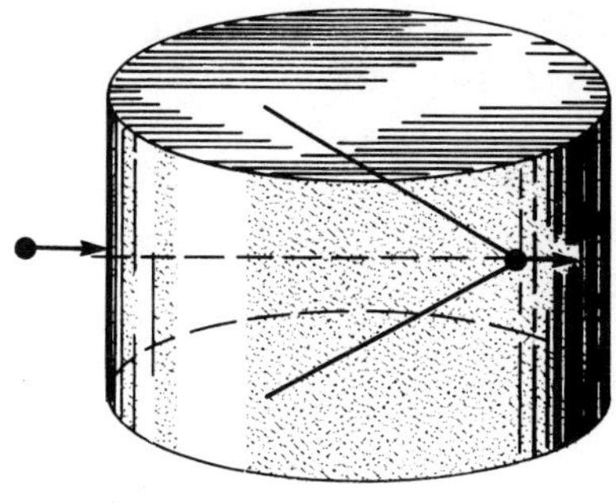

Fig. 38K. The conical wave from a high-speed atomic particle in a transparent medium; Cerenkov radiation.

Cerenkov radiation. Such a conically shaped wave is analogous to the shock wave from the nose of the airplane traveling with a speed greater than sound or to the V-shaped wave from the bow of a ship sailing faster than the speed of surface water waves.

Suitable transparent materials emitting strong Cerenkov radiation are also mounted in the end of photomultiplier tubes, and connected to an

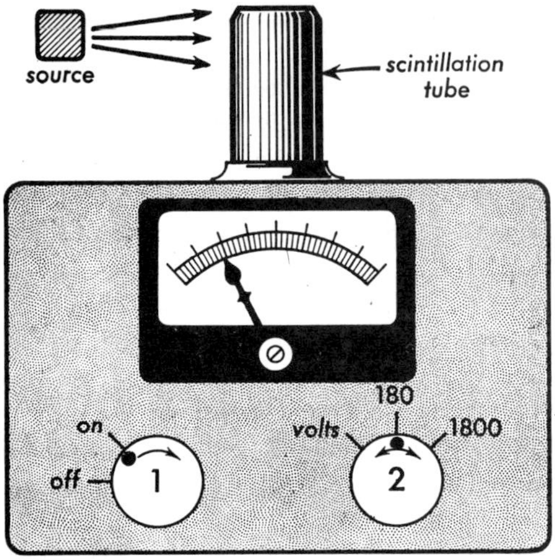

Fig. 38L. Scintillation counter arrangement.

electronic counting device as shown in Fig. 38L, form what is called a Cerenkov counter.

38.9 MEDICAL APPLICATIONS Since the time of its discovery radio-sodium has found numerous and important applications in many branches of science. It has, for example, been used as a means of tracing certain organic and inorganic chemicals passing through the human body, through plants, and through chemical experiments of one kind or another. To give a simple example, one can show, by drinking water containing radio-sodium in the form of common table salt (NaCl), that within 2 min some of it has entered the blood stream and has been distributed to all parts of the body. The presence of the salt in the finger tips or the toes can be demonstrated by detecting the electrons and γ rays from the sodium with a Geiger counter. Similar experiments can be performed with trees and plants to see how rapidly their roots take up certain plant foods and distribute them to the leaves and branches.

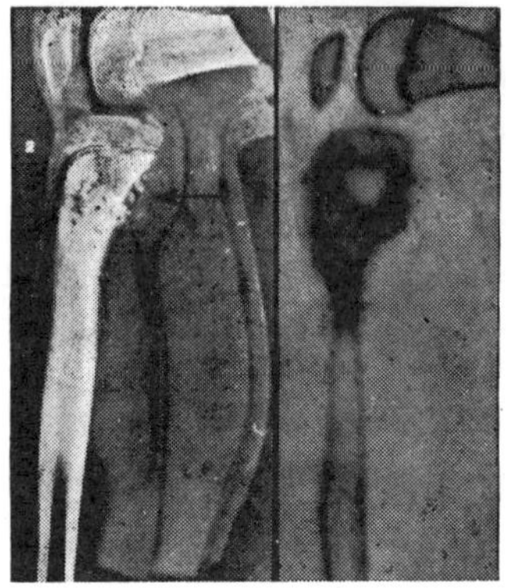

Fig. 38N. X ray and radio-autograph of a human leg. (Courtesy of Dr. J. G. Hamilton.)

At the present time radio-phosphorus 32, because of its chemical properties and its comparatively long half-life, is being used in a number of medical researches as a possible cure for certain diseases, as well as for tracing the migration of phosphorus through the body.

An important medical technique is the use of radioactive elements to obtain *radio-autographs*. A plant or animal is given a single dose of liquid containing the "tagged" atoms and at various

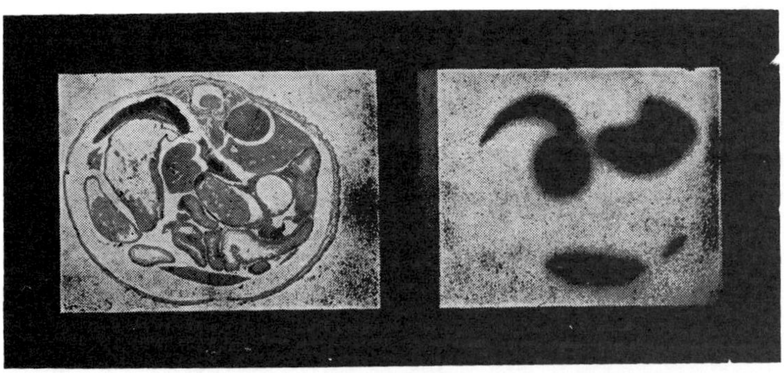

Fig. 38M. Radio-autograph of the type used in medical studies. Shows selective absorption of phosphorus by the liver and spleen.

times thereafter *thin sections of tissue* from the plant or animal are placed in direct contact with a photographic film. After several hours of exposure to any possible rays from "tagged" atoms in the section, the film is developed. Fig. 38M shows at the left a thin cross section of a mouse, previously fed radio-phosphorus, and at the right the resulting radio-autograph of this same section. The dark areas indicate the uptake of phosphorus by the *spleen* and *liver* sections.

Fig. 38N shows an *X ray* and *strontium radio-autograph* of a section of an amputated leg from a patient with osteogenic sarcoma (bone cancer). The patient received radio-strontium orally two days before removal of the extremity. The X ray shows the bone tumor at the upper end of the tibia, while the radio-autograph shows selective deposits of strontium in the tumor with small amounts in the surrounding bone and relatively little in the soft tissues. These studies indicate that radio-strontium may prove useful in clinical therapy.

QUESTIONS

1. What kinds of high-energy particles are commonly used in the bombardment of a target?

2. What kinds of high-speed particles might well be expected to be emitted as the result of various target bombardments?

3. What is induced radioactivity? What kinds of particles are usually given off by induced radioactive atoms?

4. Charged particles show their tracks in a Wilson cloud chamber. Neutrons and γ rays do not. Why?

5. How are strong beams of neutrons produced? Can neutrons be accelerated in a cyclotron?

6. What is a multiple disintegration? Give an example.

PROBLEMS

1. If a 5-MeV deuteron is captured by a N^{14} nucleus, and upon disintegrating gave out an α particle, (a) what is the reaction, and (b) what is the energy liberated? [Ans. (a) ${}_1H^2 + {}_7N^{14} = {}_2He^4 + {}_6C^{12}$. (b) 6.57 MeV.]

2. If the total available energy in Prob. 1 is divided between the two recoiling nuclei in the inverse ratio of their respective masses, what is (a) the maximum energy of the α particle, and (b) what is its range in air?

3. When α particles with a range of 7.0 cm bombard sodium-23, protons are observed coming from the target. (a) Write down the nuclear reaction, and (b) how much energy is released? [Ans. (a) ${}_2He^4 + {}_{11}Na^{23} = {}_{12}Mg^{26} + {}_1H^1$. (b) 5.8 MeV.]

4. If a 6-MeV deuteron is captured by carbon-12 nucleus and the compound nucleus disintegrates by emitting a proton, (a) what is the reaction? (b) What is the remaining nucleus? (c) What is the total energy liberated? (d) If 92% of the liberated energy goes to the proton, what will be its range in air? See Fig. 35H.

5. When 3-MeV deuterons bombard boron-10, α particles are observed among the radiated disintegration products. (a) Write down the nuclear reaction. (b) What is the total energy liberated? (c) If this total energy were divided equally among the disintegration fragments, how much energy would each one get? (d) What would be the ranges of the fragments? [Ans. (a) ${}_1H^2 + {}_5B^{10} = {}_2He^4 + {}_2He^4 + {}_2He^4$. (b) 20.91 MeV. (c) 6.97 MeV. (d) 60 cm.]

6. If a 4-MeV deuteron is captured by nitrogen-14 nucleus, and upon disintegration ejects a proton, (a) how much energy is liberated? (b) What will be the energy of the proton? (c) What will be its range in air?

7. If a neon-20 nucleus is at rest when it liberates a proton, and the total energy liberated is 12.0 MeV, find (a) the energy ratio of the two recoiling particles, (b) the energy of each particle, and (c) the range of the proton. [Ans. (a) 20:1. (b) 0.571 MeV and 11.429 MeV. (c) 148 cm.]

8. When 8 MeV neutrons are allowed to bombard magnesium-24, protons are observed being emitted from the target. (a) Write down the reaction. Find (b) the total energy liberated, (c) the energy acquired by the proton, and (d) the range of the protons.

9. When 7.5 MeV deuterons are allowed to bombard boron-10, neutrons are observed among the disintegrating fragments. (a) Write down the reaction, and (b) calculate the total energy liberated in driving the fragments apart. Calculate the energy given to the neutron. [Ans. (a) ${}_1H^2 = {}_5B^{10} = {}_6C^{11} = {}_0n^1$. (b) 13.97 MeV. (c) 12.81 MeV.]

10. If the most abundant oxygen isotopes are bombarded by neutrons, protons are liberated in the disintegration process. (a) Write down the nuclear reaction. If the neutrons have an energy of 17.0 MeV, find (b) the total energy liberated, (c) the energy of the protons, and (d) the range of the protons in air.

11. If oxygen-16 is bombarded with 6.0 MeV deuterons, and α particles are observed as disintegration fragments, find (a) the nuclear reaction, (b) the total energy liberated, (c) the energy acquired by the α particle, and (d) its range in air. [Ans. (a) ${}_1H^2 + {}_8O^{16} = {}_7N^{14} + {}_2He^4$. (b) 9.11 MeV. (c) 7.08 MeV. (d) 6.1 cm.]

12. When 6-MeV neutrons bombard chlorine-35, radioactive sulfur-35 is produced. Find (a) the reaction, (b) the total energy released, (c) the energy imparted to the proton, and (d) its range in air.

13. When 8.0 MeV deuterons bombard stable phosphorus they produce the most abundant isotope of sulfur. (a) Write the nuclear reaction. Calculate (b) the total energy released, and (c) the energy of the neutron. [Ans. (a) ${}_1H^2 + {}_{15}P^{31} = {}_{16}S^{32} + {}_0n^1$. (b) 1.362 MeV. (c) 1.321 MeV.]

14. There are seven known isobars with an atomic mass number of 87. Their accurately known masses are as follows: ${}_{35}Br^{87} = 86.92731$, ${}_{36}Kr^{87} = 86.91931$, ${}_{37}Rb^{87} = 86.91541$, ${}_{38}Sr^{87} = 86.91513$, ${}_{39}Y^{87} = 86.91682$, ${}_{40}Zr^{87} = 86.92033$, and ${}_{41}Nb^{87} = 86.92638$ atomic mass units, respectively. The first three are β active, the fourth one is stable, the fifth undergoes electron capture, and the last two undergo positron emission. Plot a graph of these masses against their atomic numbers. Use 15 vertical divisions labeled 86.914, 96.915, 86.916, up to 86.929 amu, and eight horizontal divisions labeled 34, 35, 36, up to 42. Draw a smooth curve through your plotted points. What can you say about the one stable isotope and its significance?

39 | THE ATOMIC NUCLEUS AND FISSION

One of the great unsolved mysteries of science is the nature and origin of the forces that bind the nucleus together. Why is it that certain combinations of neutrons and protons form stable nuclei, while others do not? Why does the addition of a single neutron or proton to a stable nucleus cause it to become so unstable that it disintegrates by ejecting some particle to become stable? While some progress has been made in trying to find answers to these questions, much of what is in the nucleus and what goes on there is still a tantalizing puzzle.

39.1 WHAT HOLDS THE NUCLEUS TOGETHER?

Although the disintegrations of different nuclei give rise to many different kinds of particles with different amounts of energy, we will begin by assuming only two kinds of particles exist within the nucleus—neutrons and protons. If this is correct, our task becomes the difficult one of explaining not only the disintegration mechanism of an unstable nucleus but the binding forces that hold a stable nucleus together. An answer to the latter question will serve as a starting point for the following presentations.

According to the neutron-proton theory of the atomic nucleus the deuteron nucleus contains but one neutron and one proton. See Fig. 39A. Let us compare, therefore, the mass of one free proton plus one free neutron, with their mass

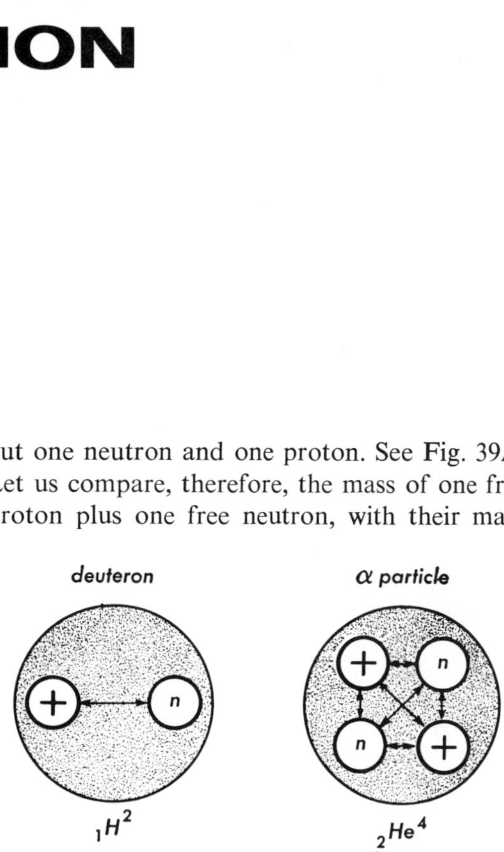

Fig. 39A. Schematic diagram of the nucleus for (a) a deuterium atom and (b) a helium atom.

when combined as a deuteron (for masses see Appendix III).

Neutron mass	$_0n^1$ =	1.008665
Hydrogen mass	$_1H^1$ =	1.007825
	Sum =	2.016490
Deuteron mass	$_1H^2$ =	2.014103

The difference in mass of 0.002387 atomic mass unit (abbr. amu) is not due to inaccurate measurements of mass but is a real difference to be accounted for as the annihilation energy that binds the two particles together. When a neutron and proton come together to form a deuteron, a small part of their mass—namely, 0.002387 amu (equivalent to 2.22-MeV energy)—is radiated from the newly formed nucleus. At close approach, in other words, the two particles attract each other so strongly that once together it takes the equivalent of a little more than two million volts of energy to pull them apart. This has been confirmed by a nuclear photoelectric effect, an experiment in which α rays of 2.22 MeV energy or greater are found to break up deuterium nuclei into their constituent parts, while γ rays of lower energy have no effect. How neutrons and protons attract each other when very close together is a question of great importance, for we now realize that the stability of all the universe as we know it depends upon these forces.

Consider as a second example the attractive forces between the four nucleons of a helium nucleus, i.e., the two neutrons and two protons of an α particle as shown in Fig. 39A. By combining the masses of the four free particles and comparing them with the mass of the helium atom, we obtain

$$\begin{aligned} 2{_0}n^1 + 2{_1}H^1 &= 4.032980 \text{ amu} \\ He^4 &= \underline{4.002604 \text{ amu}} \\ \text{Mass difference} &= 0.030376 \text{ amu} \end{aligned}$$

By multiplying this mass difference by 931, we obtain

$$E = 28.3 \text{ MeV}$$

This value of 28.3 MeV indicates a binding energy of 7 MeV per nucleon, a value considerably higher than 2.22 MeV for the deuteron. This is also the amount of the energy we must expend to break all six of the attractive bonds shown in the diagram.

If we make similar calculations for other nuclides near the beginning of the periodic table, and plot a graph of the binding energy per nucleon, E/A, where A is the mass number, we obtain Fig. 39B. If this same procedure is carried

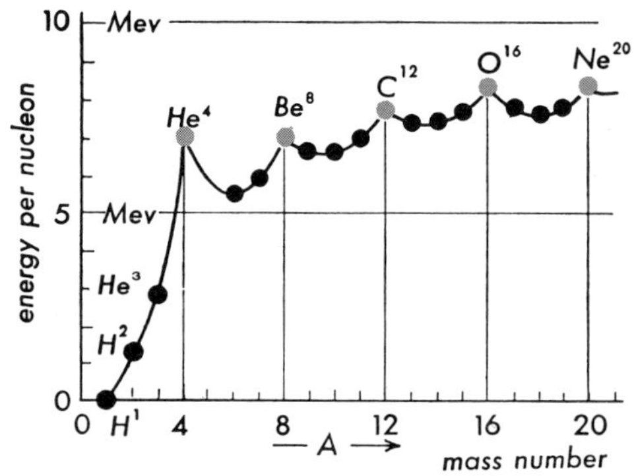

Fig. 39B. Binding energy per nucleon for the lightest elements in the periodic table.

out for the entire periodic table, a graph like the one shown in Fig. 39C is obtained. By drawing a horizontal line across the upper part of the graph at 8 MeV, as shown by the red line,

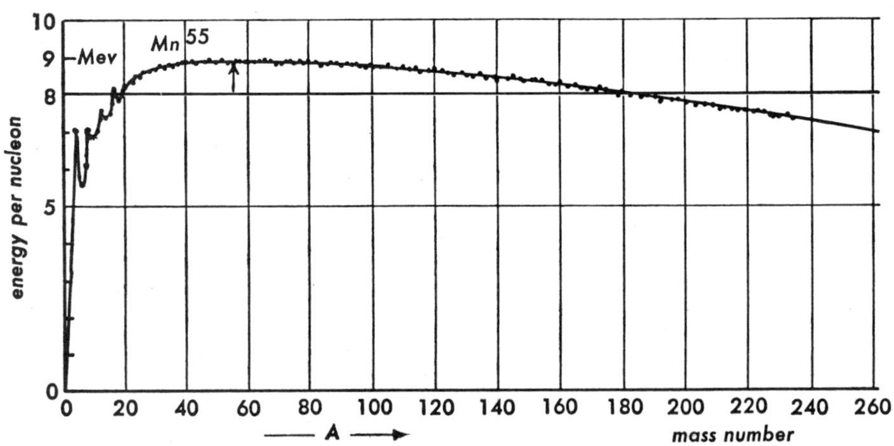

Fig. 39C. Binding energy per nucleon for the stable isotopes of the periodic table.

we obtain a kind of average binding energy per nucleon for nearly all the elements. This value is approximately the difference between unit atomic mass and the mass of a free neutron or proton.

EXAMPLE 1. Calculate the binding energy per nucleon for sodium-23.

SOLUTION. Using Appendix III, we see that $_{11}Na^{23}$ has 11 protons, and 12 neutrons, and a mass excess of −9.526 MeV. Each free proton has a mass excess of 7.288 MeV, and each neutron a mass excess of 8.071 MeV. Therefore,

$$\begin{array}{r} 7.288 \\ \times 11 \\ \hline 80.168 \text{ MeV} \end{array} \qquad \begin{array}{r} 8.071 \\ \times 12 \\ \hline 96.852 \text{ MeV} \end{array}$$

The total energy required to pull all 23 nucleons apart is, therefore, the sum 80.168 + 96.852 + 9.526 = 186.546 MeV. Dividing by 23, the number of nucleons, gives

8.11 MeV per nucleon

One process not described thus far, the nuclear photoelectric effect, is shown in Fig. 39D. A photon of energy $h\nu$, passing close to a

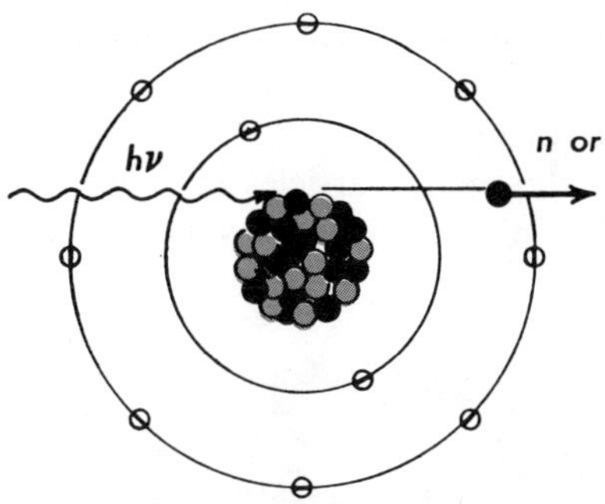

Fig. 39D. Diagram illustrating the nuclear photoelectric effect.

nucleus, is absorbed; part of the energy is used to remove a neutron or proton from the nucleus, and the remainder given to the particle as kinetic energy.

$$h\nu = W + \tfrac{1}{2} mv^2 \tag{39a}$$

The work function W, as we have seen above, is about 8 MeV and is essentially the energy required to create the extra mass needed by the nucleon to set it free.

39.2 THE NUCLEAR POTENTIAL BARRIER Early in the development of ideas concerning nuclear disintegration, Gamow proposed a model by which one might represent the atomic nucleus. This model is based upon the forces acting between two positive charges.

Picture a proton or an α particle, with its positive charge, approaching a positively charged nucleus. As the two charges come closer and closer together, they repel each other with a greater and greater force as given by Coulomb's law, Eq. (26a).

$$F = k\frac{QQ'}{d^2} \tag{39b}$$

This repulsion cannot continue to increase all the way to zero separation, however, for as the two charges come very close together, we know that there must be an attraction. Gamow proposed, therefore, that at close approach there is another law of force that comes into play and that this force is one of attraction for neutrons as well as protons and is very strong.

Since Coulomb's law is known to hold accurately at large distances, this added factor is called a short-range force. Graphs of Coulomb's law for repulsion and the short-range force for attraction are shown in Fig. 39E. Note that when

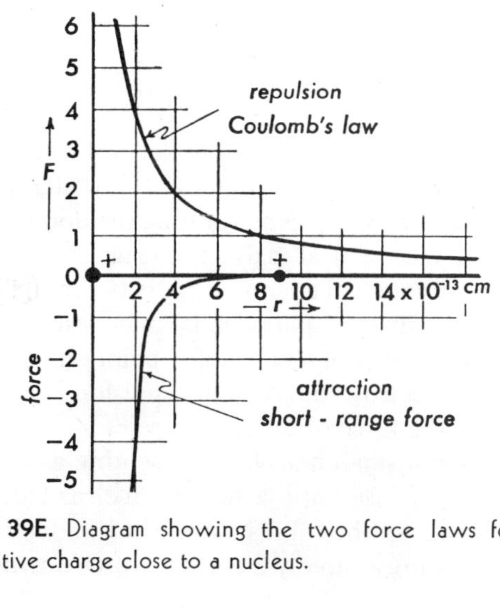

Fig. 39E. Diagram showing the two force laws for a positive charge close to a nucleus.

the particle is at the distance shown, 9×10^{-13} cm, Coulomb's law of repulsion is predominant, whereas at a distance less than 2×10^{-13} cm, the short-range force of attraction predominates. Since both of these forces are effective for all distances, they should be combined into one graph as shown by the F curve in Fig. 39F. If

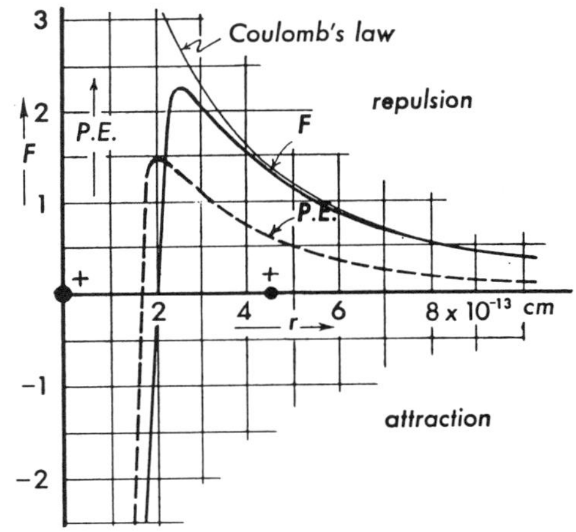

Fig. 39F. Graphs of the force and potential energy of a positively charged particle close to a nucleus.

instead of the force F we plot the potential energy stored between the two particles, we obtain the P.E. curve shown by the dotted line. Such a curve represents what is called the potential barrier of the nucleus. The highest point of the barrier is frequently called the edge of the nucleus which, for heavy atoms in the periodic table, occurs at, and gives a nuclear radius of, from 1 to 6×10^{-13} cm.

A very good mechanical model of the nucleus, having the form of a well or the crater of a volcano, can be made by rotating the dotted line around the vertical axis. See Fig. 39G. The electrical potential energy, P.E., between the two positively charged particles is analogous to the potential energy of a ball at any point on the hill, and the electrostatic force of repulsion is analogous to the force of gravity.

If now a small marble, representing a proton or an α particle, approaches the nuclear barrier, it will roll up the hill as shown in the diagram. Experiencing a rapidly increasing force down the

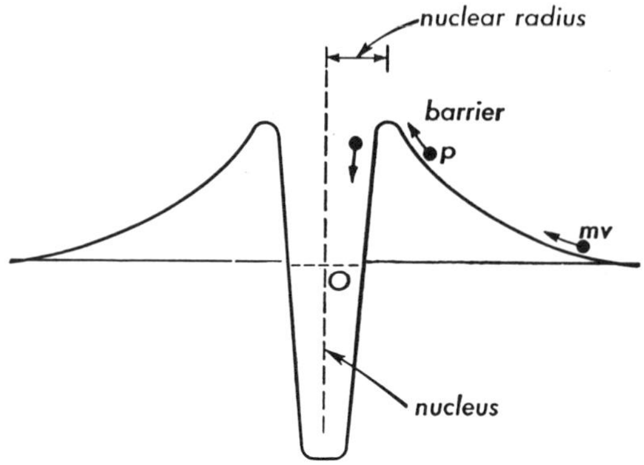

Fig. 39G. A graphical model of the atomic nucleus as proposed by Gamow. The potential barrier of a nucleus to an approaching positive charge is analogous to the crater of a volcano.

hill, the ball may be turned back, or off to one side, and we have what is called elastic scattering. If the initial velocity is high enough, however, the ball may go over the top of the barrier and drop down inside, representing capture.

39.3 BOHR'S NUCLEAR MODEL In 1937 Niels Bohr, the famous Danish physicist, made another outstanding contribution to modern physics when he improved Gamow's model of the nucleus by extending what is sometimes called the waterdrop model of the nucleus. Bohr imagines the many particles in a heavy nucleus as moving about within a spherical enclosure

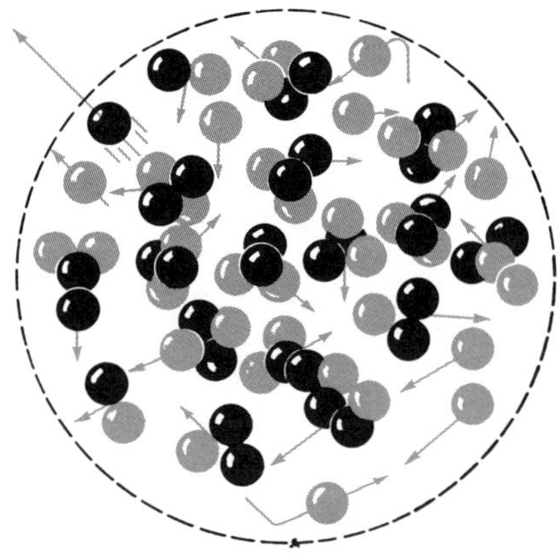

Fig. 39H. Nucleus in the act of ejecting a neutron, based upon the Bohr-Gamow waterdrop model.

with motions analogous to the molecules in a drop of water. See Fig. 39H. The surface of the spherical enclosure, which is the top of the potential barrier as represented in Fig. 39G, is analogous to the surface tension that holds a small water drop to its spherical form.

Just as the rapid motion of the molecules in water is a measure of the temperature, so Bohr speaks of the rapid motion of the neutrons and protons within the spherical boundary of the nucleus as a sort of pseudo-temperature. To explain disintegration, the analogy is drawn that the ejection of a particle from the nucleus is like the evaporation of a water molecule from a drop of water. Just as a rise in temperature brings about a more rapid evaporation of water, so an increase in the motions within the nucleus gives rise to a higher probability of disintegration.

In a stable nucleus, the particles within are moving about with very little kinetic energy, and are in the analogous state of a relatively low temperature. When a high-speed particle from outside penetrates through the potential barrier, it is accelerated toward the center of the nucleus and acquires a very high kinetic energy before it collides with one or more of the particles inside. Soon the energy becomes divided among the many particles and the nucleus takes on a higher temperature state. The potential well model of this same condition is shown in Fig. 39I.

As the particles move about inside there is a certain probability or chance that, within a given interval of time, some one particle will be hit by several particles, giving it a sufficiently high velocity in an outward direction to permit an escape through the potential barrier. The more rapid the internal motions, that is, the higher the temperature, the greater is this chance of escape.

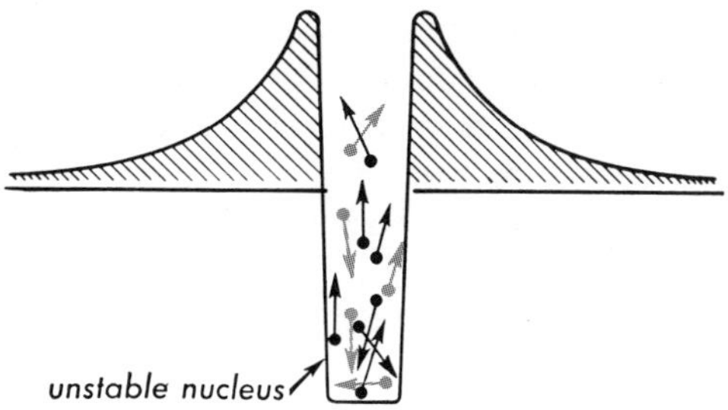

Fig. 39I. Well-model diagram representing an unstable nucleus.

A direct disintegration may be described in this way: if upon entering the nucleus a high-speed particle like a proton adds sufficient energy to give the nucleus a high temperature, another particle like a neutron or an α particle may be ejected immediately. Since such an ejected particle has to be supplied with a certain minimum energy to get free, the remaining particles will be slowed down, and the nucleus will have a lower temperature.

39.4 NUCLEAR DEMONSTRATION MODELS A demonstration model illustrating the capture of a high-speed proton or an α particle by a nucleus, prior to disintegration, is shown in Fig. 39J. Marbles rolled down the incline represent the speeding up of atomic projectiles by an accelerator like the cyclotron. Approaching the potential barrier a marble may roll part way up and then be deflected off to one side, illus-

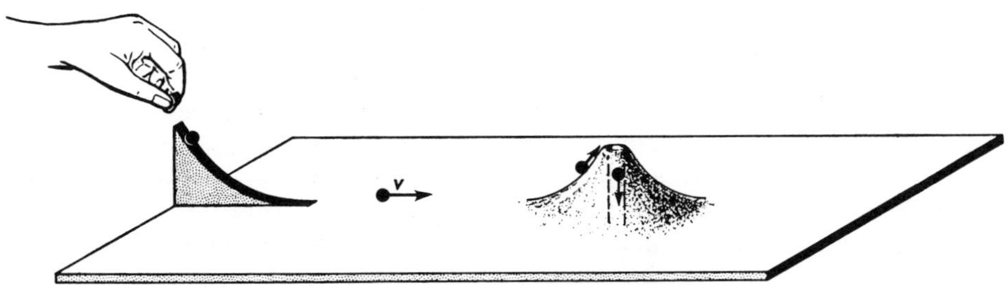

Fig. 39J. Mechanical model of a nucleus for demonstrating the capture of a high-speed proton, deuteron, or α particle, prior to disintegration.

trating an elastic collision without capture; or it may roll up the side and drop into the crater opening at the top, representing a capture prior to disintegration.

A demonstration of what happens inside the nucleus is illustrated by another model as shown in Fig. 39K. In this case the vertical scale of the barrier has, of necessity, been reduced, i.e., flattened out. When a marble is rolled down the incline and into the group of marbles at the center of the barrier, there may be several collisions before another particle usually goes bouncing out on the other side. This corresponds to a direct disintegration where one particle like a proton goes in and a neutron comes out.

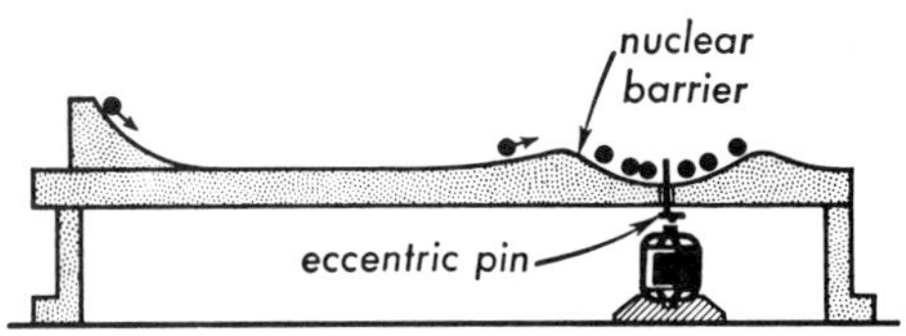

Fig. 39K. Mechanical model of a nucleus for demonstrating (a) the increased kinetic energy of nuclear particles after a capture, and (b) the chance probability of radioactive decay or disintegration by the ejection of a particle.

If a single particle does not emerge, most of the particles inside take on random motions, colliding with each other much the same as do the molecules or atoms in a gas or liquid. To prevent friction from stopping them (there is no friction in an atom), the marbles are continually agitated by a small pin protruding from underneath the barrier. This pin is mounted slightly off center at the end of the shaft of a small electric motor. If the motor is left running for some time, a single marble will eventually be hit by several particles moving in the same direction and will recoil with sufficient speed to carry it over the barrier and out. This corresponds to a disintegration or radioactive decay, which takes place according to the laws of chance, and to the resultant drop in temperature of the nucleus.

The faster the motor runs, the greater is the internal agitation and chance of ejection, and the shorter is the so-called half-life of the element.

39.5 NUCLEAR MODEL FOR NEUTRON DISINTEGRATIONS When a neutron approaches a nucleus prior to a disintegration, it does not encounter a potential barrier of the type already described for protons and α particles. A neutron has no charge so that at large distances it is not repelled by the positively charged nucleus. It may, therefore, approach a nucleus with very little speed of its own and be captured when it comes too close. At very close range the short-range attractive force shown in Fig. 39E sets in and draws the two together.

To an approaching neutron the nucleus acts as though it were a well into which the particle will fall. This is illustrated by the flat potential curve in Fig. 39L. The marble rolling along the horizontal plane toward the pit represents the influence of the nucleus upon the neutron's motion, whereas the marble rolling up the hill (dotted line) represents the influence of the same nucleus upon the motion of a proton. A mechanical model patterned after Fig. 39I, and made with the neutron barrier shape, makes an excellent demonstration of neutron scattering and capture.

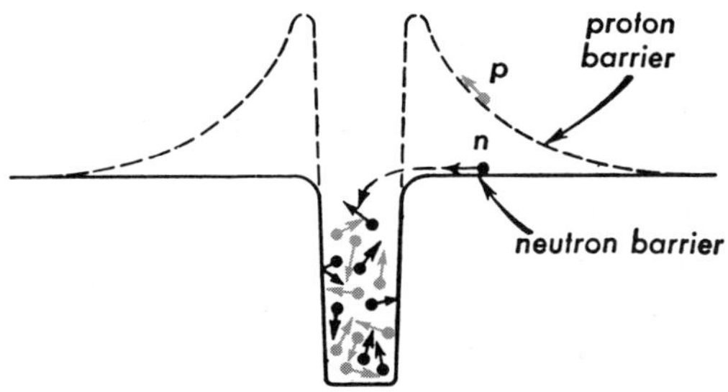

Fig. 39L. A graphical model of the nucleus as it is presented to an approaching neutron or proton.

39.6 NUCLEAR SPIN Approximately one-third of all known stable nuclides, and many of the radioactive ones, are known to be spinning, that is, to have an angular momentum. The earliest evidence for a nuclear spin, as it is called, was found in atomic spectra. Many spectrum lines show a fine structure, that is, they are made up of a number of closely spaced lines called hyperfine structure.

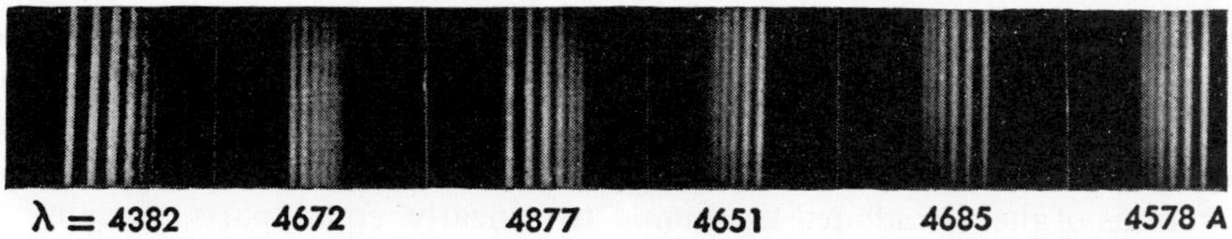

Fig. 39M. Photographs of the hyperfine structure of a few of the spectrum lines of the element praseodymium, $Z = 59$. Wavelengths are in Angstrom units. Structure produced by nuclear spin.

As an example of this hyperfine structure several of the spectrum lines of the element praseodymium are shown in Fig. 39M. In a normal spectrogram these look like single lines, but under the high dispersion and magnification shown here, each line reveals six hyperfine components. The six components signify that the nucleus of praseodymium has a spin angular momentum p_I of $\frac{5}{2} \cdot \frac{h}{2\pi}$.

TABLE 39A. Nuclear Spins of Several Nuclides

Element	Nuclide (Isotope)	Spin J
Hydrogen........	$_1H^1$	1/2
Hydrogen........	$_1H^2$	1
Lithium.........	$_3Li^6$	1
Lithium.........	$_3Li^7$	3/2
Beryllium........	$_4Be^9$	3/2
Boron...........	$_5B^{10}$	3
Oxygen..........	$_8O^{16}$	0
Oxygen..........	$_8O^{17}$	5/2
Vanadium.......	$_{23}V^{51}$	7/2
Manganese......	$_{25}Mn^{55}$	7/2
Mercury.........	$_{80}Hg^{199}$	1/2
Mercury.........	$_{80}Hg^{200}$	0
Bismuth.........	$_{83}Bi^{209}$	9/2

The angular momenta of nuclei in general are given by

$$p_I = J\frac{h}{2\pi} \qquad (39c)$$

where J is the nuclear spin quantum number, and h represents Planck's constant. See Eq. (32b). Quantitative experiments show that this quantum number J can take half-integral as well as whole number values. See Table 39A.

39.7 A NEW DISCOVERY In 1939, Hahn, Strassman, and Meitner, in Germany, subjected uranium metal to the bombardment of neutrons

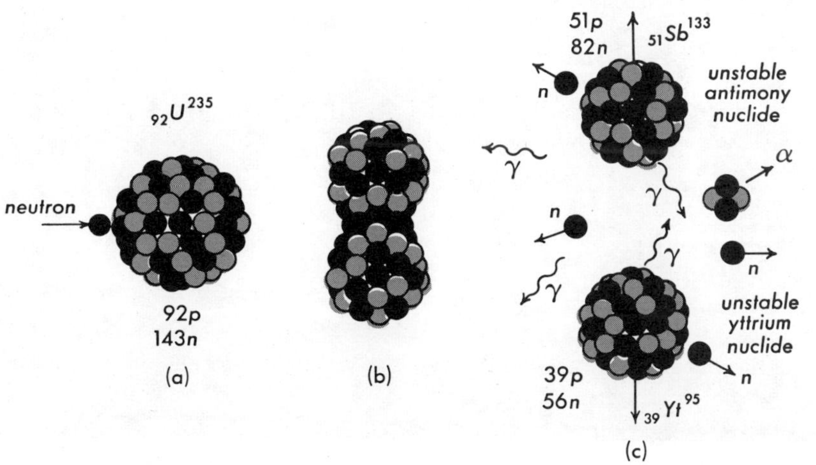

Fig. 39N. Diagrams of the fission of a uranium nucleus into two unstable nuclides, four neutrons, four gamma rays, and one alpha ray.

and observed radioactivities with several different half-lives. To find what nuclides or elements were responsible for these different activities they carefully performed a series of chemical separations of their irradiated uranium sample. To their amazement they found the radioactive atoms to be identical chemically to a number of elements near the center of the periodic table. In other words, a uranium nucleus, after the capture of a single neutron, seemed to be splitting apart into two nearly equal fragments as illustrated in Fig. 39N.

In the few weeks that followed this discovery, many observers in different laboratories the world over not only confirmed the results but extended the observations by studying in detail the products of the disintegrations. To explain the phenomenon in simple words, consider the details of the process illustrated in Fig. 39N. An original uranium nucleus, ${}_{92}U^{235}$, with its 92 protons and 143 neutrons is shown at the left as it captures a slowly moving neutron.

In the center diagram (b) the newly formed nucleus is unstable and starts to separate into two nearly equal parts. Because this process resembles cell division in the science of biology, the phenomenon is called fission. In coming apart the uranium nucleus, behaving like the analogous waterdrop, splashes out small drops, that is, neutrons and γ rays. So great is the energy liberated by this explosion of the nucleus that each of the two heavy nuclei fly apart in opposite directions. That they do so has been confirmed by many Wilson cloud-chamber photographs, one of which is reproduced in Fig. 39O.

For this photograph a thin film of uranium was placed near the center of the cloud chamber, and then irradiated. The subsequent fission of one uranium nucleus reveals two tracks of the

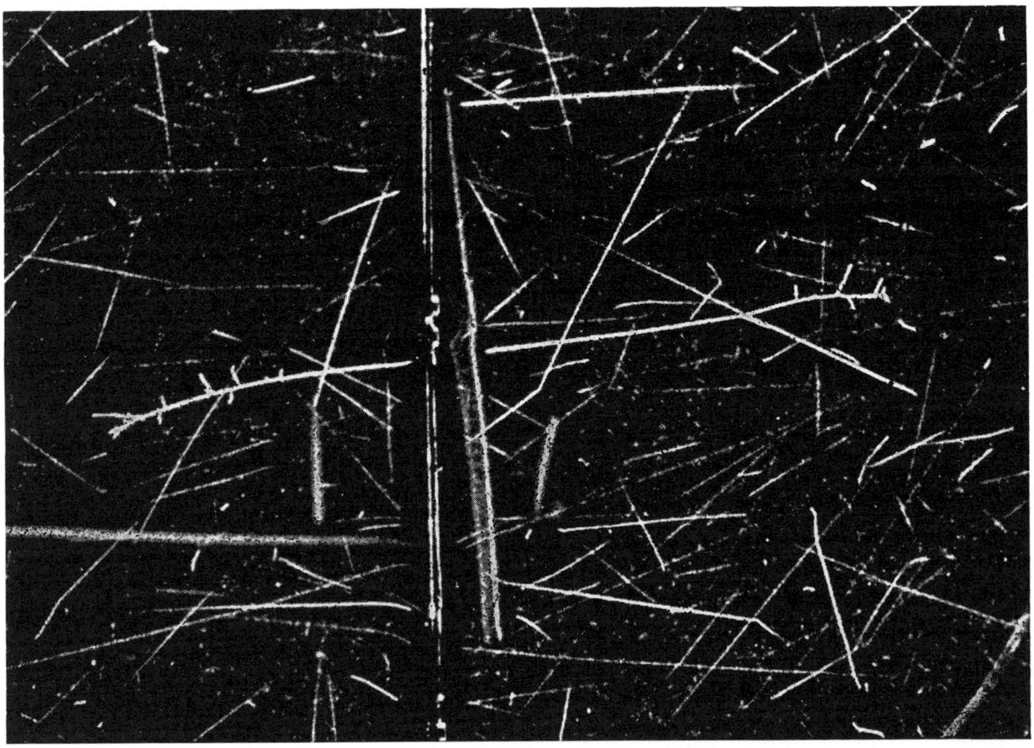

Fig. 39O. Wilson cloud-chamber photograph showing a pair of fission fragments recoiling in opposite directions from a foil containing uranium at the very center of the picture. Note the delta ray forks near the ends. Fission was produced by a neutron beam from a cyclotron. (Courtesy, I. K. Boggild.)

same density, showing clearly that the two heavy particles traveled outward in opposite directions. The heavy forks near the ends of the tracks are characteristic of highly charged fission fragments that have made several collisions with other nuclei before coming to rest.

Not all of the uranium nuclei divide into antimony and niobium as shown in Fig. 39N, but into any one of many pairs of fragments corresponding to elements near the center of the periodic table. The experimental evidence seems to favor pairs of unequal mass, accompanied by from one to five or more neutrons as shown in diagram (c).

Numerous measurements of the masses of fission fragments have made it possible to construct the graph shown in Fig. 39P. The fission fragment yield is plotted vertically to a logarithmic scale, and the mass number A is plotted horizontally to a uniform scale. The curve is seen to rise sharply between $A = 75$ and 80 and drop equally fast between $A = 155$ and 160. The most probable values for the mass numbers of the two fragments are 95 and 139 with a minimum of one-tenth of one percent at $A = 117$.

In general, fission fragments are not stable nuclei but contain an excess number of neutrons. What happens to most fragments like those shown in Fig. 39N (c) is detailed in Fig. 39Q. After a series of β emissions, in which neutrons are converted into protons in the nucleus, a stable nuclide results.

Starting at the left in the diagram with the unstable antimony nuclide of charge +51 and mass 133, the successive emission of four elec-

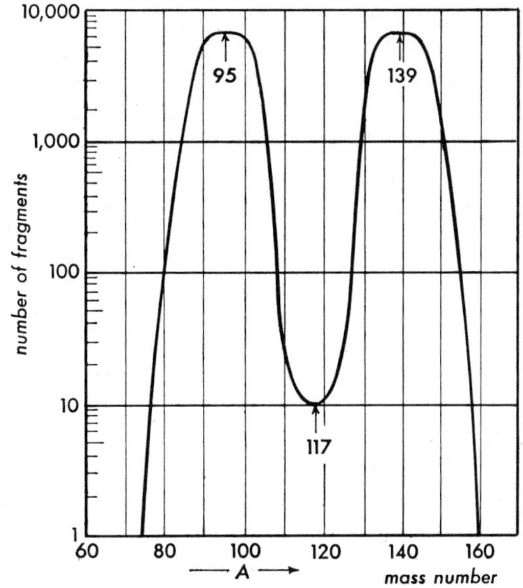

Fig. 39P. Semi-log graph showing the fission fragment yield for slow neutrons on uranium-235.

trons raises the nuclear charge by four unit steps ending with a stable cesium nuclide, ${}_{55}Cs^{133}$. The other fragment, ${}_{41}Nb^{99}$, in Fig. 39N (c), carries out a similar series of β-ray emissions ending up with ${}_{44}Ru^{99}$.

As proof that the above series are produced by fission, previously bombarded uranium has been chemically analyzed for elements near the center of the periodic table. After each chemical separation is performed, a test of the β-ray activity is made by a measurement of the half-life. A comparison of this measured half-life with the values already known for the same element from

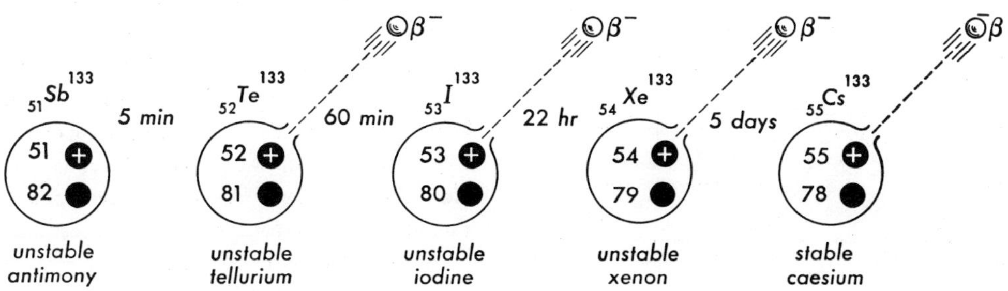

Fig. 39Q. Disintegration series starting with unstable antimony, one of the fragments of the fission of a uranium-235 nucleus.

other disintegration experiments has made it possible to identify some of the radioactive nuclei produced. Such tests, for example, have been made by Miss C. S. Wu* for the series of four elements in Fig. 39Q. Note the increasing half-lives she identified for this series, indicating increased stability as the stable nucleus cesium is approached.

Approximately 99% of the neutrons resulting from the fission of uranium occur within a small fraction of a second and are called prompt neutrons, or secondary neutrons. About one out of a hundred neutrons are emitted one or more seconds later, and these are called delayed neutrons. Delayed neutrons originate from unstable fragments that decay by neutron emission on their way to becoming stable nuclei.

39.8 FISSION ENERGY The energy liberated in the fission of uranium is due largely to the U-235 isotope. Uranium found in the earth's crust has three principal isotopes with the following relative abundance:

U-238	99.280%	4.51×10^9 y
U-235	.714%	7.10×10^8 y
U-234	.006%	2.48×10^5 y

All three of these nuclides are radioactive and decay by α emission. See Tables 39A and 39B.

When a slow or fast neutron is captured by a U-235 nucleus, the two fission fragments as well as the neutrons fly apart with a tremendous amount of kinetic energy. This energy release can best be illustrated by graphs. In Fig. 39R the number of fission fragments is plotted vertically and their kinetic energy in MeV is plotted horizontally. The result is a double peaked curve with maxima at 67 MeV and 100 MeV.

* C. S. Wu, *Physical Review*, Vol. 58, 1940, p. 925.

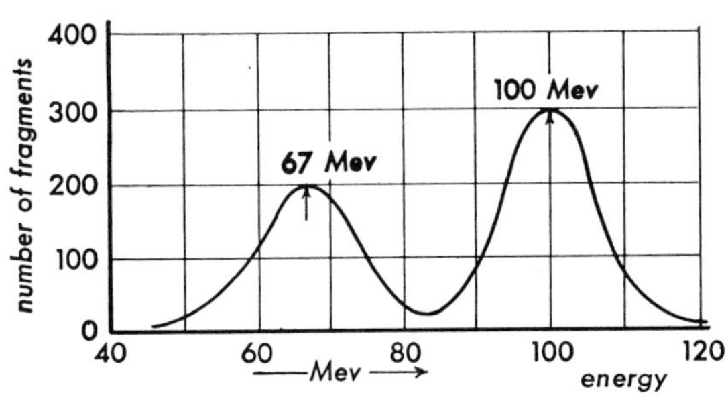

Fig. 39R. Graph of the energies of the fission fragments from uranium-235.

When U-235 nuclei undergo fission as the result of slow neutron capture, the average number of neutrons liberated is found to be 2.5 neutrons per fission. Some may yield as many as 5 neutrons, but 2 or 3 are the most probable numbers.

The average γ ray energy emitted in the fission of a single atom of U-235 is approximately 23 MeV, while the total energy is approximately 200 MeV.

39.9 THE TRANSURANIC ELEMENTS The first transuranic element, neptunium, atomic number 93, was identified in 1939 by McMillan† and Abelson. A beam of neutrons incident on a target of uranium metal gave rise to several known nuclear reactions. A neutron captured by one of the most abundant nuclei U-238 forms U-239 and a γ ray, followed by β emission, to give neptunium, element 93.

$$ {}_0n^1 + {}_{92}U^{238} = {}_{92}U^{239} + h\nu $$
$$ {}_{92}U^{239} = {}_{93}Np^{239} + {}_{-1}e^0 + h\nu $$

Ten different radioactive isotopes of neptunium are now known. Ranging in mass number from 231 to 240, isotope ${}_{93}Np^{237}$ emits α particles and has the longest half-life, 2.2-million y. While

† EDWIN M. MCMILLAN (1907–), American physicist, was educated at the California Institute of Technology, and Princeton University where he obtained his Ph.D. in 1932. His science career began as Research Associate in 1934, and as Professor of Physics in 1946 at the University of California, Berkeley. During World War II, he carried on research at the Massachusetts Institute of Technology, the Radio and Sound Laboratory of the U. S. Navy, and the Los Alamos Scientific Laboratory, where the first atomic bomb was developed. With the death of E. O. Lawrence in 1958, he became Director of the Lawrence Radiation Laboratory. He has received many awards including the Atoms for Peace Award, and the Nobel Prize in 1951 for Chemistry.

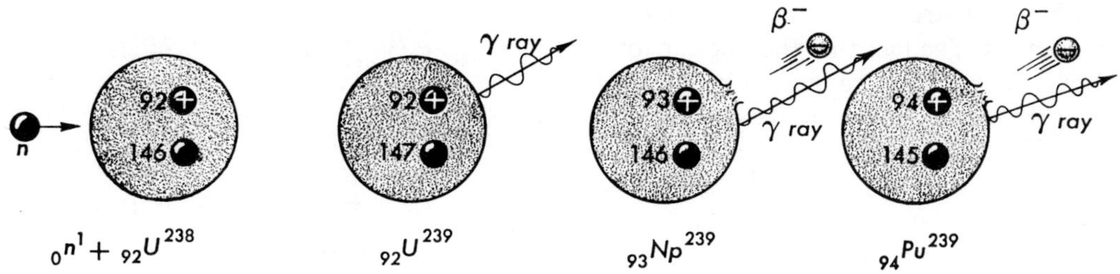

Fig. 39S. Neutron capture by uranium-238 produces, by radioactive β-decay, neptunium and plutonium.

slow or fast neutron capture by U-238 is almost always followed by β emission, about one out of a hundred neutrons will cause fission.

Plutonium (Pu), element 94, was first identified by Kennedy, McMillan, Seaborg,* Segré, and Wahl, as arising from the spontaneous emission of β particles from $_{93}Np^{239}$.

$$_{93}Np^{239} = {}_{94}Pu^{239} + {}_{-1}e^0 + h\nu$$

Schematic diagrams of the above processes are given in Fig. 39S. Fifteen different radioactive isotopes of plutonium are now known. Ranging in mass numbers from 232 to 246, several have long half-lives.

$_{94}Pu^{239}$ α active 24,300 y
$_{94}Pu^{240}$ α active 6000 y
$_{94}Pu^{242}$ α active 380,000 y
$_{94}Pu^{244}$ α active 7.6×10^7 y

Of additional transuranium elements, there are ten known radioactive isotopes of americium (Am) element 95, thirteen known isotopes of curium (Cm) element 96, eight known isotopes of berkelium (Bk) element 97, eleven known isotopes of californium (Cf) element 98, nine known isotopes of einsteinium (Es) element 99, nine known isotopes of fermium (Fm) element 100, two known isotopes of mendelevium (Md) element 101, four known isotopes of nobelium (No) element 102, and only one known isotope of lawrencium (Lw) element 103.

It is interesting to trace out the genetic relationships of the entire group of transuranium elements and see how they link up with the natural radioactive elements. All of the known nuclides fall naturally into four main series as shown in Appendix VI. The four principle activities common to all series are alpha emission (α), beta emission (β^-), electron capture (E.C.), and spontaneous fission (S.F.). While small percentages of some transuranium nuclides spontaneously fission, note that practically all californium-254 and fermium-256 nuclei decay in this way.

The discovery that some nuclei undergo fission spontaneously was made by Petrzhak and Flerov in 1940. Many of the heavier isotopes of the transuranic elements show spontaneous fission.

39.10 THE MESON THEORY OF THE NUCLEUS

The possible existence of mesons was first proposed by the Japanese mathematical physicist Yukawa in 1935. He proposed that the short-range forces between protons and neutrons inside the nucleus are to be attributed to relatively smaller particles.

According to Yukawa, nucleons emit and absorb mass-quanta, called *mesons*, just as elec-

* GLENN T. SEABORG (1912–), American chemist. He obtained his Ph.D. at the University of California, Berkeley, in 1937. He has been awarded many honorary D.Sc. degrees since then from other universities. He became Research Associate at California in 1937, Professor in 1941, and Chairman of the United States Atomic Energy Commission in 1961. From 1946 to 1958, he served as Director of Nuclear Chemistry at the Lawrence Radiation Laboratory where he and his colleagues discovered ten or more of the transuranium elements. From 1958–1961, he served as Chancellor of the Berkeley campus of the University of California. He has received numerous national and international awards; the Fermi Award in 1959, the Nobel Prize in Chemistry in 1951.

trons in the outer structure of the atom emit and absorb photons. The fact that the nuclear forces extend over only a short range can be shown to mean that the meson, unlike the massless photon, would have a finite rest mass. Furthermore, some mesons are charged, and some are neutral (see Fig. 36M).

The present concept of nucleons is that they consist of some sort of common core surrounded by a pulsating cloud of π mesons called *pions*.

Since pions are charged $+$, 0, or $-$, the rapid jumping back and forth between nucleons changes the nucleon identity equally fast and at the same time binds the two nucleons together (see Fig. 39T). This diagram might well represent a deuteron, with both spins in the same direction to give $J = 1$, and the negative pion charge at a greater distance from the center of the neutron to produce its negative magnetic moment.

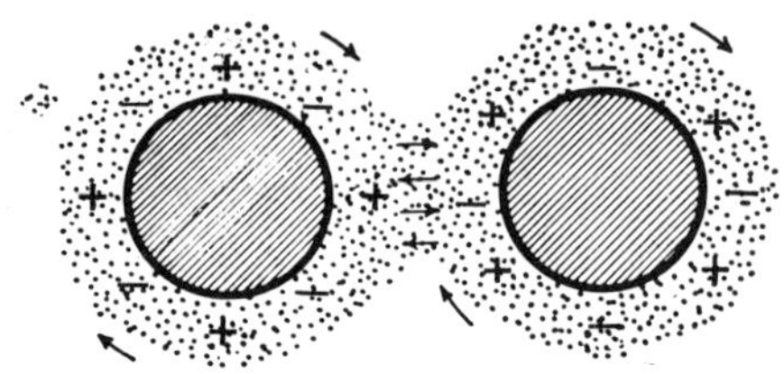

Fig. 39T. Schematic diagram of the meson clouds around each of two nucleon cores, the charge exchange accounting for the strong attractive force U.

To liberate two pions as free particles, sufficient energy must be imparted to the nucleus to create their rest mass of 270 m_e each, or a total of 0.503 BeV (see Fig. 69N).

The exchange of a single pion between two nucleons may be written

$$n \rightleftarrows p + \pi^-$$
$$p \rightleftarrows n + \pi^+$$

QUESTIONS

1. By what other name is each of the following known: (a) electron, (b) proton, (c) deuteron, (d) α particle, (e) positron, (f) X ray, (g) γ ray, and (h) photon?

2. Calculate the mass energy liberated when three neutrons and three protons combine to form lithium-6.

3. Draw a potential energy curve for a proton approaching a nucleus.

4. Draw a potential energy curve for a neutron approaching a nucleus.

5. Does Coulomb's law of repulsion between like charges hold within the nucleus of an atom?

6. About how much energy is required to remove a nucleon from a nucleus?

7. What is a potential barrier? What equivalent mechanical model is sometimes used to represent such a barrier?

8. What is the waterdrop model of the nucleus? How is it used to explain stable nuclei on the one hand and radioactive nuclei on the other?

9. What mechanical model can be used to illustrate the conditions for the nuclear capture of a high-speed proton or α particle?

10. What mechanical model can be used to illustrate disintegration of a radioactive nucleus?

11. In what way does a radioactive nucleus of short half-life compare with one of long half-life?

PROBLEMS

1. Find the total energy required to separate a beryllium-9 nucleus into its nine nuclear particles. [Ans. 58.1 MeV.]

2. Calculate the total energy required to separate a lithium-6 nucleus into three protons and three neutrons.

3. Calculate the total energy required to separate a boron-11 nucleus into its individual neutrons and protons. [Ans. 75.9 MeV.]

4. How much energy would be required to separate a sulfur-32 nucleus into its individual neutrons and protons?

5. Using the atomic masses given in Appendix III, calculate the binding energy per nucleon for (a) nitrogen-14, and (b) oxygen-16. [Ans. (a) 7.47 MeV. (b) 7.97 MeV.]

6. Using the atomic masses given in Appendix III, calculate the binding energy per nucleon for (a) fluorine-19, and (b) neon-22.

7. If one of the fission fragments of U-235 is the radioactive isotope iodine-139, what stable isotope will it become if a series of β particles are emitted? [Ans. $_{57}La^{139}$.]

8. Make a diagram like Fig. 39N, showing a plutonium-239 nucleus capturing a slow neutron and undergoing fission. Assume five prompt neutrons are emitted, and that the fragments are cesium-140, and yttrium-94, respectively.

9. If one of the fission fragments of U-235 is the radioactive isotope strontium-97, what stable isotope will it become if a series of β particles are emitted? Make a diagram like Fig. 39Q. [Ans. $_{42}Mo^{97}$.]

10. If one of the fission fragments of U-235 is the radioactive isotope krypton-95, what stable isotope will it become if a series of β particles are emitted?

11. If a small sample of plutonium-243 is bombarded by deuterons, neutrons are observed coming off. After bombardment ceases β particles are seen coming from the target. Write down the two reactions.

12. When americium-243 is bombarded by deuterons from a cyclotron, α particles are emitted. What is the reaction?

40 NUCLEAR ENERGY AND ELEMENTARY PARTICLES

For many years scientists have been searching for the ultimate particles from which all atoms are made. From the atom and its electron structure, the search has recently extended into the nucleus, and from there to the structure of the nuclear particles themselves.

When a nucleus is hit by a very high-energy particle from a giant accelerator or by a cosmic ray, a variety of particles may be produced. The questions concerning the origin of these particles present challenging problems. Do these particles exist within the nucleus or are they created from the energy liberated at impact? Many of them exist for less than a hundred-millionth of a second and disappear by a nuclear capture process, or annihilation with an anti-particle.

In this chapter we will take a look at atomic energy, brought about through the process of nuclear fission and fusion, and then consider some of the many elementary particles recently discovered in the research laboratory.

40.1 A CHAIN REACTION Not long after the discovery of fission in 1939 it became evident to many scientific groups in America and in Europe that if a sufficient quantity of pure uranium-235 (U-235) could be isolated from its more abundant isotope uranium-238 (U-238), it might have explosive powers many times greater than anything heretofore known. The reasons for believing this appeared at the time to be somewhat as follows. Suppose that a given mass of

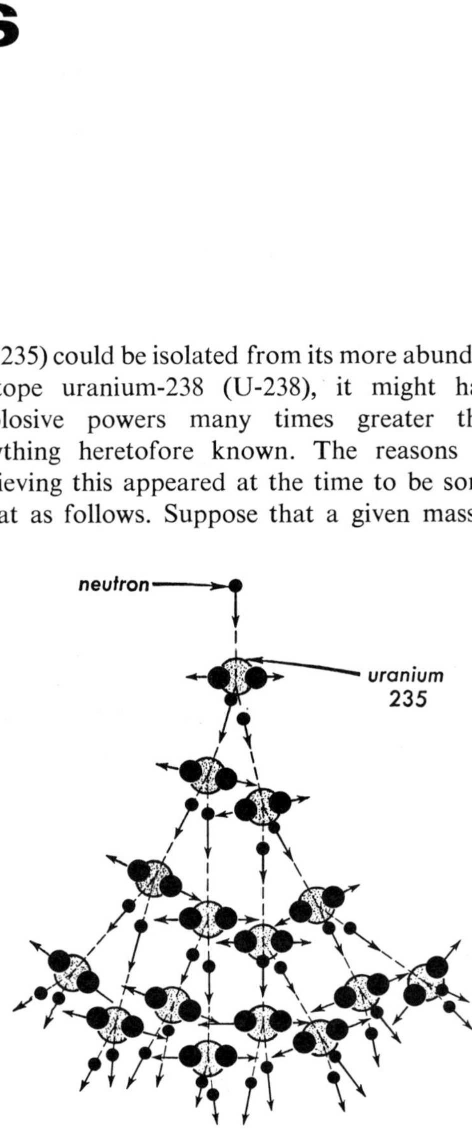

Fig. 40A. Schematic diagram of a chain reaction in pure uranium-235.

uranium metal, all composed of U-235 atoms, was brought together into one lump. The first cosmic ray that penetrated this mass and produced a neutron might well set off the chain reaction shown schematically in Fig. 40A. A U-235 nucleus would capture the neutron and, in splitting apart with great violence, liberate one or more additional neutrons. These in turn would be quickly absorbed by other nearby atoms which in turn would split up and at the same time liberate other neutrons. Hence a rapidly growing kind of avalanche might occur, a kind which, if fast enough, would have the characteristics of an explosion. A graph showing the rate of growth of such a chain process is given in Fig. 40B. Since even the slowest of neutrons in

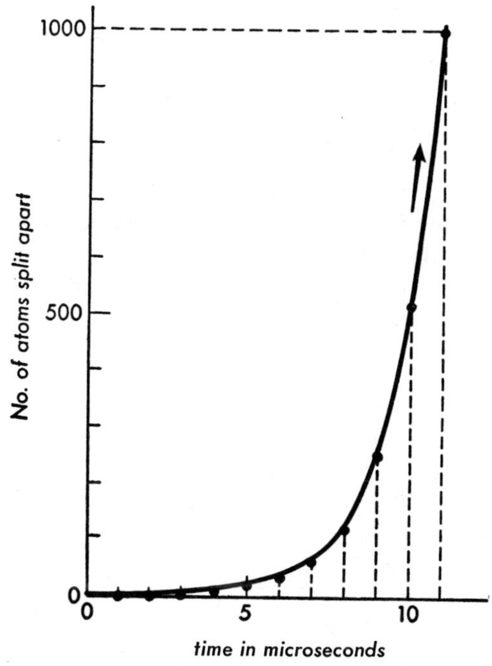

Fig. 40B. Growth curve of fission in pure uranium-235.

solid matter will have average speeds of about 200 m/s, and since many neutron collisions will, on the average, occur within several centimeters, the graph shows how quickly the growth reaches gigantic proportions. (The time scale is in microseconds.)

The escape of neutrons from any quantity of uranium is a surface effect depending on the area of the surface, whereas fission capture occurs throughout the body and is therefore a volume effect. If the assembled mass of uranium is too small, the probability that most neutrons liberated by fission would escape through the surface before being captured might well be so large that a growing chain reaction cannot occur. Since the volume of a sphere increases with the cube of the radius while the surface area increases with the square of the radius, the probability of escape would decrease with increasing size. In other words, if the uranium mass were too small, the growth process shown in Figs. 40A and 40B would be cut off before it became very large, and only if the mass were greater than some critical value would an explosion take place.

Consider, therefore, a large quantity of U-235 in two or more units, each smaller than the critical size, and separated a short distance. See Fig. 40C. Because of the relatively large surface

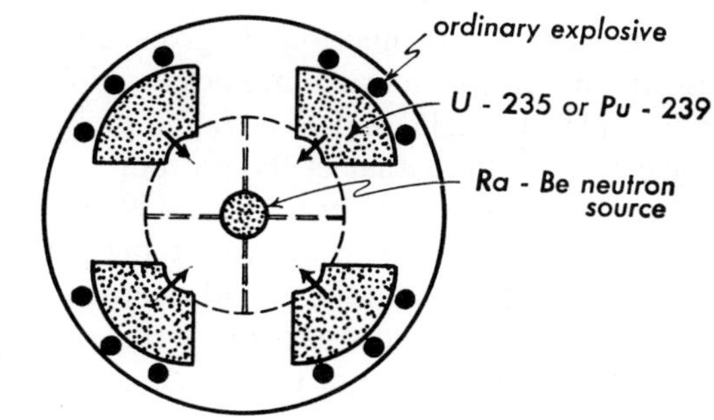

Fig. 40C. A schematic diagram of a hypothetical atomic explosion device, based upon a fast chain reaction in pure U-235 or Pu-239.

area of each unit, neutrons readily escape and a chain reaction cannot develop. Suddenly an explosive like TNT is detonated behind the separated blocks, driving them together as indicated at the center of the diagram. Neutrons entering this greater-than-critical mass from a Ra-Be source will now initiate a rapid chain reaction resulting in a violent explosion. The tremendous energy release of such a device can only be fully realized by those who have actually seen and heard one detonated.

The first atomic bomb ever produced was assembled by scientists at the Los Alamos Laboratory of the University of California, and was successfully detonated at Almagordo, New Mexico, on July 16, 1945. This device was composed of practically pure U-235. Many of the atomic weapons exploded since that time have used plutonium-239 as the fission material.

40.2 NUCLEAR RADIUS AND GEOMETRICAL CROSS SECTION Many experiments concerned with the collisions between atomic particles indicate that each of the heavy nuclei may be considered as a conglomerate of closely packed spheres. The average nucleon radius is now believed to be

$$r_0 = 1.2 \times 10^{-15}\text{ m} \tag{40a}$$

If spheres of this size are packed together into a ball-like structure as shown in Fig. 39N (a), the approximate radius of the system will be given by

$$R = r_0 \sqrt[3]{A} \tag{40b}$$

where the number of nucleons is given by the mass number A. For a uranium nucleus, U-235, for example, $A = 235$, and $R = 7.4 \times 10^{-15}$ m, a value only six times that of the proton.

The concept of *nuclear cross section* is one of considerable importance in nuclear studies. If a nucleus is set up as a target for other atoms to hit, its geometrical cross section serves as a reasonably good measure of the target size, and is given by

$$\sigma_g = \pi R^2 \tag{40c}$$

For U-238 this cross section is approximately

$$\sigma_g = 1.74 \times 10^{-28}\text{ m}^2 \tag{40d}$$

Unit area for nuclei has been arbitrarily set at 1×10^{-28} m², and is called the *barn*. The geometrical cross section for U-238, for example, would be written $\sigma_g = 1.74$ barns.

Calculated cross sections for a few nuclides are given in Table 40A, along with the heights of the nuclear barrier B for incident α particles. See Fig. 39G.

TABLE 40A. Nuclear Radii, Geometrical Cross Sections, and Potential Barrier Heights

Nuclide	R (10^{-15} m)	σ_g (barns)	B (MeV)
$_2He^4$	1.90	0.11	2.4
$_8O^{16}$	3.03	0.29	6.0
$_{30}Zn^{64}$	4.80	0.72	14.0
$_{48}Cd^{113}$	5.85	1.07	19.0
$_{100}Fm^{252}$	7.58	1.78	29.0

40.3 CROSS SECTIONS When neutrons are incident upon some material, the cross-sectional or target areas of the nuclei within seem to be different for different neutron velocities. For example, for fast neutrons a nuclear cross section might be relatively small, while for slow neutrons it may be many times larger. With this explanation we see that cross section is not necessarily an area in the literal sense but is a figurascopic cross-section σ, and for simplicity of an interaction.

The probability of interaction between a neutron and a nucleus is called the microscopic cross-section σ, and for simplicity of concept may be thought of as the effective target area of that nucleus. If a beam of neutrons is incident on one cubic centimeter of material, as shown in Fig. 40D, the total effective target area will

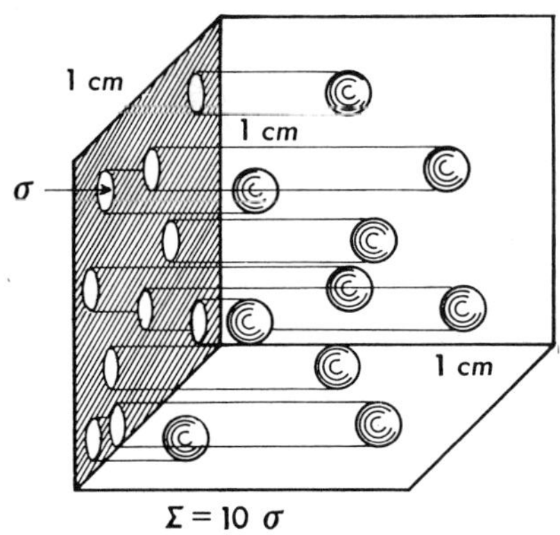

Fig. 40D. Diagram showing the microscopic cross section and the meaning of the macroscopic cross section.

be $n\sigma$, where n is the total number of atoms per cm³. This product is called the *macroscopic cross section*, and is designated Σ.

$$\Sigma = n\sigma \tag{40e}$$

The total cross-section σ_t of a single nuclide is the sum of several cross sections, one for each different kind of process that can alter the bombarding particle's energy or momentum.

$$\sigma_t = \sigma_s + \sigma_c + \sigma_f \quad (40f)$$

where σ_s is the effective nuclear area that produces measurable scattering, σ_c is the effective area for capture prior to radioactive emission, and σ_f is the effective area for capture prior to fission.

In the scattering process the incident particle is not captured by the nucleus but bounces off in some new direction.

For nuclides that do not fission, $\sigma_f = 0$.

As illustrations of these various σ's, the cross sections for several of the fissionable nuclides are given in Tables 40B and 40C. Table 40B gives the

TABLE 40B. Cross Sections in Barns for Pure U-235

	σ_s	σ_c	σ_f
Fast neutrons.........	5.0	0.25	1.27
Thermal neutrons (Slow neutrons).....	10.0	107	580

TABLE 40C. Thermal Neutron Cross Section in Barns for Fissionable Materials

Nuclide	σ_s	σ_c	σ_f
U-235.......	10	107	580
Pu-239......	9.6	315	750
U-233.......	....	52	533
U-238.......	8.3	3.50	0

cross sections for U-235 only, and shows the difference in target area for fast and slow neutrons. Two MeV neutrons have a speed of 20,000,000 m/s, while thermal neutrons (0.0253 ev) have a speed of 2200 m/s.

The whole subject of nuclear energy, and the design of nuclear reactors and explosives, is vitally dependent upon the smallness of the cross sections of some elements and the largeness of others.

40.4 CRITICAL MASS FACTORS For a mass of fissionable material like pure U-235 to be explosive, the time between fission events must be very small, and relatively few neutrons must escape through the surface, or be lost by capture by material other than the uranium. A schematic diagram of possible paths of neutrons between two consecutive fission events is shown in Fig. 40E.

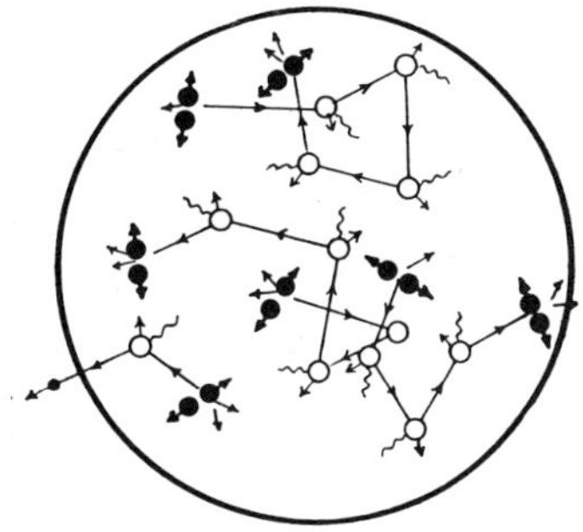

Fig. 40E. Schematic diagram of inelastic scattering between fission events in U-235 or Pu-239

Starting out as fast neutrons considerable energy is lost with each inelastic impact. As the neutron velocity decreases, the U-235 fission cross section increases and fission capture becomes more and more probable.

Whether or not any mass will sustain a chain reaction at all is determined by what is called the *reproduction factor*. The reproduction factor k is given by the ratio,

$$k = \frac{\text{rate of neutron production}}{\text{rate of neutron disappearance}} \quad (40g)$$

If the rate of neutron production equals the rate at which neutrons disappear, the mass is said to be critical, and $k = 1$.

Curves showing the growth in an assembly where k is slightly greater, and slightly smaller, than unity are given in Fig. 40F. Since one generation of neutrons requires about one microsecond, the horizontal scale can represent time in microseconds.

In 5 generations the curve $k = 1.1$ rises to

$$1.1 \times 1.1 \times 1.1 \times 1.1 \times 1.1 = 1.61 \text{ neutrons}$$

In 10 generations this same curve rises to $(1.1)^{10}$ or 2.59 neutrons, and in 100 generations

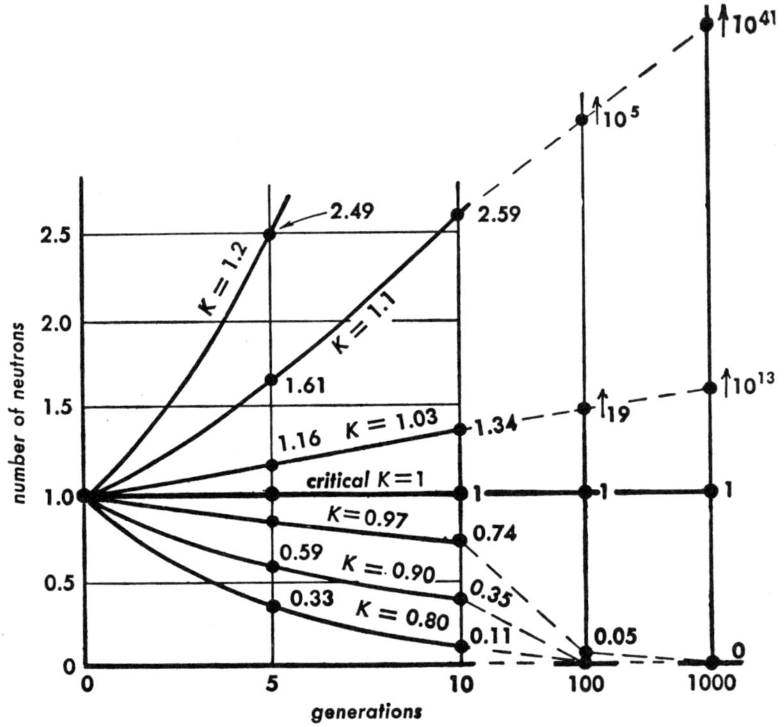

Fig. 40F. Growth curves for neutrons with reproduction factors barely above and below critical, $K = 1$.

to $(1.1)^{100}$ or 13,770 neutrons. In 1000 generations, or approximately one-thousandth of a second, the number rises to approximately 1×10^{41} neutrons. Since this represents more atoms than there are in any mass of uranium yet assembled, 1000 generations would not materialize.

The critical mass for a nuclear explosive device should lie somewhere between the size of a marble (2-cm dia) and the size of a basketball (24-cm dia). Visualize, therefore, two spheres of pure U-235 or Pu-239, one twelve times the diameter of the other. It is clear that if the average straight-line-distance between fission events is 6 cm, few neutrons will be captured in the small sphere, while nearly all of them will be captured in a mass the size of a basketball. In pure U-235, or Pu-239 the size of a marble, the reproduction factor is approximately 0.1, while for a sphere the size of a basketball $k = 2.4$.

40.5 NUCLEAR REACTORS A nuclear reactor, sometimes called an atomic pile, is an apparatus in which nuclear fission can be maintained as a self-supporting, controlled, chain reaction. It is a kind of furnace in which uranium is the "fuel burned" and many useful products such as heat, neutrons, and radioactive isotopes are liberated.

Reactors are of many kinds, sizes, and shapes, the two principal ingredients being a quantity of fissionable material, and a moderating substance for slowing down the neutrons to thermal velocities. A reactor is often designated according to the moderator, or coolant, used within it. Because of the immensity of the subject, only the simplest elements of these devices will be described here, and these as illustrations of the basic principles of many others.

Uranium metal as it is normally refined is composed of three radioactive isotopes.

U-238	99.28%
U-235	0.714%
U-234	0.006%

A self-sustaining chain reaction cannot be maintained in pure uranium alone no matter how large the mass. By properly combining or surrounding the metal with a moderator, however, the 2-MeV neutrons produced in the fission of U-235 can be slowed down by elastic collisions in the moderator to thermal energies of 0.0253 ev. At these relatively low velocities of 2200 m/s, the fission cross section of U-235 has the enormous value of 580 barns; see Table 40B.

The first self-sustaining chain reaction ever assembled by man was put into operation at the University of Chicago on December 2, 1942. This device consisted of a huge "pile" of small carbon blocks, carefully laid together to form one solid mass about the size of a normal school room. During construction, lumps of pure uranium metal were inserted at regular intervals throughout the mass.

A schematic diagram of a pile constructed of large carbon blocks is shown in Fig. 40G. Long

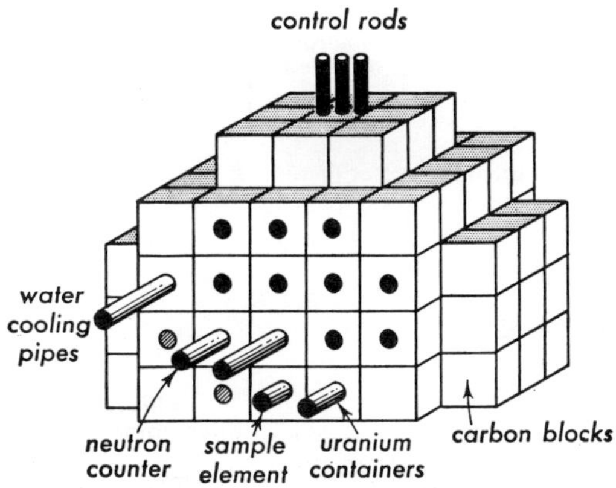

Fig. 40G. Uranium pile of carbon blocks used to produce plutonium-239 and many other radioactive atomic nuclei. (Concrete protective walls are not shown.)

cylindrical holes through the blocks provide for the insertion or removal of fuel elements, control rods, detecting devices, samples to be irradiated, etc. The fuel elements consist of pure uranium metal sealed in thin-walled aluminum cylinders.

The distance between uranium fuel elements in the moderator material is of importance in reactor design. The slowing down distances for three commonly used moderators are as follows:

ordinary water	H_2O	5.7 cm
heavy water	D_2O	11.0 cm
carbon blocks	C	19.0 cm

When, within the uranium metal, a few U-235 nuclei undergo fission, fast neutrons are liberated. Most of these enter the surrounding carbon (the moderator) where they collide elastically with carbon nuclei and slow down. Eventually many of them enter the uranium metal as thermal neutrons and are captured by U-235 nuclei to cause fission. A diagram showing this process is given in Fig. 40H.

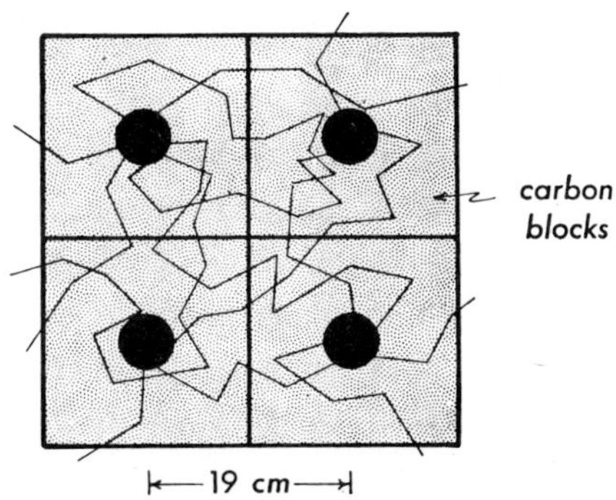

Fig. 40H. Diagram showing moderator action in an atomic reactor. Elastic collisions of neutrons with carbon nuclei slow the neutrons to thermal energies.

Not all neutrons produced within a reactor result in capture by U-235. Some are lost by escape through the surface, some by radiative capture by U-235 as well as U-238, and some by capture by structural materials and fission products. Radiative capture is a term applied to the capture of a neutron by a nucleus that results in the emission of a gamma ray from the nucleus.

The neutron balance in a natural uranium reactor, operating at the critical rate $k = 1$, is shown in Fig. 40I. Of the millions of fast neutrons produced in the reactor each microsecond, the diagram starts with 1000 fast neutrons at the center, and shows what might reasonably happen to them in regenerating 1000 more fast neutrons. The average number of neutrons produced per fission is 2.5. This factor is directly

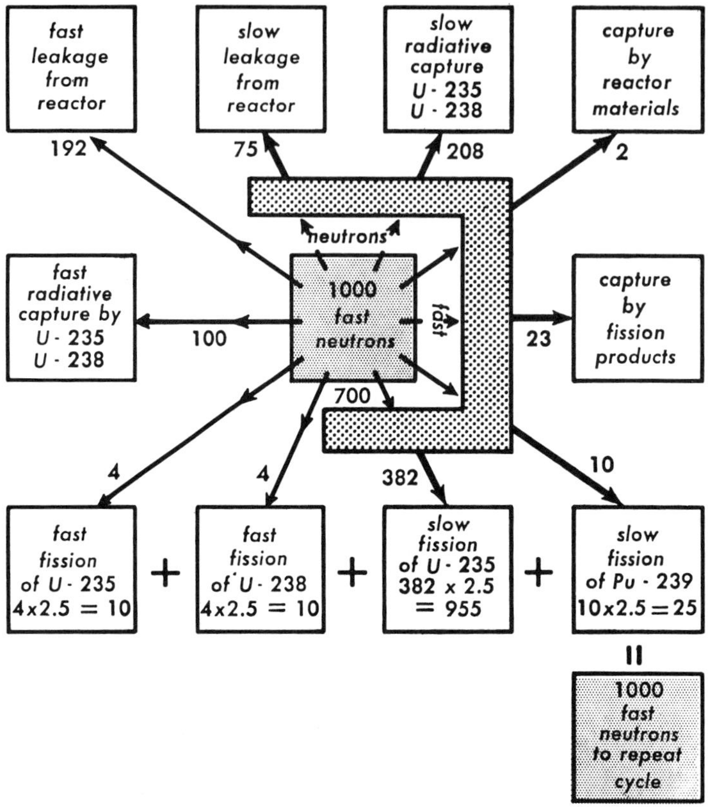

Fig. 40I. A block diagram showing the balance of neutrons in a natural uranium reactor that is just critical. (Calculated numbers by courtesy of Westinghouse Electric Corp.)

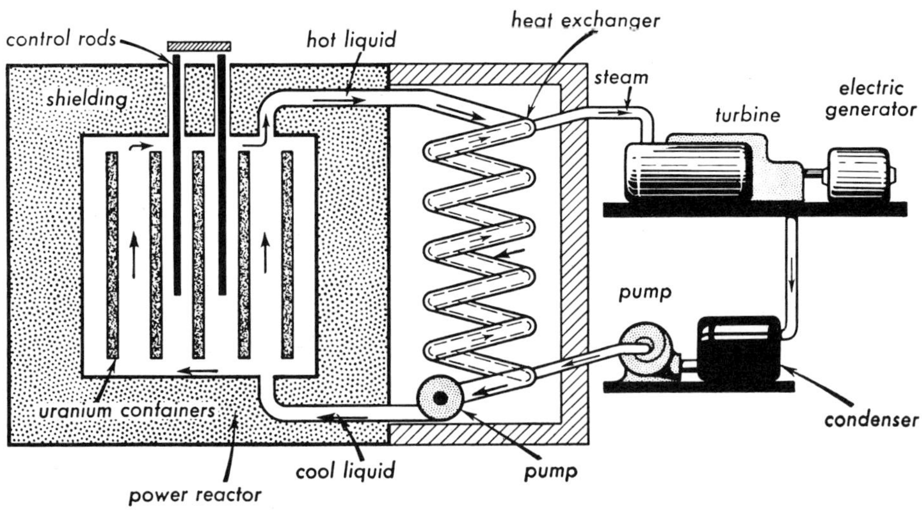

Fig. 40J. Schematic diagram of one type of nuclear power plant.

involved in the bottom four squares where it is used in accounting for reproduction. It should be noted that 60% of the neutrons disappear by other than fission processes.

If the reproduction factor k of the reactor assembly is greater than unity [see Eq. (40g)] the total number of neutrons will rise, and along with it the temperature. To prevent the temperature from rising too high, control rods, which are strong neutron absorbers, are lowered into the central core.

The most widely used control rods, or plates, contain boron or cadmium. Both these elements have enormous capture cross sections for slow neutrons. When normal cadmium metal is subjected to thermal neutrons, the average cross section is 2500 barns.

40.6 POWER REACTORS OR NUCLEAR POWER PLANTS The idea that the natural heat developed in a uranium reactor might be utilized as a source of great power has long been recognized as a feasible enterprise. The basic principles of one type of "power reactor" are shown in Fig. 40J. A quantity of enriched uranium, in the form of a pure metal, or in the form of a solution of soluble salt in water, forms the center of the heat energy source.

The energy released by fission produces great quantities of heat and the rising temperature is regulated to a predetermined value by cadmium rods. To reduce the fission rate, and thereby lower the temperature, the control rods are pushed in a little to absorb more neutrons, while to raise the temperature they are pulled out a little.

Because of the harmful effects of the intense neutron radiation to men and equipment it is not reasonable to vaporize a liquid directly as in a steam boiler; it is better to circulate a fluid through the shielded reactor and heat exchanger as shown in the diagram.

The hot liquid flowing through the heat exchanger vaporizes a more volatile liquid like water and the resulting hot gas or steam under pressure drives a turbine of special design. The turbine in turn drives an electric generator developing power that can be used to light our cities and factories, or to drive ships and submarines through the water and large planes through the air.

One of the problems connected with such power reactors concerns the effect of the intense neutron radiation on the metal structures. The neutrons change some atoms and permanently displace others from their normal positions in the crystal lattice of the solids and as a result weaken certain crucial mechanical parts. Intensive studies of the properties of various materials under conditions likely to be encountered in power reactors are continually carried on in present day research laboratories.

40.7 THE SUN'S ENERGY Measurements of solar radiation reaching the earth each day not only make it possible to calculate the surface temperature of the sun but also to determine its total radiation. The fact that the sun, over a period of many years, shows no signs of cooling off, has long been an unsolved mystery. With the discovery of nuclear disintegration and the development of methods of producing many new types of atoms, this mystery has in a measure been solved.

Although there is no direct way known of observing the interior of a star like our sun, mathematical calculations based upon well-established physical laws show that down deep within such a mass the temperature is so extremely high that matter must be a conglomeration of atoms, electrons, and light waves all moving about at tremendously high speeds.

Near the center of the sun where the temperature is about 20-million degrees, the atoms are stripped of their electrons and the light waves produced there are of such high frequencies that they should be classified as γ rays and X rays. Here, where the average particle velocity is so high, nuclear reactions must be taking place on a large scale and the liberated energy must be filtering up through to cooler and cooler layers as light waves of lower and lower frequency. At the surface most of the radiations escaping are of sufficiently low frequency to be classified as visible, ultraviolet, and infrared.

A careful study of all known nuclear reactions led Bethe* in 1938 to propose the following set of chain processes as those most probably responsible for the generation of energy at the sun's central core.

$$\begin{aligned}
&(1)\quad {}_1H^1 + {}_6C^{12} = {}_7N^{13} + \gamma \text{ ray}\\
&(2)\quad {}_7N^{13} = {}_6C^{13} + {}_1e^0\\
&(3)\quad {}_1H^1 + {}_6C^{13} = {}_7N^{14} + \gamma \text{ ray}\\
&(4)\quad {}_1H^1 + {}_7N^{14} = {}_8O^{15} + \gamma \text{ ray}\\
&(5)\quad {}_8O^{15} = {}_7N^{15} + {}_1e^0\\
&(6)\quad {}_1H^1 + {}_7N^{15} = {}_6C^{12} + {}_2He^4
\end{aligned} \qquad (40h)$$

By summing up the equations it will be seen that four hydrogen atoms are consumed and that two positrons, three γ rays, and one helium nucleus are created. The other nuclei cancel out since the original carbon atom in the first reaction is returned intact in the last reaction. Hence hydrogen is burned and helium is liberated. The loss in mass for each such cycle of reactions is, therefore, as follows:

$$4{}_1H^1 = 4.031300$$
$${}_2He^4 = 4.002604$$
$$2{}_1e^0 = 0.001098$$

Subtracting gives 4.031300 − 4.002604 − 0.001098 = 0.027598 amu. This is equivalent to 25.7 MeV energy.

The combining of hydrogen atoms to form helium is called *fusion*.

More recent experiments and calculations indicate that the following proton-proton cycle is of even greater importance in the creation of solar and stellar energy than the carbon cycle.

$${}_1H^1 + {}_1H^1 = {}_1H^2 + {}_1e^0 + 0.93 \text{ MeV}$$
$${}_1H^1 + {}_1H^2 = {}_2H^3 + \gamma \text{ ray} + 5.5 \text{ MeV}$$
$${}_2He^3 + {}_2He^3 = {}_2He^4 + 2\,{}_1H^1 + 12.8 \text{ MeV}$$

The net result is the same as before: four hydrogen atoms have been fused into one helium atom. Note that since two ${}_2He^3$ nuclei are involved in the third reaction, two proton-proton reactions and two proton-deuteron reactions are required to form one ${}_2He^4$ nucleus. Six protons are consumed but two are returned with the helium.

The rates at which these reactions should take place are not only consistent with the temperature of 20-million degrees, calculated from other considerations, but hydrogen and helium are known to be the most abundant elements of which stars are made.

In order for the sun to radiate 3.8×10^{26} joules of energy per second, Einstein's equation $E = mc^2$ shows that mass must be annihilated at the rate of 4.2×10^9 Kg/s (or 4,500,000 tons/s). While this result indicates that the sun is losing mass at a tremendous rate, the amount is small when compared with the sun's total mass of 1.98×10^{30} Kg. To illustrate, in one million years the sun is losing one ten-millionth of its total mass.

40.8 ELEMENTARY PARTICLES In these last sections of this book we will take a brief look at some of the recent discoveries and observations in "high energy physics," in the hopes that such

* HANS A. BETHE (1906–), American physicist, was born in Strassburg, Germany. He received his Ph.D. in physics at Munich in 1928. When Hitler came to power in 1933, he fled Germany and remained in England for two years before coming to America. He became Professor of Physics at Cornell University in 1937, and obtained his United States citizenship in 1939. During World War II, he served as a staff member at the Radiation Laboratory of the Massachusetts Institute of Technology. In 1943–46, he served as Chief Theoretical Physicist of the Physics Division of the Los Alamos Scientific Laboratory in New Mexico. He is a consultant for various industrial and government laboratories and has received many honorary degrees. Among others, he received the 1946 U. S. Medal of Merit, the Fermi Award of the U. S. Atomic Energy Commission in 1961, and the Nobel Prize in Physics in 1967. He is best known for his work on the source of solar energy through nuclear reactions taking place deep within the sun.

glimpses may give some clue as to what new experiments might be performed that would shed new light on the "elementary particles."

With the discovery of the neutron by Chadwick in 1932, the number of elementary particles became four in number: the *electron*, the *proton*, the *neutron*, and the *photon*. The first three are the atomic particles of which atoms are built, while the photon is the quantum unit of radiation emitted or absorbed by the electrons in the outer structure of atoms or by the particles within the nucleus.

The photon can only exist when traveling with the speed of light, and because of its motion possesses energy $h\nu$. By the mass–energy relation, $E = mc^2$, a photon also has mass $h\nu/c^2$. It possesses mass by virtue of its motion, for at rest it would have no energy and no mass.

The electron, proton, and neutron, on the other hand, have a definite rest mass m_0 and a rest energy m_0c^2. When they are set into motion, their mass increases, and their total energy is given by mc^2.

40.9 ANTIPARTICLES The positron, discovered by Anderson in 1932, is a positively charged electron. We have seen in Sec. 36.8 how electron pairs can be created and how a free positron in coming together with an electron is annihilated with the energy appearing in the form of two γ rays. It is this very property that gives the positron the name *antiparticle*—it destroys itself along with an electron and becomes another form of energy.

It is equally true that, when a high-energy photon comes close to the nucleus, mesons may be produced. If a pair of mesons is produced, one positive and one negative, one is the antiparticle of the other. These π-mesons may react with other nuclei or they may decay into muons as shown in Fig. 36N. The muons in turn may decay into electrons and positrons, or they may be captured by some other nucleus.

The discovery of the positron, three pions π^+, π^0, and π^-, the two muons μ^+ and μ^-, along with the two neutrinos ν and $\bar{\nu}$, raised the number of elementary particles by 1947 to twelve.

The number of known particles expanded rapidly between 1947 and 1954, and there were added four new groups of heavier particles that became known collectively as "strange particles." In the order of increasing mass these were called K particles, Λ particles, Σ particles, and Ξ particles.

Although the proton carries a positive charge its symbol is usually written p. A bar over any symbol represents an antiparticle. For example, $\bar{e}^+$ represents an antielectron, or positron.

40.10 *K* PARTICLES These elementary particles belong to the meson class, have a mass of about 970 electron masses, yet have no spin angular momentum. Being heavier by far than π mesons they have more possible methods, or modes, of decay. K^+ particles have an average life, or mean life, of 1.2×10^{-8} s, and decay with the following modes:

$$
\begin{aligned}
&K^+ \to \bar{\mu}^+ + \nu_\mu && 59\% \\
&K^+ \to \bar{\pi}^+ + \pi^0 && 26\%\ (\theta \text{ mode}) \\
&K^+ \to \bar{\pi}^+ + \bar{\pi}^+ + \pi^- && 5\%\ (\tau \text{ mode}) \\
&K^+ \to \bar{\mu}^+ + \nu_\mu + \pi^0 && 4\% \\
&K^+ \to \bar{e}^+ + \nu_e + \pi^0 && 4\% \\
&K^+ \to \bar{\pi}^+ + \pi^0 + \pi^0 && 2\%
\end{aligned}
\tag{40i}
$$

The K^0 and $\bar{K}^-$ particles have similar decay modes, and the K^0 has a mean life of 1×10^{-10} s. These particles are frequently called *kaons*.

40.11 Λ^0 PARTICLES All Λ^0 particles are neutral; that is, they have no charge. Their mass is approximately 2185 electron masses, which means they are heavier than nucleons. Since their mass is about 75 electron masses heavier than a proton plus a pion, it is not surprising to find that when they decay they carry out the following reactions:

$$
\begin{aligned}
&\Lambda^0 \to p + \pi^- \\
&\Lambda^0 \to n + \pi^0 \\
&\Lambda^0 \to p^- + \bar{\pi}^+
\end{aligned}
\tag{40j}
$$

Λ^0 particles have a mean life of 2.5×10^{-10} s. *All elementary particles heavier than nucleons are called* hyperons, *and by including protons and neutrons they are called* baryons.

40.12 Σ PARTICLES These baryons are of three kinds, Σ^+, Σ^0, and Σ^-, and have a mass of approximately 2335 m_e. They have a spin angular momentum of $\frac{1}{2}\frac{h}{2\pi}$, and decay with a mean life of 10^{-10}, by the following modes:

$$\begin{aligned} \Sigma^+ &\rightarrow p + \pi^0 \\ \Sigma^+ &\rightarrow n + \bar{\pi}^+ \\ \Sigma^- &\rightarrow n + \pi^- \\ \Sigma^0 &\rightarrow \Lambda^0 + \gamma \end{aligned} \tag{40k}$$

40.13 Ξ PARTICLES These baryons are about 130 electron masses heavier than $\Lambda + \pi$ and belong to the baryon class. They also have a spin of $\frac{1}{2}\frac{h}{2\pi}$ and decay principally by the following modes:

$$\begin{aligned} \Xi^- &\rightarrow \Lambda^0 + \pi^- \\ \Xi^0 &\rightarrow \Lambda^0 + \pi^0 \end{aligned} \tag{40l}$$

40.14 THE BUBBLE CHAMBER The bubble chamber, invented in 1952 by D. H. Glaser, has become one of the most valuable instruments for studying the minute details of high energy nuclear events. In a cloud chamber, fog drops form on the ions produced by charged atomic particles that have just previously traversed the gas-filled chamber. In the bubble chamber tiny bubbles form on the ions produced by a charged particle that has just previously traversed a liquid.

The basic principles of the bubble chamber involve the super heating of a liquid and the bubbles that form in the process of boiling. Water, for example, boils at 100° C at standard atmospheric pressure. If the pressure is increased boiling will not begin until a higher temperature is reached. The pressure is therefore increased and the temperature of the water raised to almost the new boiling point. If the pressure is then suddenly reduced water boils at a lower temperature; therefore, boiling begins with the sudden formation of tiny bubbles that grow quickly in size.

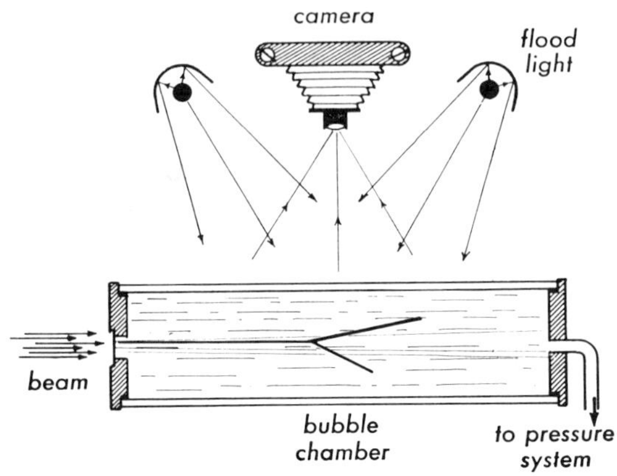

Fig. 40K. Cross-section diagram of a bubble chamber showing track illumination, camera, and incident particle beam.

A simplified diagram of a bubble chamber is shown in Fig. 40K. A box with thick glass walls, and filled with a liquid, is connected to a pressure system and then superheated to some predetermined temperature.

High energy particles enter the liquid through a thin window. A sudden release of a valve in the pressure system is quickly followed by the flash of flood lights and the snap of a camera shutter. If the chamber is timed properly, trails of bubbles that have formed on the paths of ions made by the traversing particles will be photographed. See Fig. 40L.

The extensive use of liquid hydrogen in bubble chambers has been particularly effective in the study of elementary and strange particle events. The precautions that must be taken in the handling of liquid hydrogen at −253° C in large bubble chambers have been developed by L. Alvarez and his colleagues over a period of several years. Liquid hydrogen is particularly useful in that it provides a high concentration of target protons, the simplest of atomic nuclei, and at the same time greatly shortens the distance between events that would be required in the gas-filled space of a cloud chamber.

It is customary to locate a bubble chamber at the center of a large electromagnet so that particle charge and momentum can be obtained from curvature.

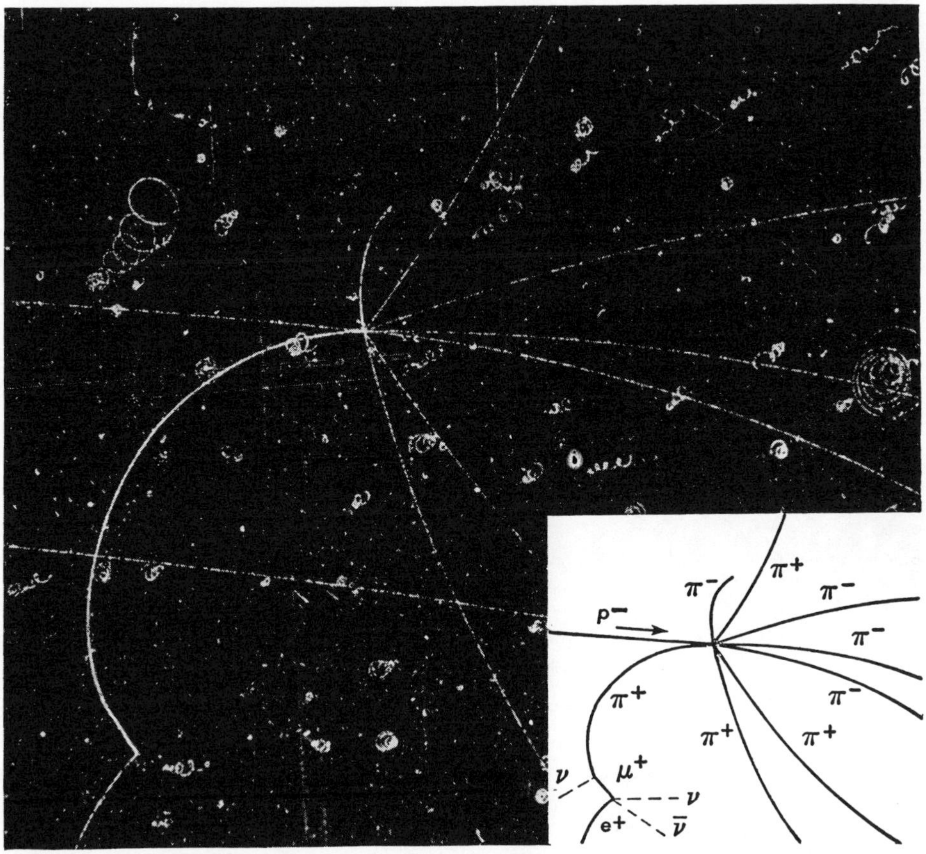

Fig. 40L. Pi-meson star produced by the proton capture of an antiproton in a liquid hydrogen bubble chamber. (Courtesy, L. Alvarez and colleagues.)

40.15 ANTIPROTONS The existence of antiprotons was discovered in 1955 by Chamberlain, Segrè,* Wiegand, and Ypsilantis at the University of California. This discovery came as the result of long range plans laid for the purpose of answering the question, "Is there a negatively charged particle with the mass of a proton?" One of the principal objectives in building the 6.2-Bev proton accelerator in 1953 was to find an answer to this question. Protons accelerated to a high enough energy should, in colliding with heavier nuclei, impart sufficient energy to create, if such were possible, a pair of protons, one plus, the other minus.

The first antiprotons were discovered as high-energy negatively charged particles emerging from a copper target in the proton beam of the bevatron, and having all of the anticipated properties. By means of a strong magnet the antiprotons were bent away from the protons and into a scintillation counter. Today antiproton

* EMILIO GINO SEGRÈ (1905–), American physicist, was born in Tivoli, Italy. He obtained his Ph.D. in physics under Enrico Fermi at Rome in 1928. After holding several professorships in Italian universities, he came to the United States to join the Radiation Laboratory and the Department of Physics at the University of California, Berkeley, in 1938. As a United States citizen, he served as a group leader during World War II at the Los Alamos Scientific Laboratory from 1943 to 1946. He has received many honorary awards, and the Nobel Prize in Physics, jointly with Owen Chamberlain, in 1959, for the discovery of the antiproton. He is best known for his work in spectroscopy and nuclear physics.

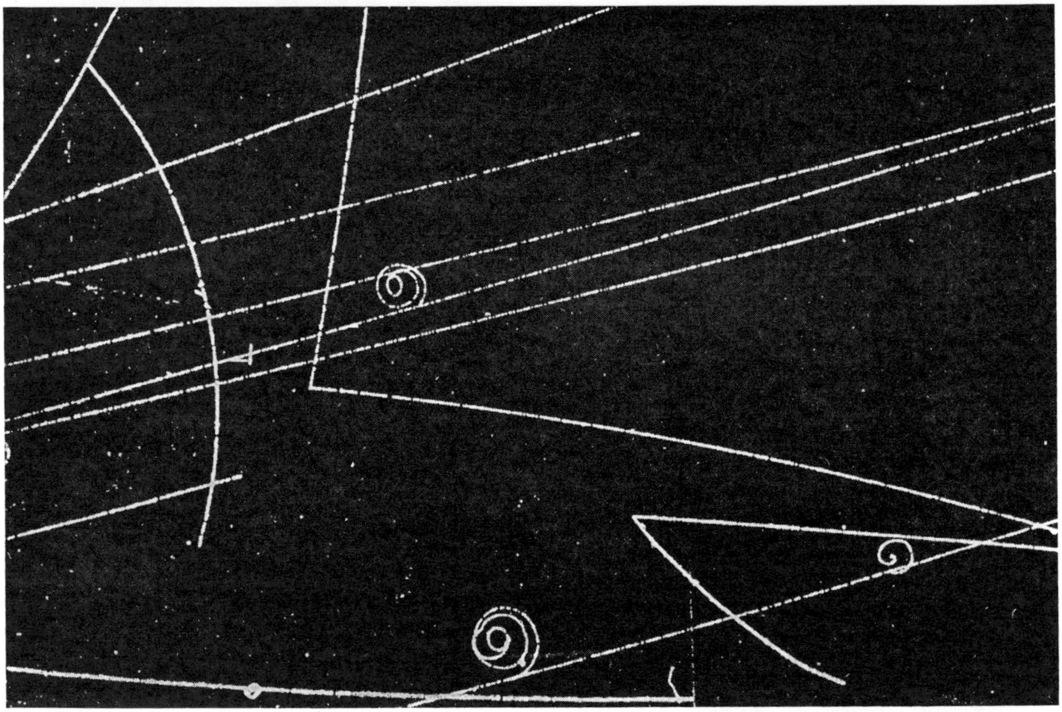

Fig. 40M. Bubble chamber photograph with π^-, p, K^0, Λ^0, π^+ and π^- tracks, all in one related event. For legend, see Fig. 40N. (Courtesy, L. Alvarez and Colleagues.)

events are commonly observed and are conveniently studied by means of a bubble chamber.

Fig. 40L is a photograph in which an antiproton entered the liquid hydrogen bubble chamber from the left. Near the center of the picture it collided with a proton, and both particles were annihilated in the production of four pairs of π mesons. Since the bubble chamber operated in the uniform field of a strong electromagnet, the π^- meson tracks curve clockwise, and the π^+ meson tracks curve counterclockwise. The π^+ meson that curved back to the left and down decayed into a μ^+ meson and an e^+ positive electron as shown in the accompanying legend diagram.

The reactions of this event are written as follows:

$$p^- + p \rightarrow 4\pi^- + 4\pi^+$$

$$\pi^+ \rightarrow \mu^+ + \bar{\nu} \qquad (40m)$$

$$\mu^+ \rightarrow e^+ + \nu + \bar{\nu}$$

Fig. 40M is a photograph in which a π^- meson enters a liquid hydrogen bubble chamber from the left. As shown in the legend diagram of Fig. 40N, interaction with a proton cancels the

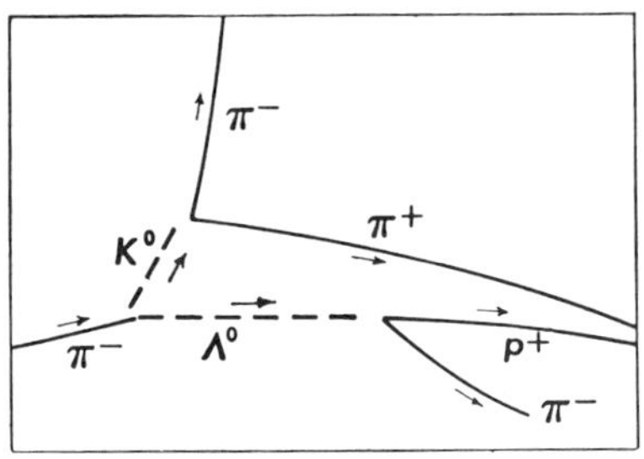

Fig. 40N. Labeled track pattern for bubble chamber photograph, Fig. 40M.

charges, and ends the track by creating two neutral particles.

These two recoiling particles, K^0 and Λ^0, do not produce tracks in the chamber but, having short lifetimes, quickly disintegrate. The neutral K^0 particle decays into a pair of pions, π^+ and

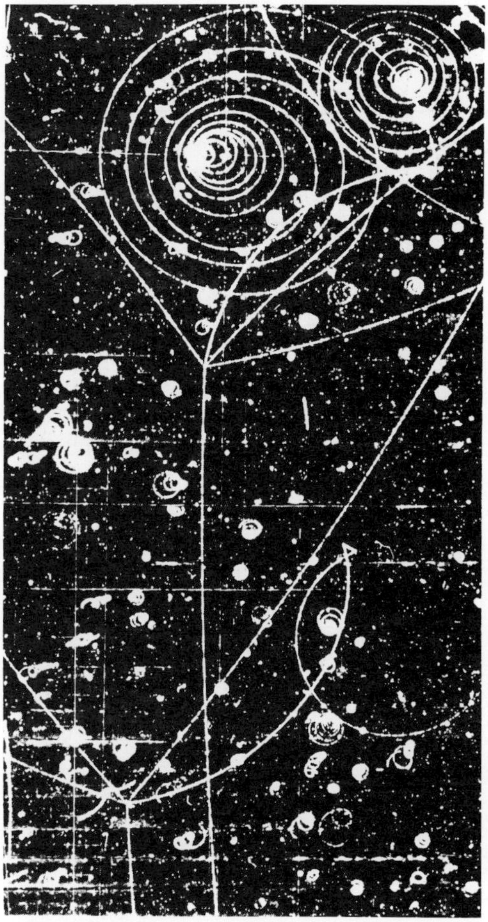

Fig. 40O(a) Photograph of antiproton reactions in a hydrogen bubble chamber showing meson pairs in the kind of reaction that first led to the identification of the ω meson. (Courtesy of the Lawrence Radiation Laboratory, and L. W. Alvarez and associates.)

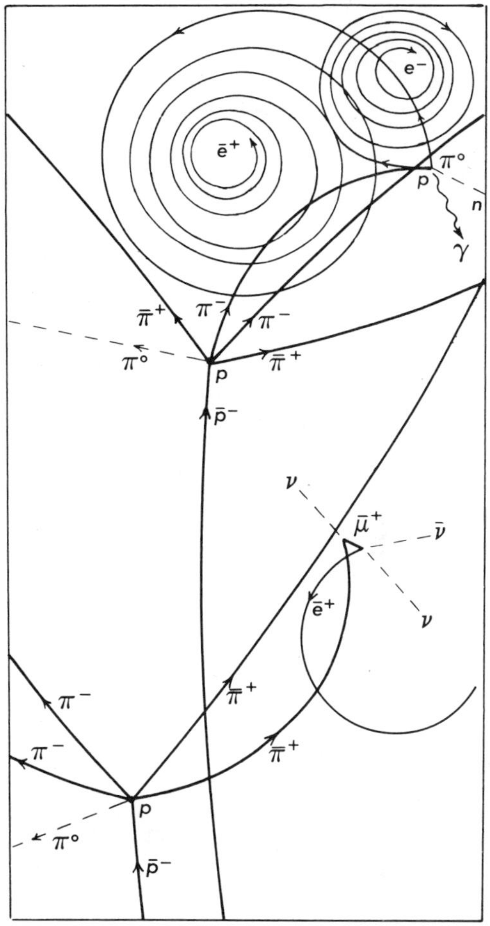

Fig. 40O(b) Legend diagram for Fig. 40O(a) showing the uncharged trackless particles as dotted lines.

π^-, while the Λ^0 particle decays into a proton and a π^-. As reactions,

$$\pi^- + p \rightarrow K^0 + \Lambda^0$$
$$K^0 \rightarrow \pi^+ + \pi^- \qquad (40n)$$
$$\Lambda^0 \rightarrow p + \pi^-$$

An excellent illustration of the decay of a $\bar{\pi}^+$ into a $\bar{\mu}^+$ and $\bar{e}^+$ is shown at the lower right in Fig. 40O (see Fig. 36N).

In the upper right in Fig. 40O, a π^- is captured by a proton to produce a neutron and a π^0. Normally a π^0 decays into γ rays, but about one out of eighty such neutral particles converts one of the γ rays into an electron pair. Such a *Dalitz pair*, as it is called, is shown by the oppositely directed spirals. The reactions for this event are

$$\pi^- + p \rightarrow \pi^0 + n$$
$$\pi^0 \rightarrow \bar{e}^+ + e^- + \gamma \text{ ray}$$

40.16 PARTICLE CLASSIFICATION The energy-level diagram in Fig. 40P represents an orderly array of 89 *strongly interacting* and *elementary particles* known today. Some of these particles were predicted before they were identified in laboratory experiments.

The vertical column of symbols and numbers

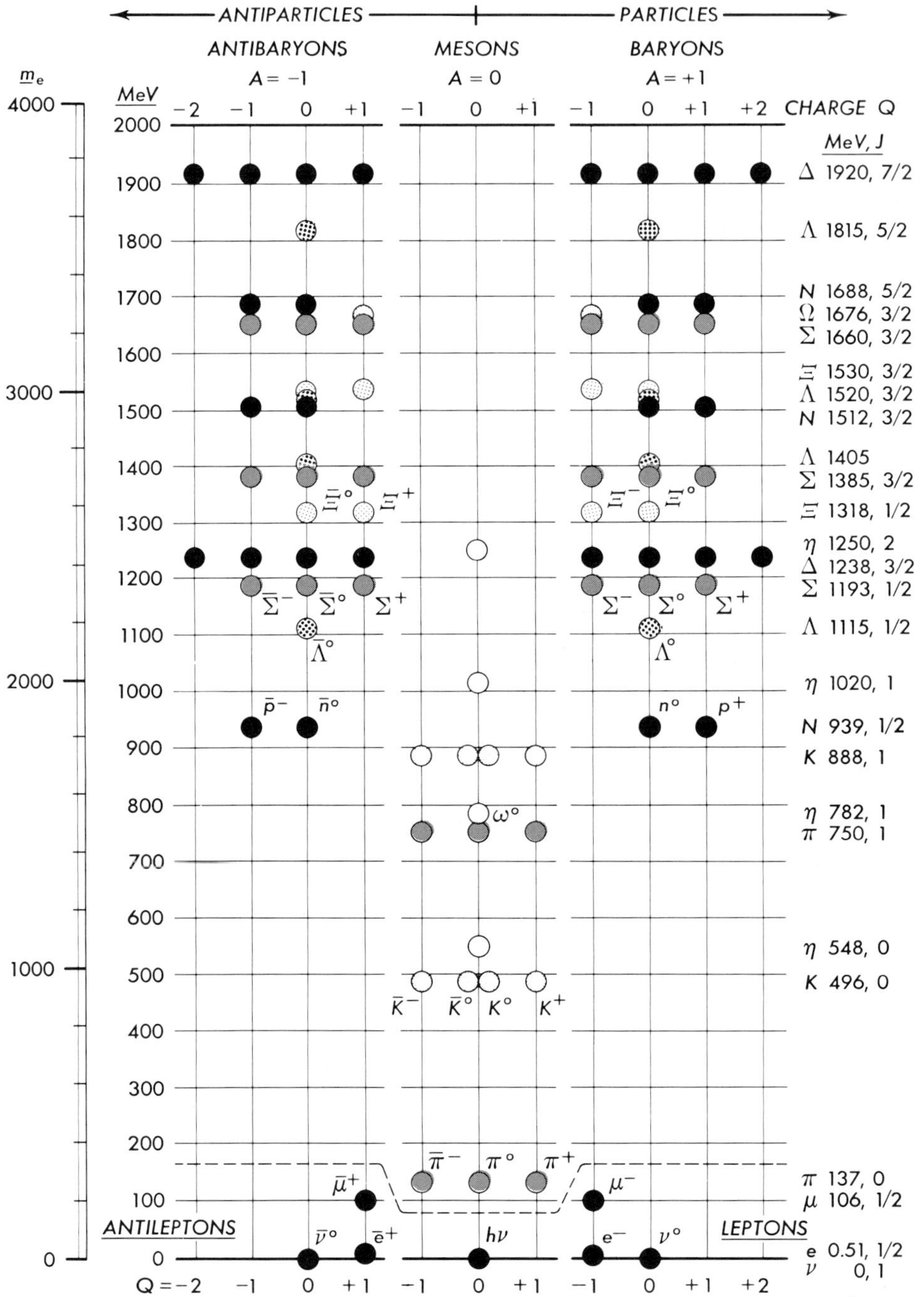

Fig. 40P. Energy-level diagram of strongly interacting and elementary particles as compiled by Geoffrey F. Chew, Murray Gell-Mann, and Arthur H. Rosenfeld. (Scientific American, February 1964)

on the right gives the *particle designations* suggested for general adoption by Chew,* Gell-Mann,† and Rosenfeld,‡ along with the *rest-mass energy* in millions of electron volts, MeV, and the particles *spin angular momentum J* in units of $\hbar$. The numbers in the third row across the top give the *baryon number*, or *atomic mass, A*, and the numbers in the fourth row give the *particle charge Q* in units of the electronic charge e. All mesons (center) and leptons (bottom) have mass or baryon number $A = 0$.

One scale on the left gives the rest mass in multiples of the electronic mass m_e, and the other scale the rest mass in MeV. The specifically labeled dots, with the exception of ω^0, represent the 30 so-called elementary particles identified prior to 1957. Note that the left half of the table is a mirror image of the right half.

The very interesting idea put forward that some distant galaxies in the heavens might be made up of antimatter might well be true. In such a galaxy, all hydrogen atoms would be made of antiprotons and positrons, and the protons and electrons would be their antiparticles. Thus Fig. 40P carries a double meaning, and *either half represents the antiparticles for the other half.*

40.17 PARTICLE SYMBOLS AND QUANTUM NUMBERS It will be observed in Fig. 40P that all particles occur in families, or charge multiplets: *singlets, doublets, triplets,* and *quadruplets.* Within each multiplet, particles differ only in electrical charge Q.

Only ten different patterns of baryons and mesons are known or predicted and each is represented by a different set of three quantum numbers, A, Y, and I.

Quantum number A, called the *baryon number*, has the value +1, 0, or −1 for all particles, and represents the familiar atomic mass number long used for nuclei. For uranium-238, $A = 238$, etc.

Quantum number Y, called the *hypercharge*, is assigned the whole number value given by twice the average charge of a multiplet.

$$Y = 2\bar{Q} \tag{40o}$$

For the nucleon doublet p^+ and n^0, for example, $\bar{Q} = +1/2$. This gives $Y = 2\bar{Q} = +1$.

Quantum number I, called *isotopic spin*, gives the number of particles in a multiplet. The multiplicity is given by

$$M = 2I + 1 \tag{40p}$$

For Δ particles, for example, $I = \frac{3}{2}$ and the multiplicity $M = 2 \times \frac{3}{2} + 1 = 4$.

For π particles $I = 1$, and $M = 2 \times 1 + 1 = 3$.

The quantum numbers assigned each of the ten known family patterns are given in Table 40D.

As an example of the proposed symbolism, a particle designated N^0, 939, $\frac{1}{2}$, is a *neutron* with

---

* GEOFFREY F. CHEW (1924–), American physicist. He obtained his Ph.D. in physics at the University of Chicago in 1948, and served during World War II as a theoretical physicist at Los Alamos Scientific Laboratory, New Mexico. He became a professor of physics at the University of California, Berkeley, in 1949. He received the Hughes Prize in 1962 and is best known for his work on strong interactions and their relation to elementary atomic particles.

† MURRAY GELL-MANN (1929–), American physicist. He obtained his Ph.D. in theoretical physics in 1951 at the Massachusetts Institute of Technology. He became Professor of Physics at the California Institute of Technology in 1953, was elected to the National Academy of Sciences, and is best known for his work on quantum field theory and elementary atomic particles, for which he received the 1969 Nobel Prize in physics.

‡ ARTHUR H. ROSENFELD (1926–), American physicist. He obtained his Ph.D. in physics at the University of Chicago in 1954. He joined the Lawrence Radiation Laboratory at the University of California in 1956 and became Professor of Physics in 1963. He is an expert on the use of high speed digital computers and is best known for his work on the classification of elementary atomic particles.

quantum numbers $A = 1$, $Y = 1$, $I = \frac{1}{2}$, $J = \frac{1}{2}$, and a rest-mass energy of 939 MeV. Similarly a particle designated $\bar{\Sigma}^-$, 1193, $\frac{1}{2}$, is an *antisigma baryon* with quantum numbers $A = -1$, $Y = 0$, $I = 1$, $J = \frac{1}{2}$, and a rest mass of 1193 MeV. A particle designated $\bar{K}^0$, 496, 0, is an antiparticle *kappa* with quantum numbers $A = 0$, $Y = -1$, $I = \frac{1}{2}$, $J = 0$, and rest-mass energy of 496 MeV.

TABLE 40D. Elementary Particle Quantum Numbers

Baryons	Baryon number A	Hypercharge Y	Isotopic spin I
Lambda Λ	1	0	0
Omega Ω	1	-2	0
Nucleon N	1	1	$\frac{1}{2}$
Xi Ξ	1	-1	$\frac{1}{2}$
Sigma Σ	1	0	1
Delta Δ	1	1	$\frac{3}{2}$

Mesons	A	Y	I
Eta H	0	0	0
Kappa K	0	1	$\frac{1}{2}$
Pi π	0	0	1

Antibaryons	A	Y	I
Lambda $\bar{\Lambda}$	-1	0	0
Omega $\bar{\Omega}$	-1	$+2$	0
Nucleon $\bar{N}$	-1	-1	$\frac{1}{2}$
Xi $\bar{\Xi}$	-1	$+1$	$\frac{1}{2}$
Sigma $\bar{\Sigma}$	-1	0	1
Delta $\bar{\Delta}$	-1	-1	$\frac{3}{2}$

Antimesons	A	Y	I
Eta $\bar{H}$	0	0	0
Kappa $\bar{K}$	0	-1	$\frac{1}{2}$
Pi $\bar{\pi}$	0	0	1

40.18 CONSERVATION LAWS Thus far in the history of science man has come up with a number of laws he considers basic. Of these fundamental laws, those classified by the word *conservation* should be listed at the top. Some quantities in nature seem to be conserved in some kinds of processes and not in others, while other quantities seem to be conserved in all situations and under all circumstances

The following is a list of the seven conservation laws found to hold in all particle interactions:

Law 1. *The total energy, including mass, remains constant.*

Law 2. *The total linear momentum remains constant.*

Law 3. *The total angular momentum remains constant.*

Law 4. *The total charge remains constant.*

Law 5. *The total electron family number remains constant.*

Law 6. *The total muon family number remains constant.*

Law 7. *The total baryon number remains constant.*

Everything that can happen without violating these laws does happen. The first four laws have been employed in establishing all the reactions given in the preceding chapters. The last three laws will now be explained and we will see how they apply to elementary particle interactions.

The electron family is composed of four elementary particles, the electron e^-, the electron neutrino ν_e, the positron $\bar{e}^+$, and the positron neutrino $\bar{\nu}_e$. The first two are particles and each has the electron family number $+1$, while the second two are antiparticles and each has the electron family number -1. All other particles have electron family number 0.

The muon family is composed of four elementary particles, the negative muon μ^-, the muon neutrino ν_μ, the antimuon $\bar{\mu}^+$, and the antimuon neutrino $\bar{\nu}_\mu$. The first two are particles and each has the muon family number $+1$, while the second two are antiparticles and each has the muon family number -1. All other particles have muon family number 0.

EXAMPLE 1. Apply the conservation laws of charge, muon number, and electron number to the natural decay of a negative pion.

SOLUTION. For this reaction, see Eq. (35a).

	$\pi^- \longrightarrow \mu^- + \bar{\nu}_\mu$
Charge No.	$-1 = -1 + 0$
Muon No.	$0 = +1 - 1$
Electron No.	$0 = \ \ 0 + 0$

The π^- and μ^- are both particles with numbers as shown. With these as fixed numbers, the muon family law shows that the neutrino must be a mu neutrino, and it must be an antiparticle. Since none of the particles in this decay are members of the electron family, all electron numbers are zero.

EXAMPLE 2. Apply the charge, muon, and electron number conservation laws to the positive muon decay.

SOLUTION. For this reaction see Eq. (35e).

	$\bar{\mu}^+ \rightarrow \bar{e}^+ + \bar{\nu}_\mu + \nu_e$
Charge No.	$+1 = +1 + 0 + 0$
Muon No.	$-1 = \ \ 0 - 1 + 0$
Electron No.	$0 = -1 + 0 + 1$

The balancing of charge is straightforward and self-explanatory. Since the positive muon is an antiparticle it has a muon family number -1. To balance this, one of the neutrinos must be a muon antineutrino with muon number -1. Since the positron is an antiparticle, its electron number of -1 must be balanced by an electron neutrino with an electron number $+1$.

We have seen in the preceding chapter that all particles having a mass equal to or greater than the nucleon are called *baryons*. Furthermore the baryon number is none other than what we have previously called the *atomic mass number A*. For most of the elementary particles known today A has the value $+1$, 0, or -1.

In applying the baryon law to interactions we count all baryons as having baryon number $+1$, and their corresponding antibaryons as having baryon number -1.

EXAMPLE 3. Apply the charge, electron, muon, and baryon conservation laws to the decay of the neutron.

SOLUTION. By Eq. (38h) the decay of the neutron gives rise to a proton, an electron, and and an antineutrino.

	$n \longrightarrow p^+ + e^- + \bar{\nu}_e$
Charge No.	$0 = +1 - 1 + 0$
Electron No.	$0 = \ \ 0 + 1 - 1$
Muon No.	$0 = \ \ 0 + 0 + 0$
Baryon No.	$1 = \ \ 1 + 0 + 0$

Note that both the neutron and proton have baryon number $+1$, but appear on opposite sides of the equation. The electron number law shows that the neutrino must belong to the electron, and it must be an antineutrino to balance the equation.

40.19 WORLD LINES

The method of depicting elementary-particle events by means of space–time graphs was introduced by R. Feynman* in 1949. As an introduction to the meaning of these useful graphical pictures, consider the flight of an airplane from one city to another as shown in Fig. 40Q. The time t is plotted vertically, and the distance x is plotted horizontally.

We now represent the plane's history by the solid line ABCD. Parked at the ramp at the Washington Airport, time t is advancing in the direction A to B. The plane takes off, and moving both in time and space, traces out the world line B to C. Landing in New York and coming to rest on the ramp, the plane again moves only in time, and its world line CD is straight upward.

Note there is no such thing as standing still

---

* RICHARD F. FEYNMAN (1918–), American physicist. He received his Ph.D. at Princeton University in 1940. He served at the Los Alamos Scientific Laboratory, New Mexico during World War II, and became Professor of Physics at Cornell University in 1945. In 1951, he became Professor of Physics at the California Institute of Technology. He won the Einstein Award in 1954. He was elected to the National Academy of Sciences and the Royal Society of London. He is best known for his theoretical work in quantum electrodynamics and his outstanding teaching ability. For his researches in quantum electrodynamics he was awarded the Nobel Prize in physics in 1965.

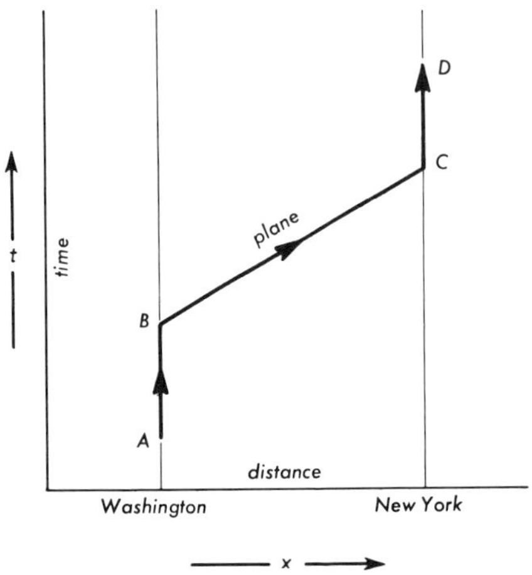

Fig. 40Q. World-line graph of a plane flight from Washington to New York.

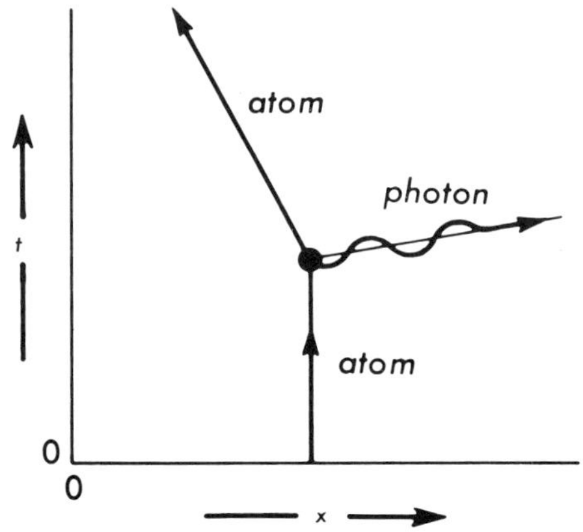

Fig. 40R. World-line graph for the emission of a photon by an atom.

in a space–time graph. A city like Washington moves through time and continually traces out its vertical world line. If one places a ruler horizontally across the bottom of the diagram and then pushes it upward with uniform speed across the diagram, the intersection of the world lines with the ruler edge gives a description of the plane's motion with respect to the two cities.

A narrow horizontal slot in a strip of cardboard is more effective than a ruler. If the strip is moved downward with uniform speed, the dots seen in the slot represent an equally valid motion, namely, that of the plane on its return flight.

Turning now to the world of elementary particles, we may trace the world lines of a simple and well-known event (see Fig. 40R). The diagram represents the emission of a photon by an excited atom.

Initially at rest the atom traces out a vertically straight world line. Upon the emission of a photon with the speed of light, the atom recoils to the left as shown. The slower a particle moves, the more nearly vertical is its world line, while the faster it moves the more nearly horizontal is its world line.

Since no particle can travel from one place to another with infinite speed, no world line can be horizontal. Photons and neutrinos provide the most nearly horizontal world lines since they travel with the speed of light c.

Figure 40S represents the decay of a negative pion

$$\pi^- \longrightarrow \mu^- + \bar{\nu}_\mu \qquad (40q)$$

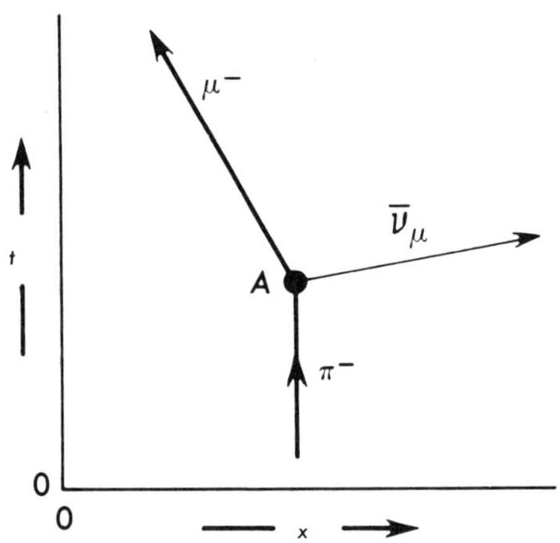

Fig. 40S. World-line graph for the decay of a pion into a muon and a neutrino.

The dot at the vertex A indicates the time at which the world line of the pion ends and the world lines of the muon and antineutrino begin.

Consider the well-known decay of a free neutron. The neutron with its half-life of 11.7 min decays into a proton, an electron, and an antineutrino (see Fig. 40T).

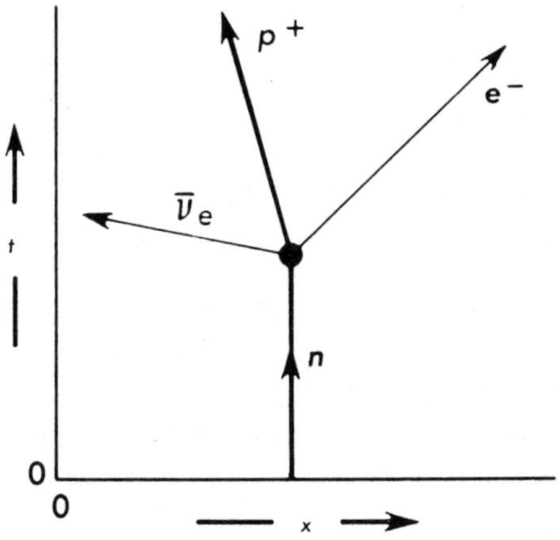

Fig. 40T. World-line graph for the decay of a free neutron into a proton, electron, and an antineutrino.

The neutron initially at rest suddenly explodes into three particles as shown. The reaction is

$$n \rightarrow p^{+} + e^{-} + \bar{\nu}_e \tag{40r}$$

Note that the world line of the antineutrino has the same slope as a photon.

If the antineutrino later encounters a proton, there is a small probability that it will be captured in what is called the *inverse beta-decay process*.

$$\bar{\nu}_e + p^{+} \rightarrow n + \bar{e}^{+} \tag{40s}$$

While the electron appears in the neutron decay process, the laws of charge conservation and electron family conservation require the creation of a positron in the capture process.

40.20 PARITY Parity is a mathematical treatment of what is best described as a mirror symmetry of many natural phenomena. Until recently it was believed that the mirror image of any physical phenomenon, or laboratory experiment, is just as true to nature as the direct image itself. This was called conservation of parity.

According to the conservation of parity it was believed that if one observed an experiment by looking in a mirror and was not told he was looking in a mirror, there would be no way in which he would know it. As an example of nature's tendency toward symmetry, many crystalline structures show mirror symmetry. With many cubic crystals, for example, there is no question since the image and mirror image have identical structures.

Diagrams of two kinds of quartz crystals found in nature are shown in Fig. 40U. One

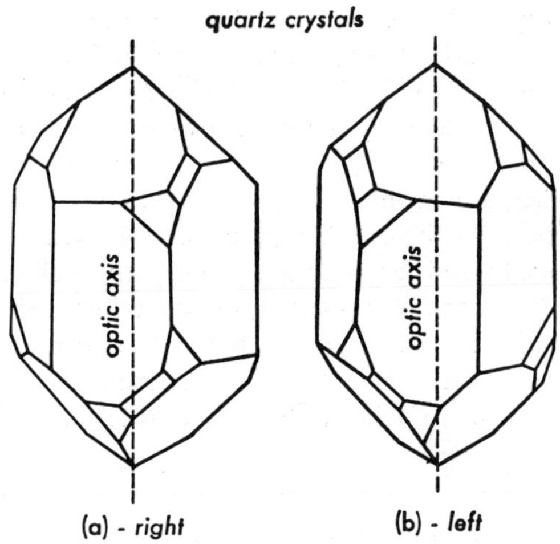

Fig. 40U. Diagrams of right- and left-handed quartz crystals. Each is a mirror image of the other.

called right-handed quartz has its silicon and oxygen atoms lined up in clockwise helixes around the axis, while the left-handed quartz is a mirror image of the first and has its molecules lined up in counterclockwise direction. One will rotate plane polarized light clockwise as it traverses the crystal along the optic axis, while the other will rotate it counterclockwise. These crystals are consistent with the concept of the conservation of parity.

The first serious question regarding the conservation of parity arose in the minds of two young theoretical physicists, Yang and Lee, when in 1956 they pointed out that some K^+ mesons decay into two pions while others decay into three. As a crucial test of parity an experiment was set up at the National Bureau of Standards under the direction of Miss C. S. Wu.* This experiment showed that when radioactive cobalt-60 nuclei were lined up with their spin axes parallel to each other, more β particles were emitted along this axis in one direction than in the other.

To realize the significance of this result consider the common decay of a π^+ meson into a μ^+ and a ν. The π^+ has zero spin, while the μ^+ and ν have spins of $\frac{1}{2}$. To conserve angular momentum the two particles must fly apart spinning in opposite directions as shown above in Fig. 40V. Moving in the direction of the arrows, both are advancing as would left-handed screws. The mirror image below, which is apparently forbidden by nature, shows both particles advancing as right-handed screws. A spinning particle advancing as a right-handed screw is said to have positive helicity; advancing as a left-handed screw it is said to have negative helicity.

From the observed spin directions of β particles it is now known that all neutrinos, ν, have negative helicity while antineutrinos, $\bar{\nu}$, have positive helicity. With this rule and spin

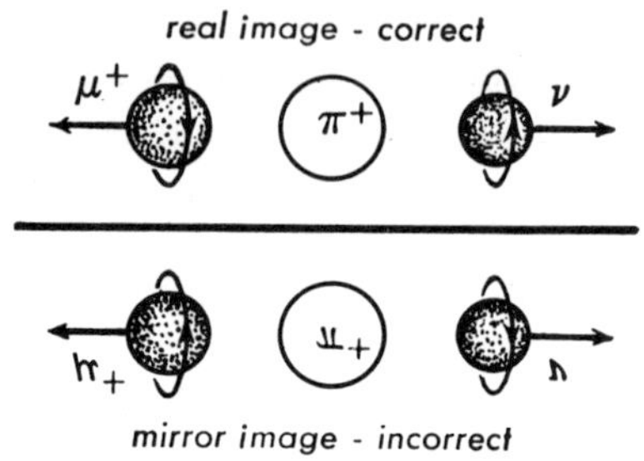

Fig. 40V. Schematic diagram showing the left-handed, or negative, helicity of μ^+ and ν-particles in the decay of π^+ mesons.

conservation, we see from Fig. 40W that in π^- decay both μ^- and $\bar{\nu}$ have positive helicity. This is an intrinsic property of the neutrino only, since it moves with the speed of light. To an observer moving faster than the muon, the μ^- would appear to spin in the opposite direction and to have negative helicity. *Muons* are μ mesons.

If all the particles in the mirror image of Fig. 40V are changed to their corresponding antiparticles, the resulting interaction becomes the real interaction of Fig. 40W. Similarly if all the particles in the mirror image of Fig. 40W are

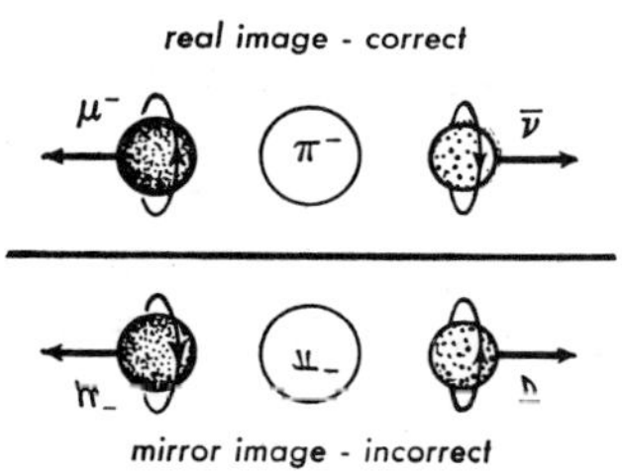

Fig. 40W. Schematic diagram showing the right-handed, or positive, helicity of a μ^- and $\bar{\nu}$ particles in the decay of π-mesons.

changed to their corresponding antiparticles, we obtain the real interaction of Fig. 40V. Thus there is an over-all symmetry between both decay reactions.

If a distant world is made of antimatter, there is no known way whereby we on the earth can determine this. Furthermore, although μ^- and μ^+ have positive and negative helicity, respectively, there is no information we could impart to a being on a distant world that would tell him which is which.

* CHIEN-SHIUNG WU (1913–   ), American physicist, was born in Shanghai, China. She became a United States citizen and was married in 1942. She obtained her Ph.D. at the University of California, Berkeley, in 1940, and received a number of honorary degrees from other Universities. She has held the position of Professor of Physics at Columbia University since 1957, and was elected to the National Academy of Sciences, and the Chinese Academy of Sciences. She received the American Association of University Women award in 1960, and is best known for her experimental work on the nonconservation of parity in beta-decay. The author had the pleasure of knowing Miss Wu in her graduate school days at the University of California.

QUESTIONS

1. Define or briefly explain each of the following: (a) geometrical cross section, (b) scattering cross section, (c) capture cross section, and (d) fission cross section.

2. In what units are cross sections measured?

3. (a) What is the average velocity of thermal neutrons? (b) What is their average energy in volts?

4. Briefly explain the meaning of (a) the reproduction factor, (b) nuclear pile, and (c) reactor moderator.

5. (a) Write down the carbon cycle believed to be the source of solar energy. (b) Write down the proton-proton cycle.

6. How many elementary particles are there? Make a list of them. What are antiparticles?

7. How would you define (a) lepton, (b) nucleon, (c) baryon, and (d) hyperon?

8. What advantages does a bubble chamber have over a cloud chamber?

PROBLEMS

1. Find (a) the radius, and (b) the geometrical cross section for the nucleus of barium-138. [Ans. (a) 6.20×10^{-15} m. (b) 1.208×10^{-28} m², or 1.208 barns.]

2. Find (a) the radius, and (b) the cross section of gold-197 nucleus.

3. Find (a) the radius, and (b) the geometrical cross section of a lead-208 nucleus. [Ans. (a) 7.11×10^{-15} m. (b) 1.588 barns.]

4. Find (a) the radius, and (b) the geometrical cross section of iodine-127.

5. Since the critical size for a U-235 explosion lies between a marble and a basketball, find (a) the ratio of these sphere diameters, (b) the ratio of their surface areas, (c) the ratio of their volumes. [Ans. (a) 12:1. (b) 144:1. (c) 1728:1.]

6. Find the total cross section for thermal neutrons in pure U-235.

7. If the reproduction factor for fission in a mass of U-235 is 1.15, how many secondary neutrons will be produced in (a) 5 generations, and (b) 10 generations? [Ans. (a) 2.011. (b) 4.05.]

8. If the reproduction factor for fission in a mass of Pu-239 is 1.35, how many secondary neutrons will be produced in (a) 5 generations, and (b) 10 generations.

9. An atomic particle is designated Σ^-, 1385, $\frac{3}{2}$. Write down the values of the quantum numbers Y, A, I, J, and M, and specify the mass in units of the electronic mass m_e. [Ans. $Y = 0$, $A = 1$, $I = 1$, $J = \frac{3}{2}$, $M = 3$, $m = 2710\ m_e$.]

10. What are the quantum numbers Y, A, I, J, and M, for an atomic particle designated $\bar{\Delta}$, 1238, $\frac{1}{2}$. Write down the particle mass in m_e.

11. Write down the minimum energy necessary to produce four π mesons. See Fig. 40L. [Ans. 559 MeV.]

12. Calculate the energy necessary to produce a pair of protons.

13. Apply the conservation laws of charge, muon number, electron number, and baryon number to the natural decay of a positive pion.

14. Apply the conservation laws of charge, baryon number, muon number, and electron number to the first reaction in Eq. (40n).

15. Apply the conservation laws of charge, baryon number, muon number, and electron number to the first reaction in Eq. (40k).

16. Apply the conservation laws of charge, muon number, and baryon number, and electron number to the natural decay of a negative muon.

17. Construct a world-line diagram for the decay of a negative muon.

18. Construct a world-line diagram for the decay of a positive muon.

19. Write down the first reaction in Eq. (40n), and then replace each particle by its antiparticle. Does this reaction obey the baryon law? [Ans. yes.]

20. Apply the law of conservation of baryons to show that the reaction shown in Fig. 40N could not have resulted in the reaction $\pi^- + p = \Sigma^0 + \Lambda^0$.

21. Apply the law of conservation of baryon number and show why the reaction in Fig. 40N could not have been $\bar{p}^- + p = \Lambda^0 + K^0$.

22. Apply the conservation laws for charge, electron number, and baryon number to Eq. (40s).

Appendix I. Abbreviations Used in This Text

A	ampere(s)	hp	horsepower	μW	microwatt		
A-turn	ampere turn(s)	hr	hour	mi	mile		
Å	angstrom unit(s)	in.	inch	mm	millimeter		
A	atomic mass number	J	joule	MeV	million electron volts		
amu	atomic mass unit(s)	Kcal	kilocalorie	ms	millisecond		
Z	atomic number	Kc/s	kilocycles/second	mV	millivolt		
b	barn	Km	kilometers	mW	milliwatt		
BeV	billion electron volts	Kg	kilogram	min	minute		
GeV	or giga electron volts	KHz	kilohertz	ns	nanosecond		
Btu	British thermal unit	KV	kilovolt	N	newton		
cal	calorie	KW	kilowatt	Ω	ohm		
cm	centimeter	MHz	megahertz	rad	radian		
C	coulomb	MV	megavolt	rps	revolutions/second		
db	decibel	MW	megawatt	s	second		
F	farad	m	meter	sl	slug		
ft	foot	μA	microampere	KeV	thousand electron volts		
G	gauss	μC	microcurie	V	volt		
g	gram	μF	microfarad	W	watt		
H	henry	μH	microhenry	Wb	weber		
Hz	hertz	$\mu\mu$F	micromicrofarad	y	year		
		μ	micron				

Appendix II. Periodic Table of the Chemical Elements

IA	IIA	IIIB	IVB	VB	VIB	VIIB	VIIIB	VIIIB	VIIIB	IB	IIB	IIIA	IVA	VA	VIA	VIIA	
		Metals										Nonmetals					
1 H 1.0080																	2 He 4.00260
3 Li 6.941	4 Be 9.01218											5 B 10.81	6 C 12.011	7 N 14.0067	8 O 15.9994	9 F 18.9984	10 Ne 20.179
11 Na 22.9898	12 Mg 24.305											13 Al 26.9815	14 Si 28.086	15 P 30.9738	16 S 32.06	17 Cl 35.453	18 Ar 39.948
19 K 39.102	20 Ca 40.08	21 Sc 44.9559	22 Ti 47.90	23 V 50.9414	24 Cr 51.996	25 Mn 54.9380	26 Fe 55.847	27 Co 58.9332	28 Ni 58.71	29 Cu 63.546	30 Zn 65.37	31 Ga 69.72	32 Ge 72.59	33 As 74.9216	34 Se 78.96	35 Br 79.904	36 Kr 83.80
37 Rb 85.4678	38 Sr 87.62	39 Y 88.9059	40 Zr 91.22	41 Nb 92.9064	42 Mo 95.94	43 Tc 98.9062[a]	44 Ru 101.07	45 Rh 102.9055	46 Pd 106.4	47 Ag 107.868	48 Cd 112.40	49 In 114.82	50 Sn 118.69	51 Sb 121.75	52 Te 127.60	53 I 126.9045	54 Xe 131.30
55 Cs 132.9055	56 Ba 137.34	57–71 Lanthanide Elements	72 Hf 178.49	73 Ta 180.9479	74 W 183.85	75 Re 186.2	76 Os 190.2	77 Ir 192.22	78 Pt 195.09	79 Au 196.9665	80 Hg 200.59	81 Tl 204.37	82 Pb 207.2	83 Bi 208.9806	84 Po (210)[b]	85 At (210)[b]	86 Rn (222)[b]
87 Fr (223)[b]	88 Ra 226.0254[a]	89–103 Actinide Elements	104 Rf (258)[b]	105 Ha (260)[b]													

Lanthanide Elements (Rare Earth)	57 La 138.9055	58 Ce 140.12	59 Pr 140.9077	60 Nd 144.24	61 Pm (147)[b]	62 Sm 150.4	63 Eu 151.96	64 Gd 157.25	65 Tb 158.9254	66 Dy 162.50	67 Ho 164.9303	68 Er 167.26	69 Tm 168.9342	70 Yb 173.04	71 Lu 174.97
Actinide Elements	89 Ac (227)[b]	90 Th 232.0381[a]	91 Pa 231.0359[a]	92 U 238.029	93 Np 237.0482[a]	94 Pu (242)[b]	95 Am (243)[b]	96 Cm (247)[b]	97 Bk (247)[b]	98 Cf (251)[b]	99 Es (254)[b]	100 Fm (253)[b]	101 Md (256)[b]	102 No (254)[b]	103 Lw (257)[b]

[a] Mass of most commonly available, long-lived isotope. [b] Mass number of most stable or best known isotope.

Based upon carbon-12 as 12 even.
(**s = second; m = minute; h = hour; d = day; y = year**)

SYMBOL	ATOMIC MASS (amu)	MASS EXCESS MeV	ABUND. OR ACTIVITY	HALF-LIFE
${}_{-1}e^0$	0.000 549	0.511	—	stable
${}_1p^1$	1.007 276	6.774	—	stable
${}_0n^1$	1.008 665	8.071	β^-	13 m
${}_1H^1$	1.007 825	7.288	**99.985**	stable
H^2	2.014 103	13.135	**0.015**	stable
H^3	3.016 049	14.949	β^-	12.2 y
${}_2He^3$	3.016 030	14.931	**0.00013**	stable
He^4	4.002 604	2.425	**99.9999**	stable
He^5	5.012 296	11.453	$\alpha + n$	inst.
He^6	6.018 900	17.604	β^-	0.82 s
${}_3Li^5$	5.012 541	11.681	$\alpha + p$	inst.
Li^6	6.015 126	14.089	**7.52**	stable
Li^7	7.016 005	14.908	**92.48**	stable
Li^8	8.022 488	20.947	β^-	0.86 s
Li^9	9.027 300	25.400	β^-	0.17 s
${}_4Be^6$	6.019 780	18.430	β^+	0.4 s
Be^7	7.016 931	15.770	E.C.	53.6 d
Be^8	8.005 308	4.944	2α	inst.
Be^9	9.012 186	11.350	**100**	stable
Be^{10}	10.013 535	12.607	β^-	2.5×10^6 y
${}_5B^8$	8.024 612	22.924	β^+	0.6 s
B^9	9.013 335	12.420	$2\alpha + p$	inst.
B^{10}	10.012 939	12.052	**18.6**	stable
B^{11}	11.009 305	8.667	**81.4**	stable
B^{12}	12.014 353	13.369	β^-	0.022 s
${}_6C^{10}$	10.016 830	15.670	β^+	19.1 s
C^{11}	11.011 433	10.649	β^+	20.5 m
C^{12}	12.000 000	0	**98.892**	stable
C^{13}	13.003 354	3.124	**1.108**	stable
C^{14}	14.003 242	3.020	β^-	5.6×10^3 y
C^{15}	15.010 600	9.873	β^-	2.3 s
${}_7N^{12}$	12.018 709	17.426	β^+	0.012 s
N^{13}	13.005 739	5.345	β^+	10.1 m
N^{14}	14.003 074	2.864	**99.635**	stable
N^{15}	15.000 108	0.101	**0.365**	stable
N^{16}	16.006 089	5.672	β^-	7.36 s
N^{17}	17.008 449	7.869	β^-	4.14 s

SYMBOL	ATOMIC MASS (amu)	MASS EXCESS MeV	ABUND. OR ACTIVITY	HALF-LIFE
$_8O^{14}$	14.008 597	8.008	β^+	74 s
O^{15}	15.003 072	2.861	β^+	2.0 m
O^{16}	15.994 915	−4.736	**99.76**	stable
O^{17}	16.999 133	−0.807	**0.04**	stable
O^{18}	17.999 160	−0.782	**0.20**	stable
O^{19}	19.003 577	3.332	β^-	29.4 s
O^{20}	20.004 071	3.792	n	13.6 s
$_9F^{17}$	17.002 098	1.954	β^+	66 s
F^{18}	18.000 950	0.884	β^+	1.87 h
F^{19}	18.998 405	−1.486	**100**	stable
F^{20}	19.999 986	−0.013	β^-	11.2 s
F^{21}	20.999 972	−0.026	β^-	5 s
$_{10}Ne^{18}$	18.005 715	5.323	β^+	1.6 s
Ne^{19}	19.001 892	1.762	β^+	19 s
Ne^{20}	19.992 440	−7.041	**90.92**	stable
Ne^{21}	20.993 849	−5.729	**0.26**	stable
Ne^{22}	21.991 385	−8.025	**8.82**	stable
Ne^{23}	22.994 475	−5.146	β^-	40.2 s
Ne^{24}	23.993 597	−5.964	β^-	3.4 m
$_{11}Na^{20}$	20.008 890	8.280	β^+	0.3 s
Na^{21}	20.997 638	−2.200	β^+	23 s
Na^{22}	21.994 435	−5.183	β^+, E.C.	2.6 y
Na^{23}	22.989 773	−9.526	**100**	stable
Na^{24}	23.990 967	−8.414	β^-	15.0 h
Na^{25}	24.989 920	−9.390	β^-	60 s
$_{12}Mg^{23}$	22.994 135	−5.463	β^+, E.C.	11 s
Mg^{24}	23.985 045	−13.930	**78.60**	stable
Mg^{25}	24.985 840	−13.189	**10.11**	stable
Mg^{26}	25.982 591	−16.215	**11.29**	stable
Mg^{27}	26.998 346	−14.581	β^-	9.45 m
Mg^{28}	27.983 880	−15.015	β^-	21.4 h
$_{13}Al^{24}$	24.000 090	0.090	β^+	2.1 s
Al^{25}	24.990 414	−8.928	β^+	7.6 s
Al^{26}	25.986 900	−12.201	β^+, E.C.	10^5 y
Al^{27}	26.981 535	−17.199	**100**	stable
Al^{28}	27.981 908	−16.851	β^-	2.3 m
Al^{29}	28.980 442	−18.217	β^-	6.6 m
$_{14}Si^{27}$	26.986 701	−12.387	β^+, E.C.	4.9 s
Si^{28}	27.976 927	−21.491	**92.27**	stable
Si^{29}	28.976 491	−21.897	**4.68**	stable
Si^{30}	29.973 761	−24.440	**3.05**	stable
Si^{31}	30.975 349	−22.961	β^-	2.6 h
Si^{32}	31.974 020	−24.200	β^-	7×10^2 y

SYMBOL	ATOMIC MASS (amu)	MASS EXCESS MeV	ABUND. OR ACTIVITY	HALF-LIFE
$_{15}P^{28}$	27.991 740	−7.690	β^+	0.28 s
P^{29}	28.981 816	−16.937	β^+	4.4 s
P^{30}	29.978 320	−20.193	β^+	2.5 m
P^{31}	30.973 763	−24.438	**100**	stable
P^{32}	31.973 908	−24.303	β^-	14.3 d
P^{33}	32.971 728	−26.334	β^-	24.4 d
P^{34}	33.973 340	−24.830	β^-	12.4 s
$_{16}S^{31}$	30.979 599	−19.002	β^+	2.6 s
S^{32}	31.972 074	−26.012	**95.054**	stable
S^{33}	32.971 461	−26.583	**0.740**	stable
S^{34}	33.967 865	−29.932	**4.190**	stable
S^{35}	34.969 034	−28.843	β^-	87.1 d
S^{36}	35.967 091	−30.653	**0.016**	stable
S^{37}	36.971 040	−26.980	β^-	5.0 m
S^{38}	37.971 220	−26.800	β^-	6 m
$_{17}Cl^{32}$	31.986 030	−13.010	β^+	0.31 s
Cl^{33}	32.977 446	−21.008	β^+	2.4 s
Cl^{34}	33.973 764	−24.437	β^+	32.4 m
Cl^{35}	34.968 855	−29.010	**75.4**	stable
Cl^{36}	35.968 312	−25.516	β^-, E.C.	3.2 y
Cl^{37}	36.965 898	−31.766	**24.6**	stable
Cl^{38}	37.968 002	−29.804	β^-	37.5 m
Cl^{39}	38.968 003	−29.803	β^-	55.5 m
Cl^{40}	39.970 400	−27.500	β^-	1.4 m
$_{18}Ar^{35}$	34.975 275	−23.030	β^+	1.83 s
Ar^{36}	35.963 548	−30.227	**0.337**	stable
Ar^{37}	36.966 772	−30.950	E.C.	34.1 d
Ar^{38}	37.962 725	−34.720	**0.063**	stable
Ar^{39}	38.964 321	−33.233	β^-	2.6×10^2 y
Ar^{40}	39.962 384	−35.037	**99.600**	stable
Ar^{41}	40.964 508	−33.058	β^-	1.83 h
Ar^{42}	41.963 043	−34.423	β^-	3.5 y
$_{19}K^{38}$	37.969 090	−28.791	β^+	7.7 m
K^{39}	38.963 714	−33.798	**93.08**	stable
K^{40}	39.964 008	−33.524	**0.0119**	10^9 y
K^{41}	40.961 835	−35.548	**6.91**	stable
K^{42}	41.962 417	−35.006	β^-	12.5 h
K^{43}	42.960 731	−36.577	β^-	22.4 h
K^{44}	43.962 040	−35.360	β^-	22.0 m
$_{20}Ca^{39}$	38.970 706	−27.286	β^+	1.0 s
Ca^{40}	39.962 589	−34.846	**96.97**	stable
Ca^{41}	40.962 279	−35.135	E.C.	2×10^5 y
Ca^{42}	41.958 628	−38.536	**0.64**	stable
Ca^{43}	42.958 780	−38.394	**0.145**	stable
Ca^{44}	43.955 490	−41.458	**2.06**	stable
Ca^{45}	44.956 189	−40.807	β^-	164 d
Ca^{46}	45.953 689	−43.136	**0.0033**	stable
Ca^{47}	46.954 512	−42.370	β^-	4.9 d
Ca^{48}	47.952 363	−44.371	**0.185**	10^{16} y
Ca^{49}	48.955 662	−41.298	β^-	8.8 m

For atomic masses in *amu* and mass excess energies in *Mev*, for isotopes in the remainder of the periodic table, see *Nuclear Physics*, March-April, pp. 28–42, 1962, by L. A. König, J. H. E. Mattauch, and A. H. Wapstra.

Appendix IV. Relative Masses of One Stable Isotope for Each of the Elements

Based on $C^{12} = 12$.

These values are taken from Everling, König, Mattauch, and Wapstra, *Nucl. Phys.* **18,** 529 (1960); and König, Mattauch, and Wapstra, *Nucl. Phys.* **31,** 18 (1962).

Space is left between third and fourth decimal place to facilitate reading.

SYMBOL	MASS	SYMBOL	MASS
$_{1}H^{1}$	1.007 8252	$_{51}Sb^{121}$	120.903 750
$_{2}He^{4}$	4.002 6036	$_{52}Te^{128}$	127.904 710
$_{3}Li^{7}$	7.016 005	$_{53}I^{127}$	126.904 352
$_{4}Be^{9}$	9.012 186	$_{54}Xe^{129}$	128.904 784
$_{5}B^{11}$	11.009 3051	$_{55}Cs^{133}$	132.905 090
$_{6}C^{12}$	12 even	$_{56}Ba^{138}$	137.905 010
$_{7}N^{14}$	14,003 0744	$_{57}La^{139}$	138.906 060
$_{8}O^{16}$	15.995 9149	$_{58}Ce^{140}$	139.905 280
$_{9}F^{19}$	18.998 4046	$_{59}Pr^{141}$	140.907 390
$_{10}Ne^{20}$	19.992 4404	$_{60}Nd^{142}$	141.907 478
$_{11}Na^{23}$	22.989 773	$_{61}Pm^{143}$	143.910 800
$_{12}Mg^{24}$	23.985 045	$_{62}Sm^{144}$	143.911 650
$_{13}Al^{27}$	26.981 535	$_{63}Eu^{153}$	152.920 720
$_{14}Si^{28}$	27.976 927	$_{64}Gd^{156}$	155.922 240
$_{15}P^{31}$	30.973 763	$_{65}Tb^{159}$	158.924 300
$_{16}S^{32}$	31.972 074	$_{66}Dy^{164}$	163.928 100
$_{17}Cl^{35}$	34.968 855	$_{67}Ho^{165}$	164.929 600
$_{18}Ar^{40}$	39.962 384	$_{68}Er^{166}$	165.931 900
$_{19}K^{39}$	38.963 714	$_{69}Tm^{169}$	165.932 700
$_{20}Ca^{40}$	39.962 589	$_{70}Yb^{177}$	176.943 940
$_{21}Sc^{45}$	44.955 916	$_{71}Lu^{177}$	176.942 450
$_{22}Ti^{48}$	47.947 948	$_{72}Hf^{180}$	179.945 120
$_{23}V^{51}$	50.943 978	$_{73}Ta^{181}$	180.646 180
$_{24}Cr^{52}$	51.940 514	$_{74}W^{184}$	183.949 150
$_{25}Mn^{55}$	54.941 293	$_{75}Re^{187}$	186.954 980
$_{26}Fe^{56}$	55.934 932	$_{76}Os^{189}$	188.957 220
$_{27}Co^{59}$	58.933 189	$_{77}Ir^{193}$	192.962 340
$_{28}Ni^{58}$	57.935 342	$_{78}Pt^{195}$	194.964 460
$_{29}Cu^{63}$	62.929 594	$_{79}Au^{197}$	196.966 552
$_{30}Zn^{64}$	63.929 145	$_{80}Hg^{202}$	201.970 630
$_{31}Ga^{69}$	68.925 682	$_{81}Tl^{205}$	204.974 462
$_{32}Ge^{74}$	73.921 150	$_{82}Pb^{206}$	205.974 460
$_{33}As^{75}$	74.921 580	$_{83}Bi^{209}$	208.980 417
$_{34}Se^{80}$	79.916 512	$_{84}Po^{214}$	213.995 192
$_{35}Br^{81}$	80.916 344	$_{85}At^{217}$	217.002 405
$_{36}Kr^{84}$	83.911 504	$_{86}Rn^{222}$	222.015 365
$_{37}Rb^{85}$	84.911 710	$_{87}Fr^{221}$	221.014 176
$_{38}Sr^{88}$	87.905 610	$_{88}Ra^{226}$	226.025 360
$_{39}Y^{89}$	88.905 430	$_{89}Ac^{227}$	227.027 814
$_{40}Zr^{90}$	89.904 320	$_{90}Th^{232}$	232.038 211
$_{41}Nb^{93}$	92.906 020	$_{91}Pa^{234}$	234.043 370
$_{42}Mo^{98}$	97.905 510	$_{92}U^{238}$	238.050 760
$_{43}Tc^{99}$	98.907 300	$_{93}Np^{239}$	239.052 938
$_{44}Ru^{102}$	101.903 720	$_{94}Pu^{240}$	240.053 974
$_{45}Rh^{103}$	102.903 800	$_{95}Am^{241}$	241.056 689
$_{46}Pd^{106}$	105.903 200	$_{96}Cm^{242}$	242.058 800
$_{47}Ag^{107}$	106.904 970	$_{97}Bk^{245}$	245.066 240
$_{48}Ca^{112}$	111.902 840	$_{98}Cf^{246}$	246.068 780
$_{49}In^{115}$	114.904 070	$_{99}Es^{249}$	249.076 220
$_{50}Sn^{120}$	119.902 130	$_{100}Fm^{252}$	252.082 650

Appendix V. Complete List of the Stable Isotopes of the Chemical Elements

Atomic weights are based on the atomic mass of C^{12} = 12 even.

AT. NO.	ELEMENT	SYM.	ISOTOPES, MASS NO.	AT. WT.
1	hydrogen	H	1, (2)	1.00797[a]
2	helium	He	**4,** (3)	4.0026
3	lithium	Li	6, **7**	6.939
4	beryllium	Be	**9**	9.0122
5	boron	B	10, **11**	10.811[a]
6	carbon	C	**12,** (13)	12.01115[a]
7	nitrogen	N	**14,** (15)	14.0067
8	oxygen	O	**16,** (18), (17)	15.9994[a]
9	fluorine	F	**19**	18.9984
10	neon	Ne	**20,** (21), **22**	20.183
11	sodium	Na	**23**	22.9898
12	magnesium	Mg	**24,** 25, 26	24.312
13	aluminum	Al	**27**	26.9815
14	silicon	Si	**28,** 29, 30	28.086[a]
15	phosphorus	P	**31**	30.9738
16	sulfur	S	**32,** 33, 34	32.064[a]
17	chlorine	Cl	**35,** 37	35.453[b]
18	argon	Ar	(36), (38), **40**	39.948
19	potassium	K	**39,** (40), 41	39.102
20	calcium	Ca	**40,** (42), (43), 44	40.08
21	scandium	Sc	**45**	44.956
22	titanium	Ti	46, 47, **48,** 49, 50	47.90
23	vanadium	V	**51**	50.942
24	chromium	Cr	50, **52,** 53, 54	51.996[b]
25	manganese	Mn	**55**	54.9380
26	iron	Fe	54, **56,** 57, (58)	55.847[b]
27	cobalt	Co	**59**	58.9332
28	nickel	Ni	**58,** 60, 61, 62 **(64)**	58.71
29	copper	Cu	**63,** 65	63.54
30	zinc	Zn	**64,** 66, 67, 68, (70)	65.37
31	gallium	Ga	**69,** 71	69.72
32	germanium	Ge	70, 72, 73, **74,** 76	72.59
33	arsenic	As	**75**	74.9216
34	selenium	Se	(74), 76, 77, 78, **80,** 82	78.96
35	bromine	Br	**79, 81**	79.909[b]
36	krypton	Kr	(78), 80, 82, 83, **84,** 86	83.80
37	rubidium	Rb	**85,** 87	85.47
38	strontium	Sr	(84), 86, 87, **88**	87.62
39	yttrium	Yt	**89**	88.905
40	zirconium	Zr	**90,** 91, 92, 94, 96	91.22
41	niobium	Nb	**93**	92.906
42	molybdenum	Mo	92, 94, 95, 96, 97, **98,** 100, 102	95.94
43	technetium	Tc	99	97.2
44	ruthenium	Ru	96, 98, 99, 100, 101, **102,** 104	101.07
45	rhodium	Rh	**103**	102.905
46	palladium	Pd	(102), 104, 105, **106,** 108, 110	106.4
47	silver	Ag	107, **109**	107.870[b]
48	cadmium	Cd	106, (108), 110, 111, **112,** 113, 114, 116	112.40
49	indium	In	113, **115**	114.82

[a] Atomic weights so designated are known to be variable because of natural variations in isotopic composition. The observed ranges are: Hydrogen ±0.00001; Boron ±0.003; Carbon ±0.00005; Oxygen ±0.0001; Silicon ±0.001; Sulfur ±0.003.

[b] Atomic weights so designated are believed to have the following experimental uncertainties: Chlorine ±0.001; Chromium ±0.001; Iron ±0.003; Bromine ±0.002; Silver ±0.003.

50	tin.	Sn	112, (114), (115), 116, 117, 118, 119, **120,** 122, 124	118.69
51	antimony.	Sb	**121, 123**	121.75
52	tellurium	Te	(120), 122, 123, 124, 125, 126, **128,** 130	127.60
53	iodine.	I	**127**	126.9044
54	xenon.	Xe	(124), (126), 128, **129,** 130, 131, 132, 134, 136	131.30
55	cesium	Cs	**133**	132.905
56	barium.	Ba	(130), (132), 134, 135, 136, 137, **138**	137.34
57	lanthanum.	La	**139**	138.91
58	cerium	Ce	(136), (138), **140,** 142	140.12
59	praseodymium	Pr	**141**	140.906
60	neodymium.	Nd	**142,** 143, **144,** 145, 146, (148), (150)	144.24
61	prometeum	Pm		146.0
62	samarium.	Sa	144, 147, 148, 149, 150, **152,** 154	150.35
63	europium.	Eu	**151, 153**	151.96
64	gadolinium	Gd	155, **156,** 157, **158,** 160	157.25
65	terbium	Tb	**159**	158.924
66	dysprosium	Dy	161, 162, 163, **164**	162.50
67	holmium	Ho	**165**	164.930
68	erbium.	Er	**166,** 167, 168, 170	167.26
69	thulium	Tm	**169**	168.934
70	ytterbium.	Yb	171, 172, 173, **174,** 176	173.04
71	lutecium.	Lu	**175**	174.97
72	hafnium.	Hf	176, 177, 178, 179, **180**	178.49
73	tantalum	Ta	**181**	180.948
74	tungsten.	W	182, 183, **184, 186**	183.85
75	rhenium.	Re	185, **187**	186.2
76	osmium	Os	186, (187), 188, 189, 190, 192	190.2
77	iridium.	Ir	191, **193**	192.2
78	platinum	Pt	(192), 194, **195,** 196, 198	195.09
79	gold	Au	**197**	196.967
80	mercury.	Hg	(196), 198, 199, 200, 201, **202,** 204	200.59
81	thallium.	Tl	203, **205**	204.37
82	lead..	Pb	204, 206, 207, **208**	207.19
83	bismuth	Bi	**209**	208.98

Although some of them are found in nature, elements 84 and up are unstable. For their isotopes see Appendix VI.

Appendix VI. The Radioactive Nuclides of the Heavy Elements Showing Decay Schemes.

81	thallium	Tl
82	lead	Pb
83	bismuth	Bi
84	polonium	Po
85	astatine	At
86	radon	Rn
87	francium	Fr
88	radium	Ra
89	actinium	Ac
90	thorium	Th
91	protactinium	Pa
92	uranium	U
93	neptunium	Np
94	plutonium	Pu
95	americium	Am
96	curium	Cm
97	berkelium	Bk
98	californium	Cf
99	einsteinium	Es
100	fermium	Fm
101	mendelevium	Md
102	nobelium	No
103	lawrencium	Lw
104	rutherfordium	Rf
105	hahnium	Ha

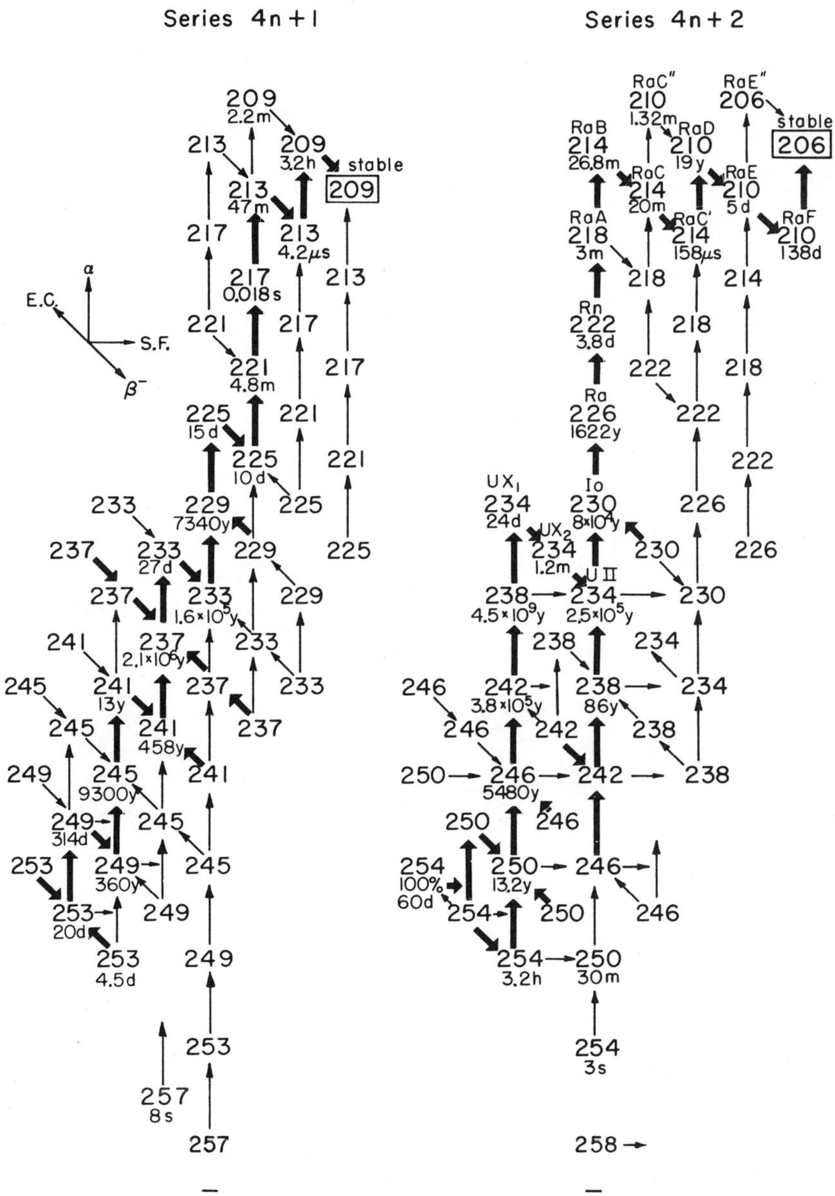

81	thallium	Tl
82	lead	Pb
83	bismuth	Bi
84	polonium	Po
85	astatine	At
86	radon	Rn
87	francium	Fr
88	radium	Ra
89	actinium	Ac
90	thorium	Th
91	protactinium	Pa
92	uranium	U
93	neptunium	Np
94	plutonium	Pu
95	americium	Am
96	curium	Cm
97	berkelium	Bk
98	californium	Cf
99	einsteinium	Es
100	fermium	Fm
101	mendelevium	Md
102	nobelium	No
103	lawrencium	Lw
104	rutherfordium	Rf
105	hahnium	Ha

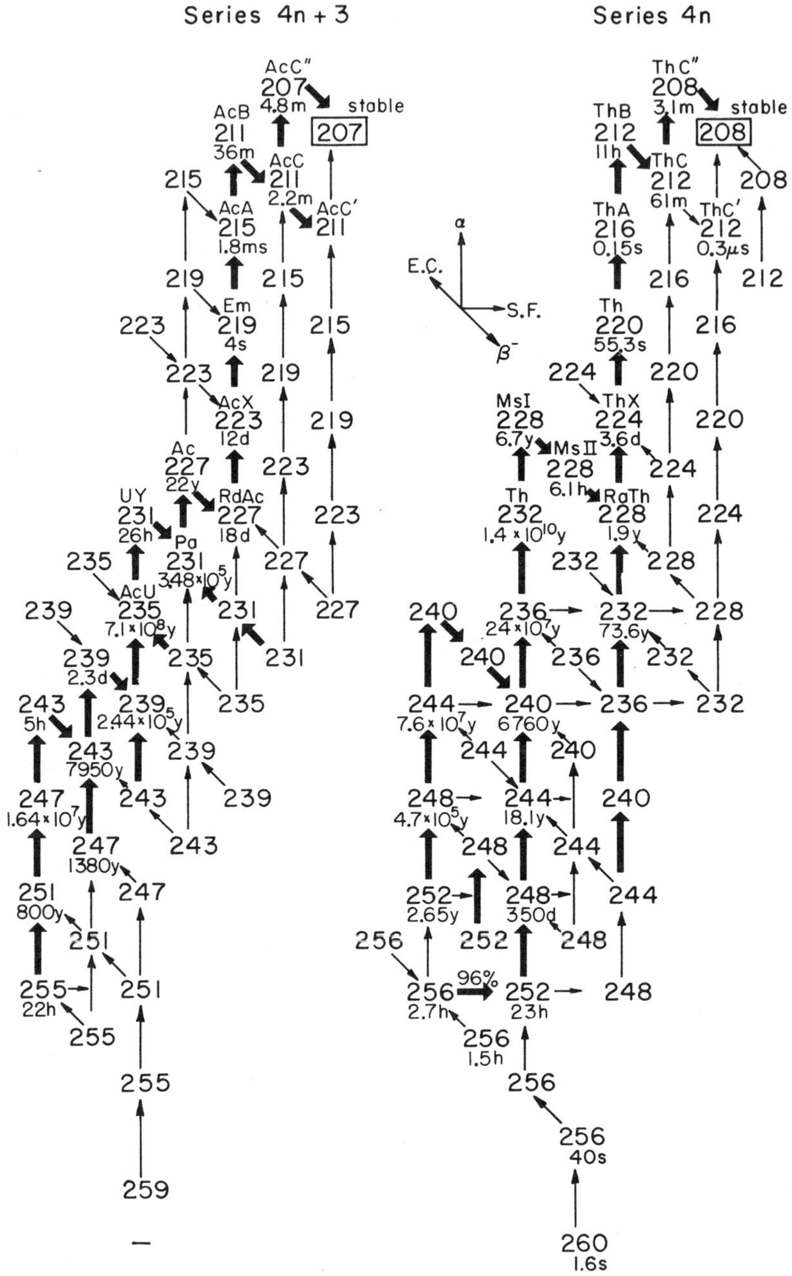

Appendix VII. Electron Subshells

The table shows the order in which the electron subshells are filled in, in the building up of the elements of the periodic table. Atomic weights are given with respect to carbon-12 as 12 even.

n + 1	Sub-Shells	1	2	3	4	5	6	7	8	9	10	11	12	13	14
1	1s	1.0080 H 1	4.003 He 2												
2	2s	6.939 Li 3	9.012 Be 4												
3	2P	10.81 B 5	12.000 C 6	14.007 N 7	15.999 O 8	18.998 F 9	20.183 Ne 10								
	3s	22.990 Na 11	24.312 Mg 12												
4	3p	26.98 Al 13	28.08 Si 14	30.974 P 15	32.064 S 16	35.453 Cl 17	39.948 Ar 18								
	4s	39.102 K 19	40.08 Ca 20												
5	3d	44.96 Sc 21	47.90 Ti 22	50.94 V 23	51.996 Cr 24	54.94 Mn 25	55.85 Fe 26	58.93 Co 27	58.71 Ni 28	63.54 Cu 29	65.37 Zn 30				
	4p	69.72 Ga 31	72.59 Ge 32	74.92 As 33	78.96 Se 34	79.91 Br 35	83.80 Kr 36								
	5s	85.47 Rb 37	87.62 Sr 38												
6	4d	88.90 Y 39	91.22 Zr 40	92.91 Nb 41	95.94 Mo 42	(99) Tc 43	101.1 Ru 44	102.91 Rh 45	106.4 Pd 46	107.870 Ag 47	112.40 Cd 48				
	5p	114.82 In 49	118.69 Sn 50	121.75 Sb 51	127.60 Te 52	126.90 I 53	131.3 Xe 54								
	6s	132.90 Cs 55	137.34 Ba 56												
7	4f	138.91 La 57	140.12 Ce 58	140.91 Pr 59	144.24 Nd 60	(145) Pm 61	150.35 Sm 62	151.96 Eu 63	157.25 Gd 64	158.92 Tb 65	162.50 Dy 66	164.93 Ho 67	167.2 Er 68	168.93 Tm 69	173.04 Yb 70
	5d	174.97 Lu 71	178.5 Hf 72	180.95 Ta 73	183.92 W 74	186.2 Re 75	190.2 Os 76	192.2 Ir 77	195.09 Pt 78	196.97 Au 79	200.59 Hg 80				
	6p	204.37 Tl 81	207.19 Pb 82	208.98 Bi 83	210 Po 84	(210) At 85	222 Rn 86								
	7s	(223) Fr 87	226.05 Ra 88												
8	5f	227 Ac 89	232.04 Th 90	231 Pa 91	238.03 U 92	(237) Np 93	(242) Pu 94	(243) Am 95	(245) Cm 96	(245) Bk 97	(248) Cf 98	(253) Es 99	(254) Fm 100	(256) Md 101	No 102
	6d	Lw 103	(258) Rf 104	(260) Ha 105	106	107	108	109	110	111	112				

In speaking of the size and shape of an object or the time interval between the occurrence of two events it is convenient to express very large numbers and very small decimals in an abbreviated form. This is done principally to conserve time and space. It is convenient for the astronomer in the study of stars, on the one hand, and the physicist and chemist in the study of atoms, on the other. The abbreviations in common use are based upon powers of ten as follows:

$$10 = 10^1 \qquad 1 = 10^0$$

$$100 = 10^2 \qquad 0.1 = 10^{-1}$$

$$1000 = 10^3 \qquad 0.01 = 10^{-2}$$

$$10{,}000 = 10^4 \qquad 0.001 = 10^{-3}$$

$$100{,}000 = 10^5 \qquad 0.0001 = 10^{-4}$$

$$1{,}000{,}000 = 10^6 \qquad 0.00001 = 10^{-5}$$

The abbreviated form on the right side of each equation is mathematically correct. For example

$$10^3 = 10 \times 10 \times 10 = 1000$$

and

$$10^{-3} = \frac{1}{10^3} = \frac{1}{1000} = 0.001$$

In every case *the exponent is seen to give directly the number of digits the decimal is moved from unity*, positive integers specifying the number of places the decimal point is moved to the right to make large numbers, and negative integers specifying the number of places it is moved to the left to make small fractions. To illustrate the use of this system suppose we say that a stone weighs three million kilograms. This can be written

$$3{,}000{,}000 \text{ Kg} = 3 \times 1{,}000{,}000 \text{ Kg} = 3 \times 10^6 \text{ Kg}$$

In the abbreviated notation the mass is therefore written 3×10^6 Kg. If more than one numeral occurs any of several abbreviations might be written. For example, in the case of large numbers,

$$840{,}000{,}000 = 84 \times 10{,}000{,}000 = 84 \times 10^7$$

or

$$840{,}000{,}000 = 8.4 \times 100{,}000{,}000 = 8.4 \times 10^8$$

In the case of small numbers, on the other hand,

$$0.0024 = 2.4 \times 10^{-3}, \quad \text{or} \quad 24 \times 10^{-4}$$

To illustrate the advantage of this abbreviated notation the mass of the *earth* and the mass of an *electron* are found by experiment to be as follows:

$$\text{mass of the earth,} \quad m = 5.97 \times 10^{24} \text{ Kg}$$

$$\text{mass of an electron,} \quad m = 9.11 \times 10^{-31} \text{ Kg}$$

If these are written down in complete decimal form they would appear as follows:

mass of the earth $= 5{,}970{,}000{,}000{,}000{,}000{,}000{,}000{,}000$ Kg

mass of the electron $= 0.000{,}000{,}000{,}000{,}000{,}000{,}000{,}000{,}000{,}000{,}911$ Kg

The multiplication and division of large and small numbers in the abbreviated notation involve the addition and subtraction of exponents.

Rule 1. When a power number is changed from numerator to denominator, or vice versa, the sign of the exponent is changed. For example,

$$\frac{5}{2 \times 10^{-6}} \text{ equals } \frac{5 \times 10^6}{2}$$

Rule 2. When two power numbers are multiplied their exponents are added. For example

$$3 \times 10^5 \times 2 \times 10^4 = 3 \times 2 \times 10^{5+4} = 6 \times 10^9$$

Again:

$$3 \times 10^{17} \times 2 \times 10^{-12} = 3 \times 2 \times 10^{17-12} = 6 \times 10^5$$

Rule 3. When two power numbers are divided, their exponents are subtracted. For example

$$\frac{8 \times 10^9}{2 \times 10^4} = \frac{8 \times 10^{9-4}}{2} = 4 \times 10^5$$

Again:

$$\frac{6 \times 10^{-7}}{3 \times 10^{-2}} = \frac{6 \times 10^{-7+2}}{3} = 2 \times 10^{-5}$$

At a meeting held by the International Union for Pure and Applied Physics a few years ago, the following symbolism was adopted for general use:

10^3	Kilo	K	10^{-3}	milli	m
10^6	Mega	M	10^{-6}	micro	μ
10^9	Giga	G	10^{-9}	nano	n
10^{12}	Terra	T	10^{-12}	pico	p

One of the reasons for adopting these notations was that B, for billion, stands for 10^9 in the United States, and for 10^{12} in Great Britain. Because the term "billion" is so thoroughly entrenched in the minds of scientists in the United States, only a few have adopted G and T, while the change from *k* to *K* for *one thousand* is becoming more common.

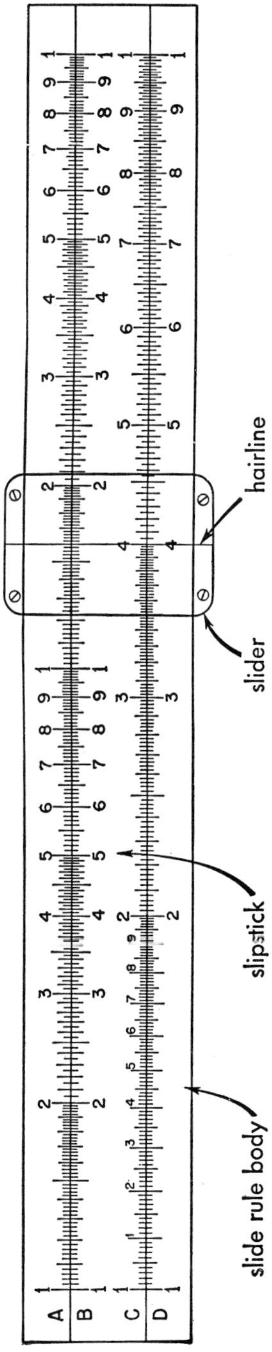

Fig. A.

A slide rule is a simple mechanical device used for carrying out the arithmetic processes of *multiplication* and *division*. Since slide rules are easy to use, and inexpensive ones are quite adequate for most purposes, every physics student should acquire a slide rule and learn how to use it.

The beginner should select a straight inexpensive rule about ten inches in length and one that contains at least four but not more than six scales. The scales most commonly used are the *A*, *B*, *C*, and *D* scales as shown in Fig. A. Note that the slide rule consists of the slide rule *body*, and two movable parts which we will call the *slipstick* and the *slider*.

Significant Figures

Before learning to use a slide rule you should clearly understand the meaning of the term *significant figures*. The three lists of numbers given below will help to illustrate its meaning.

1. All digits other than zero are significant. The measurement 42.65 Kg contains four significant figures.

2. The first significant figure of a number is the first digit that is not zero. The measurement 0.0132 grams contains three significant figures.

3. Zeros to the right of a decimal point, and to the right of a nonzero digit are significant. The reading 46.270 kilometers contains five significant figures.

In powers of ten notation, a simple rule may be applied to specify significant figures. This assumes that at least one nonzero digit appears in front of the decimal point.

4. All zeros that appear in the base number are significant. The reading 2.40×10^5 meters contains three significant figures. If this number is desired to four or more significant figures it may be assumed that zeros follow the last zero shown.

A Three significant figures	B Four significant figures	C Five significant figures
374	5279	24794
21.5	63.08	6.9428
6.05	0.1062	0.37625
0.00328	0.04503	0.053177
546000	692700	46009

If numbers like those in column C are to be used in any slide rule calculation, they should be reduced to four significant figures if the first figure begins with 1, 2, or 3, and three significant figures if the first figure begins with 4, 5, 6, 7, 8, or 9. These particular numbers would, therefore, be assumed to be 2479, 694, 3762, 532, and 460 respectively. Since many of the measurements made in the science laboratory are accurate to only three or four significant figures, the use of a slide rule is usually, but not always, sufficient and justified.

Most slide rules are capable of handling the multiplication and division of numbers to three or four significant figures only. Furthermore, the answers are correct to three or four significant figures only.

When illustrating basic scientific principles by experiment, or by mathematical problems, some measurements may be specified by small whole numbers and others to several significant figures. Suppose for example that a small car is said to travel a distance of 9 meters in 2.15 seconds, and we wish to calculate the average speed. If the answer is to be expressed to *three significant figures* it is common practice to assume both quantities are known to at least three figures.

$$v = \frac{9\text{ m}}{2.15\text{ s}} = 4.19\,\frac{\text{m}}{\text{s}}$$

What has been done to obtain this answer is to assume that the numerator has the value 9.00000 m, and upon dividing by 2.150 s, obtain 4.186 m/s, which to three significant figures is 4.19 m/s.

If a calculator or computer is used in working problems, it is quite proper to carry each figure out as far as it is specified. When the final answer is obtained it is customary to express it to one more figure than the significant figures found in the number containing the least number of significant figures. Usually the rules given above for slide rule accuracy are adequate.

The C and D Scales. In making your first critical examination of your slide rule, note that there are four fundamental scales, and that they are marked at the left end of the rule by *A*, *B*, *C*, and *D*. (See Fig. A.) Since multiplication and division are usually carried out with the two identical *C* and *D* scales, we shall examine the entire length of these scales in three sections and see how to determine any position of the slider hairline.

Consider the left-hand section of the *C* and *D* scales from the left index 1 to the principal number 2, as shown in Fig. B. Note that the interval is divided into ten divisions representing tenths, and each of these into ten divisions representing hundredths.

Fig. B.

If the slider hairline were located in the position shown by the arrow at the left, its position would be written down as 1.140. Similarly, the second arrow position represents the reading 1.370. The third arrow position appears to be halfway between 1.53 and 1.54; hence we do what is called *interpolate* and write 1.535. Interpolation is a process of inserting or finding intermediate values by making a reasonable estimate as to an exact value. Since interpolation is nothing more than a good guess, the reading 1.535 is a little better than three significant figures. Applying the interpolation process to the fourth arrow position, we could write 1.950, with the understanding that the last figure 0 may not be exact.

Although a decimal point has been inserted after the first digit of each arrow position in Fig. B, it could just as readily be placed before the first digit, after the second digit, or after the third digit, etc. In other words the left index 1 can represent any of the numbers 1, 10, 100, 1000, etc. If we let this index represent 10, the right-hand number 2 will represent 20, and the arrow positions would be read as 11.40, 13.70, 15.35, and 19.50. If we let the left index represent 100, the right-hand number 2 will represent 200, and the arrow positions would read as 114.0, 137.0, 153.5, and 195.0

Set the slider of your rule at random between the principal numbers 1 and 2, and then practice reading its position until you feel confident you are reading it correctly.

We are now ready to examine the center section of the *C* and *D* scales between the index numbers 2, 3, and 4. Note in Fig. C, and on your own rule, that each of the two intervals are divided into tenths, and each subdivision into fifths.

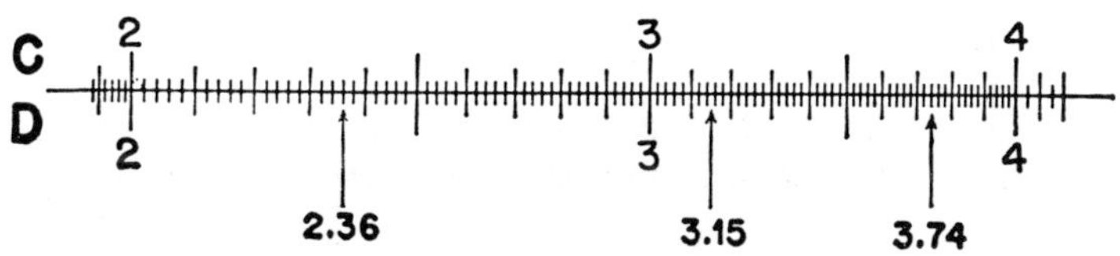

Fig. C.

In other words, the smallest intervals, as marked, each represent 2/10 of a subdivision. Suppose we want to locate the position 2.360 on these scales. The first significant figure indicates that the position lies between 2 and 3. Set the hairline of your rule on 2. The second significant figure 3 indicates that the position is between the third and fourth subdivision, while the third significant figure 6 shows it is 6/10 or 3/5 of the distance between these two subdivisions. Hence the left arrow is properly located at 2.360.

Since the index numbers 3, 4, and 5 can also represent the numbers 30, 40, and 50, or 300, 400, and 500, the arrow positions could represent 23.60, 31.50, and 37.40, or 236.0, 315.0, and 374.0, respectively.

Finally we can examine the right-hand section of the *C* and *D* scales between the index numbers 4, 5, 6, 7, 8, 9, and 1. (See Fig. D.) Between each pair of numbers are ten subdivisions and each subdivision is divided into halves or 5/10. Note carefully the positions of the arrows and be sure you can read any position of your slider hairline when it is located any place on the *C* and *D* scales.

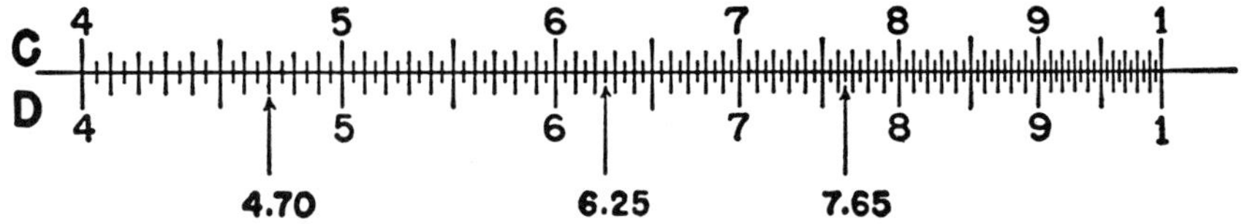

Fig. D.

Multiplication. To multiply two factors, carry out the four following steps: (1) Set the slider hairline on your first number on the *D* scale. (2) Move the slipstick so the left-hand index 1 of the *C* scale is exactly under the hairline. (3) Move the slider hairline to your second number on the *C* scale. (4) Read your answer on the *D* scale under the hairline.

Suppose for example you want to find the product 3 × 2. While you don't need the slide rule to find the answer, the performance of such a simple operation will clarify the method. (See Fig. E.) (1) Set the slider hairline on 3 of the *D* scale (2) Move the slipstick and bring the left index 1 of the *C* scale to the hairline. (3) Move the slider to the second number 2 on the *C* scale. (4) Read the answer as 6 on the *D* scale.

Fig. E.

A little study will show the slide rule settings for the products 30 × 20, 3 × 20, 30 × 2300 × 20, etc., would all be the same as the above. The numeral 6 is the correct first numeral for all answers, but the number of zeros to be added must be determined by inspection.

Consider the procedure for obtaining the product of 24.0 × 1.5. (1) Set the slider hairline exactly on the fourth subdivision line beyond the 2 of the *D* scale. This position represents 24.0. (2) Move the left-hand index 1 of the *C* scale to the hairline. (3) Move the slider to the fifth subdivision between 1 and 2 on the *C* scale. (4) The answer is under the hairline on the *D* scale and appears exactly at the sixth subdivision beyond 3 or 360. To locate the decimal point, inspect the two original numbers. One-and-a-half times 24 cannot be as small as 3.60 nor as large as 360. Therefore it must be 36.0. Similarly the product 24.0 × 2 is 48.0 as shown by the hairline position at the right in Fig. F.

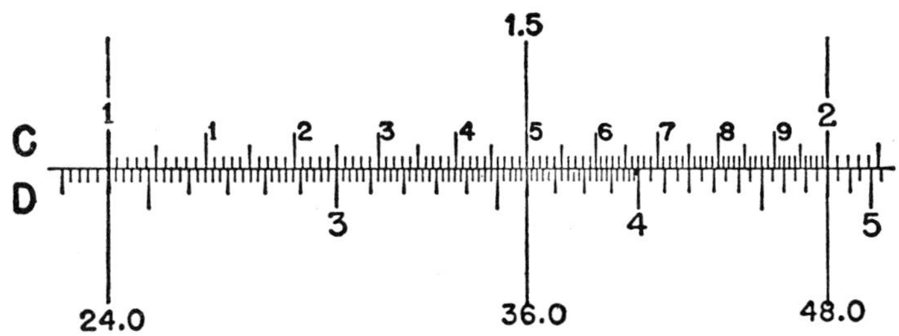

Fig. F.

Find the product 3.8 × 56.5. When the above steps are carried out for this product, the answer will be found to lie beyond the right-hand end of the *D* scale. In such cases one moves the right-hand index 1 of the *C* scale to the first number 3.8 on the *D* scale. (See Fig. G.) Moving the slider to the left, until the hairline is on the second number 56.5 on the *C* scale, the answer is read off the *D* scale as 215.

The Decimal Point. To find the decimal point in any problem, substitute round numbers for those appearing in the problem and determine the position of the decimal point by approximation. For example, in the last problem round off the number 3.8 to 4, and the number 56.5 to 60. The product 4 × 60 by mental arithmetic is 240, so the answer is 215 and not 21.5 or 2150.

Division. Division on the slide rule is just the reverse of multiplication.

Rule. To divide one number by the other, carry out the following steps: (1) Set the slider hairline on the dividend, or numerator, on the D *scale. (2) Move the slipstick until the divisor, or denominator, on the* C *scale lines up with the hairline. (3) The quotient, or answer, is under the* C *index at one end of the slipstick or the other.* In simple terms, line up the numerator on the *D* scale with the denominator on the *C* scale, and the answer is under the *C* scale index.

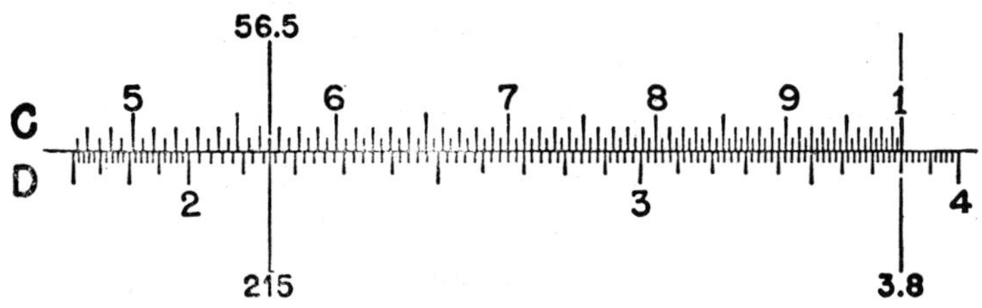

Fig. G.

For example, find the quotient for 6/2. (1) Set the slider hairline on 6 of the *D* scale. (See Fig. E.) (2) Bring the 2 of the *C* scale to the hairline. (3) The answer 3 is now under the *C* index on the *D* scale. Note that the fraction 6/2 is just upside down on the rule.

As a second example find the quotient of 215/56.5. (See Fig. G.) Set the hairline on 215 of the *D* scale and then line up 56.5 of the *C* scale. The answer 3.8 is on the *D* scale under the *C* index at the right. Note again that the fraction is lined up but inverted at the hairline position.

To find the decimal point in the last problem, round off the numbers and use mental arithmetic. For the number 215 write 200 and for 3.8 write 4. The quotient for 200/4 is 50. The answer, therefore, is 56.5 and not 5.65 or 565.

Square and Square Roots. To find the square of a number, one multiplies that number by itself. To find the square root of a number requires the carrying out of a complicated arithmetic process. With the slide rule these operations are relatively simple.

To find the square of a number, use the following rules: (1) Set the slider hairline to the number on the D *scale. (2) Read the answer under the hairline on the* A *scale.*

Inspection of Fig. A shows that directly above 2 on the *D* scale we find its square, or 4, on the *A* scale. Directly above 3 on the *D* scale is its square, or 9, on the *A* scale. Directly above 4 on the *D* scale (the position of the slider hairline shown) we find its square, or 16, on the *A* scale.

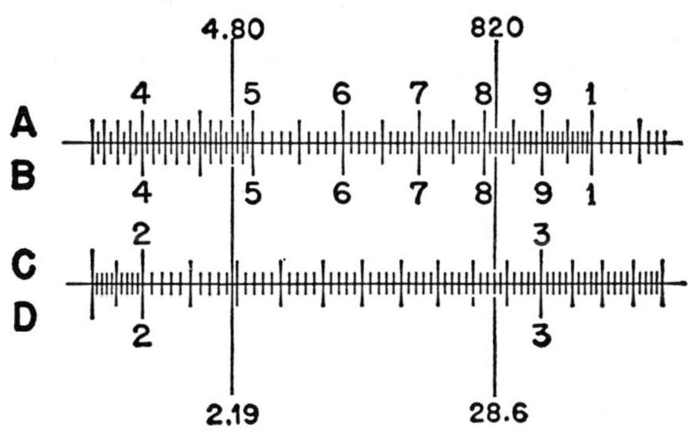

Fig. H.

To find the square root of a number, the reverse process is used: (1) Set the hairline at the number on the A *scale. (2) Read the square root on the* D *scale under the hairline.* Always use the left half of the *A* scale for numbers with an odd number of figures in front of the decimal point, and the right half for those with an even number of figures before the decimal point.

In Fig. H the hairline settings are shown for finding the square roots of 4.80 and 820. The two answers are shown on the *D* scale as 2.190 and 28.60.

Summary

The slide rule is most frequently and conveniently used for multiplication, division, squares, and square roots. A ten-in. rule is capable of an accuracy to three or four significant figures. The first significant figure of a number is the first numeral that is not zero. The last significant figure is the last numeral that is not zero.

Questions

1. Which two scales of the slide rule should be used for (a) multiplication and (b) division?
2. Which two scales of the slide rule should be used for finding (a) the squares and (b) the square roots of numbers?
3. To how many significant figures can slide rule calculations be depended upon?
4. Name the three principal parts of the slide rule.
5. Why should a long slide rule be more accurate than a shorter one?
6. Which end of the slide rule has the greatest accuracy as regards (a) significant figures and (b) percentage error? (*Note:* Can you interpolate any more accurately between 98 and 99 than between 101 and 102?)

Problems

1. Express the following figures to four significant figures only: (a) 23591, (b) 14.3264, (c) 0.39527, (d) 0.0076472, (e) 187.728.

2. Express the following to four significant figures only: (a) 59,346, (b) 8.59762, (c) 15.1697, (d) 0.19652, (e) 765.28.

3. Express the following numbers to three significant figures only: (a) 6497, (b) 78.27, (c) 0.43927, (d) 0.06008, (e) 349.82.

4. Express the following numbers to three significant figures only: (a) 59,330, (b) 6.586, (c) 26.243, (d) 0.08327, (e) 8337.

5. Using the C and D scales of a slide rule, find the product of each of the following: (a) 1.5×3.5, (b) 1.8×2.5, (c) 2.2×4.0, (d) 8.5×6.0, (e) 5.8×5.0.

6. Using the C and D scales find the following products: (a) 18×3, (b) 55×4.2, (c) 38×2.5, (d) 5.8×25, (e) 8.6×4.4.

7. Using the C and D scales of the slide rule, find the value of the following products: (a) 3.5×2.46, (b) 2.74×4.5, (c) 1.75×4.6, (d) 7.8×1.35, (e) $4.95 \times \pi$.

8. Find the product of each of the following: (a) 1.36×2.74, (b) 2.28×2.50, (c) 3.42×2.16, (d) 9.5×4.85, (e) $7.45 \times \pi$.

9. Find the quotient of each of the following by using the C and D scales of a slide rule: (a) 5/4, (b) 75/5, (c) 45/25, (d) 38/4, (e) 36/8.

10. Find the quotient of each of the following: (a) 32/25, (b) 48/15, (c) 4.8/2.2, (d) 81.5/3.6, (e) 36.4/1.90.

11. Using the A and D scales of the slide rule find the square roots of each of the following: (a) 9, (b) 16, (c) 18, (d) 24.5, (e) 87.5.

12. Using the A and D scales of a slide rule find the square roots of the following: (a) 25, (b) 49, (c) 2.50, (d) 2.40, (e) 62.5.

13. Using the slide rule find the squares of the following: (a) 4, (b) 2.5, (c) π, (d) 4.2, (e) 7.5.

Appendix X. The Nobel Prize Winners in Physics

1901 Röntgen, Wilhelm Conrad (1845–1923), German. *Discovery of X rays.*
1902 Lorentz, Hendrik Antoon (1853–1928), Dutch, and
Zeeman, Pieter (1865–1943), Dutch *Zeeman Effect.*
1903 Becquerel, Henri Antoine (1852–1908), French. *Discovery of Radioactivity.*
Curie, Pierre (1859–1906), French, and
Curie, Marie Sklodowska (1867–1934), Polish chemist in France. *Studies of Radioactivity.*
1904 Lord Rayleigh (John William Strutt)(1842–1919), English. *Studies of Gases.*
1905 Lenard, Philipp (1862–1947), German. *Studies of Cathode Rays.*
1906 Thomson, Sir Joseph John (1856–1940), English. *Discharges through Gases.*
1907 Michelson, Albert A. (1852–1931), American. *Precision Optical Instruments.*
1908 Lippmann, Gabriel (1845–1921), French. *Interference Color Photography.*
1909 Marconi, Guglielmo (1874–1937), Italian, and
Braun, Ferdinand (1850–1918), German. *Development of Wireless.*
1910 Van der Waals, Johannes D. (1837–1923), Dutch. *Gas Laws.*
1911 Wien, Wilhelm (1864–1928), German. *Heat Radiation Laws.*
1912 Dalen, Gustaf (1869–1937), Swedish. *Automatic Lighting of Lighthouses.*
1913 Kamerlingh-Onnes, Heike (1853–1926), Dutch. *Liquid Helium and Low Temperatures.*
1914 Von Laue, Max Theodor Felix (1879–1960), German. *Diffraction of X rays.*
1915 Bragg, Sir W. H. (1862–1942), English, and his son
Bragg, W. L. (1890–), English. *Crystal Structure.*
1917 Barkla, Charles G. (1877–1944), English. *Characteristic X rays of the Elements.*
1918 Planck, Max (1858–1947), German. *Quantum Theory of Radiation.*
1919 Stark, Johannes (1874–1957), German. *The Stark Effect of Spectrum Lines.*
1920 Guillaume, Charles E. (1861–1938), Spanish. *Study of Nickel-Steel Alloys.*
1921 Einstein, Albert (1879–1955), German. *Theory of Relativity and the Photoelectric Effect.*
1922 Bohr, Niels (1885–1962), Danish. *Theory of Atomic Structure.*
1923 Millikan, Robert A. (1868–1953), American. *Charge on Electron and Photoelectric Effect.*
1924 Siegbahn, Karl M. (1886–), Swedish. *X-ray Spectroscopy.*
1925 Franck, James (1882–1964), German, and
Hertz, Gustav (1887–), German. *Electron Impact on Atoms.*
1926 Perrin, Jean B. (1870–1942), French. *Discovery of the Equilibrium of Sedimentation.*
1927 Compton, Arthur H. (1892–1962), American. *Compton Effect.*
Wilson, Charles T. R. (1869–1959), English. *Wilson Cloud Chamber.*
1928 Richardson, Owen Willans (1879–1959), English. *Studies of Thermal Ions.*
1929 De Broglie, Louis V. (1892–), French. *Wave Character of Electrons.*
1930 Raman, Sir Chandrasekhara V. (1888–), Indian, *Raman Effect.*
1932 Heisenberg, Werner (1901–), German. *Creation of Quantum Mechanics.*
1933 Schrödinger, Edwin (1887–1961), German, and
Dirac, P. A. M. (1902–), English. *Atomic Theory.*
1935 Chadwick, James (1891–), English. *Discovery of the Neutron.*
1936 Hess, Victor F. (1883–1964), Austrian. *Discovery of Cosmic Rays.*
Anderson, Carl D. (1905–), American. *Discovery of Positron.*
1937 Davisson, Clinton J. (1881–1958), American and
Thompson, George P. (1892–), English. *Electron Diffraction by Crystals.*
1938 Fermi, Enrico (1901–1954), Italian. *Slow-Neutron Reactions.*
1939 Lawrence, Ernest O. (1901–1958), American. *Development of Cyclotron.*
1940, 1941, 1942, not awarded
1943 Stern, O. (1888–1969), German. *Magnetic Moment of Proton.*
1944 Rabi, I. I. (1898–), American. *Magnetic Moments of Nuclei.*
1945 Pauli, W. (1900–1958), Swiss. *Pauli Exclusion Principle.*

1946 Bridgman, P. W. (1882–1961), American. *Physical Effects of High Pressures.*
1947 Appleton, Sir Ed. V. (1892–), English. *Exploration of the Ionosphere.*
1948 Blackett, P. M. S. (1897–), English. *Discoveries in Cosmic Radiation.*
1949 Yukawa, H. (1907–), Japanese. *Theoretical Prediction of Mesons.*
1950 Powell, C. F. (1903–1969), English. *Photographic Cosmic-Ray Studies.*
1951 Cockcroft, Sir J. D. (1897–1967), English, and
Walton, E. T. S. (1903–), English. *First Transmutation of Atomic Nuclei.*
1952 Bloch, Felix (1905–), American. *Nuclear Magnetic Moments.*
Purcell, Ed. M. (1912–), American. *Radio Astronomy.*
1953 Zernike, Frits (1888–1966), Dutch. *Phase Contrast Microscope.*
1954 Born, Max (1882–), German. *Quantum Mechanics, and Wave Functions.*
Bothe, Walther (1891–1957), German. *Quantum Mechanics, and Wave Functions.*
1955 Kusch, P. (1911–), American, and
Lamb, W. E. (1913–), American. *Microwave Spectroscopy and Atomic Structure.*
1956 Shockley, W. (1910–), American,
Brattain, W. H. (1902–), American, and
Bardeen, J. (1908–), American. *Semiconductors, and Their Application to Transistors.*
1957 Yang, C. N. (1922–), American, and
Lee, T. D. (1926–), American. *Studies of the Concept of Parity in Atomic Physics.*
1958 Cerenkov, P. A. (1904–), Russian,
Tamm, I. E. (1895–), Russian, and
Frank, I. M. (1908–), Russian. *Discovery and Study of Cerenkov Radiation.*
1959 Segrè, E. (1905–), American, and
Chamberlain, O. (1920–), American. *Discovery of the Antiproton.*
1960 Glaser, D. A. (1926–), American. *The Bubble Chamber.*
1961 Hofstadter, Robt. (1916–), American, and
Mossbauer, R. L. (1930–), German. *Nuclear Radiation and Absorption.*
1962 Landau, L. D. (1908–1968), Russian. *Mathematical Explanation of Very Low Temperature Phenomena.*
1963 Goeppert-Mayer, Maria (1906–), American, and
Jensen, J. Hans D. (1906–), German. *Nuclear Shell Model.*
Wigner, E. (1902–), American. *Theoretical Work on Symmetry and Parity.*
1964 Townes, C. H. (1915–), American,
Basov, N. (1922–), Russian, and
Prokhorov, A. (1916–), Russian. *Development of Lasers and Masers.*
1965 Feynman, R. P. (1918–), American,
Schwinger, J. S. (1918–), American, and
Tomonaga, S. (1906–), Japanese. *Research in Quantum Electrodynamics.*
1966 Kastler, Alfred (1902–), French. *Theoretical Atomic Investigations.*
1967 Bethe, H. A. (1906–), American. *Sources of Solar and Stellar Energy.*
1968 Alvarez, L. W. (1911–), American. *Elementary Atomic Particle Research.*
1969 Gell-Mann, M. (1929–), American. *Nuclear Theory.*
1970 Alfvén, H. (1908–), Swedish. *Study of Plasmas.*
Néel, L. (1905–), French. *Research in Ferrimagnetism and Antiferromagnetism.*

INDEX

Renolds NUMBER = IF/VF

IF = MASS x ACCEL

INERTIAL FORCE = Dencity x Speed x Length

VISCOUS FORCE = Voscocity of FLUID x speed x Length x

VF = SPEED x VOSCOCITY FLUID x Lenth

VF = VSL

IF = DSL

$\frac{IF}{DL} = S$

$VF = V\frac{IF}{DL}L$

$VF = \frac{V(IF)}{D}$

$VF = \frac{\text{visocity}}{\text{dencity}} IF$

VF = (VIS/DEN)(MA)

$VF = \left(\frac{S^2}{T} \div M/S^3\right) MA$

$VF = \left(\frac{S^2}{T} \times \frac{S^3}{M}\right) MA$

$VF = \left(\frac{S^5}{T}\right) A = \frac{S^5}{T} \frac{S}{T^2} = S^6/T^3$